Chemistry of Plant Phosphorus Compounds

Chemistry of Plant Phosphorus Compounds

Arlen W. Frank

ELSEVIER

AMSTERDAM • BOSTON • HEIDELBERG • LONDON • NEW YORK • OXFORD • PARIS
SAN DIEGO • SAN FRANCISCO • SYDNEY • TOKYO

Elsevier
225, Wyman Street, Waltham, MA 02451, USA
The Boulevard, Langford Lane, Kidlington, Oxford OX5 1 GB, UK
Radarweg 29, PO Box 211, 1000 AE Amsterdam, The Netherlands

Library of Congress Cataloging-in-Publication Data
Application Submitted

British Library Cataloguing in Publication Data
A catalogue record for this book is available from the British Library

For information on all **Elsevier** publications
visit our web site at store.elsevier.com

Printed and bound in USA
13 14 15 16 17 10 9 8 7 6 5 4 3 2 1

ISBN: **978-0-12-407194-0**

This book has been manufactured using Print On Demand technology. Each copy is produced to order and is limited to black ink. The online version of this book will show color Figures where appropriate

Contents

About the Author

Born Nov 22, 1928, in Lima, Peru; married May 24, 1958, in Galena, MD, Marcia D. Craddock; three children, eight grandchildren.

Education: Ginásio diploma, St. Paul's School, São Paulo, Brazil, 1943; H. S. diploma, São Paulo Graded School, 1945; B. Sc. (Chemistry and Mathematics), Acadia University, Wolfville, NS, Canada, 1950; Ph.D. (Chemistry), McGill University, Montreal, PQ, Canada, 1954.

Employment: Postdoctoral Fellow, National Research Council, Ottawa, Canada, 1954–1955; Research Chemist, Du Pont Experimental Station, Wilmington, DE, 1956–1959; Research Chemist, Research Center, Hooker Chemical Corp., Grand Island, NY, 1959–1966; Research Chemist, Southern Regional Research Center, U.S. Department of Agriculture, New Orleans, LA, 1967–1990; Technical Editor, Gmelin Institute, Frankfurt-am-Main, Germany, 1990–1992.

Publications

Books: Frank, Arlen W., "The Cotton Gazetteer," self-published, © 1985, xii + 193 pp.; Frank, W. A. [*sic*], "Phosphonous Acids (Thio-, Seleno- Analogs) and Derivatives," Chapter 10 in Kosolapoff, G. M.; Maier, L. "Organic Phosphorus Compounds," Wiley-Interscience, New York, 1972, Vol. 4 (of 7), 531 pp.

Reviews: Frank, Arlen W., "The Phosphonous Acids and Their Derivatives," *Chem. Rev.* 1961, 30(4), 389–424; Frank, Arlen W., "Synthesis and Properties of N-, O- and S- Phospho Derivatives of Amino Acids, Peptides, and Proteins," *Crit. Rev. Biochem. Molec. Biol.* 1984, 16(1), 51–101.

Other Publications: About 50 articles and about 25 patents, mostly in the field of organophosphorus chemistry.

Preface

This book had its inception in 1990 when I was beginning a 2-year tour of duty as a technical editor for the Gmelin Institute in Frankfurt am Main, Germany, one of the institutes of the Max Planck Society. The multivolume *GMELIN Handbook of Inorganic Chemistry* series, published by Springer-Verlag since 1924, was published in German during its early years but increasingly in English in recent years. Our task as editors was to render manuscripts written by authors not fully versed in the English language into a form suitable for publication. The presence of the joint Gmelin/Beilstein Library onsite and the University of Frankfurt's Senckenberg Library a short distance away prompted me to start work on a long-cherished dream—a comprehensive book on the chemistry of phosphorus compounds in plants—during periods of inactivity between editorial assignments. Alas, the *Handbook* ceased publication in 1998 and the Gmelin Institute itself was closed in 1999, a victim of budget constraints following the merger of the former East German Science Institutes into the Max Planck Society in the early 1990s.

My original intent was to cover all types of phosphorus compounds found in plants, both organic and inorganic, but I soon realized that some measures would have to be taken to keep the book a manageable size. The first to go were the nucleic acids RNA and DNA, despite their importance—they comprise, for example, 11.1% and 1.44%, respectively, of the total phosphorus in the duckweed *Spirodela oligorrhiza* (pp 442–443). Next to go were the inorganic phosphates—phosphoric acid (P_i), di- or pyrophosphoric acid (PP_i), and the polyphosphoric acids—70.7%, 0.17%, and undetected,[a] respectively, in *S. oligorrhiza.* Phosphoric acid presents a problem because it is difficult (but not impossible) to distinguish P_i released during workup from P_i already present in the plant. Last to go were the phosphoproteins because the information on them is too old and sparse to be relevant. The coverage in this book is therefore restricted to monomeric phosphorus-containing compounds with the sole exception of the starch phosphates.

A substantial setback occurred on August 29, 2005 when hurricane Katrina, with a single violent surge of flood water, destroyed all of the reprints I had assembled over the years along with most of my notes.

Little did I know that this book would take 20 years to complete, but here it is at last, my *magnum opus.*

Slidell, LA, USA
July 7, 2012

[a]The lower plants store phosphorus in the form of polyphosphates but the higher plants use phytic acid (Chapter 2) for this function.[1]

Acknowledgments

I am grateful to the librarians and their staff of the Earl K. Long Library at the University of New Orleans and the Howard-Tilton Library at Tulane University, New Orleans, where most of my research for this book was carried out; to the editors and staff of Elsevier, in particular to Mr. Adrian Shell, Ph.D., Senior Acquisitions Editor, Oxford, UK, to Jessica Vaughan, Editorial Project Manager, Waltham, MA, and to my two Production Managers in Gurgaon, India,* Mr. Hemachander Timiri Sundaram and his staff and Mr. Mohana Priyan Rajendran and his staff, for carrying the project through to completion.

Introduction

Arlen W. Frank

This book is intended to be a reference source for scientists engaged in any aspect of plant research—chemistry, biochemistry, or physiology—with primary focus on the chemistry of phosphorus-containing compounds that occur naturally in the plant kingdom and specifically in the higher plants (Plantae). Algae, fungi, and other lower plant forms are not covered.

In general, the subject matter is treated in order of increasing complexity, compounds of C, H, and O in Chapters 1–3 and compounds of C, H, O, and N in Chapters 4–7. Chapter 8 covers the various methods developed in early studies to determine the overall distribution of phosphorus-containing compounds in plants. Appendix A covers the changes in phosphorus-containing compounds during germination and Appendix B their accumulation during growth and senescence. The final sections of the book comprise a comprehensive list of references and a combined index of the plants and phosphorus compounds.

The Phosphorus Compounds

Nomenclature

The naming of the phosphorus compounds in this book conforms to the recommendations of the Drake Committee on the Nomenclature of Organic Phosphorus Compounds published in 1952.[2] *Chemical Abstracts* followed closely these recommendations until the early 1970s but then began to move away from them toward the present system, which in effect subordinates P, O, N, and other atoms to C and indexes the compounds as derivatives of the organic compound having the longest carbon chain. Thus, D-glyceraldehyde 3-phosphate, formerly indexed as a derivative of D-glyceraldehyde, is now indexed as a derivative of propanal. The rationale for doing this was to facilitate the searching and indexing of the chemical literature and was never intended to replace the IUC nomenclature:

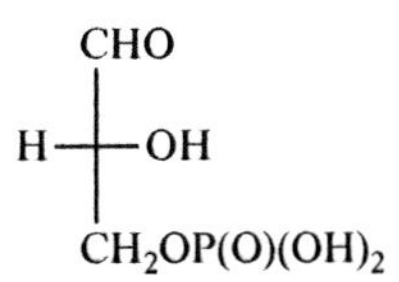

IUC Name: D-Glyceraldehyde, 3-phosphate

CAS Name: Propanal, R-(2)-hydroxy-3-phosphonooxy)-

The order of presentation for each compound in this book is as follows: (1) its IUC name; (2) its commonly used abbreviation in parentheses; (3) its molecular formula;

(4) its *Beilstein* reference if available; (5) its *Chemical Abstracts* name in italics; and (6) the year it was first reported to be a plant constituent.

Occurrence

If the phosphorus compound is a common plant metabolite, as is often the case, its place in the metabolic sequence is illustrated by a figure. The figures in this book are composites drawn from various sources, not reproductions from a single source, except for the few instances in which a reference is cited. The shadowed boxes in the figures show metabolites inside the box and precursors and products outside the box. The box itself does not represent a cell wall or other physical barrier. If the phosphorus compound is unique for a particular plant species, which sometimes occurs owing to some metabolic quirk peculiar to that species, the original paper should be consulted for further details. A few of the plant species in which a particular phosphorus compound was first mentioned are also cited under this heading, documenting the date given under (6) above.

Physical Properties

The physical appearance and solubility characteristics of each compound are given if the information is available. A few salts are included for each compound if useful for characterization but esters, amides, and other derivatives are not. References to spectra, chromtography and electrophoresis are selective rather than comprehensive—usually a single citation selected from those that are available.

Synthesis

A brief survey is given of the methods that have been used to prepare each compound in the laboratory. The coverage is comprehensive with respect to the variety of methods but selective in the number of references cited, often just a single citation.

Hydrolysis

The sensitivity of the compounds to cleavage of the phosphate group varies significantly from one compound to another. Particular attention is given to hydrolysis under acid or alkaline conditions or in the presence of heavy metal catalysts. Rate constants are given whenever possible.

Phosphorylation

Studies under this heading are restricted to the phosphorylation of monophosphates to di-, tri-, or polyphosphates by means of known phosphorylating reagents. Some enzymatic phosphorylations appear in the metabolic pathways discussed under the section "Occurrence" and some instances of the use of phosphorylating reagents to introduce a phosphate group elsewhere in the molecule appear under the section "Synthesis."

Structure

Studies under this heading are concerned with problems such as keto/enol tautomerism or pyranose/furanose conformation in phosphorylated sugars or preferred conformation in phytic acid.

Extraction

No single extraction procedure is optimal for all phosphorus compounds. Attention here is focused on the methods that are most suitable for a particular compound or class of compounds. Methods that are generally applicable to a class of compounds are described in detail at the beginning of each chapter, with selected examples from the literature for methods that are widely accepted and in general use.

Separation and Analysis

The same principles apply here. Methods that are generally applicable to the separation and analysis of the individual phosphorus compounds present in an extract are described in detail at the beginning of each chapter.

Analysis Without Separation

Two methods in particular fall under this heading: the ^{31}P NMR method and the enzymatic method. The first method provides a characteristic signal for each phosphorus compound, some of which are distinct enough to be useful for positive identification, and the second method may be specific enough to permit measurement of a particular phosphorus compound in a complex mixture.

Index of Compounds

The phosphorus compounds are listed in the Index alphabetically according to their IUC names with cross-referencing to their CAS names and commonly used abbreviations.

The Plants

Tables

The plants are listed alphabetically in the various tables according to the binary scientific names adopted by the International Code of Botanical Nomenclature (ICBN). Their common English names are also given if known. The measurements are presented in nmol/g as the preferred format unless stated otherwise.

Index of Plants

The plants are listed in the Index alphabetically according to their ICBN names with cross-referencing to their common names in English or other language that may be provided by the authors.

References

In order to avoid duplication, all of the references are assembled in a single group near the end of the book.

1 Sugar Phosphates

This chapter covers the sugar phosphates and their derivatives with the exception of phytic acid and the inositol phosphates, which are discussed in separate chapters. The following is a complete list of the compounds treated in this chapter: Their full names are given in the introduction to each section.

C_n	Sugar phosphate	Oxidized	Reduced
C_2		GAP	
C_3	G3P, DHAP	PEP, 2-PGA, 3-PGA, 1,3-DPGA, 2,3-DPGA	α-GP
C_4	E4P		
C_5	R5P, Ru5P, Ru1,5P, Xu5P		
C_6	G1P, G6P, G1,6P, F1P, F6P, F1,6P, F2,6P, Gal1P, Gal6P, Man1P, Man6P	6-PG	So1P, So6P
C_7	Se7P, Se1,7P, Ma7P		
C_8	Id1,8P		
C_{12}	Sucrose 6′-P		
C_x	Starch-P		

For general references on this subject, see LePage and Umbreit,[3] MacDonald,[4] or Bergmeyer and Grassl[5].

Extraction

Many plant tissues contain phosphatases that can rapidly hydrolyze the sugar phosphates during isolation. These phosphatases must be quickly and permanently destroyed.[6] The procedures most widely used in current practice employ ice-cold trichloroacetic acid (TCA) or perchloric acid (PCA), both of which kill the plant tissue and extract the sugar phosphates without degrading them. After extraction, the excess TCA is removed by extraction with ether and the excess PCA by precipitation as the water-insoluble potassium perchlorate.

> *Fully expanded leaf blades (about 0.5 g) were frozen with liquid nitrogen and ground in a mortar and pestle with 2.5 ml of 0.75*M *$HClO_4$. After 20 min, the homogenate was transferred to centrifuge tubes and the pH of the homogenate was adjusted to 5.0 with 1*M *KOH. The homogenate was centrifuged at 10,000 × g for 10 min and then the supernatant was made up to 10 ml with distilled water and immediately subjected to analysis. All the extraction procedures were carried out at 4 °C.*
>
> Murata, S.; Sekiya, J., "Effects of Sodium on Photosynthesis in *Panicum coloratum*", *Plant Cell Physiol.* **1992**, 33(8), 1239–1242,[7] reprinted by permission of Oxford University Press, © 1992.

Several procedures have been developed that employ different extractants. In contrast to the phospholipids or nucleotides, there is no leading reference for the extraction of sugar phosphates. Isherwood and Barrett[8] recommended the addition of 8-hydroxyquinoline to TCA as a chelating agent to prevent the loss of sugar phosphate by adsorption. Bieleski and Young[9] preferred extraction with methanol/chloroform/water (12:5:3) or methanol/chloroform/2*M* formic acid (12:5:3).[10] Others have suggested formic acid alone[11] or formic acid/ethanol.[12] Boiling 80% ethanol, once considered a useful solvent for the extraction of sugar phosphates from plant tissues,[6] is no longer recommended as it does not completely extract the soluble phosphate esters nor does it completely inactivate the phosphatases.[10,12]

Separation and Analysis

(1) LePage and Umbreit Method

This method (procedure B in Umbreit's book[3]) uses fractional precipitation with a 25% barium acetate solution of pH 8.2 and 95% ethanol to separate the TCA extract into three fractions: (1) a barium-soluble fraction containing triose-P, PEP, pentose-P, G1P, G6P, and F6P; (2) a barium-insoluble fraction containing 3-PGA and F1,6P; and (3) an ethanol-soluble fraction containing unidentified phosphorus compounds. The first two fractions are then analyzed for each component by appropriate methods.

This procedure was used to determine the levels of triose-P in potato tuber,[13] of 3-PGA in oat embryo[14] and sugar beet leaf and root,[15] of G1P, G6P, F6P, and F1,6P in potato tuber,[16] oat embryo,[14] sugar beet leaf, root and petiole,[15,17] and tomato flower ovary,[18] all carried out between 1943 and 1959. This method has been superseded by those which follow.

(2) Ion Exchange Chromatography (IEC) Method

The use of IEC to separate the sugar phosphates was introduced by Benson[6] in 1950. Sugar phosphates of widely differing structure are well separated on columns containing strongly basic anion resins such as Dowex 1-X8 by stepwise or gradient elution with $0.01 \rightarrow 0.1N$ HCl[19] or $0 \rightarrow 8.2M$ formic acid/1.7*M* ammonium formate,[20] but the pentose, hexose, and heptose monophosphates tend to elute together because of their similar structures and dissociation constants. This problem has been solved by borate complexing. The resin can be used in the borate form, but Khym and Cohn[21] found it preferable to use the resin in the chloride form and elute the sugar phosphates stepwise with 0.025*M* NH_4Cl/0.0025*M* NH_4OH buffer containing $0.01 \rightarrow 0M$ borate. This procedure was used to separate the hexose monophosphates (G1P, G6P, and F6P) of kidney beans[22] and carrots.[23] The IEC method has been automated.[24]

(3) High-Performance Liquid Chromatography (HPLC) Method

The behavior of the sugar phosphates toward ion exchange under HPLC conditions is similar to that of conventional IEC. Good resolution is obtained for all but the pentose and hexose monophosphates with LiChrosorb AN (a strongly basic anion exchanger) and gradient elution with $0.005 \rightarrow 0.4M$ KH_2PO_4,[25] whereas all of the sugar phosphates, including G1P, G6P, and F6P, are well separated if the mobile phase contains a $0.6 \rightarrow 0M$ borate gradient.[26] The use of a solid-phase extraction (SPE) cartridge for sample cleanup is now recommended.[27]

(4) Paper Chromatography (PC) Method

Two-dimensional PC has been the method of choice for the separation of plant sugar phosphates since the early 1950s. Adequate resolution is achieved with the ternary solvents used for the separation of sugars, organic acids, and the like, provided that precautions are taken to remove cations such as calcium, magnesium, or iron by the use of acid-washed paper,[6] by treatment of the plant extract with an ion exchange resin,[9] or by the addition of EDTA to the solvent.[28,29]

The ternary solvent often contains an acid for one dimension and a base, usually NH_4OH, for the second. The combinations of solvents listed in Table 1-1 have been found to be useful. R_f values for some of the sugar phosphates of this chapter in selected pairs of ternary solvents are given in Table 1-2. Some authors prefer to report their data in the form of travel constants[9,12] or P values[28,31] relative to P_i.

The spray reagent most often used for the visualization of the sugar phosphates is that of Hanes and Isherwood, a solution of 60% perchloric acid, 4% ammonium molybdate, and $1N$ HCl in water (5:25:10:60 v:v).[34] After UV exposure to improved color development, the sugar phosphates appear as blue spots on a white background whereas inorganic phosphate has a yellow-green color.[35]

Table 1-1 Solvent Systems for Paper Chromatography

	Solvent Combinations	References
Acidic solvents		
A1	Methanol/formic acid/water (80:15:5)	30
A2	*n*-Propyl acetate/formic acid/water (11:5:3)	9
A3	*n*-Butanol/propionic acid/water (375:180:245)	28
A4	Ethanol/*n*-BuOH/picric acid/water (32:52.5:2:15.5)	29
A5	Ethyl acetate/acetic acid/water (3:3:1)	31
Basic solvents		
B1	Methanol/NH_4OH/water (60:10:30)	30
B2	*n*-Propanol/NH_4OH/water (6:3:1)	9
B3	2-Methoxyethanol/2-butanone/3*N* NH_4OH (7:2:3)	31
Neutral solvents		
N1	Methanol/1*M* ammonium acetate (7:3)	12
N2	Ethanol/1*M* ammonium acetate (70:30)	29
N3	Isobutyric acid/NH_4OH/water (57:4:39)	12,28
N4	Phenol/water (72:28)	6

Table 1-2 Travel Constants for Some Sugar Phosphates

Compound	PC (R_f)[a]		PE[b] (M_P)	TLC[c] (R_P)	
	1st	2nd		1st	2nd
P_i	0.47	0.74	100	100	100
G3P	0.45		74		
DHAP	0.43	0.49	81	104	87
α-GP			78	119	118
2-PGA	0.35	0.69			
3-PGA	0.31	0.69	107	80	88
2,3-DPGA	0.16	0.69	132		
PEP			120	88	111
R5P	0.46	0.41	70	109	62
Xu5P	0.44	0.34			
Ru1,5P	0.19	0.34		58	38
G1P	0.46	0.26	64		
G6P	0.34	0.23	64	77	61
6-PG			80	65	62
F6P	0.43	0.34	62	81	66
F1,6P	0.15	0.34	100	57	32
Gal1P	0.43	0.23			
Man6P			64		
Se7P	0.39	0.28	64		

[a]Wawszkiewicz.[29]
[b]Wade and Morgan.[32]
[c]Feige *et al.*[33] Solvents (see Table 1-1): PC, N2 and A4; PE, 0.025*M* sodium butyrate buffer; TLC, N3 and A3.

Paper containing ^{14}C- or ^{32}P-labeled phosphate esters is placed in direct contact with X-ray film for 2-7 days or determined directly with the aid of a thin window gas-flow counter.[9]

(5) Thin-Layer Chromatography (TLC) Method

TLC was first applied to sugar phosphates by Bieleski in 1965, but has not displaced PC despite the promise of a 20-fold increase in sensitivity in a fifth the time with no loss of resolution.[36] Two-dimensional TLC is almost always performed on TLC plates coated with 250-μ thick layers of Macherey–Nagel unbound cellulose powder, using the same ternary solvents as those used for PC. The relative positions of the phosphate esters and the patterns obtained by the two techniques are essentially the same. Travel constants relative to P_i are reported in Table 1-2 for several of the sugar phosphates of this chapter; see also Cole and Ross.[12]

(6) Electrophoresis Methods

The sugar phosphates can be separated from each other by paper electrophoresis (PE)[9,32] or thin layer electrophoresis (TLE),[36] but little use has been made of either method. PE was used to distinguish F1P from F6P in a potato tuber extract. The hexose monophosphates migrate as a single band in acetate buffer[37,38] but are separated by borate buffers.[39] PE and TLE are more often used as the second dimension in two-dimensional PC/PE[38] or TLC/TLE[36,40] separations.

Buffers that have been used for this method are 0.25*M* ammonium acetate, pH 3.6 (PE, TLE),[9] 0.025*M* sodium butyrate, pH 3.2 (PE),[32] and 0.2*M* sodium borate, pH 9.5 (PE).[39] Electrophoresis is carried out at voltages of 400 V (6 h)[41] to 1000 V (16 min).[36] The paper electrophoretic mobilities (M_P) of some of the sugar phosphates relative to P_i are given in Table 1-2; see also Bieleski and Young.[9] The mobilities on paper and thin layer plates are almost identical.[36]

Analysis Without Separation

(1) ^{31}P Nuclear Magnetic Resonance (NMR) Method

Several plant species have been examined for the presence of sugar phosphate signals since the advent of high-field ^{31}P FT NMR spectroscopy in the early 1980s. Roberts *et al.*[42] identified G6P in the ^{31}P NMR spectrum of maize root tips and established that the G6P is in a cytoplasmic (pH ~7.1) rather than vacuolar (pH ~5.5) environment. The G6P signal is clearly visible in the ^{31}P NMR spectra of maize[42,43] and sorghum[44] root tips, Jerusalem artichoke stolons,[45] and sycamore, soybean, and *Catharanthus roseus* suspension cells.[46] No other sugar phosphate has been unequivocally identified in living plant tissue by this method with the possible exception of F6P in sycamore cells.[46]

As might be expected, better resolution is attained with TCA or PCA extracts of plant tissues than with living plant tissues, and other signals can now be identified. In addition to G6P, signals for F6P,[46,47] F1,6P,[48,49] α-GP,[49–51] and 3-PGA[50,51] as well as DHAP, PEP, and 6-PG are identifiable. Quantitative data were reported for G6P and α-GP.[50]

(2) Enzymatic Methods

The enzyme-coupled assay method, first modified for plant use by Rowan[52,53] in 1957, has become the method of choice for the analysis of sugar phosphates for a variety of reasons. It enables the individual sugar phosphates in a plant extract to be quantitated rapidly and accurately without prior separation and, with few exceptions, each assay is specific for the compound being analyzed. The basis of the method is the change in absorbance of NAD (or NADP) at 340 nm, which increases as NAD(P) is reduced to NAD(P)H and decreases as NAD(P)H is oxidized to NAD(P). For those reactions in which a sugar phosphate interacts directly with NAD(P) or NAD(P)H in an enzyme-catalyzed reaction, this change in absorbance provides a direct measure of the amount of sugar phosphate that causes the change. The seven reactions pertinent to the compounds of this chapter are listed in Table 1-3.

Accordingly, the substrates G3P, DHAP, α-GP, 1,3-PGA, G6P, and 6-PG are assayed directly by reaction with NAD(P) or NAD(P)H in the presence of the indicated enzyme. The other sugar phosphates listed in Table 1-3 may only be assayed after conversion to one or more of these substrates, usually by methods taken from well-known metabolic sequences; in practice, they are assayed immediately after the substrates by adding the

Table 1-3 NAD(P) Coupled Reactions

	Reaction	For analysis of
1	G3P + NAD → 3-PGA + NADH	G3P, E4P, R5P, Ru5P, Xu5P, F1,6P
2	DHAP + NADH → α-GP + NAD	G3P, DHAP, 3-PGA, E4P, F1P, F6P, F1,6P, F2,6P, Se1,7P
3	α-GP + NAD → DHAP + NADH	α-GP
4	1,3-DPGA + NADH → G3P + NAD	3-PGA, 1,3-DPGA, R5P, Ru5P, Xu5P, Ru1,5P
5	Pyruvate + NADH → Lactate + NAD	2-PGA, 3-PGA, PEP
6	G6P + NADP → 6-PG + NADPH	E4P, G1P, G6P, G1,6P, F6P, F1,6P, Se7P
7	6-PG + NADP → Ru5P + CO_2 + NADPH	R5P, 6-PG

Enzymes: (1,4) G3P dehydrogenase [E.C. 1.2.1.12]; (2,3) α-GP dehydrogenase [E.C. 1.1.1.8]; (5) lactate dehydrogenase [E.C. 1.1.1.27]; (6) G6P dehydrogenase [E.C. 1.1.1.49]; (7) 6-PG dehydrogenase [E.C. 1.1.1.44].

appropriate enzyme (mutase, aldolase, kinase, etc.) and taking another reading at 340 nm. This process can be repeated, with the result that two or more compounds can be assayed in sequence in a single run. Further details are dealt with in later sections of this chapter since the procedures are different for each compound.

The lower limit of the spectrophotometric assay is about 10 nmol/mL. If the concentration of a metabolite is below this, the assay can be made more sensitive by fluorimetry or enzyme cycling. According to Passonneau and Lowry, these modifications reduce the lower limits of the assay to 50 and 0.01 pmol/mL, respectively.[54]

Diose Phosphates

The parent diose (glycolaldehyde) is not a plant constituent but its oxidation product (glycolic acid) is

$$HO_2CCH_2OP(O)(OH)_2$$

Glycolic Acid Phosphate (GAP), $C_2H_5O_6P$; Beilstein **3**, E_{III} 377, E_{IV} 578. *Phosphonooxyacetic Acid* [1952].

Occurrence

GAP is an important metabolite of the photorespiratory carbon oxidation cycle (C_2 cycle) of higher plants. It is formed in the chloroplast, together with 3-PGA, by oxidative cleavage of Ru1,5P and is subsequently hydrolyzed to glycolate by GAP phosphatase (Fig. 1-1). The participation of GAP in the C_2 cycle has been confirmed by numerous ^{14}C studies (Table 1-4), but the actual level of the metabolite in higher plants is unknown.

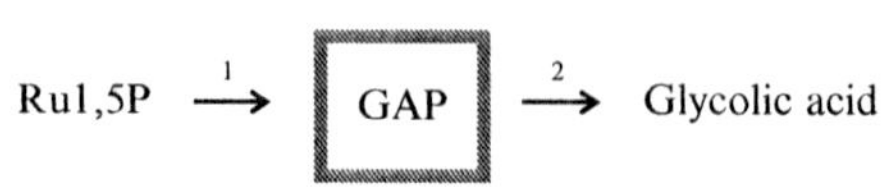

FIGURE 1-1 Glycolic Acid Phosphate. (1) Ru1,5P carboxylase-oxygenase [Rubisco, E.C. 4.1.1.39], O_2, H_2O, Mg^{2+}. (2) GAP phosphatase [E.C. 3.1.3.18].

Table 1-4 Incorporation of ^{14}C Label into Glycolic Acid Phosphate

No.	Plant part	Plant		References
1	Gametophyte	*Concephalum* spp.	Liverwort	55
1	Young leaf	*Cucurbita* spp.	Squash	55
1	Young leaf	*Glycine max* (L.) Merr.	Soybean	56
1	Leaf	*Hordeum vulgare* L.	Barley	57
1	Shoot	*Isopterygium distichaceum* (Mitt.) Jaeg.	A moss	58
1	Young leaf	*Juniperus* spp.	Juniper	55
1	Young leaf	*Lycopersicon esculentum* L.	Tomato	55
1	Leaf	*Nicotiana tabacum* L.	Tobacco	59
1	Leaf	*Oryza sativa* L.	Rice	60,61
1	Young leaf	*Persea gratissima* Gaertn. f.	Avocado	55
1	Leaf	*Pisum sativum* L.	Pea	62
1	Prothalli	*Polystichum* spp.	A fern	55
1	Leaf	*Sorghum bicolor* L.	Sorghum	63
1	Chloroplast	*Spinacia oleracea* L.	Spinach	20,64
1	–, [^{32}P] label			20
1	Leaf	*Zea mays* L.	Maize	57

The concentration of GAP in pea leaf (*Pisum sativum*) is below 2 nmol, the level of sensitivity of the enzymatic assay.[65]

Physical Properties

Glycolic Acid Phosphate (GAP)

GAP [R.N. 13147-57-4]: Monoclinic colorless needles, mp 110-112 °C,[59] pK_a, 3.15, 6.40.[66] Spectra: FTIR (fig.),[67] ^{31}P NMR,[68] X-ray.[69] Chromatography: PC,[32] TLC,[33] IEC,[59] HPLC.[70] Electrophoresis: PE,[32] TLE.[20]

$NaC_2H_4O_6P$ [R.N. 147735-68-0]: Monoclinic crystals, d 1.98. Spectra: X-ray.[69]

$KC_2H_4O_6P$ [R.N. 147735-67-9]: Monoclinic crystals, d 2.07. Spectra: X-ray.[69]

$Ba_3(C_2H_2O_6P)_2{\cdot}4H_2O$: Crystalline solid.[71]

$BaC_2H_3O_6P{\cdot}H_2O$ [R.N. 58389-61-0]: Leaflets, sol. H_2O 1.42 g/L at 20 °C,[72] insol. alcohol.[66]

$Ba_3(C_2H_2O_6P)_2{\cdot}4H_2O$: Light pearly tablets, sol. H_2O 0.10 g/L at 26 °C.[73]

$(C_6H_{11}NH_3)_3C_2H_2O_5P$ [R.N. 95648-83-2]: Colorless monoclinic crystals, d 1.490. Spectra: X-ray,[74] MS.[75]

$(C_{20}H_{24}N_2O_2)_2C_2H_5O_6P{\cdot}2H_2O$ (quinine salt): White tablets, mp 148-149 °C, $[\alpha]_D^{20}$ −182.6° (50% EtOH), sparingly sol. H_2O, slightly sol. EtOH.[72]

$(C_{21}H_{22}N_2O_2)_2C_2H_5O_6P{\cdot}2H_2O$ (strychnine salt): White tablets, mp 221-222 °C, $[\alpha]_D^{20}$ −24.2° (50% EtOH), sparingly sol. H_2O, sol. EtOH.[72]

See also X-ray structure determinations on the Ca,[76] Zn,[76] cyclohexylamine,[77] and other salts.[69,76,77]

Synthesis

GAP may be prepared (a) by phosphorylation of glycolic acid with phosphorochloridic acid[59] or of ethyl glycolate with polyphosphoric acid[78] or (b) by periodate cleavage of α-glycerophosphoric acid,[73] F_6P,[79] or F1,6P.[80]

Hydrolysis

The half-life for hydrolysis of GAP to glycolic acid and P_i is 90 min at 125 °C (*N* HCl)[66] or 35 min in the presence of molybdate at pH 4.0, 100 °C.[81] The rate is higher in *N* acetic acid than in *N* H_2SO_4.[72] The acid is much more resistant to alkaline hydrolysis: half-life ~800 h in *N* NaOH at 100 °C.[82]

Extraction, Separation, and Analysis

GAP can be separated from 3-PGA, Ru1,5P, and other metabolites by IEC on Dowex 1 (OAc) resin,[59] by ion-pairing HPLC,[70] and other methods, but its levels in plant tissues are too low to be detected by any of these methods. Evidence for the participation of GAP in plant metabolism is based on the $^{14}CO_2$ labeling studies listed in Table 1-4. A typical experiment follows:

> *Barley (*Hordeum vulgare *L.) [was] grown by water culture...in a dark room at 25 °C for one week ... After exposure to light (ca. 1,000 ft-c) [for periods ranging from 1 to 24 h], leaves were detached by cutting under water... The detached first foliage leaves were placed vertically in a fixation chamber and the bases of leaves were immersed into water ... $^{14}CO_2$ was generated in the CO_2 fixation chamber by pouring 50 % (v/v) lactic acid into $NaH^{14}CO_3$ solution (65 μCi/ml) and CO_2 concentration was adjusted to about 0.03 %. After feeding of $^{14}CO_2$ for 3 min., the leaves were transferred to boiling ethanol (70 %, v/v) [and] extracted in sequence with boiling ethanol (50 %, v/v) and boiling water. The extracts were concentrated under reduced pressure ..., passed through a column of Dowex 50 (H^+) resin [to remove amino acids, and then chromatographed on] a column of Dowex 1 (CH_3COO^-) resin for organic acid by Zelitch's method [59] ... Radioactivity was measured by a liquid scintillation counter (Beckman Instruments Ltd., LS-250). After 3 hr illumination, 9.0 % of the $^{14}CO_2$ was incorporated into glycolic acid and 4.8 % into GAP.*
>
> Lee, S. H.; Ikeda, M.; Kang, Y. H.; Yamada, Y., "Studies on Carbon Fixation in Barley and Maize Leaves", *J. Fac. Agric. Kyushu Univ.* **1979**, *24*(1), 1–9,[57] reprinted by permission of the Editor-in-Chief, © 1979.

GAP was readily identified by PC in extracts of 27 plants (representing 9 phyla) that were subjected to short-term photosynthesis using $^{14}CO_2$.[55] In other experiments, a phosphate tentatively identified as [^{32}P]-GAP was isolated from spinach chloroplasts subjected to photosynthesis in the presence of $^{32}P_i$.[20] In all of these experiments, the CO_2 content of the air had to be kept low, conditions that favor photorespiration rather than photosynthesis. If the CO_2 content is high, the Ru1,5P is all converted to 3-PGA (see Fig. 1-2) and no GAP is formed.

Triose Phosphates

This group comprises two triose phosphates (G3P and DHAP), one reduced triose phosphate (α-GP), and five oxidized triose phosphates (2-PGA, 3-PGA, 1,3-DPGA, 2,3-DPGA, and PEP).

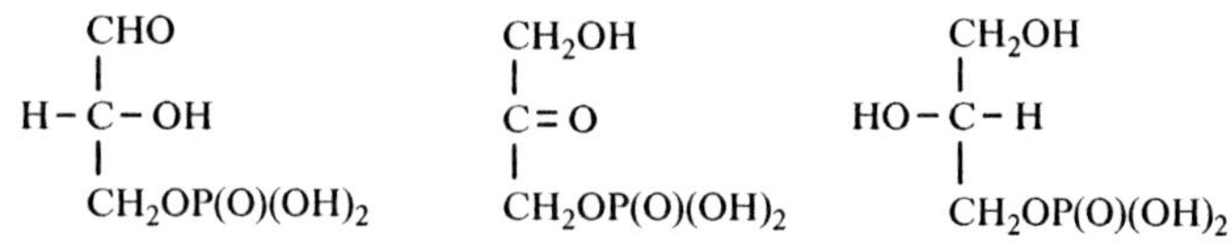

G3P **DHAP** **α-GP**

D-Glyceraldehyde 3-Phosphate (G3P), $C_3H_7O_6P$; Beilstein **1**, E_{III} 3290, E_{IV} 4117. *(R)-2-Hydroxy-3-(phosphonooxy)propanal*, R.N. 591-57-1 [1955].

Dihydroxyacetone Phosphate (DHAP), $C_3H_7O_6P$; Beilstein **1**, E_I 429, E_{III} 3297, E_{IV} 4120. 1,3-Dihydroxy-2-propanone Phosphate; *1-Hydroxy-3-(phosphonooxy)-2-propanone*, R.N. 57-04-5 [1952].

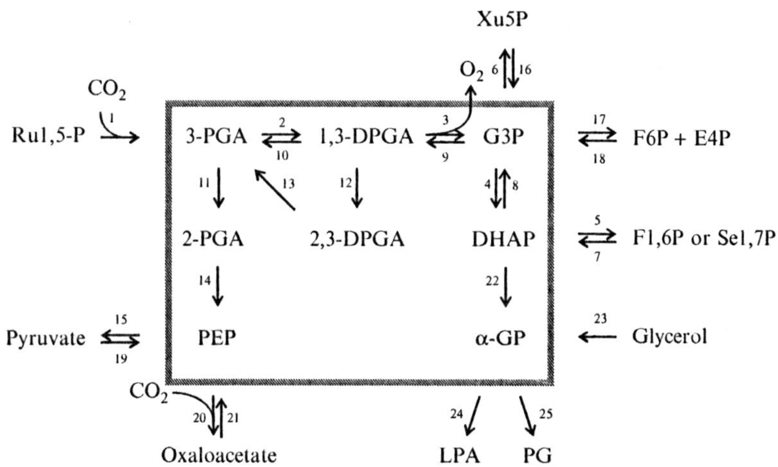

FIGURE 1-2 Triose Phosphates. (1) Ru1,5P carboxylase-oxygenase [Rubisco, E.C. 4.1.1.39], CO_2 or O_2, H_2O, Mg^{2+}. (2) 3-PGA kinase [E.C. 2.7.2.3], ATP, Mg^{2+}. (3) G3P dehydrogenase [E.C. 1.2.1.12], NADH. (4,8) Triose phosphate isomerase [E.C. 5.3.1.1]. (5,7) F1,6P aldolase [E.C. 4.1.2.13]. (6,16) Transketolase [E.C. 2.2.1.1]. (9) G3P dehydrogenase [E.C. 1.2.1.12], NAD, Mg^{2+}. (10) 3-PGA kinase [E.C. 2.7.2.3], ADP, Mg^{2+}. (11) PGA mutase [E.C. 2.7.5.3], Mg^{2+}. (12) DPGA mutase [E.C. 2.7.5.4]. (13) 2,3-DPGA phosphatase [E.C. 3.1.3.13]. (14) Enolase [E.C. 4.2.1.11], Mg^{2+}. (15,19) Pyruvate kinase [E.C. 2.7.1.40], ATP, Mg^{2+}, K^+. (17,18) Enolase [E.C. 4.2.1.11], Mg^{+2}. (20) PEP carboxylase [E.C. 4.1.1.31], CO_2, H_2O. (21) PEP carboxykinase [E.C. 4.1.1.49], ATP. (22) α-GP dehydrogenase [E.C. 1.1.1.8], NADH. (23) Glycerokinase [E.C. 2.7.1.30], ATP. (24) α-GP 1-O-acyltransferase [E.C. 2.3.1.15], acyl-CoA. (25) α-GP 3-phosphatidyltransferase [E.C. 2.7.8.5], CDP-DG.

α-Glycerophosphoric Acid (α-GP), $C_3H_9O_6P$. Beilstein **1**, H 517, E_I 517, E_{II} 275, E_{III} 592, E_{IV} 2333. *sn*-Glycero-3-phosphoric Acid; D-1-Glycerophosphoric Acid; *(R)-1,2,3-Propanetriol 1-(Dihydrogen Phosphate),* R.N. 17989-41-2 [1955].

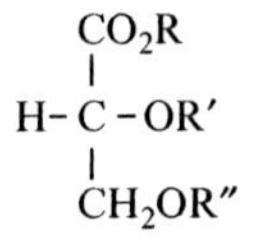

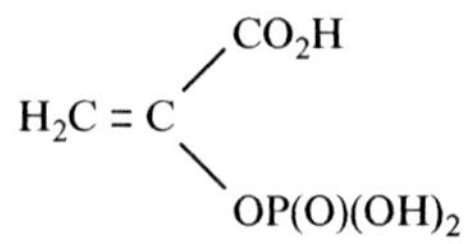

2-PGA, R = R″ = H, R = $P(O)(OH)_2$,
3-PGA, R = R′ = H, R′ = $P(O)(OH)_2$
1,3-DPGA, R′ = H, R = R″ = $P(O)(OH)_2$
2,3-DPGA, R = H, R′ = R″ = $P(O)(OH)_2$

PEP

D-2-Phosphoglyceric Acid (2-PGA), $C_3H_7O_7P$; Beilstein **3**, E_{II} 262, E_{III} 846, E_{IV} 1050. Glyceric Acid 2-(Dihydrogen Phosphate); *(R)-3-Hydroxy-2-(phosphonooxy)propanoic Acid,* R.N. 3443-57-0; see also 2553-59-5 [1934].

D-3-Phosphoglyceric Acid (3-PGA), $C_3H_7O_7P$; Beilstein **3**, E_{II} 262, E_{III} 847, E_{IV} 1051. Glyceric Acid 3-(Dihydrogen Phosphate); Nilsson ester; *(R)-2-Hydroxy-3-(phosphonooxy) propanoic Acid,* R.N. 3443-58-1; see also 820-11-1 [1930].

D-1,3-Diphosphoglyceric Acid (1,3-DPGA), $C_3H_8O_{10}P_2$; Beilstein **3**, E_{III} 851. Negelein ester; *(R)-1,3-Bis(phosphonooxy)propanoic Acid,* R.N. 38163-82-0; see also 1981-49-3 [1988].

D-2,3-Diphosphoglyceric Acid (2,3-DPGA), $C_3H_8O_{10}P_2$; Beilstein **3**, E_{II} 262, E_{III} 850, E_{IV} 1052. Greenwald ester; *(R)-2,3-Bis(phosphonooxy)propanoic Acid,* R.N. 14438-19-8; see also 138-81-8 [1925].

Phosphoenolpyruvic Acid (PEP), $C_3H_5O_6P$; Beilstein **3**, E_{III} 682, E_{IV} 977. 2-Hydroxyacrylic Acid Dihydrogen Phosphate; *2-Phosphonooxy-2-propenoic Acid,* R.N. 138-08-9 [1934].

Occurrence

G3P, DHAP, 3-PGA, and 1,3-DPGA are metabolites of the photosynthetic carbon reduction cycle (Calvin cycle) of higher plants (Fig. 1-2, reading clockwise). In step 1, Ru1,5P combines with CO_2 and water to form an intermediate which is immediately cleaved to two molecules of 3-PGA. In step 2, 3-PGA is phosphorylated by ATP to 1,3-DPGA. In step 3, 1,3-DPGA is reduced by NADH to G3P with the loss of the phosphate group at C_1. In step 4, G3P is converted into its isomer DHAP. In step 5, DHAP condenses with G3P forming F1,6P or with E4P forming Se1,7P. In step 6, G3P reacts with F6P forming Xu5P and E4P or with Se7P forming Xu5P and R5P. These reactions all occur within the chloroplast, found only in the green leaf.

G3P, DHAP, 3-PGA, and 1,3-DPGA are also metabolites of the Embden-Meyerhof-Parnas glycolysis pathway (Fig. 1-2, reading counterclockwise), in which F1,6P is degraded to 3-PGA (steps 7-10). In step 11, 3-PGA is converted to its isomer 2-PGA; 2,3-DPGA, a cofactor for the enzyme involved in this conversion, is formed from 1,3-DPGA and degraded to 3-PGA as shown in steps 12 and 13.[5] In step 14, 2-PGA is dehydrated to PEP and in step 15 the PEP is hydrolyzed to pyruvate. These reactions all take place in the cytoplasm outside any organelle.

G3P is also involved in the pentose phosphate respiratory pathway. It is formed from Xu5P by the transfer of the terminal two-carbon segment to either E4P or R5P (step 16) and reacts with Se7P to form E4P and F6P (step 17). These reactions take place in the cytoplasm, and in the chloroplast in periods of darkness. The metabolite levels are highest when the respiration rate is high, which occurs in the young plant before flowering and in the young leaves and roots of the mature plant.

PEP is a metabolite of C_4 plants, which differ from the usual C_3 plants in that the initial product of CO_2 fixation is malate or aspartate rather than 3-PGA. In step 19, pyruvate is phosphorylated to PEP by ATP. In step 20, PEP combines with CO_2 and H_2O to form oxaloacetate and P_i. Step 19 occurs within the chloroplast whereas step 20 occurs outside, in the mesophyll of the green leaf. In one C_4 subgroup, PEP is formed from oxaloacetate and ATP in a reaction catalyzed by PEP carboxykinase (step 21). This enzyme is located in most cases in the cytoplasm of green leaf tissue.

α-GP is a precursor of lysophosphatidic acid (LPA, step 24) and of the terminal glycerol unit of phosphatidylglycerol (PG, step 25). It is formed from DHAP by reduction with NADH (step 22) or from glycerol by phosphorylation with ATP (step 23).

Physical Properties

D-Glyceraldehyde 3-Phosphate

G3P [R.N. 591-57-1]: $[\alpha]_D^{25}$ +14.5° (0.1*N* HCl)[83] $\Delta G°$ −105.9 kcal/mol.[84] Spectra: ^{13}C NMR.[85] Chromatography: PC,[9,32] TLC,[12] IEC,[86,87] HPLC.[26] Electrophoresis: PE,[32] zone,[88] CZE,[89] CE,[90] isotachophoresis.[88]

$CaC_3H_5O_6P{\cdot}2H_2O$: White crystals.[91]

More extensive data are available on synthetic G3P, which also contains the L-isomer: D,L-G3P [R.N. 142-10-9]: Very hygroscopic granular powder, begins to melt to a glass above 75 °C[92]; readily soluble in water; pK_a 2.10, 6.75.[93] Spectra: UV, IR (fig.),[94] FTIR (fig.),[95] GC-MS,[96] ^1H and ^{13}C NMR,[97] ^{31}P NMR (fig.).[98] Chromatography: HPLC,[99] HPLC-MS-MS.[100]

Dihydroxyacetone Phosphate

DHAP [R.N. 57-04-5]: pK_a 1.77, 6.45[93]; unstable at pH > 6[101]; $\Delta G°$ −106.1 kcal/mol.[84] Spectra: UV,[102] IR (fig.),[103] FTIR (fig.),[95] GC-MS,[96] ^1H NMR (fig.),[103] ^{13}C NMR (fig.),[85] ^{31}P NMR (fig.).[104] Chromatography: PC,[32] TLC,[105] IEC,[86] HPLC,[106] HPLC-MS-MS,[100] IC.[107] Electrophoresis: PE,[32] TLE,[105] zone,[88] CZE,[89] CE,[90] isotachophoresis.[88]

$KC_3H_6O_6P$: Solid, sol. H_2O, insol. EtOH.[101]

$CaC_3H_5O_6P{\cdot}1/2H_2O$: Amorphous solid, often with irregular leaflets; unstable even at 0 °C.[108]

$BaC_3H_5O_6P{\cdot}1/2H_2O$: Like the calcium salt.[108]

α-Glycerophosphoric Acid

α-GP [R.N. 17989-41-2]: Thick, colorless syrup, readily sol. H_2O, methanol and ethanol, insol. ether; $[\alpha]_D$ −1.45° (2*N* HCl).[109] Spectra: FTIR (fig.),[110] GC-MS,[96] ^{31}P NMR.[68] Chromatography: TLC.[111]

$Na_2C_3H_7O_6P$: $[\alpha]_D$ 0 (H_2O).[112]

$BaC_3H_7O_6P$: Crystals, $[\alpha]_D$ 0° (H_2O)[109]; solubility in water, 2.1% at 18 °C,[112] less soluble in hot water.[109]

$Ag_2C_3H_7O_6P$: Colorless needles, $[\alpha]_D^{20}$ +0.8° (H_2O).[113]

$(C_{20}H_{24}N_2O_2)_2C_3H_9O_6P$ (quinine salt): White needles, mp 155 °C, $[\alpha]_D$ −150.3° (ethanol)[112]; sol. ethanol, hot water, insol. cold water.[114]

$C_{21}H_{22}N_2O_2 \cdot C_3H_9O_6P$ (strychnine salt): Prisms, mp 22 °C, $[\alpha]_D$ −21.99° (H_2O).[112]

More extensive data is available on synthetic α-GP, which also contains the L-isomer:

DL-α-GP [R.N. 57-03-4]: pK_1 1.40,[93] pK_2 6.648, $\Delta G°$ 9081 cal, $\Delta H°$ −749 cal, $\Delta S°$ −33.0 cal/deg at 25 °C[115]; $\Delta G°$ −114.2 kcal/mol[84]; LCAO-MO.[116] Spectra: Far-UV,[117] ^1H NMR,[118] ^{13}C NMR (fig.).[118] Chromatography: PC,[9] TLC,[119] IEC,[120] HPLC,[121] cellulose column,[122] IC,[107] GC (fig.).[123] Electrophoresis: PE.[9]

DL-$Na_2C_3H_7O_6P \cdot 6H_2O$ [R.N. 34363-28-5]: Nonhygroscopic, very hard twinned crystals[124] or monoclinic truncated trapezoids, d 1.606[125]; heat of combustion Q_M 1645 kJ/mol.[126] Spectra: X-ray.[125]

DL-$CaC_3H_7O_6P \cdot 2H_2O$ [R.N. 55128-84-2]: X-ray (fig.).[127] Anhydrous salt: Square leaflets, solubility in water 1.9% at 13 °C.[128]

D-2-Phosphoglyceric Acid

2-PGA [R.N. 3443-57-0; see also 2553-59-5]: $[\alpha]_D^{21}$ +13.0° (1*N* HCl),[129] pK_a 3.62, 6.97,[130] and 1.8[129]; conformation[131,132]; $\Delta G°$ −158.0 kcal/mol.[84] Spectra: GC-MS,[96] ^1H NMR,[131] ^{13}C NMR,[132] ^{31}P NMR.[68] Chromatography: PC,[9] TLC,[133] IEC,[134] HPLC,[135] IC.[107] Electrophoresis: PE,[9] TLE,[36] zone,[88] CZE,[136] CE.[90]

$Na_3C_3H_4O_7P \cdot 5H_2O$: long needles, $[\alpha]_D^{22}$ +3.6° (H_2O), sol. H_2O, insol. MeOH, ether.[129]

Hexahydrate [R.N. 99470-03-8]: Orthorhombic plates, d 1.76; X-ray.[137]

$Ag_3C_3H_4O_7P$: White cryst.[138]

$BaC_3H_5O_7P \cdot 1.5H_2O$: Leaflets.[139]

D-3-Phosphoglyceric Acid

3-PGA [R.N. 3443-58-1; see also 820-11-1]: $[\alpha]_D^{20}$ −13.8°,[140] pK_a 3.59, 6.76,[130] and 1.42[93]; countercurrent distribution[141]; conformation[132]; $\Delta G°$ −160.1 kcal/mol,[104] −3.1 kcal/mol.[142] Spectra: FTIR (fig.),[67] GC-MS,[96] ^1H NMR (fig.),[132] ^{13}C NMR,[132] ^{31}P NMR.[68] Chromatography: PC,[9,32] TLC,[133] GC,[143] IEC,[134] HPLC,[70,144] IC (fig.).[107] Electrophoresis: PE,[32] TLE,[36] zone,[88] isotachophoresis,[88] CZE,[136] CE.[89]

$Na_3C_3H_4O_7P$ [R.N. 61546-67-6]: $[\alpha]_D$ +11.39° (H_2O).[145]

$Na_2C_3H_5O_7P$ [R.N. 80731-10-8]: Monoclinic cryst., d 2.143,[146] $[\alpha]_D$ +7.71° (H_2O)[145]; X-ray.[146]

$KC_3H_6O_7P$ [R.N. 164034-33-7]: Colorless orthorhombic prisms, d. 1.85. Spectra: X-ray.[147]

$BaC_3H_5O_7P{\cdot}2H_2O$ [R.N. 86879-11-0]: Monoclinic plates, d 2.53,[148] $[\alpha]_D$ +5.10° (H_2O).[145] Spectra: X-ray,[148] XPS.[149]

$CdC_3H_5O_7P{\cdot}3H_2O$ [R.N. 32690-78-1]: Monoclinic prisms, d 2.203; X-ray.[150]

See also X-ray structure determinations on the calcium and cyclohexylamine salts.[147,151,152]

1,3-Diphosphoglyceric Acid

1,3-DPGA [R.N. 38163-82-0; see also 1981-49-3]: Unstable, decomposing spontaneously into 3-PGA and phosphoric acid; ΔG° −160.0 kcal/mol.[84] Spectra: UV,[153] ^{31}P NMR.[154] Chromatography: HPLC.[99]

$(C_{21}H_{22}N_2O_2)_4C_3H_8O_{10}P_2$ (strychnine salt): Crystals, unstable in the dry state, insol. H_2O.[153]

D-2,3-Diphosphoglyceric Acid

2,3-DPGA [R.N. 14438-19-8; see also 138-81-8]: Hygroscopic sirup, readily sol. alcohol, $[\alpha]_D^{20}$ −2.29°,[155] pK_a 6.39, 7.39 (apparent), 6.99, 7.28 (intrinsic)[156]; conformation[157]; enthalpy of ionization 0.0 ± 0.3 kcal/mol[158]; ΔG° −158.2 kcal/mol,[84] −11.3 kcal/mol.[142] Spectra: FTIR (fig.),[67] GC-MS,[96] ^1H NMR (fig.),[159] ^{31}P NMR.[68] Chromatography: PC,[32] TLC,[160] IEC,[86] HPLC.[135,161] Electrophoresis: PE,[32] zone,[88] isotachophoresis.[88]

$Na_5C_3H_3O_{10}P_2$: $[\alpha]_D$ −3.5° (H_2O).[162]

Ca salt: Amorph., insol. H_2O.[155]

Ba salt: Amorph., insol. H_2O[155] Spectra: IR.[159]

Phosphoenolpyruvic Acid

PEP [R.N. 138-08-9]: Slightly hygroscopic triclinic plates, mp 103 °C, d 1.72,[163] pK_a 1.41, 3.56, 6.22[164]; ΔG° −15.2 kcal/mol, ΔH° −14.3 kcal/mol, ΔS° +3 cal/deg/mol[165]; MO,[166] LCAO-MO,[116] ΔE (*ab initio* calcn.).[167] Spectra: UV (fig.),[168] GC-MS,[96] ^1H NMR,[169] X-ray.[163] Chromatography: PC,[9,32] TLC,[36] IEC,[86] HPLC,[135] IC.[107] Electrophoresis: PE,[32] TLE,[36] zone,[88] CZE,[89] isotachophoresis.[88] It is best handled and stored in the form of its stable, crystalline monopotassium salt.[170]

$NaC_3H_4O_6P{\cdot}H_2O$ [R.N. 77617-15-3]: Monoclinic prisms, d 1.89[171]; Spectra: IR,[172] X-ray.[171]

$AgBaC_3H_2O_6P{\cdot}2H_2O$ [R.N. 129176-93-8]: Colorless prisms (photo.).[173]

$(C_6H_{11}NH_2)C_3H_5O_6P$ [R.N. 10526-80-4]: Orthorhombic needles, d 1.36 or oblique monoclinic parallelepipeds, d 1.36 (two polymorphic forms). Spectra: X-ray.[174]

$(C_6H_{11}NH_2)_3C_3H_5O_6P$ [R.N. 35556-70-8]: Cryst., dec. 155-180 °C.[168] Monohydrate: Colorless orthorhombic crystals, d 1.22. Spectra: X-ray.[175]

See also X-ray structure determinations on the K,[176] Ca,[177] ammonium,[178] and other[179,180] salts.

Synthesis

G3P may be prepared (a) by phosphorylation of 2-*O*-benzyl D-glyceraldehyde dimethyl acetal with diphenyl phosphorochloridate[83] or of (*R*)-glycidaldehyde diethyl acetal with Na_2HPO_4,[181] (b) by cleavage of R5P, F6P, or F1,6P with lead tetraacetate[85,91] or periodate,[71] or (c) by enzymatic cleavage of F1,6P with zymohexase in the presence of hydrazine.[182]

DHAP may be prepared (a) by phosphorylation of 1,3-dihydroxyacetone with $POCl_3$[108] or ethyl metaphosphate,[183] (b) by phosphorylation of *O*-acetyl-1,3-dihydroxyacetone[103,184] or 2,5-diethoxy-*p*-dioxane-2,5-dimethanol[185,186] with diphenyl phosphorochloridate or cyanoethyl phosphate, (c) by hydrolysis of bromoacetone phosphate,[187] (d) by enzymatic phosphorylation of 1,3-dihydroxyacetone with glycerokinase and ATP in the presence of PEP or acetyl phosphate,[101] (e) by enzymatic cleavage of F1,6P with zymohexase in the presence of sodium bisulfite,[188] or (f) by enzymatic oxidation of α-GP with α-GP oxidase in the presence of catalase.[189] Because of its instability, DHAP is best stored as a ketal from which it can be prepared when needed.

α-GP may be prepared (a) by phosphorylation of *O,O*-isopropylidene-D(+)-glycerol with phosphorus oxychloride, (b) by fermentation of glucose in the presence of P_i and fluoride,[113] (c) by resolution of synthetic DL-α-GP,[112] or (d) by phosphorylation of glycerol with glycerol kinase, preferably immobilized on polyacrylamide gel, in the presence of acetyl fluoride and ATP.[190] Hydrolysis of lecithin with either acid or alkaline catalysts is unsuitable because of migration of the phosphate group, resulting in a product contaminated with β-GP and D-α-GP.[191]

2-PGA may be prepared (a) by phosphorylation of methyl 3-*O*-benzyl D-glycerate with diphenyl phosphorochloridate,[129] (b) by acid-catalyzed isomerization of 3-PGA,[192] (c) by periodate cleavage of D-gluconic acid 2-phosphate,[193] or by selective degradation of the L-isomer of DL-2-PGA with *Leuconostoc mesenteroides*.[194]

3-PGA may be prepared (a) by phosphorylation of D-(−)-glyceric acid with ethyl metaphosphate,[138] (b) by the action of baker's yeast on sugar phosphate (sucrose and NaH_2PO_4) in the presence of acetaldehyde and NaF,[140] or (c) by selective degradation of the L-isomer of DL-3-PGA with *L. mesenteroides*.[194]

1,3-DPGA may be prepared by enzymatic phosphorylation of G3P with phosphoric acid in the presence of acetaldehyde[153] or sodium pyruvate.[195] 2,3-DPGA may be prepared by phosphorylation of methyl D-glycerate with diphenyl phosphorochloridate.[162]

PEP may be prepared (a) by phosphorylation of pyruvic acid[173] or 3-chlorolactic acid[196] with $POCl_3$ or of methyl pyruvate with polyphosphoric acid,[78] (b) by the reaction of bromopyruvic acid with trimethyl phosphite,[197] tribenzyl phosphite,[172] or dimethyl trimethylsilyl phosphite,[198] or (c) by enzymatic rearrangement of 2-PGA with enolase.[129]

Structure

G3P exists in aqueous solution as a 29:1 mixture of the *gem*-diol **1** and free aldehyde, but only the latter is active as a substrate for G3P dehydrogenase [E.C. 1.2.1.12]. The first-order rate constant k for the conversion of diol to aldehyde is $0.087\ s^{-1}$ at 20 °C.[199] In aqueous solutions of DHAP, however, the ratio of keto to *gem*-diol **2** is 3:2.[103,104] About 1.0% of the DHAP is in the enol form at pH 7 and 20 °C[200]. The preferred conformation of 2-PGA in aqueous solution is the *gauche* form **3** in which the alcoholic OH is hydrogen-bonded to the carboxyl group.[131]

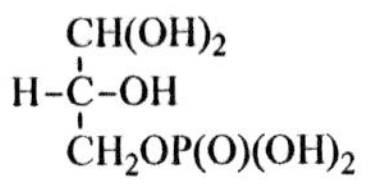

1 (G3P)

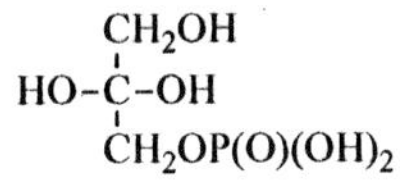

2 (DHAP)

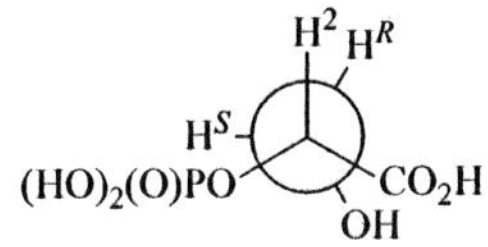

3 (2-PGA)

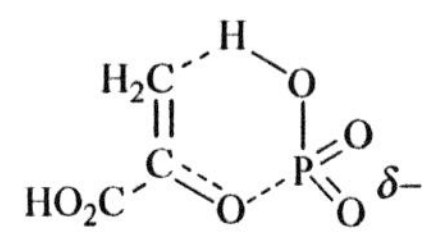

4 (PEP)

Hydrolysis

G3P is completely hydrolyzed by 1*N* acid to methylglyoxal and P_i in 60 min at 100 °C.[83] The rate constant k is $6.25 \times 10^{-4}\ s^{-1}$ in 1*N* HCl at 100 °C.[108] Exposure to 2*N* NaOH for 20 min completely hydrolyzes G3P to P_i and lactic acid.[201] In alkaline media, methylglyoxal is rapidly converted to lactic acid. For the kinetics of hydrolysis of DL-G3P in the neutral region (pH 5-9), see Kozlova *et al.*[202]

DHAP, which is isomeric with G3P, exhibits similar behavior. The rate constant k is $5.62 \times 10^{-4}\ s^{-1}$ for hydrolysis in 1*N* HCl at 100 °C.[108] Exposure to 2*N* NaOH at room temperature hydrolyzes DHAP completely to P_i and lactic acid.[201] The rate increases with KOH concentration up to 0.5 mol/L and remains the same at higher concentrations.[203]

The hydrolysis of DL-α-GP is catalyzed by molybdate,[204] Pb(II),[205] Ce(III),[206] La(III),[206] and Th(IV),[207] but not by Zr(IV).[208] In the pH 3-7 range, where the monoanion HA^- is predominant, the rate constant for the hydrolysis of DL-α-GP to glycerol at 100 °C is $1.35 \times 10^{-5}\ s^{-1}$ compared to $2.77 \times 10^{-5}\ s^{-1}$ for β-GP. In solutions of pH<3, however, the rates become equal owing to acid-catalyzed rearrangement.[209] At lower pH values the rates fall to a minimum (neutral acid H_2A), then increase linearly with acid concentration.[210] Experiments with ^{18}O-enriched water show that only P—O bond fission occurs despite differences in mechanism between HA^- and H_2A.[210]

Both isomers are stable to boiling 0.025*N* NaOH.[211] α-GP is readily hydrolyzed by acid phosphatase, which is highly specific for this substrate,[212] but less rapidly than β-GP by alkaline phosphatase.[213]

2-PGA is hydrolyzed to D-glyceric acid when heated in 1*N* HCl in a sealed tube for 18 h at 125 °C, but under milder conditions (0.25*N* HCl, 75 °C) it rearranges to 3-PGA without releasing any inorganic phosphate.[129] Owing to this rearrangement, the rates of acid hydrolysis of 2-PGA and 3-PGA are identical.[66] The rate constant k is $2.58 \times 10^{-5}\ s^{-1}$.[139]

For the effect of pH over the 2-10 range on the hydrolysis of 2-PGA at 100 °C, see Wold and Ballou.[214] At pH 4 and 100 °C, in the presence of molybdate catalyst, half of the phosphate in 2-PGA was released in 35 min whereas 3-PGA was unaffected.[81] 3-PGA is unusually resistant to acids and alkalis, and also to thermal degradation.[140] Hydrolysis of 3-PGA in the neutral region is catalyzed by Th(IV),[215] Ce(III),[216] La(III),[216] and UV radiation.[217]

1,3-DPGA decomposes spontaneously in aqueous solution. The rate constant is 0.026 min^{-1} at 38 °C and pH 7.2 and is catalyzed by acids and bases.[153] Experiments with ^{18}O-labeled water show that 1,3-DPGA is an acyl donor toward GDH in the dehydrogenase reaction (Fig. 1-2, step 3) and a phosphate donor toward ADP in the 3-PGA kinase reaction (Fig. 1-2, step 10). The bond cleavage is C—O in the former and P—O in the latter.[218]

2,3-DPGA is extraordinarily resistant to acid hydrolysis; it requires boiling for days with 5% H_2SO_4 to effect complete decomposition to D(−)-glyceric acid.[219] In 1*N* HCl at 125 °C the hydrolysis proceeds stepwise, the initial rate being $5.97 \times 10^{-5}\ s^{-1}$ and the final rate $2.80 \times 10^{-5}\ s^{-1}$. If hydrolysis is interrupted after 2 h, the product is found to be a mixture of 2-PGA and 3-PGA.[139]

Hydrolysis of PEP yields only P_i and pyruvic acid over the pH 0-8 range at 75 °C. The rate constants for the dianion HA^{2-}, monoanion H_2A^-, and neutral acid H_3A are all abnormally high, in keeping with the high-energy character of PEP in relation to other monoalkyl phosphates. A mechanism was proposed in which the expulsion of metaphosphate is enhanced by the formation of cyclic transition state, **4**.[164] Hydrolysis of PEP at 75 °C is catalyzed by $>10^6$-fold by Hg(II) via addition to the enol double bond, and 10-fold by Cu(II) via a chelation or complexation mechanism. Other metal ions are ineffective.[220]

Metal Ion Complexes

DL-α-GP forms binary 1:1 complexes with metal(II) ions. The extent of complexing can be determined by potentiometric titration. Stability constants have been reported for the Mg(II) and Ca(II) complexes.[221]

Separation and Analysis

(1) LePage and Umbreit Method[3]

G3P and DHAP appear in the barium-soluble fraction of the TCA extract (see Chapter 8, pp. 459-460). They are determined together as triose phosphate (TP) by the amount of P_i released in 20 min at room temperature with 2*N* NaOH. 3-PGA appears in the barium-insoluble fraction of the TCA extract and is analyzed by a colorimetric method. The LePage and Umbreit method has been superseded by those which follow.

(2) Chromatographic Methods

The triose phosphates can be separated from other phosphate esters by IEC or HPLC after extraction with perchloric acid or trichloroacetic acid, but their separation from each other requires two-dimensional PC/PC, PC/PE, TLC/TLC, or TLC/TLE. Basic solvents like *n*-propanol/NH_4OH/H_2O[9,29] should be avoided in the case of G3P or DHAP because of their extreme sensitivity to base, but neutral solvents that contain ammonium salts such as ammonium acetate or ammonium isobutyrate are acceptable.[12,28] 2-PGA and 3-PGA should be separable from each other by these methods but there are no examples of this in Table 1-5. α-GP can be separated from other phosphate esters by GLC after extraction with perchloric acid.[255] Quantitative data have been reported for TP, 2-PGA, and 3-PGA by the IEC method, for α-GP by the GLC method, and for DHAP, 3-PGA, and PEP by the PC/PC and PC/PE methods.

PEP is best extracted from plant tissues as the 2,4-dinitrophenylhydrazone and separated from other α-keto acid DNPs by two-dimensional PC:

> *5 g fresh weight of each plant part [was] homogenised with 0.25 ml cold sulphuric acid, 50 mg of 2,4-dinitrophenylhydrazine (2,4-DNPH) and 60 ml of 80 % ethanol. . . , kept at room temperature with occasional stirring for an hour . . . , [and] centrifuged. The residue [was extracted four times with] 20 ml of cold ethyl acetate . . . Extracts were mixed . . . evaporated to dryness . . . [and] extracted thrice with 2 ml 10 %* Na_2CO_3*. The solution of keto acid hydrazones (Na salts) was diluted to a final volume of 10-15 ml with water. The combined extracts were then reextracted twice with 2 ml aliquots of ethyl acetate [to remove] the uncombined hydrazones . . . The carbonate solution was cooled and adjusted to pH 2.0 with ice-cold conc.* H_2SO_4 *and extracted thrice with ethyl acetate. The combined extracts were evaporated to dryness under a current of cold air. Small crystals of keto acid hydrazones [were] found in the residue . . . Different DNP's were separated by two-dimensional paper chromatography using n-butanol:ammonia:water (16:3:1) and n-butanol:ethanol: water (5:1:4) for the 1st and 2nd direction, respectively. Keto acids were detected on the chromategrams by their yellow spots and identified by their* R_f *values . . . For quantitative analysis, keto acid spots . . . were eluted with 5 ml of 0.2M* Na_2CO_3 *and [read] in a photoelectric colorimeter at 420 nm.*
>
> Mukherjee, D., "Keto Acids and Amino Acids Changes in Leaves, Flowers and Fruits of *Cajanus cajan*", *J. Ind. Bot. Soc.* **1974**, *53*(1/2), 115–118,[241] reprinted by permission of the Chief Editor, © 1974.

Analysis Without Separation

(1) ^{31}*P NMR Method*

DHAP, α-GP, 3-PGA, and PEP have been identified in perchloric acid extracts of beans, spinach, and sycamore by high-field FT ^{31}P NMR spectroscopy.[49–51] A signal assigned to 3-PGA appears in the spectrum of compressed spinach leaf.[50]

(2) Enzymatic Method

In this method, G3P and DHAP are analyzed by reactions coupled to the oxidation of NADH to NAD (Table 1-3, reaction 2). In the following procedure, they are analyzed

Table 1-5 Triose Phosphates in Plants*

No.	Plant part	Method	G3P	DHAP	TP	α-GP	2-PGA	3-PGA	DPGA	PEP	References
Acer pseudoplatanus L. (Sycamore Maple)											
1	Cell suspension	^{31}P NMR				+		+			51
Actinidia deliciosa Liang and Ferg. (Kiwi)											
3	Fruit	Enzym.								5.1	222
Albizia lebbeck (L.) Benth. (Lebbek Tree)											
1	Leaf[a]	PC/PC								720	223
1	Flowering bud[a]	PC/PC								640	223
1	Flower[a]	PC/PC								720	223
Amaranthus caudatus L. (Love-Lies-Bleeding)											
1	Leaf[b]	Enzym.			714			667		147	224
Amaranthus cruentus L. (Purple Amaranth)											
1	Leaf[c,d]	Enzym.						78			225
Amaryllis vittata L'Her. (Amaryllis)											
1	Pollen[d]	Enzym.	440	130						1170	226
Apium graveolens L. var. *dulce* (Celery)											
1	Petiole vascular bundle	PC/PC						+			227
Arachis hypogaea L. (Groundnut)											
1	Leaf[a]	PC/PC								30	228
Arbutus unedo L. (Strawberry Tree)											
1	Leaf[b,d]	Enzym.			25			391			229
Arum maculatum L. (Cuckoopint)											
1	Club	Enzym.					13.2	156.8		23.2	230
1	–, cytosol and vacuole[b]	Enzym.			130			11			231
Avena sativa L. (Oats)											
1	Seed	Enzym.						0		0	232
1	1-day seed embryo[f]	LePage						102			14
1	Leaf mesophyll protoplast, chloroplast[b]	Enzym.			42			43			231
1	–, mitochondria[b]	Enzym.			28			61			231
1	–, androecium[a]	PC/PC								1040	233
1	–, gynoecium[a]	PC/PC								1760	233
1	Immature fruit[a]	PC/PC								1200	233
1	Mature fruit[a]	PC/PC								0	233
1	Root[a]	PC/PC								800	234
1	Seed, hypocotyl[a]	PC/PC								5600	234

1	–, cotyledon[a]	PC/PC								912	234
Bauhinia purpurea L. (Purple Bauhinia)											
1	Leaf[a]	PC/PC								800	233
1	Flower, calyx[a]	PC/PC								1440	233
1	–, corolla[a]	PC/PC								1600	233
Bauhinia variegata L. (Orchid Tree)											
1	Flowering bud[a]	PC/PC								1140	235
1	Leaf [a]	PC/PC								720	233
1	Flower, corolla[a]	PC/PC								1140	233
1	–, androecium[a]	PC/PC								2720	233
1	–, other parts[a]	PC/PC								0	233
1	Immature seed[a]	PC/PC								400	233
Beta vulgaris (L.) Banks (Sugar Beet)											
1	Root, stored	LePage						240			15
1	–, vegetating	LePage						260			15
1	Leaf	LePage						1040			15
1	Leaf[b,d]	Enzym.			51			432			236
Beta vulgaris L. ssp. *rapacea* (Koch) (Döll Forage Beet)											
1	Leaf[g]	Enzym.						2110			237
Brassica campestris L. (Turnip Rape)											
1	Pod, deseeded	Enzym.	14.3	17.3				830			238
Brassica napus L. (Rape)											
1	Leaf[e]	Enzym.		9.5				81.9			239
Brassica pekinensis (Lour.) Rupr. (Pe-tsai)											
1	Petiole vascular bundle	PC/PC						+			36
1	Petiole vascular bundle	TLC/TLC						+		+	36
1	Petiole vascular bundle	TLC/TLE						+		+	36
Brassica rapa L. (Turnip Rape)											
1	Vascular bundle	TLC/TLC						+			240
Cajanus cajan (L.) Mill. (Pigeon Pea)											
1	Leaf[a]	PC/PC								1920	241
1	Open flower[a]	PC/PC								1440	241
1	Young flower[a]	PC/PC								1300	241
Capsicum annuum L. (Bell Pepper)											
1	Leaf	Enzym.	266	178			88	86		341	242
Catharanthus roseus (L.) G. Don (Madagascar Periwinkle)											
1	Suspension cells	Enzym.	10.2	20.4			19.7	25.2	n.d.	28.1	243

*nmol/g fresh weight basis unless stated otherwise; [a]μg/g; [b]nmol/mg chlorophyll; [c]mg C/dm^2; [d]Basis unstated; [e]μmol/m^2; [f]nnmol/plant part; [g]Dry weight basis; [h]nmol/mL; [i]1,3-DPGA; [j]μmol/g; [k]conc., m*M*; [l]μmol/mg protein; [m]Percentage by weight; [n]Including 2-PGA.

Continued

Table 1-5 Triose Phosphates in Plants*—cont'd

No.	Plant part	Method	G3P	DHAP	TP	α-GP	2-PGA	3-PGA	DPGA	PEP	References
Centaurea moschata L. (Sweet Sultan)											
1	Leaf[a]	PC/PC								1740	244
Chloris gayana Kunth. (Rhodes Grass)											
1	Leaf	Enzym.						2800		140	245
Cicer arietinum L. (Chickpea)											
1	Seed	Enzym.							<100		246
1	48-h seedling, cotyledon[d,f]	Enzym.	5.2	34.5						41.7	247
1	–, embryonic axis[d,f]	Enzym.	1.2	9.2						4.2	247
Citrus sinensis L. (Orange)											
1	Fruit juice[d]	IEC		+		+		+	+	+	248
Coriandrum sativum L. (Coriander)											
1	Leaf[b]	Enzym.		88.8							249
Cucumis sativus L. (Cucumber)											
1	Leaf	IEC						144			250
Cucurbita maxima Duch. (Hubbard Squash)											
1	Sieve tube exudate	TLC/TLC						+			240
1	Ripe fruit exudate[d,h]	Enzym.	12	31			69	253		243	251
1	Seedling cotyledon	Enzym.			46		37	235		111	252
Cucurbita pepo L. (Pumpkin)											
1	Stem exudate[d,h]	Enzym.	38	47			71	277		104	251
1	Unripe fruit exudate[d,h]	Enzym.	15	20			27	112		36	251
Daucus carota L. (Carrot)											
1	Root	Enzym.		2.5				28			253
1	Root	Enzym.			10.4						254
Distichlis spicata (L.) Greene (Coastal Saltgrass)											
1	Leaf	GLC				420					255
Erythronium japonicum Decne (Katakuri)											
1	Leaf[e]	Enzym.						102.8			256
Fagopyrum esculentum L. (Buckwheat)											
1	Seedling	Enzym.		3		30		25		8	257
Fragaria spp. (Strawberry)											
1	Leaf	Enzym.		5.5				27.5			258
1	Leaf	Enzym.						22.9			259
Glycine max (L.) Merr. (Soybean)											
1	Seed	Enzym.								2.8	260

1	Leaf[d,e]	Enzym.			26			123			261
1	Seedling	Enzym.			85						262
Gossypium barbadense L. (Sea Island Cotton)											
1	Leaf[a]	PC/PC						2.25			263
Gossypium hirsutum L. (Upland Cotton)											
1	Leaf, glandless	Enzym.						570			264
Helianthus annuus L. (Sunflower)											
1	Leaf[e]	Enzym.		14.2				85.5			239
Helianthus tuberosus L. (Jerusalem Artichoke)											
1	Tuber	Enzym.						101		23.8	265
Hemerocallis hybrid (Daylily)											
1	Flower petal	TLC/TLC				+		+		+	266
Hordeum distichon Alef. ssp. *nutans* (Two-Rowed Barley)											
1	Sprout	Enzym.	1800	13,500							267
Hordeum vulgare L. (Barley)											
1	Leaf[b,d]	Enzym.			16		6	102		19	268
1	Leaf[b,d]	Enzym.		24					333		269
1	Leaf[b,d]	Enzym.				35					270
1	Leaf[b,d]	Enzym.						367			271
1	–, stroma[b,d]	Enzym.						217			271
1	–, cytosol[b,d]	Enzym.						151			271
1	–, vacuole[b,d]	Enzym.						0			271
2	Seedling	TLC/TLC						+		+	272
Kalanchoe blossfeldiana v. Poelln. (Flaming Katy)											
1	Leaf[g]	Enzym.			60		50	50		200	273
Kalanchoë daigremontiana Ham. and Per. (Devil's Backbone)											
1	Leaf	Enzym.			8.5			16.5		26.5	274
1	Leaf[b]	Enzym.						355		70	275
Lactuca sativa L. (Lettuce)											
1	Seed[g]	Enzym.								80	276
Lens culinaris Medik. (Lentil)											
1	Seed	Enzym.							<100		246
1	Sprout	Enzym.							<500		246
Lycopersicon esculentum Mill. (Tomato)											
1	Leaf	Enzym.						25			277
1	Fruit	Enzym.			7.5					14	278
1	Green fruit	Enzym.								17.5	258

*nmol/g fresh weight basis unless stated otherwise; [a]μg/g; [b]nmol/mg chlorophyll; [c]mg C/dm^2; [d]Basis unstated; [e]μmol/m^2; [f]nnmol/plant part; [g]Dry weight basis; [h]nmol/mL; [i]1,3-DPGA; [j]μmol/g; [k]conc., m*M*; [l]μmol/mg protein; [m]Percentage by weight; [n]Including 2-PGA.

Continued

Table 1-5 Triose Phosphates in Plants*—cont'd

No.	Plant part	Method	G3P	DHAP	TP	α-GP	2-PGA	3-PGA	DPGA	PEP	References
Malus domestica Borkh. (Apple)											
1	Fruit, immature	Enzym.		3				11		7.5	258
2	Fruit, mature	Enzym.		2-3			1-2	7-10		3-4	279
Malus sylvestris (L.) Mill (Crabapple)											
1	Stem phloem tissue	PC/PC						+			227
Medicago sativa L. (Alfalfa)											
1	Root	PC/PC	+	+				+			280
1	Root nodule	Enzym.								160	281
1	Shoot	PC/PC	+	+				+			280
Mucuna pruriens (L.) D.C. (Cowitch Bean)											
1	Leaf[e]	Enzym.			15			46			282
Musa cavendishii Lam. ex Pax. (Dwarf Banana)											
1	Fruit[g]	Enzym.			60						283
1	Fruit, green	Enzym.		6.5						10	258
Nicotiana sylvestris Speg. and Comes (Woodland Tobacco)											
1	Leaf[e]	Enzym.						214			284
Nicotiana tabacum L. (Tobacco)											
1	Leaf	PC/PC				+					285
1	Leaf[d,e]	Enzym.		35			21	153		51	286
1	–, mature	Enzym.								59.2	287
1	–, senescent	Enzym.								16.9	287
1	Shoot	Enzym.						23.4			288
1	Root	Enzym.						13.4			288
Oryza sativa L. (Rice)											
1	Grain[d]	Enzym.	n.d.	85				820		180	289
1	Grain, germ-free	Enzym.							1000		246
1	–, germ	Enzym.							<100		246
Panicum coloratum Walt. (Kleingrass)											
1	Leaf	Enzym.								400	7
Persea americana Mill. (Avocado)											
1	Fruit, mesocarp[h]	Enzym.	33				23	17	17[i]	32	290
Phaseolus vulgaris (L.) Banks (Common Beans)											
1	Bean	Enzym.							<100		246

1	Leaf	IEC					1360				22
1	Leaf[d]	Enzym.			1450			5200			291
1	Seed cotyledon	^{31}P NMR				+					49
Phragmites communis Trin. (Water Reed)											
1	Leaf[b]	PC/PC		10				96		0.010	292
Picea abies (L.) Karst. (Norway Spruce)											
1	Needle (leaf)[b,g]	Enzym.			910			640			293
Pisum sativum L. (Peas)											
1	Leaf[b]	Enzym.		58.4							249
1	Leaf chloroplast	HPLC			+			+			106
1	Root[d,j]	Enzym.			10					19	294
1	Seed	IEC						160			24
1	Seed, round	Enzym.			20						295
1	Seed, wrinkled	Enzym.			55						295
1	Seed	Enzym.							<100		246
1	Seed	Enzym.		20			50	33		112	296
Populus deltoides Bartr. ex Marsh (Eastern Cottonwood)											
1	Leaf[b]	Enzym.			30			190			297
Populus gelrica (Houtz.) Houtz.											
1	Bark[d,k]	Enzym.	0.004								298
Populus nigra L. (Black Poplar)											
1	Bark[d,k]	Enzym.	0.003								298
Portulaca grandiflora Hook. (Rose Moss)											
1	Leaf[b]	PC/PC								0	244
Prunus armeniaca L. (Apricot)											
1	Leaf	TLC/TLC						+		+	37
Pyrus communis L. (Pear)											
1	Fruit	Enzym.			+		+	+	+	+	299
Quercus rubra L. (Red Oak)											
1	Leaf[e]	Enzym.					Data not given				282
Raphanus sativus L. (Radish)											
1	Leaf	Enzym.			42			95			300
Rhododendron ponticum L. (Common Rhododendron)											
1	Detached leaf	Enzym.						54		14	301
Ricinus communis L. (Castor Bean)											
1	Seed, endosperm[d,f]	Enzym.		0.010				0.070		0.041	302

*nmol/g fresh weight basis unless stated otherwise; [a]μg/g; [b]nmol/mg chlorophyll; [c]mg C/dm^2; [d]Basis unstated; [e]μmol/m^2; [f]nnmol/plant part; [g]Dry weight basis; [h]nmol/mL; [i]1,3-DPGA; [j]μmol/g; [k]conc., m*M*; [l]μmol/mg protein; [m]Percentage by weight; [n]Including 2-PGA.

Continued

Table 1-5 Triose Phosphates in Plants*—cont'd

No.	Plant part	Method	G3P	DHAP	TP	α-GP	2-PGA	3-PGA	DPGA	PEP	References
1	Seedling, cotyledon	Enzym.			14.3		76	680		315	303
1	–, hypocotyl	Enzym.			18			25		4	304
1	–, phloem exudate[h]	Enzym.			62			63		23	304
Saccharum officinarum L. (Sugarcane)											
1	Leaf	Enzym.								820	305
1	Leaf	PC/PC						+		+	306
1	Root	PC/PC						+		+	306
1	Stalk, third internode[l]	Enzym.			10.3					6.3	307
			and variation from third to ninth internode								
Saccharum officinarum × *S. spontaneum* × *S. barberi*											
1	Leaf[e]	Enzym.			63			171		94	308
Saccharum sinense Roxb.											
1	Leaf[f]	Enzym.								59	305
Salix triandra L. (Almond Willow)											
1	Phloem exudate	TLC/TLC						+		tr	309
Sesamum alatum Thonn.											
1	Leaf[c]	Enzym.			50.7			208.5			310
Sesasum indicum L. (Sesame)											
1	Leaf[c]	Enzym.			42.8			185.6			310
1	Leaf	Enzym.			19			230			311
1	–, stroma[b]	Enzym.		8				81			312
Solanum tuberosum L. (Potato)											
1	Tuber[g,m]	LePage			0.37						13
14	Tuber	PC/PC						+			313
1	Tuber[d]	PC/PC			2.5	2.2		12[n]		1.5	314
1	Tuber	PC				+		+			315
1	Tuber	IEC						147			316
1	Tuber	Enzym.							<500		246
1	Tuber	Enzym.			4.7						317
1	Mature tuber	Enzym.		1.5				25		13	258
1	–, cytosol[b]	Enzym.		8				22			312
1	Tuber	Enzym.		6			12	40		24	318
			and variation in TP, 3-PGA, and PEP levels along the stolon-apex axis								319
Sorghum bicolor (L.) Moench (Sorghum)											
1	Leaf[b]	Enzym.		150						250	320
1	Leaf	GLC				700					255

Spartina alterniflora Lois. (Smooth Cordgrass)											
1	Leaf	GLC				1480					255
Spinacia oleracea L. (Spinach)											
1	Root[d,j]	Enzym.			13					26	294
1	Leaf	PC/PC						+	+		321
1	Leaf	^{31}P NMR		+		0.5		+		+	50
1	Leaf[b]	Enzym.			35		13	500		31	322
1	Leaf[b,d]	Enzym.			35			373		26	323
1	Leaf[b,d]	Enzym.				32					270
1	Leaf[b,d]	Enzym.		~ 20							324
1	–, stroma[b,d]	Enzym.		16				268			325
1	–, cytosol[b,d]	Enzym.		6				66			325
1	Leaf chloroplast[b,d]	IEC			2.94				70.9		20
1	Leaf, protoplast[b]	Enzym.			44.3			89.7			326
1	–, chloroplast[b]	Enzym.			17.4			64.6			326
1	–, extrachloroplast[b]	Enzym.			33.4			17.0			326
Spirodela oligorrhiza (Kurz.) Hegelm. (Duckweed)											
1	Whole plant	PC/PC			15	15		230		37	327
Tilia platyphyllos Scop. (Large-Leaved Linden)											
1	Phloem exudate	TLC/TLC						+		+	328
Trifolium subterraneum L. (Sub Clover)											
1	Whole plant	IEC						2570			329
1	Seed germ	Enzym.							<500		246
1	6-day root[g]	Enzym.	6.7	29							330
1	6-day shoot[g]	Enzym.	1.6	37							330
Triticum aestivum L. (Wheat)											
1	Root[d,j]	Enzym.			39					13	294
1	Seed[g]	Enzym.	2.6	35.1							330
1	10-day leaf	PC/PE	+					+		+	331
1	10-day leaf	TLC/TLE						+		+	331
1	Leaf[b,d]	Enzym.			30			752		34	323
1	Leaf	GLC	n.d.(< 10)								255
Vicia faba L. (Broad Bean)											
1	Leaf	PC/PC						+	+		321
Vigna mungo (L.) Hepper (Black Gram)											
1	4-day seedling, hypocotyl tip	Enzym.	0.8	1.0							332
Vigna radiata (L.) Wilcz. (Mung Bean)											
1	Seedling	Enzym.			96						262
1	Leaf[c]	Enzym.		60.2							249

*nmol/g fresh weight basis unless stated otherwise; [a]μg/g; [b]nmol/mg chlorophyll; [c]mg C/dm^2; [d]Basis unstated; [e]μmol/m^2; [f]nmmol/plant part; [g]Dry weight basis; [h]nmol/mL; [i]1,3-DPGA; [j]μmol/g; [k]conc., m*M*; [l]μmol/mg protein; [m]Percentage by weight; [n]Including 2-PGA.

Continued

Table 1-5 Triose Phosphates in Plants*—cont'd

No.	Plant part	Method	G3P	DHAP	TP	α-GP	2-PGA	3-PGA	DPGA	PEP	References
Vitis vinifera L. (Wine Grape)											
1	Seedless berry	Enzym.			7					11	333
1	Fruit, juice	Enzym.		0							334
1	Leaf	Enzym.	3-PGA/TP ratio unchanged during 4 AM-6 PM photoperiod								335
Xanthium strumarium L. (Cocklebur)											
1	Leaf[b,d]	Enzym.			32			280			336
Yucca flaccida Haw (Weak-Leaf Yucca)											
1	Phloem exudate	TLC/TLC						+		+	309
Zea mays L. (Corn)											
1	Leaf[b]	PC/PE		45				185		49.5	337
1	Leaf[b,d]	Enzym.		234				429		59	338
1	Leaf[b,d]	Enzym.			233			341		51	339
1	–, midrib[b]	Enzym.			417					104	340
1	–, lamina[b]	Enzym.			350					116	340
1	Leaf, chloroplast[b]	Enzym.			131			461			341
1	–, cytosol[b]	Enzym.			279			349			341
1	–, vacuole[b]	Enzym.			12			5			341
1	Grain, endosperm[g]	Enzym.	0.8	1.3				20.7		3.2	342
1	–, starch granule[g]	Enzym.	0.06	0.4				1.6		0.7	342
1	–, germ	Enzym.							<100		246
1	Flower, ovary[h]	Enzym.						28			343
1	Root	PC/PC	+							+	12
1	Root	TLC/TLC	+							+	12

*nmol/g fresh weight basis unless stated otherwise.
[a]µg/g.
[b]nmol/mg chlorophyll.
[c]mg C/dm^2.
[d]Basis unstated.
[e]µmol/m^2.
[f]nnmol/plant part.
[g]Dry weight basis.
[h]nmol/mL.
[i]1,3-DPGA.
[j]µmol/g.
[k]conc., m*M*.
[l]µmol/mg protein.
[m]Percentage by weight.
[n]Including 2-PGA

consecutively, together with F6P and F1,6P, in a single metabolically related sequence (Fig. 1-2):

> *DHAP was assayed in the presence of 60 μmoles of tris-chloride, pH 7.6; 0.2 μmole of NADH in 20 mM EDTA and sample. The reaction, in a total volume of 1.04 ml, was started by the addition of 20 IU of α-glycerophosphate dehydrogenase. After completion, [G3P] was assayed by the addition of 60 IU of triose-P isomerase.*
>
> Steer, B. T., "Control of Diurnal Variations in Photosynthetic Products. I. Carbon Metabolism", *Plant Physiol.* **1974** *54*(5), 758–761,[242] reprinted by permission of the American Society of Plant Biologists, © 1974.

G3P and DHAP are often assayed together as triose phosphate (TP) using a mixture of α-GP dehydrogenase and TP isomerase in the procedure above. To assay α-GP, the procedure is reversed (Table 1-3, reaction 3). α-GP is oxidized to DHAP by NAD in the presence of α-GP dehydrogenase and hydrazine; the amount of α-GP is measured by the increase in absorbance of NADH at 340 nm. PEP, 2-PGA, and 3-PGA may also be analyzed consecutively in reactions coupled to the oxidation of NADH to NAD (Table 1-3, reaction 5).

> *Pyruvate was assayed using a reaction mixture containing 60 mmoles of tris-chloride [buffer], pH 7.6; 60 mmoles of KCl; 0.1 mmole of $MnSO_4$; 0.2 mmole of ADP; 0.2 mmole of 2,3-diphosphoglycerate; 0.2 mmole of NADH in 20 mM EDTA and sample. The reaction was started by adding 86 IU of lactic dehydrogenase and recording changes in absorbance at 340 nm. The total volume was 1.13 ml. The PEP, 2PGA, and 3PGA were determined sequentially by adding 37 IU of pyruvate kinase, 13 IU of enolase, and 5 IU of phosphoglycerate mutase, respectively.*
>
> Steer, B. T., "Control of Diurnal Variations in Photosynthetic Products. I. Carbon Metabolism", *Plant Physiol.* **1974** *54*(5), 758–761,[242] reprinted by permission of the American Society of Plant Biologists, © 1974.

An alternative sequence for 3-PGA, preferred by some because two moles of NADH are oxidized per mole of 3-PGA, involves a two-step reduction of 3-PGA to α-GP via G3P and DHAP (Table 1-3, reactions 2 and 4).

1,3-DPGA and 2,3-DPGA

The record on these substances in plants is sparse. An unidentified DPGA was detected in extracts of spinach and broad bean leaves by two-dimensional PC[321] and in extracts of ^{32}P-labeled maize roots by one-dimensional PC.[344] 1,3-DPGA, a Calvin cycle metabolite, has been detected in Bartlett pear fruit[299] but not in *C. roseus* suspension cells.[243] 2,3-DPGA, a cofactor for PGA mutase, has been detected in orange juice[248] and quantitated in a variety of plant extracts.[246] Both substances were assayed by enzymatic methods: 1,3-DPGA by oxidation to G3P with NADH in the presence of G3P dehydrogenase (Table 1-3, reaction 4), and 2,3-DPGA by a colorimetric method based on its stimulation of PGA mutase (Fig. 1-2, step 11).

Tetrose Phosphates

This group is represented by a single aldose phosphate, E4P.

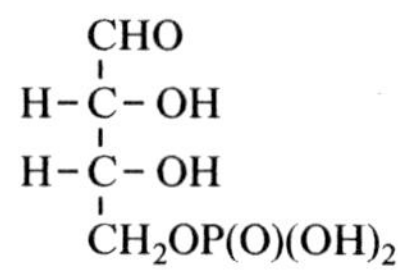

D-Erythrose 4-Phosphate (E4P), $C_4H_9O_7P$; Beilstein **1**, E_{IV} 4175. *(R*,R*)-2,3-Dihydroxy-4-(phosphonooxy)butanal* [1961].

Occurrence

E4P is a metabolite of the photosynthetic carbon reduction cycle (Calvin cycle) of higher plants (Fig. 1-3). In step 1, F6P transfers a two-carbon ketol segment to G3P, forming E4P and Xu5P. In step 2, the E4P thus formed undergoes an aldol condensation with another molecule of G3P, giving Se1,7P. These reactions occur within the chloroplast, found only in the green leaf. E4P is also a metabolite of the oxidative pentose phosphate pathway. In step 3, Se7P transfers a three-carbon ketol segment to G3P, forming E4P and F6P. In step 4, the E4P undergoes a similar reaction with Xu5P, forming F6P and G3P. These reactions take place in the cytoplasm, and in the chloroplast during periods of darkness. E4P is required for the synthesis of shikimic acid, a precursor of the phenolics and the aromatic amino acids.

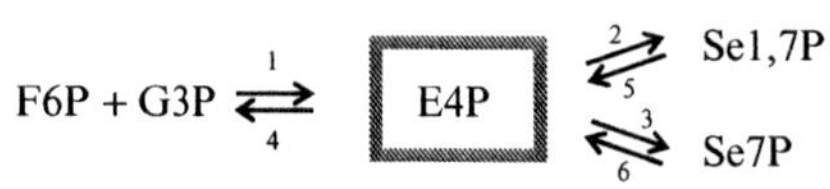

FIGURE 1-3 Tetrose Phosphates. (1, 4) Transketolase [E.C. 2.2.1.1], TPP, Mg^{2+}. (2,5) Aldolase [E.C. 4.1.2.13]. (3,6) Transaldolase [E.C. 2.2.1.2].

Physical Properties

D-Erythrose 4-Phosphate

E4P [R.N. 19234-99-2; see also 585-18-2]: $[\alpha]_D$ +0.5°[30]; $\Delta G°$ −142.7 kcal/mol.[84] Spectra: ^{1}H NMR (fig.),[97] ^{13}C NMR,[345] ^{31}P NMR (fig.)[97]; GC-MS.[96] Chromatography: PC,[346] TLC,[347] GC,[143] IEC,[347] HPLC.[106] Electrophoresis: TLE,[105] zone.[88] E4P is over 90% hydrated (to the *gem*-diol) in aqueous solution.[348] It tends to dimerize during evaporation.[349]

$K_2C_4H_7O_7P$: $[\alpha]_D$ −4.7°.[30]

Table 1-6 Tetrose Phosphates in Plants*

No.	Plant part	E4P	References
Populus gelrica (Houtz.) Houtz.			
1	Bark[a,b]	0.003	298
Populus nigra L. (Black Poplar)			
1	Bark[a,b]	0.020	298
Vigna mungo (L.) Hepper (Black Gram)			
1	4-day seedling, hypocotyl tip	2.3	332

*nmol/g fresh weight basis unless stated otherwise.
[a]Conc., m*M*.
[b]Basis unstated.

Synthesis

E4P may be prepared (a) by phosphorylation of 2,3-di-*O*-benzoyl-D-erythrose dimethyl acetal with diphenyl phosphorochloridate[350] or (b) by cleavage of G6P with lead tetraacetate.[349] It is best stored as the penultimate products of synthesis: dimethyl acetal cyclohexylamine salt[350] or hydrazide barium salt.[349]

Hydrolysis

E4P is rapidly hydrolyzed by 1*N* NaOH at room temperature,[350] but is stable to 0.1*N* HCl and to buffers of pH 3 to 5 at 37 °C.[351] Its half-life in 1*N* H_2SO_4 at 100 °C is 20 min.[350]

Analysis

The estimation of E4P is based on enzymatic methods. In the presence of F6P, E4P is converted quantitatively by transaldolase to Se7P and G3P (Fig. 1-3, reaction 6) and assayed either as G3P (Table 1-3, reaction 1)[298] or as DHAP (Table 1-3, reaction 2)[332] by measuring the change in optical density at 340 nm.

Pentose Phosphates

This group comprises one aldose phosphate (R5P) and three ketose phosphates (Ru5P, Ru1,5P, and Xu5P).

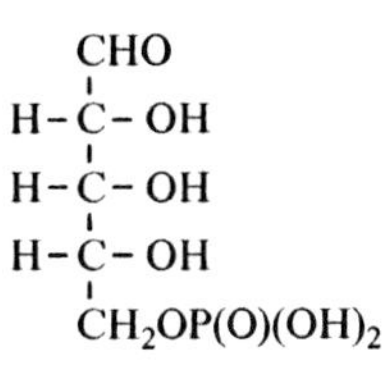

R5P

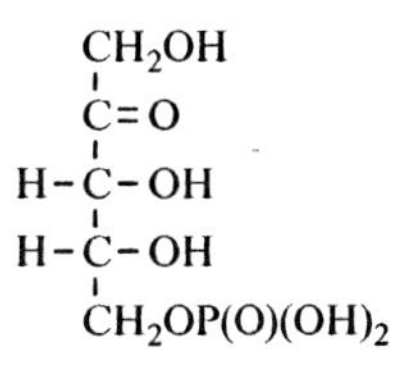

Ru5P

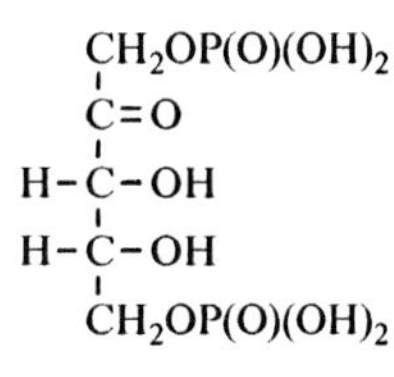

Ru1,5P

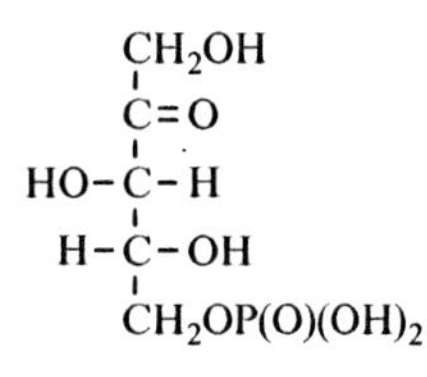

Xu5P

D-Ribose 5-Phosphate (R5P), $C_5H_{11}O_8P$; Beilstein **1**, H 859, E_I 434, E_{IV} 4245, **31**, 21. *D-Ribose 5-(Dihydrogen Phosphate)* [1953].

D-Ribulose 5-Phosphate (Ru5P) $C_5H_{11}O_8P$; not listed in Beilstein. *D*-erythro-*2-Pentulose 5-(Dihydrogen Phosphate)* [1956].

D-Ribulose 1,5-Bisphosphate (Ru1,5P), $C_5H_{12}O_{11}P_2$; not listed in Beilstein. *D*-erythro-*2-Pentulose 1,5-Bis(dihydrogen Phosphate)* [1963].

D-Xylulose 5-Phosphate (Xu5P), $C_5H_{11}O_8P$; not listed in Beilstein. *D*-threo-*2-Pentulose 5-(Dihydrogen Phosphate)* [1977].

Occurrence

R5P, Ru5P, Ru1,5P, and Xu5P are metabolites of the photosynthetic carbon reduction cycle (Calvin cycle) of higher plants (Fig. 1-4, reading clockwise). In steps 1 and 2, F6P or Se7P transfer a two-carbon ketol segment to G3P, forming E4P (or R5P) and Xu5P. In step 3, R5P is converted to its isomer Ru5P. In step 4, Ru5P is phosphorylated by ATP, completing the cycle and regenerating Ru1,5P for further CO_2 fixation (step 5). In step 6, Xu5P is converted to its epimer Ru5P by the enzyme Ru5P 3-epimerase. All of these reactions occur within the chloroplast, found only in the green leaf.

R5P, Ru5P, and Xu5P are also metabolites of the pentose phosphate respiratory pathway (Fig. 1-4, reading counterclockwise). In step 7, 6-PG is oxidized by NADP, giving Ru5P, NADPH, and CO_2. In steps 8 and 9, Ru5P is converted to its isomer R5P and epimer Xu5P, respectively. In step 10, R5P and Xu5P interact to form Se7P and G3P. In step 11, Xu5P interacts with E4P to form F6P and G3P. These reactions all take place in the leaf cytoplasm outside any organelle and in chloroplasts in periods of darkness.

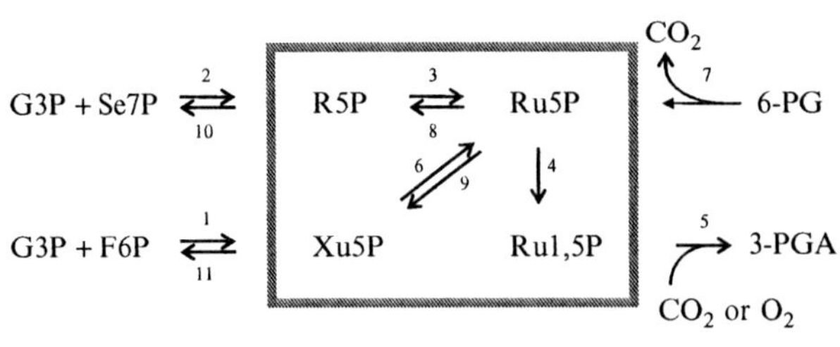

FIGURE 1-4 Pentose Phosphates. (1, 2) Transketolase [E.C. 2.2.1.1]. (3) R5P isomerase [E.C. 5.3.1.6]. (4) Ru5P kinase [E.C. 2.7.1.19], ATP. (5) Ru1,5P carboxylase-oxygenase [Rubisco, E.C. 4.1.1.39], CO_2 or O_2, H_2O, Mg^{2+}. (6) Ru5P epimerase [E.C. 5.1.3.1]. (7) 6-PG dehydrogenase [E.C. 1.1.1.44], NADP.

Physical Properties

D-Ribose 5-Phosphate

R5P [R.N. 4300-28-1; see also 3615-55-2]: White powder,[345] $[\alpha]_D^{20}$ +22.8° (0.2*N* HCl)[352]; pK_a 6.24,[353] 13.05 (thermometric, 25 °C)[354]; conformation[355]; heat of ionization ($\Delta H°$, $\Delta S°$).[354] Spectra: Far-UV (fig.),[356] FTIR,[357] Raman (fig.),[358] GC-MS,[96] ^{1}H NMR (fig.),[359] ^{13}C NMR (fig.),[85] ^{31}P NMR (fig.).[360] Chromatography: PC,[9,32] TLC,[105] GC,[143] IEC,[85,361]

HPLC,[26,144] IC.[107] Electrophoresis: PE,[32] TLE,[105] zone,[88] CZE,[136] CE,[90] isotachophoresis.[88]
$Na_2C_5H_9O_8P$ [R.N. 18265-46-8]: Colorless hygroscopic solid,[360] $[\alpha]_D^{22}$ +14.8° (H_2O).[362]
$BaC_5H_9O_8P{\cdot}5H_2O$ [R.N. 15673-79-7]: Monoclinic,[363] colorless hexagonal plates,[364] d 2.05,[363] $[\alpha]_D^{26}$ +5.99° (H_2O).[364] Spectra: Far-UV (fig.),[365] X-ray.[363]

D-Ribulose 5-Phosphate

Ru5P [R.N. 551-85-9; see also 4151-19-3]: $[\alpha]_D^{20}$ −40° (0.2*N* HCl)[352] ΔG° −179.6 kcal/mol.[84] Spectra: UV,[366] CD,[366] ORD,[366] GC-MS,[96] ^{13}C NMR (fig.),[85] ^{31}P NMR.[68] Chromatography: PC,[367] TLC,[105] IEC,[352,361] HPLC,[368] HPLC-MS-MS.[100] Electrophoresis: PE,[367] TLE,[105] zone,[88] isotachophoresis.[88]
$Li_2C_5H_9O_8P$: White nonhygroscopic powder.[369]
$BaC_5H_9O_8P$: Precipitate, sol. H_2O, insol. EtOH.[352]

D-Ribulose 1,5-Bisphosphate

Ru1,5P [R.N. 24218-00-6; see also 2002-28-0]: Spectra: UV (fig.),[370] IR (fig.),[103] GC/MS,[96] ^{1}H NMR (fig.),[103] ^{13}C NMR (fig.),[85] ^{31}P NMR.[68] Chromatography: PC,[9] TLC,[33] IEC,[371] HPLC,[26,144] IC (fig.).[107] Electrophoresis: PE,[9] TLE,[105] zone,[88] CE,[90] isotachophoresis.[88]
$Ba_2C_5H_8O_{11}P_2$ [R.N. 82130-67-4]: Solid, sol. H_2O, insol. EtOH.[371] Stable for months at −20 °C but isomerizes to xylulose 1,5-bisphosphate if stored at room temperature.[372]

D-Xylulose 5-Phosphate

Xu5P [R.N.4212-65-1; see also 28848-87-5]: Pale yellow syrup, $[\alpha]_D^{25}$ −6.5° (H_2O),[373] ΔG° −179.6 kcal/mol.[84] Spectra: UV,[366] ^{1}H NMR,[373] ^{13}C NMR,[374] GC-MS.[96] Chromatography: PC,[367] IEC.[375] Electrophoresis: PE.[367]

Synthesis

R5P may be prepared (a) by phosphorylation of methyl 2,3-isopropylidene-β-D-ribofuranoside with $POCl_3$/pyridine[364] or dibenzyl phosphorochloridate,[376] (b) by acid-catalyzed hydrolysis of AMP with Dowex 50 at 100 °C,[377] or (c) by enzymatic phosphorylation of D-ribose with ribokinase and ATP.[360] The relative merits of these methods has been discussed.[360] Ru5P may be prepared (a) by enzymatic phosphorylation of D-ribulose with isomerase-free ribulokinase and ATP,[378] (b) by isomerization of R5P with R5P isomerase,[25] or (c) by oxidation of 6-PG with crystalline 6PG dehydrogenase.[369] Ru1,5P may be prepared (a) by phosphorylation of R5P with phosphoribulokinase and ATP in the presence of phosphoriboisomerase[371] or (b) by the combined action of four immobilized enzymes on G6P.[360] Xu5P may be prepared (a) by enzymatic phosphorylation of D-xylulose with xylulokinase and ATP,[379] (b) by isomerization of Ru5P with 3-epimerase,[380] or (c) by transketolase-catalyzed condensation of hydroxypyruvate with G3P.[25]

Structure

An aqueous solution of R5P at pH 5.5 contains predominantly α-furanose (**5**, 35.6%) and β-furanose (**6**, 63.9%) forms, with only 0.1% free aldehyde and 0.5% *gem*-diol. The rates of ring-opening at pH 7.5 and 40 °C are 100 and 40 s^{-1}, respectively, and are catalyzed by acid below pH 2 and by base above pH 7[348]:

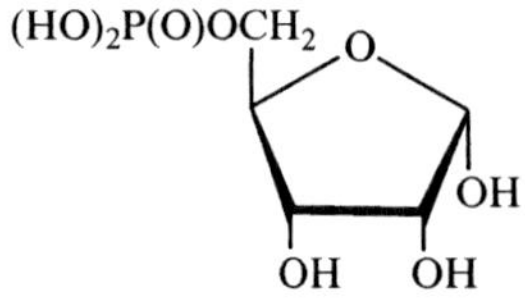

5, R.N. 34980-65-9

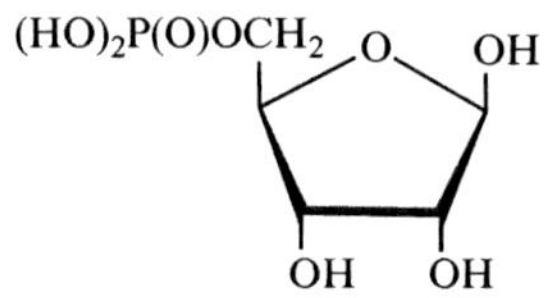

6, R.N. 34980-66-0

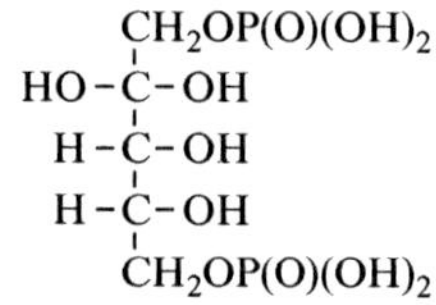

7

Aqueous solutions of Ru1,5P contain 12% (NMR)[85] to 16% (IR)[103] of *gem*-diol **7** in addition to the free ketose. Ru5P solutions also appear to contain predominantly free ketose.[85] Ru1,5P tends to isomerize to Xu1,5P when exposed to solutions of pH > 8.[360]

Xu5P is unstable in buffer solutions in the pH 5.3-8.4 range[381]; aqueous solutions should be maintained at pH 5.0 or below to avoid isomerization.[382]

Hydrolysis

Hydrolysis of R_5P with 0.01*N* HCl at 100 °C liberated 73.2% of the P after 28 h[364]; its half-life in 0.1*N* NaOH at 22 °C was about 20 h.[377] For the effect of pH over the range 1.5-12 on the hydrolysis of R5P at 100 °C, see Dimroth *et al.*[205]

Ru5P is less stable to acid than R5P; its half-life in 1*N* H_2SO_4 at 100 °C is about 60 min.[352] Ru5P is unstable above pH 7.[381] Calorimetric measurements gave $\Delta H°$ −5.69 kJ/mol (enthalpy) and $\Delta C°_P$ −63 J/(mol K) (heat capacity) for the hydrolysis of R5P with alkaline phosphatase. The corresponding data for Ru5P were −12.43 kJ/mol and −84 J/(mol K), respectively.[383]

The hydrolysis of Ru1,5P with 1*N* H_2SO_4 at 100 °C is first order with no change in slope, indicating that the two phosphate groups are equally labile. The half-life is 20 min.[371] Hydrolysis in 1*N* NaOH is also first-order ($t_{1/2}$ 2.8 min)[371] but only one phosphate group is labile.[384] The C_1 phosphate group is lost by β-elimination, followed by further degradation.[385] For the kinetics of hydrolysis over the pH 2.0-13.0 range, see Haga *et al.*[384]

Phosphorylation

Phosphorylation of R5P with phosphoric acid/DCC in the presence of pyridine gives D-ribose-5-di- and -triphosphate, separated by IEC in yields of 20-25% and 10%, respectively.[386]

Separation and Analysis

R5P and Ru1,5P can be separated from other phosphate esters by IEC, by HPLC, or by two-dimensional PC, TLC, or PC/PE after extraction with perchloric acid or trichloroacetic acid. Quantitative data were reported for Ru1,5P in some cases. These methods are applicable in principle to the pentulose phosphates Ru5P and Xu5P, but Table 1-7 shows neither.

Table 1-7 Pentose Phosphates in Plants*

No.	Plant part	Method	R5P	Ru5P	Ru1,5P	Xu5P	References
Amaranthus caudatus L. (Love-Lies-Bleeding)							
1	Leaf[a,b]	Enzym.			70		224
Arbutus unedo L. (Strawberry Tree)							
1	Leaf[a,b]	Enzym.			91		229
Beta vulgaris L. Sugar Beet							
1	Leaf[a,b]	Enzym.			140		236
Brassica campestris L. (Turnip Rape)							
1	Pod, deseeded	Enzym.			3148		238
Brassica napus L. (Rape)							
1	Leaf[c]	Enzym.			41.6		239
Erythronium japonicum Decne. (Katakuri)							
1	Leaf[c]	Enzym.			10.3		256
Glycine max (L.) Merr. (Soybean)							
1	Leaf[a,c]	Enzym.	80[d]		57		389
1	Leaf[b]	HPLC	+		210		26
Gossypium hirsutum L. (Upland Cotton)							
1	Leaf, glandless	Enzym.			210		264
Helianthus annuus L. (Sunflower)							
1	Leaf[c]	Enzym.			54.3		239
Hordeum vulgare L. Barley							
1	Leaf[a,b]	Enzym.			63		269
2	Seedling	TLC/TLC			+		272
Kalanchoë daigremontiana Ham. and Perr. (Devil's Backbone)							
1	Leaf[b]	Enzym.			50		275
Lycopersicon esculentum Mill. (Tomato)							
1	Leaf[b]	Enzym.			165		390
Mucuna pruriens (L.) DC (Cowitch Bean)							
1	Leaf[c]	Enzym.			22		282
Nicotiana sylvestris Speg. and Comes (Woodland Tobacco)							
1	Leaf[c]	Enzym.			21		284
Phaseolus vulgaris L. Beans							
1	Leaf[a]	Enzym.			800		291
Phragmites communis Trin. (Water Reed)							
1	Leaf[b]	PC/PC			12		292
Pisum sativum L. (Peas)							
1	Leaf chloroplast	HPLC			+		106
Populus gelrica (Houtz.) Houtz.							
1	Bark[a,e]	Enzym.	0				298
Populus nigra (Black Poplar)							
1	Bark[a,e]	Enzym.	0.017				298
Pyrus communis L. (Pear)							
1	Fruit skin	IEC	+				19
Quercus rubra L. Red Oak							
1	Leaf[c]	Data not given					282
Raphanus sativus L. (Radish)							
1	Leaf[a,c]	Enzym.			92		11

Table 1-7 Pentose Phosphates in Plants*—cont'd

No.	Plant part	Method	R5P	Ru5P	Ru1,5P	Xu5P	References
Secale cereale L. (Rye)							
1	Leaf chloroplast, diploid[f]	Enzym.	48.0				391
1	–, tetraploid[f]	Enzym.	34.9				391
Solanum tuberosum L. (Potato)							
1	Tuber	PC/PC	+				392
Spinacia oleracea L. (Spinach)							
1	Leaf	PC/PC			+		321
1	Leaf[a,b]	Enzym.			77.2		393
1	–, stroma[a,b]	Enzym.			183		325
1	–, chloroplast[a,b]	IEC	1.08		9.43		20
1	Leaf protoplast[b]	Enzym.			21.5		326
1	–, chloroplast[b]	Enzym.			21.9		326
1	–, extrachloroplast[b]	Enzym.			0		326
Triticum aestivum L. (Wheat)							
1	Leaf[a,b]	Enzym.			50		387
1	10-day leaf	PC/PE	+				331
Triticum durum Desf. (Durum Wheat)							
1	Leaf[b]	Enzym.			333		394
Vicia faba L. (Broad Bean)							
1	Leaf	PC/PC			+		321
Vigna mungo (L.) Hepper (Black Gram)							
1	4-day seedling, hypocotyl tip	Enzym.	5.8	0.9		1.4	332
Xanthium strumarium L. (Cocklebur)							
1	Leaf[a,b]	Enzym.			100-120		395
Zea mays L. (Corn)							
1	Leaf[b]	PC/PC			11.4		337
1	Leaf[a,b]	Enzym.			56		338

*nmol/g fresh weight basis unless stated otherwise.
[a]Basis unstated.
[b]nmol/mg chlorophyll.
[c]μmol m^2.
[d]Including Ru5P.
[e]Conc., m*M*.
[f]μg/g.

Analysis Without Separation

(1) Enzymatic Method

The pentose phosphates may be determined by enzymatic reactions coupled to either the oxidation of G3P to 3-PGA by NAD (Table 1-3, reaction 1) or the reduction of 1,3-DPGA to G3P by NADH (Table 1-3, reaction 4). In either case, the amount of pentose phosphate is calculated from the change in optical density of the NAD(H) at 340 nm.

The first method (oxidation) was used for the determination of R5P in poplar bark[298] and for the sequential determination of R5P, Ru5P, and Xu5P in *Vigna mungo* seedlings.[332]

The second method (reduction) was used for the determination of Ru1,5P in turnip rape,[238] velvet bean,[282] red oak,[282] wheat,[387] corn,[338] and cotton.[264]

A radiometric method based on the following reaction, introduced by Latzko and Gibbs[388] in 1969, has become the method of choice for the determination of Ru1,5P in plant extracts because it employs a single enzyme (Rubisco) instead of three, and is capable of measuring pools of 1 n*M* or less[5b]:

$$\mathrm{Ru1,5P} + {}^{14}\mathrm{CO_2} + \mathrm{H_2O} \xrightarrow{\text{Rubisco}} 3-\mathrm{PGA} + [{}^{14}\mathrm{C}]3-\mathrm{PGA}$$

Ribulose-1,5-bisphosphate was determined by ^{14}C incorporation into PGA using purified RuBP carboxylase [E.C. 4.1.1.39].... Assays were run in 300 ml total volume (100 mM N,N-bis(2-hydroxyethyl)glycine [Bicine], pH 8.0, 20 mM $MgCl_2$, 13 mM [^{14}C]$NaHCO_3$ [40-50 Bq $nmol^{-1}$], 50 mg carboxylase) with 100 ml of sample per assay. Duplicate assays were run for 30 min at room temperature; the reactions were terminated by adding 0.2 ml 2N *formic acid. The samples were dried in an air stream and counted by liquid scintillation spectrometry.*

Badger, M. R.; Sharkey, T. D.; von Caemmerer, S., "The Relationship between Steady-State Gas Exchange of Bean Leaves and the Levels of Carbon-Reduction-Cycle Intermediates", *Planta* **1984**, *160*(4), 306–315,[291] reprinted by permission of Springer, © 1984.

R5P, Ru5P, and Xu5P could also be determined by the $^{14}CO_2$ method with the aid of the appropriate enzymes,[5b] but this method has not yet been applied to the higher plants.

Hexose Phosphates

This large group comprises primarily the 1- and 6-phosphates of glucose and fructose and to a lesser extent the corresponding mannose and galactose phosphates. The glucose phosphates (G1P, G6P, and G1,6P), including reduced glucose phosphates (So1P and So6P) and oxidized glucose phosphates (6-GP and its lactones), are considered first:

Glucose Phosphates

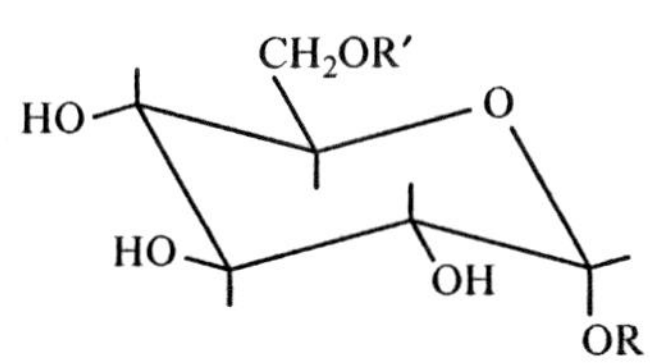

G1P, R′ = H, R = $P(O)(OH)_2$
G6P, R = H, R′ = $P(O)(OH)_2$
G1,6P, R = R′ = $P(O)(OH)_2$

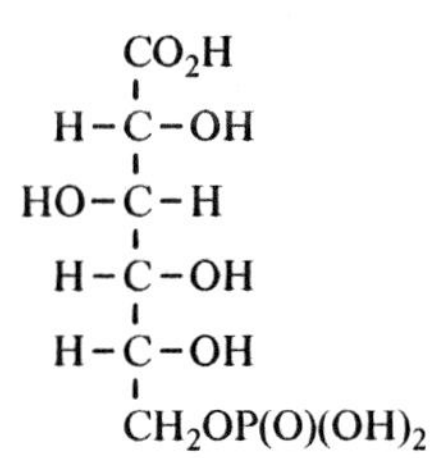

6-PG

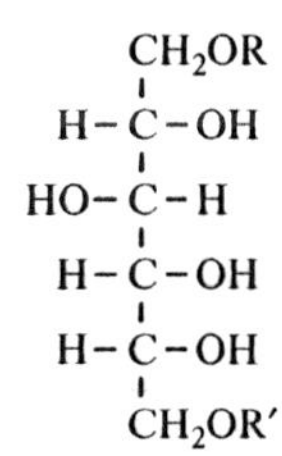

So1P, R′ = H, R = $P(O)(OH)_2$
So6P, R = H, R′ = $P(O)(OH)_2$

D-Glucose 1-Phosphate (G1P), $C_6H_{13}O_9P$; not listed in Beilstein. Cori ester, *α-D-Glucopyranose 1-(Dihydrogen Phosphate)* [1944].

D-Glucose 6-Phosphate (G6P), $C_6H_{13}O_9P$; Beilstein **1**, E_{IV} 4381. Robison ester, *D-Glucose 6-(Dihydrogen Phosphate)* [1949].

D-Glucose 1,6-Bisphosphate (G1,6P), $C_6H_{14}O_{12}P_2$; not listed in Beilstein. *α-D-Glucopyranose 1,6-Bis(dihydrogen Phosphate)* [1963].

D-6-Phosphogluconic Acid (6-PG), $C_6H_{13}O_{10}P$; Beilstein **3**, E_{III} 1063, E_{IV} 1258. *D-Gluconic Acid 6-(Dihydrogen Phosphate)* [1963].

Sorbitol 1-Phosphate (So1P), $C_6H_{15}O_9P$: not listed in Beilstein. *D-Glucitol 1-(Dihydrogen Phosphate)* [1978].

Sorbitol 6-Phosphate (So6P), $C_6H_{15}O_9P$: Beilstein **1**, E_{IV} 2849. *D-Glucitol 6-(Dihydrogen Phosphate)* [1977].

Occurrence

G1P and G6P are metabolites of the Embden-Meyerhof-Parnas glycolysis pathway, a sequence common to all of the higher plants (Fig. 1-5). G1P, formed from starch by the action of starch phosphorylase (step 1), is converted to G6P by phosphoglucomutase (step 2). The next step is the isomerization of G6P to F6P (step 3). These reactions take place in the cytoplasm outside any organelle. G6P and 6-PG are also metabolites of the hexose monophosphate shunt, a part of the pentose phosphate respiratory pathway in which G6P is oxidized via 6-PG lactone to 6-PG (step 12) and then to Ru5P with the loss of

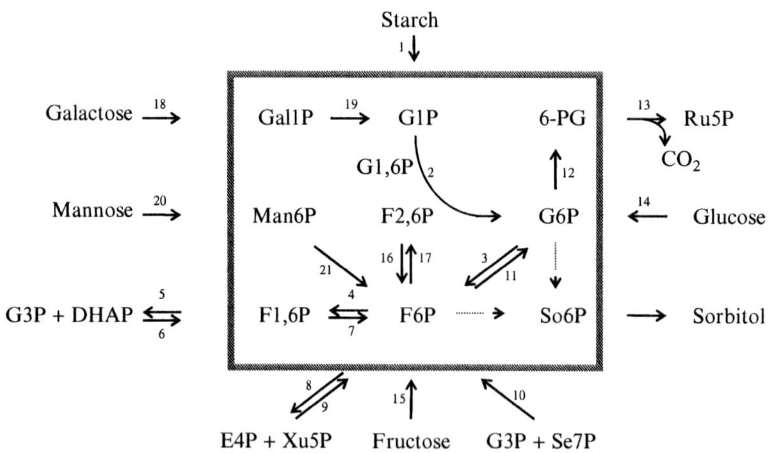

FIGURE 1-5 Hexose Phosphates. (1) Starch phosphorylase [E.C. 2.4.1.1], $H_2PO_4^-$. (2) Phosphoglucomutase [E.C. 2.7.5.1], G1,6P, Mg^{2+}. (3,11) Phosphoglucose isomerase [E.C. 5.3.1.9]. (4,15) Phosphofructokinase [PFK, E.C. 2.7.1.11], ATP, Mg^{2+}. (5,6) F1,6P aldolase [E.C. 4.1.2.13]. (7) Fructose 1,6-bisphosphatase [E.C. 3.1.3.11]. (8,9) Transketolase [E.C. 2.2.1.1]. (10) Transaldolase [E.C. 2.2.1.2]. (12) G6P dehydrogenase [E.C. 1.1.1.49], NADP. (13) 6-PG dehydrogenase [E.C. 1.1.1.44], NADP. (14,20) Hexokinase [E.C. 2.7.1.1], ATP, Mg^{2+}. (16) F6P 2-kinase [E.C. 2.7.1.105], ATP. (17) Fructose 2,6-bisphosphatase [E.C. 3.1.3.46]. (18) Galactokinase [E.C. 2.7.1.6], ATP. (19) Gal1P uridyltransferase [E.C. 2.7.7.10], UDPG. (21) Man6P isomerase [E.C. 5.3.1.8].

CO_2 (step 13). These reactions take place in the cytoplasm and (during periods of darkness) in chloroplasts. G6P is also formed from glucose by phosphorylation with ATP (step 14) and from F6P by the action of G6P isomerase (step 11).

G1,6P is a required cofactor for phosphoglucomutase [E.C. 2.7.5.1], the enzyme that catalyzes the G1P ↔ G6P interconversion (step 2). Animal tissues contain enzymes that are capable of synthesizing and degrading G1,6P, but it is not known if these enzymes are present in plants.[396]

Some plants produce sorbitol rather than sucrose as their major product of photosynthesis. So1P and So6P have both been shown to be present in apricot (*Prunus armeniaca* L.) leaf in almost equal amounts (0.0064 and 0.0072 mmol/g fwb, respectively).[397] Sorbitol phosphate was also detected in *Cotoneaster dammeri*.[37] So6P is probably formed from G6P or F6P by enzymatic reduction, but So1P is an anomaly.

Physical Properties

Glucose 1-Phosphate

G1P [59-56-3; see also 6866-69-9]: $[\alpha]_D^{25}$ +120° (H_2O),[398] pK_a 1.1[398] and 6.504[399]; $\Delta G°$ −5 kcal/mol[142]; dissociation constant ($\Delta G°$, $\Delta H°$, $\Delta S°$)[399]; countercurrent distribution[141]; MO calc.[166,400] Spectra: Far-UV (fig.),[401] ORD,[402] GC/MS,[96] HPLC-MS-MS,[100] ESI-MS-MS,[403] FAB-MS-MS,[404] ^{1}H NMR (fig.),[131] ^{13}C NMR,[405] ^{31}P NMR,[68], ESR.[406] Chromatography: PC,[9,32] TLC,[36] IEC,[86,134] HPTLC,[407] GC,[408] HPLC,[26] IC.[107] Electrophoresis: PE,[32] TLE,[36] CZE.[136]

$Na_2C_6H_{11}O_9P \cdot 3.5H_2O$ [R.N. 32972-46-6; see also 56401-20-8]: Monoclinic cryst., d 1.74[409]; diamagnetic susceptibility χ_M -135×10^{-6} cm^3/mol at 25 °C.[410] Spectra: ^{1}H, ^{23}Na, and ^{31}P NMR,[411] LD-MS,[412] X-ray.[409]

$KC_6H_{12}O_9P$ [R.N. 53823-71-5; see also 105751-36-8]: Colorless orthorhombic prisms, d 1.81. Spectra: X-ray.[413]

$K_2C_6H_{11}O_9P \cdot 2H_2O$ [R.N. 5996-14-5; see also 6736-77-2, 89705-70-4]: Fine needles, $[\alpha]_D^{26}$ +78° (H_2O),[414] d 1.85[415]; heat of combustion Q_M 2761 kJ/mol.[416] Spectra: IR (fig.),[417] ^{13}C and ^{31}P NMR (figs.),[418] X-ray.[415,419]

$CaC_6H_{11}O_9P$ [R.N. 6997-09-7]: Dissociation constants ($\Delta G°$, $\Delta H°$, $\Delta S°$).[420]

$BaC_6H_{11}O_9P \cdot 3H_2O$ [R.N. 71662-16-3]: White, nonhygroscopic powder, easily sol. H_2O, insol. EtOH, $[\alpha]_D^{25}$ +75°.[398]

$(C_6H_{11}NH_2)_2C_6H_{13}O_9P$ [R.N. 90357-92-9]: Cryst., mp 163-169 °C dec., $[\alpha]_D^{25}$ +64.0° (H_2O).[421] Spectra: ^{1}H, ^{13}C, and ^{31}P NMR.[422]

$(C_{23}H_{26}N_2O_4)_2C_6H_{13}O_9P$ (brucine salt): Colorless needles, mp 182-184 °C dec., $[\alpha]_{589}^{27} + 0.5°(H_2O)$.[423]

Glucose 6-Phosphate

G6P [R.N. 56-73-5]: Amorphous, nonhygroscopic white powder,[424] $[\alpha]_{546}^{20} + 41.4°(H_2O)$,[425] p$K_a$ 1.50, 6.22,[426] 11.71 (thermometric, 25 °C)[354]; countercurrent distribution[141]; redox potential[427]; MO calc.[166] Spectra: UV,[94] far-UV (fig.),[401] IR (fig.),[94] CD,[94,102] ORD,[402] GC/MS,[96] ESR (fig.),[406] ^{1}H NMR,[131] ^{7}C NMR,[345] ^{31}P NMR.[68] Chromatography: PC,[9,32] TLC,[36] IEC,[86,134] GC,[408] HPLC,[26,428] IC (fig.).[107] Electrophoresis: PE,[32] TLE,[36] zone,[88] CZE,[89] CE,[90] isotachophoresis.[88]

$Li_2C_6H_{11}O_9P$ [R.N. 52888-21-8]: $[\alpha]_D^{20}$ +22.5° (H_2O).[429]

$NaC_6H_{12}O_9P$ [R.N. 54010-71-8]: Monoclinic needles, d 1.83. Spectra: X-ray.[430]

$Na_2C_6H_{11}O_9P{\cdot}2H_2O$ [R.N. 3671-99-6; see also 111283-45-5]: Diamagnetic susceptibility χ_M -135×10^{-6} cm^3/mol at 25 °C.[410] Spectra: FD-MS.[431]

$K_2C_6H_{11}O_9P$ [R.N. 5996-17-8]: mp 95-97 °C dec.,[432] $[\alpha]_D^{24}$ +21.2° (H_2O), readily sol. H_2O.[433]

$BaC_6H_{11}O_9P{\cdot}7H_2O$ [R.N. 5996-16-7]: Monoclinic cryst., mp 157 °C dec.,[432] $[\alpha]_D^{25}$ +17.4° (H_2O),[434] d 2.09[435]; solubility in water, 0.002 mol/L at 25 °C, 0.001 mol/L at 0 °C.[436] Spectra: X-ray.[435]

$(C_{23}H_{26}N_2O_4)_2C_6H_{13}O_9P$ (brucine salt): Colorless needles, mp 182-184 °C, $[\alpha]_{589}^{27} + 0.5°(H_2O)$.[423]

Glucose 1,6-Bisphosphate

G1,6P [R.N. 305-58-8; see also 10139-18-1]: Spectra: ^{1}H NMR (fig.).[437] Chromatography: PC,[438] TLC,[105] IEC,[86] HPLC.[161] Electrophoresis: PE,[439] TLE,[105] CE.[90]

$Na_4C_6H_{10}O_{12}P_2$: $[\alpha]_D^{28}$ +83.0° (H_2O).[440]

$Ba_2C_6H_{10}O_{12}P_2{\cdot}H_2O$: Precipitate.[440]

$(C_6H_{11}NH_2)_4C_6H_{14}O_{12}P_2{\cdot}4H_2O$: Cryst., $[\alpha]_D^{20}$ +31.0° (H_2O).[439]

$(C_{23}H_{26}N_2O_4)_4C_6H_{14}O_{12}P_2$ (brucine salt): Elongated prisms, dec. 185 °C, $[\alpha]_D^{26}$ −16.3° (H_2O).[440]

6-Phosphogluconic Acid

6-PG [R.N. 921-62-0; see also 2464-13-3]: $[\alpha]_{546}^{21} 0°(H_2O)$.[425] Spectra: GC/MS,[96] ^{13}C NMR (fig.),[441] ^{31}P NMR.[68] Chromatography: PC,[9,32] TLC,[33] IC (fig.),[107] HPLC.[442] Electrophoresis: PE,[32] zone,[88] isotachophoresis.[88]

$Na_3C_6H_{10}O_{10}P$: Colorless, granular solid, $[\alpha]_D^{25}$ +1.9° (H_2O).[443] Dihydrate [R.N. 53411-70-4]: Monoclinic cryst., d 1.86; X-ray.[444]

$BaC_6H_{11}O_{10}P$: $[\alpha]_{546}^{21} + 0.2°(H_2O)$.[425]

$Ba_3(C_6H_{10}O_{10}P)_2$: $[\alpha]_{546}^{21} - 1.5°(H_2O)$; solubility in H_2O, 7.1 g/L at 100 °C.[425]

$(C_{23}H_{26}N_2O_4)_3C_6H_{13}O_{10}P$ (brucine salt): Needles, $[\alpha]_D^{25}$ −29.9° (H_2O).[443]

γ-Lactone [R.N. 97323-75-6]: $[\alpha]_{546}^{21} + 21°(0.1N HCl)$.[425] Spectra: ^{13}C and ^{31}P NMR.[445] Chromatography: GC.[446]

δ-Lactone [R.N. 2641-81-8]: Standard transformed Gibbs energy of formation ($\Delta_f G'°$).[447] Spectra: ^{13}C and ^{31}P NMR.[445]

Sorbitol 1-Phosphate

So1P [R.N. 67670-70-6]: Spectra: GC/MS.[397] Chromatography: IEC, PC, TLC, GC.[397] Electrophoresis: TLE.[397]

Sorbitol 6-Phosphate

So6P [R.N. 20479-58-7]: Standard transformed Gibbs energy of formation ($\Delta_f G'^\circ$).[447]
Spectra: GC/MS.[96] Chromatography: IEC,[448] PC,[448] TLC,[449] GC.[449] Electrophoresis: TLE.[449]
$BaC_6H_{13}O_9P$: Cryst.[450]

Synthesis

G1P may be prepared (a) by phosphorylation of tetraacetylglucosyl-1-bromide with Ag_3PO_4[398] or silver diphenyl phosphate[414] or of 2,3,4,6-tetraacetyl-D-glucose with dibenzyl *N,N*-diethylphosphoramidite,[451] (b) by the reaction of β-D-glucose pentaacetate[452] or MOP β-D-glucopyranoside[422] with anhydrous phosphoric acid, or (c) by enzymatic hydrolysis of starch with potato phosphorylase in pH 6.7 phosphate buffer.[453]

G6P may be prepared (a) by phosphorylation of glucose with polyphosphoric acid[454] or metaphosphoric acid,[424] (b) by phosphorylation of glucose,[432] methyl α-D-glucopyranoside,[429] or O^1,O^2-isopropylidene-α-D-glucofuranose[434] with $POCl_3$, (c) by phosphorylation of O^1,O^2,O^3,O^4-tetraacetyl-β-D-glucopyranose with diphenyl phosphorochloridate[433] or bis(2,4-dichlorophenyl) phosphorochloridite[455] or of O^1,O^2-isopropylidene-α-D-glucofuranose with dibenzyl phosphorochloridate,[456] or (d) by fermentation of glucose[425] or sucrose[457] with yeast. It can also be prepared enzymatically (e) by the action of phosphoglucomutase on G1P[458] or of hexose-phosphate isomerase on F6P.[436]

G1,6P may be prepared (a) by phosphorylation of O^2,O^3,O^4-triacetyl-O^6-(diphenylphosphono)glucosyl bromide with Ag_3PO_4 or silver diphenyl phosphate,[440] (b) by the reaction of tetraacetyl β-D-glucose 6-phosphate with anhydrous phosphoric acid,[438,439] or (c) by fermentation of sucrose with yeast.[459] 6-GP may be prepared (a) by oxidation of G6P with bromine[425] or iodine[443] or by enzymatic oxidation of G6P with G6P dehydrogenase. The use of immobilized enzymes is advantageous.[460] So1P and So6P may be prepared by borohydride reduction of F1P[397] and F6P,[448,450] respectively.

Structure

The preferred conformation of the pyranose ring of G1P in aqueous solution is $^4C_1$ with the C_5—C_6 bond in gg↔tg equilibrium.[461] G6P solutions contain no free aldehyde (<0.4%)[94]; the bulk of the G6P is a mixture of α (**8**) and β (**9**) pyranose anomers in 38.5:61.5 ratio.[462] The rate of interconversion of the anomers is 0.0658 s^{-1} (α to β) and 0.0389 s^{-1} (β to α).[463] The interconversion is catalyzed by G6P 1-epimerase.

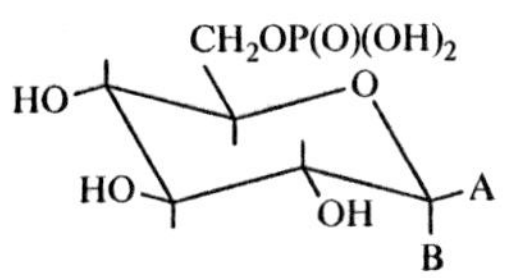

8, A = H, B = OH [R.N. 1520911-7]
9, A = OH, B = H [R.N. 15209-12-8]

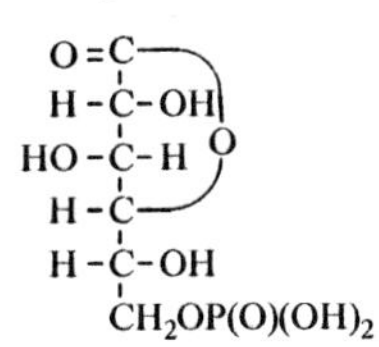

10

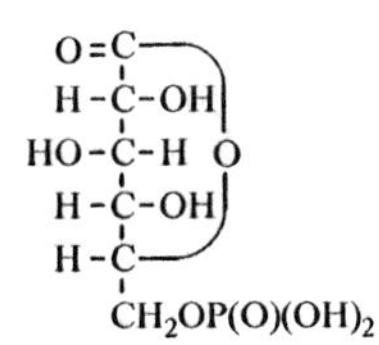

11

6-PG cyclizes spontaneously to D-gluconolactone 6-phosphate as observed by the change in optical rotation over a 4-h period.[425,443] The product, like that of D-gluconic acid itself, is the γ (1-4) lactone **10**. The initial product of enzymatic oxidation of G6P, however, is the δ (1-5) lactone **11**,[464] which is subsequently hydrolyzed to 6-PG by 6-PG lactonase. The rates of ring-opening hydrolysis of **10** and **11** at pH 7.4 and 34 °C were 3.98×10^{-3} and $7.14 \times 10^{-3}\ s^{-1}$, respectively, reflecting the greater stability of the γ-lactone.[465]

Hydrolysis

G1P is stable to base but the rate of hydrolysis at 100 °C increases slowly from pH 8 to pH 4 and more rapidly at lower pH.[466] The rate was $0.288 \times 10^{-5}\ s^{-1}$ at pH 7.46 and $161 \times 10^{-5}\ s^{-1}$ at pH 2.5.[467] This is interpreted as a normal rate for the monoanion AH^- and an abnormal rate for the neutral acid H_2A relative to simple monoalkyl phosphates. In strongly acidic media, there is a rapid acid-catalyzed reaction whose rate depends on the Hammett acidity function. This reaction, in contrast to enzymatic hydrolysis,[468] proceeds by C—O bond fission.[467] For further detailed studies, see the cited references.[469,470]

The behavior of G6P toward hydrolysis is similar to that of simple monoalkyl phosphates in the acid region (pH < 5.8) but differs in the alkaline region. At pH 4.0, where the monoanion HA^- is predominant, the rate constant for hydrolysis at 100 °C is $2.0 \times 10^{-5}\ s^{-1}$. At lower pH, the rate profile falls to about $0.5 \times 10^{-5}\ s^{-1}$ (neutral acid, H_2A), then rises steeply at pH < 0 owing to acid catalysis. At pH 7, the rate increases to a second plateau (dianion, A^{2-}), then rises steeply above pH 9 owing the base-catalyzed hydrolysis. The kinetics have been studied in detail.[426,471] Calorimetric measurements gave ΔH° 0.91 kJ/mol, ΔC_P° −48 J/(mol K) for the hydrolysis of G6P by alkaline phosphatase.[383]

Hydrolysis of both G1P and G6P in the mildly acid to neutral region is catalyzed by molybdate,[204] Th(IV),[215] Ce(III), and other M(III) salts,[216] and by UV radiation.[217] Hydrolysis of G1,6P with 1*N* HCl at 100 °C for 30 min gives G6P and P_i.[439] The rate constant k in 1*N* H_2SO_4 at 30 °C is $1.31 \times 10^{-5}\ s^{-1}$.[440] The rate of hydrolysis of 6-PG in 1*N* HCl at 100 °C falls from 1.02×10^{-5} to $0.59 \times 10^{-5}\ s^{-1}$, probably reflecting different rates for the free acid and its lactone.[472]

So6P is quite stable to hydrolysis. Less than 10% of the phosphorus is liberated by either 1*N* HCl or 0.1*N* NaOH after 1 h at 100 °C,[448] but So6P is readily hydrolyzed to sorbitol by wheat germ acid phosphatase.[397]

Metal Ion Complexes

G6P forms 1:1 complexes with Mg(II) and Ca(II). Their stability constants have been reported.[221]

Separation and Analysis

(1) LePage and Umbreit Method

G1P and G6P appear together with other phosphate esters in the barium-soluble fraction (see p 2) of the TCA extract. G1P is measured by the amount of P_i released in 7 min at

100 °C with 1*N* HCl correlated with an increase in reducing value, and G6P is measured by its reducing value corrected for those due to F6P and pentose phosphate. This method has been superseded by those which follow.

(2) Chromatographic Methods

G1P and G6P can be separated from other phosphate esters by IEC or HPLC, but not from each other unless the mobile phase contains a borate gradient (pages 2,3). An HPLC procedure was developed for the analysis of G1P and G6P in potato tuber based on selective conversion to 6-PG by PG mutase and 6-PG dehydrogenase.[473] The detection of G1,6P and 6-PG in orange juice, based on retention time and behavior on mild hydrolysis, could not be confirmed.[248] 6-PG was, however, isolated from an extract of ladino clover leaf by reverse phase HPLC on a Brownlee column with water/acetonitrile/dicyclohexylamine (84.5:14.5:1) as the mobile phase and quantitated by GC/MS after derivatization with TMS.[474] All but So1P and So6P can be separated by two-dimensional PC, TLC, PC/PE, or TLC/TLE (Table 1-8). Quantitative data were reported in a few instances for G1P, G6P, and 6-PG. The identities and relative amounts of the two sorbitol phosphates were determined by GC/MS after conversion to the tri-methysilyl derivatives.[397]

Analysis Without Separation

(1) ^{31}P NMR Method

G6P has been identified by high-field FT ^{31}P NMR in a number of plant tissues and extracts, and 6-PG in a spinach leaf extract (Table 1-8). For further details, see p 5.

(2) Enzymatic Methods

G6P may be determined by enzymatic oxidation to 6-PG by NADP in the presence of G6P dehydrogenase (Table 1-3, reaction 6). The amount of G6P is calculated from the change in optical density of the NADP at 340 nm.

> *Into a 1.0-cm absorption cell are pipetted 0.65 ml of water, 0.15 ml of 0.25*M *glycylglycine buffer, pH 7.5, 0.10 ml of 0.1*M *$MgCl_2$, 0.10 ml of triphosphopyridine nucleotide (TPN) (1.4 μ*M*/ml), and 0.02 ml of glucose-6-phosphate dehydrogenase (ca 0.02 mg). The solution is mixed, and an initial density reading at 340 mm is taken. The glucose-6-phosphate sample . . . is added, and the increase in density at 340 mm is followed until a constant value is reached. The amount of glucose-6-phosphate added is calculated from the relation G6P (micromoles) = $(d_f - d_o) \times 1.04/6.22$ where d_f and d_o are the final and initial density readings, respectively.*
>
> Horecker, B. L.; Wood, W. A., "D-Glucose-6-Phosphate", *Methods Enzymol.* **1957**, *3*, 152–154,[547] reprinted by permission of Elsevier, © 1957.

Metabolites that are converted to G6P upon treatment with the appropriate enzyme, such as G1P/PG mutase or F6P/PG isomerase, can be assayed by the same method. Others are ATP/hexokinase,[282] UDPG/UDPG pyrophosphorylase,[527] and ADPG/ADPG

Table 1-8 Glucose and Fructose Phosphates in Plants*

No.	Plant part	Method	G1P	G6P	G1,6P	6-PG	F1P	F6P	F1,6P	F2,6P	References
Acer pseudoplatanus L. (Sycamore Maple)											
1	Suspension cells	^{31}P NMR		+			+				46
Actinidia deliciosa Liang. and Ferg. (Kiwi)											
3	Fruit	Enzym.	12.9	113.5				34.8	1.5		222
1	Callus culture	Enzym.		105							475
1	Callus culture	^{31}P NMR		+							475
Allium cepa L. (Onion)											
1	Leaf[a,b]	Enzym.								1.0	476
Amaranthus caudatus L. (Love-Lies-Bleeding)											
1	Leaf[a,b]	Enzym.		183				85	24		224
Amaryllis vittata L'Her. (Amaryllis)											
1	Pollen[a]	Enzym.							600		226
Ananas comosus (L.) Merr. (Pineapple)											
1	Leaf[a]	Enzym.								+	477
1	Phylloides	Enzym.								5-13	478
Annona cherimola Mill. (Cherimoya)											
1	Fruit, mesocarp	Enzym.						102	75	0.046	479
Arabidopsis thaliana (L.) Heynh (Mouse-Ear Cress)											
1	Leaf[b]	Enzym.	20	200	2.98			50		0.52	480
Arbutus unedo L. (Strawberry Tree)											
1	Leaf[a,b]	Enzym.		57				42	6		229
Arum maculatum L. (Cuckoopint)											
1	Club	Enzym.	22.3	210.4				39.5	21.8		230
Avena sativa L. (Oats)											
1	Seed	Enzym.		0						–[c]	232
1	–, embryo[d]	LePage	141					97	7		14
Beta vulgaris (L.) Banks (Sugar Beet)											
1	Root, stored	LePage	50	490				120	90		15
1	–, vegetating	LePage	50	450				160	80		15
1	Leaf	LePage	70	630				130	180		15
1	Petiole, parenchyma	LePage		970				200			17
1	–, fibrovascular bundle	LePage		250				70			17
2	Axis[a,e]	Enzym.	3.5-6.0	6.0-9.0				12.5-17.5			481

1	Leaf[a,f]	Enzym.	3.6	8.7			1.7	1.2	0.06	482
1	Leaf	PC/PC		+		+	+		0.6	483
1	Leaf	PC/PC					+	+		484
Brassica campestris L. (Turnip Rape)										
1	Pod, deseeded[b]	Enzym.		145			13.7	9.9	0.032	238
Brassica napus L. (Rape)										
1	Leaf[g]	Enzym.		51.7			24.3	1.8		239
Brassica nigra (L.) Koch (Black Mustard)										
1	Suspension cells	Enzym.							0.051	485
Brassica pekinensis (Lour.) Rupr. (Pe-tsai)										
1	Petiole, vascular bundle	PC/PC	+	+			+			36
1	Petiole, vascular bundle	TLC/TLC	+	+			+			36
1	Petiole, vascular bundle	TLC/TLE	+	+			+			36
Brassica rapa L. (Turnip Rape)										
1	Vascular bundle	TLC/TLC		+			+			240
–, var. *rapa* (DC) Metzg. (White Turnip)										
1	Tuber	Enzym.				21.1				486
1	Tuber	PC				+				486
Bryiphyllum tubiflorum Harv. (Chandelier Plant)										
1	Leaf	Enzym.							+	477
1	Phylloides	Enzym.							0.1-4.0	478
Camellia japonica L. (Camellia)										
1	Pollen	Enzym.					11	0	0.01	487
Camellia sinensis L. (Tea)										
1	Leaf	Enzym.	6.3	0.7						488
–, var. *assamica* (Mast.) Kitam. (Assam Tea)										
1	Leaf	Enzym.	12.5		2.0					488
Capsicum annuum L. (Bell Pepper)										
1	Seed pericarp								0.60	489
1	Leaf	Enzym.					226	130		242
1	–, tip	Enzym.							3.6	490
1	–, base	Enzym.							7.8	490
Catharanthus roseus (L.) G. Don (Madagascar Periwinkle)										
1	Suspension cells	Enzym.	34.5	337.1			58.2	31.8	0.04	243
1	Suspension cells	^{31}P NMR		+			+			46

*nmol/g fresh weight basis unless stated otherwise; [a]Basis unstated; [b]nmol/mg chlorophyll; [c]Almost unmeasurable; [d]nmol/plant part; [e]nmol/axis; [f]nmol/mg protein; [g]μmol/m^2; [h]nmol/mL; [i]Dry weight basis; [j]mg C/dm^2; [k]mol/mol CABP binding site; [l]μmol/g; [m]μg/g; [n]μg/mm^2; [o]Incl. G1P; [p]μmol/mg protein; [q]Conc., m*M*.

Continued

Table 1-8 Glucose and Fructose Phosphates in Plants*—cont'd

No.	Plant part	Method	G1P	G6P	G1,6P	6-PG	F1P	F6P	F1,6P	F2,6P	References
Cicer arietinum L. (Chickpea)											
1	Seedling, cotyledon[a,d]	Enzym.		138.8				32.4	8.3		247
1	–, embryonic axis[a,d]	Enzym.		20.0				8.7	0.5		247
1	Leaf exudate[h]	HPLC		8.33							491
Citrullus lanatus (Thunb.) Mats. and Nakai (Watermelon)											
1	Cotyledon	Enzym.								0.025	492
Citrus acida Roxb. (Citron)											
1	Mature fruit[i]	IEC	0	440				380	180		493
1	Immature fruit[i]	IEC	550	720				3000	480		493
Citrus paradisi Macf. (Grapefruit)											
1	Leaf	Enzym.								0.75	494
1	Fruit juice sac	Enzym.								0.08	494
Citrus sinensis (L.) Osbeck (Orange)											
1	Fruit juice	IEC	+	+	+	+		+	+		248
1	Fruit juice sac	Enzym.								0.16	494
Coriandrum sativum L. (Coriander)											
1	Leaf[b]	Enzym.							96.4	0.072	249
Cucumis sativus L. (Cucumber)											
1	Leaf	IEC	125.4	390.0				113.1			250
Cucurbita maxima Duch. (Hubbard Squash)											
1	Sieve tub exudate	TLC/TLC	+	+				+			240
Cucurbita pepo L. (Pumpkin)											
1	Leaf[j]	Enzym.						46.7	5.4		252
1	Seedling cotyledon	Enzym.		2113							252
1	Stem exudate[a,b]	Enzym.	128	746		97		319	134		251
1	Unripe fruit exudate[a,b]	Enzym.	41	221		34		89	37		251
1	Ripe fruit exudate[a,b]	Enzym.	67	2459		118		434	25		251
Daucus carota L. (Carrot)											
1	Tuber[k]	IEC	0.14	0.23				0.16			23
1	Root	Enzym.		111						0.120	253
1	Root	Enzym.					19.5				486
1	Root	Enzym.						3.04	0.23	0.014	317
1	Root	Enzym.					+				486

Egeria densa Planch. (Brazilian Waterweed)											
1	Leaf[l]	Enzym.		105							495
Fagopyrum esculentum L. (Buckwheat)											
1	Seedling	Enzym.		120		4		15	3		257
Fragaria vesca L. (Woodland Strawberry)											
1	Leaf	Enzym.	28	247	8.3			46	4.3		259,488
Glycine max (L.) Merr. (Soybean)											
1	Leaf[a]	Enzym.	84	675				126		0.195	496
1	Leaf[b]	HPLC	31	307				175	32		26
1	Suspension cells	^{31}P NMR		+				+			46
1	Seed, cotyledon	^{31}P NMR		+				+			47
1	Seedling	Enzym.		95	18			26	39	45	262
1	Root tip[i]	Enzym.		7900		<300		<200			497
Gossypium barbadense L. (Sea Island Cotton)											
1	Leaf[m]	PC/PC	2.18				0.89				263
Gossypium hirsutum L. (Upland Cotton)											
1	Leaf[b]	Enzym.		170×10^3				100.3×10^3			498
Helianthus annuus L. (Sunflower)											
1	Leaf[g]	Enzym.		103.7				34.4	5.6		239
1	–, youngn	PC		0				2160	1000		499
1	–, budding stage[m]	PC		0				0	950		499
1	–, flowering stage[m]	PC		60				260	4950		499
Helianthus tuberosus L. (Jerusalem Artichoke)											
1	Tuber	IEC		+			+	+	+		500
1	Tuber	Enzym.		133				33		1.2	265
1	Young stolon	^{31}P NMR		+							45
Hemerocallis hybrid (Daylily)											
1	Flower petal	TLC/TLC	+	+		+		+			266
Heracleum sphondylium L. (Hogweed)											
1	Leaf	Enzym.								0.105	501
Hordeum distichon Alef. ssp. *nutans*											
1	Sprout[n]	Enzym.	1.25	77.6				37.5	3.2		267
Hordeum vulgare L. (Barley)											
2	Dehusked grain[a,d]	Enzym.	2.6-2.8	37				9.0-11.0			502
1	Leaf[b]	Enzym.	16	89	3.8						396

*nmol/g fresh weight basis unless stated otherwise; [a]Basis unstated; [b]nmol/mg chlorophyll; [c]Almost unmeasurable; [d]nmol/plant part; [e]nmol/axis; [f]nmol/mg protein; [g]μmol/m^2; [h]nmol/mL; [i]Dry weight basis; [j]mg C/dm^2; [k]mol/mol CABP binding site; [l]μmol/g; [m]μg/g; [n]μg/mm^2; [o]Incl. G1P; [p]μmol/mg protein; [q]Conc., m*M*.

Continued

Table 1-8 Glucose and Fructose Phosphates in Plants*—cont'd

No.	Plant part	Method	G1P	G6P	G1,6P	6-PG	F1P	F6P	F1,6P	F2,6P	References
1	Leaf[a,b]	Enzym.						73	6	0.048	269
1	–, tillering stage[m]	PC		0				0	0		499
1	–, ear-forming stage[m]	PC		1450				4620	1020		499
2	Seedling	TLC/TLC	+	+		+		+	+		272
1	Root tip	^{31}P NMR		+							503
Impatiens balsamina L. (Garden Balsam)											
1	Root tip[i]	Enzym.		4700							497
Kalanchoë blossfeldiana v. Poelln. (Flaming Katy)											
1	Leaf[i]	Enzym.		120				30	5		273
Kalanchoë daigremontiana Hammett et Perrier (Devil's Backbone)											
1	Leaf	Enzym.		73.7		3.0		9.8	54		274
1	Leaf[a]	Enzym.								+	477
Lactuca sativa L. (Lettuce)											
1	Leaf	^{31}P NMR		+				+			48
Larix decidua Mill. (European Larch)											
1	Cambial sap	IEC/PC		+							504
Lathyrus sativus L. (Grass Pea)											
1	Leaflet[m]	TLC/TLC	312		593						505
Lolium temulentum L. (Darnel)											
1	Leaf	Enzym.		160				63			506
1	Leaf[a,b]	Enzym.								0.100	507
Lotus corniculatus L. (Birdsfoot Trefoil)											
1	Root	Enzym.		190.7							508
Lycopersicon esculentum Mill. (Tomato)											
1	Green fruit	Enzym.		80				15	0.3		258
1	Fruit	Enzym.		240				5	20		278
1	Flower ovary[i,m]	LePage	130	420				150	360		18
Malus domestica Borkh. (Apple)											
1	Immature fruit	Enzym.		80				10	1.4		258
1	Flower bud[l]	Enzym.		610							509
2	Mature fruit	Enzym.		22-25				4-5	1		279
Medicago sativa L. (Alfalfa)											
1	Root	PC/PC		+				+	+		280

1	Shoot	PC/PC		+	+	+		280
1	Root	Enzym.		122.1				508
1	–, nodule	Enzym.		1040	700	160		281
Mesembryanthemum crystallinum L. (Ice Plant)								
1	Leaf[m]	Enzym.		32	8	1	0.08	510
Mucuna pruriens (L.) D.C. (Cowitch Bean)								
1	Leaf[g]	Enzym.		21	15	4.4		282
Musa cavendishii Pax. (Dwarf Banana)								
1	Green fruit	Enzym.		70	15	0.5		258
1	Fruit[a]	Enzym.					0.25	511
Nicotiana tabacum L. (Tobacco)								
1	Leaf[a,g]	Enzym.	56	413	84	2.7		286
1	Leaf	Enzym.					0.038	501
1	–, mature	Enzym.				25.6		287
1	–, senescent	Enzym.				7.8		287
1	Shoot	Enzym.	7.36	34.1	10.5		0.0357	288
1	Root	Enzym.		43.3			0.0114	288
Oryza sativa L. (Rice)								
1	Grain[a]	Enzym.	61	1800	360	130		289
1	Seed, milky stage	PC/PC	+					512
1	Leaf[a,b]	Enzym.					0.100	513
1	Seedling[a,f]	Enzym.					10	514
Persea americana Mill. (Avocado)								
1	Fruit	^{31}P NMR		+				515
1	Fruit	Enzym.					0.34	516
1	–, mesocarp[h]	Enzym.		9	3	29		290
Persea gratissima Gaertn. f. (Avocado)								
1	Leaf	PC/PC		+				517
Phaseolus coccineus L. (Scarlet Runner Bean)								
1	Seedling, cotyledon	Enzym.					0.9	518
1	–, leaf	Enzym.					0.60	518
1	–, root	Enzym.					0.09	518
1	–, stem	Enzym.					0.10	518
Phaseolus vulgaris L. (Beans)								
1	Leaf[a]	PC/PE		29.7[o]	15.5			38
1	Leaf[a]	Enzym.		210	110	100		291
1	Leaf[h]	Enzym.	0.87	2.86				519

*nmol/g fresh weight basis unless stated otherwise; [a]Basis unstated; [b]nmol/mg chlorophyll; [c]Almost unmeasurable; [d]nmol/plant part; [e]nmol/axis; [f]nmol/mg protein; [g]μmol/m^2; [h]nmol/mL; [i]Dry weight basis; [j]mg C/dm^2; [k]mol/mol CABP binding site; [l]μmol/g; [m]μg/g; [n]μg/mm^2; [o]Incl. G1P; [p]μmol/mg protein; [q]Conc., m*M*.

Continued

Table 1-8 Glucose and Fructose Phosphates in Plants*—cont'd

No.	Plant part	Method	G1P	G6P	G1,6P	6-PG	F1P	F6P	F1,6P	F2,6P	References
1	Leaf	IEC	1430	490				430	480		22
1	Fruit	IEC		110							520
1	Seed[a,d]	Enzym.								16	521
1	Seed, cotyledon	^{31}P NMR		+				+			49
Picea abies (L.) Karst. (Norway Spruce)											
1	Needle (leaf)	Enzym.						480		1.4	293
1	Needle (leaf)	Enzym.	Variation in F2,6P level from base to tip								522
1	Seedling endosperm[a,d]	Enzym.		2.6							523
1	–, embryo[a,d]	Enzym.		3.7							523
1	Root	Enzym.	Variation in F2,6P level from base to tip								522
Pinus silvestris L. (Scotch Pine)											
1	Seedling endosperm[a,d]	Enzym.		3.1							523
1	–, embryo[a,d]	Enzym.		1.5							523
Pisum sativum L. (Peas)											
1	Leaf	Enzym.		120				17			524
1	Leaf[b]	Enzym.							75.2	0.078	249
1	–, chloroplast	HPLC							+		106
1	Root[a,l]	Enzym.		97							294
1	Root[f]	Enzym.								0.053	525
1	Root tip	^{31}P NMR		+							526
1	Pea	IEC	+	527				93			24
1	Seed	Enzym.		8.0				14.5	8.0		296
2	–, embryo (round)	Enzym.	31	442				92	26	1.415	295
	–, embryo (wrinkled)		70	1253				283	54	0.290	
1	–, cotyledon	Enzym.	35	479				120			527
Populus deltoides Bartr. ex Marsh (Eastern Cottonwood)											
1	Leaf[b]	Enzym.	24	140				45	25	0.030	297
Populus gelrica (Hioutz.) Houtz.											
1	Twig bark[i]	Enzym.	20	400				80			509
1	–, xylem[i]	Enzym.	20	300				60			509
1	Bark[a,d]	Enzym.	0.01	0.20		0.02		0.04	0.004		298
Populus nigra L. (Black Poplar)											
1	Bark[a,d]	Enzym.	0.02	0.33		0.02		0.06	0.004		298

Populus tremula L. (European Aspen)										
1	Leaf[e]	PC/PE		83.4[o]			34.8			38
Prunus armeniaca L. (Apricot)										
1	Leaf[a]	PC/PE		26.6			7.7			37
Pyrus communis L. (Pear)										
1	Fruit	Enzym.		+			8.0	3.5	0.006	299
1	–, skin	IEC						+		19
Quercus rubra L. (Red Oak)										
1	Leaf[g]	Enzym.	Data not given							282
Raphanus sativus L. (Radish)										
1	Leaf	Enzym.		68			32	27		300
Rhododendron ponticum L. (Common Rhododendron)										
1	Detached leaf	Enzym.		78	5		16	0.8		301
Ricinus communis L. (Castor Bean)										
1	Seed, endosperm[a,d]	Enzym.		0.176			0.074	0.015		302
1	–, endosperm[a,b]	Enzym.							0.169	528
1	Seedling, cotyledon	Enzym.	151	2000			600	6.7	0.55	303
1	–, hypocotyl	Enzym.	9	395			64	3		304
1	–, phloem exudate[h]	Enzym.	80	550			171	9		304
Saccharum spp., hybrid										
1	Internode[a,d]	Enzym.						F2,6P	0.157 -0.936	529
Saccharum officinarum L. (Sugar Cane)										
1	Leaf	PC/PC	+	+			+			306
1	Root	PC/PC	+	+			+			306
1	Stalk, third internode[p]	Enzym.		13.0			7.9	1.8	0.4	307
				and variation from third to ninth internode						
Saccharum officinarum × *S. spontaneum* × *S. barberi*										
1	Leaf[g]	Enzym.	12	74			36	43		530
Salix triandra L. (Almond Willow)										
1	Phloem exudate	TLC/TLC		+		+	+	+		309
Secale cereale L. (Rye)										
1	Root tip[q]	Enzym.		0.078						531
1	Leaf chloroplast (diploid)[m]	Enzym.	64.4	37.9			65.1			391
1	Leaf chloroplast (tetraploid)[m]	Enzym.	67.3	116.3			44.5			391
Sesamum alatum Thonn.										
1	Leaf[b]	Enzym.					22.6	83.8		310
Sesamum indicum L. (Sesame)										
1	Leaf[b]	Enzym.					18.9	52.5		310

*nmol/g fresh weight basis unless stated otherwise; [a]Basis unstated; [b]nmol/mg chlorophyll; [c]Almost unmeasurable; [d]nmol/plant part; [e]nmol/axis; [f]nmol/mg protein; [g]μmol/m^2; [h]nmol/mL; [i]Dry weight basis; [j]mg C/dm^2; [k]mol/mol CABP binding site; [l]μmol/g; [m]μg/g; [n]μg/mm^2; [o]Incl. G1P; [p]μmol/mg protein; [q]Conc., m*M*.

Continued

Table 1-8 Glucose and Fructose Phosphates in Plants*—cont'd

No.	Plant part	Method	G1P	G6P	G1,6P	6-PG	F1P	F6P	F1,6P	F2,6P	References
Solanum tuberosum L. (Potato)											
1	Tuber[i,k]	LePage	0	3.50				0.17	0		13
1	Tuber	IEC	114	722				306	48		316
1	Tuber	PC/PC	+	+				+	+		9
1	Tuber[a]	PC/PC	18	60		3		22	1.0		314
14	Tuber	PC/PC	tr	+				+			313
1	Tuber	PC					+	+	+		315
1	Tuber	PE	+				+	+			39
1	Tuber	Enzym.	11	110							318
1	Tuber	Enzym.						28	1.5	0.03	317
3	Tuber	Enzym.								0.2-0.7	532
		and variation in G1P, G6P, F6P, F1,6P and F2,6P levels along the stolon-apex axis									533
1	Mature tuber	Enzym.		115				20	0.3		258
1	Developing tuber	HPLC	15.7	157.5				68.5			473
1	–, starch grain[m]	Enzym.		15							318
1	Leaf	Enzym.								0.061	501
1	Leaf	Enzym.		208				42			311
1	–, stroma[b]	Enzym.		24				23	35		312
1	–, cytosol[b]	Enzym.		85				59	3		312
Sorghum bicolor (L.) Moench (Sorghum)											
1	Leaf[b]	Enzym.								0.1	320
1	Root tip	^{31}P NMR		+							44
Spinacia oleracea L. (Spinach)											
1	Root[a,l]	Enzym.		188							294
1	Leaf	PC/PC		+	+			+	+		321
1	Leaf	^{31}P NMR		7.0		+					50
1	Leaf[a,b]	Enzym.	16.8	87.9							393
1	Leaf[b]	Enzym.	20	280				90	10	0.075	322
1	Leaf[a,b]	Enzym.						83	6	0.095	269
1	Leaf	Enzym.			1.4						396
1	–, stroma[a,b]	Enzym.		48				34	35		325
1	–, cytosol[a,b]	Enzym.		82				27	2.3		325
1	–, cytosol[a,b]	Enzym.								0.222	534
1	Leaf, protoplast[b]	Enzym.		81.2				29.7	36.9		326
1	–, chloroplast[b]	Enzym.		35.9				16.1	30.7		326
1	–, chloroplast[d]	Enzym.								0	534

1	–, extrachloroplast[b]	Enzym.		61.0				17.9	2.8		326
1	–, chloroplast[a,b]	IEC							2.20		20
Spirodela oligorrhiza (Kurz.) Hegelm. (Duckweed)											
1	Whole plant	PC/PC	50	500		18		100	40		327
Tilia platyphyllos Scop. (Large-Leaved Linden)											
1	Phloem exudate	TLC/TLC		+			+	+			328
Trifolium pratense L. (Red Clover)											
1	Stem[l]	Enzym.		0.64							535
1	Leaf[n]	HPLC				6.1					474
Trifolium repens L. (White Clover)											
1	Creeping stem[l]	Enzym.		0.41							535
Trifolium subterraneum L. (Sub Clover)											
1	Whole plant	IEC		1930				310			329
Triticosecale Witt. (Triticale)											
1	Leaf[b]	Enzym.								0.070	536
Triticum aestivum L. (Wheat)											
1	Grain[a,d]	Enzym.	1.7	34.0				5.0	4.0	0.108	537
1	–, endosperm	Enzym.	10.8	224	4.9			54.1			538
1	Root[a,l]	Enzym.		179							294
1	–, tip[q]	Enzym.		0.108							531
1	6-day root[i]	Enzym.							0.42		330
1	Leaf[a,b]	Enzym.	26	115				77	11	0.191	323
1	Leaf[i,m]	Enzym.		106.6		99.5					539
1	Seedling, leaf	PC/PE	+					+			331
1	Shoot[i]	Enzym.							2.3		330
Urtica dioica L. (Stinging Nettle)											
1	Leaf	Enzym.								0.652	501
Vicia faba L. (Broad Bean)											
1	Leaf	PC/PC		+	+			+	+		321
Vigna mungo (L.) Hepper (Black Gram)											
1	Seedling, cotyledon[a,d]	Enzym.		8.8		0.5		1.5			540
1	–, embryonic axis[a,d]	Enzym.		30.1		0.6		6.0			540
1	–, hypocotyl tip	Enzym.		43.1		0.2		10.9	1.3	0.186	332
			and variation in F2,6P content from base to tip								541
1	Root tip	^{31}P NMR		+							542,543
Vigna radiata (L.) R. Wilcz. (Mung Bean)											
1	Seedling	Enzym.		58	12			22	44	21	262

*nmol/g fresh weight basis unless stated otherwise; [a]Basis unstated; [b]nmol/mg chlorophyll; [c]Almost unmeasurable; [d]nmol/plant part; [e]nmol/axis; [f]nmol/mg protein; [g]μmol/m^2; [h]nmol/mL; [i]Dry weight basis; [j]mg C/dm^2; [k]mol/mol CABP binding site; [l]μmol/g; [m]μg/g; [n]μg/mm^2; [o]Incl. G1P; [p]μmol/mg protein; [q]Conc., m*M*.

Continued

Table 1-8 Glucose and Fructose Phosphates in Plants*—cont'd

No.	Plant part	Method	G1P	G6P	G1,6P	6-PG	F1P	F6P	F1,6P	F2,6P	References
1	Leaf[b]	Enzym.							70.3	0.080	249
Vitis vinifera (Wine Grape)											
1	Seedless berry	Enzym.		117				30	0.14		333
1	Juice							0			334
Xanthium strumarium L. (Cocklebur)											
1	Leaf[a,b]	Enzym.							33		336
Yucca flaccida Haw. (Weak-Leaf Yucca)											
1	Phloem exudate	TLC/TLC		+			+	+	+		309
Zea mays L. (Corn)											
1	Root	PC/PC	313	456							12
1	Root	TLC/TLC	+	+			+	+			12
1	Root tip	^{31}P NMR		+							42,43,544
1	Leaf[b]	PC/PC		110				61	21.4		337
1	Leaf[a,b]	Enzym.		105				58	21	0.13	545
1	–, midrib[b]	Enzym.		287				135	78		340
1	–, lamina[b]	Enzym.		121				58	46		340
1	Grain, endosperm[i]	Enzym.	0.7	38.1				5.9	2.7		342
1	–, endosperm	Enzym.	140.5	2180.0				558.9	467.2	6.1	546
1	–, starch granule[i]	Enzym.	0.2	8.6				1.8	0.5		342
1	Flower, ovary	Enzym.	0.13	2.2				0.18			343

*nmol/g fresh weight basis unless stated otherwise.
[a]Basis unstated.
[b]nmol/mg chlorophyll.
[c]Almost unmeasurable.
[d]nmol/plant part.
[e]nmol/axis.
[f]nmol/mg protein.
[g]μmol/m^2.
[h]nmol/mL.
[i]Dry weight basis.
[j]mg C/dm^2.
[k]mol/mol CABP binding site.
[l]μmol/g.
[m]μg/g.
[n]μg/mm^2.
[o]Incl. G1P.
[p]μmol/mg protein.
[q]Conc., m*M*.

pyrophosphorylase.[548] Kumar and Singh[548] were able to determine all six metabolites in a wheat extract by adding two units (5 mL) of each enzyme in sequence to the assay mixture (1.5 mL) in a cuvette and recording the absorbance after each addition. G1P and G6P may also be determined by conversion to F6P with PG mutase and PG isomerase and assayed for F6P by the luciferin-luciferase method (Fig. 1-5, steps 2 and 3).[481]

Two enzymatic methods have been reported for the determination of G1,6P in plant extracts, both of which require brief heating with dilute alkali to remove interfering substances. In one, G1,6P is hydrolyzed with 0.1*N* HCl for 7 min at 100 °C and assayed for G6P by fluorimetry.[488] In the other, G1,6P is assayed as a cofactor of the PG mutase reaction (Fig. 1-5, step 2) using G1P and dephosphorylated PG mutase to convert the G1,6P to G6P.[396]

6-PG may be determined by enzymatic oxidation to Ru5P and CO_2 by NADP in the presence of 6-PG dehydrogenase (Table 1-3, reaction 7). The amount of 6-PG is calculated from the change in absorbance at 340 nm.[257]

Fructose Phosphates

This group comprises two monophosphates (F1P and F6P) and two bisphosphates (F1,6P and F2,6P).

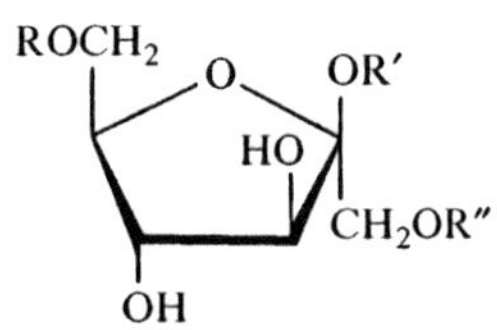

F1P, R = R′ = H, R″ = $P(O)(OH)_2$
F6P, R = $P(O)(OH)_2$, R′ = R″ = H
F1,6P, R = R″ = $P(O)(OH)_2$, R′ = H
F2,6P, R = R′ = $P(O)(OH)_2$, R″ - H

D-Fructose 1-Phosphate (F1P), $C_6H_{13}O_9P$; Beilstein **1**, E^{IV} 4421. *β-D-Fructofuranose 1-(Dihydrogen Phosphate)* [1956].

D-Fructose 6-Phosphate (F6P), $C_6H_{13}O_9P$; Beilstein **1**, E^{I} 464, E^{IV} 4423; **31**, 337. Neuberg ester, *β-D-Fructofuranose 6-(Dihydrogen Phosphate)* [1943].

D-Fructose 1,6-Bisphosphate (F1,6P), $C_6H_{14}O_{12}P_2$; Beilstein **1**, E^{I} 464, E^{IV} 4424; **31**, 337. Harden-Young ester, *β-D-Fructofuranose 1,6-Bis(dihydrogen Phosphate)* [1952].

D-Fructose 2,6-Bisphosphate (F2,6P), $C_6H_{14}O_{12}P_2$; not listed in Beilstein. *β-D-Fructofuranose 2,6-Bis(dihydrogen Phosphate)* [1981].

Occurrence

F6P and F1,6P are metabolites of the photosynthetic carbon reduction cycle (Calvin cycle) of higher plants (Fig. 1-5). In step 6, DHAP undergoes an aldol condensation with G3P forming F1,6P. In step 7, F1,6P is hydrolyzed by a specific fructose 1,6-bisphosphatase to F6P. In step 8, F6P reacts with another molecule of G3P, forming E4P and Xu5P. These reactions all take place within the chloroplast, found only in the green leaf.

F6P and F1,6P are also metabolites of the Embden-Meyerhof-Parnas glycolysis pathway, a sequence common to all of the higher plants. F6P, derived from fructose by phosphorylation with ATP (step 15) or from G6P by the action of G6P isomerase (step 3), is phosphorylated to F1,6P by ATP (step 4). The F1,6P is then split into G3P and DHAP by aldolase (step 5). These reactions take place in the cytoplasm outside any organelle.

In addition, F6P is a metabolite of the pentose phosphate respiratory pathway (hexose monophosphate shunt). F6P, derived from E4P and Xu5P by the action of transketolase (step 9) or from G3P and Se7P by the action of transaldolase (step 10), is converted to its isomer G6P by G6P isomerase (step 11). These reactions take place in the cytoplasm and (during periods of darkness) in chloroplasts.

F2,6P is a recently discovered metabolite whose function in the higher plants is to regulate the synthesis of sucrose by controlling the enzymes that catalyze the F6P $\leftrightarrow$ F1,6P interconversion. It is formed from F6P by phosphorylation with ATP (step 16) and is hydrolyzed back to F6P by a specific phosphatase (step 17). As a signal regulator, its content in plant tissue, usually at the nanomole level, can change dramatically in response to changes in metabolite level, nitrogen or phosphorus level, light level, and other factors.

The presence of F1P in potato,[39] carrot,[486] turnip,[486] and artichoke[500] tubers has been established by experiments with [^{32}P]-KH_2PO_4 and [^{14}C]-fructose, but its source could not be determined and it did not appear to be metabolized.[500] F1P was postulated to be a metabolite in the photosynthesis of sucrose in cotton[263] and sugar beet[483] leaves, but this was later retracted.[549]

Physical Properties

Fructose 1-Phosphate

F1P [R.N. 15978-08-2; see also 52050-05-2]: $[\alpha]_D^{20}$ −52.5°.[550] Spectra: UV,[102] IR (fig.),[94] CD,[102] GC/MS,[96] ^{13}C NMR,[551] ^{31}P NMR.[68] Chromatography: PC,[9] TLC,[160] IEC,[552] GC.[446] Electrophoresis: PE,[9] zone,[88] isotachophoresis.[88]

$BaC_6H_{11}O_9P{\cdot}2H_2O$ [R.N. 53823-70-4]: Amorphous solid,[550] $[\alpha]_D^{26}$ −30.4° (H_2O).[553]

$(C_6H_{11}NH_2)_2C_6H_{13}O_9P$ [R.N. 74381-67-2]: Fine long rods, sol. methanol, insol. ethyl acetate.[553]

$(C_{23}H_{26}N_2O_4)_2C_6H_{13}O_9P$ (brucine salt): $[\alpha]_{546}^{20}$ − 52.1°(H_2O).[554]

Fructose 6-Phosphate

F6P [R.N. 643-13-0; see also 41452-29-3]: $[\alpha]_D^{21}$ +2.5°,[555] pK_a 0.97[555] and 6.35[130]; redox potential,[427] countercurrent distribution.[141] Spectra: UV,[102] IR (fig.),[94] CD,[102] GC/MS,[96] ^{1}H NMR,[131] ^{13}C NMR (fig.),[551] ^{31}P NMR,[68] ESR.[406] Chromatography: PC,[9,32] TLC,[36] IEC,[86,134] HPTLC,[407] GC,[408] HPLC,[26] IC (fig.).[107] Electrophoresis: PE,[32] TLE,[36] zone,[88] CZE,[89] CE,[90] isotachophoresis.[88]

$Na_2C_6H_{11}O_9P{\cdot}2H_2O$ [R.N. 26177-86-6]: White, amorphous powder. Spectra: ^1H, ^{23}Na, and ^{31}P NMR.[556] Heptahydrate: Orthorhombic cryst., d 1.61; X-ray.[557]
$K_2C_6H_{11}O_9P{\cdot}6H_2O$ [R.N. 99481-18-2]: Monoclinic cryst., d 1.66; X-ray.[558]
$BaC_6H_{11}O_9P$ [R.N. 6035-54-7]: White powder, $[\alpha]_D^{19}$ +3.58° (H_2O).[559]

Fructose 1,6-Bisphosphate
F1,6P [R.N. 488-69-7; see also 34693-15-7]: $[\alpha]_D^{17}$ +4.04° (2*N* HBr),[559] pK_a 1.48,[560] 6.425 and 7.276[561]; redox potential,[427] countercurrent distribution.[141] Spectra: UV (fig.),[561] IR (fig.),[103] FTIR (fig.),[95] Raman,[562] CD,[102] ORD (fig.),[561] GC/MS,[96] ^1H NMR (fig.),[103,131] ^{13}C NMR (fig.),[103] ^{31}P NMR (fig.),[103] ESR.[406] Chromatography: PC,[9,32] TLC,[160] IEC,[86,134] HPTLC,[407] GC,[408] HPLC,[26] HPLC-MS-MS,[100] IC (fig.).[107] Electrophoresis: PE,[32] TLE,[105] zone,[88] CZE,[89] CE,[90] isotachophoresis.[88]
$Na_3C_6H_{11}O_{12}P_2$ [R.N. 38099-82-0]: mp 125 °C, $[\alpha]_D$ +2.6° (H_2O).[563] Octahydrate [R.N. 81028-91-3]: Monoclinic cryst., d 1.704; X-ray.[564]
$Na_4C_6H_{10}O_{12}P_2$ [R.N. 23784-19-2]: Hygroscopic, yellowish powder, $[\alpha]_D^{21}$ +2.79° (H_2O).[565]
$Mg_2C_6H_{10}O_{12}P_2$: Colorless granules; water solubility, 30%.[566]
$Ca_2C_6H_{10}O_{12}P_2$ [R.N. 6055-82-9]: Amorphous solid[567]; heat of combustion 2808 kJ/mol.[126]
$BaC_6H_{12}O_{12}P_2$: Snow-white solid, very sol. H_2O (~1:2).[559]
$Ba_2C_6H_{10}O_{12}P_2$ [R.N. 6035-52-5]: Fluffy, snow-white powder, sparingly sol. H_2O.[559]
$(C_6H_{11}NH_2)_3C_6H_{14}O_{12}P_2{\cdot}H_2O$: Rods, mp 163-166 °C dec.[568]

Fructose 2,6-Bisphosphate
F2,6P [R.N. 77164-51-3; see also 79082-92-1]: Conformation.[569] Spectra: ^1H NMR (fig.), ^{13}C NMR (fig.), ^{31}P NMR (fig.).[569] Chromatography: PC,[570] IEC,[570] HPLC.[428] Electrophoresis: CZE.[89]
$C_6H_{11}NH_2$ salt: Powder.[570]

Synthesis

F1P may be prepared (a) by phosphorylation of 2,3,4,5-diisopropylidene-D-fructopyranose with $POCl_3$,[571] P_2O_5,[553] or diphenyl phosphorochloridate,[572] (b) by partial enzymatic hydrolysis of F1,6P with bone phosphatase,[554] or (c) by aldolase-catalyzed condensation of DHAP with D-glyceraldehyde.[554] F6P may be prepared (a) by partial hydrolysis of F1,6P with HCl at 35 °C[559] or liver hexose diphosphatase at pH 9.7[573] or (b) by phosphorylation of D-fructose with hexokinase.[101] F1,6P may be prepared (a) by phosphorylation of sucrose with baker's yeast[574] or (b) by aldolase-catalyzed condensation of DHAP with G3P.[101] F2,6P may be prepared (a) from F_6P by phosphorylation with phosphoric acid[575] or (b) from F1,6P by DCC-promoted cyclization followed by alkaline hydrolysis.[576]

Hydrolysis

The rate of hydrolysis of F1P in 1*N* HCl at 100 °C is 2.68×10^{-3} s^{-1}.[554]

Hydrolysis of F6P in 1*N* HCl at 100 °C liberates orthophosphate quantitatively ($k = 1.67 \times 10^{-4}$ s^{-1}),[577] but in 0.75*M* NaOH only 80% of the orthophosphate is liberated owing to degradation of the sugar moiety.[578] For the behavior of F6P over the pH range 0-8.3 at 100 °C, see Fleury *et al.*[579] The hydrolysis is catalyzed by molybdate (fivefold at pH 5.8),[204] Th(IV),[215] Ce(III), and other M(III) salts,[216] and by UV radiation.[217] Measurements on the hydrolysis of F6P with alkaline phosphatase gave ΔG° 13.70 kJ/mol, ΔH° −7.61 kJ/mol, and ΔC°_P −28 J/(mol K).[383]

Hydrolysis of F1,6P in 1*N* HCl at 100 °C liberates P_i in two stages ($k = 2.0 \times 10^{-3}$ s^{-1} for P_1 and 1.5×10^{-4} s^{-1} for P_6),[580] but alkaline hydrolysis is complicated by degradation of the sugar moiety to saccharinic acid,[581] lactic acid,[565] and pyrotartaric acid.[578] The hydrolysis is catalyzed by molybdate,[204] Th(IV),[215] Ce(III), and other M(III) salts,[216] and by UV radiation.[217]

Hydrolysis of F2,6P with 0.5*N* HCl at 30 °C for 15 min liberates equimolar amounts of F6P and inorganic phosphate.[570] Its half-life at pH 4 is 5.2 min.[576] F2,6P is stable to base, no change being noted after 30 min in 0.01*M* NaOH at 100 °C.[576]

Structure

F1P shows no free keto form (<1.0%) in aqueous solution.[94] F6P solutions show 2.5%[94] to 4.1%[582] of the open-chain keto form **12** and 20% of the α-furanose anomer **13**.[583] The rate of ring-opening at pH 7.5 and 40 °C is 18 s^{-1} for the α anomer **13** and 21 s^{-1} for the β anomer **14**.[348]

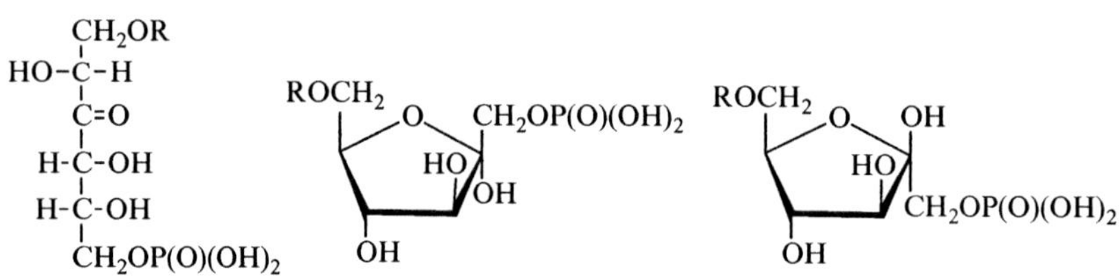

12, R = H | **13**, R = H | **14**, R = H
15, R = $P(O)(OH)_2$ | **16**, R = $P(O)(OH)_2$ | **17**, R = $P(O)(OH)_2$

The behavior of F1,6P is analogous: Aqueous solutions show 1.7%[103] to 2.0%[582] of open-chain keto form **15** (not hydrated to an appreciable extent [<0.1%] to the *gem*-diol[103]) and 10% of the α-furanose anomer **16**.[104] The rates of ring-opening at pH 7.2 and 25 °C are about 8 s^{-1} for the α anomer **16** and 35 s^{-1} for the β anomer **17**.[582]

Separation and Analysis

(1) LePage and Umbreit Method[3]

F6P and F1,6P are present in different fractions of the TCA extract, together with other phosphate esters (see p 2). They are determined by analysis for fructose, on the assumption that no other fructose-containing compounds are present. This method has been superseded by those which follow.

(2) Chromatographic Methods

F6P and F1,6P in plant extracts can be separated from other phosphate esters and from each other by IEC, by HPLC, or by two-dimensional PC, TLC, PC/PE, or TLC/TLE (Table 1-8). Quantitative data have been reported in several instances. F1P has been detected in a few plant extracts by two-dimensional PC, but F2,6P occurs at too low a level to be detected by any chromatographic method.

Analysis Without Separation

(1) ^{31}P NMR Method

F6P and F1,6P have been identified in a few plant extracts by high-field FT ^{31}P NMR (Table 1-8). F6P was also tentatively identified in the spectra of sycamore cells suspended in a P_i-free culture medium.[46]

(2) Enzymatic Methods

F1P may be determined by an indirect enzymatic method coupled to the reduction of DHAP to α-GP by NADH (Table 1-3, reaction 2) using aldolase to convert the F1P to DHAP.[486] F6P is usually determined by an indirect enzymatic method coupled to the oxidation of G6P by NADP (Table 1-3, reaction 6 and Fig. 1-5, steps 11 and 12):

> *Glc-6-P concentration was measured in 1 mL of assay mixture containing 0.2*M *triethanolamine/HCl, pH 7.6, 5* mM *$MgCl_2$, 0.5* mM *$NADP^+$, 0.5 unit glucose 6-phosphate dehydrogenase (EC 1.1.1.49) and 25 µL of plant extract. After total oxidation of Glc-6-P, Fru-6-P concentration was determined by addition of 1 unit hexose-phosphate isomerase (EC 5.3.1.9).*
>
> Sola, M. M.; Gutiérrez, M.; Vargas, A. M., "Regulation of Hexose-Phosphate Cycle Determines Glucose and Fructose Accumulation in Cherimoya (*Annona cherimola* Mill.) during Ripening", *J. Plant Physiol.* **1994**, *144*(4/5), 569–575,[479] reprinted by permission of Elsevier, © 1994.

F6P may also be determined by enzymatic phosphorylation to F1,6P with PFK kinase/ATP (Fig. 1-5, step 4) and assayed directly by measuring the consumption of ATP by the luciferin/luciferase method[481] or indirectly by the DHAP/NADH route as shown below[242]:

F1,6P may be determined by indirect enzymatic methods coupled to the reduction of DHAP to α-GP (Table 1-3, reaction 2),[242] to the oxidation of G3P to 3-PGA (Table 1-3, reaction 1),[258] or to the oxidation of G6P to 6-PG (Table 1-4, reaction 6).[236]

$$\mathrm{F6P} \xleftarrow{7} \mathrm{F1,6P} \xrightarrow{5} \begin{cases} \mathrm{DHAP} \rightarrow \alpha\text{-GP} \\ \mathrm{G3P} \rightarrow \text{3-PGA} \end{cases}$$

The first two methods employ F1,6P aldolase and triose-P isomerase to convert the F1,6P to DHAP or G3P. The DHAP reduction method, which was used in the majority of the enzymatic analyses reported in Table 1-8, is described on pp 13-14. The G3P oxidation method was employed by Barker *et al.*[258,584] for the fluorimetric analysis of F1,6P in a variety of plant extracts. The third method employs F1,6P phosphatase to convert the F1,6P to F6P and PG isomerase to convert the latter to G6P.[387]

F2,6P is acid-labile and therefore requires a sample that has been extracted from plant tissue at pH 7 or above. Most determinations are made by the procedure of van Schaftlingen *et al.*[585] which is based on the stimulation of the potato enzyme PP_i-PFK [E.C. 2.7.1.90] by F2,6P in the phosphorylation of F6P by pyrophosphate ($F6P + PP_i \rightarrow F1{,}6P + P_i$). The assay then follows the DHAP reduction method for F1,6P described above. An important modification of the original procedure is the use of an acid wash to remove traces of F2,6P from the F6P reagent used in the assay.[586]

> *The assay medium (1 ml) contained 50* mM *Tris/HCl, pH 8.0, 5* mM *$MgCl_2$, 1* mM *EDTA, 0.1* mM *DTT, 0.15* mM *NADH, 1* mM *acid-washed Fru-6-P, 0.5 unit fructose 1,6-bisphosphate aldolase (E.C. 4.1.2.13), 1.5 unit triose-phosphate isomerase (E.C. 3.3.1.1), 0.5 unit glycerol 3-phosphate dehydrogenase (EC 1.1.1.8), approximately 50 mUnit potato tuber PP_i-PFK and 0.1 mL of plant extract. The reactions were initiated with 0.5* mM *Na_4PP_i (previously adjusted to pH 8.0) after preincubation for 5 min at 30 °C. The assays were carried out for 10 min at 30 °C. Fru-2,6-P_2 concentration was calculated with respect to a standard curve. Recovery experiments (Ball and ap Rees, 1988) always gave values greater than 90%.*
>
> Sola, M. M.; Gutiérrez, M.; Vargas, A. M., "Regulation of Hexose-Phosphate Cycle Determines Glucose and Fructose Accumulation in Cherimoya (*Annona cherimola* Mill.) during Ripening", *J. Plant Physiol.* **1994**, *144*(4/5), 569–575,[479] reprinted by permission of Elsevier, © 1994.

F2,6P can also be determined by mild acid hydrolysis, provided that the reducing sugar phosphates (G6P, F6P, etc.) are first destroyed by treatment with alkali or $NaBH_4$.[262] The F6P so produced is then analyzed in the usual manner. This method has found little favor, though it is still used as the reference method for purified standards.[5a]

Galactose and Mannose Phosphates

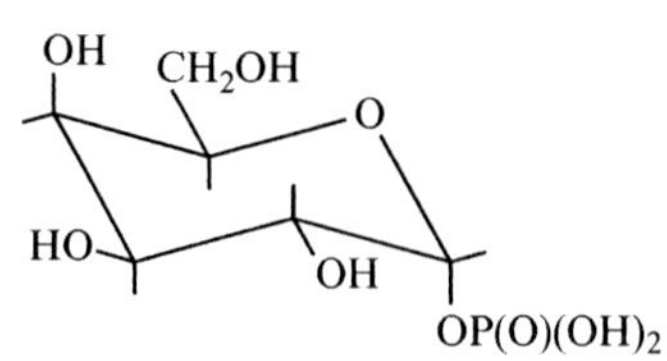

Gal1P

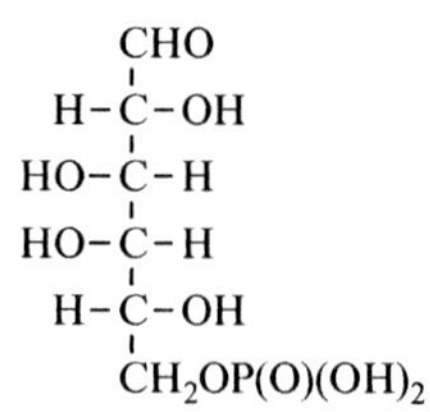

Gal6P

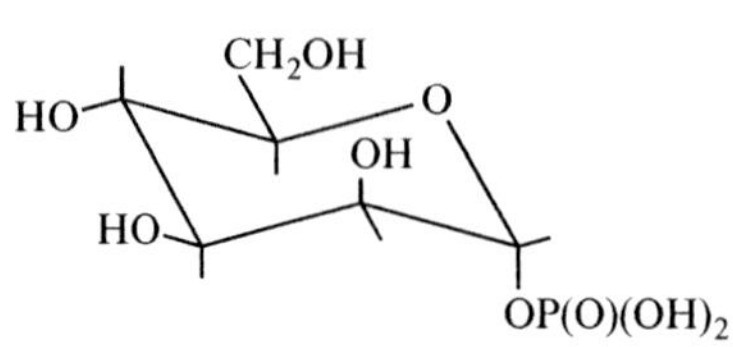

Man1P

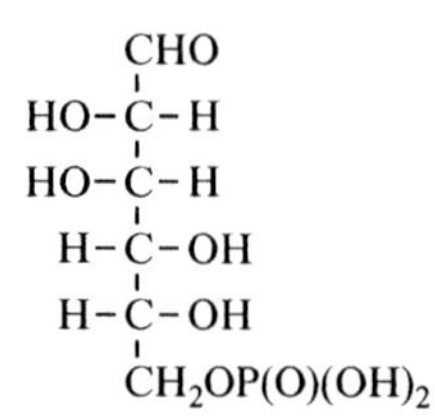

Man6P

D-Galactose 1-Phosphate (Gal1P), $C_6H_{13}O_9P$; not listed in Beilstein. *α-D-Galactopyranose 1-(Dihydrogen Phosphate)* [1938].

D-Galactose 6-Phosphate (Gal6P), $C_6H_{13}O_9P$; Beilstein **1**, E_{IV} 4383. *D-Galactose 6-(Dihydrogen Phosphate)* [1931].

D-Mannose 1-Phosphate (Man1P), $C_6H_{13}O_9P$; not listed in Beilstein. *α-D-Mannopyranose 1-(Dihydrogen Phosphate)* [1938].

D-Mannose 6-Phosphate (Man6P), $C_6H_{13}O_9P$; Beilstein **1**, E_{IV} 4382. *D-Mannose 6-(Dihydrogen Phosphate)* [1932].

Occurrence

These phosphates are less well known than those of glucose or fructose. Galactosyl and mannosyl residues are present in plant tissue as components of polysaccharides, glycolipids, or glycoproteins, but both are often toxic when supplied to the plant in the form of free galactose or mannose. Numerous plants have been shown to be capable of metabolizing D-galactose and D-mannose to Gal1P and Man6P, respectively, but few have the enzymes needed to convert these to starch or sucrose. The phosphates accumulate, depleting the supply of cytoplasmic P_i.

The known metabolic pathways whereby the plant salvages these sugars are illustrated in Fig. 1-6. In steps 1 and 3, D-galactose and D-mannose (released from their combined forms by enzymatic reactions) are phosphorylated by ATP to Gal1P and Man6P, respectively. The Gal1P reacts with UTP (step 2) to give UDPGal, which is either metabolized to galactosyl polysaccharides, lipids, etc., or isomerized to UDPG, G1P, and ultimately starch.[587] The Man6P is isomerized to F6P (step 4), leading ultimately to sucrose.[588]

Traces of a galactose monophosphate identified as Gal6P have been detected by paper chromatography in duckweed[327] and in the cambial sap of larch.[504] Its source is unknown. The galactose monophosphate formed in the tomato is definitely not the 6-phosphate.[589]

Gal1P and Man6P accumulate in corn (*Zea mays* L.) root and leaf upon treatment with D-galactose and D-mannose, respectively, based upon ^{14}C labeling experiments and changes in their high-field FT ^{31}P NMR spectra.[587,590] The Man6P is slowly metabolized by a pathway involving Man1P.[590]

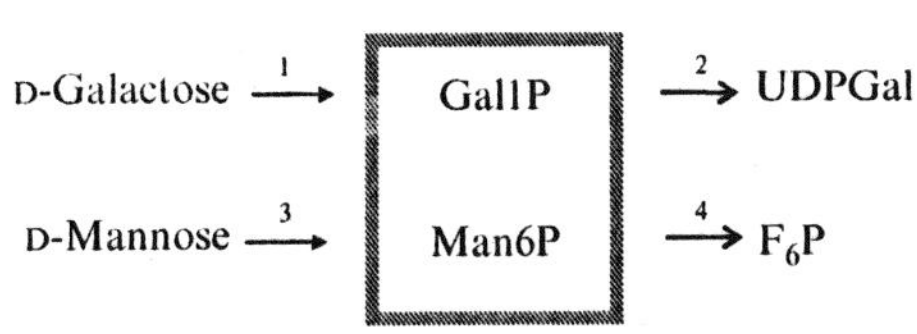

FIGURE 1-6 Galactose and Mannose Phosphates. (1) D-Galactokinase [E.C. 2.7.1.6], ATP. (2) UDP-D-galactose pyrophosphorylase [E.C. 2.7.7.10], UTP. (3) Hexokinase [E.C. 2.7.1.1], ATP. (4) Man6P isomerase [E.C. 5.3.1.8].

Physical Properties

D-Galactose 1-Phosphate

Gal1P [R.N. 2255-14-3]: $[\alpha]_D^{25}$ +143° (H_2O),[591] pK_a 1.00, 6.17,[592] standard transformed Gibbs energy of formation, $\Delta_f G'^{\circ}$.[447] Spectra: GC/MS (fig.),[96] ^{1}H NMR,[593] ^{13}C NMR,[594] ^{31}P NMR.[68] Chromatography: PC,[29] TLC,[160] IEC,[595] ESI-MS-MS,[403] HPLC-MS-MS.[100] Electrophoresis: PE.[596]

$K_2C_6H_{11}O_9P{\cdot}2H_2O$ [R.N. 19046-60-7]: Crystals, very sol. H_2O, insol. EtOH, $[\alpha]_D^{18}$ +108.2° (H_2O).[597] Pentahydrate [R.N. 132929-14-7]: Monoclinic crystals, d 1.756. Spectra: X-ray.[598]

$BaC_6H_{11}O_9P{\cdot}3H_2O$ [R.N. 103213-39-4]: Amorphous nonhygroscopic white powder, easily sol. H_2O, insol. 50% EtOH, $[\alpha]_D^{25}$ +91° (H_2O).[591]

$(Et_3N)_2C_6H_{13}O_9P$ [R.N. 139883-88-8]: $[\alpha]_D^{25}$ +66.5° (H_2O).[599]

$(C_6H_{11}NH_2)_2C_6H_{13}O_9P{\cdot}0.5H_2O$ [R.N. 38099-40-0]: Hygroscopic needles, mp 141-143 °C,[600] $[\alpha]_D^{20}$ +78.5° (H_2O).[452] Spectra: ^{1}H, ^{13}C, and ^{31}P NMR.[422]

D-Galactose 6-Phosphate

Gal6P [R.N. 6665-00-5; see also 33163-99-4]: Colorless glass, $[\alpha]_D^{20}$ +36.5° (H_2O).[601] Spectra: GC/MS (fig.),[96] ^{13}C NMR,[602] ^{31}P NMR.[68] Chromatography: PC,[9,32] GC,[446] IEC,[552] HPLC,[27] IC.[603] Electrophoresis: PE.[32]

$Li_2C_6H_{11}O_9P$ [R.N. 52888-24-1]: $[\alpha]_D^{20}$ +31.6° (H_2O).[429]

$BaC_6H_{11}O_9P$ [R.N. 4198-47-4]: $[\alpha]_D^{25}$ +24.5° (H_2O).[604]

Brucine salt: Elongated hexagonal plates.[605]

D-Mannose 1-Phosphate

Man1P [R.N. 15978-07-1]: $[\alpha]_D^{25}$ +58° (H_2O).[591] Spectra: GC/MS,[96] ^{1}H NMR (fig.),[593] ^{13}C NMR,[606] ^{31}P NMR.[68] Chromatography: PC,[29] TLC,[160] IEC,[552] HPLC-MS-MS,[100] ESI-MS-MS.[403] Electrophoresis: PE.[596]

$Li_2C_6H_{11}O_9P{\cdot}H_2O$ [R.N. 14985-29-6]: Hygroscopic solid, $[\alpha]_D^{25}$ +46.3° (H_2O).[607]

$BaC_6H_{11}O_9P{\cdot}3H_2O$ [R.N. 51306-18-4]: Amorphous nonhygroscopic white powder, easily sol. H_2O, insol. 50% EtOH; $[\alpha]_D^{25}$ +36° (H_2O).[591]

$(Et_3N)_2C_6H_{13}O_9P$ [R.N. 139883-95-7]: $[\alpha]_D^{25}$ +19.7° (H_2O).[599]

$(C_6H_{11}NH_2)_2C_6H_{13}O_9P{\cdot}2H_2O$ [R.N. 51306-17-3]: Colorless cryst., mp 137-138.5 °C, $[\alpha]_D^{20}$ +26° (H_2O).[608] Monohydrate: Cryst., mp 181-182 °C.[609]

D-Mannose 6-Phosphate

Man6P [R.N. 3672-15-9]: $[\alpha]_D^{22}$ +13.3° (0.1N HCl),[610] $[\alpha]_{546}^{20}$ + 15.1°(H_2O),[577] pK_a 6.1.[405] Spectra: GC/MS,[96] ^{1}H NMR,[611] ^{13}C NMR,[606] ^{31}P NMR.[68] Chromatography: PC,[9,32] TLC,[36] IEC,[552] GC,[446] HPLC.[442] Electrophoresis: PE,[32] TLE.[36]

$Li_2C_6H_{11}O_9P$ [52888-25-2]: $[\alpha]_D^{20}$ +7.7° (H_2O).[429]

$Na_2C_6H_{11}O_9P$ [R.N. 33068-18-7]: $[\alpha]_D^{25}$ +7.39° (H_2O).[611]

$BaC_6H_{11}O_9P{\cdot}2H_2O$: Fine needles, $[\alpha]_D^{20}$ +3.2° (H_2O).[429]

$(Et_3N)_2C_6H_{13}O_9P$: $[\alpha]_D$ +10° (MeOH).[612] Spectra: ^{1}H NMR.[612]
Brucine salt: Short needles.[577]

Synthesis

Gal1P may be prepared (a) by phosphorylation of tetra-*O*-acetyl-α-D-galactopyranoside with diphenyl phosphorochloridate[599] or dibenzyl *N,N*-diethylphosphoramidite,[451] (b) by phosphorylation of tetra-*O*-acetyl-α-D-galactosyl bromide with Ag_3PO_4[597] or silver diphenyl phosphate,[414] (c) by the reaction of penta-*O*-acetyl-β-D-galactopyranoside[452] or MOP β-D-galactopyranoside[422] with anhydrous phosphoric acid, or (d) by enzymatic phosphorylation of D-galactose with galactokinase and ATP in the presence of PEP or acetyl phosphate.[594] Gal6P may be prepared by phosphorylation of D-galactose with polyphosphoric acid,[602] of methyl α-D-galactopyranoside with $POCl_3$,[429] or of di-*O*-isopropylidene-D-galactopyranose with $POCl_3$,[604] diphenyl phosphorochloridate,[601] or bis(2,4-dichlorophenyl)phosphorochloridite.[455]

Man1P may be prepared (a) by phosphorylation of tetra-*O*-acetyl-α-D-mannopyranose with diphenyl phosphorochloridate[599] or dibenzyl *N,N*-diethylphosphoramidite,[451] (b) by phosphorylation of tetra-*O*-acetyl-α-D-mannosyl bromide with Ag_3PO_4[591] or of the corresponding chloride with silver diphenyl phosphate or silver dibenzyl phosphate,[610] (c) by the reaction of penta-*O*-acetyl-α-D-mannopyranose with anhydrous phosphoric acid,[607] or (d) by reaction of a tri-*O*-acetyl-β-D-mannopyranosyl orthoacetate with dibenzyl phosphate.[609] Man6P may be prepared by phosphorylation (a) of penta-*O*-trimethylsilyl-α-D-mannopyranose[611] or methyl α-D-mannopyranoside[429] with $POCl_3$, (b) of tetra-*O*-acetyl-β-D-mannopyranose with diphenyl phosphorochloridate,[610] (c) of D-mannose with hexokinase and ATP,[613] or (d) by fermentation of mannose with dried yeast.[614]

Hydrolysis

Gal1P is completely hydrolyzed by 0.09*N* H_2SO_4 in 2 min at 100 °C but is unaffected by 0.2*N* NaOH after 30 min at 100 °C.[597] In contrast, Gal6P is only 7% hydrolyzed after 5 h in 0.1*N* H_2SO_4 at 100 °C but 90% hydrolyzed in less than 10 min in 0.1*N* NaOH at 100 °C.[615] The hydrolysis of Gal6P is catalyzed by Ce(III) and La(III).[616] Rate constants for the hydrolysis of the four compounds are as follows:

Compound	**Conditions**	$k \times 10^{-3}$ **min**$^{-1}$	**References**
Gal1P	0.25*N* HCl, 25°C	2.05	597
Gal6P	0.01*N* HCl, 100 °C	0.87	605
Man1P	0.95*N* H_2SO_4, 30 °C	1.90	610
Man6P	1*N* HCl, 100 °C	0.67	577

The α and β anomers of Gal1P and Man1P can be distinguished by the rate of release of P_i with molybdate-malachite green at 37 °C (k_β/k_α 1.34 and 3.16, respectively).[617]

Calorimetric measurements gave ΔH° 1.40 kJ/mol (enthalpy) and ΔC_p° −46 J/(mol K) (heat capacity) for the hydrolysis of Man6P with alkaline phosphatase.[383]

Phosphorylation

Gal1P is phosphorylated by G1,6P in the presence of phosphoglucomutase, cysteine, and Mg(II), giving G6P and (presumably) Gal1,6P. The rate is somewhat slower than that for ribose 1-phosphate.[618]

Separation and Analysis

Gal1P and Man6P may be separated from other phosphate esters and from each other by two-dimensional PC, TLC, PC/PE, or TLC/TLE (Table 1-9). Gal1P was determined in sugar beets by the consumption of ATP (luciferin-luciferase method) after conversion to F6P with a mixture of Gal1P uridyltransferase, PG mutase, and PG isomerase (Fig 1-5, steps 19, 2, and 3).[481]

Structure

The preferred conformation of the pyranose ring of the 1-phosphate in aqueous solution is $^4C_1$ with the C_5-C_6 bond in gt↔tg equilibrium for Gal1P and gg↔tg for Man1P.[461] Pure Gal6P is a mixture of α and β pyranose anomers in a 32:64% ratio with no more than 4% furanose forms. The ratio of pyranose anomers is the same (1:2) in aqueous solution.[602]

Heptose Phosphates

This group comprises two sedoheptulose phosphates, Se7P and Se1,7P, and one manno-heptulose phosphate, Ma7P.

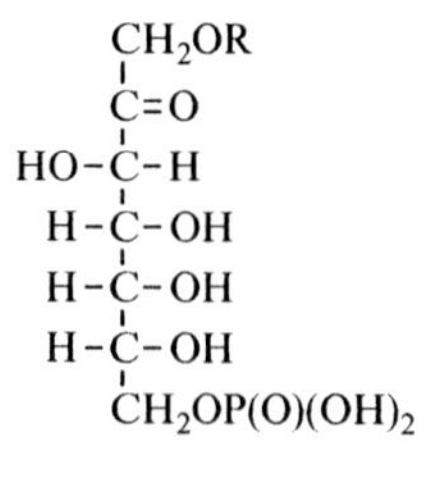

Se7P, R = H
Se1,7P, R = P(O)(OH)$_2$

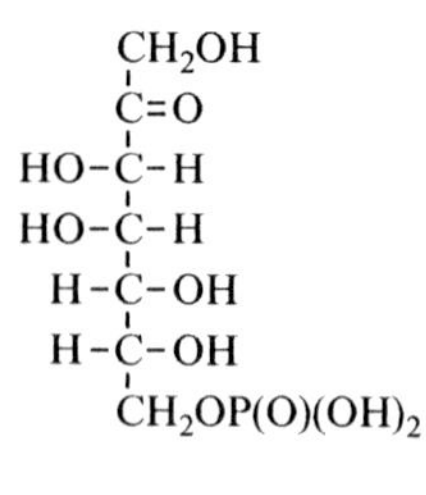

Ma7P

D-Sedoheptulose 7-Phosphate, $C_7H_{15}O_{10}P$; not listed in Beilstein. D-altro-*2-Heptulose 7-(Dihydrogen Phosphate)* [1951].

D-Sedoheptulose 1,7-Bisphosphate, $C_7H_{16}O_{13}P_2$; not listed in Beilstein. *D-altro-2-Heptulose 1,7-Bis(dihydrogen Phosphate)* [1963].

Table 1-9 Galactose and Mannose Phosphates*

No.	Plant part	Method	Gal1P	Gal6P	Man1P	Man6P	References
Beta vulgaris L. (Sugar Beet)							
1	Leaf[a]	PC/PC				+	483
2	Axis[a,b]	Enzym.	7.5-11.0				481
Brassica pekinensis (Lour.) Rupr. (Pe-tsai)							
1	Petiole vascular bundle	PC/PC				+	36
1	Petiole vascular bundle	TLC/TLC				+	36
1	Petiole vascular bundle	TLC/TLE				+	36
Brassica rapa L. (Turnip Rape)							
1	Petiole vascular bundle	TLC/TLC				+	240
Cucumis sativus L. (Cucumber)							
1	Root	PC	+				619
Cucurbita maxima Duch. (Hubbard Squash)							
1	Leaf sieve tube exudate	TLC/TLC				+	240
Hemerocallis hybrid (Day Lily)							
1	Flower petal	TLC/TLC				+	266
Larix decidua Mill. (European Larch)							
1	Cambial sap	IEC/PC		tr			504
Lycopersicon esculentum Mill. (Tomato)							
1	Fruit	TLE	+				589
Prunus armeniaca L. (Apricot)							
1	Leaf	PC/PC				17.3	397
1	Leaf[a]	PC/PE				14.2	37
Solanum tuberosum L. (Potato)							
1	Tuber[a]	PC/PC				13	314
Spirodela oligorrhiza (Kurz.) Hegelm. (Duckweed)							
1	Whole plant	PC/PC		~40		120	327
Zea mays L. (Corn)							
1	Root[c]	PC	550				619

*nmol/g fresh weight basis unless stated otherwise.
[a]Basis unstated.
[b]nmol/axis.
[c]μg/g.

D-Mannoheptulose 7-Phosphate, $C_7H_{15}O_{10}P$: not listed in Beilstein. *D-manno-2-Heptulose 7-(Dihydrogen Phosphate)*, [1954].

Occurrence

Se7P and Se1,7P are metabolites of the photosynthetic carbon reduction cycle (Calvin cycle) of higher plants (Fig. 1-7). In step 1, E4P undergoes an aldol condensation with DHAP, giving Se1,7P. In step 2, Se1,7P is hydrolyzed by sedoheptulose 1,7-bisphosphatase to Se7P and P_i. In step 3, Se7P transfers a two-carbon ketol segment to G3P, forming R5P and Xu5P. These reactions take place in the chloroplast, found only in the green leaf.

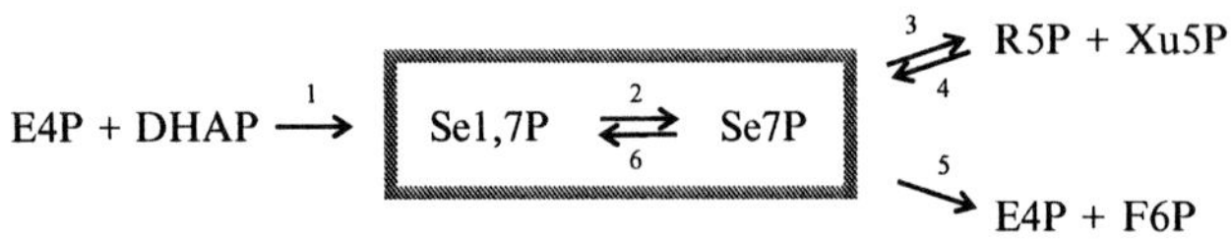

FIGURE 1-7 Heptose Phosphates. (1) Aldolase [E.C. 4.1.2.13]. (2) Sedoheptulose 1,7-bisphosphatase [E.C. 3.1.3.37]. (3, 4) Transketolase [E.C. 2.2.1.1]. (5) Transaldolase [E.C. 2.2.1.2]. (6) F6P kinase [E.C. 2.7.1.11], ATP.

Se7P is also a metabolite of the respiratory pentose phosphate pathway. It is formed, together with G3P, by the interaction of R5P and Xu5P (step 4), and reacts with G3P to form E4P and F6P (step 5). These reactions take place in the cytoplasm, and in chloroplasts in periods of darkness.

A mannoheptulose phosphate, presumably the 7-ester, has been positively identified only in the avocado leaf,[517] but may be expected to occur in other plant species that have been shown to accumulate mannoheptulose.[620] Its metabolism is unknown. Mannoheptulose and its phosphate are both labeled rapidly upon brief photosynthesis in $^{14}CO_2$ but the corresponding heptitol (perseitol), which occurs together with mannoheptulose in avocado, is not.[621]

Physical Properties

D-Sedoheptulose 7-Phosphate

Se7P [R.N. 2646-35-7]: $[\alpha]^{22}_{589}$ + 4.65°(0.3*N*HBr),[622] $\Delta G°$ −253.2 kcal/mol.[84] Spectra: ORD,[622] GC/MS,[96] ^{13}C NMR,[623] ^{31}P NMR.[68] Chromatography: PC,[32] TLC,[105] IEC,[622] HPLC,[106] IC.[107] Electrophoresis: PE,[32] TLE,[105] zone,[88] isotachophoresis.[88]
$BaC_7H_{13}O_{10}P$: Amorphous solid.[622]

D-Sedoheptulose 1,7-Bisphosphate

Se1,7P [R.N. 815-91-8]: Faintly yellow sirup,[91] $\Delta G°$ −253.4 kcal/mol.[84] Spectra: ^{13}C NMR,[623] ^{31}P NMR.[68] Chromatography: TLC,[105] IEC,[624] HPLC.[26] Electrophoresis: TLE,[105] zone,[88] isotachophoresis.[88]
$Ba_2C_7H_{12}O_{13}P_2$: White solid.[91]

D-Mannoheptulose 7-Phosphate

Ma7P [R.N. 69975-90-2]: Spectra: ^{13}C NMR, ^{31}P NMR.[374] Chromatography: PC,[621] IEC.[345]

Synthesis

Se7P may be prepared (a) by condensation of R5P with 2-nitroethanol,[622] (b) by the action of transaldolase-free enzyme preparations from spinach,[625] *Vicia faba*[626] or *Streptococcus faecalis* 10Cl[627] on R5P, (c) by the action of transketolase on a mixture of R5P and L-erythrulose, [628] or (d) by the action of Se1,7P diphosphatase on Se1,7P.[91] Enzymatic phosphorylation of sedoheptulose with hexokinase/ATP gives the 6-phosphate instead.[629] Se1,7P may be prepared by the action of aldolase and transketolase on a mixture of Se7P and F6P[625] or on a mixture of G3P and E4P.[91]

Ma7P may be prepared (a) by phosphorylation of methyl α-D-mannoheptuloside or di-*O*-isopropylidene-D-mannoheptulofuranose with diphenyl phosphorochloridate[374] or (b) by the action of transketolase on a mixture of arabinose-5-phosphate and Xu5P[345] or L-erythrulose.[628]

Separation and Analysis

Se7P and Se1,7P can be separated from other phosphate esters by IEC or HPLC and identified by the Dische method,[19] by TLC/TLE,[20] or by retention time[26,106] (Table 1-10). In many cases, the heptose phosphates have been separated by two-dimensional PC and their identities confirmed by PC of the free sugars after enzymatic dephosphorylation.

Se7P and Se1,7P have also been determined directly on PCA extracts without prior separation by enzymatic methods similar to those employed with F6P and F1,6P. Four different procedures were adopted for the compounds in Table 1-11, each capable of distinguishing sedoheptulose esters from fructose esters.

Structure

In aqueous solution, both sedoheptulose phosphates exist as equilibrium mixtures of β-furanose, α-furanose (**18** and **20**), and α-pyranose (**19** and **21**) forms. The ratios for Se7P and Se1,7P are 67:17:16 and 74:13:13, respectively.[623] Ma7P exists exclusively as the α-pyranose in aqueous solution[374]:

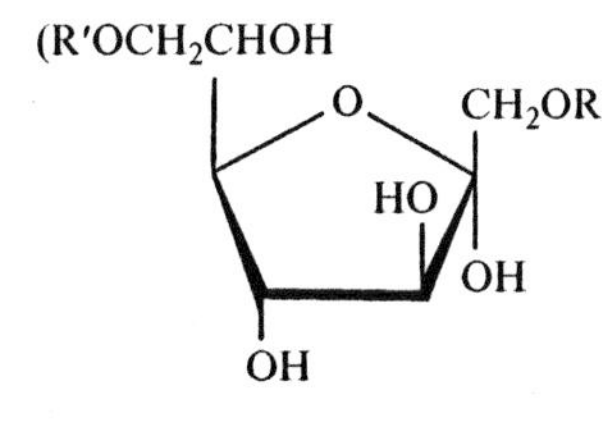

18, R = H, R′ = $P(O)(OH)_2$
20, R = R′ = $P(O)(OH)_2$

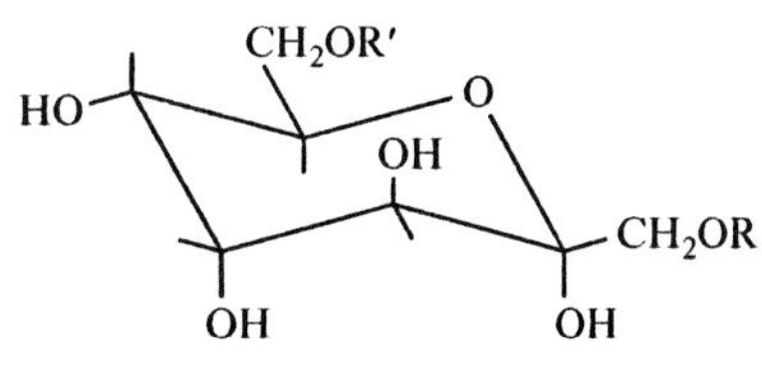

19, R = H, R′ = $P(O)(OH)_2$
21, R = R′ = $P(O)(OH)_2$

Octose Phosphates

D-*altro*-2-Idooctulose 1,8-bisphosphate (Id1,8P), a metabolite of the pentose phosphate pathway in rat liver, was reported in a meeting abstract to be formed in spinach (*Spinacia oleracea* L.) leaf during brief photosynthesis in $^{14}CO_2$; its concentration in the chloroplast was about 1.3 nmol/mg chlorophyll.[633] No further details are available.

Table 1-10 Heptose Phosphates in Plants*

No.	Plant part	Method	Ma7P	Se7P	Se1,7P	References
Beta vulgaris L. (Sugar Beet)						
1	Leaf[a]	PC/PC		+		483
Brassica pekinensis (Lour.) Rupr. (Pe-tsai)						
1	Petiole vascular bundle	PC/PC	+			36
1	Petiole vascular bundle	TLC/TLC	+			36
1	Petiole vascular bundle	TLC/TLE	+			36
Coleus blumei Benth. (Painted Nettle)						
3	Leaf	PC/PC		+		630
Glycine max (L.) Merr. (Soybean)						
1	Leaf	HPLC			+	26
Kalanchoë daigremontiana Hammett et Perrier (Devil's Backbone)						
1	Leaf[b]	Enzym.			15	275
Persea gratissima Gaertn. (Avocado)						
1	Leaf	PC/PC	+	+		517
Pisum sativum L. (Peas)						
1	Leaf chloroplast	HPLC		Low	+	106
Populus gelrica (Houtz.) Houtz.						
1	Bark[c,d]	Enzym.		0.042		298
Populus nigra L. (Black Poplar)						
1	Bark[c,d]	Enzym.		0.026		298
Pyrus communis L. (Pear)						
1	Fruit skin	IEC		+		19
Robinia pseudoacacia L. (Black Locust)						
1	Leaf	PC		+	+	631
Spinacia oleracea L. (Spinach)						
1	Leaf	PC/PC		+	+	321
1	Leaf[b]	Enzym.			12	632
1	–, chloroplast[b,d]	IEC		+	2.15	20
Triticum aestivum L. (Wheat)						
1	10-day leaf	PC/PC		+		331
Vicia faba L. (Broad Bean)						
1	Leaf	PC/PC		+	+	321
Vigna mungo (L.) Hepper (Black Gram)						
1	Seedling	PC		+		631
1	–, hypocotyl tip	Enzym.		2.2		332

*nmol/g fresh weight basis unless stated otherwise.
[a]Air-dry basis.
[b]nmol/mg chlorophyll.
[c]Conc., m*M*.
[d]Basis unstated.

Sucrose 6′-Phosphate

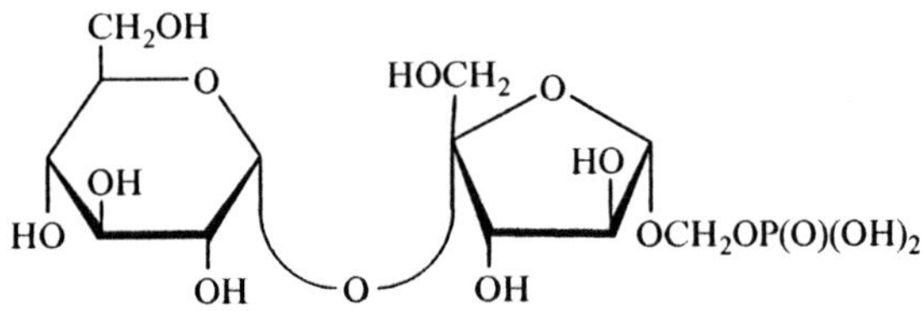

Sucrose 6′-Phosphate, $C_{12}H_{23}O_{14}P$; not listed in Beilstein. *Sucrose 6-(Dihydrogen Phosphate)* [1953].

Occurrence

Sucrose 6′-phosphate is a metabolite in the synthesis of sucrose in the higher plants. It is formed by the reaction of UDPG with F6P and is hydrolyzed to sucrose by sucrose phosphatase (Fig. 1-8).

Physical Properties

Sucrose 6′-Phosphate [R.N. 22372-29-8]: Spectra: ^{1}H NMR (fig.).[634] Chromatography: PC,[635] IEC,[636] HPLC.[637] Electrophoresis: CE.[90]

$K_2C_{12}H_{21}O_{14}P{\cdot}3H_2O$: Microcrystalline solid, $[\alpha]_D^{20}$ +34° (H_2O).[549]

$BaC_{12}H_{21}O_{14}P{\cdot}3H_2O$: Amorphous solid, $[\alpha]_D^{20}$ +35.4° (H_2O).[549]

Synthesis

Sucrose 6′-phosphate may be prepared (a) by phosphorylation of sucrose with sodium metaphosphate[634] or of sucrose 2,3,4,6,1′,3′,4′-heptaacetate with cyanoethyl phosphate/DCC[549] or (b) by reaction of UDPG with F6P catalyzed by sucrose phosphate synthase.[638]

Hydrolysis

Sucrose 6′-phosphate is stable to alkali. Mild hydrolysis in 1*N* acid (5 min at 100 °C) yields D-glucose and F6P.[638]

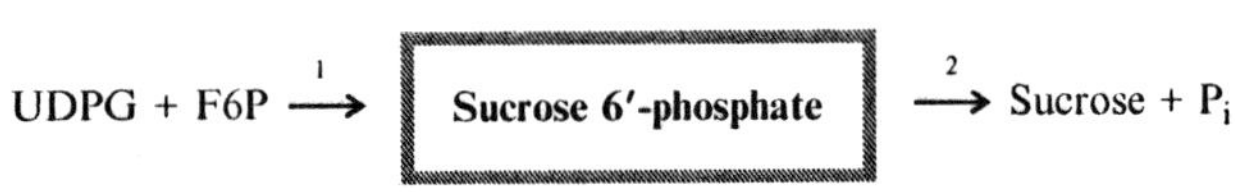

FIGURE 1-8 Sucrose 6′-Phosphate. (1) Sucrose phosphate synthase [E.C. 2.4.1.14]. (2) Sucrose phosphatase [E.C. 3.1.3.24].

Analytical Methods

The sucrose 6′-phosphate content of plant tissue is difficult to determine because of the presence of active sucrose synthase and phosphatase enzymes in the same tissue and low levels of the metabolite. In early work it was determined by two-dimensional PC and autoradiography after labeling with ^{14}C or ^{32}P (Table 1-11).[327,483] Later, it was determined by hydrolysis with 0.1*N* HCl at 100 °C and analyzed for F6P by fluorimetry.[488] In the enzymatic method, which is now preferred, the sucrose 6′-phosphate is isolated by IEC, hydrolyzed with alkaline phosphatase, and assayed for sucrose by coupled enzymatic assay.[527]

Starch Phosphate

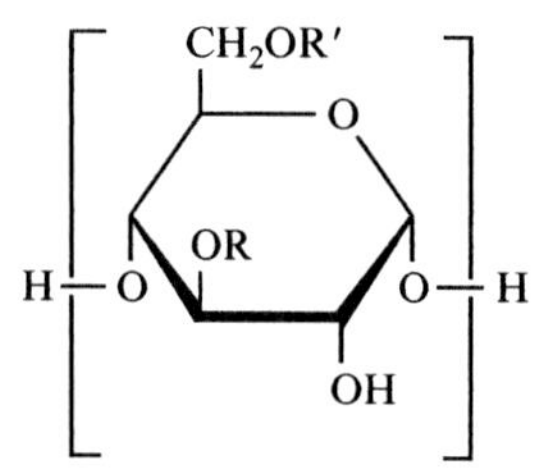

Starch, R = R′ = H
Starch 6-Phosphate, R = H, some R′ = $P(O)(OH)_2$
Starch 3-Phosphate, R′ = H, some R = $P(O)(OH)_2$

Table 1-11 Sucrose 6′-Phosphate in Plants*

No.	Plant part	Method	Sucrose-6P	References
Beta vulgaris L. (Sugar Beet)				
1	Leaf	PC/PC	+	483
Camellia sinensis L. Tea				
1	Leaf	Enzym.	0.1	488
–, var.*assamica* (Mast.) Kitam. (Assam Tea)				
1	Leaf	Enzym.	0.5	488
Fragaria vesca L. (Woodland Strawberry)				
1	Leaf	Enzym.	3.6	488
Pisum sativum L. (Peas)				
1	Cotyledon	Enzym.	23	527
Ricinus communis L. (Castor Bean)				
1	Cotyledon	Enzym.	52	303
Spinacia oleracea L. (Spinach)				
1	Leaf[a]	Enzym.	0.6-0.7	637
Spirodela oligorrhiza (Kurz.) Hegelm. (Duckweed)				
1	Whole plant	PC/PC	30	327

*nmol/g fresh weight basis unless stated otherwise.
[a]nmol/mg chlorophyll.

Occurrence

It has long been known that starch contains a small amount of residual phosphorus that cannot be removed by extraction.[639] In some plants, such as the potato, the residual phosphorus is carbohydrate-bound whereas in others, such as wheat or corn, the residual phosphorus is lipid-bound. In this chapter we are concerned only with carbohydrate-bound phosphorus.

In the potato, the residual phosphorus seems to be confined to the amylopectin component of the starch. The amount is small—about one phosphorus for every 400 D-glucosyl residues.[640] About one third of the phosphorus is bound as glucose 3-phosphate and two thirds as glucose 6-phosphate.[641] The phosphate level is independent of potato variety for C-3 but not for C-6, and as a consequence the C-6 to C-3 ratio varies from 2.0:1 to 3.4:1.[642]

The phosphate group in the amylopectin protects the $1 \rightarrow 4$ α-D-glucosidic linkages in its vicinity from attack by amylase in a manner similar to the protection afforded by the $1 \rightarrow 6$ α-D-glucosidic branch points. Both the crystallinity[643] and enthalpy of gelatinization[644] of the native starch decrease with increasing phosphorylation. These results are consistent with the fact that the C-3 carbons face the interior of the amylopectin helix whereas the C-6 carbons face the exterior, insulating the former and exposing the latter to interaction with other molecules.[643]

Hydrolysis

Enzymatic degradation of potato starch with amylase yields mixtures of phosphorylated oligosaccharides (phosphodextrins) of varying chain length, depending on the nature of the amylose and the severity of the reaction conditions. With few exceptions, the phosphodextrins are linear, contain one phosphate group per mole, and have at least one unsubstituted glucosyl residue at the reducing end.

The smallest phosphodextrins obtained by this method are maltose 6-phosphate,[641] maltotriose 6^2-phosphate,[645,646] maltotetraose 6^3-phosphate,[647,648] maltopentaose 6-phosphate,[646] and maltohexaose 6^3-phosphate,[647] where the superscript denotes the glucosyl residue bearing the phosphate substituent counting from the reducing end (right end in the formula illustrated above). All are prepared by treatment of potato starch with α-amylase to the limit of enzymatic hydrolysis, followed in some cases by further hydrolysis with β-amylase or mineral acid. Digestion of potato starch with α-amylase alone yields a series of phosphodextrins from pentaose through octaose (separated by paper chromatography) which contain progressively more phosphate on C-3.[645]

The ester linkage in starch 3-phosphate is somewhat acid-labile, which makes it difficult to quantify.[649] Tabata and Hizukuri[641] found evidence for an even more acid-labile phosphate in potato starch (approximately 1% of the phosphate in the starch), possibly as glucose 2-phosphate, but this has yet to be confirmed.

The residual phosphorus content of starch in various plants and the distribution between glucose 3-phosphate and glucose 6-phosphate if known is given in Table 1-12.

Table 1-12 Residual Phosphorus Content of Starch*

No.	Plant part	P (G_3P)	P (G_6P)	P (total)	References
Aesculus hippocastanum L. (European Horse Chestnut)					
1	Nut, amylopectin			0.015	650
Amaranthus hypochondriacus L. (Prince's Feather)					
1	Waxy seed		0.003[a]	0.003	651
Amorphophallus campanulatus (Roxb.) Blume ex Deane (Telingo Potato)					
1	Corm, amylopectin			0.044	650
Arenga pinnata (Wurmb.) Merr. (Sugar Palm)					
1	Pith, amylopectin			0.015	650
Arum maculatum L. (Cuckoopint)					
1	Root, amylopectin			0.014	650
Avena sativa L. (Oats)					
1	Seed		n.d.[a]	0.056	651
Canna indica L. (Indian Shot)					
1	Root			0.044	652
1	–, amylopectin			0.106	650
Castanea sativa Mill. (Spanish Chestnut)					
1	Seed, amylopectin			0.007	650
Colchicum autumnale L. (Autumn Crocus)					
1	Corm, amylopectin			0.007	650
Colocasia antiquorum Schott. (Culcas, Egyptian Taro)					
1	Corm, amylopectin			0.006	650
Colocasia esculenta (L.) Schott. (Taro)					
1	Corm		0.022	0.027	653
1	–, amylopectin			0.018	650
Curcuma longa L. (Turmeric)					
1	Root		0.186		654
1	–, amylopectin	Chain length distribution			654
1	Rhizome, amylopectin			0.123	650
Curcuma viridiflora Roxb.					
1	Rhizome, amylopectin			0.053	650
Cyperus esculentus L. (Yellow Nutsedge)					
1	Tuber, amylopectin			0.031	655
1	–, dextrin			0.020	655
Cyperus rotundus L. (Purple Nutsedge)					
1	Tuber, amylopectin			0.0045	655
1	–, dextrin			0.048	655
Fagopyrum esculentum L. (Buckwheat)					
1	Seed, amylopectin			0.003	650
Glycine max (L.) Merr. (Soybean)					
1	Cotyledon		0.008	0.013	653
Hordeum vulgare L. (Barley)					
1	Seed			0.059	656
1	–, amylose			0.001	656
1	–, amylopectin			0.041	656
Ipomeoa batatas (L.) Lam. (Sweet Potato)					

Table 1-12 Residual Phosphorus Content of Starch*—cont'd

No.	Plant part	P (G_3P)	P (G_6P)	P (total)	References
1	Tuber		0.008	0.012	653
1	–, amylopectin			0.0027	655
1	–, dextrin			0.0024	655
Ipomoea purga Hayne (Jalap)					
1	Tuber, amylopectin			0.006	650
Iris spp. (Iris)					
1	Rhizome, amylopectin			0.048	640
Lens culinaris Medik. (Lentil)					
1	Seed		0.002[a]	0.008	657
1	–, amylopectin			0.004	650
Malus domestica Borkh. (Apple)					
1	Fruit, amylopectin			0.025	640
Mangifera indica L. (Mango)					
1	Seed, amylopectin			0.012	640
Manihot esculenta Crantz (Cassava)					
1	Root		0.008		654
1	Tapioca		0.008[a]	0.008	651
2	Tapioca			0.005-0.018	658
1	Tapioca			0.060	659
1	–, amylose			0.018	659
1	–, amylopectin			0.026-0.030	659
Maranta arundinacea L. (Arrowroot)					
1	Root		0.021[a]	0.021	651
1	–, amylopectin			0.017	650
1	–, amylopectin	Chain length distribution			654
Metroxylon sagu Rottb. (Sago Palm)					
1	Pith			0.012	652
1	–, amylopectin			0.016	650
Musa paradisiaca L. (Plantain)					
1	Fruit, amylopectin			0.022	650
Nelumbo nucifera Gaertn. (Sacred Lotus)					
1	Seed		0.005[a]	0.005	651
Oryza sativa L. (Rice)					
6	Nonwaxy seed			0.014-0.032	660
1	Nonwaxy seed		0.013[a]	0.065	651
7	Nonwaxy seed			0.012-0.005	661
7	–, amylose			0.0024-0.0084	661
7	–, amylopectin			0.0017-0.0080	661
7	–, amylopectin	Chain length distribution			654
1	Waxy seed		0.003[a]	0.005	651
2	Waxy seed			0.001-0.002	660
1	–, amylopectin			0.0011	661
Oxalis purpurata Jacq.					
1	Tuber, amylopectin			0.0047	655
1	–, dextrin			0.0045	655

*Percentage, dry weight basis. [a]Total sugar phosphorus.

Continued

Table 1-12 Residual Phosphorus Content of Starch*—cont'd

No.	Plant part	P (G_3P)	P (G_6P)	P (total)	References
Panicum miliaceum L. (Proso Millet)					
1	Seed			0.043	662
Pastinaca sativa L. (Parsnip)					
1	Seed, amylopectin			0.054	640
Pennisetum glaucum R. Br. (Cattail Millet)					
1	Seed		n.d.[a]	0.058	651
Phalaris canariensis L. (Canary Grass)					
1	Seed			0.15	663
Phaseolus lunatus L. (Lima Bean)					
1	Bean	0.006	0.005	0.011	651
Phaseolus vulgaris L. (Bean)					
1	Bean, amylopectin			0.005	650
Pisum sativum L. (Pea)					
1	Seed	0.001	0.003	0.006	651
1	–, amylopectin			0.003	650
Pueraria hirsuta Matsum. (Kudzu)					
1	Root		0.005	0.009	653
Raphanus sativus L. (Winter Radish)					
1	Root, amylopectin			0.025	650
Sagittaria trifolia L. (Chinese Arrowhead)					
1	Root			0.009	664
Secale cereale L. (Rye)					
1	Seed			0.025	665
Solanum phureja Juz. and Buk. (Nightshade)					
1	Tuber	+	0.078		649
Solanum tuberosum L. (Potato)					
26	Tuber		0.025-0.078		666
2	Tuber	Variation in G_6P content from cortex to pith			666
2	Tuber	Variation in G_6P content with tuber diameter			319
7	Tuber			0.078-0.096	667
1	Tuber		0.053	0.075	653
1	Tuber, amylopectin		0.032	0.048	653
1	–, amylose			0.0009	668
1	–, amylopectin			0.076	668
1	–, α-limit dextrin	G_3P/G_6P ratio 1:4			669
Sorghum bicolor (L.) Moench (Sorghum)					
1	Seed		0.003		654
Trapa natans L. (Water Chestnut)					
1	Seed		0.004[a]	0.004	651
Triticosecale Witt. (Triticale)					
1	Seed			0.031	665
Triticum aestivum L. (Wheat)					
1	Seed		0.001[a]	0.054	651
1	Seed			0.118	659
1	–, amylose			0.018	659

Table 1-12 Residual Phosphorus Content of Starch*—cont'd

No.	Plant part	P (G_3P)	P (G_6P)	P (total)	References
1	–, amylopectin			0.116-0.154	659
Vicia faba L. (Broad Bean)					
1	Bean, amylopectin			0.017	640
Vicia spp. (Vetch)					
1	Tuber, amylopectin			0.033	655
1	–, dextrin			0.005	655
Vigna radiata (L.) Wilcz. (Mung Bean)					
1	Bean, amylopectin	0.004	0.007	0.012	651
Zea mays L. (Maize)					
1	Seed			0.017	670
1	Seed		tr[a]	0.016	651
1	–, amylose			0	671
1	Waxy seed		0.002	0.002	651
1	–, high amylose		0.013[a]	0.020	651

*Percentage, dry weight basis.
[a]Total sugar phosphorus.

2

Phytic Acid

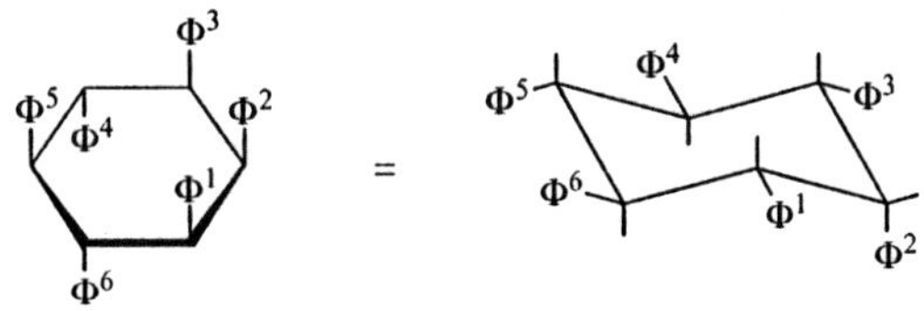

Phytic acid, $\Phi = OP(O)(OH)_2$

Phytic Acid ($InsP_6$), $C_6H_{18}O_{24}P_6$; Beilstein **6**H 1197, E_I 590, E_{II} 1160, E_{III} 6935, E_{IV} 7927. myo-*Inositol Hexakis(dihydrogen phosphate)* [1872]

Occurrence

Phytic acid, the hexaphosphate ester of *myo*-inositol, is found in all plant seeds, being most abundant in the cereals and legumes. It occurs frequently in the form of a calcium magnesium salt for which Posternak[672] coined the name "phytin" [R.N. 3615-82-5]. Some plant phytins, however, contain more potassium than calcium, and other elements may be present such as Mn, Fe, or Zn.[673] Phytin, therefore, is not a definite compound but a collective term for the various salt forms of phytic acid in its natural state.

Phytic acid was isolated from wheat seed by Pfeffer[674] in 1872 and identified as an ester of *myo*-inositol and phosphoric acid by Winterstein[675] in 1897, but its structure was not established unambiguously until much later.

For general references on this subject see Cosgrove and Irving[676] or Graf.[673]

Physical Properties

Phytic Acid [R.N. 83-86-3]: Thick, practically colorless syrup that darkens on standing or on heating to 100 °C or above[677]; d^{25} 1.369,[678] $[\alpha]_D^{20}$ 0.00 (H_2O)[679]; soluble in water and acetone, slightly soluble in methanol and ethanol, and insoluble in ether, hexane, benzene, chloroform, and glacial acetic acid[678,680]; pK_a −0.15 (P_2) to 11.76 ($P_{1,3}$ and $P_{4,6}$), all 12 pK_as assigned[681]; conductivity.[682] Spectra: UV (fig.),[683] IR,[679,684] near IR,[685] Raman,[686] ^{1}H NMR,[687] ^{13}C NMR,[686] ^{31}P NMR,[679,686,688] HMQC ^{31}P NMR (fig.),[689] FAB/MS,[690] ESR.[691] Chromatography: PC,[692] TLC,[693] IEC,[694] HPTLC,[407] GFC,[695] HPLC,[696] IC.[697] Electrophoresis: PE,[698] TLE,[693] disk gel,[699] CZE.[700]

$Na_{12}C_6H_6O_{24}P_6{\cdot}38H_2O$ [R.N. 14306-25-3]: Colorless rhombic prisms, mp 58-59 °C,[701,702] soluble in hot H_2O, insoluble cold,[702] piezoelectric positive[701]; X-ray, d 1.75[701,703]; loses $35H_2O$ at 120 °C giving a trihydrate, white amorphous solid, dec. 210 °C.[702]

$Na_{12}C_6H_6O_{24}P_6{\cdot}nH_2O$: Hydrates of composition $n=1$[679]; $n=33$, X-ray, DTA, PE[679,704]; $n=35$, X-ray, DTA[679,704]; $n=47$[702]; $n=x$ (indefinite), solubility in H_2O[705] and aqueous ethanol[706]; near IR (fig.),[685] TGA,[707] Raman,[686] FAB/MS (positive and negative).[690]

$Ca_5C_6H_8O_{24}P_6{\cdot}3H_2O$ [R.N. 78342-17-3]: White amorphous powder, slightly soluble in H_2O[708]; solubility versus pH[709]; solubility product in 0.2 and 1.0*M* NaCl[710]; X-ray (fig.).[710]

$Ca_6C_6H_6O_{24}P_6{\cdot}xH_2O$ [R.N. 7776-28-5]: IR, Raman.[686]

$(C_{23}H_{26}N_2O_4)_6C_6H_{18}O_{24}P_6{\cdot}24H_2O$ (brucine salt): Needles, mp ~210-211 °C.[711]

$(C_{21}H_{22}N_2O_4)_6C_6H_{18}O_{24}P_6$ (strychnine salt): Slender needles, mp 203-204 °C,[712] ~210-211 °C,[711] or 218-219 °C.[713]

Synthesis

Phosphorylation of *myo*-inositol with phosphoric acid/phosphorus pentoxide at 130 °C yields a mixture of *myo*-inositol polyphosphates ($n=4$-6) from which phytic acid can be isolated by fractional crystallization of the barium salts. The yield is only 8%.[702] Alternatively, *myo*-inositol may be phosphorylated with polyphosphoric acid at 160 °C and purified by ion exchange chromatography.[714] Phosphoric acid[715] and pyrophosphoric acid[708] yield only partially phosphorylated products. Phosphorylation of *myo*-inositol with triethylammonium *N*-benzoylphosphoramidate in DMF followed by removal of the protecting groups gives phytic acid ($InsP_6$) together with a small amount of Ins (1,3,4,5,6)P_5. The protective reagents usually employed in the phosphorylation of alcohols are ineffective.[716]

Hydrolysis

Hydrolysis of phytic acid under carefully controlled conditions leads to a series of *myo*-inositol derivatives ranging from pentakisphosphates to monophosphates. Owing to the unsymmetrical nature of the *myo*-inositol molecule, there are 6 possible pentakisphosphates, 15 tetrakisphosphates, 20 trisphosphates, 15 bisphosphates, and 6 monophosphates. Some of these are D,L-stereoisomer pairs, reducing the number of chemically distinct isomers to 4, 9, 12, 9, and 4, respectively.

Hydrolysis of phytic acid at 95 °C, measured by the release of P_i, follows first-order kinetics with a rate constant of 8.74×10^{-4} min^{-1} at pH 4, falling to 3.98×10^{-4} at pH 1, and 0.28×10^{-4} at pH 12. The presence of metal ions has little influence on the rate.[717]

Hydrolysis of phytic acid at pH 4.0 and 110 °C for 100 min in a sealed tube yields a pentakisphosphate fraction[718] that can be separated into its individual components by ion exchange chromatography[719] or moving paper electrophoresis.[720] The structures of the isomers (confirmed by later studies) are illustrated in Fig. 2-1.[719]

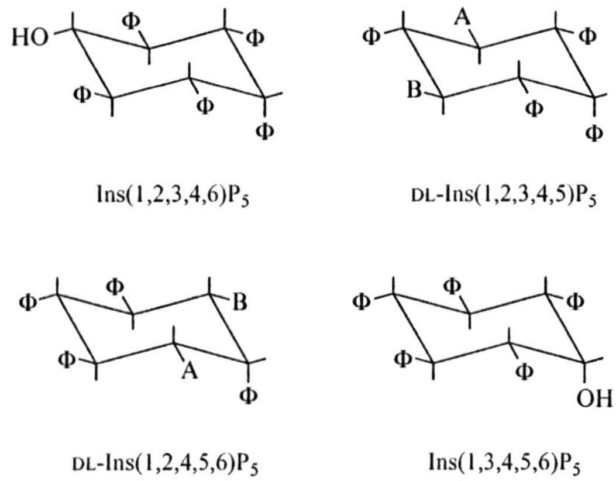

FIGURE 2-1 *myo*-Inositol Pentakisphosphates. A = OH, B = Φ for the D isomer and A = Φ, B = OH for the L isomer.

Further hydrolysis of phytic acid with acid, base, or enzyme catalysts can be made to give tetrakis-, tris-, bis-, or monophosphates depending on the severity of the reaction conditions. The complexity of the mixtures does not permit the isolation of individual isomers until the monophosphate stage is reached. The only monophosphate formed under acid or base catalysis is the *meso* isomer Ins(2)P,[721,722] but some Ins(1)P is also formed under enzyme catalysis.[723]

Complete hydrolysis of phytic acid to *myo*-inositol and phosphoric acid requires more drastic conditions such as 8 h refluxing with 35% H_2SO_4[724] or autoclaving at 220-230 °C with 20% NaOH.[675] Phytase does not remove the last phosphate group.

Conformation

The conformation of calcium phytate, like that of *myo*-inositol itself, is known to be $1_{ax}/-5_{eq}$ in the solid state, based on Raman spectroscopic analysis.[686] Sodium phytate, however, has the inverted $5_{ax}/1_{eq}$ conformation, based on X-ray diffraction measurements on the crystalline salt.[703]

The situation is much less clear for phytic acid and its salts in aqueous solution. According to ^{13}C NMR, ^{31}P NMR, and Raman spectroscopy, sodium phytate has the $1_{ax}/5_{eq}$ conformation at low pH and the $5_{ax}/1_{eq}$ conformation at high pH. The inversion occurs at pH 9.4.[686] The tetra-*n*-butylammonium salt exhibits three inversions: from $5_{ax}/1_{eq}$ to $1_{ax}/5_{eq}$ at pH 2 and at pH 12 and from $1_{ax}/5_{eq}$ to $5_{ax}/1_{eq}$ at pH 5.[688] The preferred conformation apparently depends on the nature of the counterion as well as the degree of protonation. That this matter remains unresolved is evident from two papers published in 1999 in which one group carried out a detailed study of an equatorial to axial inversion occurring in the pH 10-11.3 range[725] while another group, using the same reagents and spectroscopic methods, found no such inversion up to pH 12 (Fig. 2-2).[726]

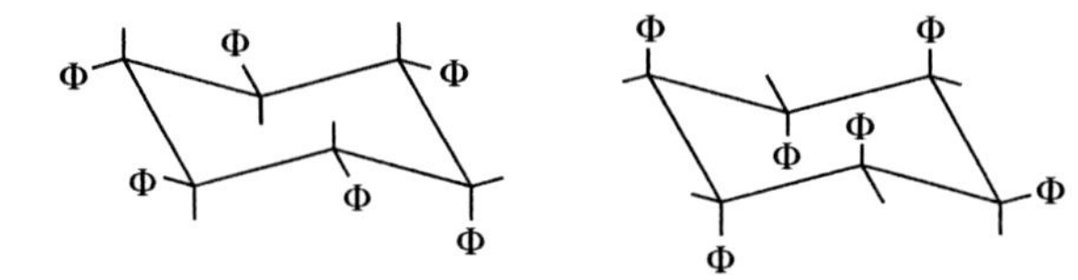

FIGURE 2-2 Conformational isomers of phytic acid. Left, $1_{ax}/5_{eq}$ conformer; right, $5_{ax}/1_{eq}$ conformer. The unique substituent is at C2 in both cases. Note the steric crowding of the phosphate (Φ) groups in the $5_{ax}/1_{eq}$ conformer.

Preparative Methods

Phytic acid can be isolated from the seeds of a variety of plant species by extraction with acid. Some of the sources for which detailed procedures have been described are listed in Table 2-1.

Despite the availability of several synthetic methods, phytic acid is seldom prepared by chemical synthesis since it is so readily available from natural sources. Commercial phytic acid is prepared from corn steep liquor in the United States[673c] and from rice bran in Japan.[678] Phytic acid solutions that contain salts can be desalted by ion exchange chromatography or more conveniently by dialysis.[729] Calcium precipitation is inefficient.[730]

Table 2-1 Phytic Acid from Various Plant Sources

Plant species	% Yield	References
Brassica campestris L. (Field Mustard)	0.38-0.57	712
Acer saccharinum L. (Silver Maple)	0.49	727
Pisum sativum L. (Peas)	0.30	728
Triticum aestivum L. (Wheat)	0.14	713

Quantitative Analysis

There are no known reagents specific for phytic acid. It does not have a characteristic UV, visible, or IR spectrum. Much attention, therefore, has been devoted to the development of suitable methods for its analysis. The more rigorous of these methods include hydrolysis of the phytic acid and separate analysis of its component parts: *myo*-inositol and phosphoric acid.

For many years, the most widely used method was that of Heubner and Stadler[731] in which phytic acid is precipitated as the sparingly soluble ferric salt (Fe:P ratio 4:6).[732] Several modifications of this method have been introduced and some have been widely adopted.[733–735] The lower limit of detection is ~3 μg/mL of phytate phosphorus.[736] In 1977, Harland and Oberleas[737] proposed an ion exchange chromatographic method that has since been adopted as an official AOAC standard.[738] In this method, an aliquot of the extract is transferred to a column containing a strong base anion exchange resin such as AG1-X4 (Cl form), eluted with NaCl in a gradient from 0.1 to 0.7*M*, digested with acid, and analyzed for %P. The use of reverse phase HPLC for the analysis of phytic acid proposed by Tangendjaja *et al.*[739] in 1980 is becoming more and more widely used as modifications have been introduced to increase its accuracy and sensitivity.[740–742] Detection limits as low as 2 ng have been reported.[740] Other methods include ^{31}P NMR,[743] polarography,[744] bioassay with phytase,[745] isotachophoresis,[746] ion chromatography,[697] near IR,[685] and kinetics of crystal growth.[747] Methods based on the analysis of the *myo*-inositol component include oxidation with periodic acid[724] and bioassay with *Schizosaccharomyces pombe.*[748]

(1) McCance-Widdowson Method

This method is a widely used modification of the ferric phytate method introduced by Heubner and Stadler in 1914, which was based on the insolubility of ferric phytate in the dilute acid used to extract the phytic acid from the plant.[731] The original method is now seldom used because of interference with other polyphosphates and with phosphoric acid, and because the endpoint of the titration is often difficult to determine precisely. In the McCance-Widdowson modification (1935), the ferric phytate is converted to sodium phytate, digested with perchloric acid, and analyzed for phosphorus:

> *5-10 g. of dried, finely-ground material were shaken in a small glass-stoppered bottle with 100 ml. N/2 HCl for 2 hours to extract the phytic acid; 10 to 40 ml. of the filtered extract were neutralized to phenolphthalein with NaOH, rendered slightly acid with HCl and made up to 50 ml. Duplicate 20 ml. aliquots were treated in 50 ml. centrifuge-tubes, with 4 ml. of the $FeCl_3$ solution [containing 0.5 mg. ferric iron per ml. of N HCl]. The tubes were heated in a boiling water bath for 15 min. to flocculate the precipitate of ferric phytate, cooled, centrifuged and the supernatant liquid poured off. The precipitate was washed with 5 ml. N/6 HCl, centrifuged again and the acid decanted. The precipitate was then stirred up with 2 ml. distilled water and heated in a boiling water bath for a few minutes. 2 ml. of 2% NaOH were then added, and the heating continued for a further 15 min. The solution containing the phytin as sodium phytate was filtered into a Kjeldahl flask, the precipitated ferric hydroxide was well washed with hot water and the washings were added to the filtrate in the flask. 1 ml. of conc. H_2SO_4 and 1 ml. of 65% $HClO_4$ were added, and the mixture incinerated very gently until completely digested. It was then heated strongly for 60 min. to drive off any residual $HClO_4$. When cool, about 20 ml. of water were added and the contents of the flask just neutralized to phenolphthalein with 40% NaOH. The solution was then made up to 100 ml. [and analyzed for phosphorus by the phosphomolybdate method].*
>
> Reproduced with permission from McCance, R. A.; Widdowson, E. M., "Phytin in Human Nutrition", *Biochem. J.* **1935**, *29*, 2694–2699,[735] © 1935 the Biochemical Society.

Some minor improvements have been described.[732,749–752] An alternative procedure is that of Young[753] in which the ferric ion is estimated calorimetrically, either in the precipitate[733,754] or in the supernatant solution.[734,753,754] As this method depends on the 6:4 stoichiometry of P:Fe in ferric phytate, much attention has been devoted to ensuring that the ratio is correct. The most widely used variants of this method are those of Makower,[754] Young,[753] and Wheeler and Ferrel.[733] The Wade reagent used by Latta and Eskin[734] can be used if the phosphoric acid concentration does not exceed 20 mg/mL.[755] The lower limit of measurement in the McCance/Widdowson method is 0.16 mmol/mL.[746] The method has been semiautomated.[756]

(2) Michel-Durand Method

This method was developed to overcome the coprecipitation of other phosphorus compounds, notably nucleic acids and sugar phosphates, with ferric phytate in the McCance/Widdowson method. In principle, the acid extract is treated with ammonia-magnesium reagent to precipitate inorganic and phytate P and the precipitate is dissolved in 0.5*N* HCl. Upon addition of calcium acetate, the phytate P precipitates but the inorganic P remains in solution.[757] In one study,[758] the calcium acetate treatment was omitted. Improvements in the Fe precipitation method (method 1) and the development of newer methods have rendered this method obsolete.

> *A volume of cold acid extract, equivalent to 200-300 g phytic P, is measured in a conical centrifuging tube; 2 ml. of magnesium reagent is added and the mixture is neutralized with 8* N *NH_4OH in the presence of neutral red, then made alkaline with 1 ml. 8* N *NH_4OH. The precipitate is a mixture of crystalline MgNH, phosphate and amorphous Mg phytate. After standing overnight in the cold, it is separated by centrifugation, dissolved in 5 ml. 0.5* N *HCl, and immersed in a boiling-water bath for 10 min.; 5 ml. 25% Ca acetate is then added and the Ca phytate precipitate is heat-coagulated for 5 min. It is recovered by centrifuging and ashed for P determination.*
>
> Bourdet, A.; Feillet, P., "Distribution of Phosphorus Compounds in the Protein Fractions of Various Types of Wheat Flour", *Cereal Chem.* **1967**, *44*(5), 457–482,[759] reprinted by permission of AACC International, © 1967.

(3) Harland-Oberleas Method

In this method, introduced in 1977, phytic acid is separated from phosphoric acid and any lower inositol phosphates by ion exchange chromatography, digested with concentrated HNO_3-H_2SO_4, and analyzed for phosphorus.[737]

> *Samples of dried material (5 g) were extracted for 2 hr at room temperature with 100 ml of 1.2 % HCl. . . . . Glass barrel Econo-columns (. . . . .0.7 x 15 cm) equipped with a 3-way Luer fit stop-cock. . . . .were clamped vertically and packed with about 0.5 g of 200-400 mesh AG1-X8 chloride form anion-exchange resin. . . . . Before use, the columns were washed first with 0.7* M *NaCl to assure chloride saturation of the resin and then with distilled water until the eluate was salt-free. The sample extract was diluted before being placed on the column so that the total anion concentration was less than 0.05* M *(about 1:25). The sample container was rinsed and the column*

was washed with 15 ml of distilled water to elute most of the inorganic phosphate. The remaining inorganic phosphate was eluted with 15 ml of 0.05 M *NaCl. The phytate was then eluted with 15 ml of 0.7* M *NaCl and collected in a microKjeldahl flask or digestion tube. To the digestion vessel was added 0.5 ml of concentrated H_2SO_4, 3.0 ml of concentrated HNO_3, and a few boiling chips. Digestion was continued until only the H_2SO_4 remained (25-30 min)... . About 10-15 ml of distilled water was added to the digestion vessels, the contents were mixed thoroughly, and the vessels were placed in a boiling water bath for 15 min to hydrolyze any polyphosphates which may have formed during digestion..... An aliquot of the digested eluate was then taken for phosphorus determination by the Fiske-Subbarow method.*

Harland, B. F.; Oberleas, D., "A Modified Method for Phytate Analysis Using an Ion Exchange Procedure: Application to Textured Vegetable Proteins", *Cereal Chem.* **1977**, *54*(4), 827–832,[737] reprinted by permission of AACC International, © 1977.

To avoid digestion, many analysts prefer the modification of Latta and Eskin[734] in which the phytate content of the column eluate is determined spectrophotometrically with a modified Wade reagent (ferric chloride and sulfosalicylic acid, absorbance at 500 nm). One newer method measures the chloranilic acid released from zinc chloranilate by phytic acid at 332 nm,[760] another measures the fluorescence of 2,2′-dipyridyl ketone hydrazone after Cu(II)-catalyzed oxidation,[761] and yet another measures the molybdenum blue/phytic acid complex by flow injection analysis.[762] Analysis for phytate phosphorus without digestion is also possible with inductively coupled plasma spectrometry.[752] Minor improvements are the use of EDTA to remove interfering cations[763] and a more reliable resin (AG1-X4) for the ion exchange.[764] These changes are incorporated in the AOAC standard.[738]

(4) High-Performance Liquid Chromatography (HPLC) Method

This method was introduced in 1978 by Gerritse,[696] who used it to measure the phytic acid content of a variety of organic materials including potatoes and annual bluegrass. Three variants of the method have evolved, one employing conventional ion exchange and the other two employing ion-pair and ion chromatography.

Cilliers and van Niekerk[765] developed a rapid, sensitive procedure that requires minimal sample preparation. It employs a low capacity anion exchange resin (IC-PAK-A), an isocratic mobile phase (0.045*M* $NaNO_3$/0.018*M* HNO_3), and postcolumn colorimetric detection with the Wade reagent ($FeCl_3$/sulfosalicylic acid). If hydrolysis of the phytic acid is suspected, the use of a linear gradient of 0-0.5*M* $NaNO_3$ in 0.01*M* 1-methylpiperazine provides a better separation of the lower inositol phosphates.[766] Strong anion exchange resins such as Aminex A-27 or PL-SAX can be employed provided that column hydrolysis is avoided by gradient elution with 0.1-0.5*M* NaCl containing Na_4EDTA.[767,768] Phytic acid can also be detected by the change in refractive index,[697] UV/vis spectrum,[697] or conductivity[769] of the eluate; by flame ionization (FID)[770]; or by evaporative light scattering (ELS).[771]

Early attempts to determine phytic acid by HPLC on reverse phase C_{18} columns such as μBondapak or Spherisorb ODS were hampered by poor resolution owing to weak retention on the column. Phytic acid eluted with the solvent front.[739–741,772] This problem was overcome by adding an ion-pairing reagent, tetrabutylammonium hydroxide, to the mobile phase[742] and by adjusting the pH to 4.3 to optimize the separation of the inositol tris-, tetrakis-, pentakis-, and hexakisphosphates.[773] In this procedure, the InsP's must be separated from the plant extract prior to injection into the HPLC column. The following procedure is typical:

> *Samples of 0.5 g freeze-dried, ground food or intestinal contents were extracted under vigorous mechanical agitation with 20 mL 0.5* M *HCl for 2 hr at 20 °C (Graf and Dintzis, 1982). The extract was centrifuged and the supernatant decanted, frozen overnight and filtered through MF-Millipore (0.22 μm) under pressure. The inositol phosphates were separated from the filtrate and concentrated by a modification of the ion-exchange procedure of Graf and Dintzis (1982).... The eluent was evaporated to dryness and diluted with 1 mL of water.... The mobile phase consisted of 0.05* M *formic acid:methanol (Fison HPLC grade) 49:51 and 1.5 mL/100 mL of TBA-OH (tetrabutylammonium hydroxide) (Fluka, 40 % in water) were added. The pH was adjusted to 4.3 by addition of 9* M *sulfuric acid. The mobile phase was filtered through a MF-Millipore (0.45 μm) under vacuum and degassed by sonication for 15 min. A reverse phase C-18 [HPLC] column (Waters Radialpak) 5 micron particle size was equilibrated with the mobile phase for 1 hr. Analysis was conducted with an HPLC pump (Waters Model 510) equipped with a Z-module. Inositol phosphates were detected by refractive index.... The optimal flow rate was 3 mL/min. Retention times and peak areas were measured....*
>
> Sandberg, A.-S.; Ahderinne, R., "HPLC Method for Determination of Inositol Tri-, Tetra-, Penta- and Hexaphosphates in Foods and Intestinal Contents", *J. Food Sci.* **1986**, *51*(3), 547–550,[773] reprinted by permission of John Wiley & Sons, © 1985.

Lehrfeld[774,775] has introduced numerous modifications to improve the speed and reliability of the method. The third variant is ion chromatography, run in the isocratic mode with Dionex AS3 as the stationary phase and 0.11*M* HNO_3 as the mobile phase.[697] Treatment of the plant extract with EDTA eliminates the interference of metal ions.[776] The lower limit of detection for the HPLC method is reported to be 5 mg[696,741] to 2 ng[742] phytic acid at sample sizes down to 50 mg.[777]

(5) Gel Filtration Chromatography (GFC) Method

Gel filtration chromatography (GFC) has been employed in a few instances to separate phytic acid from other plant phosphorus compounds. The column packing used most often is Sephadex G-25. The lower limit of detection is 0.05-0.5 mg phytate P/g.[696] Analysis of the phytate fractions for other elements in two of the cited references[778,779] (the third[696] used EDTA) revealed the presence of sodium, potassium, and magnesium but no calcium; this method therefore seems to offer a useful means for identifying the cationic components of plant phytins, a feature that should be further exploited.

(6) Near-Infrared Method

Parrish *et al.*[685] recently developed a near-IR method for the analysis of phytic acid in cottonseed, using a pair of wavelengths (1568 and 1230 nm) selected from the 1100- to 2500-nm spectrum. The phytic acid content is calculated by means of the equation (% phytic acid $= 100.625 - 0.657\ (\lambda_{1568}/\lambda_{1230})$). The method is simple, rapid, and does not require prior extraction of the sample, but does require calibration for each plant substrate by one of the standard methods of analysis. The results correlated well ($R = 0.998$) with those determined for cottonseed by the HPLC method.[697] Nevertheless, Le Boever *et al.*[780] found the method to be of limited value for predicting the phytate content of 19 Belgian foodstuffs, even when split into starchy and protein-rich groups.

(7) ^{31}P Nuclear Magnetic Resonance (NMR) Method

In the pH 4-5 range, the ^{31}P NMR spectrum of phytic acid exhibits a pattern of four signals in a 1:2:2:1 ratio that is readily distinguishable from those of inorganic phosphate and the lower inositol phosphates. In this method, introduced by O'Neill *et al.* in 1980, the peak area of one of these signals, usually the C-2 phosphate, is measured after a large number of transients by Fourier transform and compared to a calibrated standard[743]:

> *The sample (2.5 g) was stirred vigorously for 0.5 h in a solution of aqueous 3% w/v trichloroacetic acid (10 mL) containing 0.1% w/v sodium cyclotrimetaphosphate tetrahydrate at room temperature.... Additional 3% w/v trichloroacetic acid was used when necessary to ensure efficient stirring. After centrifuging, the supernatant was adjusted to pH 4.5 ± 0.5 by addition of tetrasodium EDTA before filtering (0.45-μm Millipore) and adding 2 mL to sucrose (2.5 g) with stirring to aid dissolution.... All spectra were recorded with proton noise-decoupling on a JEOL PFT-100 NMR spectrometer fitted with a 5-kHz crystal filter. The calibration and sample spectra were recorded [at 40.3 MHz, pulse width 15 ms, data acquisition time 1.02 s, deuterium lock... the] standard and phytate C-2 phosphate peak areas were calculated from the triangulated signals resulting from peak extrapolation on these spectra.*
>
> O'Neill, I. K.; Sargent, M.; Trimble, M. L., "Determination of Phytate in Foods by Fourier Transform Phosphorus-31 Nuclear Magnetic Resonance Spectrometry", *Anal. Chem.* **1980**, *52*(8), 1288–1291,[743] reprinted by permission of the American Chemical Society, © 1980.

The method was modified by Mazzola *et al.*[781] and Ersöz *et al.*,[782] who eliminated the need for sucrose as a relaxation agent by using phosphorous acid[782] or phytic acid itself[781] as the standard. Unlike other methods this method is specific for phytic acid, but it requires large sample sizes (2.5-5.0 g) and its sensitivity to samples of low phytate content (<0.1%) is poor.

(8) Inositol Method

In this method, phytic acid is determined by bioassay of the *myo*-inositol liberated after hydrolysis with 1*N* HCl in a sealed Pyrex tube at 123 °C. *myo*-Inositol itself is stable under these conditions. The results must be corrected for any free *myo*-inositol present in the plant tissue, which is determined separately by bioassay after extraction with 0.04*N* HCl at 4-5 °C[748]:

About 100 mg. of sample was heated with 2 ml. of N-*HCl in a sealed Pyrex tube at 123° for 48 h. After cooling, the tube was opened, the hydrolysate filtered with numerous washings into a measuring flask, the pH adjusted to 4.8 with NaOH and the solution made up to volume.... The method used [to assay inositol] was described by Norris & Darbre (1956). The test organism was a strain of* S. pombe.... *The method is, in general, accurate within ±3%.*

Reproduced with permission from Darbre, A.; Norris, F. W., "Vitamins in Germination. Determination of Free and Combined Inositol in Germinating Oats", *Biochem. J.* **1956**, *64*(3), 441–446,[748] © 1956 the Biochemical Society.

Free *myo*-inositol may also be determined by bioassay after prior separation from its phosphates, including phytic acid, by ion exchange chromatography on Amberlite IRA-410 (OH^-).[783] The bioassay organism may be *S. pombe*[784] or *Saccharomyces carlsbergensis.*[783] Alternatively, free *myo*-inositol may be determined by GC using scillitol as an internal standard.[785]

(9) Phytase Method

This method, used extensively in Sweden for the analysis of phytic acid in rapeseed, is based on the hydrolysis of phytic acid by the enzyme phytase (E.C. 3.1.3.26). Since this method measures the P_i released, and wheat phytase hydrolyzes only five of the six phosphate groups of phytic acid *in vitro*,[786] it would seem that the values should be low by a factor of 1/6 (16.7%), but direct comparison shows that the values are actually slightly higher than those of the Wheeler and Ferrel method.[745]

Defatted RSM [rapeseed meal] (1 g) was extracted with 25 ml 15% trichloroacetic acid (TCA) for 1 h at room temperature with mechanical shaking. After filtering, 6 ml of the solution was transferred to a 25-ml volumetric flask. Sodium acetate buffer (4 ml, 2.8 M*) containing 0.1* M *magnesium sulphate and 3 ml of [freshly prepared wheat] phytase solution was added.... The mixture was diluted to volume with distilled water, after which the pH was 5.1. Then 1 ml of this solution was taken and used for background phosphate analysis, and the remainder of the solution incubated for 16 h at room temperature. Thereafter, 1 ml of the hydrolysed solution was analyzed for phosphate by the ammonium molybdate method. The content of phytate phosphorus [was] calculated from the difference between the measurement before and after the enzymatic hydrolysis.*

Uppström, B.; Svensson, R., "Determination of Phytic Acid in Rapeseed Meal", *J. Sci. Food Agric.* **1980**, *31*(7), 651–656,[745] reprinted by permission from John Wiley & Sons, © 1980.

(10) Paper Electrophoresis (PE) Method

Phytic acid is readily separated from phosphoric acid and the lower inositol phosphates by paper electrophoresis.[787] The following method was adopted for the quantitative determination of phytic acid in pollen:

> *Dry pollen (200 mg) was ground in a mortar precooled with liquid N_2, then 2 ml 0.02* M *Na_4EDTA was added, allowed to freeze over the pollen and the whole ground into a fine powder. The powder was allowed to thaw, with further grinding, poured into a small tube and heated in a boiling water bath for 20 min. After cooling to 0° [and] centrifugation (Sorvall SSI; 6000 g for 20 min)... the first... pellet was re-extracted... with 2 ml 0.02* M *Na_4EDTA.... Both [extracts] were subjected to paper electrophoresis (0.1* M *oxalic acid, pH 1.5; 550 V for 2 hr; on Whatman 3 MM paper 10 x 43 cm). Under these conditions phytic acid migrates well ahead of other cellular phosphate esters.... Phosphate-containing areas in the dry electrophoretogram were then stained by dipping in an acid molybdate solution in acetone... dried overnight in the dark and exposed briefly to sunlight next day. The intensity of the discrete blue spot from pollen extracts with the same mobility as phytic acid, was compared with that from a series of standard phytic acid spots run under the same conditions. It was found that as little as 1 nmol phytic acid (6 nmol of phosphate) could be detected by this method.*
>
> Jackson, J. F.; Jones, G.; Linskens, H. F., "Phytic Acid in Pollen", *Phytochemistry* **1982**, *21*(6), 1255–1258,[788] reprinted by permission of Elsevier, © 1982.

Williams[789] cut out the phytate spots and digested them in Kjeldahl flasks with 70% (w/v) perchloric acid for P analysis by the Bartlett method.

(11) Capillary Zone Electrophoresis Method

In this method, phytic acid is separated from phosphoric acid and the lower inositol phosphates by capillary zone electrophoresis (CZE), using appropriate buffers. The procedure of Blatny *et al.*[700] enables $InsP_6$, $InsP_5$, and P_i to be analyzed in a single run. Others used potential gradient detection[746] or indirect UV detection with histidine as the UV absorber.[790]

> *Two to five grams of the ground sample is weighed into a 100 mL volumetric flask, and 50 mL of HCl of different concentrations (0-5%) is added. IP6 is extracted from the sample by shaking for 1 h at room temperature. After this period, the mixture is filtered, diluted, and directly analyzed by isotachophoresis. The volume-coupling instrument (Ionosep 900.1, Recman-Laboratorni Technika, Ostrava, Czech Republic) was equipped with Teflon capillaries.... Detection was carried out with a contactless high-frequency conductivity cell. The following constant current was applied:... initially 70 mA (15 mA during detection).... The electrolyte system consist[ed] of*

> *0.01 mol/L hydrochloric acid, 0.0056 mol/L bis-tris-propane (pH 6.1) as leading electrolyte, and 0.005 mol/L 2-morpholinoethanesulfonic acid as terminating electrolyte.... The detection limit was less than 10^{-6} mol/L, which corresponds to a phytic acid content of the samples of less than 0.08 %.*

Blatny, P.; Kvasnicka, F.; Kenndler, E., "Determination of Phytic Acid in Cereal Grains, Legumes, and Feeds by Capillary Isotachophoresis", *J. Agric. Food Chem.* **1995**, *43*(1), 129–133,[700] reprinted by permission of the American Chemical Society, © 1995.

(12) Other Quantitative Methods

Miscellaneous other methods used for the analysis of phytic acid are outlined in Table 2-2. In some cases insoluble salts other than Fe (McCance method) or Mg (Michel method), such as the Ca,[708] Ba,[715,806] or Cu[708,807] salts, were used for the precipitation of phytate. The highly insoluble thorium[808] and bismuth[809] salts have been suggested but have not been used for this purpose. Babakhodzhaeva *et al.*[744] titrated a lead nitrate solution with a rice flour extract and determined its phytic acid content by polarography. In some cases the method was undisclosed or could not be determined.

Table 2-2 Other Quantitative Methods

Author	Method	References
Schmalfuss (1940)	NH_4 citrate extn., Cu acetate pptn.	791,792
Mukhamedova *et al.* (1977)	HNO_3 extn., pptn. of phytin	793–796
Suzuki (1907)	P (HCl-soluble) – P_i	797–799
Ogawa *et al.* (1975)	P (total)	800
Ogawa *et al.* (1977)	P (acid-soluble)	801
Fontaine *et al.* (1946)	P (total) – P_i	802
Kazakov *et al.* (1980)	H_2O extn., pptn. of phytin	803
Morton *et al.* (1964)	P (total) – P (lipid) – P (nucleic acid)	804,805

Qualitative Analysis

(1) Chromatographic Methods

Paper chromatography (PC) and thin layer chromatography (TLC) have been employed in a few instances for the identification of phytic acid in plant extracts. Sobolev[810] and others used *n*-PrOH/NH_3/-H_2O (5:4:1) to separate phytic acid from P_i and the lower inositol phosphates (see Chapter 3, Table 3-2)[693,811,812] and from other phosphate esters such as ATP or phosphatidylinositol (PI).[813] Other solvent mixtures have been used to separate phytic acid from inorganic polyphosphates[814] and sugar phosphates.[315]

(2) Disk Gel Electrophoresis (DGE) Method

Polyacrylamide disc gel electrophoresis of aqueous seed extracts has been used to verify complex formation between phytic acid and the plant proteins. Soybean and sesame protein form phytate complexes but corn germ does not.[699] The phytate band is visualized by staining with ferric chloride, which produces a white precipitate on an orange background. A modified staining technique was found to be nonspecific.[815]

(3) Mössbauer Spectroscopy

Mössbauer spectroscopy of wheat seeds grown in a medium enriched in ^{57}Fe provides direct evidence that the iron in wheat is combined in the form of monoferric phytate in both seed and bran. Monoferric phytate exhibits a quadrupole splitting of 0.55 mm/s and an isomer shift of 0.77 mm/s relative to sodium nitroprusside at 80 K.[816] In soybean, however, the splitting is too large, suggesting some other combining form for the ferric ion. Ferrous ion is also present, especially in the immature seed.[817]

(4) Fourier Transform Infrared (FTIR) Spectroscopy

Sowa and Connor[818] used FTIR spectroscopy to follow the changes that occur in pine and spruce pollen during germination. The decrease in P—O vibrations in the 1300-900 cm^{-1} frequency range reflects the utilization of phytate.

(5) Light Microscopy

The aleurone grains of barley and other plant seeds contain globoids that stain intensely red with toluidine blue owing to the presence of phytin.[819] This metachromasia, though not specific for phytin, is useful as an aid in the localization of phytin when used in conjunction with other methods. Another useful marker is acriflavine-HCl, which imparts an intense red fluorescence to phytin globoids.[820]

> *[Aleurone] layers were... fixed... in 2.5% glutaraldehyde in 0.05* M *cacodylate buffer, pH 7.2... for 6 h at 2 °C. After fixation, the layers were washed for 12-15 h in several changes of the fixing buffer at 2 °C. The tissue was supported in a gelatin-antifreeze medium... for sectioning at −24 °C... [2 μ] sections were stained in 0.02% toluidine blue O (Harleco, C.I. No. 52040) in acetate buffer... at pH 5.0. The slides were examined immediately after rinsing and mounting in buffer and also after dehydration in 95% ethanol (10 s), ethyl cellosolve (10 s) and xylene before mounting in Eukitt (O. Kindler, Freiburg).*
>
> Jacobsen, J. V.; Knox, R. R.; Pyliotis, N. A., "The Structure and Composition of Aleurone Grains in the Barley Aleurone Layer", *Planta* **1971**, *101*(3), 189–209,[819] reprinted by permission of Springer, © 1971.

(6) Electron Microscopy

Electron microscopy has been found to be useful in determining the distribution of phytic acid in plant tissues at the cellular level, particularly in plant species whose seeds contain phytin crystals embedded in protein bodies or globoids. The phytin crystals, which are naturally electron-dense, appear white in scanning electron microscopy (SEM) but black in transmission electron microscopy (TEM).[673b]

These techniques, in combination with X-ray microanalysis, also make it possible to determine the identity and relative amounts of mineral elements contained in phytin globoids. Table 2-4 identifies studies in which electron microprobe X-ray (EMX) analysis has been used in combination with SEM and those in which energy-dispersive X-ray (EDX) analysis has been used in combination with either SEM or TEM.

> Scanning electron microscopy (SEM)—*Thick endosperm slices were fixed in half-strength Karnovsky's fixative (1965) in 50 mM phosphate buffer at pH 7.2 in the refrigerator for several days. Slices were rinsed in buffer, fractured with a dull razor blade and subsequently dehydrated in an acetone series. The tissue was critical point dried and the fractured surface was coated with a gold/palladium mixture. Specimens were viewed on a JEOL-35C microscope and photographed on Polaroid Type 55 film.*
>
> [Transmission] electron microscopy (TEM) and X-ray microprobe analysis (EDAX)—*Endosperm from seeds soaked for 2 to 7 days was cut into small pieces and fixed in 2 % glutaraldehyde in 50 mM phosphate buffer at pH 7.2 at room temperature for 4 to 6 hr and post-fixed in 1 % osmium tetroxide in the same buffer in the refrigerator overnight. The tissue was dehydrated in acetone and infiltrated in the low viscosity recipe of Spurrs' epoxy resin (Spurr, 1969) over several days. The blocks were sectioned on glass knives and stained with lead citrate and uranyl acetate. Specimens were viewed on a Phillips 400 electron microscope. For the EDAX preparations, tissue was fixed in 2 % glutaraldehyde in 50 mM cacodylate buffer and dehydrated in acetone without osmium post-fixation. Sections were collected on copper grids. Electron-opaque deposits were subjected to energy dispersive X-ray microanalysis using a Philips 400-EDAX 9100 system, an accelerating voltage of 40 or 60 KeV, and a takeoff angle of 24 degrees. A control spectrum was collected from an adjacent area of protein for each deposit analyzed.*
>
> DeMason, D. A., "Endosperm Structure and Storage Reserve Histochemistry in the Palm, *Washingtonia filifera*", *Am. J. Bot.* **1986**, *73*(9), 1332–1340,[1174]
>

Radioactive Labeling

Phytic acid labeled with ^{3}H, ^{14}C, or ^{32}P has been found to be useful in studies of the metabolism of phytic acid in birds, rats, and other animal species as well as in the plant itself. The radioisotope is usually introduced by incubating the seed with an aqueous solution of a labeled precursor such as [^{3}H]-*myo*-inositol, [^{14}C]-*myo*-inositol, or [^{32}P]-phosphoric acid, but it can also be injected into the stalks (Table 2-3).[1185,1190] In one study, the label was applied to plant cells in suspension culture.[1186]

Other Inositol Isomers

L'Annunziata and Fuller[1080] isolated *myo*-, D-*chiro*-, and *muco*-inositol from the polyphosphates of ponderosa pine needles and velvet mesquite leaves recovered from forest or desert soils. It seems likely that these isomers are the result of epimerization of *myo*-inositol catalyzed by soil enzymes as they are not present in living ponderosa pine[818] or velvet mesquite[1094] tissue.

Table 2-3 Radioactive Labeling

No.	Plant	Plant part	References
Labeled with ^{3}H			
2	Rice	*Oryza sativa* L.	1185,1186
8	Wheat	*Triticum aestivum* L.	811,1138,1185,1187
1	Wheat, club	*Triticum compactum* L.	1138
Labeled with ^{14}C			
1	Water meal	*Wolffiella floridana* (Sm.) Thomps.	695
1	Gibbous duckweed	*Lemna gibba* L.	695
1	Mustard, white	*Sinapis alba* L.	1188
1	Rice	*Oryza sativa* L.	1189
1	Wheat	*Triticum aestivum* L.	1190
Labeled with ^{32}P			
1	Subclover	*Trifolium subterraneum* L.	693
1	Corn	*Zea mays* L.	1191,1192
2	Mung bean	*Vigna radiata* (L.) Wilcz.	693,1193
2	Rice	*Oryza sativa* L.	783,1186,1187,1189
4	Wheat	*Triticum aestivum* L.	693,811,1187

Table 2-4 Phytic Acid Content of Plants

No.	Plant part	Method	% P (phytate)	% P (total)	References
Abelmoschus esculentus (L.) Moench (Okra)					
1	Vegetable	McCance	0	0.616	821
1	Boiled leaf	Harland	0.129		822
Abies nordmanniana (Stev.) Spach. (Caucasian Fir)					
1	Pollen	PE	0.04	0.56	788
Abutilon theophrasti Medik. (Velvetleaf)					
1	Defatted seed meal	Mukhamedova	0.84		796
Acacia conferta Cunn. ex Benth. (Crowded-Leaf Wattle)					
1	Seed cotyledon	TEM/EDX	+		823
Acacia decurrens (Wendl.) Willd. (Green Wattle)					
1	Pollen	PE	0	0.22	788
Acacia karoo Hayne (Karroo Thorn)					
1	Seed, axis[a]	^{31}P NMR	86		824
1	–, cotyledon[a]	^{31}P NMR	79		824
Acacia nilotica (L.) Delile (Babul)					
1	Seed	McCance	0.20	0.45	825
Acacia senegal (L.) Willd. (Gum Arabic Tree)					
1	Seed	McCance	0.29	0.48	825
Acer pseudoplatanus L. (Sycamore Maple)					
1	Culture cells	^{31}P NMR	+		51
Acer spp. (Maple)					
1	Pea[b]	McCance	0.194	0.334	826
1	Fat-free leaf powder	Michel	Trace		757

[a]Percentage of total P; [b]Species unstated; [c]Basis not given; [d]Negligible (<1% of total P); [e]As is basis; [f]Fresh weight basis; [g]Air-dry basis; [h]Mean ± standard error; [i]Mixture; [j]See also ref. 1184.

Continued

Table 2-4 Phytic Acid Content of Plants—cont'd

No.	Plant part	Method	% P (phytate)	% P (total)	References
Achras zapota L. (Chicle)					
1	Flesh	McCance	0	0.051	827
Adansonia digitata L. (Baobab)					
1	Seed	McCance	0.106		828
1	Fruit pulp (milk)	Harland	0.00011	0.0018	829
Adenanthera pavonina L. (Coralwood)					
1	Seed	McCance	0.17	0.49	825
Adonis turkestanicus (Korsh.) (Adolf)					
1	Defatted seed meal	Mukhamedova	0.79		796
Aegilops speltoides Tausch (Goatgrass)					
1	Seed	Harland	fig.	0.710	830
Aesculus hippocastanum L. (European Horse Chestnut)					
1	Fat-free leaf flour[b,c]	Michel	0	0.0037	757
Afzelia africana Sm. ex Pers. (African Mahogany)					
1	Seed	McCance	0.23	0.39	825
Agropyron desertorum (Fisch.) Schult. (Crested Wheatgrass)					
1	Seed	Harland	0.27		831
1	Seed	Inositol	+		831
Albizia julibrissin Durazz. (Mimosa)					
1	Defatted seed meal	Mukhamedova	0.39		795
Albizia lophantha (Willd.) Benth. (Plume Albizzia)					
1	Seed, axis[a]	^{31}P NMR	68		824
1	–, cotyledon[a]	^{31}P NMR	75		824
Alisma orientale Juz. (Water Plantain)					
1	Rhizome	GFC	0.098		779
1	Rhizome	Inositol, PC, PE, TLC	+		779
Allium ascalonicum L. (Shallot)					
1	Root	McCance	0.037	0.066	832
Allium cepa L. (Onion)					
1	Leaf	McCance	0.020	0.037	832
1	Tuber	McCance	0.006	0.031	832
Allium porrum L. (Leek)					
1	Leaf	McCance	0.004	0.032	832
Allium sativum L. (Garlic)					
1	Leaf	McCance	0.008	0.048	832
1	Bulb	McCance	0.064	0.196	832
Allium turcomanicum Regel					
1	Defatted seed meal	Mukhamedova	0.96		794
Alnus crispa (Ait.) Pursh (Green Alder)					
1	Leaf and stem	McCance	negl.[d]		833
Aloe excelsa A. Berger					
1	Pollen	PE	0.03	0.31	788
Alternanthera sessilis (L.) R. Br. (Siseles)					
1	Leaf	McCance	0.027	0.063	832

Table 2-4 Phytic Acid Content of Plants—cont'd

No.	Plant part	Method	% P (phytate)	% P (total)	References
Althaea nudiflora Lindl.					
1	Defatted seed meal	Mukhamedova	0.17		796
Althaea officinalis L. (Marsh Mallow)					
1	Defatted seed meal	Mukhamedova	0.68		796
Amaranthus caudatus L. (Love-Lies-Bleeding)					
4	Seed	McCance	0.37-0.52	0.46-0.62	834
1	Seed[e]	Inositol	0.42		835
1	Defatted seed meal	McCance	0.39	0.71	836
1	Leaf and stem	McCance	0.043	0.290	837
Amaranthus cruentus L. (Purple Amaranth)					
4	Whole grain	McCance	0.15-0.16		838
1	–, dehulled	McCance	0.14		838
1	–, hull	McCance	0.14		838
1	Seed[e]	Inositol	0.38		835
1	Defatted seed meal	McCance	0.34-0.47	0.61-0.80	836
Amaranthus gracilis Desf. (Spineless Amaranth)					
1	Whole seed[c]	MCance	0.12		839
1	Leaf	McCance	0.044	0.090	832
Amaranthus hybridus L. (Smooth Pigweed)					
1	Whole grain	McCance	0.17		838
Amaranthus hypochondriacus L. (Prince's Feather)					
2	Whole grain	McCance	0.16-0.17		838
1	–, dehulled	McCance	0.15		838
1	–, hull	McCance	0.12		838
1	Seed[e]	Inositol	0.42		835
1	Leaf[c]	McCance	0.93	1.1	840
A. hybridus × *A. hypochondriacus*					
1	Whole grain	McCance	0.17		838
Amaranthus mantegazzianus					
1	Defatted seed meal	McCance	0.38	0.77	836
Amaranthus polygamus L.					
1	Whole seed[c]	McCance	0.17		839
Amaranthus retroflexus L. (Redroot Pigweed)					
1	Defatted seed meal	Mukhamedova	0.39		796
Amaranthus spinosus L. (Spiny Amaranth)					
1	Whole seed[c]	McCance	0.15		839
Amaranthus standleyanus Parodi ex Cuvas (Indehiscent Pigweed)					
1	Seed flour	McCance	0.29	0.57	841
Amaranthus tenuifolius Willd.					
1	Whole seed[c]	McCance	0.16		839
Amaranthus thunbergii Moq. (Wild Blite)					
1	Boiled leaf	Harland	0.067		822

[a]Percentage of total P; [b]Species unstated; [c]Basis not given; [d]Negligible (<1% of total P); [e]As is basis; [f]Fresh weight basis; [g]Air-dry basis; [h]Mean ± standard error; [i]Mixture; [j]See also ref. 1184.

Continued

Table 2-4 Phytic Acid Content of Plants—cont'd

No.	Plant part	Method	% P (phytate)	% P (total)	References
Amaranthus tricolor L. (Tampala)					
1	Whole, except root and tough stem	McCance	0.036	0.586	827
1	Leaf[e]		0	0.024	842
Amorpha fruticosa L. (Bastard Indigo)					
1	Defatted seed meal	Mukhamedova	0.36		795
Amorphophallus campanulatus (Roxb.) Blume (Elephant Yam)					
1	Leaf	McCance	0.008	0.048	832
1	Tuber	McCance	0.022	0.224	821
Anacardium occidentale L. (Cashew)					
1	Nut	McCance	0.170	0.574	842
1	Seed cotyledon	TEM/EDX	+		843
1	Young leaf	McCance	0.067	0.122	832
Ananas comosus (L.) Merr. (Pineapple)					
1	Peeled ripe fruit	McCance	0.95		844
Angelica acutiloba Kitag.					
1	Root	GFC	0.119		779
1	Root	Inositol, PC, PE, TLC	+		779
Annona muricata L. (Soursop, Guayabano)					
1	Peeled ripe fruit	McCance	1.28		844
Annona squamosa L. (Sweetsop, Ates)					
1	Peeled fruit	McCance	0.80		844
Antirrhinum majus L. (Snapdragon)					
1	Pollen	PE	0.59	1.55	788
Apium graveolens L., var. *dulce* (Mill.) Pers. (Celery)					
1	Stalk	McCance	0.006	0.031	832
Apium graveolens L., var. *rapaceum* (Mill.) Gaud. (Celeriac)					
1	Stalk	McCance	0.018	0.570	821
Arachis hypogaea L. (Peanut)					
9	Seed	McCance	0.289-0.396	0.46-0.54	845
6	Seed, defatted flour	McCance	0.395-0.623	0.616-0.969	846
1	Seed, lipid-free cotyledon	McCance	0.48	0.90	847
1	–, protein-rich fraction	McCance	0.14-1.60	0.32-1.87	847
1	–, starch grain	McCance	0.14	0.31	847
1	–, cell wall	McCance	0.003	0.09	847
1	–, vascular tissue	McCance	0.003	0.78	847
1	Seed, protein isolate	McCance	~0.28		848
1	Seed, aleurone grain[c]	McCance	1.21	1.33	849
1	–, globoid[c]	McCance	7.88	7.4	849
1	Nut	Harland	0.279		822
1	Nut	PE	+		849
2	Seed flour[b]	Harland	0.412-0.546		738
1	Nut, defatted[b]	HPLC	0.53		850
1	Seed, axis[a]	^{31}P NMR	81		824
1	–, hypocotyl[a]	^{31}P NMR	88		824
1	Seed, globoid[c]	Inositol	7.22		849
1	Acid-washed fiber	Harland	0.047		851
1	Leaf[f]	HPLC	0.00013		852

Table 2-4 Phytic Acid Content of Plants—cont'd

No.	Plant part	Method	% P (phytate)	% P (total)	References
Araucaria angustifolia (Bert.) O. Kuntze (Paraná Pine)					
1	Defatted seed flour	McCance	0.0588	0.303	853
Araucaria araucana (Mol.) K. Koch (Monkey Puzzle Tree)					
1	Defatted seed meal	McCance	0.0695	0.349	854
Araucaria bidwillii Hook. Bunya Bunya					
1	Defatted seed flour	McCance	0.0518	0.174	853
1	Pollen	PE	0.17	0.50	788
Arctotheca calendula Wendl.					
1	Pollen	PE	0	0.53	788
Arenga pinnata (Wurmb) Merr. (Sugar Palm)					
1	Seed	McCance	0.008	0.011	832
Artocarpus altilis (Park.) Fosb. (Breadfruit)					
1	Fruit	McCance	0.21		844
1	Peeled unripe fruit	McCance	0.04	0.22	855
Artocarpus camansi Blanco (Breadnut)					
1	Unripe fruit	McCance	0.008	0.039	832
Artocarpus heterophyllus Lam. (Jackfruit)					
1	Seed	McCance	0.012	0.083	832
1	Ripe fruit	McCance	0.008	0.017	832
1	Unripe fruit	McCance	0.004	0.030	832
Astragalus bisulcatus (Hook.) A. Gray (Milk Vetch)					
1	Seed embryo	TEM/EDX	+		856
Athyrium esculentum (Rezt.) Copel. (Vegetable Fern)					
1	Leaf	McCance	0.005	0.052	832
Atriplex patula L. (Spreading Orach)					
1	Defatted seed meal	Mukhamedova	0.51		796
Avena nuda L. (Naked Oats)					
1	Seed caryopsis	HPLC	+		857
Avena sativa L. (Oats)					
9	Seed	McCance	0.222-0.284	0.368-0.454	858
6	Grain[b]	McCance	0.13-0.29	0.19-0.44	859
5	Whole grain	McCance	0.54-0.87		860
5	–, groats	McCance	0.72-1.12		860
5	–, hulls	McCance	0.21-0.44		860
6	Threshed grain	McCance	0.185-0.260	0.314-0.426	861
4	Grain, groats	McCance	0.29-0.39		862
7	Grain, groats	McCance	0.27-0.40		863
0	Whole milled grain[b]	McCance	0.185-0.227		864
1	Seed meal, pinhead[b]	McCance	0.299	0.495	826
1	Dehusked seed	McCance	0.311-0.355	0.474-0.495	865
1	–, meal	McCance	0.291	0.507	865
1	–, flakes	McCance	0.183-0.226	0.325-0.367	865
1	Grain, kernel	McCance	0.20	0.34	866
1	–, coarse bran	McCance	0.28	0.15	866

[a]Percentage of total P; [b]Species unstated; [c]Basis not given; [d]Negligible (<1% of total P); [e]As is basis; [f]Fresh weight basis; [g]Air-dry basis; [h]Mean ± standard error; [i]Mixture; [j]See also ref. 1184.

Continued

Table 2-4 Phytic Acid Content of Plants—cont'd

No.	Plant part	Method	% P (phytate)	% P (total)	References
1	–, fine bran	McCance	0.34	0.10	866
1	–, bran flour	McCance	0.17	0.05	866
1	–, white flour	McCance	0.11	0.04	866
2	Seed[b]	Harland	0.263-0.347		738
1	Seed hull	Harland	0.010		851
1	Entire panicle (early dough stage)[b]	Harland	0.028	0.32	867
2	Seed, aleurone tissue	Microscopy	+		868,869
1	Seed, aleurone cell	SEM	+		870
1	Seed aleurone protein body	Harland	0.72	0.82	871
1	Starchy seed endosperm	Harland	0.00	0.69	871
1	Seed embryo	TEM/EDX	+		872
1	Acid-washed fiber	Harland	0		851
1	Whole seed[b]	HPLC	0.21		774
1	–, groats	HPLC	0.38		774
1	–, bran	HPLC	0.39		774
1	Seed kernel flour	^{31}P NMR	0.369		873
1	Seed	Inositol	0.264		874
1	Whole seed	Phytase	+	+	14
1	Defatted seed powder	Michel	0.272		758
1	10-day seedling	Inositol	0.227		874
1	Shoot	Harland	0		875
1	Root	Harland	0		875
Avena strigosa Schreb. (Bristle Oats)					
1	Threshed grain	McCance	0.233-0.308	0.395-0.483	861
A. strigosa × *A. brevi*					
1	Threshed grain	McCance	0.272-0.322	0.452-0.489	861
Averrhoa carambola L. (Carambola)					
1	Fruit	McCance	0.022	0.044	832
Azadirachta indica A. Juss. (Neem)					
1	Seed	McCance	0.098		828
1	Leaf	McCance	0.89	0.4	840
Azolla africana Desv. (Water Fern)					
1	Leaf protein concentrate	McCance	0.169		876
1	–, pulp	McCance	0.155		876
1	–, juice	McCance	0.008		876
Bambusa spinosa Roxb. (Bamboo)					
1	Shoot	McCance	0.002	0.026	832
Banksia coccinea R.Br. (Scarlet Banksia)					
1	Seed cotyledon	Microscopy	+		877
Banksia hookeriana Meissn. (Hooker's Banksia)					
1	Seed cotyledon	Microscopy	+		877
1	Seed cotyledon	SEM/EMX	+		877
Basella rubra L. (Malabar Nightshade)					
1	Leaf	McCance	0.006	0.076	832
Bauhinia monandra Kurz (Butterfly Flower)					
1	Seed	McCance	0.25	0.41	825

Table 2-4 Phytic Acid Content of Plants—cont'd

No.	Plant part	Method	% P (phytate)	% P (total)	References
Bauhinia purpurea L. (Purple Bauhinia)					
1	Seed	McCance	0.135	0.365	878
Bauhinia variegata L. (Orchid Tree)					
1	Seed axis[a]	^{31}P NMR	79		824
1	–, hypocotyl[a]	^{31}P NMR	91		824
Benincasa hispida (Thumb.) Cogn. (Wax Gourd)					
1	Gourd	McCance	0.003	0.018	832
Berlinia auriculata Benth.					
1	Seed	McCance	0.12	0.18	825
Bertholletia excelsa Humb. and Bonpl. (Brazil Nut)					
1	Nut	McCance	0.313	0.805	842
1	Radicle	TEM/EDX	+		879
Beta vulgaris (L.) Banks (Sugar Beet)					
1	Leaf	Michel	0.0048	0.091	15
1	Root	Michel	0.0029	0.057	15
1	Tuber	McCance	0.012	0.039	832
1	Tuber[b]	HPLC	0.003		774
1	Tuber	^{31}P NMR	n.d.		43
1	Acid-washed fiber	Harland	0.006		851
Betula papyrifera (Reg.) Fern and Raup (Paper Birch)					
1	Leaf and stem	McCance	negl.[d]		833
Bixa orellana L. (Annatto)					
1	Seed	McCance	0.094		828
Blighia sapida K. König (Akee)					
1	Seed	McCance	0.290		828
Brassica campestris L. (Turnip Rape)					
2	Seed	McCance	0.69-0.72		880
4	Defatted seed flour	McCance	1.1-1.4		881
1	Defatted seed meal	Harland	1.1		882
1	Defatted seed flour[b]	Harland	1.63		883
2	Seed meal, defatted (1)	Phytase	0.99-1.11	1.08-1.18	745
	Seed meal, defatted (2)	Phytase	0.74-0.85	1.08-1.18	745
3	Seed tissue	Microscopy	+		820
1	Seed embryo	TEM/EDX	+		884
1	Leaf	McCance	0.025		885
Turnip rape, low erucic acid (Canola)					
57	Seed	Phytase	0.98-1.43		745
Brassica carinata A. Braun (Abyssinian Mustard)					
2	Seed	McCance	0.52-0.62		886
4	Seed	McCance	0.65-0.71		880
4	Defatted seed meal	McCance	0.99-1.07		880
Brassica chinensis L. (Pak-choi)					
1	Leaf	McCance	0.012	0.041	832
1	Boiled leaf	Harland	0.022		822
Brassica integrifolia (West) Rupr. (Indian Brown Mustard)					
1	Leaf	McCance	0.017	0.022	832

[a]Percentage of total P; [b]Species unstated; [c]Basis not given; [d]Negligible (<1% of total P); [e]As is basis; [f]Fresh weight basis; [g]Air-dry basis; [h]Mean ± standard error; [i]Mixture; [j]See also ref. 1184.

Continued

Table 2-4 Phytic Acid Content of Plants—cont'd

No.	Plant part	Method	% P (phytate)	% P (total)	References
Brassica juncea L. (Indian Mustard)					
4	Seed	McCance	0.72-1.01		880
4	–, defatted meal	McCance	0.99-1.07		880
1	Seed embryo	SEM, TEM/EDX	+		884
1	Seed, protein isolate	McCance	0.56		887
Brassica napobrassica (L.) Mill. (Rutabaga)					
1	Tuber[b,f]	McCance	0	0.0190	735
Brassica napus L. (Rape)					
3	Seed[c]	McCance	0.85-1.21	1.04-1.49	751
3	Seed meal, defatted	McCance	0.82-1.04	1.14-1.34	745
3	Seed meal, defatted	Phytase	0.90-1.10	1.14-1.34	745
1	Seed meal, defatted	HPLC	+		888
6	Seed, defatted flour	McCance	1.0-1.4		881
1	–, hull	McCance	0.10		881
2	Defatted seed meal	Harland	1.2-1.3		882
1	Seed flour	Harland	0.828	1.507	889
1	–, protein concentrate	Harland	0.999	1.737	889
1	Seed, protein concentrate	McCance	1.534	1.737	889
1	Seed, protein isolate	McCance	0.004-0.011	0.23-1.14	890
1	Seed, flour	Phytase	0.860	1.507	889
1	–, protein concentrate	Phytase	1.078	1.737	889
2	Seed tissue	Microscopy	+		820
1	Seed embryo	TEM/EDX	+		884
1	Root	SEM/EMX	+		891
Rape, low erucic acid (Canola)					
127	Seed meal, defatted	Phytase	0.90-1.62		745
Rape, low glucosinolate (Canola)					
4	Seed	Harland	0.35-0.65		886
4	Defatted seed meal	McCance	0.95-1.14	1.13-1.43	892
1	Defatted seed meal	Harland	0.98	1.25	892
1	Seed, protein isolate	McCance	0.01-0.02		893
Rape, low glucosinolate, low erucic acid (Canola)					
19	Defatted seed meal	Phytase	1.09-1.45		745
5	Defatted seed meal[c]	McCance	0.82-0.96	1.27-1.55	894
2	Defatted seed meal	McCance	0.82-0.90		887
2	–, protein isolate	McCance	0.62-0.73		887
Brassica oleracea L. ssp. *acephala* (Collards)					
1	Leaf[f]	HPLC	0.0011		852
Brassica oleracea L. ssp. *botrytis* L. (Cauliflower)					
1	Flower	McCance	0.007	0.030	832
Brassica oleracea L. ssp. *capitata* L. (Cabbage)					
1	Leaf	McCance	0.011	0.499	821
1	Leaf[f]	HPLC	0.00028		852
1	Root	SEM/EMX	+		891

Table 2-4 Phytic Acid Content of Plants—cont'd

No.	Plant part	Method	% P (phytate)	% P (total)	References
Brassica oleracea L. ssp. *gemmifera* Zenk. (Brussels Sprouts)					
1	Inner leaf	McCance	0.037	0.630	827
Brassica oleracea L. ssp. *italica* Plenck. (Broccoli)					
1	Sprout[b]	HPLC	0		741
Brassica rapa L. (Turnip)					
1	Leaf[f]	HPLC	0.00009		852
1	Tuber[b,f]	McCance	0	0.0275	735
Bromus pratensis Lam. (Brome Grass)					
1	Defatted seed powder	Michel	0.218		758
Buchanania latifolia Roxb. (Chirauli Nut)					
1	Nut	McCance	0.163	0.716	827
Bulnesia bonariensis Griseb. (Jaboncillo)					
1	Seed cotyledon	TEM	+		895
Bulnesia schickendantzii Hieron					
1	Seed cotyledon	TEM	+		895
Cajanus cajan (L.) Huth (Pigeon Pea)					
22	Pea[c]	McCance	0.8-2.0	0.230-0.350	896
10	Pea	McCance	0.00003-0.00011		897
6	Whole seed	McCance	0.265±0.020	0.390±0.032	898
6	–, decorticated	McCance	0.267±0.017	0.411±0.029	898
18	Decorticated seed	McCance	0.184-0.476	0.292-0.560	899
1	Pea	Harland	0.194-0.222		822
2	Seed	HPLC	0.035-0.038		900
1	Seed, axis[a]	^{31}P NMR	43		824
1	–, cotyledon[a]	^{31}P NMR	55		824
1	3-day sprout	McCance	0.043		901
Calendula officinalis L. (Pot Marigold)					
1	Defatted seed meal	Mukhamedova	0.73		794
Canarium ovatum Engl. (Pili Nut)					
1	Nut	McCance	0.097	0.555	832
Canavalia ensiformis (L.) DC (Jack Bean)					
1	Bean	McCance	0.156		902
1	Seed	HPLC	0.033		900
Cannabis sativa L. (Hemp)					
1	Seed[b]	McCance	0.285	0.692	826
1	Decorticated seed[b,c]	Posternak	1.330	1.460	807
1	–, aleurone grain[b,c]	Posternak	3.61	3.83	807
Capparis decidua (Forsk.) Edgew. (Ker)					
1	Unripe fruit	McCance	0.086	0.179	903
Capsella bursa-pastoris (L.) Medik. (Shepherd's Purse)					
1	Seed embryo	TEM/EDX	+		904
Capsicum annuum L. (Sweet Pepper)					
1	Pod	McCance	0.080	0.386	821
1	Fruit	McCance	0.017	0.027	832

[a]Percentage of total P; [b]Species unstated; [c]Basis not given; [d]Negligible (<1% of total P); [e]As is basis; [f]Fresh weight basis; [g]Air-dry basis; [h]Mean ± standard error; [i]Mixture; [j]See also ref. 1184.

Continued

Table 2-4 Phytic Acid Content of Plants—cont'd

No.	Plant part	Method	% P (phytate)	% P (total)	References
Capsicum frutescens L. (Chili Pepper)					
1	Leaf	McCance	0.010	0.054	832
Carapa procera DC (African Crabwood)					
1	Seed	McCance	0.093		828
Carduus acanthoides L. (Black Thistle)					
1	Defatted seed flour	McCance	0.097	0.955	905
Carica papaya L. (Papaya)					
1	Young leaf	McCance	0.019	0.100	832
1	Unripe fruit	McCance	0.005	0.030	832
1	Peeled ripe fruit	McCance	0.36		844
Carpinus betulus L. (European Hornbeam)					
1	Fat-free leaf powder[b,c]	Michel	Trace		757
Carthamus tinctorius L. (Safflower)					
1	Seed, protein isolate[c]	McCance	0.003	0.23	906
Carum copticum (L.) Benth. and Hook. (Bishop's Weed)					
1	Seed	McCance	0.332	0.346	821
Carya illinoinensis (Wangenh.) K. Koch (Pecan)					
2	Nut[b,g]	McCance	0.39-0.43		907
Carya spp. (Hickory)					
2	Nut	McCance	0.42-0.47		907
Cassia artemisiodes DC (Silver Cassia)					
1	Seed cotyledon	TEM/EDX	+		823
Cassia nodosa Buch.-Ham. ex Roxb. (Jointwood)					
1	Seed	McCance	0.12	0.39	825
Cassia sericea Sweet					
1	Seed	McCance	0		908
Cassia sieberiana DC (African Laburnum, Ratu)					
1	Seed	McCance	0.26	0.45	825
Castanea sativa Mill. (Spanish Chestnut)					
1	Fat-free leaf powder[b,c]	Michel	Trace		757
Castanea spp. (Chestnut)					
1	Nut[b,f]	McCance	0.009	0.074	735
Casuarina cunninghamiana Miq. (Cunningham Beefwood)					
1	Seed cotyledon	TEM/EDX	+		872
Casuarina littoralis Salisb. (Black She Oak)					
1	Seed cotyledon	TEM/EDX	+		872
Casuarina muelleriana Miq. (Slaty She Oak)					
1	Seed cotyledon	TEM/EDX	+		872
Cedrela odorata L. (Spanish Cedar)					
1	Seed	McCance	0.338		828
Ceiba pentandra (L.) Gaertn. (Kapok Tree)					
1	Seed meal	McCance	0.64	0.93	855
Celosia argentea L. (Cockscomb)					
1	Leaf	McCance	0.151	1.2	909

Table 2-4 Phytic Acid Content of Plants—cont'd

No.	Plant part	Method	% P (phytate)	% P (total)	References
Centaurea calcitrapa L. (Abrepuño Colorado (Star Thistle))					
1	Defatted seed flour	McCance	0.073	0.682	910
Centaurea solstitialis L. (Abrepuño Amarillo (Yellow Star Thistle))					
1	Defatted seed flour	McCance	0.084	0.995	910
Ceratonia siliqua L. (Carob)					
1	Seed, defatted flour	McCance	0.14		911
1	–, protein isolate	McCance	0.08		911
1	Seed germ	Harland	0.55		761
1	Seed, axis[a]	^{31}P NMR	92		824
1	–, cotyledon[a]	^{31}P NMR	96		824
Ceratophyllum spp. (Hornwort)					
1	Whole plant	Harland	0.0017	0.126	912
Chaenomeles japonica Lindl. (Japanese Quince)					
1	Defatted seed meal	Mukhamedova	0.39		795
Chenopodium album L. (Lambsquarters)					
1	Leaf	McCance	0.084		885
Chenopodium botrys L. (Jerusalem Oak)					
1	Defatted seed meal	Mukhamedova	0.11		796
Chenopodium quinoa Willd. (Quinoa)					
5	Grain	Undisclosed	0.27-0.37		913
1	Whole seed	McCance	0.174	0.360	914
1	–, whole meal	McCance	0.184		914
1	–, flour	McCance	0.083		914
1	–, bran	McCance	0.167		914
1	–, hulls	McCance	0.077	0.320	914
1	Seed	Harland	0.293		915
1	–, polished	Harland	0.220		915
1	Sweet, seed	HPLC	0.206		916
1	Bitter, seed	HPLC	0.160		916
Chloris gayana Kunth. (Rhodes Grass)					
1	Pollen	PE	0	0.37	788
Chromolaena odorata (L.) King and H. Rob. (Jack in the Bush)					
1	Leaf	McCance	0.034	0.10	917
Chrysophyllum albidum G. Don (African Starapple)					
1	Fruit, peel	McCance	0.0002	0.077	918
1	–, deseeded pulp	McCance	0.0004	0.075	918
Cicer arietinum L. (Chickpea)					
11	Seed	McCance	0.25-0.31	0.286-0.338	919
8	Whole seed	McCance	0.211-0.225		920
5	Whole seed	McCance	0.076-0.114	0.164-0.308	921
16	Decorticated seed	McCance	0.147-0.334	0.211-0.419	899
2	Seed	HPLC	0.107-0.110		922
5	Seed, cotyledon	McCance	0.059-0.106	0.160-0.330	921
2	Seed, dehulled	McCance	0.27-0.37		923
2	–, hull	McCance	0.02-0.056		923
1	Immature seed	McCance	0-0.082	0.380-0.506	924

[a]Percentage of total P; [b]Species unstated; [c]Basis not given; [d]Negligible (<1% of total P); [e]As is basis; [f]Fresh weight basis; [g]Air-dry basis; [h]Mean ± standard error; [i]Mixture; [j]See also ref. 1184.

Continued

Table 2-4 Phytic Acid Content of Plants—cont'd

No.	Plant part	Method	% P (phytate)	% P (total)	References
1	Pea	Harland	0.095	0.257	925
1	Defatted seed powder	Michel	0.354		758
1	Defatted seed flour	Harland	0.146		926
1	–, protein fraction	Harland	0.360		926
1	–, starch fraction	Harland	0.042		926
1	Seed, axis[a]	^{31}P NMR	41		824
1	–, hypocotyl[a]	^{31}P NMR	58		824
8	68-h sprout	McCance	0.147-0.153		920
1	Sprout	McCance	0.035	0.230	927
Cichorium intybus L. (Chicory)					
1	Seed flour	McCance	0.048	0.536	928
Citrullus vulgaris Schrad. (Watermelon)					
1	Seed	McCance	0.600	1.456	929
1	–, protein isolate	McCance	0.28	0.30	930
1	Defatted seed powder	Michel	0.120		758
1	Fruit	McCance	0.0017		885
Citrus aurantium L. (Sour Orange)					
1	Defatted seed powder	Michel	0.2945		758
Citrus limon (L.) Burm. f. (Sour Lemon)					
1	Fruit[b]	McCance	0.026		931
Citrus nobilis Lour. (Tangor)					
1	Defatted seed powder	Michel	0.1162		758
Citrus sinensis (L.) Osbeck (Orange)					
1	Fruit[b]	McCance	0.029		931
1	Acid-washed fiber	Harland	0		851
Clianthus formosus (G. Don) Ford and Vick. (Sturt Desert Pea)					
1	Seed cotyledon	TEM/EDX	+		823
Clitoria ternatea L. (Blue Pea)					
1	Leaf	McCance	0.010	0.061	832
Cocos nucifera L. (Coconut)					
1	Seed	McCance	0.067	0.336	929
1	Seed meal[b,c]	McCance	0.258	0.505	932
1	Nut[b,c]	Harland	0.67		933
1	Shoot	McCance	0.010	0.076	832
Coffea arabica L. (Coffee)					
5	Bean	McCance	0.062-0.098		934
6	Raw bean	HPLC	0.046-0.104	0.159-0.214	935
1	Roasted bean[b]	Harland	0.050		776
Cola acuminata (Beauv.) Schott and Endl. (Abata Cola)					
1	Seed	McCance	0.038	0.193	929
Colocasia antiquorum Schott. (Egyptian Taro, Culcas)					
1	Stem and leaf	McCance	0	0.562	827
Colocasia esculenta (L.) Schott (Taro)					
1	Leaf	McCance	0.009	0.042	832
1	Root	McCance	0.022	0.053	832
1	Tuber[c]	Harland	0.241		936

Table 2-4 Phytic Acid Content of Plants—cont'd

No.	Plant part	Method	% P (phytate)	% P (total)	References
Conium maculatum L. (Poison Hemlock)					
1	Defatted seed meal	Mukhamedova	0.70		796
Corchorus olitorius L. (Tossa Jute)					
1	Leaf	McCance	0.028	0.130	832
Coriandrum sativum L. (Coriander)					
1	Seed	McCance	0.350	0.454	821
1	Leaf	McCance	0.088	0.441	827
Corylus avellana L. (European Hazel)					
1	Nut, shelled[b,e]	McCance	0.099	0.230	937
1	Nut, shelled	HPLC	0.048	0.299	938
1	Seed cotyledon	TEM/EDX	+		939
Coula edulis Baill. (Gabon Nut)					
1	Seed	McCance	0.211		828
Crambe abyssinica Hochst. (Crambe)					
1	Seed	SEM/EMX	+		940
1	Seed	TEM/EDX	+		940
Crassocephalum biafrae (Oliv. and Hiern) S. Moore (Sierra Leone Bologi)					
1	Leaf	McCance	0.076	0.7	909
Crassocephalum crepidioides (Benth.) S. Moore (Hawksbead Velvetplant)					
1	Leaf	McCance	0.070	0.9	909
Cucumis melo L. (Melon)					
1	Seed, protein isolate	McCance	1.02	1.25	930
1	Fruit	McCance	0.004	0.011	832
Cucumis sativus L. (Cucumber)					
1	Cotyledon	Microscopy	+		819
1	–, aleurone grain	Microscopy	+		941
1	Fruit	McCance	0.026	0.036	832
Cucurbita andreana Naudin. (Winter Squash)					
1	Seed cotyledon	TEM/EDX	+		942
Cucurbita digitata Gray (Coyote Gourd)					
1	Defatted flour	McCance	2.4		943
1	Defatted seed embryo	Harland	0.87		944
Cucurbita foetidissima HBK (Calabazilla)					
1	Defatted flour	McCance	2.2		943
1	Defatted seed embryo	Harland	0.53		944
1	Seed cotyledon	TEM/EDX	+		942
Cucurbita maxima Duch. (Hubbard Squash)					
1	Seed embryo, defatted	Harland	0.31		944
5	Seed cotyledon	TEM/EDX	+		945,946
1	Boiled leaf	Harland		0.087	822
1	Young leaf	McCance	0.003	0.046	832
1	Fruit	McCance	0.008	0.044	832
1	Flower	McCance	0.023	0.046	832

[a]Percentage of total P; [b]Species unstated; [c]Basis not given; [d]Negligible (<1% of total P); [e]As is basis; [f]Fresh weight basis; [g]Air-dry basis; [h]Mean ± standard error; [i]Mixture; [j]See also ref. 1184.

Continued

Table 2-4 Phytic Acid Content of Plants—cont'd

No.	Plant part	Method	% P (phytate)	% P (total)	References
Cucurbita mixta Pang. (Silver-Seed Gourd)					
1	Seed cotyledon	TEM/EDX	+		942
Cucurbita moschata (Duch.) Poir. (Warty Squash)					
1	Seed embryo, defatted	Harland	0.56		944
1	Seed cotyledon	TEM/EDX	+		942
Cucurbita palmata S. Wats. (Coyote Gourd)					
1	Defatted flour	McCance	2.6		943
Cucurbita pepo L. (Pumpkin)					
1	Seed[b,c]	Harland	1.21		933
1	Seed, unshelled	Lendle	0.0031		947
1	Defatted seed embryo	Harland	0.56		944
1	Seed cotyledon	TEM/EDX	+		942
1	Defatted seed meal	McCance	0.847	1.783	948
1	–, protein concentrate	McCance	1.016	1.853	948
1	–, protein isolate	McCance	0.158	0.766	948
Cuminum cyminum L. (Cumin)					
1	Seed	McCance	0.171	0.514	821
Curcuma longa L. (Turmeric)					
1	Seed	McCance	0.112	0.327	821
Cuscuta reflexa Roxb. (Dodder)					
1	Stem	McCance	0.0103	0.0198	814
1	Stem	PC	+	0.0198	814
Cyamopsis psoralioides (Lam.) DC (Cluster Bean)					
1	Bean	McCance	0.022	0.224	821
Cynoglossum creticum Mill. (Blue Hound's Tongue)					
1	Defatted seed meal	Mukhamedova	0.70		795
Cyperus esculentus L. (Yellow Nutsedge, Chufa)					
1	Defatted seed	McCance	0.183	0.307	865
1	Tuber	Harland	0.204		949
Cyphomandra betacea (Cav.) Sendtn. (Tree Tomato)					
1	Flesh and seed	McCance	0.043	0.333	827
Dactylis glomerata L. (Orchard Grass)					
1	Defatted seed powder	Michel	0.436		758
Daniellia oliveri Hutch. and Dalz. (African Copaiba Balsam)					
1	Seed	McCance	0.13	0.19	825
Datura stramonium L. (Jimsonweed)					
1	Defatted seed meal	Mukhamedova	0.39		796
Daucus carota L. (Carrot)					
1	Tuber	McCance	0	0.253	821
1	Tuber[b]	HPLC	0.145		741
1	Tuber[c]	PE	0		950
1	Root	^{31}P NMR	n.d.		43
Delonix regia (Boj.) Rafir. (Royal Poinciana, Flamboyant)					
1	Defatted seed flour[c]	McCance	0.6		951
1	–, protein isolate	McCance	0.03		951

Table 2-4 Phytic Acid Content of Plants—cont'd

No.	Plant part	Method	% P (phytate)	% P (total)	References
Dendrobium lowii Lindl.					
1	Pollen	PE	0.04	0.56	776
Deschampsia caespitosa (L.) Beauv. (Tufted Hairgrass)					
1	Root	SEM/EMX	+		952
Detarium microcarpum Guill. and Perr.					
1	Seed	McCance	0.06	0.56	825
Dioclea reflexa Hook. f. (Sea Purse)					
1	Seed	McCance	0.0896	0.420	953
Dioscorea alata L. (Winged Yam)					
6	Tuber	McCance	0.038 ± 0.013	0.171 ± 0.014	954
1	Boiled root	Harland	0.007		955
Dioscorea batatas Decne. (Chinese Yam)					
1	Tuber mucilage	PC, PE	+	3-4	956
Dioscorea esculenta (Lour.) Burk. (Lesser Yam)					
3	Tuber	McCance	0.032 ± 0.003	0.117 ± 0.009	954
Dioscorea rotundata Poir. (White Guinea Yam)					
1	Tuber	McCance	0.055	0.092	929
1	Tuber[b]	Harland	0.179		936
Dioscorea versicolor Wall.					
1	Skinned tuber	McCance	0.024	0.250	827
Dioscoreophyllum cumminsii (Stapf.) Diels (Serendipity Berry)					
1	Seed	McCance	0.248		828
Diospyros discolor Willd. (Persimmon)					
1	Peeled ripe fruit	McCance	0.92		844
Dipsacus fullonum L. var. *sylvestris* (Teasel)					
1	Defatted seed flour	McCance	0.148	1.376	957
Dipteryx alata Vog. (Baru)					
1	Seed		No data given in abstract		958
Diptychocarpus strictus (Fisch.) Trautv.					
1	Defatted seed meal	McCance	1.01	3.3	959
Dryandra formosa Meissn. (Showy Dryandra)					
1	Seed cotyledon	Microscopy	+		877
Echinochloa frumentacea Link. (Japanese Millet)					
1	Seed bran[c]	Suzuki	0.344	0.765	797
Echium italicum L. (Italian Bugloss)					
1	Defatted seed meal	Mukhamedova	0.79		795
Elaeis guineensis Jacq. (Oil Palm)					
1	Seed meal	McCance	0.481	0.576	932
Elaeocarpus oblongus (Wild Olive)					
1	Fruit flesh and skin	McCance	0.012	0.064	827
Eleocharis dulcis (Burm. f.) Trin. (Water Chestnut)					
1	Nut	McCance	0.067	0.146	832
Eleusine coracana (L.) Gaertn. (Ragi)					
12	Colored seed[h]	McCance	0.188 ± 0.0075	0.243 ± 0.0094	960
3	White seed[h]	McCance	0.154 ± 0.0060	0.214 ± 0.0020	960
1	Seed flour	HPLC	0.145	0.128	961

[a]Percentage of total P; [b]Species unstated; [c]Basis not given; [d]Negligible (<1% of total P); [e]As is basis; [f]Fresh weight basis; [g]Air-dry basis; [h]Mean ± standard error; [i]Mixture; [j]See also ref. 1184.

Continued

Table 2-4 Phytic Acid Content of Plants—cont'd

No.	Plant part	Method	% P (phytate)	% P (total)	References
Emilia sonchifolia (L.) DC (Red Tasselflower)					
1	Leaf	McCance	0.072	0.16	917
Enterolobium cyclocarpum Griseb. (Elephant's Ear)					
1	Seed	McCance	0.29	0.35	825
Erodium cicutarium (L.) L'Hér. (Red-Stemmed Filaree)					
1	Defatted seed meal	Mukhamedova	0.28		796
Erythrina coralloides DC (Naked Coral Tree)					
1	Seed, axis[a]	^{31}P NMR	63		824
1	–, cotyledon[a]	^{31}P NMR	79		824
Eucalyptus deglupta Bl. (Mindanao Gum)					
1	Seed	McCance	0.275		828
Eucalyptus erythrocorys F. Muell. (Illyarie)					
1	Seed cotyledon	TEM/EDX	+		962
Eucalyptus youngiana F. Muell. (Ooldea Mallee)					
1	Pollen	PE	0.06	0.56	788
Euphorbia hirta L. (Kaka Wie Adwe)					
1	Leaf	McCance	0.184	0.004	963
Fagopyrum esculentum Moench (Buckwheat)					
1	Dehusked grain	McCance	0.322	0.410	827
1	Seed, flour[b,g]	McCance	0.52		907
1	–, hull[b,g]	McCance	0.28		907
Ferula karataviensis (Regel and Schmalh.) Korov.					
1	Defatted seed meal	Mukhamedova	0.34		794
Ferula kelleri K.-Pol.					
1	Defatted seed meal	Mukhamedova	1.21		794
Ficus carica L. (Fig)					
1	Fruit[b]	McCance	0.0119	0.0915	735
Foeniculum vulgare Mill. (Fennel)					
1	Mature seed	McCance	0.369		964
1	Immature seed	McCance	0.328		964
Fragaria spp. (Strawberry)					
1	Fruit[b]	McCance	0.010		931
Fuchsia spp. hort. (Fuchsia)					
1	Pollen	PE	0.04	0.59	788
Garcinia kola Heck. (Efiari)					
1	Fruit, mesocarp	McCance	0.00025		965
1	–, pericarp	McCance	0.00031		965
Garcinia mangostana L. (Mangosteen)					
1	Peeled ripe fruit	McCance	0.19		844
Gevuina avellana Mol. (Chile Nut)					
1	Defatted meal	McCance	0.157	0.419	966
Ginkgo biloba L. (Ginkgo)					
1	Nut	Harland	0.172		967
Glaucium fimbrilligerum Boiss. (Sea Poppy)					
1	Defatted seed meal	Mukhamedova	2.25		794

Table 2-4 Phytic Acid Content of Plants—cont'd

No.	Plant part	Method	% P (phytate)	% P (total)	References
Gleditsia triacanthos L. (Honey Locust)					
1	Defatted seed meal	Mukhamedova	0.62		795
1	Seed, axis[a]	^{31}P NMR	82		824
1	–, hypocotyl[a]	^{31}P NMR	90		824
Glycine max (L.) Merr. (Soybean)					
38	Seed	McCance	0.391-0.648	0.58-0.91	968
~15	Seed	McCance	0.231-0.575	0.442-0.838	969
2	Full-fat seed flour[b]	McCance	0.424-0.509	0.563-0.676	970
7	–, defatted[b]	McCance	0.457-0.522	0.665-0.753	970
12	Defatted seed meal	McCance	0.16-0.20	0.73-0.86	971
7	Bean[g]	McCance	0.48-0.73		907
15	Seed	McCance	0.282-0.414	0.532-0.745	858
5	Bean	McCance	0.42-0.53	0.341-0.410	919
17	Defatted seed meal	Harland	0.529-1.145	1.277-2.833	972
1	Seed oil[b]	Harland	n.d.		973
1	Seed bran	Harland	0.029		851
1	Acid-washed fiber	Harland	0.019		851
4	Bean	HPLC	0.31-0.48		765
1	Seed	CZE	0.771		700
1	Seed	Mössbauer	(–)		817
1	Seed cotyledon	TEM/EDX	+		974
1	Seed embryo axis	TEM/EDX	+		823
1	Seed flake, defatted	DGE	+		699
1	Flake, defatted[b]	HPLC	0.50-0.52		850
1	Hull, defatted[b]	HPLC	0.034		850
1	Seed, axis[a]	^{31}P NMR	54		824
1	–, hypocotyl[a]	^{31}P NMR	87		824
1	Seed protein isolate[b,c]	^{31}P NMR	0.31		781
1	Seed meal[b]	Inositol	0.581		784
1	Defatted seed flour[b]	Harland	0.328		926
1	–, protein concentrate[b]	Harland	0.381		926
2	Seed, protein concentrate[b]	McCance	0.351-0.612	0.654-0.899	970
2	Seed, protein isolate[b]	McCance	0.56-0.58	0.847-0.885	975
1	Seed, 7S protein	McCance	0.39	0.63-0.68	976
1	–, 11S protein	McCance	0.02	0.08	976
1	–, whey isolate	McCance	12.77	10.49-15.20	976
2	Seed protein isolate	Harland	0.404-0.513		738
1	Seed protein isolate	DGE	+		815
1	Seed protein isolate[b]	Harland	0.20-0.69	0.38-0.84	977
1	–, 7S protein	Harland	0.15	0.61	977
1	–, 11S protein	Harland	0.007	0.060	977
3	Protein isolate[b,c]	HPLC	0.45-0.56	0.79-0.89	978
1	–, 7S protein	HPLC	0.02	0.18	978
1	–, 11S protein	HPLC	0.003	0.06	978
1	Seed protein body	Harland	0.53	0.78	979

[a]Percentage of total P; [b]Species unstated; [c]Basis not given; [d]Negligible (<1% of total P); [e]As is basis; [f]Fresh weight basis; [g]Air-dry basis; [h]Mean ± standard error; [i]Mixture; [j]See also ref. 1184.

Continued

Table 2-4 Phytic Acid Content of Plants—cont'd

No.	Plant part	Method	% P (phytate)	% P (total)	References
1	–, globoid	Harland	6.70	7.65	979
1	Seed protein body[b]	Morton	0.38	0.43	804
1	Root	SEM/EMX	+		891
2	3-day sprout	McCance	0.15-0.18	0.44-0.47	980
1	10-day seedling	Inositol	0.464		784
1	Whole plant (early flowering stage)[b]	Harland	0.009	0.28	867
1	–, stem	Harland	0.000	0.30	867
1	–, young leaf	Harland	0.000	0.41	867
1	–, first trifoliate leaf	Harland	0.014	0.28	867
Glycine soja Sieb and Zucc. (Soybean)					
20	Seed Several sources say *Glycine soja* is the same as *Glycine max*, but in this reference Raboy *et al.* provide data for 38 varieties of *G. max* and 20 varieties of *G. soja*	McCance	0.529-0.780	0.81-1.07	968
Glycyrrhiza glabra L. (Licorice)					
1	Defatted seed meal	Mukhamedova	0.14		795
Gnetum africanum Wlew. (Afang)					
1	Leaf	McCance	0.070	0.300	981
Gnetum buchholzianum Engl. (Joint Fir)					
1	Leaf	McCance	0.087	0.1	909
Gnetum gnemon L. (Buko, Spinach Joint Fir)					
1	Leaf	McCance	0.017	0.088	832
Gossypium barbadense L. (Sea Island Cotton)					
1	Meal	McCance	1.02		982
Gossypium herbaceum L. (Levant Cotton)					
1	Seed	McCance	0.322-0.410		865
Gossypium hirsutum L. (Upland Cotton)					
8	Ground seed kernel	McCance	0.429-1.069	0.692-1.315	983
8	–, defatted	McCance	0.686-1.768	1.078-2.200	983
3	Acid-delinted seed	McCance	0.649-0.958	0.796-1.148	984
1	–, cotyledon	McCance	0.046-0.654	0.716-1.361	984
1	–, hypocotyl and root	McCance	0.00	0.568-0.878	984
1	Glandless defatted seed flour	McCance	1.59	1.85	985
1	Glandless seed, aleurone grain[c]	PE	2.25	2.39	986
1	–, globoid[c]	PE	13.85	14.21	986
1	–, globoid[c]	Inositol	13.62	14.21	986
Gossypium spp. (Cotton)					
14	Seed kernel, dehulled[c]	HPLC	0.462-0.842		685
14	Seed kernel, dehulled[c]	Harland, near IR			685

Table 2-4 Phytic Acid Content of Plants—cont'd

No.	Plant part	Method	% P (phytate)	% P (total)	References
1	Seed protein fraction[b]	Fontaine	1.151	1.164	802
1	Cotyledon	Microscopy	+		819
Grevillea annulifera F. Muell. (Prickly Plume Grevillea)					
1	Seed cotyledon	Microscopy	+		877
1	Seed cotyledon	SEM/EMX	+		877
Grevillea leucopteris Meissn.					
1	Seed cotyledon	Microscopy	+		877
1	Seed cotyledon	SEM/EMX, TEM	+		877
Grewia asiatica L. (Phalsa)					
1	Fruit, flesh, and skin	McCance	0.068	0.203	827
Guizotia abyssinica (L.f.) Cass. (Nigerseed)					
1	Lipid-protein seed concentrate	McCance	0.835	2.75	987
Hakea platysperma Hook. (Cricket Ball Hakea)					
1	Seed cotyledon	Microscopy	+		877
1	Seed cotyledon	SEM/EMX, TEM	+		877
Hakea victoriae J. Drumm.					
1	Seed cotyledon	Microscopy	+		877
1	Seed cotyledon	SEM/EMX	+		877
Hedera helix L. (English Ivy)					
1	Fat-free leaf powder[b,c]	Michel	Trace		757
Helianthus annuus L. (Sunflower)					
2	Whole seed	McCance	0.45-0.48		988
1	Mature seed	PC	+		810
1	Whole seed[b,c]	McCance	0.22		989
1	–, dehulled	McCance	0.30		989
1	–, hull	McCance	0.05		989
1	–, defatted flour	McCance	0.54-0.56		989
1	–, protein isolate	McCance	0.08-0.23		989
1	Whole seed	Harland	0.43	0.53	990
1	–, meal	Harland	1.2	1.50	990
1	–, kernel	Harland	0.62	0.69	990
1	–, hull	Harland	0.03	0.04	990
1	–, protein concentrate	Harland	0.9	1.21	990
1	Whole seed	HPLC	0.48	0.53	990
1	–, kernel	HPLC	0.62	0.69	990
1	–, hull	HPLC	0	0.04	990
1	–, meal	HPLC	1.3	1.50	990
1	–, protein concentrate	HPLC	1.0	1.21	990
1	Seed, cotyledon	TEM/EDX	+		843
1	Seed protein isolate[b,c]	Harland	0.28-1.04		755
1	Seed, protein concentrate	McCance	0.94		988
1	Lipid-protein concentrate[c]	McCance	0.880	1.67	987

[a]Percentage of total P; [b]Species unstated; [c]Basis not given; [d]Negligible (<1% of total P); [e]As is basis; [f]Fresh weight basis; [g]Air-dry basis; [h]Mean ± standard error; [i]Mixture; [j]See also ref. 1184.

Continued

Table 2-4 Phytic Acid Content of Plants—cont'd

No.	Plant part	Method	% P (phytate)	% P (total)	References
1	Decorticated seed[b,c]	Posternak	0.723	0.830	807
1	–, aleurone grain[b,c]	Posternak	2.71	2.78	807
1	Germinated seed	^{31}P NMR	+		43
1	Root	SEM/EMX	+		891
1	Pollen	PE	0	0.90	788
Helianthus tuberosus L. (Jerusalem Artichoke)					
1	Tuber	^{31}P NMR	+		43
1	Tuber, boiled[b,f]	McCance	0.0092	0.0370	735
1	Other parts	^{31}P NMR	0		45
Helichrysum bracteatum (Vent.) Andr. (Golden Everlasting Daisy)					
1	Seed cotyledon	TEM/EDX	+		843
Heliotropium lasiocarpum Fisch. and Mey. (Heliotrope)					
1	Defatted seed meal	Mukhamedova	0.56		794
Heliotropium olgae Bunge					
1	Defatted seed meal	Mukhamedova	0.56		795
Hevea brasiliensis (Willd.) Müll. (Pará Rubber)					
1	Defatted seed kernel	McCance	0.090	0.24	991
1	Seed meal	McCance	0.35	0.58	855
Hibiscus cannabinus L. (Kenaf)					
1	Defatted seed meal	Mukhamedova	1.3		793
Hibiscus huegelii Endl. (Lilac Hibiscus)					
1	Pollen	PE	0.07	0.56	788
Hibiscus sabdariffa L. (Roselle)					
3	Seed	McCance	0.26-0.33		992
1	Leaf and stem	McCance	0.004	0.228	837
Hippophae rhamnoides L. (Sea Buckthorn)					
1	Defatted seed meal	Mukhamedova	0.96		796
Hordeum bulbosum L. (Bulbous Barley)					
1	Pollen	PE	0	0.40	788
Hordeum vulgare L. (Barley)					
18	Seed	McCance	0.273-0.327	0.397-0.494	858
9	Seed	McCance	0.27-0.31		993
1	Meal	McCance	0.216	0.355	826
1	High protein seed flour	McCance	0.68-0.81		765
1	High bran seed flour	McCance	0.90-0.98		765
1	Seed	Harland	0.34		994
3	Seed, 2-rowed	Harland	0.250-0.276		995
3	Seed, 6-rowed	Harland	0.267-0.273		995
1	Seed[b]	HPLC	0.33		774
1	Seed[b]	Inositol	0.314		784
1	Seed	Michel	0.143	0.253	996
2	Seed aleurone tissue	Microscopy	+		819,868
2	Seed aleurone cell	SEM/EDX	+		870
1	Seed aleurone cell	TEM/EDX	+		870

Table 2-4 Phytic Acid Content of Plants—cont'd

No.	Plant part	Method	% P (phytate)	% P (total)	References
1	Grain	CZE	0.250		746
1	Seed bran[b,c]	Suzuki	0.238	0.541	797
1	Acid-washed fiber	Harland	0.127		851
1	Defatted seed powder	Michel	0.338		758
1	Germinated seed	HPLC	+		857
1	84-h sprout	McCance	0.058	0.56	997
Hyoscyamus niger L. (Henbane)					
1	Defatted seed meal	Mukhamedova	1.01		796
Hyphaene thebaica (L.) Mart. (Doum Palm)					
1	Seed	McCance	0.093		828
Icacina manni (Efik Isun)					
1	Peeled tuber	McCance	0.042		998
Ipomoea aquatica Forssk. (Swamp Cabbage)					
1	Leaf	McCance	0.010	0.058	832
1	Stalk	McCance	0	0.398	821
Ipomoea asarifolia (Desr.) Roem. and Schult. (Ginger-Leaf Morning-Glory)					
1	Leaf	McCance	0.025	0.13	917
Ipomoea batatas (L.) Lam. (Sweet Potato)					
1	Young leaf	McCance	0.019	0.068	832
1	Leaf[f]	HPLC	0.00059		852
49	Whole tuber, milled	McCance	0.0014-0.0043	0.038-0.182	999
1	Boiled tuber	Harland	0.011		822
Ipomoea involucrata Beauv. (Iehowa Dua)					
1	Leaf	McCance	0.050	0.034	963
Isoetes muelleri A. Braun					
1	Corm	Microscopy	+		1000
1	Corm	SEM, TEM	+		1000
Jatropha curcas L. (Barbados Nut)					
18	Seed kernel	McCance	1.85-2.84		1001
Jatropha dulcis J. F. Gmel.					
1	Root[b,c]	Undisclosed	+		1002
Jatropha utilissima					
1	Root[b,c]	Undisclosed	+		1002
Juglans nigra L. (Black Walnut)					
1	Nut[b,g]	McCance	0.57		907
1	Pollen	PE	0.22	0.46	788
Juglans regia L. (English Walnut)					
1	Nut	McCance	0.179	0.403	842
1	Seed cotyledon	TEM/EDX	+		939
Kerstingiella geocarpa Harms. (Kersting's Groundnut)					
1	Seed		0.014-0.084		1003
Lablab purpureus (L.) Sweet (Hyacinth Bean)					
5	Seed		0.282-0.380	0.385-0.483	1004
1	Seed, axis[a]	^{31}P NMR	67		824

[a]Percentage of total P; [b]Species unstated; [c]Basis not given; [d]Negligible (<1% of total P); [e]As is basis; [f]Fresh weight basis; [g]Air-dry basis; [h]Mean ± standard error; [i]Mixture; [j]See also ref. 1184.

Continued

Table 2-4 Phytic Acid Content of Plants—cont'd

No.	Plant part	Method	% P (phytate)	% P (total)	References
1	Seed, hypocotyl[a]	^{31}P NMR	74		824
1	Boiled bean	Harland	0.070		822
Lactuca sativa L. (Lettuce)					
1	Seed	McCance	0.39		1005
1	–, cotyledon	McCance	0.31		1005
1	–, axis	McCannce	0.08		1005
1	–, endosperm	McCance	0.007		1005
1	Leaf[b]	HPLC	0		741
1	Leaf[c]	^{31}P NMR	+		48
Lagenaria siceraria (Mol.) Standl. (Bottle Gourd)					
1	Gourd	McCance	0.003	0.017	832
Lansium domesticum Corr. Serr. (Lansones, Langsat)					
1	Peeled ripe fruit	McCance	0.31		844
Larix laricina (Du Roi) K. Koch (American Larch)					
1	Needle and seed	McCance	negl.[d]		833
Lasianthera africana P.B. Mundja (Editan)					
1	Leaf	McCance	0.038	0.425	981
Lathyrus sativus L. (Grass Pea)					
1	Seed	McCance	0.082	0.262	827
1	Seed	Harland	0.325		1006
Lathyrus tingitanus L. (Tangier Pea)					
2	Seed	McCance	0.13-0.15	0.32-0.35	859
Launaea taraxacifolia (Willd.) Amin (Wild Lettuce)					
1	Leaf	McCance	2.61 [*sic*]	0.012	963
Lemna gibba L. (Gibbous Duckweed)					
1	Whole plant	PE	+		695
Lemna minor L. (Lesser Duckweed)					
1	Whole plant	GFC	+		683
1	Frond	SEM/EMX	+		1007
1	Frond	TEM	+		1007
Lemna spp. (Duckweed)					
1	Whole plant	Harland	0.0069	0.384	912
Lens culinaris Medic. (Lentil)					
~15	Seed	McCance	0.044-0.339	0.234-0.544	969
2	Seed	McCance	0.120-0.195	0.153-0.247	1008
4	Seed	McCance	0.043-0.051	0.261-0.331	1009
5	Seed	McCance	0.08-0.11	0.190-0.217	919
3	Seed	Harland	0.236-0.295		1006
2	Seed	Harland	0.042-0.434	0.23-0.67	1010
1	Seed	HPLC	0.069		922
1	Defatted seed powder	Michel	0.3685		758
1	Flour	Harland	0.180		926
1	–, protein fraction	Harland	0.528		926
1	–, starch fraction	Harland	0.042		926

Table 2-4 Phytic Acid Content of Plants—cont'd

No.	Plant part	Method	% P (phytate)	% P (total)	References
Lepidium sativum L. (Garden Cress)					
1	Seed	McCance	0.457	0.747	827
Leucaena leucocephala (Lam.) De Wit (Ipil Ipil (Leadtree))					
2	Whole seed	McCance	0.40-0.61	0.298-0.300	1011
2	–, kernel	McCance	0.42-0.73	0.607-0.615	1011
2	–, hull	McCance	0.33-0.35	0.052-0.054	1011
1	Seed, axis[a]	^{31}P NMR	78		824
1	–, hypocotyl[a]	^{31}P NMR	79		824
1	Leaf meal	McCance	0.02	0.22	855
Lilium henryi Bak. (Lily)					
1	Pollen	PE	0.31	0.84	788
Lilium longiflorum Thunb. (Trumpet Lily)					
1	Pollen	McCance	0.229	0.889	1012
1	Pollen	Harland	0.244	0.889	1012
Lindelofia macrostyla (Bunge) M. Pop.					
1	Defatted seed meal	Mukhamedova	0.39		794
Lindelofia tschimganica (Lipsky) Pazij.					
1	Defatted seed meal	Mukhamedova	0.45		794
Linum usitatissimum L. (Flax)					
1	Whole seed, defatted	McCance	0.606-0.785	0.818-1.077	865
1	–, meal	McCance	1.351	1.907	865
1	–, shell	McCance	0.270	0.350	865
1	Defatted flour[b,c]	McCance	1.7-1.8		943
8	Defatted meal	McCance	0.374-1.071		335
10	Defatted meal	Harland	0.68-0.76	0.99-1.07	1013
10	Seed, protein isolate[i]	McCance	0.36	0.45	1014
1	Aleurone grain	Microscopy	+		941
Lolium cuneatum Nevski (Poison Ryegrass)					
1	Defatted seed meal	Mukhamedova	0.45		794
Lolium multiflorum Lam. (Italian Ryegrass)					
1	Root	SEM/EMX	+		891
Lolium spp. (Ryegrass)					
3	Seed[b]	McCance	0.19-0.27	0.35-0.43	1015
Lonicera nummulariifolia Jaub. and Spach. (Honeysuckle)					
1	Defatted seed meal	Mukhamedova	0.42		796
Lophira alata Banks (Niam Tree)					
1	Seed	McCance	0.204		828
Lotus corniculatus L. (Birdsfoot Trefoil)					
1	Root	McCance	0.023	0.55	1016
Luffa acutangula (L.) Roxb. (Angled Luffa)					
1	Gourd[e]	McCance	0	0.037	842
Luffa aegyptiaca Mill. (Sponge Gourd)					
1	Gourd	McCance	0.009	0.031	832
1	Seed, protein isolate	McCance	0.40	0.47	930

[a]Percentage of total P; [b]Species unstated; [c]Basis not given; [d]Negligible (<1% of total P); [e]As is basis; [f]Fresh weight basis; [g]Air-dry basis; [h]Mean ± standard error; [i]Mixture; [j]See also ref. 1184.

Continued

Table 2-4 Phytic Acid Content of Plants—cont'd

No.	Plant part	Method	% P (phytate)	% P (total)	References
Lupinus albus L. (White Lupine)					
4	Seed[c]	McCance	0.40 ± 0.18		1017
2	Seed	McCance	0.085-0.099	0.302-0.378	1018
2	Seed	Harland	0.14-0.17		1019
1	Seed	HPLC	0.0087		1020
1	Seed flour	Harland	0.230		926
1	Seed, axis[a]	^{31}P NMR	0.630	0.714	824
1	–, cotyledon[a]	^{31}P NMR	0.309	0.334	824
Lupinus angustifolius L. (Blue Lupine)					
4	Seed[c]	McCance	0.41 ± 0.12		1017
6	Seed	Harland	0.20-0.31		1019
1	Dehulled seed	McCance	0.132	0.29	1021
1	6-day sprout	McCance	0.045	0.35	1021
Lupinus atlanticus Gladst.					
2	Seed	Harland	0.31-0.34		1019
Lupinus cosentinii Guss.					
1	Seed	Harland	0.11-0.17		1019
Lupinus digitatus Forsk.					
1	Seed	Harland	0.22		1019
Lupinus hispanicus Boiss. and Reut. (Spanish Lupine)					
1	Seed	Harland	0.34		1019
Lupinus hispanicus ssp. *bicolor* (Merino) Gladst.					
18	Seed[c]	McCance	0.40 ± 0.07		1017
Lupinus hispanicus ssp. *hispanicus*					
16	Seed[c]	McCance	0.42 ± 0.09		1017
Lupinus luteus L. (Yellow Lupine)					
4	Seed[c]	McCance	0.76 ± 0.27		1017
2	Whole seed flour	Harland	0.85-0.87		1022
1	Seed	HPLC	0.201		1020
1	Seed aleurone grain	McCance	1.07		1023
1	–, light	McCance	0.53		1024
1	–, heavy	McCance	3.58		1024
Lupinus micranthus Guss.					
1	Seed	Harland	0.25		1019
Lupinus mutabilis Sweet (Tarwi)					
4	Seed[c]	McCance	0.77 ± 0.10		1017
1	Seed axis[a]	^{31}P NMR	78		824
1	–, cotyledon[a]	^{31}P NMR	81		824
Lupinus palestinus Boiss.					
1	Seed	Harland	0.34		1019
Lupinus pilosus L. (Blue Lupine)					
2	Seed	McCance	Trace-0.13	0.26-0.39	859
3	Seed	Harland	0.22-0.25		1019
Lupinus polyphyllus Lindl. (Washington Lupine)					
1	Seed	Harland	0.20		1019

Table 2-4 Phytic Acid Content of Plants—cont'd

No.	Plant part	Method	% P (phytate)	% P (total)	References
Lupinus spp. (Lupine)					
1	Root	SEM/EMX	+		891
Lycopersicon esculentum Mill. (Tomato)					
1	Seed	SEM	+		1025
1	Seed cotyledon	TEM/EDX	+		1025
1	Seed, defatted groats	McCance	0.1-0.4		1026
1	Acid-washed fiber	Harland	0.240		851
1	Root	SEM/EMX	+		891
1	Whole plant[f]	McCance	5.0		1027
1	Fruit	McCance	0.010	0.021	832
–, ssp. *cerasiforme* Alef. (Cherry Tomato)					
1	Seed	TEM/EDX	+		904
Lysiloma watsonii Rose (Feathertree)					
1	Seed	Harland	0.28		1028
Macadamia integrifolia Maid. and Bet. (Macadamia Nut)					
1	Seed cotyledon	TEM/EDX	+		939
Macrotyloma uniflorum (Lam.) Verdc. (Horse Gram)					
1	Seed	McCance	0.184	0.320	1029
Malus domestica Borkh. (Apple)					
1	Stem cortex, 1 year old	McCance	0.034		1030
1	Acid-washed fiber	Harland	0.038		851
Malus sylvestris (L.) Mill. (Crabapple)					
1	Cored fruit	McCance	0	0.129	827
Mangifera indica L. (Mango)					
1	Seed kernel meal	McCance	0.082	0.168	1031
2	Fruit	McCance	0.010-0.015	0.016-0.029	832
1	Fruit	Harland	0.039		822
Manihot esculenta Crantz (Cassava)					
1	Leaf, very young	McCance	0.028	0.23	1032
1	–, mature	McCance	0.047	0.18	1032
1	Leaf meal	McCance	0.04	0.42	855
1	Boiled leaf	Harland	0.053		822
1	Tuber	McCance	0.019	0.080	827
7	Peeled tuber, edible portion	McCance	0.26-0.56		844
1	Tuber flour	McCance	0.0033	0.0327	1033
1	Root	Harland	0.045		822
Maranta arundinacea L. (Arrowroot)					
1	Root flour	McCance	0	0.020	821
Marsdenia latifolium Benth. (Utasi)					
1	Leaf	McCance	0.031	0.395	981
Matricaria chamomilla L. (Sweet False Chamomile)					
1	Defatted seed meal	Mukhamedova	0.11		794
Medicago sativa L. (Alfalfa)					
1	Defatted seed[b]	McCance	0.006	0.283	1034
1	Seed meal[b]	McCance	0.027	0.131	826
1	Leaf protein concentrate	McCance	0.00		1035

[a]Percentage of total P; [b]Species unstated; [c]Basis not given; [d]Negligible (<1% of total P); [e]As is basis; [f]Fresh weight basis; [g]Air-dry basis; [h]Mean ± standard error; [i]Mixture; [j]See also ref. 1184.

Continued

Table 2-4 Phytic Acid Content of Plants—cont'd

No.	Plant part	Method	% P (phytate)	% P (total)	References
1	Root	McCance	0.009	0.43	1016
1	Root	SEM/EMX	+		891
8	Root and crown	McCance	0.022-0.029	0.191-0.254	1036
3	Root and crown	Harland	0.006-0.016	0.191-0.221	1036
3	Root and crown	PE	+	0.191-0.221	1036
1	Whole plant (1/4 bloom)[b]	Harland	0.006	0.31	867
Melia azedarach L. (Chinaberry)					
1	Defatted seed meal	Mukhamedova	0.34		796
Melilotus officinalis (L.) Lam. (Yellow Sweet Clover)					
1	Root	McCance	0.018	0.52	1016
Mentha spicata L. (Spearmint)					
1	Leaf	McCance	0.022	0.268	821
Metroxylon sagu Rottb. (Sago Palm)					
1	Palm	McCance	0	0.007	821
1	Pith starch[b,f]	McCance	0.019	0.038	735
Momordica charantia L. (Balsam Pear)					
1	Leaf	McCance	0.042	0.108	832
1	Fruit	McCance	0.004	0.050	832
Monodora ternifolia					
1	Seed	McCance	0.183		828
Moringa oleifera Lam. (Horseradish Tree)					
1	Leaf	McCance	0.009	0.192	832
1	Pod	McCance	0.026	0.048	832
Moringa peregrina (Forssk) Fion (Yassar)					
1	Seed, defatted flour	McCance	0.732		1037
1	–, protein concentrate	McCance	1.188		1037
1	–, protein isolate	McCance	0.278		1037
Mucuna deeringiana (Bort.) Merr. (Velvet Bean)					
1	Bean	McCance	0.17	0.29	855
Mucuna monosperma DC ex Wight (Negro Bean)					
2	Bean	McCance	0.142-0.178		1038
Mucuna utilis (Wight) Burck. (Mucuna Bean)					
1	Whole seed	McCance	0.087	0.22	1039
Murraya koenigii (L.) Spreng. (Curry Leaf)					
1	Leaf	McCance	0.010	0.170	821
Musa errans (Blanco) T.B. Teodoro (Botoan)					
1	Inflorescence	McCance	0.004	0.038	832
Musa × paradisiaca L. (Banana)					
4	Fruit	McCance	0.008-0.029	0.012-0.071	832
1	Fruit	Harland	0.022		822
2	Boiled ripe fruit	Harland	0		955
2	Boiled unripe fruit	Harland	0.0003-0.010		955
Myriophyllum spp. (Milfoil)					
1	Whole plant	Harland	0.0058	0.149	912
Myristica fragrans Houtt. (Nutmeg)					
1	Seed	McCance	0.189	0.308	821

Table 2-4 Phytic Acid Content of Plants—cont'd

No.	Plant part	Method	% P (phytate)	% P (total)	References
Nasturtium officinale R. Br. (Watercress)					
1	Leaf	McCance	0.016	0.057	832
Nelumbo spp. (Lotus)					
1	Stem[b]	McCance	0	0.141	827
1	Fruit	PE	+		779
Nitraria komarovii Iljin and Lava					
1	Defatted seed meal	Mukhamedova	0.22		794
Nitraria schoberi L.					
1	Defatted seed meal	Mukhamedova	0.36		794
Ocimum basilicum L. (Basil)					
1	Leaf	McCance	0.073	0.503	981
Ocimum viride Willd. (Iko)					
1	Leaf	McCance	0.025	0.440	981
Oenothera striata Link.					
1	Pollen	PE	0.11	0.40	788
Onopordon acanthium L. (Scotch Thistle)					
1	Defatted flour	McCance	0.107	1.053	905
Onopordon olgae Bunge and Schmalh.					
1	Defatted seed meal	Mukhamedova	0.28		794
Oryza sativa L. (Rice)					
1	Whole grain[g]	McCance	0.25	0.31	1040
1	–, germ	McCance	0.98	1.30	1040
1	–, endosperm	McCance	0.004	0.11	1040
1	–, pericarp	McCance	0.95	1.04	1040
9	Grain	McCance	0.029-0.233	0.095-0.435	832
2	Grain	CZE	0.249-0.380		746
1	Polished grain, long[b]	McCance	0.097	0.189	1041
2	–, medium[b]	McCance	0.039-0.054	0.080-0.128	1042
1	–, short[b]	McCance	0.038	0.114	1042
8	White grain	McCance	0.035-0.055	0.096-0.121	1043
8	–, true bran	McCance	2.231-2.637	2.476-2.868	1043
3	Brown grain	McCance	0.155-0.201	0.320-0.348	1044
3	–, lightly milled	McCance	0.100-0.118	0.153-0.162	1044
3	–, well milled	McCance	0.037-0.053	0.080-0.128	1044
3	–, whole kernel	McCance	0.239-0.253	0.239-0.272	1044
3	–, bran	McCance	0.660-0.810	1.955-2.264	1044
3	Brown grain[b]	Michel	0.10-0.28	0.16-0.38	1045
3	–, milled	Michel	0.02-0.07	0.06-0.16	1045
2	–, endosperm, subaleurone	Michel	0.21-0.24	0.21-0.35	1045
2	–, endosperm, middle	Michel	0.010-0.025	0.06-0.09	1045
2	–, endosperm, inner	Michel	0.003-0.005	0.05	1045
2	–, bran	Michel	0.84-2.28	1.00-2.60	1045
2	–, polish	Michel	0.60-1.58	0.67-1.61	1045
1	Brown seed, unpolished	Inositol	0.222		785
1	Brown seed, parboiled[b]	HPLC	0.438		741

[a]Percentage of total P; [b]Species unstated; [c]Basis not given; [d]Negligible (<1% of total P); [e]As is basis; [f]Fresh weight basis; [g]Air-dry basis; [h]Mean ± standard error; [i]Mixture; [j]See also ref. 1184.

Continued

Table 2-4 Phytic Acid Content of Plants—cont'd

No.	Plant part	Method	% P (phytate)	% P (total)	References
2	Enriched polished seed	Harland	0.067-0.084		738
1	Enriched seed, defatted[b]	HPLC	0.064		850
1	Polished seed[b]	^{31}P NMR	0.12		743
1	Wild seed[b]	Harland	0.62		994
12	Seed kernel	Harland	0.214-0.284	0.269-0.375	1046
1	Seed, aleurone particle[c]	Inositol	9.71	11.61	1047
1	Seed, aleurone particle[c]	PE	11.3	11.6	1047
1	Seed, aleurone particle[c]	PC	+		1047
1	Seed, aleurone grain	TEM/EDX	+		1048
1	High-phytic seed particles	Ogawa	18.77-18.93	18.77-18.93	800
1	Scutellum particles	Ogawa	19.79		801
1	Scutellar cell	TEM/EDX	+		1049
1	Scutellum	SEM/EMX	+		1050
1	Seed bran	GFC	+		778
1	Seed bran	Harland	0.431		851
1	Bran, stabilized[b]	HPLC	2.20		772
1	Nonwaxy seed kernel	Harland	0.266	0.340	1046
1	–, embryo	Harland	1.918	2.167	1046
1	–, endosperm	Harland	0.038	0.100	1046
1	–, bran	Harland	1.984	2.114	1046
1	Immature grain[b,e]	Inositol	0.0233		783
1	–, leaf, and stem	Inositol	0.0039		783
1	–, root	Inositol	0.00057		783
1	Sprout	McCance	0.230	0.318	1051
2	Leaf (straw)	Michel	0.02	0.49-0.74	1052
Oxytropis lambertii Pursh. (Locoweed)					
1	Seed embryo	TEM/EDX	+		856
Pachyrhizus erosus (L.) Urb. (Yambean)					
1	Pod	McCance	0.026	0.045	832
1	Tuber	McCance	0.015	0.029	832
Padus racemosa Lam. (Bird Cherry)					
1	Stem cortex, 1 year old	McCance	0.041		1030
Panicum miliaceum L. (Proso Millet)					
24	Whole seed[e]	McCance	0.05-0.13		1053
5	–, dehulled	McCance	0.05-0.09		1053
5	–, hull	McCance	0.14-0.45		1053
6	Dehulled seed	McCance	0.14-0.19	0.31-0.34	1054
1	Seed[b]	^{31}P NMR	0.21		743
Panicum sumatrense Roem. and Schult. (Little Millet)					
1	Dehusked grain	McCance	0.058	0.174	827
Papaver somniferum L. (Poppy)					
1	Mature seed	PC	+		810
1	Defatted seed powder	Michel	0.316		758
1	Press-cake flour[c]	McCance	0.808	1.58	987
1	Defatted flour [c]	McCance	0.96		1055
1	–, protein isolate	McCance	0.00		1055

Table 2-4 Phytic Acid Content of Plants—cont'd

No.	Plant part	Method	% P (phytate)	% P (total)	References
Parkia clappertoniana Keay (West African Locust Bean)					
1	Seed	McCance	0.16	0.43	825
Parkinsonia aculeata L. (Mexican Palo Verde)					
1	Seed	Harland	0.20-0.22		1028
Paspalum scrobiculatum L. (Kodo Millet)					
1	Dehusked grain	McCance	0.135	0.136	827
Pastinaca sativa L. (Parsnip)					
1	Tuber	McCance	0.016	0.245	821
1	Tuber[b]	HPLC	0.23		741
1	Tuber	^{31}P NMR	+		43
Peganum harmala L. (African Rue)					
1	Defatted seed meal	Mukhamedova	0.87		794
Pennisetum glaucum (L.) R.Br. (Pearl Millet)					
14	Whole grain	McCance	0.035-0.221	0.245-0.348	1056
8	Whole grain	McCance	0.167-0.293	0.300-0.390	1057
2	Whole seed	McCance	0.179-0.306		1058
1	–, decorticated	McCance	0.312		1058
1	–, germ	McCance	0.752		1058
1	–, endosperm	McCance	0.086		1058
1	–, bran	McCance	0.278		1058
1	Whole seed meal	McCance	0.245		1058
1	Grain flour	McCance	0.177	0.311	1059
1	84-h sprout	McCance	0.013		1060
Pennisetum spp. (Millet)					
1	Seed, flour[b]	HPLC	0.165		1061
1	–, 48-h sprout[b]	HPLC	0.099		1061
Pentaclethra macrophylla Benth. (African Oil Bean)					
1	Dehulled seed	McCance	0.320	0.300	1062
1	Hull	McCance	–	0.040	1062
Perilla frutescens (L.) Britt. (Perilla Mint)					
1	Whole flour	McCance	0.68		1063
1	–, defatted	McCance	1.15		1063
Persea americana Mill. (Avocado)					
2	Peeled ripe fruit	McCance	0.35-0.69		844
1	Fruit	Harland	0.014		822
Petroselinum crispum (Mill.) Nym. (Parsley)					
1	Leaf	McCance	0.053	0.477	821
1	Root[b]	Schmalfuss	0	1.04	792
Petunia hybrida hort. (Petunia)					
1	Pollen	^{31}P NMR	n.d.		788
3	Flower, pollen	PE	0.28-0.56		1064
3	–, stigma	PE	0.08-0.17		1064
3	–, style	PE	0-0.014		1064
3	–, ovary, filament, or petal	PE	0		1064

[a]Percentage of total P; [b]Species unstated; [c]Basis not given; [d]Negligible (<1% of total P); [e]As is basis; [f]Fresh weight basis; [g]Air-dry basis; [h]Mean ± standard error; [i]Mixture; [j]See also ref. 1184.

Continued

Table 2-4 Phytic Acid Content of Plants—cont'd

No.	Plant part	Method	% P (phytate)	% P (total)	References
Phaseolus acutifolius Gray (Tepary Bean)					
2	Domestic bean	Harland	0.17-0.21		1028
1	Wild bean	Harland	0.29		1028
Phaseolus coccineus L. (Runner Bean)					
1	Bean	Harland	0.321		1065
Phaseolus lunatus L. (Lima Bean)					
18	Bean	McCance	0.150-0.211	0.316-0.480	1066
1	Bean	Harland	0.197		822
2	Seed	HPLC	0.041-0.052		900
1	Defatted seed powder	Michel	0.1962		758
1	Seed flour	Harland	0.146		926
1	–, protein fraction	Harland	0.384		926
1	–, starch fraction	Harland	0.052		926
2	Seed protein isolate	McCance	0.80-1.00		815
1	Seed protein isolate	DGE	+		815
1	Bean crystalline protein	McCance	0.47-1.25	0.46-1.20	1067
3	3-day sprout	McCance	0.10-0.13	0.28-0.35	980
Phaseolus vulgaris L. (Bean (Common))					
10	Whole bean	McCance	0.33-0.82		1068
10	–, dehulled	McCance	0.46-1.03		1068
50	Whole bean	McCance	0.15-0.44	0.259-0.556	1069
~15	Bean	McCance	0.139-0.373	0.295-0.514	969
1	Bean flour	McCance	0.75		1070
1	–, CDF fraction	McCance	1.55		1070
1	Immature bean	McCance	0.0014-0.0120	0.0843-0.2688	1071
2	Bean[b]	HPLC	0.25-0.48		772
1	Bean, defatted[b]	HPLC	0.39		850
1	Bean	Inositol	0.484		784
1	Bean, axis	^{31}P NMR	0.335	0.675	824
1	–, cotyledon	^{31}P NMR	0.383	0.453	824
3	Dehulled bean	Harland	0.392-0.467	0.493-0.535	1072
3	–, protein fraction	Harland	0.668-0.851	0.844-0.916	1072
3	–, starch fraction	Harland	0.121-0.246	0.274-0.400	1072
3	–, hull	Harland	0.288-0.382	0.447-0.507	1072
5	Bean protein concentrate	McCance	0.132-0.428		1073
2	Bean protein isolate	McCance	1.11-1.36		815
2	Bean protein isolate	DGE	+		815
2	Bean crystalline protein	McCance	0.79-2.16	0.74-2.16	1067
1	Pod	McCance	0.005-0.012		754
1	Seedling, embryonic axis[f]	McCance	0.0030		888
1	–, vacuole[f]	McCance	0		888
Philydrella pygmaea (R.Br.) Caruel					
1	Corm, seed	Microscopy	+		1000
1	Corm	SEM/EMX	+		1000
1	Corm	TEM	+		1000

Table 2-4 Phytic Acid Content of Plants—cont'd

No.	Plant part	Method	% P (phytate)	% P (total)	References
Phoenix dactylifera L. (Date Palm)					
1	Seed	McCance	0.023		865
1	Defatted seed powder	Michel	0.0232		758
1	Haustoria	Microscopy	+		1074
1	Haustoria	TEM/EDX	+		1074
1	Deseeded fruit	McCance	0.04	0.18	855
Phylloglossum drummondii Kunze					
1	Stem tuber	Microscopy	+		1000
1	Stem tuber	SEM, TEM	+		1000
Physalis ixocarpa Brot. (Tomatillo)					
1	Whole fruit	McCance	0.084	0.482	827
Physalis peruviana L. (Cape Gooseberry)					
1	Whole fruit	McCance	0.107	0.440	827
Picea glauca (Moench) Voss (White Spruce)					
1	Whole seed	Harland	0.688	0.782	1075
1	–, kernel	Harland	0.936	1.151	1075
1	–, zygotic embryo	Harland	0.347	1.070	1075
1	–, female gametophyte	Harland	1.014	1.163	1075
1	Seed, cotyledon	TEM/EDX		+	1076
1	–, hypocotyl	TEM/EDX		+	1076
1	–, radicle	TEM/EDX		+	1076
Picea mariana (Mill.) Brit., St. and Pog. (Black Spruce)					
1	Needle and stem	McCance	negl.[j]		833
Picea pungens Engelm. (Colorado Spruce)					
1	Pollen	FTIR	+	0.16	818
1	Pollen	McCance	0.1075	0.16	818
Picea spp. (Spruce)					
1	Seed[b,c]	Posternak	0.600	0.656	807
1	–, aleurone grain[b,c]	Posternak	2.58	2.67	807
Pinus banksiana Lamb. (Jack Pine)					
1	Seed cotyledon	TEM/EDX	+		1077
1	Seedling cotyledon	TEM/EDX	+		1077
Pinus canariensis C. Sm. (Canary Island Pine)					
1	Pollen	PE	0.07	0.50	1078
Pinus cembra L. (Swiss Stone Pine)					
1	Pollen	PE	0.05	0.46	1078
Pinus contorta Douglas ex Louden (Lodgepole Pine)					
1	Seed cotyledon, hypocotyl, and root	TEM/EDX	+		1079
Pinus coulteri D. Don (Coulter Pine)					
1	Seed cotyledon, hypocotyl, and root	TEM/EDX	+		1079
Pinus flexilis James (Limber Pine)					
1	Pollen	PE	0.03	0.40	788
Pinus griffithii McClelland (Himalayan Pine)					
1	Pollen	PE	0.03	0.46	1078

[a]Percentage of total P; [b]Species unstated; [c]Basis not given; [d]Negligible (<1% of total P); [e]As is basis; [f]Fresh weight basis; [g]Air-dry basis; [h]Mean ± standard error; [i]Mixture; [j]See also ref. 1184.

Continued

Table 2-4 Phytic Acid Content of Plants—cont'd

No.	Plant part	Method	% P (phytate)	% P (total)	References
Pinus koraiensis Sieb. and Zucc. (Korean Pine)					
1	Seed cotyledon, hypocotyl, and root	TEM/EDX	+		1079
Pinus mugo Turra (Mugo Pine)					
1	Seed cotyledon, hypocotyl, and root	TEM/EDX	+		1079
1	Pollen	PE	<0.01	0.54	1078
Pinus mugo ssp. *mughus* Scop.					
1	Pollen	PE	<0.01	0.50	1078
Pinus mugo ssp. *rostrata* (Ant.) Hoopes					
1	Pollen	PE	0.11	0.43	1078
Pinus nigra Black Pine					
1	Seed cotyledon, hypocotyl, and root	TEM/EDX	+		1079
Pinus pinea L. (Italian Stone Pine)					
1	Pollen	PE	<0.01	0.31	1078
Pinus ponderosa (Dougl. ex) Laws. (Ponderosa Pine)					
1	Seed cotyledon, hypocotyl, and root	TEM/EDX	+		1079
1	Pollen	McCance	0.0076	0.017	818
1	Pollen	FTIR	+	0.017	818
1	Fallen needle	Inositol	+		1080
Pinus radiata D. Don (Monterey Pine)					
1	Pollen	PE	0.02	0.37	1078
1	Needle (leaf)	PE	<0.0008	0.22	1078
1	Seed	PE	0.79	0.78	1078
Pinus resinosa Ait. (Red Pine)					
1	Seed cotyledon, hypocotyl, and root	TEM/EDX	+		1079
Pinus sabiniana Douglas					
1	Seed cotyledon, hypocotyl, and root	TEM/EDX	+		1079
Pinus strobus L. (Eastern White Pine)					
1	Seed cotyledon, hypocotyl, and root	TEM/EDX	+		1079
Pinus sylvestris L. (Scotch Pine)					
1	Seed cotyledon, hypocotyl, and root	TEM/EDX	+		1079
Pinus yunnanensis Franchet (Yunnan Pine)					
1	Pollen	PE	0.04	0.40	1078
Pinus spp. (Pine)					
1	Nut	McCance	0.318	0.514	1081
Piper clusii C. DC (African Cubeb)					
1	Seed	McCance	0.064	0.422	821

Table 2-4 Phytic Acid Content of Plants—cont'd

No.	Plant part	Method	% P (phytate)	% P (total)	References
Piper guineense Schum. and Thonn. (Etinkene (Guinea Cubeb))					
1	Leaf	McCance	0.055	0.215	
Piper nigrum L. (Black Pepper)					
1	Seed	McCance	0.132	0.226	821
Pisonia alba Span. (Lettuce Tree)					
1	Leaf	McCance	0.007	0.010	832
Pistacia atlantica Desf. (Mt. Atlas Mastic Tree)					
1	Defatted seed powder	Michel	0.176		758
Pistacia chinensis Bunge (Chinese Pistachio)					
1	Pollen	PE	0.07	0.56	788
Pistacia vera L. (Pistachio)					
1	Shelled nut[b,e]	McCance	0.164	0.529	937
1	Seed cotyledon	TEM/EDX	+		843
Pisum sativum L. (Peas)					
~15	Seed	McCance	0.092-0.311	0.234-0.528	969
15	Seed	Harland	0.178-0.376		1082
1	Seed	CZE	0.619		700
2	Seed	PC	+		1083
1	Whole pea	McCance	0.221		1084
1	–, cotyledon	McCance	0.220		1084
1	–, germ	McCance	0.345		1084
1	–, hull	McCance	0.003		1084
2	Pea, interior region[b,c]	McCance	0.20-0.50		1085
2	–, exterior region[b,c]	McCannce	0.30-0.64		1085
1	Pea flour	McCance	0.601	0.883	1086
3	–, protein flour	McCance	0.56-0.63	1.05-1.20	1087
1	Pea, protein fraction	McCance	0.53	0.86	1088
1	–, starch fraction	McCance	0.06	0.09	1088
1	Seed flour	Harland	0.209		926
1	–, protein fraction	Harland	0.532		926
1	–, starch fraction	Harland	0.051		926
1	Immature pea	McCance	0.045-0.048		1089
2	Individual peas[c]	McCance	0.35-0.43		1090
3	Whole seed	Harland	0.21-0.26		1091
2	–, protein concentrate	Harland	0.37-0.53		1091
1	Acid-washed fiber	Harland	0.014		851
1	Pea, split[b]	HPLC	0.473		741
1	Seed, axis[a]	^{31}P NMR	60		824
1	–, cotyledon[a]	^{31}P NMR	82		824
1	Seed cotyledon	SEM/EDX	+		706
1	Seed cotyledon	TEM/EDX	+		706
1	Root	SEM/EMX	+		891
Pithecellobium dulce (Roxb.) Benth. (Korukapalli (Manila Tamarind))					
1	Leaf	McCance	0	0.394	821
1	Peeled deseeded fruit	McCance	0	0.309	827

[a]Percentage of total P; [b]Species unstated; [c]Basis not given; [d]Negligible (<1% of total P); [e]As is basis; [f]Fresh weight basis; [g]Air-dry basis; [h]Mean ± standard error; [i]Mixture; [j]See also ref. 1184.

Continued

Table 2-4 Phytic Acid Content of Plants—cont'd

No.	Plant part	Method	% P (phytate)	% P (total)	References
Pithecellobium flexicaule (Benth.) Coult. (Ébano)					
1	Seed cotyledon, immature	McCance	0.22		1092
1	–, mature	McCance	0.76		1092
Poa annua L. (Annual Bluegrass)					
1	Root	GFC	0.05	0.210	696
1	Root	HPLC	n.d.	0.210	696
Portulaca oleracea L. (Purslane)					
1	Stem and leaf	McCance	0.036	0.482	827
1	Leaf	McCance	0.009	0.053	832
Potentilla asiatica Juz. (Cinquefoil)					
1	Defatted seed meal	Mukhamedova	0.65		794
Poterium polygamum Waldst. and Kit.					
1	Defatted seed meal	Mukhamedova	0.56		795
Premna odorata Blanco (Premna)					
1	Leaf	McCance	0.035	0.068	832
Prosopis africana (Gui., Per., and Rich.) Taub.					
1	Seed	McCance	0.25	0.47	825
Prosopis chilensis (Mol.) Stuntz. (Algarrobo)					
1	Seed	McCance	0.13	0.48	825
Prosopis glandulosa Torr. (Honey Mesquite)					
1	–, seed	Harland	0.282	0.31	1093
1	–, pericarp	Harland	0.006	0.08	1093
1	Seed[c]	Inositol	0.24		1094
1	–, pericarp	Inositol	0		1094
1	Green pod	Harland	0.044	0.11	1093
1	Mature pod	Harland	0.046	0.13	1093
Prosopis velutina Woot. (Velvet Mesquite)					
1	Seed[c]	Inositol	0.25		1094
1	–, pericarp	Inositol	0		1094
1	Fallen leaf	Inositol	+		1080
Protea compacta R.Br.					
1	Cotyledon	SEM/EDX	+		1095
Prunus armeniaca L. (Apricot)					
1	Defatted kernel flour	McCance	0.0039	0.667	1096
1	Kernel protein isolate[b,c]	McCance	0.51	0.65	1097
Prunus avium (L.) L. (Cherry)					
1	Fruit[b]	McCance	0.013		931
Prunus bucharica Korsh.					
1	Defatted seed meal	Mukhamedova	1.04		795
Prunus domestica L. (Plum)					
1	Flesh and skin	McCance	0	0.265	827
1	Kernel protein isolate[b,c]	McCance	0.14	0.32	1097
Prunus dulcis (Mill.) D.A. Webb (Almond)					
1	Nut	McCance	0.242	0.645	842
1	Seed	Harland	0.60		761
1	Seed cotyledon	TEM/EDX	+		879

Table 2-4 Phytic Acid Content of Plants—cont'd

No.	Plant part	Method	% P (phytate)	% P (total)	References
Prunus laurocerasus L. (Cherry Laurel)					
1	Fat-free leaf flour[b,c]	Michel	0.023-0.045		757
Prunus mahaleb L. (Mahaleb Cherry)					
1	Defatted seed meal	Mukhamedova	0.62		795
Prunus persica (L.) Batsch. (Peach)					
1	Flesh and skin	McCance	0.008	0.325	827
1	Defatted kernel flour[c]	McCance	0.0149		1098
1	Kernel protein isolate[b,c]	McCance	0.30	0.33	1097
Prunus petunnikowii Litv.					
1	Defatted seed meal	Mukhamedova	0.98		795
Psidium guajava L. (Guava)					
1	Peeled ripe fruit	McCance	0.76		844
Psophocarpus tetragonolobus (L.) DC (Winged Bean)					
16	Whole seed[e]	McCance	0.172-0.212		1099
12	Seed	Harland	0.219-0.339	0.433-0.632	1100
3	Whole seed	McCance	0.179-0.202	0.263-0.278	1101
12	Defatted seed flour[e]	McCance	0.28-0.48	0.36-0.64	1102
1	–, protein concentrate[e]	McCance	0.13		1102
1	Hull[e]	McCance	<0.03		1102
Psoralea drupacea Bunge					
1	Defatted seed meal	Mukhamedova	0.36		795
Pterocarpus mildbraedii Harms. (Mkpa)					
1	Leaf	McCance	0.072	0.410	981
Pterocarpus osun Craib. (African Padauk)					
1	Seed	McCance	0.28	0.46	825
Punica granatum L. (Pomegranate)					
1	Defatted seed meal	Mukhamedova	0.53		796
Quercus lusitanica Lam. (Portuguese Oak)					
1	Pollen	PE	0.15	0.56	788
Quercus robur L. (English Oak)					
1	Pollen	PE	0.18	0.56	788
Quercus spp. (Oak)					
1	Fat-free acorn flour[b,c]	Michel	0.01		757
Ranunculus spp. (Buttercup)					
1	Seed[b]	McCance	0.32	0.46	1015
Raphanus sativus L. (Radish)					
1	Defatted seed powder	Michel	0.071		758
2	Tuber	McCance	0	0.208-0.447	821
1	Defatted tuber flour	McCance	0.2047	1.870	1103
1	Root	SEM/EMX	+		891
Rauvolfia vomitoria Afzel. (Swizzlestick)					
1	Seed	McCance	0.099		828
Rhaponticum integrifolium C. Winkl.					
1	Defatted seed meal	Mukhamedova	0.28		794
Rhodomyrtus tomentosa (Ait.) Hassk. (Downy Myrtle)					
1	Cored fruit	McCance	0.058	0.079	827

[a]Percentage of total P; [b]Species unstated; [c]Basis not given; [d]Negligible (<1% of total P); [e]As is basis; [f]Fresh weight basis; [g]Air-dry basis; [h]Mean ± standard error; [i]Mixture; [j]See also ref. 1184.

Continued

Table 2-4 Phytic Acid Content of Plants—cont'd

No.	Plant part	Method	% P (phytate)	% P (total)	References
Rhus hirta (L.) Sudw. (Staghorn Sumac)					
1	Defatted seed meal	Mukhamedova	0.79		796
Ricinus communis L. (Castor Bean)					
1	Immature seed	McCance	10.2		1104
1	Bean	TLC	+		813
1	Defatted flour[b,c]	McCance	1.7		943
1	Seed, aleurone grain	McCance	2.90		1023
1	–, globoid	McCance	17.2		1023
1	Defatted seed powder	Michel	0.5086		758
2	Seed endosperm	Microscopy	+		819,1105,1106
1	Seed endosperm	SEM/EDX	+		1105
Roemeria refracta DC (Roemer Poppy)					
1	Defatted seed meal	Mukhamedova	1.21		794
Rosa canina L. (Dog Rose)					
1	Defatted seed meal	Mukhamedova	0.70		794
Rosa rubiginosa L. (Sweetbrier Rose)					
1	Defatted seed flour	McCance	0.046	0.167	1107
1	Deseeded fruit	McCance	Trace	0.034	1108
Rubia tinctorum L. (Madder)					
1	Defatted seed meal	Mukhamedova	0.17		796
Rubus spp. (Blackberry)					
1	Fruit[d,f]	McCance	0.0042	0.0259	735
Rumex acetosa L. (Sorrel)					
1	Leaf	McCance	0.063		1109
Rumex halaczii Rech.					
1	Defatted seed meal	Mukhamedova	0.17		796
Rumex vesicarius L. (Chukai Shak)					
1	Leaf	McCance	0.0015		902
Saccharum officinarum L. (Sugarcane)					
1	Acid-washed fiber	Harland	0		851
Salvia macrosiphon Boiss. (Sage)					
1	Defatted seed meal	Mukhamedova	1.07		796
Sandoricum koetjape (Burm. f.) Merr. (Santol)					
1	Fruit		0.008	0.026	832
Santalum acuminatum R.Br. (Quandong)					
1	Endosperm	TEM/EDX	+		879
Saposhnikovium spp.					
1	Root	PE	+		779
Secale cereale L. (Rye)					
3	Grain[b,g]	McCance	0.29-0.53		907
2	Grain	CZE	0.151-0.160		746
1	Seed	Michel	0.197	0.342	996
1	Seed	HPLC	0.28		774
1	Whole grain[b,c]	Harland	0.41		933
4	Whole grain[g]	McCance	0.188-0.284		1110
1	–, kernel, upper half	McCance	0.226		1110

Table 2-4 Phytic Acid Content of Plants—cont'd

No.	Plant part	Method	% P (phytate)	% P (total)	References
1	–, kernel, germ half	McCance	0.296		1110
2	–, coarse bran	McCance	1.020-1.217		1110
2	–, fine bran	McCance	0.227-0.809		1110
4	Seed flour[g]	McCance	0.080-0.118		1110
1	Seed flour, 100% extraction	McCance	0.23	0.37	1111
1	–, 65% extraction	McCance	0.01	0.15	1111
1	Seed aleurone cell	SEM	+		870
1	Pollen	PE	0	0.62	788
Senecio cruentus DC (Cineraria)					
1	Pollen	PE	0	0.59	788
Sesamum indicum L. (Sesame)					
1	Seed	McCance	0.239	0.461	832
1	Seed meal[b]	McCance	1.00	1.4	1112
1	Defatted seed meal[b]	McCance	1.44		699
1	Defatted seed meal[b]	DGE	+		699
1	Seed, dehulled, and defatted[b]	HPLC	1.51		850
1	Seed protein concentrate	McCance	0.181		1113
5	Seed protein isolate	McCance	0.21	0.38	1114
1	Seed β-globulin[c]	Harland	n.d.	0.06	1115
Sesbania grandiflora (L.) Poir. (Agati (Scarlet Wistaria Tree))					
1	Leaf	McCance	0	0.355	821
1	Flower	McCance	0.021	0.024	832
Sesuvium portulacastrum L. (Sea Purslane)					
1	Leaf	McCance	0.013	0.038	832
Setaria italica (L.) Beauv. (Italian (Foxtail) Millet)					
1	Seed	McCance	0.237	0.356	821
4	Dehulled seed	McCance	0.18-0.20	0.17-0.36	1054
Simmondsia chinensis (Link.) Schneid. (Jojoba)					
1	Seed cotyledon	TEM/EDX	+		962
1	Defatted meal[c]	McCance	0.002		1116
1	Decorticated meal[c]	McCance	0.003		1116
1	Seed protein concentrate[c]	McCance	n.d.		1116
1	Germinated seed	^{31}P NMR	+		43
Sinapis alba L. (White Mustard)					
1	Defatted seed flour	McCance	0.90		881
1	Defatted seed powder	Michel	0.3924		758
1	Seed, aleurone grain	McCance	1.49		1023
Solanum africanum Mill.					
1	Leaf	McCance	0.050	0.8	909
Solanum gnomalum					
1	Leaf and stem	McCance	0.033	0.192	837
Solanum mauritianum Scop. (Earleaf Nightshade)					
1	Pollen	PE	0.31	0.40	788
Solanum melongena L. (Eggplant)					
1	Vegetable	McCance	0.055	0.504	821

[a]Percentage of total P; [b]Species unstated; [c]Basis not given; [d]Negligible (<1% of total P); [e]As is basis; [f]Fresh weight basis; [g]Air-dry basis; [h]Mean ± standard error; [i]Mixture; [j]See also ref. 1184.

Continued

Table 2-4 Phytic Acid Content of Plants—cont'd

No.	Plant part	Method	% P (phytate)	% P (total)	References
Solanum nigrum L. (Black Nightshade)					
1	Leaf	McCance	0.045	0.9	909
Solanum tuberosum L. (Potato)					
1	Leaf[b]	Suzuki	0.096	0.433	798
5	Tuber	McCance	0.055-0.068	0.205-0.266	667
1	Tuber[b]	HPLC	0.10	0.230	696
1	Tuber	GFC	0.04	0.230	696
1	Tuber[b]	^{31}P NMR	0.03		743
1	Tuber, stem end[e]	McCance	0.0043	0.0281	1117
1	–, rose end[e]	McCance	0.0091	0.0400	1117
2	Tuber	PC	+	0.062-0.079	315
2	Tuber, mature[b,c]	Harland	0.0149-0.0160	0.0616-0.0717	1118
2	–, immature[b,c]	Harland	0.0090-0.0103	0.053-0.062	1118
1	Tuber, pith	^{31}P NMR	+		43
1	–, other parts	^{31}P NMR	n.d.		43
Sophora japonica L. (Japanese Pagoda Tree)					
1	Seed meal, defatted	Mukhamedova	0.42		795
1	Seed, axis[a]	^{31}P NMR	79		824
1	–, hypocotyl[a]	^{31}P NMR	90		824
Sorghum bicolor (L.) Moench (Sorghum)					
16	Whole grain	McCance	0.12-0.18	0.28-0.50	1119
30	Whole grain	McCance	0.172-0.407	0.224-0.480	756
6	–, dehulled grain	McCance	0.05-0.18	0.09-0.22	756
6	–, bran	McCance	0.73-1.16	0.84-1.31	756
7	Whole grain[b]	McCance	0.20-0.37		1120
7	–, germ	McCance	0.54-1.91		1120
7	–, bran	McCance	0.19-0.49		1120
7	–, grits (endosperm)	McCance	0.02-0.18		1120
2	Grain	McCance	0.160-0.273	0.223-0.307	1121
1	Grain	Harland	0.249		936
1	Whole grain[b]	Inositol	+		1120
4	Pearled grain	HPLC	0.06-0.28		765
2	Grain pearling	HPLC	0.31-0.34		765
1	Seed aleurone tissue	Microscopy	+		868
1	5 ft plant, entire panicle[b]	Harland	0.040	0.30	867
1	–, 11th leaf blade	Harland	0.009	0.22	867
1	–, root	Harland	0.009	0.17	867
Sorghum halepense (L.) Pers. (Johnson Grass)					
1	Defatted seed meal	Mukhamedova	0.45		796
Sphenostylis stenocarpa (Hochst ex A. Rich) Harms (African Yambean)					
1	Seed	McCance	0.174	0.248	1122
3	Bean	Harland	0.178-0.215	0.267-0.303	1123
1	Seed	HPLC	0.016		900
Spinacia oleracea L. (Spinach)					
1	Stalk	McCance	0	0.229	821

Table 2-4 Phytic Acid Content of Plants—cont'd

No.	Plant part	Method	% P (phytate)	% P (total)	References
1	Leaf	McCance	0.010	0.360	821
1	Mature leaf, cytosol	^{31}P NMR	+		50
1	–, chloroplast	^{31}P NMR	0		50
Spinacia turkestanica Iljin					
1	Defatted seed meal	Mukhamedova	0.25		796
Spirodela polyrrhiza (L.) Schleid. (Giant Duckweed)					
1	Pulp	McCance	0.146		876
1	Juice	McCance	0.098		876
1	Leaf protein concentrate	McCance	0.163		876
Spondias mombin L. (Yellow Mombin)					
1	Seed	McCance	0.234		828
Spondias purpurea L. (Red Mombin)					
1	Young leaf	McCance	0.056	0.099	832
1	Fruit	McCance	0.008	0.036	832
Sterculia foetida L. (Indian Almond, Jangli Badam)					
1	Kernel	McCance	0.306	0.644	827
Stillingia spp. (Toothleaf)					
1	Defatted seed flour[b,c]	McCance	1.8		943
Stizolobium aterrimum Piper and Tracy (Bengal Bean)					
1	Boiled bean	Harland	0.146		822
Strelitzia reginae (Banks.) Ait. (Bird-of-Paradise)					
1	Pollen	PE	0.15	0.74	788
Stylidium petiolare Sond.					
1	Corm, seed	Microscopy	+		1000
1	Corm	SEM/EMX	+		1000
1	Corm	TEM	+		1000
Talinum triangulare (Jacq.) Willd. (Fameflower, Philippine Spinach)					
1	Leaf	McCance	0.004	0.018	832
Tamarindus indica L. (Tamarind)					
1	Seed	McCance	0.013	0.165	1124
1	–, pulp	McCance	n.d.	0.091	1124
1	Unripe seed[e]	McCance	0	0.101	842
1	Leaf and stem	McCance	0	0.475	827
1	Young leaf	McCance	0.008	0.023	832
1	Unripe fruit	McCance	0.016	0.034	832
1	Flower	McCance	0.004	0.050	832
Telfairia occidentalis Hook. f. (Fluted Pumpkin)					
1	Leaf	McCance	0.15	1.2	840
Terminalia catappa L. (Indian Almond)					
1	Seed	McCance	0.449		828
Tetracarpidium conophorum (Müll. Arg.) Hutch. and Dalz. (Conophor)					
1	Boiled seed		0.063	0.774	1125
Tetrapleura tetraptera (Schum. and Thonn.) Taub					
1	Seed	McCance	0.16	0.21	825

[a]Percentage of total P; [b]Species unstated; [c]Basis not given; [d]Negligible (<1% of total P); [e]As is basis; [f]Fresh weight basis; [g]Air-dry basis; [h]Mean ± standard error; [i]Mixture; [j]See also ref. 1184.

Continued

Table 2-4 Phytic Acid Content of Plants—cont'd

No.	Plant part	Method	% P (phytate)	% P (total)	References
Tetrastigma harmandii Planch. (Ayo)					
1	Leaf	McCance	0.020	0.080	832
Theobroma cacao L. (Cocoa)					
1	Bean[b,c]	Suzuki	0.24	0.37	799
1	–, shell[b,c]	Suzuki	0.29	0.52	799
1	–, powder[b,c]	Suzuki	0.47	0.63	799
1	Husk	HPLC	0.28	0.69	1126
Treculia africana Decne. (African Breadfruit)					
1	Seed	McCance	0.002	0.382	1127
Trichodesma incanum (Bunge) DC					
1	Defatted seed meal	Mukhamedova	0.34		794
Trichosanthes anguina L. (Snake Gourd)					
1	Gourd[e]	McCance	0	0.029	842
Trifolium pratense L. (Red Clover)					
1	Defatted meal[b]	Schmalfuss	0	2.23	791
1	Herbage[g]	Kazakov	0.13		803
1	Root	McCance	0.009	0.44	1016
Trifolium repens L. (White (Ladino) Clover)					
1	Whole plant (full bloom)[b]	Harland	0.009	0.28	867
1	Radicle	Microscopy	+		1128
Trifolium spp. (Clover)					
1	Seed[b,g]	McCance	0.29	0.56	1129
Trigonella foenum-graecum L. (Fenugreek)					
2	Seed	McCance	0.173-0.207	0.321-0.418	1018
1	Seed	Harland	0.047	0.284	1130
1	Seed, cotyledon	McCance	0.08		1131
1	–, axis	McCance	0.03		1131
1	–, endosperm	McCance	0.01		1131
1	Leaf	McCance	0.90	1.0	840
2	Sprout	McCance	0.087-0.112	0.174-0.280	1018
Triticum aestivum L. (Wheat)					
38	Seed	McCance	0.174-0.380	0.289-0.524	858
21	Whole grain[c]	McCance	0.152-0.328	0.233-0.378	1132
19	Whole grain	McCance	0.148-0.298	0.310-0.490	1133
8	Seed flour[b,g]	McCance	0.19-0.36		907
10	Grain, dryland crop[j]	McCance	0.233-0.286	0.323-0.359	1134
10	–, irrigated crop[j]	McCance	0.270-0.342	0.342-0.401	1134
24	Grain	HPLC	0.092-0.280	0.225-0.424	1135
12	Grain[c]	McCance	0.109-0.318	0.190-0.449	1136
5	Grain[b]	McCance	0.12-0.32	0.19-0.44	859
1	Grain	CZE	0.109		746
1	Grain	PC	+		811
1	Seed	Mössbauer	+		816
1	Seed protein body	Morton	1.2	0.96	805
1	Seed protein concentrate[b,e]	McCance	0.347-0.696	0.486-1.383	733
1	Pollard seed[b]	McCance	0.277-0.962	0.350-1.238	826
1	Soft, whole kernel[g]	McCance	0.32	0.42	1040

Table 2-4 Phytic Acid Content of Plants—cont'd

No.	Plant part	Method	% P (phytate)	% P (total)	References
1	–, germ	McCance	1.10	1.66	1040
1	–, endosperm	McCance	0.001	0.11	1040
1	–, hull	McCance	0	0.08	1040
1	–, aleurone	McCance	1.16	1.39	1040
3	Aleurone tissue	Microscopy	+		819,868,1137
1	Aleurone cell	SEM	+		870
2	Aleurone grain	TEM	+		868
6	Defatted seed powder	Michel	0.145-0.200		758
8	Whole seed[b]	Michel	0.252-0.371	0.341-0.498	759
8	–, flour[b]	Michel	0.024-0.060	0.092-0.134	759
8	–, gliadin[b]	Michel	0.008-0.036	0.081-0.184	759
8	–, glutenin[b]	Michel	0.044-0.083	0.208-0.314	759
8	–, gluten[b]	Michel	0.003-0.087	0.213-0.281	759
9	Seed	Harland	fig.	0.463-0.550	830
5	Seed	GFC, PE	+		1138
1	Seed flour[b]	Harland	0.06-0.07		734
2	Seed Waldron bran[b]	Harland	0.96-1.31		738
1	Seed germ	Harland	0.63		761
1	Seed embryo	TEM/EDX	+		1139
1	Whole grain[b]	HPLC	0.30		774
1	Break flour	HPLC	<0.02		777
1	Germ, defatted[b]	HPLC	0.68		850
1	Whole grain[b]	^{31}P NMR	0.876		873
3	Grain, 100% extraction[c]	^{31}P NMR	0.24-0.27		743
1	–, flour[b]	^{31}P NMR	0.036		743
1	–, bran[b]	^{31}P NMR	0.96		743
1	Seed flour[b]	Inositol	0.077		784
1	–, germ tailings	Inositol	0.896		784
1	Seed flour	Phytase	0.498		1140
1	Seed gluten[b]	McCance	0.060		1141
15	Bran	HPLC	0.72-1.67		1142
1	Bran[b]	Harland	1.04		737
1	Bran[g]	McCance	1.105	1.20	1143
	See also refs. 740–742,772 for other data on soft white and hard red wheat brans				
1	Bran[b]	Inositol	0.949		785
1	Bran	CZE	2.964		700
1	Bran	Mössbauer	+		816
1	Root	SEM/EMX	+		891
1	Pollen	PE	0	0.43	788
T. aestivum × *Secale cereale* (× *Triticosecale* Wittm.) (Triticale)					
12	Whole grain	McCance	0.137-0.531	0.57-1.00	1144
12	–, flour	McCance	0.052-0.188	0.16-0.29	1144
12	–, bran	McCance	0.057-0.219	0.74-1.08	1144

[a]Percentage of total P; [b]Species unstated; [c]Basis not given; [d]Negligible (<1% of total P); [e]As is basis; [f]Fresh weight basis; [g]Air-dry basis; [h]Mean ± standard error; [i]Mixture; [j]See also ref. 1184.

Continued

Table 2-4 Phytic Acid Content of Plants—cont'd

No.	Plant part	Method	% P (phytate)	% P (total)	References
13	Whole grain	McCance	0.075-0.118		1145
13	–, flour	McCance	0.041-0.068		1145
13	–, bran	McCance	0.023-0.095		1145
5	Seed	Harland	0.28-0.33		1146
1	Grain[b]	HPLC	0.28		774
See also ref. 1147 for various *Triticale* × *Agropyron* crosses					
Triticum compactum Host. (Club Wheat)					
1	Seed	GFC, PE	+		1138
Triticum durum Desf. (Durum Wheat)					
6	Whole kernel[b]	McCance	0.25-0.40		1148
6	–, bran[b]	McCance	0.65-1.22		1148
1	Seed germ[b,c]	McCance	0.597	1.244	1149
6	Seed flour[b]	McCance	0.13-0.20		1148
6	Semolina[b]	McCance	0.04-0.10		1148
1	Seed protein body	Morton	0.68	0.75	805
1	Waldron bran[b]	McCance	1.31		749
1	Bran[c]	HPLC	0.79		740
Triticum monococcum L. (Einkorn)					
1	Seed	Harland	fig.	0.640	830
Triticum spelta L. (Spelt)					
1	Whole seed meal	McCance	0.29	0.48	1150
1	Unripe seed	McCance	0.27		1151
Triticum turgidum L. (Poulard Wheat)					
6	Seed	Harland	fig.	0.470-0.610	830
Vaccaria segetalis (Neck.) Garcke (Cow Herb)					
1	Defatted seed meal	Mukhamedova	0.36		796
Vallisneria spp. (Eelgrass)					
1	Whole plant	Harland	0.0056	0.130	912
Vernonia amygdalina Delile (Bitter Leaf)					
1	Leaf and stem	McCance	0.018	0.240	837
1	Leaf	McCance	0.061	0.7	909
Vernonia galamensis (Cass.) Less. (Ironweed)					
1	Defatted seed meal[c]	Harland	0.716	64.4 (*sic*)	1152
Vicia faba L. (Broad (Faba) Bean)					
23	Whole bean	McCance	0.21-0.47	0.405-0.894	1153
5	Seed, cotyledon	McCance	0.49-0.53		1154
7	Bean	McCance	0.13-0.21		1155
5	Bean	McCance	0.39-0.42	0.315-0.384	919
2	Decoated seed	McCance	0.089-0.290	0.394-0.786	1156
2	Seed coat	McCance	0.010-0.040	0.031-0.073	1156
1	Mature seed	PC	+		810
1	–, immature seed	PC	+		810
1	Seed aleurone grain	McCance	0.89	1.4	812
1	Seed aleurone grain	PC	+	1.4	812
1	Seed protein fraction	McCance	1.18		1088
1	Seed starch fraction	McCance	0.25		1088
1	Defatted seed powder	Michel	0.5275		758

Table 2-4 Phytic Acid Content of Plants—cont'd

No.	Plant part	Method	% P (phytate)	% P (total)	References
4	Whole seed	Harland	0.20-0.32		1091
2	–, dehulled	Harland	0.27-0.33		1091
2	–, hull	Harland	0.017-0.056		1091
2	–, protein concentrate	Harland	0.58-0.59		1091
1	–, protein isolate	Harland	0.68		1091
2	Seed	HPLC	0.097-0.172		922
1	Seed flour	Harland	0.103		926
1	–, protein fraction	Harland	0.243		926
1	–, starch fraction	Harland	0.045		926
1	Seed, axis	^{31}P NMR	0.422	0.728	824
1	–, cotyledon	^{31}P NMR	0.336	0.400	824
1	Seed cotyledon	TEM/EDX	+		823
2	Sprout	McCance	0.202-0.203	0.416-0.518	1018
Vicia faba L. ssp. *minor* L. (Small Horsebean)					
1	Bean	Undisclosed	0.402		1157
1	Germinated bean	Undisclosed	0.486		1157
Vigna aconitifolia (Jacq.) Mar. (Moth Bean)					
1	Seed	MCance	0.32	0.45	1158
1	Seed flour	McCance	0.210	0.340	1159
4	Whole seed	McCance	0.240-0.253		1160
4	60-h sprout	McCance	0.108-0.129		1160
Vigna ambacensis Welw. ex Baker					
3	Seed	Harland	0.121-0.200		1161
Vigna angularis (Willd.) Ohwi and Ohashi (Adzuki Bean)					
1	Seed	McCance	0.042		1162
Vigna catjang (Burm. f.) Walp. (Catjang)					
1	Bean	McCance	0.167	0.447	827
Vigna luteola (Jacq.) Benth. (Hairypod Cowpea)					
3	Seed	Harland	0.187-0.269		1161
Vigna mungo (L.) Hepper (Black Gram)					
6	Whole seed	McCance	0.203±0.009	0.415±0.027	898
6	–, decorticated	McCance	0.228±0.004	0.455±0.011	898
4	Whole seed	McCance	0.182-0.188		920
1	Seed meal	McCance	0.48	0.54	1163
1	–, whey	McCance	0.30		1163
1	Defatted seed	McCance	0.17		1164
4	48-h sprout	McCance	0.102-0.115		920
Vigna oblongifolia A. Rich.					
3	Seed	Harland	0.240-0.414		1161
Vigna racemosa G. Don					
3	Seed	Harland	0.174-0.242		1161
Vigna radiata (L.) Wilcz. (Mung Bean)					
5	Whole seed	McCance	0.311±0.018	0.464±0.005	898
5	–, decorticated	McCance	0.347±0.005	0.489±0.008	898
1	Seed	McCance	0.188-0.203	0.410	1051
1	Defatted seed	McCance	0.18		1164

[a]Percentage of total P; [b]Species unstated; [c]Basis not given; [d]Negligible (<1% of total P); [e]As is basis; [f]Fresh weight basis; [g]Air-dry basis; [h]Mean ± standard error; [i]Mixture; [j]See also ref. 1184.

Continued

Table 2-4 Phytic Acid Content of Plants—cont'd

No.	Plant part	Method	% P (phytate)	% P (total)	References
1	Bean	Harland	0.177	0.376	925
1	Bean	TLC, TLE	+		693
1	Seed flour	Harland	0.144		926
1	–, protein fraction	Harland	0.404		926
1	–, starch fraction	Harland	0.044		926
1	Sprout	McCance	0.161	0.540	1051
V. radiata × *V. mungo*					
4	Seed	McCance	0.196-0.211		1165
Vigna reticulata Hook. f.					
3	Seed	Harland	0.134-0.457		1161
Vigna sesquipedalis (L.) Fruw. (Yard Long Bean)					
1	Seed	McCance	0.072	0.135	832
1	Green pod	McCance	0.034	0.045	832
Vigna subterranea (L.) Verdc. (Bambara Groundnut)					
1	Seed	McCance	0.083		901
1	Boiled bean	Harland	0.197		822
1	3-day sprout	McCance	0.064		901
Vigna umbellata (Thunb.) Ohwi and Ohashi (Rice Bean)					
5	Whole bean	McCance	0.528-0.639	0.234-0.249	1166
5	–, defatted	McCance	0.10-0.14		1164
1	Sprout	McCance	0.120	0.400	1167
Vigna unguiculata (L.) Walp. (Cowpea)					
15	Seed	McCance	0.452-0.706		1168
13	Seed	Harland	0.14-0.29		1169
13	Pea	McCance	0.079-0.223	0.382-0.473	1170
10	Whole bean	McCance	0.079-0.093		1171
4	Whole seed	McCance	0.28-0.39		1172
4	–, dehulled	McCance	0.28-0.39		1172
1	Defatted bean	McCance	0.14		1164
1	Seed flour	Harland	0.214		926
1	–, protein fraction	Harland	0.460		926
1	–, starch fraction	Harland	0.070		926
1	Seed flour	HPLC	0.156		769
1	Seed, axis	^{31}P NMR	0.187	0.538	824
1	–, cotyledon	^{31}P NMR	0.301	0.360	824
1	Young pod	McCance	0.016	0.067	832
10	3-day sprout	McCance	0.040-0.050		1171
3	3-day sprout	McCance	0.10-0.12	0.29-0.37	980
Vigna unguiculata ssp. *dekindtiana* (Harms.) Verdc.					
4	Seed	Harland	0.175-0.429		1161
Vigna vexillata (L.) A. Rich. (Zombi Pea)					
5	Seed	Harland	0.185-0.428		1161
Vigna vexillata ssp. *macrosperma*					
3	Seed	Harland	0.183-0.384		1161
Vinca erecta Regel and Schmalh. (Periwinkle)					
1	Defatted seed meal	Mukhamedova	0.01		796

Table 2-4 Phytic Acid Content of Plants—cont'd

No.	Plant part	Method	% P (phytate)	% P (total)	References
Viscum album L. (Mistletoe)					
1	Shoot	McCance	0.39-0.43		1173
1	Fat-free leaf flour[b,c]	Michel	0.110		757
Vitex doniana (Sweet)					
1	Seed	McCance	0.177		828
Vitis vinifera L. (Wine Grape)					
1	Seed	Harland	0.06		761
Voandzeia poissonii A. Chev. (Dougoufoulo)					
1	Nut[b,c]	Undisclosed	+		1002
Washingtonia filifera (Lindl.) Wendl. (California Fan-Palm)					
1	Seed endosperm	Microscopy	+		1174
1	Seed endosperm	SEM	+		1174
1	Seed endosperm	TEM/EDX	+		1174
1	Haustoria	Microscopy	+		1074
1	Haustoria	TEM/EDX	+		1074
Wolffiella floridana (Sm.) Thomps. (Water Meal)					
1	Whole plant	PE	+		695
Xanthosoma mafaffa (L.) Schott. (Lime Zinger)					
1	Leaf	McCance	0.037	0.017	963
Xanthosoma sagittifolium (L.) Schott. (Tannia)					
1	Leaf	McCance	0.048	1.3	909
Xanthosoma spp. (Cocoyam)					
1	Boiled leaf	Harland	0.005		955
1	Boiled root	Harland	0.010		955
Xylia xylocarpa (Roxb.) Theob. (Burma Pyingado)					
1	Seed flour	McCance	0.102	0.342	1175
Xylomelum angustifolium Kipp.					
1	Seed cotyledon	Microscopy	+		877
1	Seed cotyledon	SEM/EMX	+		877
Xylopia aethiopica (Dun.) A.Rich. (Guinea Pepper)					
1	Seed	McCance	0.149		828
Zantedeschia aethiopica (L.) Spreng. (Calla Lily)					
1	Pollen	PE	0.07	0.62	788
Zea mays L. (Corn)					
9	Grain	McCance	0.313-0.419		736
2	Yellow grain	McCance	0.149-0.276	0.199-0.368	826
1	–, meal	McCance	0.277-0.314	0.364-0.407	826
1	Mature seed	PC	+		810
1	Seed germ, defatted	McCance	1.35		699
2	Seed germ, defatted	DGE	+		699
1	Seed flour[b]	McCance	0.045	0.103	1176
1	Whole kernel	McCance	0.144-0.234	0.20-0.30	1177
1	–, germ	McCance	1.208-1.782	1.47-2.10	1177
1	–, floury endosperm	McCance	0.0244-0.0520	0.05-0.09	1177
1	–, horny endosperm	McCance	0.0031-0.0148	0.02-0.04	1177
1	–, hull	McCance	0.0053-0.0197	0.02-0.03	1177

[a]Percentage of total P; [b]Species unstated; [c]Basis not given; [d]Negligible (<1% of total P); [e]As is basis; [f]Fresh weight basis; [g]Air-dry basis; [h]Mean ± standard error; [i]Mixture; [j]See also ref. 1184.

Continued

Table 2-4 Phytic Acid Content of Plants—cont'd

No.	Plant part	Method	% P (phytate)	% P (total)	References
1	Seed, milky stage	McCance	0.107	0.393	842
1	High protein seed (65th to 81st generation)	McCance			1178
1	–, germ	McCance	3.0	3.6	1178
1	–, endosperm	McCance	0.022	0.037	1178
		Also data for low protein seed (65th to 81st) and both (83rd)			
2	Seed gluten meal	McCance	0.35-0.36	0.58	1179
1	Degermed seed meal	McCance	0.07	0.10	1179
1	Seed zein	McCance	0.00		1141
1	Defatted seed powder	Michel	0.294		758
2	Grain	Harland	0.193-0.207		936
1	Grain	CZE	0.363		700
1	Whole seed, embryo	Microscopy	+		1180
1	Seed flour, 65% extraction	Harland	0.066		822
1	–, bran	Harland	0.341		822
1	Whole grain[b]	HPLC	0.30		774
2	–, bran[b]	HPLC	0.01-0.02		774
1	Seed bran[b]	McCance	0		740
1	Whole seed meal[b]	^{31}P NMR	0.17		743
1	Bran[b,c]	^{31}P NMR	0		781
1	Root	^{31}P NMR	n.d.		43
1	Root	SEM/EMX	+		891
1	Germinated seed	SEM/EMX	+		43
1	Flower ovule	McCance	0.013	0.608	1181
1	Pollen	McCance	0.043		1181
1	Silks	McCance	0.017		1181
		No phytic acid in stalks, leaves, tassels, shank, or cob.			1182
1	Pollen	PE	0.07	0.53	788
High lysine corn					
1	Grain[b,e]	McCance	0.28		1183
1	Whole kernel[g]	McCance	0.27	0.31	1178
1	–, germ[g]	McCance	1.61	1.78	1178
1	–, endosperm[g]	McCance	0.01	0.03	1178
1	–, hull[g]	McCance	0.07	0.10	1178
Zingiber officinale Rosc. (Ginger)					
1	Root[e]	McCance	0.0048	0.062	842

In this table, unlike most of the others in this book, the coverage for each entry is selective rather than comprehensive because of extensive duplication in the published literature. Preference is given to studies that report data for two or more varieties or cultivars, but for the remainder, the author has selected the one that in his opinion is the most reliable.

[a]Percentage of total P.

[b]Species unstated.

[c]Basis not given.

[d]Negligible (<1% of total P).

[e]As is basis.

[f]Fresh weight basis.

[g]Air-dry basis.

[h]Mean ± standard error.

[i]Mixture.

[j]See also ref. 1184.

3

Lower *myo*-Inositol Phosphates

This chapter covers the *myo*-inositol phosphates and their glyceryl esters with the exception of phytic acid, which is covered in Chapter 2, and of phosphatidylinositol (PI) and its derivatives, which are covered in Chapter 4. The following is a complete list of the compounds described in this chapter:

$InsP_n$	Inositol phosphate	Glyceryl ester
InsP	Ins(1)P, Ins(2)P, Ins(3)P, Ins(4)P	GroPIns
$InsP_2$	Ins(1,2)P_2, Ins(1,4)P_2, Ins(3,4)P_2, Ins(4,5)P_2	GroPIns(3)P, GroPIns(4)P
$InsP_3$	Ins(1,2,3)P_3, Ins(1,2,6)P_3, Ins(1,4,5)P_3, Ins(1,5,6)P_3, Ins(3,4,6)P_3	GroPIns(3,4)P_2, GroPInsP(4,5)P_2
$InsP_4$	Ins(1,2,3,6)P_4, Ins(1,2,5,6)P_4, Ins(1,3,4,5)P_4, Ins(1,4,5,6)P_4, Ins(3,4,5,6)P_4	
$InsP_5$	Ins(1,2,3,4,6)P_5, Ins(1,2,3,5,6)P_5, Ins(1,2,4,5,6)P_5, Ins(1,3,4,5,6)P_5	

The full names and structures of the lower *myo*-inositol phosphates are given in the introduction to each section. The nomenclature used here conforms to the recommendations of the Joint IUPAC-IUB Commission on the Nomenclature of Cyclitols issued in 1973.[1194] Caution should be exercised when reading older literature as the inositol ring numbering may be different and the sign (+ or −) may be reversed.

For general references on the subject of this chapter see Cosgrove and Irving,[676] Irvine,[1195] Morré *et al.*,[1196] or Biswas and Biswas.[1197]

Occurrence

No *myo*-inositol phosphate with fewer than six phosphate groups can be detected in rice or wheat seedlings during the first 14 days of germination, during which *myo*-inositol and inorganic phosphate are released and the phytate content falls to zero.[1187,1198] Short-term feeding experiments with [^{3}H]-*myo*-inositol or $^{32}P_i$ produced InsP and phytate but no other intermediates.[1187] Many of the lower *myo*-inositol phosphates reported in Table 3-4 may be artifacts of the hydrolysis of phytic acid by endogenous phytase[1208] or by chemical reagents during the course of extraction, isolation, and analysis.

Synthesis

The *myo*-inositol phosphates ($InsP_1$ through $InsP_6$) may be prepared (a) by phosphorylation of *myo*-inositol with polyphosphoric acid[714] or ATP,[1209] (b) by acid- or base-catalyzed hydrolysis of phytic acid,[721] or (c) by enzymatic hydrolysis of phytic acid with 3-phytase

[E.C. 3.1.3.8], 6-phytase [E.C. 3.1.3.26],[1210] or acid phosphatase [E.C. 3.1.3.2].[1210] The polyphosphates are readily separated by ion exchange chromatography.

Hydrolysis

Hydrolysis of phytic acid ($InsP_6$) *in vitro* with acid, base, or enzyme catalysts produces a mixture of *myo*-inositol polyphosphates whose relative proportions change during the course of the reaction. The polyphosphates are readily separated into pentakis-, tetrakis-, tris-, bis-, and monophosphate fractions by chromatography or electrophoresis. Each fraction can (and often does) contain more than one stereoisomer (Table 3-1).

Table 3-1 Isomer Distribution in *myo*-Inositol Polyphosphates

	Maximum number of isomers				Hydrolysis of phytic acid by phytase		
$InsP_n$	**D**	**L**	**Meso**	**Total**	**$[\alpha]_D$ [a]**	**$K_M \times 10^{-3}$ [a]**	**P_{50} [b]**
$InsP_5$	2	2	2	6	n.d.	1.7	–
$InsP_4$	6	6	3	15	−4.09	5.0	17.8
$InsP_3$	8	8	4	20	−8.26	4.4	16.8
$InsP_2$	6	6	3	15	−5.95	1.2	15.4
$InsP_1$	2	2	2	6	0	1.5	–

[a] Posternak and Posternak.[1211]
[b] Percentage of total P at 50% hydrolysis.[1212]

In vitro hydrolysis of phytic acid by wheat bran phytase at pH 5.2 and 55 °C produces a mixture of polyphosphates whose middle fractions—$InsP_4$, $InsP_3$, and $InsP_2$—are optically active (Table 3-1).[723,1211] The initial rates of hydrolysis and activation energy (11 kcal/mol) of each fraction are nearly the same, even though the more highly phosphorylated members are attacked preferentially.[1212] The heats of hydrolysis of the pentakis-, tetrakis-, and trisphosphates are not significantly different (ΔH_{app} −2820 to −3040 cal/mol P_i), but that of the bisphosphate is higher (~4000 cal/mol P_i).[1213]

Phytic acid is scarcely attacked by acids or bases at temperatures below 100 °C. In aqueous solution at 100 °C, the rate of hydrolysis is negligible at pH 12, rises to a maximum at pH 4, falls to a minimum at pH −1, and then rises steeply with increasing acidity.[1214] This behavior is typical of monoalkyl esters of phosphoric acid.

Hydrolysis at pH 4, monitored by PC,[692] PE,[698] or IEC,[719] produces a mixture of polyphosphates whose distribution changes from pentakis- to monophosphate with time of heating. Prolonged hydrolysis at pH 9 yields mainly the tetrakis- and trisphosphates.[721] If the pH is not allowed to fall below 7.5, the monophosphate Ins(2)P is obtained in 55% yield after 60 h of refluxing.[1215] For accumulation of the tris- and bisphosphates, autoclaving at 120 °C is more efficient than enzymatic hydrolysis.[1210]

Extraction

The inositol polyphosphates are usually extracted from plant tissues with ice-cold 0.5*M* perchloric acid or trichloroacetic acid. After extraction, it is good practice to remove the excess acid promptly by neutralization with KOH to precipitate the perchloric acid as its highly insoluble potassium salt[1216] or by extraction with ether to remove the trichloroacetic acid.[950] Other extractants that have been employed are 2*N* formic acid,[1217] 0.4*N* HCl,[1218] and phenol.[683] If the plant tissue is to be monitored for phosphatidylinositol (PI), it may be extracted with $CHCl_3$/methanol/3*N* HCl (6:6:1) and the extracts analyzed separately for PI (organic phase) and InsPs (aqueous phase).[1219,1220] In one case *n*-Bu_4NHSO_4 was used in place of HCl.[1221]

Separation and Analysis

(1) Precipitation Method

Before the advent of chromatography, the inositol polyphosphates were separated from each other by selective precipitation as barium salts whose solubility increases in the order $InsP_5$ (and $InsP_6$) > $InsP_4$ > $InsP_3$ > $InsP_2$ and $InsP_1$. The barium inositol monophosphates are water soluble.[1211,1222] This method is still employed occasionally in conjunction with other methods of isolation.

(2) Chromatographic Methods

The use of chromatographic methods to separate the *myo*-inositol polyphosphates into fractions of equal phosphorus content is illustrated in Table 3-2. $InsP_6$ and $InsP_5$ are poorly separated with solvents or buffers of pH 3.6 or above[692,698] but can be separated by lowering the pH to 1.5.[787,1226] Paper chromatography (PC) and paper electrophoresis (PE) have been employed much more often than thin-layer methods (TLC or TLE) for qualitative identification of InsP fractions. Gel filtration chromatography (GFC) is not as useful. The InsP spots are usually detected by spraying the paper or thin-layer plate with the Hanes/Isherwood reagent followed by UV exposure to improve color development; this treatment hydrolyzes the InsPs leaving blue spots (phosphomolybdate) on a white background.[35]

Ion exchange chromatography (IEC) is the method of choice for either analytical or preparative resolution of InsP mixtures. In early work, Dowex 2 (Cl^-) resin was employed, but

Table 3-2 Chromatography of *myo*-Inositol Polyphosphates

Fraction	PC[a]	TLC[b]	PE[c,d]	TLE[d,e]	GFC[f]	HPLC[g]
$InsP_6$	0.09	1.00	2.05	1.68	0.19	58
$InsP_5$	0.10	1.22	1.95	1.60	0.19	53.8
$InsP_4$	0.17	1.32		1.47	0.25	48.9
$InsP_3$	0.22		1.45	1.30	0.33	43.1
$InsP_2$	0.27		1.20	1.10	0.44	36.4
$InsP_1$	0.38		0.92	0.86	0.59	27

[a]R_f in *n*-PrOH/NH_3/H_2O (5:4:1).[692]
[b]Travel relative to phytic acid = 1.00 on silica gel H after methylation with diazomethane and elution with methanol.[1223]
[c]800 V, 0.1*M* oxalate buffer, pH 1.5, 0 °C.[1224]
[d]Travel relative to picric acid = 1.00.
[e]1000 V, 0.1*M* oxalate buffer, pH 1.5.[693]
[f]Distribution coefficient K_D on Sephadex G-25, 0.2*M* LiCl.[1225]
[g]Elution time (min.) on Aminex A-27 with a 0.10*M*→0.50*M* NaCl gradient.[767]

Dowex 1 (Cl^- or HCO_2^-) is now preferred. The InsPs are eluted in the order $InsP_1 \rightarrow InsP_6$ with gradients of HCl[721] or LiCl.[723] In some cases, the $InsP_2$, $InsP_3$, $InsP_4$, or $InsP_5$ fractions split into two or more discrete peaks owing to differences in elution behavior between the isomers.[719,1210,1227] The lower InsPs ($InsP_1$ to $InsP_3$) are often eluted from Dowex 1 (HCO_3^-) with an ammonium formate gradient followed by an HCl gradient for the higher InsPs.

Since 1985, high-performance liquid chromatography (HPLC) methods have been developed for separating the *myo*-inositol polyphosphates into fractions of different phosphorus content. Many of these methods employ strong (Partisil SAX, Aminex A-27, Mono Q) or weak (IC-PAK-A) anion exchange stationary phases, and mobile phases similar to those of conventional IEC. Separation of a partial hydrolyzate of phytic acid into six fractions on Aminex A-27 (Bio-Rad) with a $0.1M \rightarrow 0.5M$ linear NaCl gradient at pH 10 is shown in Table 3-2.[767] Some HPLC methods are powerful enough to separate the InsP fractions into their individual isomers.[1228]

One such method is gradient ion chromatography (IC) using a Dionex AS3 column installed in the HPLC unit.[1229]

Analysis Without Separation

(3) Protein-Binding Assay (PBA) Method

Intense interest in D-*myo*-inositol 1,4,5-trisphosphate as a calcium stimulant in animal cells prompted the development of the PBA method, which is specific for this compound. Similar in principle to the Gilman protein-binding assay for c-AMP described in Chapter 5, page 375, the PBA method employs a binding protein of bovine origin and 2-[^{3}H]Ins(1,4,5)P_3 as the tracer. All the data given in Table 3-3 were obtained using commercial kits available from Amersham, UK.

> *Ins(1,4,5)P_3 levels in [wheat] root tissues were determined using an IP_3 assay kit (TRK 1000; Amersham; Toyoda et al., 1993; Challis, 1995). Briefly, root tips were frozen in liquid N_2 and then ground in 0.5 mL of 20% perchloric acid at 4 °C for 5 min; the samples were then centrifuged at 2000g for 15 min. The supernatant was then neutralized with 1.5 M KOH–60 mM Hepes, and the precipitated $KClO_4$ was removed by centrifugation (2000g, 15 min). Aliquots of the supernatant were used for analysis of IP_3 as described in the assay kit.*
>
> Jones, D. L.; Kochian, L. V., "Aluminum Inhibition of the Inositol 1,4,5-Trisphosphate Signal Transduction Pathway in Wheat Roots: A Role in Aluminum Toxicity?", *Plant Cell* **1995**, *7*(11), 1913–1922,[1235] reprinted by permission of the American Society of Plant Biologists, © 1995.

(4) Other Methods

Johnson *et al.*[689] have described a method whereby the major components of a complex mixture, such as a partial phytic acid hydrolyzate, can be analyzed without prior separation. The method uses total correlation spectroscopy (TOCSY) to identify the ^{1}H NMR resonance of the six *myo*-inositol hydrogens in each component.

Table 3-3 D-Ins(1,4,5)P_3 in Plants by the PBA Method*

No.	Plant part	Ins(1,4,5)P_3	References
Beta vulgaris L. (Red beet)			
1	Tuber	0.0088	1230
Lycopersicon esculentum Mill. (Tomato)			
1	Seedling, root	3.550	1231
1	–, cotyledon	0.275	1231
1	–, epicotyl	0.180	1231
Medicago sativa L. (Lucerne)			
1	Cell suspension culture	0.030	1232
Papaver rhoeas L. (Corn Poppy)			
1	Pollen	0.5	1233
Pisum sativum L. (Peas)			
1	Plasma membrane[a]	0.8	1234
1	Epicotyl	0.013	1234
Triticum aestivum L. (Wheat)			
1	Root	0.4	1235
Zea mays L. (Maize)			
1	Seedling, root	0.23	1236
1	Coleoptile	0.82	1236

*nmol/g fresh weight basis unless stated otherwise.
[a]nmol/mg protein.

myo-Inositol Monophosphates

There are six *myo*-inositol monophosphates, of which four are important plant metabolites ($\Phi = OP(O)(OH)_2$) (Fig. 3-1):

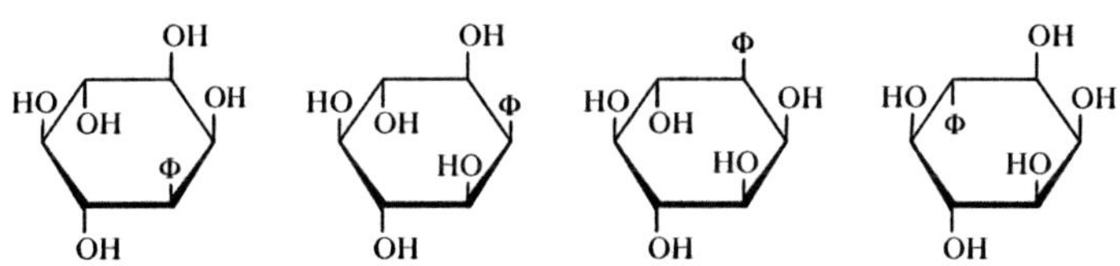

D-Ins(1)P L-Ins(3)P | **Ins(2)P** *meso* | **D-Ins(3)P** L-Ins(1)P | **D-Ins(4)P** L-Ins(6)P

D-*myo*-Inositol 1-Phosphate (Ins(1)P), $C_6H_{13}O_9P$; Beilstein **6**, E_{III} 6931-6932. D-*(or 1D)-myo-Inositol 1-(Dihydrogen Phosphate)* [1959].

myo-Inositol 2-Phosphate (Ins(2)P), $C_6H_{13}O_9P$; Beilstein **6**, E_I 590, E_{II} 1159, E_{III} 6932, E_{IV} 7925. *myo-Inositol 2-(Dihydrogen Phosphate)* [1914].

D-*myo*-Inositol 3-Phosphate (Ins(3)P), $C_6H_{13}O_9P$; Beilstein **6**, E_{III} 6931. D-*myo-Inositol 3-(Dihydrogen Phosphate)* [1960].

D-*myo*-Inositol 4-Phosphate (Ins(4)P), $C_6H_{13}O_9P$; Beilstein **6**, E_{III} 6932. D-*myo-Inositol 4-(Dihydrogen Phosphate)* [1961].

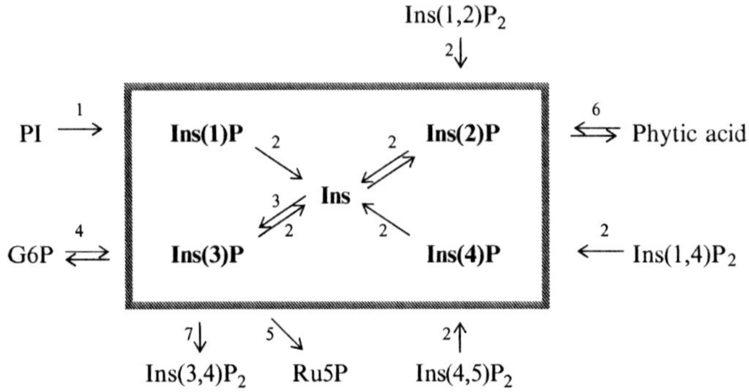

FIGURE 3-1 ***myo*-Inositol Monophosphates.** (1) PI phospholipase C [E.C. 3.1.4.10]. (2) InsP phosphatase [E.C. 3.1.3.25], Mg(II). (3) Ins(3)P kinase [E.C. 2.7.1.64], ATP, Mg(II).[1237] (4) Ins(3)P synthase [E.C. 5.5.1.4], NAD.[1237] (5) Ins(3)P dehydrogenase, NAD. (6) 6-Phytase [E.C. 3.1.3.26]. (7) Phytate synthase.[1238]

Occurrence

Ins(1)P has been identified in carrot,[950] lucerne,[1239] tobacco,[1200] and Madagascar periwinkle[1200] cell suspension cultures, in apple buds,[1240] in pea flour,[1241] in trumpet lily pollen,[1217] and in rain tree pulvini.[1219] Ins(2)P occurs in rice,[1186] tobacco,[1200] and Madagascar periwinkle[1200] cell suspension cultures, in wheat ear stalks,[1242] and in ripening rice grain.[1242] Ins(3)P has been identified in barley aleurone tissue[1208] and in giant duckweed.[1243] Ins(4)P occurs in tobacco and Madagascar periwinkle cell suspension cultures,[1200] in cauliflower hypocotyl,[1244] and in pea flour.[1241]

Physical Properties

D-*myo*-Inositol 1-Phosphate

Ins(1)P [R.N. 15421-51-9]: $[\alpha]_D^{25}$ −9.8° (H_2O)[1245]; pK_a 6.16 (potentiometric),[1246] 6.08 (^{31}P NMR).[68] Spectra: ORD,[1247] ^{31}P NMR.[68] Chromatography: PC,[1245] TLC,[813] IEC,[1248] GC,[1249] HPLC,[1250] IC,[107] LC/MS.[1251] Electrophoresis: PE.[1239] Additional data for DL-Ins (1)P: pK_a 6.50,[1252] ^{1}H NMR (fig.),[687,1253] ^{13}C NMR (fig.),[1253] ^{31}P NMR (fig.),[1252] HPTLC,[407] capillary electrophoresis.[1254]

$(C_6H_{11}NH_2)_3C_6H_{13}O_9P$: Needles,[1255] mp 197-209 °C (variable),[1245] $[\alpha]_D^{25}$ +3.4° (H_2O),[1245] $[M]_D$ +8.8° (calc.).[1256] Spectra: IR (fig.),[1257] ORD (fig.),[1257] ^{1}H NMR,[1255] ^{13}C and ^{31}P NMR,[1258] MS (FAB),[1259] MS/MS (FAB, fig.).[1260] Chromatography: PC.[1257]

myo-Inositol 2-Phosphate

Ins(2)P [R.N. 7336-80-3]: Large colorless prisms[1222] or hexagonal plates[1261] with a sharp, somewhat astringent taste, mp 201-202 °C dec., $[\alpha]_D$ 0° (H_2O), very sol. H_2O, insol. organic solvents[1222]; pK_a 1.21, 5.73 (potentiometric),[1215] 5.83 (^{31}P NMR),[1252] *ab initio* calculations (HF/6-31G* method).[1262] Spectra: IR (fig.),[1263] ^{1}H NMR,[687,1264] 2d-^{1}H NMR (fig.),[689] ^{13}C NMR (fig.),[1253] ^{31}P NMR,[68] MS (FAB),[690] ORD.[1242] Chromatography: PC,[1265] TLC,[1266] IEC,[1267] cellulose powder,[1265] HPLC,[1268] IC,[107]

GC,[1269] GC/MS (EI,CI),[690] LC/MS.[1251] Electrophoresis: PE,[1239] capillary,[1254] CZE/MS.[1270] *Monohydrate* [R.N. 54275-72-8]: White acicular monoclinic crystals, mp ≅ 190 °C dec., d 1.791; conformation $1_{ax}(C^2)/5_{eq}$, X-ray.[1271]
$Na_2C_6H_{11}O_9P$: Fine white amorphous powder, slightly hygroscopic.[1261]
$BaC_6H_{11}O_9P$: Amorphous solid, sol. H_2O.[1261] Tetrahydrate.[1211] Other salts are water soluble except for the lead salt.[1222]
$(C_6H_{11}NH_2)C_6H_{13}O_9P{\cdot}H_2O$: Needles, mp 203-205 °C.[722]
$(C_6H_{11}NH_2)_2C_6H_{13}O_9P{\cdot}4H_2O$: Rosettes of needles, mp 211-213 °C.[722] Spectra: IR,[1245] MS(FAB),[690] MS/MS (FAB, fig.),[1260] X-ray (powder photo).[722] Chromatography: PC.[1245]
$(C_{23}H_{26}N_2O_4)C_6H_{13}O_9P{\cdot}1.5H_2O$ (brucine salt): Crystalline, mp 244-247 °C.[722]

D-*myo*-Inositol 3-Phosphate

Ins(3)P [R.N. 2831-74-5]: $[\alpha]_D^{25}$ +9.3° (H_2O).[1272] Spectra: ^1H NMR (fig.).[1264] Chromatography: PC,[1272] column,[1265] HPLC,[1268] GC,[1249] GC/MS (EI,CI).[690]
$(C_6H_{11}NH_2)_2C_6H_{13}O_9P$: Needles, mp 190-192 °C,[1273] $[\alpha]_D^{25}$ −3.2° (H_2O).[1272] Spectra: IR,[1272] MS (FAB).[1259]

D-*myo*-Inositol 4-Phosphate

Ins(4)P [R.N. 69256-52-6]: Spectra: GC/MS (EI,CI).[690] Chromatography: PC,[1274] IEC,[1274] HPLC.[1268]
$(C_6H_{11}NH_2)_2C_6H_{13}O_9P{\cdot}1.5H_2O$: Crystals, mp 183-188 °C dec.,[1259] $[\alpha]_D^{25}$ −1.3° (H_2O).[1274] Spectra: IR,[1274] ^1H NMR,[1259] ^{13}C NMR,[1259] ^{31}P NMR,[1259] MS (FAB).[1259] Chromatography: HPLC.[1275]

Synthesis

Ins(1)P may be prepared (a) by phosphorylation of D-2,3,4,5,6-penta-*O*-benzyl-*myo*-inositol with diphenyl phosphorochloridate,[1257] of D-2,3,4,5,6-penta-*O*-benzoyl-*myo*-inositol with *o*-xylene-α,α-diyl *N,N*-diethylphosphoramidite,[1255] of 4,6-dibenzyl-1,2-*O*-isopropylidene-5-tetrahydro-2*H*-pyran-2-yl-1L-*myo*-inositol with 2-diethylamino-1,3,2-benzodioxaphosphepane,[1258] or of D-2,3,4,5,6-penta-*O*-acetyl-*myo*-inositol with $POCl_3$ and benzyl alcohol,[1276] or (b) by hydrolysis of soybean phosphatidylinositol with 2*N* KOH.[1245]

Ins(2)P may be prepared (a) by phosphorylation of 1,3,4,5,6-penta-*O*-acetyl-*myo*-inositol with diphenyl phosphorochloridate[1261] or of 4,6-di-*O*-benzyl-*myo*-inositol orthoformate with tetrabenzyl pyrophosphate,[1277] (b) by hydrolysis of phytic acid with wheat bran phytase [E.C. 3.1.3.26],[1278] or (c) by prolonged heating of phytic acid in aqueous solution at pH 7.5 and 100 °C.[1215]

Ins(3)P may be prepared (a) by phosphorylation of D-1,2,4,5,6-penta-*O*-benzyl (or *O*-benzoyl)-*myo*-inositol with diphenyl phosphorochloridate[1272,1273] or of D-1,2,4,5,6-penta-*O*-acetyl-*myo*-inositol with $POCl_3$ and benzyl alcohol[1276] or (b) by the action of Ins(3)P synthase [E.C. 5.5.1.4] on G6P in the presence of DPN.[1279]

Ins(4)P may be prepared by phosphorylation of D-3-*O*-benzyl-1,2:4,5-*O*-dicyclohexylidene-*myo*-inositol with diphenyl phosphorochloridate,[1275] of

D-2,3-*O*-camphanylidene-5,6-*O*-isopropylidene-*myo*-inositol with 2-dimethylamino-5,6-benzo-1,3,2-dioxaphosphepane,[1259] or of 3-(1-(*S*)-camphanyl)-1,2:5,6-dicyclohexylidene-D-*myo*-inositol with dibenzyl *N,N*-diethylphosphoramidite.[1280]

Hydrolysis

Ins(1)P and Ins(3)P are both hydrolyzed to *myo*-inositol and P_i by yeast inositol 1-phosphatase [E.C. 3.1.3.27]. Their rates of hydrolysis at 29 °C and pH 7.7 are identical.[1279] Ins(2)P is hydrolyzed to *myo*-inositol and P_i by 3% H_2SO_4 at 120-125 °C or by 10% NH_4OH at 150 °C.[1222]

Conformation

Ins(2)P adopts the $1_{ax}/5_{eq}$ conformation over the entire pH range 1.0-13.0, according to ^1H NMR analysis. There is no evidence for the $5_{ax}/1_{eq}$ form.[1281]

Analytical Methods

Ins(1)P and Ins(3)P may be assayed by enzymatic hydrolysis with InsP phosphatase [E.C. 3.1.3.25] from rat testes; Ins(2)P does not interfere.[1282] Ins(2)P may be assayed by oxidation with periodic acid and colorimetric measurement of the liberated P_i, but this method does not distinguish Ins(2)P from the other monophosphates.[4221]

myo-Inositol Bisphosphates

There are 15 *myo*-inositol bisphosphates, of which 4 are important plant metabolites ($\Phi = OP(O)(OH)_2$):

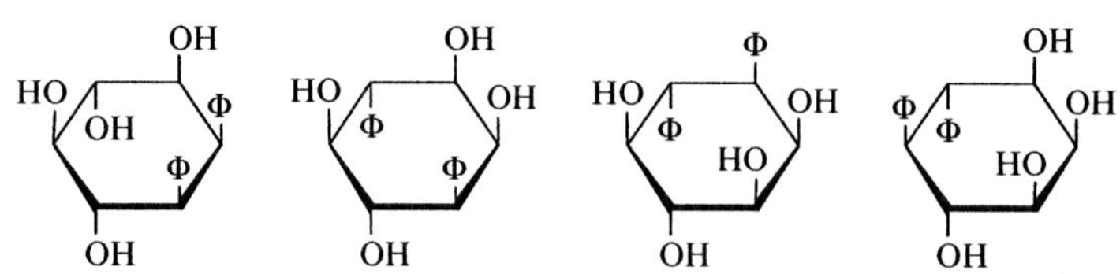

D-Ins(1,2)P$_2$	**D-Ins(1,4)P$_2$**	**D-Ins(3,4)P$_2$**	**D-Ins(4,5)P$_2$**
L-Ins(2,3)P$_2$	L-Ins(3,6)P$_2$	L-Ins(1,6)P$_2$	L-Ins(5,6)P$_2$

D-*myo*-Inositol 1,2-Bisphosphate (Ins(1,2)P_2), $C_6H_{14}O_{12}P_2$; Beilstein **6**, E_{III} 6933. D-*myo-Inositol 1,2-Bis(Dihydrogen Phosphate)* [1962].

D-*myo*-Inositol 1,4-Bisphosphate (Ins(1,4)P_2), $C_6H_{14}O_{12}P_2$; unlisted in Beilstein. D-*myo-Inositol 1,4-Bis(Dihydrogen Phosphate)* [1961].

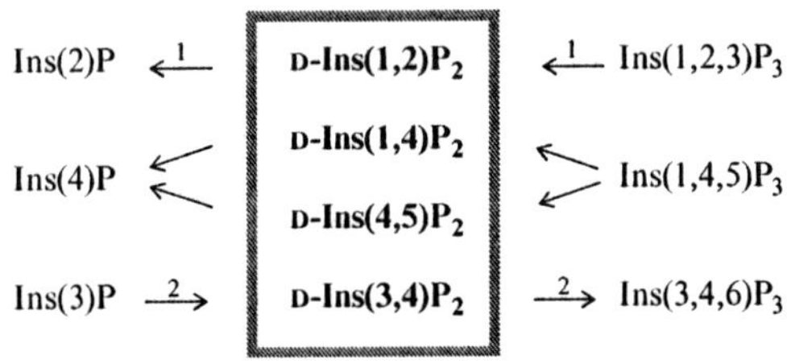

FIGURE 3-2 ***myo*-Inositol Bisphosphates.** (1) 6-Phytase [E.C. 3.1.3.26]. (2) Phytic acid synthase.[1238]

D-*myo*-Inositol 3,4-Bisphosphate ($Ins(3,4)P_2$), $C_6H_{14}O_{12}P_2$; unlisted in Beilstein. D-*myo-Inositol 3,4-Bis(Dihydrogen Phosphate)* [1996].

D-*myo*-Inositol 4,5-Bisphosphate ($Ins(4,5)P_2$), $C_6H_{14}O_{12}P_2$; Beilstein **6**, E_{III} 6934. D-*myo-Inositol 4,5-Bis(Dihydrogen Phosphate)* [1961].

Occurrence

$Ins(1,2)P_2$ has been identified only in barley aleurone tissue.[1208] $Ins(1,4)P_2$ occurs in barley aleurone tissue[1208] and mesophyll cells,[1283] in apple buds,[1240] in giant duckweed,[1243] in rain tree pulvini,[1219] in cauliflower hypocotyl,[1244] and in tobacco,[1200] Madagascar periwinkle[1200] and lucerne[1239] cell suspension cultures. $Ins(1,5)P_2$ has been reported in pea flour.[1241] $Ins(2,4)P_2$ has been reported in tobacco and Madagascar periwinkle cell suspension cultures[1200] and in pea flour.[1241] $Ins(3,4)P_2$ has been identified in giant duckweed.[1243] $Ins(4,5)P_2$ occurs in barley mesophyll cells (Fig. 3-2).[1283]

Physical Properties

D-*myo*-Inositol 1,2-Bisphosphate

$Ins(1,2)P_2$ [R.N. 23410-61-9]: $[\alpha]_D^{25}$ −4.4° (H_2O).[1284] PE,[723,1284] HPLC.[161] Additional data for DL-$Ins(1,2)P_2$ [R.N. 131321-70-5]: pK_a 5.74, 7.97; microconstants for the first four dissociation equilibria.[1252] Spectra: 2d-^1H NMR,[687] ^{13}C NMR,[1195] ^{31}P NMR.[1285] Chromatography: IEC.[1227] Electrophoresis: CE,[1254] CZE/MS.[1270]

$Ba_2C_6H_{10}O_{12}P_2$: Slightly soluble in H_2O.[676]

$(NH_3)_4C_6H_{14}O_{12}P_2$: $[\alpha]_D^{25}$ −1.2° (H_2O).[1284]

$(C_6H_{11}NH_2)_4C_6H_{14}O_{12}P_2$: $[\alpha]_D^{25}$ −4.1° (H_2O).[723]

D-*myo*-Inositol 1,4-Bisphosphate

$Ins(1,4)P_2$ [R.N. 74465-19-3]: Spectra: ^1H NMR (fig.),[1264] ^{31}P NMR,[1264] MS(FAB, fig.),[690] GC/MS.[690] Chromatography: IEC,[1248] HPLC,[1250] IC,[107] GC.[1269] Electrophoresis: PE.[1239] Additional data for DL-$Ins(1,4)P_2$ [R.N. 107494-29-1]: pK_a 6.22, 6.59 (potentiometric).[1286] Spectra: 2d-^1H NMR,[687] ^{13}C NMR.[1285]

$Ba_2C_6H_{10}O_{12}P_2$: Slightly soluble in H_2O.[676]

$(NH_3)_4C_6H_{14}O_{12}P_2$: Spectra: FAB-MS/MS (fig.).[1260]

$(C_6H_{11}NH_2)_4C_6H_{14}O_{12}P_2$ [R.N. 126722-35-8]: Crystals, mp 174-176 °C,[1287] $[\alpha]_D^{25}$ +3.4° (H_2O).[1274] Spectra: IR,[1287] MS (FAB).[1275] Chromatography: PC.[1274]

D-*myo*-Inositol 3,4-Bisphosphate

Ins(3,4)P_2 [R.N. 69256-53-7]: HPLC.[1250] Additional data for DL-Ins(3,4)P_2: ^1H and ^{31}P NMR,[1285] IEC.[1227]

$(C_6H_{11}NH_2)_4C_6H_{14}O_{12}P_2$: Crystals, mp 158-161 °C, $[\alpha]_D^{20}$ +3.9° (H_2O), IR, PC.[1287]

D-*myo*-Inositol 4,5-Bisphosphate

Ins(4,5)P_2 [R.N. 69256-54-8]: Conformation.[1288] Chromatography: HPLC,[161] GC.[1269] Additional data for DL-Ins(4,5)P_2 [R.N. 109584-32-9]: pK_a 8.01, 8.15 (potentiometric).[1286] Spectra: ^1H NMR,[1285] 2d-^1H NMR (fig.),[687] ^{13}C NMR,[1195] ^{31}P NMR.[1286] Chromatography: IEC.[1227]

$(NH_3)_4C_6H_{14}O_{12}P_2$: Spectra: FAB-MS/MS (fig.).[1260]

$(C_6H_{11}NH_2)_4C_6H_{14}O_{12}P_2$: Crystals, mp 138-140 °C,[1287] $[\alpha]_D^{25}$ −15.2° (H_2O).[1274] Spectra: IR.[1287] Chromatography: PC.[1274]

Synthesis

Ins(1,4)P_2 may be prepared by phosphorylation of D-2,3:5,6-di-*O*-cyclohexylidene-*myo*-inositol with diphenyl phosphorochloridate[1287] or of D-2,3-*O*-camphanylidene-5,6-*O*-isopropylidene-*myo*-inositol with 2-dimethylamino-5,6-benzo-1,3,2-dioxaphosphepane.[1259] Ins (3,4)P_2 may be prepared (a) by phosphorylation of D-1,2:5,6-di-*O*-cyclohexylidene-*myo*-inositol with diphenyl phosphorochloridate[1287] or (b) by partial dephosphorylation of D-Ins (1,3,4)P_3 with alkaline phosphatase.[1250] Ins(4,5)P_2 may be prepared by phosphorylation of D-1,6:2,3-di-*O*-cyclohexylidene-*myo*-inositol with diphenyl phosphorochloridate in pyridine.[1287] No synthesis of Ins(1,2)P_2 has been published.

Conformation

Ins(1,2)P adopts the $1_{ax}/5_{eq}$ conformation over the entire pH range 1.0-13.0, according to ^1H NMR analysis. There is no evidence for the $5_{ax}/1_{eq}$ form.[1281]

myo-Inositol Trisphosphates

There are 20 *myo*-inositol trisphosphates, of which 5 are important plant metabolites ($\Phi = OP(O)(OH)_2$):

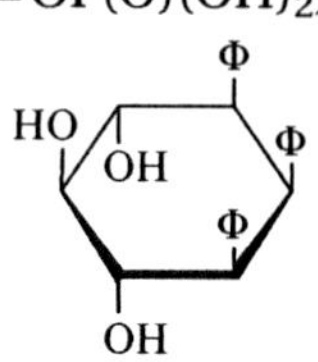

Ins(1,2,3)P_3
(*meso*)

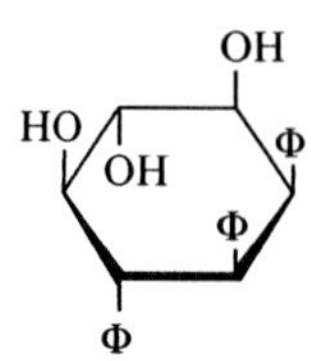

D-Ins(1,2,6)P_3
L-Ins(2,3,4)P_3

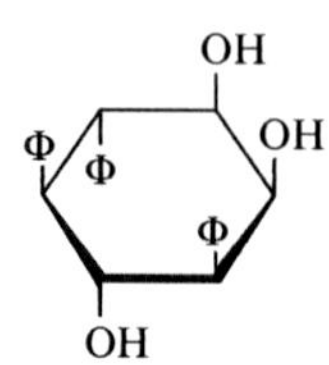

D-Ins(1,4,5)P_3
L-Ins(3,5,6)P_3

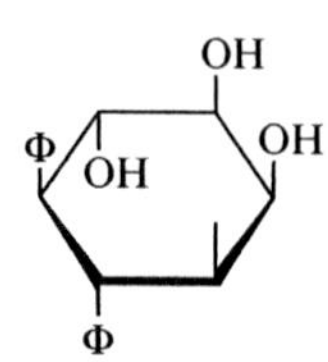

D-Ins(1,5,6)P_3
L-Ins(3,4,5)P_3

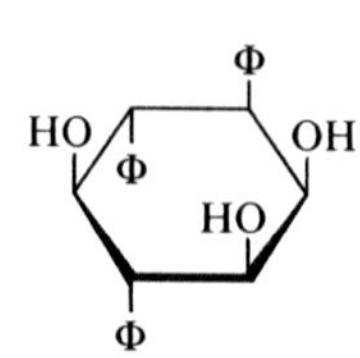

D-Ins(3,4,6)P_3
L-Ins(1,4,6)P_3

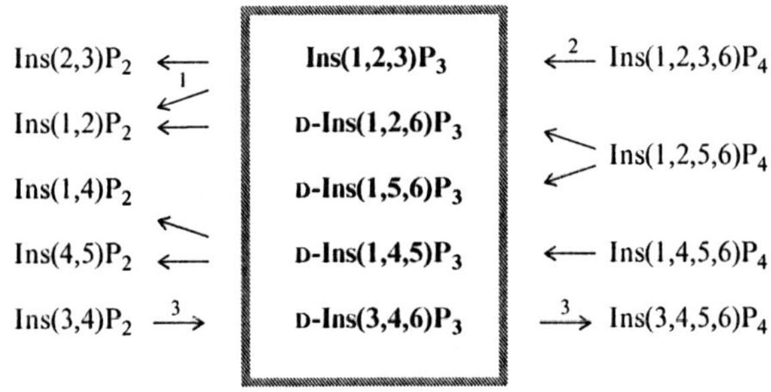

FIGURE 3-3 ***myo*-Inositol Trisphosphates.** (1) 6-Phytase [E.C. 3.1.3.26]. (2) 5-Phytase.[1291] (3) Phytic acid synthase.[1238]

myo-Inositol 1,2,3-trisphosphate (Ins(1,2,3)P_3), $C_6H_{15}O_{15}P_3$; unlisted in Beilstein. *myo-Inositol 1,2,3-Tris(Dihydrogen Phosphate)* [1962].

D-*myo*-Inositol 1,2,6-Trisphosphate (Ins(1,2,6)P_3), $C_6H_{15}O_{15}P_3$; unlisted in Beilstein. D-*myo-Inositol 1,2,6-Tris(Dihydrogen Phosphate)* [1962].

D-*myo*-Inositol 1,4,5-Trisphosphate (Ins(1,4,5)P_3), $C_6H_{15}O_{15}P_3$; unlisted in Beilstein. D-*myo-Inositol 1,4,5-Tris(Dihydrogen Phosphate)* [1961].

D-*myo*-Inositol 1,5,6-Trisphosphate (Ins(1,5,6)P_3), $C_6H_{15}O_{15}P_3$; unlisted in Beilstein. D-*myo-Inositol 1,5,6-Tris(Dihydrogen Phosphate)* [1988].

D-*myo*-Inositol 3,4,6-Trisphosphate (Ins(3,4,6)P_3), $C_6H_{15}O_{15}P_3$; unlisted in Beilstein. D-*myo-Inositol 3,4,6-Tris(Dihydrogen Phosphate)* [1996].

Occurrence

Ins(1,2,3)P_3 and Ins(1,2,6)P_3 have been identified in barley aleurone tissue[1208] and Ins (1,3,4)P_3 in pea flour.[1289] Ins(1,4,5)P_3 occurs in cauliflower hypocotyl,[1244] in pea flour,[1289] in apple buds,[1240] in carnation flower petals,[1290] in rain tree pulvini,[1219] and in Madagascar periwinkle,[1200] tobacco,[1200] and lucerne[1239] cell suspension cultures. Ins(2,4,5)P_3 has been identified in pea flour[1289] and Ins(3,4,6)P_3 in giant duckweed (Fig. 3-3).[1243]

Physical Properties

myo-Inositol 1,2,3-Trisphosphate

Ins(1,2,3)P_3 [R.N. 41613-02-9]: $[\alpha]_D^{25}$ 0° (H_2O).[1284] Spectra: ^1H NMR (fig.),[1292] 2d-^1H NMR (fig.),[1292] ^{13}C NMR (calc.).[1195] Chromatography: IEC,[1227] HPLC.[161] Electrophoresis: PE.[1284]

$Ba_3C_6H_9O_{15}P_3$: Slightly soluble in H_2O.[676]

$(NH_4)_6C_6H_9O_{15}P_3$: $[\alpha]_D^{25}$ 0° (H_2O).[1284]

$(C_6H_{11}NH_3)_6C_6H_9O_{15}P_3$: Crystals. Spectra: ^1H NMR, ^{31}P NMR.[1293]

D-*myo*-Inositol 1,2,6-Trisphosphate

Ins(1,2,6)P_3 [R.N. 28841-62-5]: $[\alpha]_D^{22}$ −3.2° (H_2O)[1294]; pK_a 5.28, 6.54, and 8.44 (potentiometric, 37 °C); microscopic protonation constants given for the 12 dissociation equilibria.[1295] Spectra: ^{1}H NMR (fig.),[1292] 2d-^{1}H NMR (fig.),[1292] ^{13}C NMR.[1195] Chromatography: HPLC,[161] LC/MS.[1251] Electrophoresis: PE.[1291] Additional data for DL-Ins(1,2,6)P_3 [R.N. 121321-74-9]: ^{31}P NMR (fig.),[1252] IEC,[1227] CZE/MS.[1270]

D-*myo*-Inositol 1,4,5-Trisphosphate

Ins(1,4,5)P_3 [R.N. 85166-31-0]: Spectra: ^{1}H NMR (fig.),[1296] ^{13}C NMR (fig.),[1296] ^{31}P NMR (fig.),[1296] MS (FAB, fig.).[690] Chromatography: PC,[1274] IEC,[1248] TLC,[1297] HPTLC,[407] HPLC,[1250] IC,[107] GC/MS,[690] SFC/MS.[1298] Electrophoresis: PE.[1239] Additional data for DL-Ins(1,4,5)P_3 [R.N. 109583-09-7]: Potentiometric titration,[1286] high-pH IEC,[730] CE,[1254] CZE.[1299]

$Ba_3C_6H_9O_{15}P_3$: Slightly soluble in H_2O.[676]

$(NH_4)_6C_6H_9O_{15}P_3$: $[\alpha]_D^{25}$ −9.4° (H_2O).[1297] Spectra: FAB-MS/MS (fig.).[1260]

$(C_6H_{11}NH_3)_6C_6H_9O_{15}P_3$ [R.N. 120710-76-1]: $[\alpha]_D^{25}$ −27.4° (H_2O).[1274] PC.[1274]

D-*myo*-Inositol 1,5,6-Trisphosphate

Ins(1,5,6)P_3 [R.N. 121251-13-6]: Spectra: ^{1}H NMR (fig.),[1292] 2d-^{1}H NMR (fig.),[1292] ^{13}C NMR and ^{31}P NMR.[1300] Chromatography: HPLC.[161] Additional data for DL-Ins(1,5,6)P_3 [R.N. 118354-68-0]: ^{1}H NMR,[687,1229] IEC,[1227] high-pH IEC,[730] IC.[1229]

$Na_6C_6H_9O_{15}P_3$: $[\alpha]_D$ −2.8° (H_2O).[1300]

D-*myo*-Inositol 3,4,6-Trisphosphate

Ins(3,4,6)P_3 [R.N. 114418-84-7]: Chromatography: IEC,[1301] HPLC.[1243] Additional data for DL-Ins(3,4,6)P_3 [R.N. 131321-75-0]: ^{1}H NMR,[687] IEC.[1227]

$Na_3C_6H_{12}O_{15}P_3$: $[\alpha]_D^{20}$ +9.4° (H_2O).[1302] Additional data for DL-$Na_3C_6H_{12}O_{15}P_3$: Fine crystals, mp > 240 °C, Spectra: ^{1}H, ^{13}C, and ^{31}P NMR.[1302]

$(Et_3N)_xC_6H_9O_{15}P_3$: $[\alpha]_D$ +25.0° (TEAB).[1301] Spectra: ^{1}H NMR, ^{31}P NMR, MS (FAB).[1301]

Synthesis

Ins(1,2,3)P_3 may be prepared by phosphorylation of 4,5,6-tri-*O*-benzoyl-*myo*-inositol with dibenzyl *N,N*-diisopropylphosphoramidite.[1293] Ins(1,4,5)P_3 may be prepared (a) by phosphorylation of D-2,3,6-tri-*O*-benzyl-*myo*-inositol or other fully protected *myo*-inositol with *N,N'*-diphenylphosphorodiamidic chloride,[1303] tetrabenzyl pyrophosphate,[1275,1303] dibenzyl *N,N*-diisopropylphosphoramidite,[1304] dibenzyl *N,N*-diethylphosphoramidite,[1305] *O*-xylylene *N,N*-diethylphosphoramidite,[1297,1303] or bis(2-cyanoethyl) *N,N*-diethylphosphoramidite[1306] or (b) by selective phosphorylation of a partially protected *myo*-inositol with *N,N'*-diphenylphosphorodiamidic chloride.[1307] Ins(1,5,6)P_3 may be prepared by phosphorylation of D-2,3-*O*-camphanylidene-4-(*t*-butyldimethylsilyl)-*myo*-inositol with 2-dimethylamino-5.6-benzo-1,3,2-dioxaphosphepane.[1293] Ins(3,4,6)P_3 may be prepared (a) by phosphorylation of 1D-5-*O*-acetyl-1,2-*O*-cyclohexylidene-*myo*-inositol with *O*-xylylene *N,N*-diethylphosphoramidite in the presence of 1*H*-tetrazole[1302] or (b) of D-1,2,5-tri-*O*-benzyl-*myo*-inositol with di-*O*-benzyl *N,N*-diisopropylphosphoramidite in the presence of 1*H*-tetrazole.[1301]

Conformation

The 1,2,3-, 1,2,6-, and 1,4,5-*myo*-inositol trisphosphates all adopt the $1_{ax}/5_{eq}$ conformation over the entire pH range from 1.0 to 13.0, with no evidence for the $5_{ax}/1_{eq}$ conformer.[1281]

Ins(2,4,5)P_3

Maitra *et al.*[1308] isolated an InsP_3 from the products of hydrolysis of phytic acid by mung bean phytase tentatively identified as Ins(2,4,5)P_3, implying the existence of an alternative pathway for phytate metabolism. The status of this finding remains unresolved.

myo-Inositol Tetrakisphosphates

There are 15 *myo*-inositol tetrakisphosphates, of which 5 are important plant metabolites (Φ=OP(O)$(OH)_2$) (Fig. 3-4):

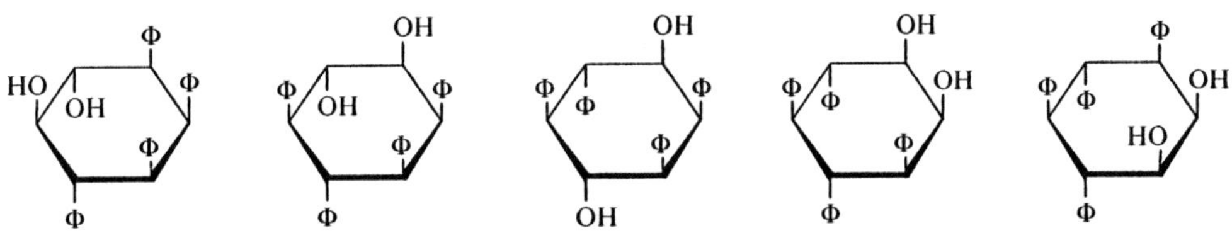

D-Ins(1,2,3,6)P_4 L-Ins(1,2,3,4)P_4 | **D-Ins(1,2,5,6)P_4** L-Ins(2,3,4,5)P_4 | **D-Ins(1,3,4,5)P_4** L-Ins(1,3,5,6)P_4 | **D-Ins(1,4,5,6)P_4** L-Ins(3,4,5.6)P_4 | **D-Ins(3,4,5,6)P_4** L-Ins(1,4,5,6)P_4

D-*myo*-Inositol 1,2,3,6-tetrakisphosphate (Ins(1,2,3,6)P_4), $C_6H_{16}O_{18}P_4$; not listed in Beilstein. *D-myo-Inositol 1,2,3,6-Tetrakis(Dihydrogen Phosphate)* [1969].

D-*myo*-Inositol 1,2,5,6-Tetrakisphosphate (Ins(1,2,5,6)P_4), $C_6H_{16}O_{18}P_4$; not listed in Beilstein. *D-myo-Inositol 1,2,5,6-Tetrakis(Dihydrogen Phosphate)* [1962].

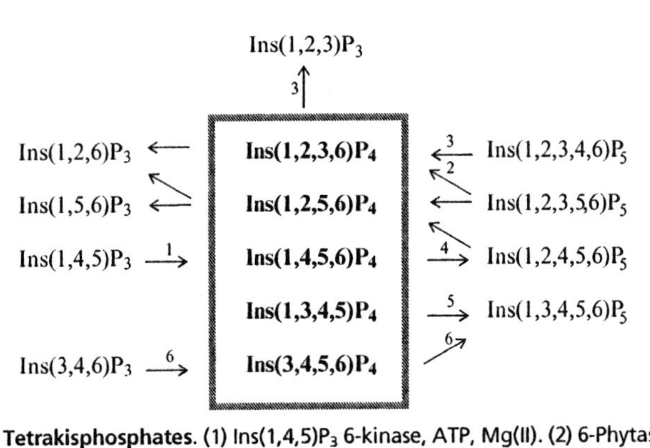

FIGURE 3-4 *myo*-Inositol Tetrakisphosphates. (1) Ins(1,4,5)P_3 6-kinase, ATP, Mg(II). (2) 6-Phytase [E.C. 3.1.3.26]. (3) 5-Phytase.[1291] (4) Ins(1,3,4,5,6)P_5 6-kinase, ATP, Mg(II).[1309] (5) Ins(1,3,4,5)P_4 6-kinase.[1310] (6) Phytic acid synthase.[1238]

D-*myo*-Inositol 1,3,4,5-Tetrakisphosphate (Ins(1,3,4,5)P_4), $C_6H_{16}O_{18}P_4$; not listed in Beilstein. D-*myo-Inositol 1,3,4,5-Tetrakis(Dihydrogen Phosphate)* [1988].

D-*myo*-Inositol 1,4,5,6-Tetrakisphosphate (Ins(1,4,5,6)P_4), $C_6H_{16}O_{18}P_4$; not listed in Beilstein. D-*myo-Inositol 1,4,5,6-Tetrakis(Dihydrogen Phosphate)* [1988].

D-*myo*-Inositol 3,4,5,6-Tetrakisphosphate (Ins(3,4,5,6)P_4), $C_6H_{16}O_{18}P_4$; not listed in Beilstein. D-*myo-Inositol 3,4,5,6-Tetrakis(Dihydrogen Phosphate)* [1996].

Occurrence

Ins(1,2,3,6)P_4 and Ins(1,2,5,6)P_4 were identified in barley aleurone tissue,[1208] Ins(1,2,5,6)P_4 in pea flour,[1289] and Ins(3,4,5,6)P_4 in giant duckweed.[1243]

Physical Properties

D-*myo*-Inositol 1,2,3,6-Tetrakisphosphate

Ins(1,2,3,6)P_4 [R.N. 41613-03-0]: $[\alpha]_D^{25}$ +6.0° (H_2O).[1284] Spectra: 2d-^{1}H NMR (fig.).[1292] Chromatography: IEC.[719] Electrophoresis: PE.[1284] Additional data for DL-Ins(1,2,3,6)P_4 [R.N. 118354-66-8]: ^{1}H NMR,[1229] IC.[1229]

Ba salt: Amorphous solid, insol. in aqueous ethanol.[676]

$(NH_4)_8C_6H_{12}O_{18}P_4$: $[\alpha]_D^{25}$ −11.3° (H_2O).[1284]

D-*myo*-Inositol 1,2,5,6-Tetrakisphosphate

Ins(1,2,5,6)P_4 [R.N. 28841-60-3]: $[\alpha]_D^{25}$ −3.5° (H_2O),[719] $[M]_{300}^{25}$ −233° (H_2O).[1311] Spectra: ORD,[1311] 2d-^{1}H NMR (fig.),[1292] FAB-MS.[690] Chromatography: IEC,[719] HPLC.[161] Electrophoresis: PE.[1291] Additional data for DL-Ins(1,2,5,6)P_4 [R.N. 118354-64-6]: pK_a (potentiometric, 37 °C) 5.23, 6.06, 7.65, and 9.39,[1312] ^{1}H NMR,[687,1229] ^{13}C NMR,[1195] IC,[1229] high-pH IEC,[730] CZE/MS.[1270]

Ba salt: Amorphous solid, insol. in aqueous ethanol.[676]

$(C_6H_{11}NH_3)_8C_6H_{12}O_{18}P_4$: $[\alpha]_D^{25}$ −14.8° (H_2O); may contain some 1,2,3,5 or 1,2,3,6 isomer.[723]

D-*myo*-Inositol 1,3,4,5-Tetrakisphosphate

Ins(1,3,4,5)P_4 [R.N. 99610-76-1]: pK_a (potentiometric, 37 °C): 5.23, 6.36, 7.66, 9.92.[1312] Spectra: ^{1}H (fig.), ^{13}C (fig.), ^{31}P (fig.), and 2d-^{1}H NMR (fig.).[1313] Chromatography: IEC,[1267] HPLC.[161] Additional data for DL-Ins(1,3,4,5)P_4 [R.N. 115184-69-5]: 2d-^{1}H NMR (fig.),[1228] IC.[1229]

$K_4C_6H_{12}O_{18}P_4$: $[\alpha]_D^{20}$ −3.85° (H_2O).[1314] Spectra: IR, ^{1}H, ^{13}C, and ^{31}P NMR, ES-MS.[1314]

$(NH_3)_8C_6H_{18}O_{18}P_4$: $[\alpha]_D^{24}$ −13° (H_2O).[1315]

$(C_6H_{11}NH_2)_{4.5}C_6H_{18}O_{18}P_4$: mp 173.7-175.7 °C, $[\alpha]_D^{25}$ −2.5° (H_2O).[1316] Spectra: ^{1}H, ^{13}C, and ^{31}P NMR, FAB-MS.[1316]

D-*myo*-Inositol 1,4,5,6-Tetrakisphosphate

Ins(1,4,5,6)P_4 [R.N. 109837-24-3]: Spectra: ^{1}H, ^{13}C, ^{31}P NMR (figs.).[1317] Chromatography: IEC,[1317] HPLC.[161] Additional data for DL-Ins(1,4,5,6)P_4 [R.N. 113666-11-8]: IC,[1229] high-pH IEC.[730]

D-*myo*-Inositol 3,4,5,6-Tetrakisphosphate

Ins(3,4,5,6)P_4: Spectra: ^{1}H and ^{31}P NMR.[1259] Chromatography: HPTLC,[407] HPLC.[1243] $Na_8C_6H_{18}O_{18}P_4$: Solid, mp > 270 °C, $[\alpha]_D^{24}$ +9.8° (H_2O).[1318]

Synthesis

Ins(1,2,5,6)P_4 is formed together with some Ins(1,2,3,6)P_4 by partial dephosphorylation of phytic acid with phytase.[723,1284] Ins(1,4,5,6)P_4 and Ins(3,4,5,6)P_4 may be prepared by phosphorylation of D-2,3- and D-1,2-*O*-cyclohexylidene-*myo*-inositol, respectively, with dibenzyl *N,N*-diisopropylphosphoramidite in the presence of 1*H*-tetrazole.[1318] Ins(1,3,4,5)P_4 may be prepared (a) by phosphorylation of D(−)-2,6-di-*O*-benzyl-*myo*-inositol with dibenzyl *N,N′*-diisopropylphosphoramidite,[1319] 2-diethylamino-1,3,2-benzodioxaphosphepane,[1314] or tetrabenzyl pyrophosphate[1315] or (b) by enzymic phosphorylation of D-Ins(1,4,5)P_3 with Ins(1,4,5)P_3 kinase.[1264] No synthesis of Ins(1,2,3,6)P_4 has been published.

Conformation

Both Ins(1,2,3,6)P_4 and Ins(1,2,5,6)P_4 adopt the $1_{ax}/5_{eq}$ conformation over the entire pH range from 1.0 to 13.0, with no evidence for the $5_{ax}/1_{eq}$conformer.[1281]

Phosphorylation

DL-Ins(1,4,5,6)P_4 is phosphorylated by Ins(1,3,4,5,6)P_5 2-kinase from immature soybeans giving a pentakisphosphate identified by IEC as DL-Ins(1,2,4,5,6)P_5.[1309]

myo-Inositol Pentakisphosphates

There are six *myo*-inositol pentakisphosphates, of which four are important plant metabolites ($\Phi = OP(O)(OH)_2$):

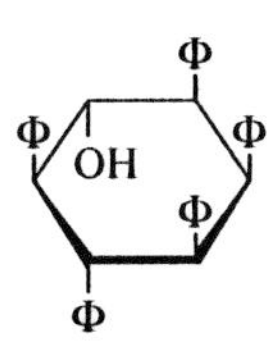

L-Ins(1,2,3,4,5)P₅
D-Ins(1,2,3,5,6)P_4

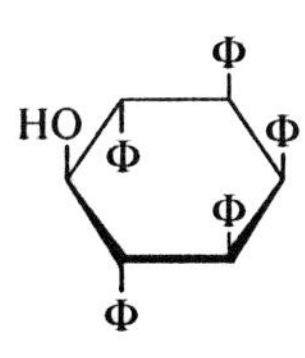

Ins(1,2,3,4,6)P_5
(*meso*)

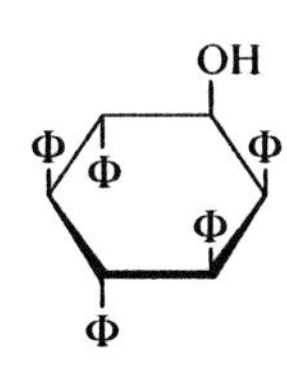

D-Ins(1,2,4,5,6)P_5
L-Ins(2,3,4,5,6)P_5

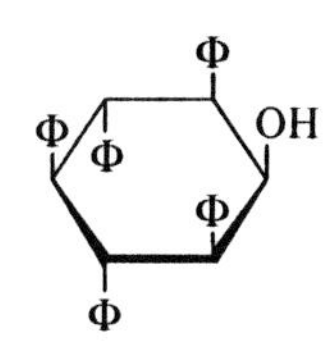

Ins(1,3,4,5,6)P_5
(*meso*)

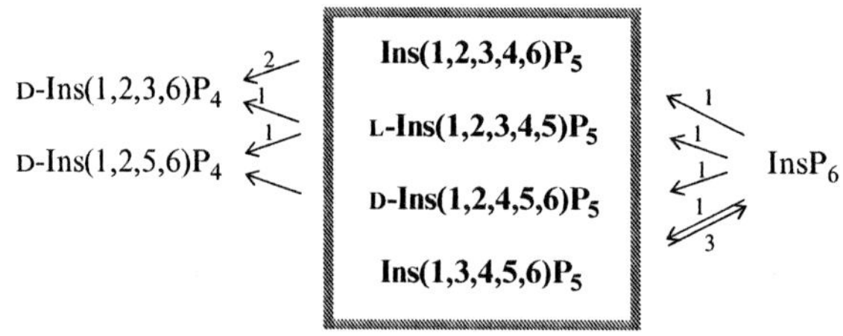

FIGURE 3-5 ***myo*-Inositol Pentakisphosphates.** (1) 6-Phytase [E.C. 3.1.3.26]. (2) 5-Phytase.[1291] (3) 2-Kinase, ATP, Mg(II).[1309]

L-*myo*-Inositol 1,2,3,4,5-Pentakisphosphate (Ins(1,2,3,4,5)P_5), $C_6H_{17}O_{21}P_5$; not listed in Beilstein. *L-myo-Inositol 1,2,3,4,5-Pentakis(Dihydrogen Phosphate)* [1962].

myo-Inositol 1,2,3,4,6-Pentakisphosphate (Ins(1,2,3,4,6)P_5), $C_6H_{17}O_{21}P_5$; not listed in Beilstein. *myo-Inositol 1,2,3,4,6-Pentakis(Dihydrogen Phosphate)* [1968].

D-*myo*-Inositol 1,2,4,5,6-Pentakisphosphate (Ins(1,2,4,5,6)P_5), $C_6H_{17}O_{21}P_5$; not listed in Beilstein. *D-myo-Inositol 1,2,4,5,6-Pentakis(Dihydrogen Phosphate)* [1968].

myo-Inositol 1,3,4,5,6-Pentakisphosphate (Ins(1,3,4,5,6)P_5), $C_6H_{17}O_{21}P_5$; not listed in Beilstein. *myo-Inositol 1,3,4,5,6-Pentakis(Dihydrogen Phosphate).*

Occurrence

Ins(1,2,3,4,6)P_5 (one of the *meso* compounds) was identified in barley aleurone tissue[1208] and in soybean seed,[1229] and Ins(1,3,4,5,6)P_5 (the other *meso* compound) was identified in barley aleurone tissue,[1208] in soybean seed[1229] and in giant duckweed.[1243] D- and/or L-Ins (1,2,3,4,5)P_5 was identified in barley aleurone tissue,[1208] in soybean seed,[1229] and in mung bean sprout.[1320] D- and/or L-Ins(1,2,4,5,6)P_5 was identified in soybean seed,[1229] in mung bean sprout,[1320] and in giant duckweed (Fig. 3-5).[1243]

Physical Properties

L-*myo*-Inositol 1,2,3,4,5-Pentakisphosphate

Ins(1,2,3,5,6)P_5 [R.N. 26326-85-2]: Solid,[1321] $[\alpha]_D^{25}$ +1.6° (H_2O).[1284] Spectra: ^1H NMR, ^{31}P NMR,[1321] FAB-MS (fig.).[690] Chromatography: IEC,[718] HPLC,[1320] IC.[1229] Electrophoresis: PE.[720] Additional data on DL-(1,2,3,5,6)InsP_5 [R.N. 23486-93-3]: ^{13}C NMR.[1195]
$Na_8C_6H_9O_{21}P_5$: White powder, FAB-MS.[1210]
Ba salt: Amorphous solid, insol. in H_2O.[676]
$(NH_4)_{10}C_6H_{12}O_{21}P_5$: $[\alpha]_D^{25}$ +3.7° (H_2O).[1284]

myo-Inositol 1,2,3,4,6-Pentakisphosphate

Ins(1,2,3,4,6)P_5 [R.N. 20298-97-9]: $[\alpha]_D^{25}$ 0° (H_2O).[1284] Spectra: ^1H NMR,[1229] 2d-^1H NMR (fig.),[1291] ^{13}C NMR,[1195] ^{31}P NMR,[1291] FAB-MS.[690] Chromatography: IEC,[719] HPLC,[1320] IC.[1229] Electrophoresis: PE.[720]
Ba salt: Amorphous solid, insol. in H_2O.[676]
$(NH_4)_{10}C_6H_{12}O_{21}P_5$: $[\alpha]_D^{25}$ 0° (H_2O).[1284]

D-*myo*-Inositol 1,2,4,5,6-Pentakisphosphate

Ins(1,2,4,5,6)P_5 [R.N. 26326-86-3]: $[\alpha]_D$ −7.4° (H_2O),[1322] $[M]^{25}_{300}$ −115° to −160° (H_2O).[1311] Spectra: ^{1}H NMR,[1320] 2d-^{1}H NMR (fig.),[1320] ^{31}P NMR (fig.),[1320] FAB-MS,[690] ORD.[1311] Chromatography: IEC,[1323] HPLC.[1320] Additional data for DL-Ins(1,2,4,5,6)P_5 [R.N. 23393-17-1]: ^{13}C NMR,[1195] IC,[1229] PE.[720]

Ba salt: Amorphous solid, insol. in H_2O.[676]

$(NH_4)_{10}C_6H_{12}O_{21}P_5$: $[\alpha]_D$ −4.0° (H_2O).[1322]

myo-Inositol 1,3,4,5,6-Pentakisphosphate

Ins(1,3,4,5,6)P_5 [R.N. 20298-95-7]: $[M]^{25}_D$ 0°.[1284] Spectra: ^{1}H NMR (fig.),[1317] ^{13}C NMR (fig.),[1317] ^{31}P NMR (fig.),[679] FAB-MS.[690] Chromatography: PC,[679] TLC,[1324] IEC,[1323] high-pH IEC,[730] HPTLC,[407] HPLC,[1320] IC,[1229] GFC.[1225] Electrophoresis: PE.[720]

Ba salt: Amorphous solid, insol. H_2O.[676] Spectra: IR.[679]

$Na_{10}C_6H_7O_{21}P_5$ [23103-36-8]: Amorphous hygroscopic solid, $[\alpha]^{20}_D$ 0.0° (H_2O).[679] Spectra: IR (Table 3-4).[679]

Synthesis

Ins(1,2,3,4,5)P_5 may be prepared by enzymatic hydrolysis of phytic acid with wheat bran phytase [E.C. 3.1.3.26] or wheat germ phosphatase [E.C. 3.1.3.2].[1210] DL-Ins(1,2,3,4,5)P_5 and DL-Ins(1,2,4,5,6)P_5 have been prepared by the reaction of *N*-benzoylphosphoramidic acid with DL-4-*O*-benzyl-*myo*-inositol and DL-3-*O*-benzyl-*myo*-inositol, respectively, in DMF.[716]

No synthesis has yet been published for either of the other two pentakisphosphates.

Phosphorylation

All four pentakisphosphates are phosphorylated by mung bean kinase/Mg(II) to phytic acid.[1320,1337] The relative first-order rate constants for DL-Ins(1,2,4,5,6)P_5, DL-Ins (1,2,3,4,5)P_5, Ins(1,2,3,4,6)P_5, and Ins(1,3,4,5,6)P_5 are 1:11:12:33, respectively.[1320]

Analytical Methods

The four InsP_5's can be separated from each other by HPLC on a Mono Q anion exchange column upon gradient elution with 0.0002→0.4*M* HCl[161] or by IC on a Dionex AS-3 column on gradient elution with 0→0.155*N* HNO_3.[1229] The order of elution in both cases is Ins(1,2,3,4,6)P_5 > Ins(1,2,3,4,5)P_5 > Ins(1,2,4,5,6)P_5 > Ins (1,3,4,5,6)P_5. They can also be separated on a Partisphere HPLC column upon gradient elution with 0-0.5*M* ammonium phosphate, pH 3.2, but require sequential runs on a weak (WAX) and a strong (SAX) anion exchanger (in any order) for complete separation.[1320]

Table 3-4 Lower *myo*-Inositol Phosphates in Plants*

No.	Plant part	Method	$InsP_1$	$InsP_2$	$InsP_3$	$InsP_4$	$InsP_5$	$InsP_6$	References
Agropyron desertorum (Fisch.) Schult. (Crested Wheatgrass)									
1	Seed[a]	IEC		0.10	0.21	0.47			831
Brassica napus L. (Rape)									
8	Whole seed[b]	HPLC			0	0	0.63-0.86	24.2-34.5	1325
8	–, dehulled[b]	HPLC			0	0	0.94-1.24	30.5-38.9	1325
8	–, hull[b]	HPLC			0	0	0.08-0.45	2.4-10.1	1325
Brassica napus L. ssp. *oleifera* (Winter Rape)									
1	Leaf	IEC	++	+	+				1326
Brassica oleracea L. ssp. *botrytis* Cauliflower									
1	Hypocotyl	$^{32}P_i$	+	+	+				1244
Cajanus cajan (L.) Mill. (Pigeon Pea)									
1	Seed	HPLC			0	0.04	2.41	7.96	1327
Catharanthus roseus (L.) G. Don (Madagascar Periwinkle)									
1	Cell suspension culture	IEC	12.8	0.29	0.16				1200
Ceratophyllum spp. (Hornwort)									
1	Whole plant[a]	IEC			0.028		+		912
Chenopodium quinoa Willd. (Quinoa)									
1	Seed, sweet	HPLC			0	0.002	0.002	0.206	916
1	Seed, bitter	HPLC			0	0.002	0.004	0.160	916
Cicer arietinum L. (Chickpea)									
1	Seed	HPLC			0	0.04	1.76	6.00	1327
2	Seed[c]	HPLC			0.025-0.027	0-0.011	0.031-0.068	0.379-0.389	922
Cocos nucifera L. (Coconut)									
1	Copra cake	HPLC			1.76	0.91	1.65	9.71	1328
Coffea arabica L. (Coffee)									
3	Bean[c]	HPLC			0	0	0.025-0.039	0.255-0.292	935
Coffea robusta (Coffee)									
2	Bean[c]	HPLC			0	0-0.040	0.031-0.110	0.306-0.368	935
Corylus avellana L. (Hazel)									
1	Shelled nut[a]	HPLC			0	0	0	2.58	938
Daucus carota L. (Carrot)									
1	Cell suspension culture	^{3}H-Ins	+						950
Dianthus caryophyllus L. (Carnation)									
1	Flower petal	^{3}H-Ins			+				1290

Glycine max (L.) Merr. (Soybean)									
1	Seed	IC			+	+	+	+	1229
1	Seed, immature[d]	IEC	1.9	Trace	0	Trace	4.9		1329
4	Seed meal	HPLC			0	0-0.60	1.09-3.98	11.78-16.84	1328
Gossypium hirsutum L. (Upland Cotton)									
1	Glandless seed, defatted	IEC, PE		0	0	+	+		1330
Helianthus annuus L. (Sunflower)									
1	Seed meal	HPLC			0.20	0.48	5.63	40.43	1328
Hordeum vulgare L. (Barley)									
4	Grain	HPLC			0	0	0	3.78-5.73	1328
1	Grain[c]	HPLC			0	0	0.087	0.887	1331
1	Unripe caryopsis	HPLC					+		857
1	Aleurone tissue	HPLC	+	+	+	+	+	+	1208
1	Mesophyll cells	^{3}H-Ins		+					1283
Lemna minor L. (Lesser Duckweed)									
1	Whole plant	IEC, GFC	+	+	+	+	+		683
Lemna spp. (Duckweed)									
1	Whole plant [a]	IEC			0.119		+		912
Lens culinaris Medic. (Lentil)									
1	Seed	HPLC			0.32	0.21	1.39	8.37	1327
1	Seed[c]	HPLC			0.016	0.019	0.048	0.244	922
1	Raw flour[a]	HPLC			0.37	0.50	1.07	10.77	1332
Lilium longiflorum Thunb. (Trumpet Lily)									
1	Pollen	^{3}H-Ins	+						1217
Lupinus albus L. (White Lupine)									
1	Seed[c]	HPLC			0	0	0	0.314	922
Lupinus luteus L. (Yellow Lupine)									
1	Seed [a]	HPLC					0.09	21.72	1333
1	Seed [c]	HPLC			0	0	0.06	0.717	922
Malus domestica Borkh. (Apple)									
1	Bud, dormant	HPLC	0.72	0.00035	0.0033				1240
Medicago sativa L. (Alfalfa)									
1	Seed meal	HPLC			0	0.26	0	0.80	1328
1	Cell suspension culture	^{3}H-Ins		+	+				1239

*nmol/g fresh weight basis unless stated otherwise; [a]μmol/g; [b]mg/g; [c]Percentage of plant tissue; [d]Percentage of total P.

Continued

Table 3-4 Lower *myo*-Inositol Phosphates in Plants*—cont'd

No.	Plant part	Method	$InsP_1$	$InsP_2$	$InsP_3$	$InsP_4$	$InsP_5$	$InsP_6$	References
Myriophyllum spp. (Water Milfoil)									
1	Whole plant[a]	IEC			0.071		+		912
Nicotiana tabacum L. (Tobacco)									
1	Cell suspension culture	^3H-Ins	+	+	+				1200
Oryza sativa L. (Rice)									
1	Seed	IEC	0	0	0	0	0		1218
1	Brown seed, nonwaxy	IEC				+	+		1046
1	–, embryo[d]	IEC				0.019	0.020		1046
1	–, endosperm	IEC				0	0		1046
1	–, bran	IEC				+	+		1046
3	Seed, milled	HPLC			0-0.93	0-0.49	1.23-3.08	13.83-28.00	1328
1	Bran	HPLC			0	0.65	7.75	59.99	1328
1	Ripening grain	^3H-Ins	+						1242
1	Cell suspension culture	^3H-Ins	+						1186
Phaseolus lunatus L. (Lima Bean)									
1	Bean	HPLC			0	0.23	2.13	9.96	1327
Phaseolus vulgaris L. (Beans (Common))									
6	Bean	HPLC			0	0.02-0.19	1.80-2.19	11.7-14.2	1327
2	Bean[c]	HPLC			0.009	0	0.019-0.020	0.564-0.628	922
1	Root[a]	IEC		3.70	3.80	0.74			1334
Pisum sativum L. (Peas)									
1	Soaked flour	HPLC		+	+	+			1289
1	Soaked flour	HPLC	+	+					1241
2	Seed[b]	HPLC			0	0.12-0.17	1.49-1.36	8.82-6.48	1327
Samanea saman Merr. (Rain Tree)									
1	Pulvini	^3H-Ins	+	+	+				1219
Secale cereale L. (Oats)									
3	Grain	HPLC			0	0	0	4.18-8.97	1328
Sorghum bicolor (L.) Moench (Sorghum)									
2	Grain	HPLC			0	0	0-0.54	8.77-9.68	1328
Spirodela polyrhiza (L.) Schleid. (Giant Duckweed)									
1	Axenic culture	^3H-Ins	+	+	+	+	+	+	1243

Theobroma cacao L. (Cocoa)									
1	Husk	HPLC			0	0	0	0.28[c]	1126
Triticum aestivum L. (Wheat)									
10	Grain	HPLC			0	0	0	4.90-9.27	1328
1	Bran	HPLC			0	1.24	6.92	39.47	1328
1	Ear stalk	$^{32}P_i$	+						1242
Vallisneria spp. (Eelgrass)									
1	Whole plant[a]	IEC			0.068		+		912
Vicia faba L. (Broad or Faba Bean)									
2	Bean[c]	HPLC			0.016-0.020	0.043-0.047	0.075-0.212	0.345-0.885	922
Vicia faba L. ssp. *minor* L. (Small Horsebean)									
1	Bean[a]	HPLC					0.14	17.45	1333
Vigna radiata (L.) Wilcz. (Mung Bean)									
1	Seed[a]	IEC		0.08	0.19	0.39	0.69		1335
1	Sprout	$^{32}P_i$					+		1320
Vigna unguiculata (L.) Walp. (Cowpea)									
1	Seed	HPLC			0.01	0.26	2.52	12.6	1327
Zea mays L. (Corn)									
1	Seed	IEC				+	+		1216
2	Grain	HPLC			0	0	0.22-0.32	7.29-11.43	1328
1	1-week seedling	IEC	+	+	+	+	+		1216
1	Root, apical segment	IEC	++	+	+				1336

*nmol/g fresh weight basis unless stated otherwise.
[a]μmol/g.
[b]mg/g.
[c]Percent of plant tissue.
[d]Percent of total P.

Glyceroinositol Phosphates

This group comprises monoglyceryl esters of three *myo*-inositol phosphates:

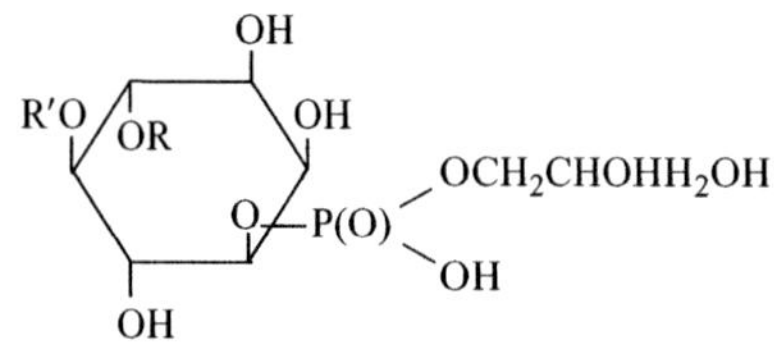

GroPIns, R = R′ = H
GroPInsP, R = P(O)(OH)$_2$, R′ = H
GroPInsP$_2$, R = R′ = P(O)(OH)$_2$

Glycerophosphoinositol (GroPIns), $C_9H_{19}O_{11}P$; Beilstein **6**, E_{III} 6933, E_{IV} 7925. D-*myo-Inositol 1-(R-2,3-Dihydroxypropyl Hydrogen Phosphate)* [1958].

Glycerophosphoinositol 4-Phosphate (GroPInsP), $C_9H_{20}O_{14}P_2$; Beilstein **6**, E_{III} 6934. D-*myo-Inositol 4-(Dihydrogen Phosphate) 1-(R-2,3-Dihydroxypropyl Hydrogen Phosphate)* [1961].

Glycerophosphoinositol 4,5-Bisphosphate (GroPInsP$_2$), $C_9H_{21}O_{17}P$; Beilstein **6**, E_{III} 6934. D-*myo-Inositol 4,5-Bis(Dihydrogen Phosphate) 1-(R-2,3-Dihydroxypropyl Hydrogen Phosphate)* [1961].

Occurrence

The presence of these compounds in the higher plants has not been rigorously established, but they are included here as they are likely products of the action of phospholipase A_1 and A_2 on PI, PIP, and PIP$_2$, all of which (including the enzymes) are known to occur in higher plants. Heim and Wagner[1200] identified GroPIns, GroPIns(4)P, and GroPIns(4,5)P$_2$ in Madagascar periwinkle and tobacco culture cell extracts by their retention times on HPLC. Brearly and Hanke[1338] identified GroPIns(3)P and GroPIns(3,4)P$_2$ in addition to the 4- and 4,5-analogs as products of deacylation of PIP$_2$ in *Spirodela polyrrhiza.* All the other compounds listed in Fig. 3-6 were prepared from the corresponding inositol phosphatides by base-catalyzed methanolysis.[1203]

Physical Properties

Glycerophosphoinositol

GroPIns [R.N. 129830-95-1; see also 2816-11-7, 27082-28-6, 37449-91-5, and 40529-46-2]: $[\alpha]_D^{20}$ −18° (H_2O).[1339] Spectra: GC/MS (fig.),[1340] ^{31}P NMR.[68,1341] Chromatography: PC,[1339] TLC,[1266] IEC,[1207] GC,[123] HPLC.[1268]

$C_6H_{11}NH_3.C_9H_{18}O_{11}P$: Crystals from alcohol, $[\alpha]_D^{25}$ −14.0° (H_2O).[1207] *Hemihydrate*: mp 124-127 °C dec., $[\alpha]_D^{20}$ −13.2° (H_2O).[1247] Spectra: IR.[1247] Chromatography: PC.[1247]

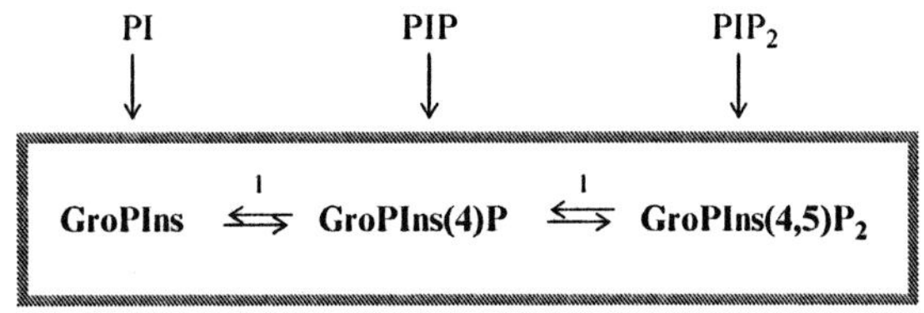

FIGURE 3-6 Glycerophosphoinositols. (1) Alkaline phosphatase [E.C. 3.1.3.1].[1205]

Glycerophosphoinositol 4-Phosphate

GroPInsP [R.N. 129830-96-2; see also 29595-51-5 and 77055-41-5]: $[\alpha]_D$ −4.8° (acid solution), −5.5° (basic solution)[1342]; pK_a 5.99.[1343] Spectra: GC/MS (fig.).[1340]
Chromatography: PC,[1342] IEC,[1342] GC,[123] HPLC.[1268]
Ca(II), Mg(II), Ni(II) complexes: Stability constants.[1343]

Glycerophosphoinositol 4,5-Bisphosphate

GroPInsP_2 [R.N. 111188-72-8; see also 92216-46-1 and 94161-15-6]: $[\alpha]_D$ −1.1° (acid solution), −3.3° (basic solution)[1342]; pK_a 5.70, 8.05.[1344] Spectra: GC/MS (fig.).[1340]
Chromatography: PC,[1342] IEC,[1342] HPLC.[1268]
Ca(II), Mg(II), Ni(II) complexes: Stability constants.[1343,1344]

Synthesis

None of the naturally occurring compounds have been synthesized. DL-GroPIns has been prepared by the reaction of DL-*myo*-inositol 1-phosphate with glycidol.[1345]

Hydrolysis

Hydrolysis of DL-GroPIns with 1*N* NaOH for 30 min at 60 °C yields a 65:35 mixture of glyceryl phosphate and *myo*-inositol phosphate in addition to glycerol and *myo*-inositol (but no P_i). PC shows that both phosphates are partly isomerized, the ratios being 44:56 for Gro(1)P:Gro(2)P and 70:30 for Ins(1)P:Ins(2)P.[1345] Similarly, hydrolysis of GroPInsP or GroPInsP_2 with 1*N* NaOH at 100 °C yields a mixture of glyceryl phosphate and InsP_2 (48:52) or InsP_3 (26:74), respectively, both partially isomerized (ratios not determined).[1205]

Hydrolysis of GroPInsP or GroPInsP_2 with bacterial phosphatase at pH 8.2 and 40 °C cleaves only the primary phosphate groups giving GroPIns and P_i.[1205]

Extraction and Separation

The three GroPIns's can be isolated from plant tissue by extraction with $CHCl_3$/methanol (1:2) containing 1% HCl and separated from the InsP's by IEC or HPLC.[1200] In most cases, however, they are prepared from the corresponding phosphatidylinositol (PI, PIP, or PIP_2) by base-catalyzed methanolysis and separated from each other by IEC, HPLC, PE,[1204] or 2d-PC.[1203] The IEC separation may be performed on a Dowex-1 column with 0→0.4*M* LiCl gradient elution[1342] and the HPLC separation on a Partisil-10 SAX anion exchange column with 0→1.0*M* ammonium formate gradient at pH 3.7[1201] or a Supelco LC-8 reversed-phase column with $0 \rightarrow 0.8M$ KH_2PO_4/n-propanol gradient containing a surfactant, hexadecyltrimethylammonium hydroxide.[1346] The order of elution is GroPIns > GroPInsP > GroPInsP$_2$ in all four methods (Table 3-5).

Table 3-5 Glyceroinositol Phosphates in Plants*

No.	Plant part		GroPIns	GroPInsP	GroPInsP$_2$	References
Brassica oleracea L. var. *botrytis* (Cauliflower)						
1	Callus	IEC	+	+	+	1199
1	Hypocotyl	IEC	+	+	+	1199
Catharanthus roseus (L.) G. Don (Madagascar Periwinkle)						
1	Cell suspension culture	IEC	8.5			1200
1	Cell suspension culture	HPLC	+	+		1200
Daucus carota L. (Carrot)						
1	Culture cells	HPLC			+	1201
Dianthus caryophyllus L. (Carnation)						
1	Flower petal	HPLC	+	+		1202
Hordeum vulgare L. (Barley)						
1	Leaf[a]	PC/PC	+			1203
1	Aleurone layer[b]	PE	+	+	+	1204
Nicotiana tabacum L. (Tobacco)						
1	Leaf [a]	PC/PC	+			1203
1	Culture cells	HPLC	+	+	+	1200
Pisum sativum L. (Peas)						
1	Leaf, 12 days old	HPLC		+	+	1201
Spirodela polyrrhiza (L.) Schleid. (Giany Duckweed)						
1	Whole plant	HPLC	+	+	+	1205
Trifolium repens L. (White Clover)						
1	Leaf	PC/PC	+			1203
Triticum aestivum L. (Wheat)						
1	Flour [a]	PC/PC	+			1206
Zea mays L. (Corn)						
1	–[a]	IEC	+			1207

*nmol/g fresh weight basis unless stated otherwise.
[a]Species not given.
[b]Ratio of GroPIns:GroPInsP:GroPInsP$_2$ is 100:0.5:0.1.

4 Phospholipids

This group comprises the phospholipids PC, PE, PG, PI, PS, PA, and DPG, and the phospholipid derivatives PIP, PIP_2, and NAPE:

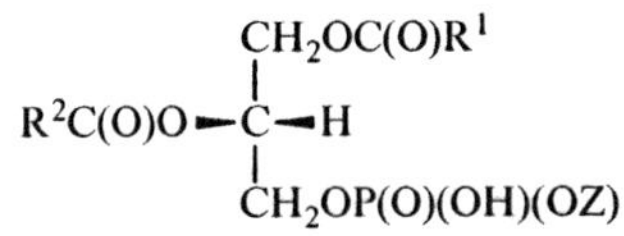

PL	ZOH component	RCO_2H component	
PA	H	16:0	Palmitic acid
PC	Choline	18:0	Stearic acid
PE	Ethanolamine	18:1	Oleic acid
PG	Glycerol	18:2	Linoleic acid
PS	L-Serine	18:3	Linolenic acid
PI	*myo*-Inositol	and other long-chain fatty	
PIP	*myo*-Inositol 4-phosphate	acids in various combinations	
PIP_2	*myo*-Inositol 4,5-bisphosphate		
DPG	PA		
NAPE	*N*-Acylethanolamine		

The plant phospholipids are all esters of phosphatidic acid (PA, Z=H). These esters are commonly called phosphatidylcholine (PC), phosphatidylethanolamine (PE), and so on, names that are adequate for this part of the chapter, which treats the phospholipids in a general way. A different nomenclature is introduced later in this chapter that enables individual members of each phospholipid class to be identified unambiguously.

For general references on the phospholipids, see Butt and Beevers,[1347] Kates,[1348] Kates,[1349] Stumpf,[1350] Moore,[1351] Coté and Crain,[1352] or Cevc.[4297]

Occurrence

The occurrence of lecithin in plant seeds was first reported by Schulze and Steiger[1353] in 1889. Its identity as phosphatidylcholine (PC) was established in a series of research studies by Schulze's group over a 20-year period. Phosphatidylethanolamine (PE) was detected in soybeans in 1925, but the "cephalin" fraction that contained it was later shown to also contain *myo*-inositol[1354] and serine[1355] residues, among others. The situation was not clarified until the advent of chromatographic methods for their separation in the 1950s. The number of phospholipid classes then expanded rapidly with the discovery of PI (1957) in runner beans,[1356] PG (1958) in tobacco,[1203] DPG (1960) in corn and clover,[1357] NAPE (1965) in wheat,[1358] and PIP and PIP_2 (1985) in wild carrot.[1359] All of these phospholipids are now known to be normal constituents of the higher plants as well as of animals and lower plants, but they are not all present in all plant tissues nor in all plant tissues at different stages of development.

The metabolic pathways for the biosynthesis of the phospholipids are outlined in Fig. 4-1. The key precursor for all of the phospholipids is phosphatidic acid (PA), formed from *sn*-glycero-3-phosphate (α-GP) by acylation with acyl-CoA (step 1). The phospholipids are seen to be split into two groups depending on which metabolite is activated by cytidine triphosphate (CTP). In one pathway, phosphocholine and phosphoethanolamine are activated by CTP forming CDP-choline and CDP-ethanolamine, respectively, which then react with diacylglycerol (DG), formed from PA by hydrolysis with phosphatase (step 2), giving PC (step 4) or PE (step 5). In the other pathway, the phosphate group of PA is activated by CTP giving cytidine diphosphate-diacylglycerol (CDP-DG, step 3), which then reacts with L-serine, *myo*-inositol, or α-GP giving PS (step 6), PI (step 7), or PG (step 8a), respectively. Interconversion of PE and PS can occur by exchange of PE with L-serine (step 9) or by decarboxylation of PS (step 10). NAPE is formed from PE in seed microsomes (step 11) by acylation.[1361] PIP and PIP_2 are formed from PI by successive phosphorylation with ATP (steps 12 and 13).[1362] DPG is formed from PG in plant mitochondria (step 14) by reaction with CDP-DG.[1363]

Stereochemistry

The phospholipids all contain a chiral center at the C_2 position of the glyceryl moiety and are therefore capable of existing in two configurations, which by convention are labeled R and S. All naturally occurring plant phospholipids have the R configuration, as illustrated later, and can therefore display optical activity. Some examples are given in Table 4-1.

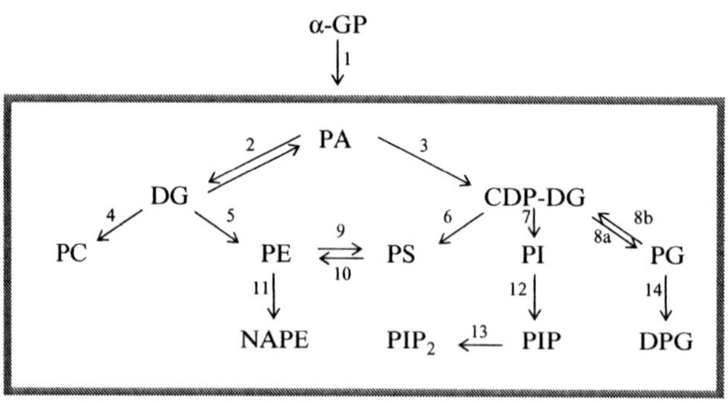

FIGURE 4-1 Phospholipid Metabolism. (1) Acyl-CoA:*sn*-glycerol-3-phosphate 1-*O*-acyltransferase [E.C. 2.3.1.15]. (2) 3-*sn*-Phosphatidate phosphohydrolase [E.C. 3.1.3.4]. (3) CTP:Phosphatidate cytidylyltransferase [E.C. 2.7.7.41], Mn(II). (4) CDP-Choline:-DG cholinephosphotransferase [E.C. 2.7.8.2], Mg(II). (5) CDP-Ethanolamine:DG ethanolaminephosphotransferase [E.C. 2.7.8.1], Mg(II). (6) CDP-DG:L-serine 3-*O*-phosphatidyltransferase [E.C. 2.7.8.8], Mn (II). (7) CDP-DG:*myo*-Inositol 3-phosphatidyltransferase [E.C. 2.7.8.11], Mn(II). (8a) CDP-DG:*sn*-glycerol 3-phosphate 3-phosphatidyltransferase [E.C. 2.7.8.5]. (8b) Phosphatidylglycerophosphate phosphohydrolase [E.C. 3.1.3.27]. (11) Phosphatidylethanolamine *N*-acyltransferase.[1360]

Table 4-1 Physical Properties of Some Plant Phospholipids

Arachis hypogaea L. (Peanut), seed	
PI	Ba pptn. sodium salt: crystalline solid[1364]
Beta vulgaris L. (sugarbeet), tuber	
PG	^{13}C NMR (fig.)[1365]
Glycine max (L.) Merr. (soybean), bean	
PC	^{31}P NMR (fig.),[1366] TGA.[1367] Prep. HPLC, FT-IR (fig.)[1368]
PE	^{31}P NMR (fig.),[1366] TGA.[1367]
PI	^{31}P NMR (fig.),[1366] FAB-MS (fig.),[1369] FTICR-MS/MS.[1370] DEAE cellulose and silicic acid chromatography, IR (fig.), PC, TLC.[1371] Ammonium salt (1:1): powder, $[\alpha]_D^{27}$ +6.0° ($CHCl_3$), silicic acid column chromatography.[1372] Na salt (1:1); small colorless needles[1372]
PA	^{31}P NMR (fig.)[1366]
NAPE	Triethylamine salt: light pale viscous oil, $[\alpha]_D^{22}$ +4.7° ($CHCl_3$), IR, TLC[1373]
LPC, LPE, LPI, LPA	^{31}P NMR (fig.)[1366]
Gossypium hirsutum L. (Cotton), seed kernel	
PC	$[\alpha]_D$ +7.7° ($CHCl_3$), IR, prep. TLC[1374]
PE	$[\alpha]_D$ +6.1° ($CHCl_3$), IR, prep. TLC[1374]
PI	$[\alpha]_D$ +3° ($CHCl_3$),[2403] IR, prep. TLC[1374]
Helianthus annuus L. (Sunflower), seed kernel	
PC	Light yellow rubberlike substance, prep. TLC,[1375] $[\alpha]_D^{27}$ +7.45° ($CHCl_3$), IR[1376]
PE	Light yellow waxy substance, prep. TLC,[1375] $[\alpha]_D{}^{27}$ +6.1° ($CHCl_3$), IR[1376]
PI	Almost colorless pulverulent substance, prep. TLC,[1375] $[\alpha]_D^{27}$ +9.7° ($CHCl_3$), IR[1376]
PA	Prep. TLC,[1375] IR[1376]
Hibiscus cannabinus L. (Kenaf), seed kernel	
PC	$[\alpha]_D^{20}$ +6.3° ($CHCl_3$), prep. TLC[1377]
Oryza sativa L. (Rice), polished grain	
LPC	$CdCl_2$ pptn., colorless needles, s. ~100 °C, dec. 262-264° C with gassing, $[\alpha]_D^{20}$ −4.52° ($CHCl_3$), sol. water, methanol, ethanol, glacial acetic acid, $CHCl_3$, pyridine, insol. ether, petr. ether, acetone. Aqueous solution, neutral with an irritating astringent taste; poisonous[1378]
Spinacia oleracea L. (Spinach), leaf	
PC	IR (fig.), DEAE cellulose column chromatography, countercurrent distribution[1379]
PE	IR (fig.), DEAE cellulose column chromatography, countercurrent distribution[1379]
PG	IR (fig.), DEAE cellulose column chromatography, countercurrent distribution.[1379] Ammonium salt: $[\alpha]_D^{25}$ +1.0° ($CHCl_3$)[1380]
PI	IR (fig.), DEAE cellulose column chromatography, countercurrent distribution[1379]
Triticum aestivum L. (Wheat), seed germ	
PI	Ba pptn., insol. ethanol. Sodium salt: White needles, mp 145 °C (s. 125-130 °C), sol. $CHCl_3$, CCl_4, benzene, slightly sol. methanol, insol. ether, petr. ether, acetone; viscous colloidal dispersion in water[1109]; photoelectron spectrum (fig.)[1381]
PA	Monopotassium salt: $[\alpha]_D^{20}$ +9.4° (CH_3OH), very sol. ether, $CHCl_3$, sol. methanol (~2%), insol. ethanol; PC, TLC[1382]
DPG	Oil, very sol. ethanol; PC, TLC. Na salt: solid, slightly sol. ethanol. Ba salt, gelatinous soln.in dry ether[1383]
Vigna mungo (L.) Hep. (Black gram), seed	
NAPE	IR (fig.), TLC, HPTLC[1384]
NALPE	IR (fig.), TLC, HPTLC[1384]

The phosphatidylglycerols (PG) contain two such chiral centers, one in each glycerol moiety, and are therefore capable of existing in four configurations, which by convention are labeled *R,S*; *R,R*; *S,R*; and *S,S*. In recent work in which the stereoisomers were separated by chiral-phase HPLC, the PG from cabbage leaf and soybean was found to consist solely of R,S isomers, whereas those of *Escherichia coli* and egg yolk consisted of 89:11% and 45:55% mixtures of R,S and R,R isomers, respectively[1385]:

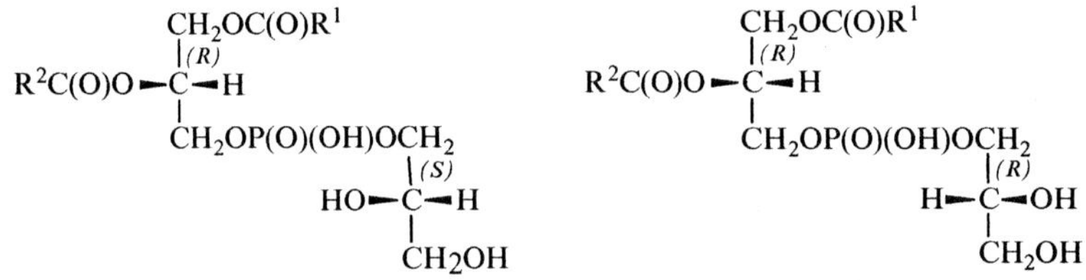

R,S (plant, microbial, animal) *R,R* (microbial, animal)

Physical Properties

Physical properties have been reported for a few phospholipids derived from plant sources without regard to the presence therein of acyl groups of different chain lengths or degrees of unsaturation. Some of these are given in Table 4-1.

Extraction

The most widely used procedures for the extraction of lipids are those of Folch *et al.*[1386] and Bligh and Dyer,[1387] modified for use with plant tissues. Both procedures employ chloroform-methanol (2:1 or 1:1) solvent mixtures, followed by backwashing with water or salt solution to remove nonlipid contaminants. A preliminary heat treatment with boiling water, boiling 2-propanol[1356] or the like is recommended to inactivate enzymes such as phospholipase D, which can hydrolyze phospholipids to phosphatidic acid or react with methanol to form phosphatidylmethanol as an unwanted artifact.[1388,1389]

> *For quantitative analyses of the constituent phospholipids, weighed amounts of cotyledons or seed coats were steamed for 15 min prior to comminution in 20 vol of chloroform-methanol (C/M 2:1, v/v) using a Polytron PT 20 homogenizer (Kinematica, Lucerne, Switzerland). Filtrates from these homogenates were partitioned against 0.2 vol of 0.9 % (w/v) NaCl .... Filtered homogenates plus washings were brought to 2:1 (v/v) or 1:1 (v/v) with respect to chloroform and methanol and were washed according to Folch* et al.[1386] *and Bligh and Dyer,*[1387] *respectively. Washed lipid extracts were dried on a rotary evaporator and the dry lipid redissolved in chloroform so that 1 ml of lipid solution was equivalent to 1 g fresh weight of cotyledons.*

Roughan, P. G.; Slack, C. R.; Holland, R., "Generation of Phospholipid Artefacts during Extraction of Developing Soybean Seeds with Methanolic Solvents", *Lipids* **1978**, *13*(7), 497–503,[1388] reprinted by permission of Springer, © 1978.

The combined lipids may be subjected to further purification by column chromatography on Sephadex G-25, which removes any carbohydrate or protein that may have coextracted with the lipid,[1390] or by thin-layer chromatography on silica gel, which removes any pigments that may be present.[1391] In some cases, more vigorous conditions are required. Lipids of seed tissues such as wheat bran, embryo, or whole grain, for example, are extracted more efficiently by boiling water-saturated n-butanol than by methanol/chloroform or 2-propanol/chloroform.[1392,1393] Supercritical fluid extraction of oilseeds with CO_2 has received much attention in recent years because the extract is free of solvent and low in free fatty acids or pigments. It can be applied to the extraction of phospholipids if a cosolvent is used (such as ethanol) because the phospholipids are not soluble in neat CO_2.[1394,1395]

Separation and Analysis

Plant tissues, in contrast to animal or microbial tissues, tend to have much lower amounts of phospholipids in relation to neutral or nonpolar lipids. For this reason, it is usually advisable to do a preliminary fractionation to remove the bulk of the nonpolar lipids from the extract. A widely used procedure is column chromatography on silicic acid. Successive elution with chloroform, acetone, and then methanol separates the lipid extract into neutral lipids, glycolipids, and phospholipids, respectively. Some of the PA and DPG, however, may be carried into the acetone fraction.[1349a]

> *The lipid preparations, dissolved in diethyl ether, were applied to columns of Mallinckrodt silicic acid (250 parts by weight; activated at 110° overnight) and the "neutral" lipids (hydrocarbons, sterol esters, triglycerides, free fatty acids, diglycerides and sterols) and most of the pigments were eluted with diethyl ether. Polar lipids (glycolipids, phospholipids, sterol glycosides) and residual pigment were then displaced with ether-methanol (1:1, v/v) and methanol. Positive N_2 pressure was generally applied to the columns and the separation of neutral and polar lipids from a total of about 250 mg lipid took less than 3 h.*
>
> Nichols, B. W., "Separation of the Lipids of Photosynthetic Tissues: Improvements in Analysis by Thin-Layer Chromatography", *Biochim. Biophys. Acta* **1963**, *70*(4), 417–422,[1396] reprinted by permission of Elsevier, © 1963.

Older methods that are now seldom used are acetone precipitation,[1349b] dialysis,[1397] countercurrent distribution,[1379] and the use of other column supports such as DEAE cellulose,[1379] Sephadex LH-20,[1398] or alumina.

(1) Thin-Layer Chromatography (TLC) Method

Two-dimensional TLC has been the method of choice for the separation and quantitation of plant phospholipids since the early 1960s. The R_f values for the major phospholipids (PL) and lysophospholipids (LPL) are given in Table 4-2.

Table 4-2 Separation of Phospholipid Classes by TLC [1393]

	Dimension			Dimension	
PL	**1st**	**2nd**	**LPL**	**1st**	**2nd**
PC	0.37	0.28	LPC	0.13	0.09
PE	0.49	0.59	LPE	0.19	0.34
PG	0.46	0.71			
PI	0.09	0.39			
PS	0.19	0.34			
PA	0.01	0.77			
DPG	0.50	0.86			
NAPE	0.84	0.96	NALPE	0.66	0.82

The polar lipid fraction [from a lipid extract of soft winter wheat grain after silicic acid column chromatography] was evaporated to dryness under N_2 and redissolved in a small vol. of $CHCl_3$. Separation of the polar lipids into classes was achieved by 2-D TLC on Si gel G [using] $CHCl_3$-MeOH-ammonia (30 %, w/v) (13:5:1) for the 1st dimension and $CHCl_3$-Me_2CO-MeOH-HOAc-H_2O (6:8:2:2:1) for the 2nd dimension. The 1st solvent was removed under vac. Atm pres was reinstated with N_2 and chromatograms were immediately run in the 2nd solvent. Chromatograms were then dried in air and the spots were detected using I_2 vapour ... Phospholipids were tentatively identified by comparing their TLC R_f values with published data, by co-chromatography with authentic samples, and by their behaviour with various spray reagents. For the latter purpose, the developed chromatograms were sprayed with 0.3 % ninhydrin–5 % lutidine in H_2O satd. n-BuOH for amino groups, a modified Dragendorff reagent for choline, and a modified Zinzadze reagent for phosphate. The use of 0.6 % $K_2Cr_2O_7$ in 55 % H_2SO_4 as a general detection reagent also allowed the identification of some nonphosphorus polar lipids. The quantities of individual phospholipids were determined [as follows]. Chromatogram spots that had been located using I_2 vapour and Si gel blanks were aspired directly into 50 ml Kjeldahl flasks containing a liquid trap of 0.7 ml 70 % $HClO_4$. The contents of the flask were refluxed at 200° until clear and then cooled. 3.3 ml H_2O, 0.5 ml 2.5 % ammonium molybdate, and 0.5 ml 10 % ascorbic acid were then added. The mixture was transferred to a reductase tube, and heated for 5 min. at 100°. After cooling, the Si gel was removed by a brief centrifugation and the A at 800 nm of the supernatant was measured.

Colborne, A. J.; Laidman, D. L., "The Extraction and Analysis of Wheat Phospholipids", *Phytochemistry* **1975**, *14*(12), 2639–2645,[1393]

The reproducibility of the method was tested in a seven-laboratory collaborative study.[1399] The method was later adopted as an AOCS standard for phospholipids in animal or vegetable lecithins.[1400] Some other methods that have been employed for the quantitation of the phospholipid classes after TLC include (a) photodensitometry after charring with 50% H_2SO_4, cupric acetate, and the like, (b) automated scanning densitometry,[1401]

(c) scintillation counting for PLs labeled with [^{3}H],[1402] [^{14}C],[1403] or [^{32}P],[1389] (d) measurement of the absorbance at 720 nm after spraying with the Dittmer-Lester reagent,[1404] and (e) fluorescence detection.[1405] Automation of the TLC method is possible with the use of Chromarods in the Iatroscan TLC/FID system.[1406]

(2) High-Performance Liquid Chromatography (HPLC) Method

The HPLC method, introduced in 1981,[1407,1408] has become a useful alternative to TLC for the separation and quantitation of plant phospholipids. With few exceptions, the column packing is silica gel (5μ Porasil, LiChrosorb Si-60, etc.) and the mobile phase is *n*-hexane/2-propanol/water or acetonitrile/methanol/water. The order of elution is PE > PG > PI > PA > PS > PC with the former[1407] and PI > PS > PE > PC with the latter,[1368] but it can deviate considerably with changes in solvent ratios or gradients. In most cases, the phospholipids are detected by UV absorption at 206 nm and quantitated by phosphorus analysis,[1407] which can be brought online by means of an automatic phosphorus analyzer,[1409] or by transmethylation with MeOH/NaOMe and GC analysis of the FAME,[1410] as described later in this chapter. The reproducibility of the method for analysis of phospholipids in commercial soybean lecithin has been tested in two multilaboratory collaborative studies.[1411,1412]

> *The lipid sample was filtered through a 0.4μm pore size polycarbonate membrane . . .; the solvent was evaporated with a stream of nitrogen gas, and the sample was dissolved in 0.5 mL of chloroform. An aliquot containing ~1 mg of phospholipid was injected into the HPLC. Liquid chromatographic analysis was performed with a Waters' Associates liquid chromatograph [consisting of] two Model 6000A solvent pumps . . . a Model 660 solvent programmer for gradient elution . . . a U6K injector . . . a 30 cm × 3.9 mm Porasil column . . . packed with 10 μm diameter silica gel particles was used. The solvent system selected was a mixture of hexane, 2-propanol, and water. . . . Solvent composition was initially . . . 6:8:0.5 v/v/v. A linear increase in the water content during a 25-min period resulted in a final solvent composition of 6:8:1.5. Solvent flow was maintained at 1.5 mL/min. Detection of eluting components was by absorbance at 206 nm . . . . The phosphorus content of the fractions collected from the HPLC was determined by using Bartlett's modification (Bartlett, 1959) of the Fiske-SubbaRow method (Fiske and SubbaRow, 1925).*
>
> Reprinted with permission from Ashworth, E. N.; St. John, J. B.; Christiansen, M. N.; Patterson, G. W., "Characterization of the Phospholipid Composition of Wheat Roots Using High-Performance Liquid Chromatography", *J. Agric. Food Chem.* **1981**, *29*(4), 879–881,[1407] © American Chemical Society.

Direct quantitation of the phospholipids is possible if the UV detection is replaced by flame ionization (FID)[1413] or evaporative light scattering (ELS)[1414] detection. Other methods of detection that have been proposed are FTIR,[1368] plasmaspray ionization MS,[1415] and postcolumn derivatization.[1416]

Moreau *et al.*[1413] developed a protocol that separates all seven of the major phospholipids—DPG, PE, PG, PI, PA, PS, and PG—in order of elution as well as LPC in one 60-min run using a Lichrosorb Si-60 column and FID detection. NAPE and NALPE can be separated from each other and from other phospholipids by HPLC on Lichrosorb Si-60 with ELS detection.[1417] The phosphoinositides PI, PIP, and PIP_2 can be separated by HPLC on an amino-NP silica column with liquid scintillation detection.[1418]

Methods have been developed for the preparation of gram-scale quantities of PC, PE, and PI of high purity (>99%) from soybean lecithin using preparative HPLC columns of 4.6- to 50-mm inner diameter.[1419,1420] Flash chromatography of rapeseed lecithin also provides gram-scale separation of PC, PE, and PI. but only the PC is 99% pure.[1421]

The normal-phase HPLC procedures described here serve to separate the phospholipids by class. Later we shall see that the phospholipids of each class can be further separated into their component molecular species by reversed-phase HPLC.

(3) Micellar Electrokinetic Chromatography (MEKC) Method

This method, which is still in the developmental stage, provides an alternative to HPLC that has the advantage of much higher peak capacity resulting in improved resolution. It separates the phospholipids based on their partition coefficients between a mobile phase and a pseudostationary micellar phase. The mobile phase is usually 30% *n*-propanol and the micellar phase a buffered solution of cholic acid[1422] or deoxycholic acid.[1423] The method has been applied to commercial soybean and rapeseed lecithins.[1422,1423]

Analysis Without Separation

(1) ^{31}P Nuclear Magnetic Resonance (NMR) Method

This method, adopted in 1988, is useful in that it quantitates phospholipids directly by their phosphorus content. A crude soybean extract was separated into 11 phospholipid classes (7 PLs and 4 LPLs) in a single 24-h run.[1366] The ^{31}P chemical shift δ for each phospholipid (PL) and lysophospholipid (LPL) class is given in Table 4-3.

For example, seven phospholipid classes and four lysophospholipid classes were identified in a crude soybean extract in a single 24-h run.[1366]

Table 4-3 ^{31}P NMR of Phospholipid Classes[1366]

PL	δ	LPL	δ
PC	−0.84	LPC	−0.28
PE	−0.01	LPE	0.43
PG	0.52	LPG	1.09
PI	−0.37	LPI	0.10
PA	0.25	LPA	0.83
DPG	0.18		

Fatty Acid Distribution

Each class of phospholipid in a typical plant extract comprises long-chain fatty acids of variable chain lengths and degrees of unsaturation. The distribution of fatty acid residues within each class is readily determined by transesterification with methanol and analysis of the resulting mixture of fatty acid methyl esters (FAME) by gas chromatography. The following procedure is typical:

> *Bands were scraped off [the TLC plates] on to small chromatographic columns from which the ... phospholipids [were eluted] with methanol containing 5 per cent. of anhydrous hydrogen chloride; 1 ml of the standard solution of methyl heptadecanoate was added to each fraction .... The phospholipids [were] transesterified by refluxing for 2 hours in the methanolic hydrogen chloride .... Methyl esters were dissolved in hexane for ... gas liquid chromatographic analyses [which] were carried out on columns of 15 per cent w/v poly(ethyleneglycol adipate) on Chromosorb W ... in a Pye 104 chromatograph. The amount of each ester present (per cent w/w) was calculated from the product of the peak height and retention time, and results obtained were converted into mole per cent ....*
>
> Christie, W. W.; Noble, R. C.; Moore, J. H., "Determination of Lipid Classes by a Gas Chromatographic Procedure", *Analyst* **1970**, *95*(11), 940–944,[1424]
> reproduced by permission of the Royal Society of Chemistry, © 1970.

Morrison *et al.*[1425] have provided a list of factors that can be used to convert % P or % FAME to % phospholipid or % lysophospholipid (Table 4-4):

The FAME method has been applied to the majority of the phospholipids listed in Table 4-15. The transesterification is often carried out with acidic catalysts such as anhydrous HCl or sulfuric acid or with basic catalysts such as sodium methoxide or anhydrous potassium hydroxide; see Hammond[2091] for a discussion of the relative merits of each method.

Table 4-4 Conversion Factors for FAME

PL	Factor		LPL	Factor	
	P[a]	FAME[b]		P[a]	FAME[b]
PC	24.60	1.33	LPC	16.47	1.79
PE	23.22	1.26	LPE	15.08	1.64
PG	24.22	1.31	LPG	16.08	1.74
PI	27.09	1.48	LPI	18.95	2.05
PS	24.67	1.34	LPS	16.53	1.79
PA	21.83	1.18			
DPG	22.75	1.23			
NAPE	31.35	1.70	NALPE	23.22	2.52

[a]Weight P × factor = weight PL or LPL.
[b]Weight FAME × factor = weight PL or LPL.

Table 4-5 Fatty Acid Groups in Tomato Leaf PG (mol%)[2092]

	Acyl Group	Composition	Distribution C_1	Distribution C_2
16:0	Palmitoyl	17	9	8
16:1	*trans*-Hexadecenoyl	37	0	37
18:0	Stearoyl	1	1	0
18:1	Oleoyl	4	4	0
18:2	Linoleoyl	23	19	4
18:3	Linolenoyl	*18*	*16*	*2*
		100	49	51

Table 4-5 illustrates the method as applied to tomato leaf phosphatidylglycerol.[2092] The predominant fatty acids are *trans*-hexadecenoic acid, located exclusively on C_2; linoleic acid and linolenic acid, located mainly on C_1; and palmitic acid, evenly divided between C_1 and C_2. The distribution between C_1 and C_2 was determined by separate experiments in which the PG was hydrolyzed by lipase from the mold *Rhizopus delemar*, which cleaves the C_1 acyl group selectively. From these results it can be inferred that the main molecular species in tomato leaf PG are 1-linoleoyl- and 1-linolenoyl-2-*trans*-hexadecenoyl-*sn*-glycero-3-phospho-1′-glycerol, but the actual combinations must be determined by methods discussed later in this chapter.

The C_{16} and C_{18} fatty acids listed on the first page of this chapter comprise the bulk of the acyl groups found in higher plants, but they are by no means the only ones. Longer chain fatty acids such as arachidonic acid (20:4) are found in algae but are rare in the higher plants with the exception of phosphatidylserine, which contains significant amounts of C_{20} to C_{26} fatty acids.[1450] Shorter chain fatty acids such as lauric acid (12:0) or myristic acid (14:0) are found in significant amounts in mint leaf,[1470] cucurbit seed,[1476] and dill leaf PI,[1470] and in trace amounts in many other plant phospholipids.

Separation into Molecular Species

Each of the foregoing phospholipid classes comprises a range of individual compounds, the so-called molecular species, which contain fatty acid residues of variable chain lengths and degrees of unsaturation. In the molecular formula given at the beginning of this chapter, the nature of the "Z" group determines the class into which a particular phospholipid falls. In this section, we are concerned with the nature of the "R" groups, R^1 and R^2, which can be the same or different. Each individual compound in a phospholipid is defined by three variables, R^1, R^2, and Z.

(1) Argentation TLC Method

Silver ion has a retarding effect on the TLC of unsaturated compounds owing to the formation of a reversible complex between the silver ion and the double bond. This effect has

Table 4-6 Argentation TLC

CC	PE (intact)[a]	PC (MADG)[b]	Molecular species
0			
1	0.60	0.88	16:0/16:1, 16:0/18:1
2	0.48	0.76, 0.82	16:0/18:2, 18:1/16:1, 18:1/18:1
3	0.30, 0.40	0.38, 0.65	16:0/18:3, 18:1/18:2
4	0.22	0.23	18:1/18:3, 18:2/18:2
5	0.12		18:2/18:3
6	0.05		18:3/18:3

[a]PE in perilla seed.[2094]
[b]PC in rape cell suspension culture.[2099]

been used to advantage in the separation of intact PC,[2093] PE,[2093,2094] or PG[1566,2095] of plant origin into four to six subclasses of different saturation levels. More often it has been applied to the separation of the 1,2-diacylglycerols (DG) or their acetates (MADG) remaining after removal of the phosphate head group by acetolysis[1845] or by hydrolysis with phospholipase C[2023,2096] followed in some cases by acetylation with acetic anhydride[2097] or acetyl chloride[2098] to prevent isomerization of the DG. Regardless of which variant is used, the fractions separate in order of increasing saturation (Table 4-6). If the subclass comprises DGs or intact PLs, its composition is almost always determined by GC of the fatty acid methyl esters (FAME) after transesterification with methanol/HCl.[2098] If the subclass comprises MADGs, its composition can also be determined by GC[2097] or GC/MS.[1904]

This procedure determines the pairings of the acyl groups in the phospholipids but not their relative positions. For this an additional step is needed, such as selective cleavage of the *sn*-1 acyl group of the DG or MADG with phospholipase A_2 followed by a second GC/FAME analysis.[2023] This method has been used in about 20 molecular species determinations (Table 4-6).

(2) High-Performance Liquid Chromatography (HPLC) Method

Reversed-phase HPLC offers a method for separating the phospholipids of each class (PC, PE, etc.) into its individual molecular species, often with baseline resolution, short columns, and isocratic solvents. In general, the retention time increases with increasing chain length and decreases with increasing unsaturation regardless of phospholipid class.[1410,1983] The following procedure is typical:

PC, PE, MGDG and DGDG were separated into molecular species on a 3.9 × 300 mm μ Bondapak C_{18} column in MeOH-H_2O-MeCN (90.5:7:2.5) at a flow rate of 1.5 ml/min .... Phospholipids and galactolipids were solubilized in 0.1 to 0.2 ml of EtOH or MeOH and portions of 10-25 μl (50-150 μg) were applied to the column. After elution, lipids were detected at 205 nm as described above.... Lipids from each fraction collected at the exit of HPLC column were extracted according to [Bligh and Dyer[1387]]. For quantification on a mass basis, an appropriate amount of an intl. standard

> *(heptadecanoic acid) was added to the pooled eluate representing each peak. Fatty acid methyl esters from each lipid then were prepared by* trans-*methylation ... and analyzed by GC ....*
>
> Demandre, C.; Tremolieres, A.; Justin, A.-M.; Mazliak, P., "Analysis of Molecular Species of Plant Polar Lipids by High Performance and Gas Liquid Chromatography", *Phytochemistry* **1985**, *24*(3), 481–485,[1410] reprinted by permission of Elsevier, © 1985.

Resolution is improved considerably if an ion-pairing reagent such as tetrabutylammonium phosphate is added to the mobile phase; this procedure is applicable to PC and PE,[2100] PI,[2101] PG and DPG,[2102] and PA.[2103] Direct quantitation of the molecular species is possible if the UV detector is replaced by a flame ionization detector (FID)[2104,2105] or by thermospray MS.[2106] An RI detector was used in the original paper on this method.[2107] Derivatization of the phospholipid prior to HPLC, for example by esterification of the P-OH group of PA[2108] or PG[2109] with methanol or of the C-NH_2 group of PE or PS with 2,4,6-trinitrobenzenesulfonic acid[2110] is possible but unnecessary.

(3) Gas-Liquid Chromatography (GLC) Method

This method, introduced in 1982,[2111] offers an alternative to HPLC and argentation TLC for separating and quantitating plant phospholipids (PC, PE, etc.) into their individual molecular species. Prior to separation, the phosphate group must be removed by hydrolysis with phospholipase C and the resulting diacylglycerol (DG) converted to its acetyl (MADG),[2112] trimethylsilyl (TMSDG)[2111] or *t*-butyldimethylsilyl (tBDMSDG)[2113] derivative. GLC chromatography on a capillary column separates the DG derivatives by carbon number, degree of unsaturation. or both, depending on whether the liquid phase is polar or nonpolar. For example, chromatography of TMS-DG from soybean PI on a 10-m capillary column coated with a polar liquid phase (Supelco SP-2330) at 250 °C gave 13 well-resolved peaks ranging from 16:0/16:0 to 18:2/18:3, whereas an 8-m fused capillary silica column coated with a nonpolar liquid phase (Hewlett-Packard SE-54) at temperatures programmed from 40 to 340 °C gave just two major peaks in the C_{32} to C_{42} range corresponding to various pairings of C_{16}/C_{18} and C_{18}/C_{18} acyl groups, where C_{16}=palmitic acid and C_{18}=stearic, oleic, vaccenic, linoleic or linolenic acid.[2114]

The molecular species are identified in most cases by comparison with the retention times of standards or with the relative retention time factors published by Myher and Kuksis,[2111] confirmed in some cases by GC/MS.[2112,2115] This method, like the first two, does not distinguish isomeric pairs such as 16:0/18:2 and 18:2/16:0. The relative positions of the acyl groups must be determined independently, for example by hydrolysis with phospholipase A_2 and GC(FAME) analysis of the fragments. The GLC method has been used in about 20 molecular species determinantions (Table 4-8).

(4) Mass Spectrometry (MS) Method

Conventional electron impact (EI) ionization techniques are unsuitable for the present purpose, but the development of "soft" ionization modes such as field desorption (FD), fast atom bombardment (FAB), or chemical ionization (CI) has enabled MS techniques to be applied to intact phospholipids such as PC, PE, PG, or PI. FD-MS of commercial soybean PC or PE gives molecular species profiles that agree with those of the HPLC or GLC methods. The formation of $[M+Na]^+$ ions can be suppressed by the addition of *p*-toluenesulfonic acid[2122] or enhanced by the addition of NaCl.[2134]

FAB-MS in either the positive or negative mode gives the $[M+H]^+$ or $[M-H]^-$ molecular ions of PC, PE, or PI, but the elevated background levels and increased level of surface reactions make it unsuitable for quantitative measurement.[2134] The background level can be reduced by a special technique that deposits most of the phospholipid on a matrix surface.[2135] Combination of negative-ion FAB-MS with collisional activation of daughter ions (CAD), which requires a triple sector mass spectrometer such as VG ZAB-3F or Kratos MS-50, provides much more complete compositional information. Over 95% of the molecular species reported by GLC for soybean PI were identified and characterized unambiguously by FAB-MS-MS, including low-level isomeric pairs such as 18:0/18:3 and 18:1/18:2. However, this method is also unsuitable for quantitative analysis.[1369,2136]

Isobutane CI-MS has been used to identify the molecular species of soybean PC, PE, and PI[2123,2137] and black mangrove PG.[2138] The $[MH]^+$ peaks are weak or nonexistent, but a method was developed for absolute quantitation based on the diglyceride ion DG^+ and internal standards. If two molecular species have the same molecular weight, they can be distinguished by tandem MS-MS daughter analysis.[2123] Other MS methods that have been applied to soybean PI are secondary ion mass spectrometry (SIMS)[2139] and ^{252}Cf plasma desorption mass spectrometry (PD-MS).[2140] In one instance, the MS method was applied to spinach PC or PG after removal of the phosphate group by hydrolysis with phospholipase C and conversion to the acetyl derivative MADG.[2141] In all, MS methods have been used in about 12 molecular species determinations (Table 4-8).

(5) Random Distribution (1R,2R) Method

If the acyl group distributions at the *sn*-1 and *sn*-2 positions of a phospholipid are known, the molecular species pairings can be calculated by the 1-random-2-random method, which assumes that each pairing is determined solely by the relative abundance of its partners. The validity of this assumption can be tested by comparing the calculated values with those found experimentally by any of the foregoing methods. Some authors[1845,2023,2094,2114] have reported good agreement while others[2096] have not.

The procedure is best illustrated by a concrete example (Table 4-7, taken from Myher and Kuksis[2114]). Analysis of a purified commercial soybean PI by the GLC/FAME method together with selective hydrolysis of the PI with phospholipase C and independently with phospholipase A_2 provide the data shown in Table 4-7, columns 2 and 3. The data shown in columns 4-7 were calculated from various pairings of those of columns 2 and 3 by using the expression mol% DG = mol% (*sn*-1)(*sn*-2)/100; for example, mol% 16:0:18:2 = (67.4)(75.2)/100 = 50.7.

The data in parentheses represent the corresponding values obtained experimentally by the GLC/TMS method.

Table 4-7 Molecular Species in Soybean PI

Fatty acid	Position		1-random-2-random distribution[a]			
	sn-1	*sn*-2	16:0	18:1	18:2	18:3
16:0	67.4	5.9	4.0 (0.1)	4.8 (4.3)	50.7 (57.3)	5.9 (6.8)
18:0	15.8	2.1	0.9	1.1 (1.0)	11.9 (13.1)	1.4 (1.3)
18:1	3.7	7.1	0.2	0.3	2.8 (3.6)	0.3
18:2	7.7	75.2	0.4	0.5	5.8 (6.5)[b]	0.7 (1.0)
18.3	0.8	8.7	0.05	0.06	0.6	0.07

[a]Experimental data in parentheses. Data include reverse isomers.
[b]Including 18:1/18:3 and 18:3/18:1, indistinguishable by this method from 18:2/18:2.

Lysophospholipids

The terms *lysolecithin* (LPC) and *lysocephalin* (LPE) were coined by Levene and Rolf[2142] in 1923 to describe the products obtained by the action of cobra venom on egg yolk PC and PE, respectively. This group now comprises the lysophospholipids LPC, LPE, LPG, LPI, LPS, and LPA and the lysophospholipid derivatives LPIP and NALPE. Each of these compounds can occur in one of two forms, depending on which acyl group is removed from the parent phospholipid. If the terminal acyl group is split off, the product is a 1-lysophospholipid, but if the middle acyl group is split off, the product is a 2-lysophospholipid:

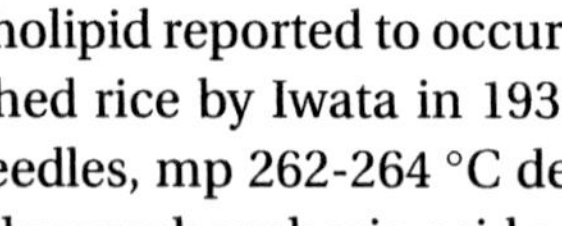

1-LPL

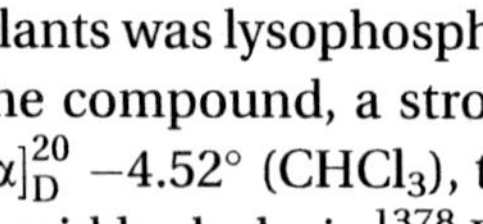

2-LPL

Occurrence

The first lysophospholipid reported to occur in plants was lysophosphatidylcholine (LPC), isolated from polished rice by Iwata in 1930. The compound, a strong hemolytic agent, formed colorless needles, mp 262-264 °C dec., $[\alpha]_D^{20}$ −4.52° ($CHCl_3$), that yielded palmitic acid, choline, and glycerophosphoric acid upon acid hydrolysis.[1378] LPC was soon shown to be widely distributed among plant seeds,[2143] but the other LPLs were not reported until much later—NALPE (1965) in wheat flour,[1358] LPE and LPS (1967) in wheat starch,[2077] LPI (1971) in several cereal starches,[1499] and LPG (1979) in corn starch.[2074] The occurrence of LPLs in cereal seeds is now well documented,[2144] but their occurrence in leaves, stems, or roots (listed in Table 4-8) remains questionable. Plant cells contain the enzymes phospholipase A_1 and A_2, which, if not inactivated, can degrade any of the phospholipids to 1-lyso- and 2-lysophospholipids, respectively, during extraction and isolation. These enzymes can be inactivated by steaming or by

LPL	ZOH component	RCO$_2$H component	
LPA	H	16:0	Palmitic acid
LPC	Choline	18:0	Stearic acid
LPE	Ethanolamine	18:1	Oleic acid
LPG	Glycerol	18:2	Linoleic acid
LPS	L-Serine	18:3	Linolenic acid
LPI	*myo*-Inositol		
LPIP	*myo*-Inositol 4-phosphate	and other long-chain fatty acids	
NALPE	*N*-Acylethanolamine		

extraction with hot 2-propanol, but in some cases, as for example leaf plasma membrane preparations, which require fractionation prior to extraction, some phospholipid degradation is inevitable.

Physical Properties

Physical properties have been reported for a few lysophospholipids derived from plant sources without regard to the presence therein of acyl groups of different chain lengths or degrees of unsaturation. Some of these are reported in Table 4-1.

Extraction

The lysophospholipids are incompletely extracted by the solvents usually employed to extract phospholipids, such as chloroform/methanol, but are efficiently extracted by water-saturated *n*-butanol (WSB) at 90-100 °C. There is no perceptible lipid degradation if the reaction is carried out under nitrogen, and care is taken to inactivate the phospholipases.[1393] The preferred procedure for plant tissues such as wheat starch, which is rich in lysophospholipids, is to remove the PLs by extraction with WSB or other solvent mixture for 10 min at 20 °C, followed by a 3-h extraction at 90-100 °C to remove the LPLs.[1392]

Separation and Analysis

(1) Thin-Layer Chromatography (TLC) Method

The method of Colborne and Laidman[1393] described on p 155 separates both Pls and LPLs in a single run. The lysophospholipids have much lower R_f values than the phospholipids in both dimensions (Table 4-2).

(2) High-Performance Liquid Chromatography (HPLC) Method

The separation of wheat LPLs on a 7μ Lichrosorb Si-60 column, hexane/THF/2-propanol/water mobile phase, and ELS detection gives, in order of elution, NALPE > LPE > LPC for the nonstarch lipids and LPE > LPG > LPC for the starch lipids.[1417] LPC, LPI, and LPS can be separated on an aminopropyl silica column with ethanol/acetonitrile/10m*M* ammonium phosphate buffer as the mobile phase.[1416]

PHOSPHATIDYLCHOLINES

This group comprises the phosphatidylcholines (PC) and the lysophosphatidylcholines (LPC).

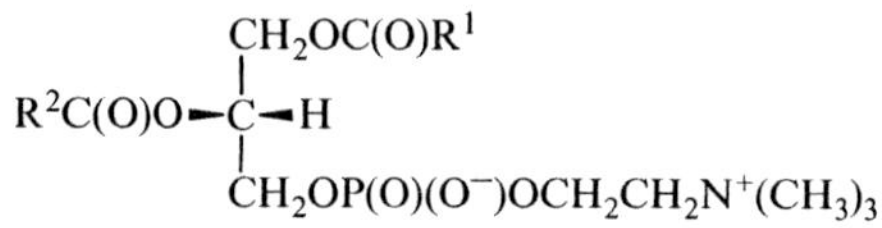

PC, $R^1 = R^2$ = alkyl
1-LPC, R^1 = H, R^2 = alkyl
2-LPC, R^1 = alkyl, R^2 = H

Phosphatidylcholines

Physical Properties

1,2-Dipalmitoyl-*sn*-glycero-3-phosphocholine (DPPC), 16:0/16:0 PC, $C_{40}H_{80}NO_8P$. *(R)-4-Hydroxy-N,N,N-trimethyl-10-oxo-7-[(1-oxohexadecyl)oxy]-3,5,9-trioxa-4-phosphapentacosan-1-aminium Hydroxide, inner salt, 4-Oxide.*

DPPC [R.N. 63-89-8]: Plates, X-ray[2145]; electron diffraction,[2146] ORD and CD (figs.)[2147]; conformation.[2148] Spectra: IR,[2149] Raman (fig.).[2150] EI-MS (fig.),[2151] FAB-MS (fig.),[2152] FD-MS (fig.),[2122] CIMS,[2153] SIMS (fig.),[2152] LSIMS,[2154] FAB MS/MS,[2155] and MALDI FT-ICR-MS.[1370] Chromatography: SFC.[2156] The anhydrous salt is over 10 times more soluble than its hydrate in solvents such as benzene.[2145] *Hydrate* [R.N. 60422-19-7]: Crystals, mp 235-236 °C, $[\alpha]_D^{25}$ +7.0° ($CHCl_3$),[2157] soluble in chloroform and hot dioxane, very slightly soluble in other solvents[2158]; countercurrent distribution[2159]; fluorescence polarization[1631]; DSC.[2160] Spectra: IR (fig.),[2161] FTIR,[2162] ^1H and ^{13}C NMR (figs.),[118] ^{31}P NMR (fig.),[2163] ^{14}N NMR (fig.),[2164] X-ray (photo.),[2158] PES.[1381] Chromatography: PC,[2165] TLC,[2166] HPTLC,[1405] HPLC.[2107] *Dihydrate* [R.N. 68573-30-8]: Orthorhombic crystals, X-ray,[2167] DTA.[2167] Spectra: Raman,[2168] ^{13}C and ^{31}P CP-MAS NMR (figs.).[2169]

$[CdCl_2]_3[DPPC{\cdot}H_2O]_2$: Amorphous solid.[2158]

1-Palmitoyl-2-oleoyl-*sn*-glycero-3-phosphocholine (POPC), 16:0/18:1 PC, $C_{42}H_{82}NO_8P$. *[R-(Z)]-4-Hydroxy-N,N,N-trimethyl-9-oxo-7-[[(1-oxohexadecyl)oxy]methyl]-3,5,8-trioxa-4-phosphahexacos-17-en-1-aminium Hydroxide, inner salt, 4-Oxide.*

POPC [R.N.26853-31-6]: $[\alpha]_D^{25}$ +6.3° ($CHCl_3$/MeOH),[2170] CD.[2147] Spectra: FTIR,[2171] Raman,[2172] ^1H and ^{13}C NMR (figs.),[2173] 2d-[^1H][^{13}C] NMR (fig.),[2174] EI-MS (fig.),[2175] CI-MS (fig.), [2153] FAB-MS (fig.),[2176] FD-MS,[2177] PD-MS (fig.),[2140] MALDI-MS (fig.),[2178] LSI-MS (fig.),[2139] FAB-MS/MS (fig.),[2176] MALDI FT-ICR MS/MS (fig.),[1370] CID-MS/MS.[2179] Chromatography: Silicic acid column,[2170] HPLC,[2107] HPLC-CIMS.[2180] Electrophoresis: CE (fig.).[2181] *Hemihydrate*: mp 225-230 °C, $[\alpha]_D^{20}$ −7.57° [*sic*] ($CHCl_3$/MeOH),[2182] DSC.[2183] Spectra: ESR.[2184] Chromatography: TLC,[2166] silica gel column.[2182]

1-Palmitoyl-2-linoleoyl-*sn*-glycero-3-phosphocholine (PLPC), 16:0/18:2 PC, $C_{42}H_{80}NO_8P$. *[R-(Z,Z)]-4-Hydroxy-N,N,N-trimethyl-9-oxo-7-[[(1-oxohexadecyl)oxy]methyl]-3,5,8-trioxa-4-phosphahexacosa-17,20-dien-1-aminium Hydroxide, inner salt, 4-Oxide.*

PLPC [R.N. 17708-90-6; see also 6931-84-6]: Pale yellow oil.[1788] Spectra: ^{13}C and ^{31}P NMR,[2083] EI-MS,[2185] FD-MS,[2177] FAB-MS and FAB-MS/MS,[2176] SIMS,[2139] CID-MS-MS.[2179] Chromatography: HPLC.[2107]

1-Palmitoyl-2-linolenoyl-*sn*-glycero-3-phosphocholine (PLnPC), 16:0/18:3 PC, $C_{42}H_{78}NO_8P$. *[R-(Z,Z,Z)]-4-Hydroxy-N,N,N-trimethyl-9-oxo-7-[[(1-oxohexadecyl)oxy]methyl]-3,5,8-trioxa-4-phosphahexacosa-17,20,23-trien-1-aminium Hydroxide, inner salt, 4-Oxide.*

PLnPC [R.N. 35418-56-5 or 95268-94-3]: $[\alpha]_D^{20}$ −2.9° ($CHCl_3$).[2186] Spectra: ^1H and ^{31}P NMR.[2186] Chromatography: Ag-TLC,[2099] HPLC.[2100]

1-Stearoyl-2-oleoyl-*sn*-glycero-3-phosphocholine (SOPC), 18:0/18:1 PC, $C_{44}H_{86}NO_8P$. *[R-(Z)]-4-Hydroxy-N,N,N-trimethyl-10-oxo-7-[(1-oxo-9-octadecenyl)oxy]-3,5,9-trioxa-4-phosphaheptacosan-1-aminium Hydroxide, inner salt, 4-Oxide.*

SOPC [R.N. 56421-10-4]: EI-MS (fig.),[2151] FD-MS,[2122] FAB-MS and FAB-MS/MS,[2176] CID-MS-MS.[2179] Chromatography: LC-MS (fig.).[2187] *Monohydrate*: White solid, mp 230-231 °C, $[\alpha]_D^{20}$ +6.0° ($CHCl_3$), soluble in $CHCl_3$, methanol, ethanol, and ether, fairly soluble in acetone, insoluble in water[2188]; countercurrent distribution[2159]; DTA[2189]; DSC[2190]; monolayer film properties.[2191] Spectra: IR (fig.),[2188] ^{13}C NMR.[2192] Chromatography: PC, TLC, silicic acid column,[2188] HPLC.[2107]
$[CdCl_2]_3[SOPC{\cdot}H_2O]_2$: Solid.[2188]

1-Stearoyl-2-linoleoyl-*sn*-glycero-3-phosphocholine (SLPC), 18:0/18:2 PC, $C_{44}H_{84}NO_8P$. *[R-(Z,Z)]-4-Hydroxy-N,N,N-trimethyl-10-oxo-7-[(1-oxo-9,12-octadecadienyl)oxy]-3,5,9-trioxa-4-phosphaheptacosan-1-aminium Hydroxide, inner salt, 4-Oxide.*

SLPC [R.N. 27098-24-4]: Spectra: FD-MS,[2122] LIMS (fig.),[2154] CID-MS-MS.[2179] Chromatography: HPLC.[2100] *Hydrate*: mp 163.0-165.0 °C, $[\alpha]_D^{20}$ +5.02° ($CHCl_3$),[2193] monolayer film properties,[2191] DSC.[2194] Spectra: UV,[2194] IR (fig.),[2193] ^{13}C NMR (fig.).[2195] Chromatography: PC,[2193] TLC,[2196] silicic acid column.[2195]

1-Stearoyl-2-linolenoyl-*sn*-glycero-3-phosphocholine (SLnPC), 18:0/18:3 PC, $C_{44}H_{82}NO_8P$. *[R-(Z,Z,Z)]-4-Hydroxy-N,N,N-trimethyl-10-oxo-7-[(1-oxo-9,12,15-octadecatrienyl)oxy]-3,5,9-trioxa-4-phosphaheptacosan-1-aminium Hydroxide, inner salt, 4-Oxide.*

SLnPC [R.N. 35418-57-6; see also 40811-59-4]: $[\alpha]_D^{20}$ −6.4° ($CHCl_3$).[2186] Spectra: ^1H and ^{31}P NMR.[2186] Chromatography: HPLC.[2051] *Hydrate*: Monolayer film properties,[2191] DSC.[2194] Spectra: UV.[2194] Chromatography: TLC, silicic acid column.[2194]

1,2-Dioleoyl-*sn*-glycero-3-phosphocholine (DOPC), 18:1/18:1 PC, $C_{44}H_{84}NO_8P$. *[R-(Z,Z)]-4-Hydroxy-N,N,N-trimethyl-10-oxo-7-[(1-oxo-9-octadecenyl)oxy]-3,5,9-trioxa-4-phosphaheptacos-18-en-1-aminium Hydroxide, inner salt, 4-Oxide.*

DOPC [R.N. 4235-95-4; see also 10015-85-7]: Colorless waxlike substance, soluble in chloroform and ethanol, less soluble in acetone, insoluble in ether, water[2197]; $[\alpha]_D^{22}$ +5.4° (1:1 $CHCl_3$/MeOH),[2197] countercurrent distribution.[2159] CD (fig.).[2147] Spectra: IR (fig.),[2198] Raman (fig.).[2199] PES,[1381] EI-MS (fig.),[2151] FAB-MS,[2200] FD-MS,[2177] CIMS,[2153]

LSIMS (fig.).[2154] Chromatography: PC,[2197] HPTLC,[1405] HPLC,[2107] silicic acid column,[2197] silica gel column.[2201] *Hydrate*: Waxlike substance, readily soluble in chloroform, methanol, ethanol, ether, and 90% acetone, moderately soluble in warm petroleum ether[2202]; $[\alpha]_D^{25}$ +6.2° (1:1 $CHCl_3$/MeOH),[2202] monolayer film properties,[2191] DSC.[2190] Spectra: IR (fig.), [2202] ^{1}H NMR (fig.),[2203] ^{13}C NMR,[2083] ^{31}P NMR (fig.),[2163] ^{14}N NMR.[2204] Chromatography: TLC,[2166] silicic acid column.[2202]
$[CdCl_2]_3[DOPC.H_2O]_2$: Solid,[2197] $[\alpha]_D$ +3.7° ($CHCl_3$).[2202]

1-Oleoyl-2-linoleoyl-*sn*-glycero-3-phosphocholine (OLPC), 18:1/18:2 PC, $C_{44}H_{82}NO_8P$. *[R-(Z,Z,Z)]-4-Hydroxy-N,N,N-trimethyl-10-oxo-7-[(1-oxo-9,12-octadecadienyl)oxy]-3,5,9-trioxa-4-phosphaheptacos-18-en-1-aminium Hydroxide, inner salt, 4-Oxide.*
OLPC [R.N. 17041-44-0; see also 99166-28-6]: FD-MS,[2122] HPLC.[2100]

1-Oleoyl-2-linolenoyl-*sn*-glycero-3-phosphocholine (OLnPC), 18:1/18:3 PC, $C_{44}H_{80}NO_8P$. *[R-(all Z)]-4-Hydroxy-N,N,N-trimethyl-10-oxo-7-[(1-oxo-9,12,15-octadecatrienyl)oxy]-3,5,9-trioxa-4-phosphaheptacos-18-en-1-aminium Hydroxide, inner salt, 4-Oxide.*
OLnPC [R.N. 105017-24-1; see also 74936-57-5, 96497-78-8, 120502-93-4]: HPLC.[2105]

1,2-Dilinoleoyl-*sn*-glycero-3-phosphocholine (DLPC), 18:2/18:2 PC, $C_{44}H_{80}NO_8P$. *[R-(all Z)]-4-Hydroxy-N,N,N-trimethyl-10-oxo-7-[(1-oxo-9,12-octadecadienyl)oxy]-3,5,9-trioxa-4-phosphaheptacosa-18,21-dien-1-aminium Hydroxide, inner salt, 4-Oxide.*
DLPC [R.N. 998-06-1; see also 6542-05-8, 81902-03-3]: Glass,[2201] solubility in $CHCl_3$ 0.8 mol/L at 20 °C.[2205] Spectra: IR (fig.),[2205] FD-MS (fig.),[2177] CIMS,[2153] LIMS.[2154] Chromatography: Silica gel column.[2201] *Hydrate*: $[\alpha]_D^{25}$ +6.02° ($CHCl_3$),[2206] DSC.[2207] Spectra: UV, [2207] IR (fig.),[2206] ^{1}H NMR,[2208] ^{13}C and ^{31}P NMR.[2083] Chromatography: PC,[2206] TLC,[2166] silicic acid column,[2206] HPTLC.[1405]

1-Linoleoyl-2-linolenoyl-*sn*-glycero-3-phosphocholine (LLnPC), 18:2/18:3 PC, $C_{44}H_{78}NO_8P$. *[R-(all Z)]-4-Hydroxy-N,N,N-trimethyl-10-oxo-7-[(1-oxo-9,12,15-octadecatrienyl)oxy]-3,5,9-trioxa-4-phosphaheptacosa-18,21-dien-1-aminium Hydroxide, inner salt, 4-Oxide.*
LLnPC [R.N. 85046-19-1; see also 105120-83-0]: FD-MS,[2122] CIMS,[2153] HPLC.[2100]

1,2-Dilinolenoyl-*sn*-glycero-3-phosphocholine (DLnPC), 18:3/18:3 PC, $C_{44}H_{76}NO_8P$. *[R-(all Z)]-4-Hydroxy-N,N,N-trimethyl-10-oxo-7-[(1-oxo-9,12,15-octadecatrienyl)oxy]-3,5,9-trioxa-4-phosphaheptacosa-18,21,24-trien-1-aminium Hydroxide, inner salt, 4-Oxide.*
DLnPC [R.N. 2701-19-1; see also 40811-59-4]: DSC.[2207] Spectra: UV,[2209] ^{1}H NMR (fig.),[2209] FD-MS,[2177] CIMS.[2153] Chromatography: TLC,[2209] HPTLC,[1405] HPLC.[2100]

Molecular Species Composition

The molecular species composition of phosphatidylcholine in plants, determined by any of the methods described on pages 168–171, is given in Table 4-8.

Synthesis

DPPC may be prepared (a) by phosphorylation of 1,2-dipalmitoyl-*sn*-glycerol with phenyl phosphorodichloridate,[2158] methyl *N,N*-diisopropylaminophosphorochloridite,[2210] or

PCl_3/imidazole[2211] followed by choline chloride or choline tosylate, (b) by base-catalyzed acylation of *sn*-glycero-3-phosphocholine or its $CdCl_2$ complex with palmitoyl chloride,[2212] palmitic anhydride,[2198] or *N*-palmitoylimidazole,[2213] or (c) by stepwise reaction of 1,2-dipalmitoyl-*sn*-glycerol with 2-bromoethyl phosphorodichloridate[2214] or 2-chloro-2-oxo-1,3,2-dioxaphospholane[2215] and trimethylamine. POPC may be prepared (a) by acylation of 1-palmitoyl-*sn*-glycero-3-phosphocholine with oleoyl chloride[2170] or oleic anhydride,[2173] or with 2-thiopyridyl oleate in the presence of silver perchlorate,[2216] (b) by acylation of 1-*O*-trityl-2-oleoyl-*sn*-glycero-3-phosphocholine with palmitic anhydride or *N*-palmitoyltriazolide,[2182] or (c) by reaction of the methyl ester of 1-palmitoyl-2-oleoyl-*sn*-glycero-3-phosphoethylene bromohydrin with trimethylamine.[2217] PLPC may be prepared by acylation of 1-palmitoyl-*sn*-glycero-3-phosphocholine with linoleic acid in the presence of *N,N'*-carbonyldiimidazole[289] or with 2-thiopyridyl linoleate in the presence of silver perchlorate.[2216] PLnPC may be prepared by acylation of 1-palmitoyl-*sn*-glycero-3-phosphocholine with linolenic anhydride[2186] or of its $CdCl_2$ complex with linolenoyl chloride.[2218]

SOPC may be prepared by the reaction of silver benzyl 1-stearoyl-2-oleoyl-*sn*-glycero-phosphate with 2-bromoethyltrimethylammonium picrate.[2188] SLPC may be prepared (a) by phosphorylation of 1-stearoyl-2-linoleoyl-*sn*-glycerol with $POCl_3$ and choline chloride,[2193] (b) by condensation of 1-stearoyl-2-linoleoyl-*sn*-glycerol 3-iodohydrin with silver 2-chloroethyl *p*-nitrobenzyl phosphate,[2196] or (c) by acylation of 1-stearoyl-*sn*-glycero-3-phosphocholine (SHPC) with linoleic anhydride.[2194] SLnPC may be prepared by acylation of 1-stearoyl-*sn*-glycero-3-phosphocholine with linolenic anhydride[2186] or of its $CdCl_2$ complex with linolenoyl chloride[2218] or linolenic anhydride.[2194]

DOPC may be prepared (a) by acylation of *sn*-glycero-3-phosphocholine with oleoyl chloride in the presence of $CdCl_2$,[2219] with oleic anhydride in the presence of tetraethylammonium oleate,[2220] sodium oleate,[2198] or 4-pyrrolidinopyridine,[2221] or with oleic imidazolide in the presence[2209] or absence[2201] of a strong base, (b) by phosphorylation of 1,2-dioleoyl-*sn*-glycerol with $POCl_3$ and choline chloride (or iodide),[2222] or with 2-chloro-2-oxo-1,3,2-dioxaphospholane and trimethylamine,[2223] or (c) by the reaction of 1,2-dioleoyl-*sn*-glycerol-3-ethylenechlorohydrin with trimethylamine.[2202] OLPC may be prepared by stepwise acylation of the barium salt of *sn*-glycero-3-phosphoethylene chlorohydrin with oleoyl chloride and linoleoyl chloride followed by reaction with trimethylamine.[2224]

DLPC may be prepared (a) by acylation of *sn*-glycero-3-phosphocholine or its $CdCl_2$ complex with linoleoyl chloride in the presence of 4-pyrrolidinopyridine[2221] or potassium linoleate,[2225] or with *N*-linoleoylimidazole in the presence[2209] or absence[2201] of a strong base, (b) by phosphorylation of 1,2-dilinoleoyl *sn*-glycerol with $POCl_3$ and choline chloride,[2193] (c) by the reaction of 1,2-dilinoleoyl-*sn*-glycero-3-phosphoethylene chlorohydrin with trimethylamine,[2224] or (d) by condensation of 1-linoleoyl-*sn*-glycero-3-phosphocholine (LHPC) with linoleic acid in the presence of DCC.[2208] DLnPC may be prepared by acylation of *sn*-glycero-3-phosphocholine or its $CdCl_2$ complex with linolenic anhydride[2226] or *N*-linolenoylimidazole.[2209,2227]

Table 4-8 Molecular Species Composition: Phosphatidylcholines (mol%)

No.	Plant part	Method	16:0 16:0	16:0 18:1	16:0 18:2	16:0 18:3	18:0 18:1	18:0 18:2	18:0 18:3	18:1 18:1	18:1 18:2	18:1 18:3	18:2 18:2	18:2 18:3	18:3 18:3	References
Arabidopsis thaliana (L.) Heynh (Mouse-Ear Cress)																
1	Leaf plasma membrane	GLC	0.4[a]	2.9	32.1	23.6	0.6	0.4	0.8	2.3	2.7	14.8[b]		13.1	4.0	1478
Arachis hypogaea L. (Peanut)																
1	Seed	HPLC									59.3		40.7			2116
Avena sativa L. (Oats)																
1	Leaf plasma membrane	GLC	1.8[a]	5.6	34.3	19.9	0.7		0.4	0.8	4.4	14.9[b]		9.6	3.7	1502
1	Root plasma membrane	GLC			55	14							23	8		2113
Brassica napus L. (Rape)																
1	Cell susp. culture	AgTLC		19[c]	8	29				16[d]	10	18				2099
Brassica napus L. (Rape, low erucic acid (Canola))																
1	Leaf	HPLC		1.8	11.5	17.9			3.9		4.0		10.1	26.1	24.6	2117
Brassica oleracea L. ssp. *botrytis* L. (Cauliflower)																
1	Inflorescence	AgTLC		1.2	7.5	37.2				0.4	3.3	18.6		6.8	21.6	2096
Camellia sinensis L. (Tea)																
1	Pollen	HPLC				60									25	1558
Carthamus tinctorius L. (Safflower)																
1	Seed	AgTLC	0.5	4.2	34.7		(18:2/18:1, 5.0)			1.4	8.0		44.8			2118
Crambe amabilis Butkov and Majlin																
1	Seed	1R,2R	0.5	5.5	4.9	3.4	1.0	1.0		12.6	12.9	7.5	7.5	5.3	7.2	2119
Daucus carota L. (Carrot)																
1	Cell culture	AgTLC		2.2	43.3						11.8		25.1	17.6		1648
Dianthus caryophyllus L. (Carnation)																
1	Flower microsomal membr.	GLC		2.5	50.7	2.1		0.8	(16:0/18:0, 0.5)		4.4	1.4	34.0	2.9		1652
Diptychocarpus strictus (Fisch.) Trautv.																
1	Seed	1R,2R	0.3	5	6.1	9	1	1.2	1.7	7.5	8.3	13.2	5	8.1	9	959

Echium plantagineum L. (Salvation Jane)																
1	Leaf	AgTLC			18	10					5[e]	6.5[e]	5	14[e]	2.5	2120
Glycine max (L.) Merr. (Soybean)																
1	Seed	Br_2			19.1					4.7	63.0		13.5			2121
1	Seed	AgTLC		3.1	24.0		(18:3/18:2, 6.9)		3.5	1.8	8.9		39.7	1.8		2118
1	Seed	GLC		2.6	25.3	2.4	0.5	6.6	0.4	0.7	8.0		42.4	8.3	0.5	2111
1	Seed	FD-MS	1.0	6.0	26.7		1.6			8.9[f]	13.8		34.0	6.9		2122
1	Seed	CI-MS		2.7	25.7	2.5	0.9	8.5	0.6	1.0	11.9		38.9	6.9	0.4	2123
1	Seed	HPLC	0.5	8.2[f]	24.8	3.3	0.6		0.3	0.7	7.8	3.5	38.7	10.7	0.9	1693
1	–, hypocotyl	AgTLC			24	9		8			7		46	2	Trace	1690
1	–, cotyledon	AgTLC			27	1		10			15		36	3	Trace	1690
1	–, cotyledon	HPLC			19.24	32.46				9.38	7.85	3.43	4.70	4.45	10.46	2124
1	–, axis	HPLC			23.9	15.8				5.4	Trace	Trace	37.5	17.4	Trace	2124
1	Leaf	HPLC			8.75	39.27				0.91	2.23	1.96	17.21	10.58	15.44	2124
1	Stem	HPLC			11.46	43.26				3.21	2.87	2.98	14.23	6.62	12.55	2124
1	Root	HPLC			22.97	38.62				6.62	6.70	4.40	2.20	3.97	8.52	2124
1	Seedling, microsomes	HPLC	0.4	3.2	26.8	5.9	0.4	5.4	1.0	0.5	9.2	1.2	35.1	9.9[e]	0.9	2125
Gossypium barbadense L. (Sea Island Cotton)																
1	Seed kernel	1R,2R	2.0	9.8	18.0		0.8	1.5		9.2	16.2		20.1			1709
Gossypium hirsutum L. (Upland Cotton)																
1	Seed	1R,2R	3.1	14.3	22.8		1.2	1.4		5.6	12.7		23.7			2098
Helianthus annuus L. (Sunflower)																
1	Seed	1R,2R	0.4	3.5	14.4		1.7	7.0		2.3	9.3		44.6			2126
Hibiscus cannabinus L (Kenaf)																
1	Seed	1R,2R	1.2	12.5	19.2	0.7	0.8	1.2		10.3	16.4	0.6	16.9	0.6		1377
Hordeum vulgare L. (Barley)																
1	Leaf	HPLC			25.2	46.1[b]								21.5	7.1	2127

[a]Including 14:0/18:0; [b]Including 18:2/18:2 (LL); [c]Including 16:0/16:1; [d]Including 18:1/16:1 PG; [e]*sn*-1/*sn*-2 positions reversed; [f]Including 18:0/18:2 (SL); [g]Including 18:3/18:2 (LnL); [h]Including 18:0/18:3 (SLn).

Continued

Table 4-8 Molecular Species Composition: Phosphatidylcholines (mol%)—cont'd

No.	Plant part	Method	16:0 16:0	16:0 18:1	16:0 18:2	16:0 18:3	18:0 18:1	18:0 18:2	18:0 18:3	18:1 18:1	18:1 18:2	18:1 18:3	18:2 18:2	18:2 18:3	18:3 18:3	References
Lavandula vera DC (English Lavender)																
1	Cell culture	AgTLC		12.2	11.7	17.3	7.1	4.6	15.8	9.6	7.1	6.1	0.5	(16:0/18:0, 8.1)		1765
Lilium longiflorum Thunb. (Trumpet Lily)																
1	Pollen	HPLC	1.7		18.5	36.0							6.5	17.6[g]	11.4	1769
Lycopersicon esculentum Mill. (Tomato)																
1	Fruit, pericarp	GLC	0	8.0	48.9	8.0		(16:1/18:2, 13.4)					13.6	3.9	0	2128
Lycoris radiata Herb. (Red Spiderlily)																
1	Flower	Prep. TLC	88.7		11.3											1788
Malus domestica Borkh. (Apple)																
1	Fruit, embryo	HPLC		17	14			3		19	20		27			2129
Medicago media Pers. (Alfalfa)																
1	Leaf	AgTLC		2.8	31.0[f]	32.7[h]							6.9	13.1	10.1	2130
Nicotiana tabacum L. (Tobacco)																
1	Leaf	HPLC			17.3	42.2							8.9	17.9	14.5	1410
1	Cell culture	AgTLC		11.5	48.0	15.6			8.1		3.0	3.2	8.2	2.4		1495
Oryza sativa L. (Rice)																
1	Bran, glutinous	AgTLC		26.8	14.3		(18:2/18:1, 9.3)			29.3	12.5		7.8			1845
1	Bran, glutinous	1R,2R		21.3	12.8		(18:2/18:1, 13.7)			23.6	14.2		8.3			1845
1	Bran, nonglutinous	AgTLC		18.5	20.3		(18:2/18:1, 10.1)			19.6	18.8		12.7			1845
1	Bran, nonglutinous	1R,2R		16.4	18.0		(18:2/18:1, 11.9)			15.5	17.1		13.1			1845
Phaseolus coccineus L. (Scarlet Runner Bean)																
1	Seed	HPLC	0.5	6.1[f]	25.2	15.3	<0.1		1.9	1.3	6.9	4.9	15.9	15.9	5.9	1693

Phaseolus vulgaris L. (Kidney Bean)																
1	Seed	HPLC	1.5	11.2[f]	20.6	19.6	1.0		1.6	3.2	9.2	8.6	6.7	11.5	5.2	1693
Picea abies (L.) Karst. (Spruce)																
1	Shoot	GLC		11.9	26.2	1.8		6.6		12.2	16.3		19.4			2115
Pinus nigra Arnold (Black Pine)																
1	Needle (leaf)	HPLC	<1	1	25	6		4			6	8	15	5	1	1892
Pisum sativum L. (Pea)																
1	Seed	AgTLC		7.9	15.2		4.8			38.6	3.4		30.1			1904
1	Seed	HPLC	0.1	17.0[f]	14.8	1.7	4.4		0.5	15.8	26.4	2.1	14.7	2.3	0.2	1693
1	Leaf	HPLC		Trace	28.0	16.0	1.5	1.3	2.6		3.2	2.2	21.2	17.8[e]	6.2	2125
Raphanus sativus L. (Radish)																
1	Seedling	HPLC	11.7	8.8	14.4	16.1		8.8	3.8		11.2	11.1	5.8	12.0	5.1	2131
Rauvolfia serpentina (L.) Benth. ex Kurz (Indian Snakeroot)																
1	Cell culture	AgTLC			73.7	1.6	Trace	9.9	0.6	2.1				12.1		1495
Secale cereale L. (Rye)																
2	Grain (1)	HPLC		7	28	5				3	13	5	30	8	1	1957
	Grain (2)	HPLC		5	23	2				2	18	3	38	8	1	1957
1	Leaf plasma membrane	GLC	0.5[a]	2.4	27.4	24.9	0.3		1.4	1.8	1.7	14.8[b]		14.0	8.1	1502
Solanum tuberosum L. (Potato)																
1	Tuber	HPLC			36.3	10.4		7.1					32.5	13.7	Trace	1410
1	–, microsomes	HPLC		35.5	10.1		6.9	2.1				31.8	13.4		Trace	2132
Spinacia oleracea L. (Spinach)																
1	Leaf	HPLC			5.4	54.1	(18:3/18:2, 0.9)			3.7	5.5	2.9	16.1	7.9	3.5	2105
1	Leaf	HPLC		3.6	4.7	15.5				6.9	6.5	9.0	11.6	13.8	28.3	2133
Triticum aestivum L. (Wheat)																
1	Germ oil	AgTLC		6.9	28.4	(18:3/18:2, 3.6)			1.7	2.5	11.8	0.4	36.3	2.0		2118
1	Flour	AgTLC		6.1	27.8	1.3	0.2	2.0	0.1	1.2	6.2	0.1	39.8			2023
1	Flour	1R,2R	1.0	6.1	29.3	1.3	0.4	2.1	0.1	1.3	6.0	0.3	38.1	1.7	0.1	2023
1	Flour	HPLC		13.44	42.13	1.12				2.80	11.36	0.84	24.88	1.17		2021
Vicia faba L. (Broad Bean)																
1	Seed	HPLC	0.2	15.8[f]	16.6	0.2	3.3		0.5	17.0	26.8	1.3	17.4	0.9	0.1	1693

[a]Including 14:0/18:0; [b]Including 18:2/18:2 (LL); [c]Including 16:0/16:1; [d]Including 18:1/16:1 PG; [e]*sn*-1/*sn*-2 positions reversed; [f]Including 18:0/18:2 (SL); [g]Including 18:3/18:2 (LnL); [h]Including 18:0/18:3 (SLn).

Continued

Table 4-8 Molecular Species Composition: Phosphatidylcholines (mol%)—cont'd

No.	Plant part	Method	16:0 16:0	16:0 18:1	16:0 18:2	16:0 18:3	18:0 18:1	18:0 18:2	18:0 18:3	18:1 18:1	18:1 18:2	18:1 18:3	18:2 18:2	18:2 18:3	18:3 18:3	References
Vigna angularis (Willd.) Ohwi and Ohashi (Adzuki Bean)																
1	Seed	HPLC		4.9[f]	31.1	10.0			1.3		6.5	2.2	27.4	14.8	1.9	2051
1	Leaf	HPLC		7.3[f]	21.2	25.1			6.5		1.5	1.9	13.7	17.4	5.4	2051
1	Root	HPLC		6.4[f]	51.2	11.9			3.3		8.3	3.2	16.2	13.1	2.6	2051
	Similar data reported for PC molecular species in cotyledon, leaf, hypocotyl, and root tissue of adzuki bean cell culture															2052
Vigna unguiculata (L.) Walp. (Cowpea)																
2	Leaf, drought-resistant	HPLC			26	24			4		7		15	18	6	2064
	Leaf, drought-sensitive	HPLC			9	41			16		5		3	12	13	2064
Zea mays L. ssp. *saccharata* Sturt (Sweet Corn)																
1	Seed	HPLC			11	18				14	32		24	1		2086

[a]Including 14:0/18:0.
[b]Including 18:2/18:2 (LL).
[c]Including 16:0/16:1.
[d]Including 18:1/16:1 PG.
[e]*sn*-1/*sn*-2 positions reversed.
[f]Including 18:0/18:2 (SL).
[g]Including 18:3/18:2 (LnL).
[h]Including 18:0/18:3 (SLn).

Hydrolysis

Hydrolysis of DPPC with either HCl or NaOH in refluxing ethanol yields palmitic acid and a mixture of α- and β-glycerophosphoric acids owing to migration of the phosphate group. Under milder conditions (0.2*N* methanolic NaOH at 37 °C), the first-order rate constant for the liberation of palmitic acid is 1.0 h^{-1}. Most, if not all, of the free choline is formed by hydrolysis of the glycerophosphocholine ($k = 0.069\ h^{-1}$) rather than the intact DPPC.[191] Hydrolysis of DPPC in acetate buffer at pH 4.0 follows pseudo-first-order kinetics ($k_{obs.} = 1.0 \times 10^{-2}\ day^{-1}$ at 30 °C). The Arrhenius plot shows a break at the gel-to-liquid phase transition temperature T_m (42 °C) with activation energy $E_a = 91$ kJ/mol below the T_m and 67 kJ/mol above it.[2228]

Enzymatic hydrolysis of DPPC, DSPC, DOPC, or POPC with *Crotalus adamanteus* snake venom phospholipase [E.C. 3.1.1.4] gives exclusively the corresponding 2-lysophospholipids PHPC,[2134] SHPC,[2229] OHPC,[2230] and PHPC,[2230] whereas enzymatic hydrolysis of DPPC, POPC, DOPC, or DLPC with lipase from the mold *Rhizopus arrhizus* [E.C. 3.1.1.3] gives exclusively the corresponding 1-lysophospholipids HPPC,[2231] HOPC,[2232] HOPC,[2233] and HLPC,[2233] respectively. Both acyl groups of POPC are cleaved in 30 min and 37 °C with 1.0*N* aqueous methanolic NaOH.[2234]

Lysophosphatidylcholines

Physical Properties

1-Palmitoyl-*sn*-glycero-3-phosphocholine (PHPC), 16:0 LPC, $C_{24}H_{50}NO_7P$. *(R)-4,7-Dihydroxy-N,N,N-trimethyl-10-oxo-3,5,9-trioxa-4-phosphapentacosan-1-aminium Hydroxide, inner salt, 4-Oxide.*

PHPC [R.N. 17364-16-8]: *Hydrate*: mp 195-196 °C, $[\alpha]_D^{25}$ −2.2° ($CHCl_3$),[2231] readily soluble in methanol,[2216] critical micelle concentration.[2235] Spectra: Raman,[2236] ^{1}H NMR and ^{31}P NMR (figs.),[2237] ^{14}N NMR,[2204] EI-MS (fig.),[2238] FAB-MS,[2135] FD-MS,[2177] CIMS (fig.),[2238] SIMS.[2239] Chromatography: PC,[2212] TLC,[2240] HPTLC,[1405] HPLC,[2216] silicic acid column.[2212] *rac*-PHPC [R.N. 17364-17-0]: Dec. 260-262 °C,[2241] IR (fig.),[2242] ESI-MS and ESI-MS/MS (figs.).[2243] $[CdCl_2]_3[PHPC.H_2O]_2$: Solid, insoluble in ether or ethanol.[2244]

2-Palmitoyl-*sn*-glycero-3-phosphocholine (HPPC), 16:0 LPC, $C_{24}H_{50}NO_7P$. *(R)-4-Hydroxy-7-hydroxymethyl-N,N,N-trimethyl-9-oxo-3,5,8-trioxa-4-phosphatetracosan-1-aminium Hydroxide, inner salt, 4-Oxide.*

HPPC [R.N. 66757-27-5; see also 80547-97-5]: Soluble in benzene, $CHCl_3$, DMSO and methanol. Chromatography: HPLC.[2216] *Hydrate*: Soluble in chloroform. Spectra: ^{1}H and ^{31}P NMR (figs.),[2237] ESI-MS/MS (fig.).[2243] Chromatography: PC,[2212] TLC,[2237] gel filtration.[2237]

1-Stearoyl-*sn*-glycero-3-phosphocholine (SHPC), 18:0 LPC, $C_{26}H_{54}NO_7P$. *(R)-4,7-Dihydroxy-N,N,N-trimethyl-10-oxo-3,5,9-trioxa-4-phosphaheptacosan-1-aminium Hydroxide, inner salt, 4-Oxide*

SHPC [R.N. 19420-57-6]: Readily soluble in methanol,[2216] insoluble in water.[2235] Spectra: IR,[2201] ^{1}H NMR,[2201] FAB-MS,[2135] FD-MS (fig.),[2177] CI-MS/MS.[2245]

Chromatography: PC,[2197] HPLC,[2216] silicic acid column.[2197] *Hydrate*: Needles, mp 257.5-258.5 °C, $[\alpha]_D^{25}$ −2.87° ($CHCl_3$/MeOH).[2246] Spectra: ^{13}C and ^{31}P NMR,[2247] CI-MS.[2248] Chromatography: TLC.[2240]
$[CdCl_2]_3[SHPC.H_2O]_2$: Solid.[2230]
$[CdCl_2]_3[SHPC.HCl]_2$: Solid.[2197]

1-Oleoyl-*sn*-glycero-3-phosphocholine (OHPC), 18:1 LPC, $C_{26}H_{52}NO_7P$. *(R)-4,7-Dihydroxy-N,N,N-trimethyl-10-oxo-3,5,9-trioxa-4-phosphaheptacos-18-en-1-aminium Hydroxide, inner salt, 4-Oxide*
OHPC [R.N. 19420-56-5]: *Hydrate*, mp 240 °C, $[\alpha]_D^{20}$ −3.2 ($CHCl_3$/ethanol).[2230] Spectra: ^1H and ^{14}N NMR,[2204] FAB-MS,[2249] FD-MS,[2177] CI-MS,[2248] ESI-MS/MS (fig.).[2243] Chromatography: PC,[2250] TLC,[2250] HPLC,[2251] silicic acid column.[2197]
$[CdCl_2]_3[OHPC{\cdot}H_2O]_2$: Solid.[2197]

2-Oleoyl-*sn*-glycero-3-phosphocholine (HOPC), 18:1 LPC, $C_{26}H_{52}NO_7P$. *(R)-4-Hydroxy-7-hydroxymethyl-N,N,N-trimethyl-9-oxo-3,5,8-trioxa-4-phosphahexacos-17-en-1-aminium Hydroxide, inner salt, 4-Oxide.*
HOPC [R.N. 22248-65-3]: $[\alpha]_D^{25}$ +3.1° ($CHCl_3$/methanol).[2233] Spectra: ESI-MS/MS (fig.).[2243] Chromatography: TLC,[2233] gel filtration column.[2233]

1-Linoleoyl-*sn*-glycero-3-phosphocholine (LHPC), 18:2 LPC, $C_{26}H_{50}NO_7P$. *(R)-4,7-Dihydroxy-N,N,N-trimethyl-10-oxo-3,5,9-trioxa-4-phosphaheptacosa-18,21-dien-1-aminium Hydroxide, inner salt, 4-Oxide*
LHPC [R.N. 22252-07-9; see also 5655-12-9]. *Hydrate*: Spectra: IR,[2252] ^1H NMR,[2204] ^{13}C NMR,[2252] FD-MS.[2122] Chromatography: TLC, HPLC, silica gel column.[2252]

2-Linoleoyl-*sn*-glycero-3-phosphocholine (HLPC), 18:2 LPC, $C_{26}H_{50}NO_7P$. *(R)-4-Hydroxy-7-hydroxymethyl-N,N,N-trimethyl-9-oxo-3,5,8-trioxa-4-phosphahexacosa-17,20-dien-1-aminium Hydroxide, inner salt, 4-Oxide.*
HLPC [R.N. 27304-12-7; see also 3973-94-2]: Spectra: IR, ^1H, and ^{13}C NMR, FAB-MS.[2252] Chromatography: TLC,[2233] HPLC,[2252] silicic acid column.[2252]

1-Linolenoyl-*sn*-glycero-3-phosphocholine (LnHPC), 18:3 LPC, $C_{26}H_{48}NO_7P$. *(R)-4,7-Dihydroxy-N,N,N-trimethyl-10-oxo-3,5,9-trioxa-4-phosphaheptacosa-18,21,24-trien-1-aminium Hydroxide, inner salt, 4-Oxide.*
LnHPC [R.N. 62512-91-8; see also 63163-01-9]: *Hydrate*: FD-MS.[2177]

2-Linolenoyl-*sn*-glycero-3-phosphocholine (HLnPC), 18:3 LPC, $C_{26}H_{48}NO_7P$. *(R)-4-Hydroxy-7-hydroxymethyl-N,N,N-trimethyl-9-oxo-3,5,8-trioxa-4-phosphahexacosa-17,20,23-trien-1-aminium Hydroxide, inner salt, 4-Oxide.*
HLnPC [R.N. 77667-11-9]: No data.

Synthesis

PHPC may be prepared (a) by catalytic hydrogenation of 1-palmitoleyl-*sn*-glycero-3-phosphocholine,[2231] (b) by partial methanolysis of DPPC with octylamine in methanol,[2253] (c) by enzymatic hydrolysis of DPPC with snake venom phospholipase A_2,[2231]

(d) by acylation of *sn*-glycero-3-phosphocholine with palmitoyl chloride,[2212] or (e) by stepwise reaction of 2-benzyl-1-palmitoyl-*sn*-glycerol with 2-bromoethyl phosphorodichloridate and trimethylamine.[2240]

HPPC may be prepared (a) by enzymatic hydrolysis of DPPC with lipase from the mold *R. arrhizus*[2237] or (b) by oxidation of a 1-(alk-1′-enyl)-2-palmitoyl-*sn*-glycero-3-phosphocholine with iodine.[2254]

SHPC may be prepared (a) by partial acylation of *sn*-glycero-3-phosphocholine with stearoyl chloride,[2197] (b) by enzymatic hydrolysis of DSPC with snake venom phospholipase A_2,[2246] (c) by catalytic hydrogenation of OHPC,[2230] or (d) by stepwise reaction of 2-benzyl-1-stearoyl-*sn*-glycerol with 2-bromoethyl phosphorodichloridate and trimethylamine[2240] or with $POCl_3$ and choline *p*-toluene-sulfonate.[2255]

OHPC may be prepared (a) by partial acylation of *sn*-glycero-3-phosphocholine with oleoyl chloride,[2197] (b) by partial methanolysis of 2-acetyl-1-oleoyl-*sn*-glycero-3-phosphocholine with *t*-BuOK in methanol,[2240] or (c) by enzymatic hydrolysis of DOPC with snake venom phospholipase A_2.[2230]

HOPC may be prepared by enzymatic hydrolysis of POPC[2232] or DOPC[2233] with lipase from the mold *R. arrhizus*.

LHPC may be prepared (a) by stepwise reaction of 1-linoleoyl-2-methoxyethoxymethyl-*sn*-glycerol with $POCl_3$ and choline *p*-toluenesulfonate[2208] or (b) by migration of the linoleoyl group during the preparation of HLPC from soybean PC in phosphate buffer.[2252]

HLPC may be prepared by hydrolysis of DLPC with lipase from the mold *R. arrhizus*[2233] or of soybean PC with lipase from the fungus *Mucor javanicus*.[2252]

Hydrolysis

The second-order rate constant for base-catalyzed hydrolysis of PHPC is about $1.8 \times 10^{-2}\ M^{-1}\ s^{-1}$ about 8500 times slower than the rate of acyl migration.[2237] PHPC, SHPC, and OHPC are rapidly hydrolyzed to *sn*-glycero-3-phosphocholine by *Penicillium notatum* phospholipase B.[2256] The rate of hydrolysis of HPPC to *sn*-glycero-3-phosphocholine by this phospholipase is about 15 times slower than that of PHPC.[2254] PHPC and SHPC are rapidly hydrolyzed to PHPA and SHPA, respectively, by Savoy cabbage phospholipase D.[2257]

Migration

PHPC gradually rearranges in aqueous solution to an equilibrium mixture containing 90% PHPC and 10% HPPC. The second-order rate constant for the migration of the palmitoyl group from *sn*-2 to *sn*-1 is about 160 M^{-1}/s for base-catalyzed migration (pH 6.2-9) and $4 \times 10^{-4}\ M^{-1}/s$ for acid-catalyzed migration (pH 1-2). Migration of the phosphoryl group is observed only at pH 1 or below.[2237]

Migration of the linoleoyl group from *sn*-2 to *sn*-1 occurs during the preparation of HLPC in phosphate buffer resulting in a mixture containing 94% HLPC and 6% LHPC, but can be prevented by the use of borate buffer.[2252]

PHOSPHATIDYLETHANOLAMINES

This group comprises the phosphatidylethanolamines (PE), the lysophosphatidylethanolamines (LPE), the *N*-acylphosphatidylethanolamines (NAPE), and the *N*-acyllysophosphatidylethanolamines (NALPE):

$$\begin{array}{l} \qquad\qquad\quad CH_2OC(O)R^1 \\ \qquad\qquad\qquad | \\ R^2C(O)O\!-\!C\!-\!H \\ \qquad\qquad\qquad | \\ \qquad\qquad\quad CH_2OP(O)(OH)OCH_2CH_2NHR^3 \end{array}$$

PE, $R^1 = R^2$ = alkyl, R^3 = H
1-LPE, $R^1 = R^3$ = H, R^2 = alkyl
2-LPE, R^1 = alkyl, $R^2 = R^3$ = H
NAPE, $R^1 = R^2 = R^3$ = alkyl
1-NALPE, R^1 = H, $R^2 = R^3$ ′ alkyl
2-NALPE, $R^1 = R^3$ = alkyl, R^2 = H

Phosphatidylethanolamines

Physical Properties

1,2-Dipalmitoyl-*sn*-glycero-3-phospho-1′-ethanolamine (DPPE), 16:0/16:0 PE, $C_{37}H_{74}NO_8P$. *(R)-Hexadecanoic Acid, 1-[[[(2-aminoethoxy)hydroxyphosphinyl]oxy] methyl]-1,2-ethanediyl Ester.*

DPPE [R.N. 923-61-5]: Microscopic spherulites (photo), mp 172.5-175 °C, $[\alpha]_D^{26}$ +6.4° ($CHCl_3$), soluble in $CHCl_3$, moderately soluble in ethanol, insoluble in acetone, ether, benzene, petroleum ether[2258]; ORD,[2147] DSC,[2259] TGA,[1367] IR dichroism.[2260] Spectra: IR (fig.),[2258] FTIR,[2261] Raman (fig.),[2262] ^{1}H NMR,[2223] ^{13}C and ^{31}P NMR (figs.),[2210] FAB-MS,[2249] FD-MS,[2263] CI-MS,[2264] LSIMS,[2154] FAB-MS/MS,[2136] CID-MS/MS (fig.).[2265] X-ray.[2258] Chromatography: TLC,[2266] HPTLC,[1405] HPLC,[2239] LC-MS,[2154] GFC,[2267] SFC.[2156] *Dihydrate*: Crystals, ^{13}C and ^{31}P CP-MAS-NMR (figs.), two conformations.[2169]

1-Palmitoyl-2-oleoyl-*sn*-glycero-3-phospho-1′-ethanolamine (POPE), 16:0/18:1 PE, $C_{39}H_{76}NO_8P$. *[R-(Z)]-9-Octadecenoic Acid, 1-[[[(2-Aminoethoxy)hydroxyphosphinyl]oxy] methyl]-2-[(1-oxo-hexadecyl)oxy]ethyl Ester.*

POPE [R.N. 39749-44-5]: Colorless waxy solid, mp 184-185 °C[2268]; CE (fig.),[2181] MALDI FT-ICR-MS/MS (fig.).[1370] *Monohydrate*: Hard fatty wax, mp 194.6 °C, $[\alpha]_D^{22}$ +6.35° ($CHCl_3$), soluble in $CHCl_3$ and methanol, insoluble in ether or acetone,[2269] DSC.[2227] Spectra: FTIR,[2270] ^{31}P NMR (fig.),[2227] 2D-[^{1}H][^{31}P]-NMR,[2271] FAB-MS,[2135] LSIMS (fig.),[2139] FAB-MS/MS,[2272] CID-MS/MS (fig.),[2265] Chromatography: HPLC,[2100] LC-MS.[2273] Electrophoresis: PE.[2100]

1-Palmitoyl-2-linoleoyl-*sn*-glycero-3-phospho-1′-ethanolamine (PLPE), 16:0/18:2 PE, $C_{39}H_{74}NO_8P$. *[R-(Z,Z)]-9,12-Octadecadienoic Acid, 1-[[[(2-aminoethoxy) hydroxyphosphinyl]oxy]methyl]-2-[(1-oxohexadecyl)oxy]ethyl Ester.*

PLPE [R.N.26662-95-3; see also 14018-09-8]: Pale yellow oil. Spectra: IR,[1787] ^{1}H and ^{13}C NMR,[2274] EI-MS,[1787] FAB-MS,[1787] CI-MS,[2264] CID-MS/MS (fig.).[2265] Chromatography: TLC,[1787] HPLC,[2100] HPLC/CI-MS,[2180] LC/MS,[2273] LC/MS/MS (fig.).[1415]

PLPE.HCl [R.N. 106498-49-1]: Colorless waxy solid, mp 155 °C, $[\alpha]_D^{20}$ +6.0° ($CHCl_3$), TLC.[2275]

1-Palmitoyl-2-linolenolyl-*sn*-glycero-3-phospho-1′-ethanolamine (PLnPE), 16:0/18:3 PE, $C_{39}H_{72}NO_8P$. *[R-(Z,Z,Z)]-9,12,15-Octadecatrienoic Acid, 1-[[[(2-Aminoethoxy) hydroxyphosphinyl]oxy]methyl]-2-[(1-oxohexadecyl)oxy]ethyl Ester.*

PLnPE [R.N. 4009-43-2]: White powder, $[\alpha]_D^{20}$ +6.0° ($CHCl_3$), IR, TLC,[2276] HPLC.[2132]

1-Stearoyl-2-oleoyl-*sn*-glycero-3-phospho-1′-ethanolamine (SOPE), 18:0/18:1 PE, $C_{41}H_{80}NO_8P$. *[R-(Z)]-9-Octadecenoic Acid, 1-[[[(2-Aminoethoxy)hydroxyphosphinyl]oxy] methyl]-2-[(1-oxooctadecyl)oxy]ethyl Ester.*

SOPE [R.N. 6418-95-7]: White solid,[2277] mp 189-190 °C, $[\alpha]_D^{20}$ +6.1° ($CHCl_3$), soluble in chloroform and ether, fairly soluble in acetone, less soluble in ethanol, insoluble in water[2188]; monolayer film properties.[2191] Spectra: IR (fig.),[2188] FAB-MS,[2135] FD-MS.[2122] Chromatography: PC,[2277] TLC,[2278] HPLC,[2110] MPLC,[2278] silicic acid column,[2188] LC-MS (fig.).[2187] *rac*-SOPE, Na salt [R.N. 78524-86-4]: DTA (fig), X-ray,[2279] ^{1}H NMR.[2280]

1-Stearoyl-2-linoleoyl-*sn*-glycero-3-phospho-1′-ethanolamine (SLPE), 18:0/18:2 PE, $C_{41}H_{78}NO_8P$. *[R-(Z,Z)]-9,12-Octadecadienoic Acid, 1-[[[(2-Aminoethoxy) hydroxyphosphinyl]oxy]methyl]-2-[(1-oxooctadecyl)oxy]ethyl Ester.*

SLPE [R.N. 7266-53-7]: DSC (fig.),[2281] monolayer film properties,[2191] FAB-MS,[2135] HPLC.[2100]
S-SLPE: $[\alpha]_D^{20}$ −6.0° ($CHCl_3$), IR, PC, TLC, silicic acid chromatography.[2282]

1-Stearoyl-2-linolenoyl-*sn*-glycero-3-phospho-1′-ethanolamine (SLnPE), 18:0/18:3 PE, $C_{41}H_{76}NO_8P$. *[R-(Z,Z,Z)]-9,12,15-Octadecatrienoic Acid, 1-[[[(2-Aminoethoxy) hydroxyphosphinyl]oxy]methyl]-2-[(1-oxooctadecyl)oxy]ethyl Ester.*

SLnPE [R.N. 61177-33-1]: HPLC,[2110] monolayer film properties.[2191]

1,2-Dioleoyl-*sn*-glycero-3-phospho-1′-ethanolamine (DOPE), 18:1/18:1 PE, $C_{41}H_{78}NO_8P$. *[R-(Z,Z)]-9-Octadecenoic Acid, 1-[[[(2-aminoethoxy)hydroxyphosphinyl]oxy]methyl]-1,2-ethanediyl Ester.*

DOPE [R.N. 4005-05-1]: Almost white, waxy, slightly hygroscopic solid,[2283] mp 195-200 °C,[2270] soluble in $CHCl_3$, ether and petroleum ether, somewhat soluble in methanol (20.2 g/100 mL) or ethanol (13.2 g/100 mL), and sparingly soluble in acetone (2.6 g/100 mL),[2283] $[\alpha]_D^{20}$ +6.0° ($CHCl_3$),[2283] DSC,[2284] TGA,[1367] monolayer film properties.[2191] Spectra: IR (fig.),[2283] FTIR,[2270] ^{1}H NMR,[2285] ^{13}C NMR,[2201] ^{31}P NMR (fig.),[2286] 2d-[^{1}H] [^{13}C]-NMR (fig.),[2174] FD-MS,[2122] LSIMS,[2239] MALDI FT-ICR/MS (fig.).[1370] Chromatography: TLC,[2166] HPLC,[2100] LC-MS,[1415] silicic acid column.[2283]

1-Oleoyl-2-linoleoyl-*sn*-glycero-3-phospho-1′-ethanolamine (OLPE), 18:1/18:2 PE, $C_{41}H_{76}NO_8P$. *[R-(Z,Z,Z)]-9,12-Octadecadienoic Acid, 1-[[[(2-Aminoethoxy) hydroxyphosphinyl]oxy]methyl]-2-[(1-oxo-9-octadecenyl)oxy]ethyl Ester.*

OLPE [R.N. 1188-56-3]: $[\alpha]_D^{25}$ +4.1° ($CHCl_3$), soluble in chloroform and ether.[2287] Chromatography: TLC,[2287], HPLC,[2100] silicic acid column.[2287] *rac-OLPE*: IR, PC.[2288]

1-Oleoyl-2-linolenoyl-*sn*-glycero-3-phospho-1′-ethanolamine (OLnPE), 18:1/18:3 PE, $C_{41}H_{74}NO_8P$. *[R-(all Z)]-9,12,15-Octadecatrienoic Acid, 1-[[[(2-Aminoethoxy) hydroxyphosphi-nyl]oxy]methyl]-2-[(1-oxo-9-octadecenyl)oxy]ethyl Ester.*

OLnPE [R.N. 124265-78-7] HPLC.[1892]

1,2-Dilinoleoyl-*sn*-glycero-3-phospho-1′-ethanolamine (DLPE), 18:2/18:2 PE, $C_{41}H_{74}NO_8P$. *[R-(all Z)]-9,12-Octadecadienoic Acid, 1-[[[(2-aminoethoxy) hydroxyphosphinyl]oxy]methyl]-1,2-ethanediyl Ester.*

DLPE [R.N. 20707-71-5]: Soft wax, mp 25 °C,[2268] soluble in chloroform, ether, and petroleum ether, less soluble in ethanol, poorly soluble in acetone; forms a stable emulsion with water,[2289] $[\alpha]_D^{20}$ +6.0° ($CHCl_3$),[2290] DSC,[2227] TGA.[1367] Spectra: IR (fig.),[2289] ^{31}P NMR.[2227] Chromatography: PC,[2290] TLC,[2289] HPLC,[2100] DEAE-celullose column,[2291] silicic acid column,[2290] LC-MS (fig.).[2273]

1-Linoleoyl-2-linolenoyl-*sn*-glycero-3-phospho-1′-ethanolamine (LLnPE), 18:2/18:3 PE, $C_{41}H_{72}NO_8P$. *[R-(all Z)]-9,12,15-Octadecatrienoic Acid, 1-[[[(2-Aminoethoxy) hydroxyphosphinyl]oxy]methyl]-2-[(1-oxo-9,12-octadecadienyl)oxy]ethyl Ester.*

LLnPE [R.N. 124264-80-8]: HPLC.[2100]

1,2-Dilinolenoyl-*sn*-glycero-3-phospho-1′-ethanolamine (DLnPE), 18:3/18:3 PE, $C_{41}H_{70}NO_8P$. *[R-(all Z)]-9,12,15-Octadecatrienoic Acid, 1-[[[(2-Aminoethoxy) hydroxyphosphinyl]oxy]methyl]-1,2-ethanediyl Ester.*

DLnPE [R.N. 34813-40-6]: $[\alpha]_D$ +6.1° ($CHCl_3$),[2292] DSC.[2227] Spectra: IR,[2292] ^{31}P NMR.[2227] Chromatography: HPLC,[2100] LC-MS.[2273] *rac*-DLnPE: PC.[2288]

Molecular Species Composition

The molecular species composition of phosphatidylethanolamine in plants, determined by any of the methods described on pages 168–171, is given in Table 4-9.

Synthesis

DPPE may be prepared (a) by phosphorylation of 1,2-dipalmitoyl-*sn*-glycerol with phenyl phosphorodichloridate,[2258] PCl_3/imidazole,[2211] or methyl *N,N*-diisopropylaminophosphorochloridite,[2210] followed by reaction with *N*-BOC-ethanolamine or *N*-tritylethanolamine, (b) by phosphorylation of 1,2-dipalmitoyl-*sn*-glycerol with 2-phthalimidoethyl phosphorodichloridate,[2266] 2-phthalimidoethyl dihydrogen phosphate,[2266] or 2-chloro-2-oxo-1,3,2-dioxaphospholane and trimethylamine,[2223] (c) by acylation of *sn*-glycero-3-phospho-*N*-phthaloylethanolamine[2293,2294] or polymer-supported *sn*-glycero-3-phospho-*N*-tritylethanolamine[2295] with palmitoyl chloride or palmitic anhydride, or (d) by condensation of 1,2-dipalmitoyl-*sn*-glycero-3-bromohydrin[2296] or -3-iodohydrin[2297] with silver phenyl 2-BOC-aminoethyl phosphate or silver benzyl 2-tritylaminoethyl phosphate.

POPE may be prepared (a) by condensation of 1-palmitoyl-2-oleoyl-*sn*-glycero-3-iodohydrin with silver *t*-butyl 2-(*N*-*t*-butoxycarbonyl)aminoethyl phosphate[2269] or of 2-tritylaminoethyl iodide with silver 1-palmitoyl-2-oleoyl-*sn*-glycero-3-(*p*-nitrobenzyl phosphate),[2268] (b) by condensation of 1-palmitoyl-2-oleoyl-*sn*-glycero-3-(dihydrogen phosphate) with *N*-BOC-ethanolamine in the presence of 2,4,6-triisopropylbenzenesulfonyl chloride,[2298] (c) by stepwise acylation of 1-trityl-*sn*-glycero-3-phospho-*N*-tritylethanolamine with oleoyl imidazolide and palmitic

Table 4-9 Molecular Species Composition: Phosphatidylethanolamines (mol%)

No.	Plant part	Method	16:0 16:0	16:0 18:1	16:0 18:2	16:0 18:3	18:0 18:1	18:0 18:2	18:0 18:3	18:1 18:1	18:1 18:2	18:1 18:3	18:2 18:2	18:2 18:3	18:3 18:3	References
Arabidopsis thaliana (L.) Heynh (Mouse-Ear Cress)																
1	Leaf plasma membrane	GLC	0.6[a]	0.8	43.0	21.0	0.4	1.5	Trace	1.2	2.4	14.8[b]		8.6	1.6	1478
Avena sativa L. (Oats)																
1	Leaf plasma membrane	GLC	1.2[a]	2.1	48.4	18.0	0.3		0.2	0.7	1.8	14.8[b]		8.5	2.6	1502
1	Root plasma membrane	GLC			63	14							17	6		2113
Brassica napus L. (Rape, low erucic acid (Canola))																
1	Leaf	HPLC			21.1	23.3			4.2				14.5	25.3	11.6	2117
Camellia sinensis L. (Tea)																
1	Pollen	HPLC				66									23	1558
Daucus carota L. (Carrot)																
1	Cell culture	AgTLC		2.2	48.0						9.7		33.4	6.7		1648
Dianthus caryophyllus L. (Carnation)																
1	Flower microsomal membr.	GLC		2.4	73.5			0.2			1.0	3.2	19.6			1652
Diptychocarpus strictus (Fisch.) Trautv.																
1	Seed	1R,2R	1.1	4.5	8.0	10.0	0.4	0.7	1.0	5.6	9.0	16.0	6.5	9.0	6.4	959
Echium plantagineum L. (Salvation Jane)																
1	Leaf	AgTLC			18	18					2[c]	5[c]	9.5	5.5[c]	2	2120
Glycine max (L.) Merr. (Soybean)																
1	Seed	HPLC	0.2	7.4[d]	32.9	4.2	0.7		0.8	1.8	8.1	2.0	33.1	8.2	0.6	1693
1	Seed	FD-MS		4.3	38.3	5.8				4.6[d]	9.9[e]		29.0	8.1		2134
1	Seed	CI-MS		5.6	44.0	2.5	0.7			7.0[d]	11.3[e]		24.1	3.3	1.4	2123
1	–, hypocotyl	AgTLC			42	7		6			4		23	16	Trace	1690
1	–, cotyledon	AgTLC			52	1		8			10		20	4	Trace	1690
1	Seedling, microsomes	HPLC		3.4	31.6	4.6		5.8	Trace	Trace	6.5	1.7	38.0	8.4[c]		2125

[a]Including 14:0/18:0; [b]Including 18:2/18:2 (LL); [c]*sn*-1/*sn*-2 positions reversed; [d]Including 18:0/18:2 (SL); [e]Including 18:0/18:3 (SIn); [f]Including 18:3/18:2 (LnL); [g]Including 18:1/18:1 (OO).

Continued

Table 4-9 Molecular Species Composition: Phosphatidylethanolamines (mol%)—cont'd

No.	Plant part	Method	16:0 16:0	16:0 18:1	16:0 18:2	16:0 18:3	18:0 18:1	18:0 18:2	18:0 18:3	18:1 18:1	18:1 18:2	18:1 18:3	18:2 18:2	18:2 18:3	18:3 18:3	References
Gossypium barbadense L. (Sea Island Cotton)																
1	Seed kernel	1R,2R	0.9	9.2	39.0		0.5	2.1		1.5	6.0		28.0			1709
Gossypium hirsutum L. (Upland Cotton)																
1	Seed	FAB-MS/MS			+++								+			2311
1	Seed kernel	1R,2R	2.2	7.2	38.3		0.3	1.9		0.7	3.5		26.4			1374
Hibiscus cannabinus L. (Kenaf)																
1	Seed	1R,2R	1.6	11.4	25.5	0.4	0.8	1.7		3.6	8.1	0.1	24.1	0.4		1377
Lavandula vera DC (English Lavender)																
1	Cell culture	AgTLC		10.4	12.4	18.7	6.9	4.9	15.7	3.5	2.0	11.4	1.5	2.0	4.4	1765
Lilium longiflorum Thunb. (Trumpet Lily)																
1	Pollen	HPLC	1.0		20.3	39.3							6.8	16.2[f]	9.1	1769
Lycopersicon esculentum Mill. (Tomato)																
1	Fruit, pericarp	GLC	1.8	1.8	53.4	4.5							19.4	0	4.1	2128
Nicotiana tabacum L. (Tobacco)																
1	Cell culture	AgTLC		Trace	67.4	13.2			10.5		Trace	Trace	8.9	Trace		1495
Oryza sativa L. (Rice)																
1	Bran, glutinous	AgTLC		21.1	33.4		(18:2/18:1, 7.0)			5.6	13.6		19.3			1845
1	Bran, glutinous	1R,2R		17.6	30.8		(18:2/18:1, 10.0)			6.6	11.5		17.4			1845
1	Bran, nonglutinous	AgTLC		16.1	38.3		(18:2/18:1, 7.6)			6.5	14.2		17.3			1845
1	Bran, nonglutinous	1R,2R		13.8	38.4		(18:2/18:1, 7.0)			4.1	11.3		19.5			1845
Perilla frutescens (L.) Britt. (Perilla Mint)																
1	Seed	AgTLC					1.7	20.9[g]			26.3[e]	14.8[b]		22.6	13.7	2094
1	Seed	1R,2R					2.3	17.6[g]			28.5[e]	16.0[b]		19.8	15.6	2094
Phaseolus coccineus L. (Scarlet Runner Bean)																
1	Seed	HPLC	0.3	6.9[d]	34.6	20.1	<0.1		0.6	0.8	5.3	3.7	12.6	11.8	3.3	1693
Phaseolus vulgaris L. (Kidney Bean)																
1	Seed	HPLC	0.7	10.4[d]	23.9	23.7	0.5		1.5	2.0	7.3	6.1	7.6	11.0	5.3	1693

Picea abies (L.) Karst. (Spruce)																
1	Shoot	GLC	6.0	3.6	35.6	0.6		6.0		3.6	2.7		28.5			2115
Pinus nigra Arnold (Black Pine)																
1	Needle (leaf)	HPLC	1	<1	14	10		1			1	11	5	2	1	1892
Pisum sativum L. (Pea)																
1	Seed	AgTLC		28.6	14.1		8.3			10.3	14.4		24.3			1904
1	Seed	HPLC	0.1	16.4[d]	21.7	1.4	2.6		0.6	8.1	21.9	1.8	22.1	2.6	0.5	1693
1	Leaf	HPLC		1.3	36.8	13.1	Trace	5.2	4.7		2.1		19.6	13.4[c]	2.4	2125
Raphanus sativus L. (Radish)																
1	Seedling	HPLC	6.1	3.1	21.8	27.0		3.1	3.5		6.5	7.3	9.2	11.0	4.5	2131
Rauvolfia serpentina Benth ex Kurz (Indian Snakeroot)																
1	Cell culture	AgTLC			59.1	2.7	Trace	4.6	0.5	0.4			32.7			1495
Secale cereale L. (Rye)																
2	Grain (1)	HPLC		4	34	3				2	11	2	33	10	1	1957
	Grain (2)			4	28	4				2	11	3	36	11	1	1957
1	Leaf plasma membrane	GLC	1.2[a]	2.2	39.1	22.0	0.4		Trace	0.4	0.9	14.6[b]		11.7	4.7	1502
Solanum tuberosum L. (Potato)																
1	Tuber	HPLC			38.2	9.1		7.1					34.4	11.1		1410
Triticum aestivum L. (Wheat)																
1	Flour	AgTLC		4.5	22.9	1.4	0.1	2.2		1.1	7.3	0.4	44.3	(18:2/18:1, 6.3)		2023
1	Flour	1R,2R	1.3	3.2	23.3	1.4	0.3	2.1	0.1	1.0	7.1	0.4	43.9	2.6	0.1	2023
Vicia faba L. (Broad Bean)																
1	Seed	HPLC	0.4	13.3[d]	26.0	3.6	2.6		1.0	7.5	20.3	1.2	21.5	2.1	0.4	1693
Vigna angularis (Willd.) Ohwi and Ohashi (Adzuki Bean)																
1	Seed	HPLC		3.8[d]	46.3	14.6			1.5		3.2	0.8	19.2	9.6	1.1	2051
1	Leaf	HPLC		5.0[d]	33.4	24.7			3.6		1.0	0.7	13.4	14.8	3.3	2051
1	Root	HPLC		2.1[d]	71.2	9.1			1.9		2.3	0.9	6.3	5.7	0.5	2051
	Similar data reported for PE molecular species in cotyledon, leaf, hypocotyl and root tissue of adzuki bean cell culture															2052
Zea mays L. ssp. *saccharata* Sturt (Sweet Corn)																
1	Seed	HPLC		6	23					5	26		39	1		2086

[a]Including 14:0/18:0.
[b]Including 18:2/18:2 (LL).
[c]*sn*-1/*sn*-2 positions reversed.
[d]Including 18:0/18:2 (SL).
[e]Including 18:0/18:3 (SLn).
[f]Including 18:3/18:2 (LnL).
[g]Including 18:1/18:1 (OO).

anhydride,[2278] (d) by the reaction of 1-palmitoyl-2-oleoyl-*sn*-glycero-3-(2-bromoethyl hydrogen phosphate) with ammonia,[2299] or (e) by reaction of POPC with ethanolamine in the presence of phospholipase D.[2227]

PLPE may be prepared (a) by condensation of 1-palmitoyl-2-linoleoyl-*sn*-glycero-3-iodohydrin with silver *p*-nitrobenzyl *N*-BOC-2-aminoethyl phosphate,[2275] (b) by acylation of *O*-methyl-1-palmitoyl-*sn*-glycero-3-phospho-(*N*-BOC)ethanolamine with linoleic acid in the presence of DCC,[2274] or (c) by reaction of PLPC with ethanolamine in the presence of phospholipase D.[2264] PLnPE may be prepared by condensation of 1-palmitoyl-2-linolenoyl-*sn*-glycerol 3-iodohydrin with silver benzyl 2-phthalimidoethyl phosphate.[2276]

SOPE may be prepared (a) by phosphorylation of 1-stearoyl-2-oleoyl-*sn*-glycerol with $POCl_3$ and quinoline followed by reaction with *N*-(2-hydroxyethyl)phthalimide,[2277] (b) by condensation of silver benzyl 1-stearoyl-2-oleoyl-*sn*-glycerol-3-phosphate with 2-iodoethylphthalimide,[2188] or (c) by stepwise acylation of 1-trityl-*sn*-glycero-3-phospho-*N*-tritylethanolamine with oleoyl imidazolide and stearic anhydride.[2278]

SLPE: None reported, but syntheses of the corresponding D- and DL-isomers are available.[2282,2288]

SLnPE may be prepared by condensation of 1-stearoyl-2-linolenoyl-*sn*-glycerol 3-iodohydrin with silver benzyl 2-tritylaminoethyl phosphate.[2300]

DOPE may be prepared (a) by phosphorylation of 1,2-dioleoyl-*sn*-glycerol with $POCl_3$/quinoline followed by reaction with *N*-(2-hydroxyethyl)phthalimide,[2283] (b) by acylation of *N*-phthaloyl-*sn*-glycero-3-phosphoethanolamine with oleoyl chloride[2301] or oleic anhydride[2302] or of *N*-trityl-*sn*-glycero-3-phosphoethanolamine with *N*-oleoylimidazole,[2201] (c) by condensation of 1,2-dioleoyl-*sn*-glycero-3-(2,2,2-tribromoethyl hydrogen phosphate) with *N*-tritylethanolamine in the presence of 2,4,6-triisopropylbenzenesulfonyl chloride,[2285] (d) by condensation of silver benzyl 1,2-dioleoyl-*sn*-glycero-3-phosphate with 2-tritylaminoethyl iodide,[2297] or (e) by reaction of DOPC with ethanolamine in the presence of phospholipase D.[2286]

OLPE may be prepared (a) by phosphorylation of 1-oleoyl-2-linoleoyl-*sn*-glycerol with $POCl_3$/quinoline followed by *N*-(2-hydroxyethyl) phthalimide[2287] or (b) by stepwise acylation of *N*-phthaloyl-*sn*-glycero-3-phosphoethanolamine with oleoyl chloride and linoleoyl chloride in the presence of pyridine.[2303]

DLPE may be prepared (a) by phosphorylation of 1,2-dilinoleoyl-*sn*-glycerol with 2-phthalimidoethyl phosphorodichloridate[2290] or *N*-(2,2,2-trichloroethoxycarbonyl)-2-aminoethyl phosphorodichloridate,[2291] (b) by acylation of *N*-phthaloyl-*sn*-glycero-3-phosphoethanolamine with linoleoyl chloride,[2289] (c) by condensation of 1,2-dilinoleoyl-*sn*-glycero-3-iodohydrin with silver benzyl 2-(*N*-tritylaminoethyl) phosphate[2268] or (e) by reaction of DLPC with ethanolamine in the presence of phospholipase D.[2227]

DLnPE may be prepared (a) by phosphorylation of 1,2,-dilinolenoyl-*sn*-glycerol with $POCl_3$/quinoline followed by *N*-BOC-ethanolamine,[2292] (b) by condensation of 1,2-dilinolenoyl-*sn*-glycero-3-phosphate with *N*-BOC-ethanolamine in the presence of trichloroacetonitrile,[2292] or (c) by reaction of DLnPC with ethanolamine in the presence of phospholipase D.[2227]

Hydrolysis

Hydrolysis of DPPE in acetate buffer at pH 4.0 follows pseudo-first-order kinetics ($k_{obs} = 0.7 \times 10^{-2}$ day^{-1} at 30 °C). There is a tendency toward faster degradation in the order DPPC < DPPC < DPPG.[2228] Upon saponification, DOPE yields 95% of the theoretical amount of oleic acid, showing that no conversion to elaidic acid has occurred.[2283] Methanolysis with HCl yields methyl oleate of 95-97% purity.[2302] Methanolysis of DLPE with BF_3/methanol yields about 98% methyl linoleate based on GC analysis.[2291]

Lysophosphatidylethanolamines

Physical Properties

1-Palmitoyl-*sn*-glycero-3-phospho-1′-ethanolamine (PHPE), 16:0 LPE, $C_{21}H_{44}NO_7P$. *(R)-Hexadecanoic Acid, 3-[[(2-aminoethoxy)hydroxyphosphinyl]oxy]-2-hydroxypropyl Ester.*
PHPE [R.N. 53862-35-4]: DSC.[2304] Spectra: Raman (fig.),[2304] ^{1}H NMR,[2305] ^{31}P NMR (fig.),[2304] FAB-MS,[2135] FD-MS (fig.),[2177] CI-MS (fig.),[2248] ESI-MS/MS (fig.),[2243] CID-MS/MS.[2265] Chromatography: TLC,[2240] HPTLC,[1405] silica gel column.[2305] DL-PHPE [R.N. 19805-23-3]: Monoclinic crystals, mp 212 °C, X-ray,[2306] PC.[2297]

2-Palmitoyl-*sn*-glycero-3-phospho-1′-ethanolamine (HPPE), 16:0 LPE, $C_{21}H_{44}NO_7P$. *(R)-Hexadecanoic Acid, 2-[[(2-aminoethoxy)hydroxyphosphinyl]oxy]-1-(hydroxymethyl)ethyl Ester.*
HPPE [R.N. 19805-25-5]: Crystals, mp 220 °C, soluble in chloroform and hot ethanol. Chromatography: PC, TLC.[2297]

1-Stearoyl-*sn*-glycero-3-phospho-1′-ethanolamine (SHPE), 18:0 LPE, $C_{23}H_{48}NO_7P$. *(R)-Octadecanoic Acid, 3-[[(2-aminoethoxy)hydroxyphosphinyl]oxy]-2-hydroxypropyl Ester.*
SHPE [R.N. 69747-55-3]: DSC.[2304] Spectra: IR,[2201] ^{1}H NMR,[2201] FAB-MS.[2135] Chromatography: TLC, silicic acid column.[2201]

1-Oleoyl-*sn*-glycero-3-phospho-1′-ethanolamine (OHPE), 18:1 LPE, $C_{23}H_{46}NO_7P$. *[R-(Z)]-9-Octadecenoic Acid, 3-[[(2-aminoethoxy)hydroxyphosphinyl]oxy]-2-hydroxypropyl Ester.*
OHPE [R.N. 89576-29-4]: White waxy plates, mp 202-204 °C dec., $[\alpha]_D^{20}$ −2.5° ($CHCl_3$).[2307] Spectra: IR (fig.),[2307] ^{1}H and ^{31}P NMR (fig.),[2308] FAB-MS,[2135] FD-MS,[2177] X-ray.[2308] ESI-MS/MS (fig.),[2243] Chromatography: TLC,[2307] silica column,[2308] CM-cellulose column.[2308]

2-Oleoyl-*sn*-glycero-3-phospho-1′-ethanolamine (HOPE), 18:1 LPE, $C_{23}H_{46}NO_7P$. *[R-(Z)]-9-Octadecenoic Acid, 2-[[(2-aminoethoxy)hydroxyphosphinyl]oxy]-1-(hydroxymethyl)ethyl Ester.*
HOPE [R.N. 60701-97-5]: ESI-MS/MS.[2243]

1-Linoleoyl-*sn*-glycero-3-phospho-1′-ethanolamine (LHPE), 18:2 LPE, $C_{23}H_{44}NO_7P$. *[R-(Z,Z)]-9,12-Octadecadienoic Acid, 3-[[(2-aminoethoxy)hydroxyphosphinyl]oxy]-2-hydroxypropyl Ester.*

LHPE [R.N. 85046-18-0]: Spectra: IR,[2252] ^{1}H and ^{13}C NMR,[2252] ^{31}P NMR (fig.),[2308] FAB-MS,[2252] FD-MS,[2122] X-ray.[2308] Chromatography: TLC,[2252] HPLC,[2252] silica column,[2252] CM-cellulose column.[2308]

2-Linoleoyl-*sn*-glycero-3-phospho-1′-ethanolamine (HLPE), 18:2 LPE, $C_{23}H_{44}NO_7P$. *[R-(Z, Z)]-9,12-Octadecadienoic Acid, 2-[[(2-aminoethoxy)hydroxyphosphinyl]oxy]-1-(hydroxymethyl)ethyl Ester.*

HLPE [R.N. 60701-98-6] Spectra: IR, ^{1}H NMR, ^{13}C NMR, FAB-MS.[2252] Chromatography: TLC, HPLC, silica column.[2252]

1-Linolenoyl-*sn*-glycero-3-phospho-1′-ethanolamine (LnHPE), 18:3 LPE, $C_{23}H_{42}NO_7P$. *[R-(Z,Z,Z)]-9,12,15-Octadecatrienoic Acid, 3-[[(2-aminoethoxy)hydroxyphosphinyl]oxy]-2-hydroxypropyl Ester.*

LnHPE [R.N. 99881-83-1]: Spectra: ^{1}H NMR, ^{31}P NMR (fig.), X-ray.[2308] Chromatography: TLC, silica column, CM-cellulose column.[2308]

Synthesis

PHPE may be prepared (a) by phosphorylation of 2-benzyl-1-palmitoyl-*sn*-glycerol with $POCl_3$ followed by reaction with *N*-BOC-ethanolamine,[2240] (b) by acylation of *N*-trityl-*sn*-glycero-3-phosphoethanolamine with palmitic anhydride or *N*-palmitoylimidazole,[2305] or (c) by enzymatic hydrolysis of DPPE with snake venom phospholipase A_2.[2309] HPPE may be prepared by condensation of 2-palmitoyl-1-trityl-*sn*-glycero-3-iodohydrin with silver benzyl 2-tritylaminoethyl phosphate.[2297]

SHPE may be prepared by acylation of *N*-trityl-*sn*-glycero-3-phosphoethanolamine with stearic anhydride[2201] or *N*-stearoylimidazole.[2305]

OHPE may be prepared (a) by condensation of 1-oleoyl-*sn*-glycero-3-iodohydrin with silver benzyl 2-phthalimidoethyl phosphate[2307] or (b) by enzymatic hydrolysis of DOPE with snake venom phospholipase A_2.[2308] HOPE may be prepared by enzymatic hydrolysis of DOPE with lipase from the mold *R. arrhizus*.[2309]

LHPE may be prepared (a) by migration of the linoleoyl group during the preparation of HLPE from soybean PE in phosphate buffer[2252] or (b) by enzymatic hydrolysis of DLPE with snake venom phospholipase A_2.[2308] HLPE may be prepared by enzymatic hydrolysis of soybean PE with lipase from the fungus *M. javanicus*.[2252]

LnHPE may be prepared by enzymatic hydrolysis of DLnPE with snake venom phospholipase A_2.[2308]

Migration

Migration of the linoleoyl group from *sn*-2 to *sn*-1 occurs during the preparation of HLPE from DLPE in phosphate buffer, resulting in a 91:9 mixture of HLPE and LHPE, but this can be prevented by the use of a borate buffer.[2252]

N-Acylphosphatidylethanolamines

Occurrence

The occurrence of NAPE and NALPE in plants was first reported by Bomstein[1358] in 1965, and subsequently shown to be widespread in plant tissues. The metabolic pathways for the biosynthesis and turnover of these phospholipids are outlined in Fig. 4-2.

Physical Properties

1,2,*N*-Tripalmitoyl-*sn*-glycero-3-phospho-1′-ethanolamine (*N*-palmitoyl-DPPE), 16:0-16:0/16:0 NAPE, $C_{53}H_{104}NO_9P$. *(R)-Hexadecanoic Acid, 5-Hydroxy-5-oxido-10-oxo-4,6-dioxa-9-aza-5-phosphapentacosane-1,2-diyl Ester*

N-Palmitoyl-DPPE [R.N. 57984-41-5]: mp 52-53 °C, $[\alpha]_D^{20}$ +5.37° ($CHCl_3$),[2313] DSC (fig.).[2314] Spectra: IR (fig.),[2315] Raman (fig.),[2315] FTIR (fig.),[2316] ATR-IR (fig.),[2317] ^{1}H NMR (fig.),[1361] FAB-MS.[1361] Chromatography: TLC.[2313]

1,*N*-Dipalmitoyl-2-linoleoyl-*sn*-glycero-3-phospho-1′-ethanolamine (*N*-palmitoyl-PLPE), 16:0-16:0/18:2 NAPE, $C_{55}H_{104}NO_9P$. *[R-(Z,Z)]-9,12-Octadecadienoic Acid, [[[Hydroxy[2-[(1-oxohexadecyl)amino]ethoxy]phosphinyl]oxy]methyl]-2-[(1-oxohexadecyl) oxy]ethyl Ester.*

N-Palmitoyl-PLPE [R.N. 168434-64-8]: FAB-MS, CAD-MS/MS.[2311]

2,*N*-Dilinoleoyl-1-palmitoyl-*sn*-glycero-3-phospho-1′-ethanolamine (*N*-linoleoyl-PLPE), 18:2-16:0/18:2 NAPE, $C_{57}H_{104}NO_9P$. *[R-(all Z)]-9,12-Octadecadienoic Acid, Ester with 1,2,3-Propanetriol Monohexadecanoate 1-[2-[(1-oxo-9,12-octadecadienyl)amino]ethyl Hydrogen Phosphate].*

N-Linoleoyl-PLPE [R.N. 168434-66-0]: FAB-MS, CAD-MS/MS.[2311]

1,2-Dilinoleoyl-*N*-palmitoyl-*sn*-glycero-3-phospho-1′-ethanolamine (*N*-palmitoyl-DLPE), 16:0-18:2/18:2 NAPE, $C_{57}H_{104}NO_9P$. *[R-(all Z)]-9,12-Octadecadienoic Acid, 1-(3-Hydroxy-3-oxido-8-oxo-2,4-dioxa-7-aza-3-phosphatricos-1-yl)-1,2-ethanediyl Ester.*

N-Palmitoyl-DLPE [R.N. 168294-52-8]: FAB-MS, CAD-MS/MS.[2311]

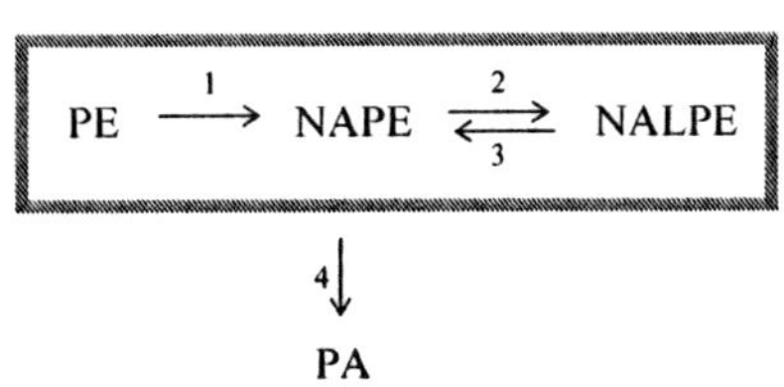

FIGURE 4-2 Metabolism of *N*-Acyl Phospholipids in Plants. (1) PE *N*-acyltransferase, fatty acid.[1361] (2) Phospholipase A_2.[2310] (3) Fatty acid, Ca(II).[2310] (4) Phospholipase D.[2310]

Table 4-10 Molecular Species Composition: *N*-Acylphosphatidylethanolamines (mol%)

No.	Plant part	Method	16:0 16:0	16:0 18:1	16:0 18:2	16:0 18:3	18:0 18:1	18:0 18:2	18:0 18:3	18:1 18:1	18:1 18:2	18:1 18:3	18:2 18:2	18:2 18:3	18:3 18:3	References
Glycine max (L.) Merr. (Soybean)																
1	Seed[a]	HPLC	0.7	7.8[b]	37.8	7.5	0.5		1.3	0.5	6.0	2.0	25.8	8.1	2.0	1693
Gossypium hirsutum L (Upland Cotton)																
1	Seed[c]				+++									+		2311
Phaseolus coccineus L. (Scarlet Runner Bean)																
1	Seed[a]	HPLC	0.9	10.2[b]	37.9	20.8	0.5		0.3	1.5	6.0	3.1	8.5	7.8	2.5	1693
Phaseolus vulgaris L. (Kidney Bean)																
1	Seed[a]	HPLC	1.2	12.1[b]	27.6	24.9	0.5		0.2	1.8	6.9	5.3	6.0	8.9	4.5	1693
Pisum sativum L. (Pea)																
1	Seed[d]	AgTLC	0.4	24.7	25.3		5.9			16.1	13.9		13.2	(16:0/18:0, 0.4)		2312
1	Seed[a]	HPLC	0.4	17.0[b]	23.9	4.1	2.6		0.6	6.9	19.2	1.7	17.9	3.4	2.5	1693
Triticum aestivum L. (Wheat)																
1	Flour[e]	HPLC		11.11	23.84	3.10	12.41			2.01	6.53	Trace	22.06	1.28		2021
Vicia faba L. (Broad Bean)																
1	Seed[a]	HPLC	1.0	14.9[b]	26.5	1.7	2.8		1.2	7.7	19.8	0.9	18.8	2.3	2.3	1693
Vigna angularis (Willd.) Ohwi and Ohashi (Adzuki Bean)																
1	Seed[a]	HPLC	1.0	4.3[b]	37.8[f]	13.2	0.7			1.2	4.8	22.6[g]		12.4	2.0	1693

[a]N-Acyl group not identified.
[b]Including 18:0/18:2 (SL).
[c]*N*-Acyl mainly 16:0 or 18:2.
[d]*N*-Acyl mainly 16:0, 18:1, or 18:2.
[e]*N*-Acyl mainly 18:2.
[f]Including 18:3/18:0 (LnS).
[g]Including 18:2/18:2 (LL).

Molecular Species Composition

The molecular species composition of *N*-acylphosphatidylethanolamine in plants, determined by any of the methods described on pages 168–171, is given in Table 4-10.

Synthesis

N-Palmitoyl-DPPE may be prepared by acylation of DPPE with palmitic anhydride in pyridine.[2313]. No synthesis of any of the other NAPEs has been reported.

Hydrolysis

Hydrolysis or methanolysis of DL-1,2,*N*-tripalmitoyl NAPE at room temperature converts it not only to the expected deacylation product, DL-glycerol 1-(2-palmitoylaminoethyl hydrogen phosphate), but also, in 20% or higher yield, to *N*-(2-hydroxyethyl)palmitamide. Subsequent heating in aqueous solution at pH 1 cleaves all of the product to *N*-(2-hydroxyethyl)palmitamide.[2318]

PHOSPHATIDYLGLYCEROLS

This group comprises the phosphatidylglycerols (PG), two kinds of lysophosphatidylglycerols (LPG), and the diphosphatidylglycerols (DPG):

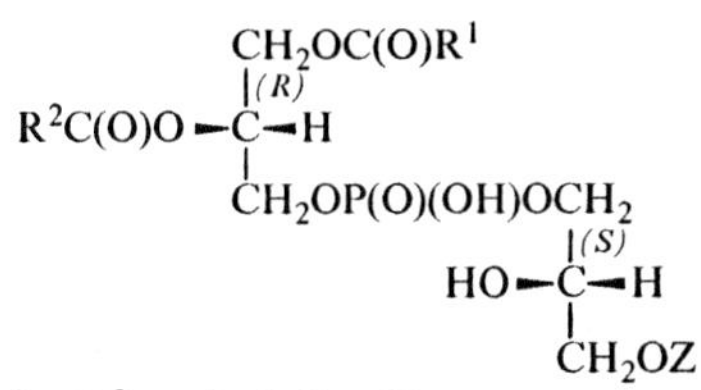

PG, $R^1 = R^2$ = alkyl, Z = H
1-LPG, R^1 = Z = H, R^2 = alkyl
2-LPG, R^1 = alkyl, R^2 = Z = H
DPG, $R^1 = R^2$ = alkyl, Z = $P(O)(OH)OCH_2CH(CH_2OC(O)R^1)OC(O)R^2$

Phosphatidylglycerols

Nomenclature

Phosphatidylglycerols containing acyl groups other than *palmitoyl* are named in the IUB system by substituting stearoyl, oleoyl, linoleoyl, linolenoyl, or 3-*trans*-hexadecanoyl for *palmitoyl* and in the Chemical Abstracts system by substituting octadecyl, 9-octadecenyl, 9,12-octadecadienyl, 9,12,15-octadecatrienyl, or 3-hexadecenyl for *hexadecyl* and the prefix [*R*-(*R**,-*S**)], [*R*-[(*R**,*S**-(*Z*))]], [*R*-[(*R**,*S**-(*Z*,*Z*))]], [*R*-[(*R**,*S**-(*Z*,*Z*,*Z*))]], or [*R*-[*R**,*S**-(*E*)]], respectively, for [*R*-(*R**,*S**)]. Chemical Abstracts Registry Numbers are given for each of the phosphatidylglycerols in the following paragraphs.

Caution: The numbering of the glycerol carbon atoms in the two systems is different and the naming and numbering of isomer pairs such as 18:0/18:3 and 18:3/18:0 in Chemical Abstracts is also different.

Stereochemistry

Because both of the glyceryl moieties are asymmetric, PG has two chiral centers—the central carbon atom of each moiety—there are four possible configurations: [*R*,*R*], [*R*,*S*], [*S*,*R*], and [*S*,*S*]. The configuration in the molecular formula illustrated above, [*R*,*S*], is the natural configuration of cabbage leaf and soybean PG[1385] and may well be the natural configuration of all plant PGs. Egg yolk PG, by contrast, contains a 45:55 mixture of [*R*,*S*] and [*R*,*R*] isomers, and *E. coli* PG contains an 89:11 mixture of [*R*,*S*] and [*R*,*R*] isomers.[1385] The orientation of the H and OH atoms in the 1′-glycerol moiety (the lower one in the formula illustrated above) is reversed in the [*R*,*R*] isomer. To date, there are no natural PGs in which the orientation of the other (upper) glycerol moiety is *S*, i.e., [*S*,*R*] or [*S*,*S*].

The absolute stereochemical configuration of PG is not easily determined. Optical rotation alone is inadequate, and enzymic methods based on phospholipases A, C, and D are not absolutely specific except for phospholipase A_2. More rigorous methods are required, such as chemical degradation,[1203] chiral phase HPLC,[1385] or synthetic methods employing reagents of known stereochemistry.

In DPG, there is only one chiral center if the acyl substituents are identical, in which case there are only two possible configurations—R and S—and the plant DPGs should contain only the R isomer.

Occurrence

The principal phosphatidylglycerols of plant tissues are listed below under *Physical Properties*. Those whose molecular species distribution is known are listed in Table 4-11. The phosphatidylglycerols differ from all the other phospholipids (PC, PE, PI, PA, and PS) in two respects.

- Many of them contain high proportions of the rare acid *trans*-3-hexadecenoic acid, located exclusively at the 2-position of the 1-glyceryl moiety (upper one in the formula given above). Roughly half of the PG comprises molecular species in which the C_2 substituent is *trans*-3-hexadecenoyl and the C_1 substituent is palmitoyl (16:0), stearoyl (18:0), oleoyl (18:1), linoleoyl (18:2), or linolenoyl (18:3). Surveys have shown that PG containing *trans*-3-hexadecenoic acid is characteristic of photosynthetic plant tissue but is not present in etiolated or nonphotosynthetic tissue, or in the PG component of blue-green algae or photosynthetic bacteria.[1348]
- The relative *sn-1* / *sn-2* position of the acyl substituents in the PG tends to have the reverse of the usual order of the phospholipids, e.g. 18:2/16:0 rather than 16:0/18:2.

A few of the molecular species of PG reported in Table 4-11 contain no *trans*-3-hexadecenoyl substituents and their profiles resemble those of the other phospholipids. They are reported separately on the last page of the table under the regular headings.

Physical Properties

1,2-Dipalmitoyl-*sn*-glycero-3-phospho-1′-*sn*-glycerol (DPPG), 16:0/16:0 PG. *[R-(R*,S*)]-Hexadecanoic Acid, 1-[[[(2,3-Dihydroxypropoxy)hydroxyphosphinyl]oxy]methyl]-1,2-ethanediyl Ester.*

DPPG [R.N. 74300-16-6]: Homogeneous solid.[2285] Spectra: ^{1}H NMR,[2285] ^{2}H and ^{31}P NMR (figs.).[2321] Chromatography: TLC,[2285] HPLC.[2064] *Na salt* (1:1) [R.N. 74300-11-1]: $[\alpha]_{546}^{30}$ +7.6° ($CHCl_3$/MeOH).[2321] *DPPG* (*R,S* unspecified) [R.N. 4537-77-3]: pK_a 2.9,[2322] DSC.[2177] Spectra: FTIR,[2323] FAB-MS,[2200] FD-MS,[2177] FAB MS/MS (fig.),[2136] fluorescence polarization (fig.).[1631] Chromatography: HPTLC.[1405] Electrophoresis: CE.[2324] *Na salt* (1:1) [R.N. 67232-81-9]: DSC,[2325] LSI-MS.[2239] *R,R-DPPG* [R.N. 74313-95-4]: $[\alpha]_D^{20}$ +6.93° ($CHCl_3$/MeOH 2:1). Spectra: ^{13}C and ^{31}P NMR.[2210] *R,rac-DPPG* [R.N. 220407-65-8]: Oil, $[\alpha]_D^{20}$ +6.1° ($CHCl_3$/MeOH 2:1).[2326] Spectra: IR.[2326] Chromatography: TLC,[2326] chiral phase HPLC,[1385] SFC.[2156] Ba salt (2:1) [R.N. 91859-08-4]: Crystals, mp 145-146 °C, $[\alpha]_D^{20}$ +9.0° (pyridine).[2326]

1-Palmitoyl-2-*trans*-3-hexadecenoyl-*sn*-glycero-3-phospho-1′-*sn*-glycerol (P(3t)PG), 16:0/16:1 PG, $C_{38}H_{73}O_{10}P$. *[R-[R*,S*-(E)]]-3-Hexadecenoic Acid, 1-[[[(2,3-Dihydroxypropoxy)hydroxyphosphinyl]oxy]methyl]-2-[(1-oxohexadecyl)oxy]-ethyl Ester.*

P(3t)PG [R.N. 85967-07-3]: Chromatography: HPLC.[2064]

1-Stearoyl-2-palmitoyl-*sn*-glycero-3-phospho-1′-*sn*-glycerol (SPPG), 18:0/16:0 PG, $C_{40}H_{75}O_{10}P$. *[R-[R*,S*]]-Octadecanoic Acid, 1-[[[(2,3-Dihydroxypropoxy)hydroxyphosphinyl]oxy]methyl]-2-[(1-oxohexadecyl)oxy]ethyl Ester.*

SPPG [R.N. 132922-64-6]: Chromatography: HPLC.[1594]

1-Stearoyl-2-*trans*-3-hexadecenoyl-*sn*-glycero-3-phospho-1′-*sn*-glycerol (S(3t)PG), 18:0/16:1 PG, $C_{40}H_{73}O_{10}P$. *[R-[R*,S*-(E)]]-Octadecanoic Acid, 3-[[(2,3-Dihydroxypropoxy)hydroxyphosphinyl]oxy]2-[(1-oxo-3-hexadecenyl)oxy]propyl Ester.*

S(3t)PG [R.N. 132943-54-5]: Chromatography: HPLC.[2064]

1-Oleoyl-2-palmitoylphosphatidylglycerol (OPPG), 18:1/16:0 PG, $C_{40}H_{77}O_{10}P$. *[R-[R*,S*-(Z)]]-9-Octadecenoic Acid, Monoester with 1,2,3-Propanetriol 1,1′-(Hydrogen phosphate) Monohexadecanoate*

OPPG [R.N. 134623-65-7]: Chromatography: HPLC.[1892] *OPPG* (*R,S* unspecified) [R.N. 81490-05-3]: Electrophoresis: CE (fig.).[2181] Na salt (1:1): White powder, mp 176-179 °C, $[\alpha]_D^{20}$ +1.02° ($CHCl_3$)[2327]; IR.[2327] Chromatography: PC, TLC, silica gel column.[2327]

1-Oleoyl-2-*trans*-3-hexadecenoyl-*sn*-glycero-3-phospho-1′-*sn*-glycerol O(3t)PG, 18:1/16:1 PG, $C_{40}H_{75}O_{10}P$. *[R-[R*,S*-(E,Z)]]-9-Octadecenoic Acid, 3-[[(2,3-Dihydroxypropoxy)hydroxyphosphinyl]oxy]-2-[(1-oxo-3-hexadecenyl)oxy]propyl Ester.*

O(3t)PG [R.N. 132922-63-5]: Chromatography: HPLC.[1594]

1,2-Dioleoyl-*sn*-glycero-3-phospho-1′-*sn*-glycerol (DOPG), 18:1/18:1 PG, $C_{42}H_{79}O_{10}P$. *[R-[R*,S*-(Z)]]-9-Octadecenoic Acid, 1-[[[(2,3-Dihydroxypropoxy)hydroxyphosphinyl]oxy]-methyl]-1,2-ethanediyl Ester*

Table 4-11 Molecular Species Composition: Phosphatidylglycerols (mol%)

No.	Plant part		16:0 16:0	18:0 16:0	18:1 16:0	18:2 16:0	18:3 16:0	16:0 16:1	18:0 16:1	18:1 16:1	18:2 16:1	18:3 16:1	References
Allium fistulosum L. (Welsh Onion)													
1	Leaf	GLC	5	1	4	27	26	2	0	1	11	24	2112
Aquilegia alpina L. (Alpine Columbine)													
1	Leaf	HPLC			3.2	16.3	10.5				25.8	42.6	2109
Arabidopsis thaliana (L.) Heynh (Mouse-Ear Cress)													
1	Leaf	HPLC	3.5		7.2	11.7	19.9	4.8	0.5	2.3	6.2	43.3	2319
Avena sativa L. (Oats)													
1	Leaf	GLC	3	0	2	11	33	2	0	0	3	45	2112
Avicennia germinans (L.) L. (Black Mangrove)													
2	Chilling-tolerant, leaf	GLC	41.6[a]	Trace	40.9[b]	6.2	4.8		2.8		3.7	Trace	2138
	–, root	GLC	34.8[a]	9.0	12.8	10.1	2.2				2.3	2.6	2138
	Chilling-sensitive, leaf	GLC	49.2[a]	Trace	31.7[b]	6.4	5.2		2.3		5.2	Trace	2138
	–, root	GLC	21.3[a]	4.2	18.6	24.5	7.9				Trace	2.5	2138
Brassica campestris L. ssp. *komatsuma* Mak. (Chinese Cabbage)													
1	Leaf	GLC	4	1	2	9	47	3	0	0	3	32	2112
Brassica napus L. (Rape, low erucic acid (Canola))													
1	Leaf	HPLC			2.0	21.0	53.0				4.0	20.0	2117
Brassica oleracea L. ssp. *capitata* L. (Cabbage)													
1	Leaf	GLC	3	0	4	45	43	1	0	0	2	2	2112
Citrus amblycarpa (Hassk.) Ochse (Djeroek Limoe)													
1	Leaf	HPLC	4	1	20	1[c]	3	10	4	46	5	7	1594
Citrus grandis (L.) Osbeck (Tosa-Buntan)													
1	Leaf	HPLC	6	1	11	6[c]	3	22	3	26	9	10	1594
Citrus junos (Sieb.) Tanaka (Yuzu)													
1	Leaf	HPLC	4	1	17	1[c]	1	14	6	43	5	6	1594
Citrus × paradisi Macf. (Pomelo)													
1	Leaf	HPLC	7	1	34	2[c]	2	11	2	32	4	4	1594
Citrus sinensis (L.) Osbeck (Orange)													
2	Leaf (1)	HPLC	9	1	23	3[c]	2	11	3	36	5	3	1594
	Leaf (2)	HPLC	4	1	21	1[c]	1	11	4	47	5	4	1594
Citrus unshiu Marc. (Mandarin)													
1	Leaf	HPLC	9	3	35	3[c]	2	10	4	29	2	3	1594

Colocasia esculenta (L.) Schott (Taro)													
1	Leaf	GLC	20	1	3	11	7	42	3	3	4	6	2112
Cucumis sativus L. (Cucumber)													
1	Leaf	HPLC	22.0	14.0	26.0	4.0	5.0	23.4			2.1	3.5	2104
Cucurbita moschata Duch. (Warty Squash)													
1	Leaf	GLC	17	0	0	13	23	24	2	5	3	23	2112
Cyclamen persicum Mill. (Cyclamen)													
1	Leaf	GLC	5	0	11	8	4	21	3	34	8	5	2112
Gossypium hirsutum L. (Upland Cotton)													
1	Leaf	HPLC	33.4	7.4	13.1	3.5	5.0	26.5			5.6	5.6	2104
Ipomoea batatas (L.) Lam. (Sweet Potato)													
1	Leaf	GLC	26	3	7	4	2	39	3	8	4	4	2112
Kalanchoë daigremontiana Ham. and Perr. (Devil's Backbone)													
1	Leaf	GLC	32	3	19	26	11	2	0	0	1	6	2112
Lactuca sativa L. (Lettuce)													
1	Leaf	GLC	13	0	1	26	26	4	1	1	4	24	2112
Luffa aegyptiaca Mill. (Sponge Gourd)													
1	Leaf	GLC	16	2	4	5	4	23	13	19	3	11	2112
Lycoris radiata Herb. (Red Spiderlily)													
1	Leaf	GLC	6	0	6	28	23	5	2	3	7	20	2112
Nerium oleander L. (Oleander)													
1	Leaf, thylakoid	HPLC	2.6	0.7	21.4	13.0	2.2	4.6	Trace	50.8	7.9	2.2	1814
Nicotiana tabacum L. (Tobacco)													
1	Leaf	GLC	19	0	1	11	7	20	5	15	7	16	2112
1	Leaf	HPLC				17.3	11.6				21.9	49.1	1410
Oriza sativa L. (Rice)													
2	Leaf thylakoid, chilling-resistant	GLC	13.3		9.8	9.6	5.9	38.7	1.4		13.4	7.9	2320
	Leaf thylakoid, chilling-sensitive	GLC	33.7	0.6	10.2	1.8	0.8	42.1	2.7	4.3	1.2	17.2	2320
		Data also given for 15 other varieties having intermediate levels of chilling resistance											2320

[a]Including about 10-15% of 16:0/16:1 PG; [b]Including 18:1/16:1 PG; [c]Including 16:0/18:3 (PLn); [d]Including 18:3/18:0 (LnS); [e]Including 18:0/18:2 (SL); [f]Including 14:0/18:0; [g]Including 18:2/18:2 (LL).

Continued

Table 4-11 Molecular Species Composition: Phosphatidylglycerols (mol%)—cont'd

No.	Plant part		16:0 16:0	18:0 16:0	18:1 16:0	18:2 16:0	18:3 16:0	16:0 16:1	18:0 16:1	18:1 16:1	18:2 16:1	18:3 16:1	References
Pinus nigra Arnold (Black Pine)													
1	Needle (leaf)	HPLC	4		<1	36	(18:1/18:3, 13)		3		13[b]	18	1892
Pisum sativum L. (Pea)													
1	Leaf	GLC	12	1	27	15	15	7	0	6	10	7	2112
1	Leaf chloroplast	HPLC		1	15	22	28				5	28	1909
1	Leaf mitochondria	HPLC			9	36	38[d]						1909
Raphanus acanthiformis Backer (Japanese Radish)													
1	Leaf	GLC	5	1	3	5	51	3	0	1	4	26	2112
Raphanus sativus L. (Radish)													
1	Seedling	HPLC	2.8		6.2	19.8	58.9	0.1				1.6	2131
Ricinus communis L. (Castor Bean)													
1	Leaf	GLC	20	7	15	3	1	22	1	29	1	1	2112
Spinacea oleracea L. (Spinach)													
1	Leaf	AgTLC				2	9	7	(16:0/18:3, 12)			56	2095
1	Leaf	GLC	1	0	1	5	16	5	0	0	0	72	2112
1	Leaf, chloroplast envelope	GC-MS					+					+++	2141
Taraxacum spp. (Dandelion)													
1	Leaf	GLC	2	0	3	16	15	2	0	12	29	20	2112
Trifolium pratense L. (Red Clover)													
1	Leaf	GLC	4	0	5	11	12	6	0	23	14	25	2112
Triticum aestivum L. (Wheat)													
1	Leaf	GLC	1	0	2	11	32	2	0	0	5	47	2112
Vigna unguiculata (L.) Walp. (Cowpea)													
2	Leaf, drought-resistant	HPLC	12		5		8	39	10		5	19	2064
	Leaf, drought-sensitive	HPLC	11		6		14	53				16	2064
Zea mays L. (Corn)													
1	Leaf	GLC	20	0	0	5	17	17	0	4	11	26	2112

No.	Plant part	Method	16:0 16:0	16:0 18:1	16:0 18:2	16:0 18:3	18:0 18:1	18:0 18:2	18:0 18:3	18:1 18:1	18:1 18:2	18:1 18:3	18:2 18:2	18:2 18:3	18:3 18:3	References
								Note changed table headings								
Daucus carota L. (Carrot)																
1	Cell culture	AgTLC		7.2	59.1						5.5		19.3	8.9		1648
Dianthus caryophyllus L. (Carnation)																
1	Flower microsomal membr.	GLC	17.6	10.2	64.6	7.3		(16:0/18:0, 0.2)								1652
Glycine max (L.) Merr. (Soybean)																
1	Seed	HPLC	2.9	14.1[e]	52.2	7.3	0.8		1.8	2.4	5.3	0.9	9.7	2.0	0.6	1693
Phaseolus coccineus L. (Scarlet Runner Bean)																
1	Seed	HPLC	7.5	13.9[e]	47.2	20.7	0.9		0.4	0.5	2.4	1.2	2.7	1.8	0.8	1693
Phaseolus vulgaris L. (Kidney Bean)																
1	Seed	HPLC	3.5	13.1[e]	30.9	34.5	0.6		2.5	1.1	2.2	2.6	2.9	3.6	2.6	1693
Pisum sativum L. (Pea)																
1	Seed	HPLC	20.1	24.6[e]	37.3	4.2	0.7		4.0	0.6	2.0	3.4	2.2	0.5	0.3	1693
Secale cereale L. (Rye)																
1	Leaf plasma membrane	GLC	5.0[f]	5.1	47.6	28.1						Trace[g]		Trace		1962
Vicia faba L. (Broad Bean)																
1	Seed	HPLC	15.3	23.6[e]	50.4	1.2	1.0		0.2	0.8	3.4	0.2	2.9	0.4	0.5	1693
Vigna angularis (Willd.) Ohwi and Ohashi (Adzuki Bean)																
1	Seed	HPLC	18.4	7.8[e]	47.0	12.5			0.5		2.5	1.2	4.8	3.8	1.6	1693
1	Cell culture, cotyledon	HPLC	Trace	11[e]	16	21			13				3	7		2052
1	–, leaf	HPLC	Trace	8[e]	21	26			9				6	9		2052
1	–, hypocotyl	HPLC	Trace	10[e]	15	29			8				8	8		2052

[a]Including about 10-15% of 16:0/16:1 PG.
[b]Including 18:1/16:1 PG.
[c]Including 16:0/18:3 (PLn).
[d]Including 18:3/18:0 (LnS).
[e]Including 18:0/18:2 (SL).
[f]Including 14:0/18:0.
[g]Including 18:2/18:2 (LL).

DOPG [R.N. 109009-74-7]: DSC.[2328] *DOPG* (*R,S* unspecified) [R.N. 62700-69-0]: Spectra: ^{1}H and ^{13}C NMR,[2329] ^{31}P NMR (fig.).[2286] Chromatography: TLC,[2166] HPLC.[2251] Na salt (1:1): Spectra: FAB-MS,[2329] LSI-MS.[2239] *R,rac DOPG*: Spectra: MS (fig.).[1385] Chromatography: Chiral phase HPLC.[1385]

1-Linoleoyl-2-palmitoyl-*sn*-glycero-3-phospho-1′-*sn*-glycerol (LPPG), 18:2/16:0 PG, $C_{40}H_{75}O_{10}P$. *[R-[R*,S*-(Z,Z)]]-9,12-Octadecadienoic Acid, 3-[[(2,3-Dihydroxypropoxy)-hydroxyphosphinyl]oxy]-2-[(1-oxohexadecyl)oxy]propyl Ester.*
LPPG [R.N. 120444-75-9]: Chromatography: HPLC.[1410]

1-Linoleoyl-2-*trans*-3-hexadecenoyl-*sn*-glycero-3-phospho-1′-*sn*-glycerol (L(3t)PG), 18:2/16:1 PG, $C_{40}H_{73}O_{10}P$. *[R-[R*,S*-(E,Z,Z)]]-9,12-Octadecadienoic Acid, 3-[[(2,3-Dihydroxypropoxy)-hydroxyphosphinyl]oxy]-2-[(1-oxo-3-hexadecenyl)oxy]propyl Ester.*
L(3t)PG [R.N. 120407-77-4]: Chromatography: HPLC.[1410,2104]

1,2-Dilinoleoyl-*sn*-glycero-3-phospho-1′-*sn*-glycerol (DLPG), 18:2/18:2 PG, $C_{42}H_{75}O_{10}P$. *[R-[(R*,S*-(all Z))]-9,12-Octadecadienoic Acid, 1-[[[(2,3-Dihydroxypropoxy) hydroxyphosphinyl]-oxy]methyl]-1,2-ethanediyl Ester.*
DLPG [R.N. 120851-76-5]: Chromatography: HPLC.[2064] *R,rac-DLPG*: Chiral phase HPLC.[1385]

1-Linolenoyl-2-palmitoyl-*sn*-glycero-3-phospho-1′-*sn*-glycerol (LnPPG), 18:3/16:0 PG, $C_{40}H_{73}O_{10}P$. *[R-[R*,S*-(Z,Z,Z)]]-9,12,15-Octadecatrienoic Acid, 3-[[(2,3-Dihydroxypropoxy)-hydroxyphosphinyl]oxy]-2-[(1-oxohexadecyl)oxy]propyl Ester.*
LnPPG [R.N. 120407-76-3]: Chromatography: HPLC.[2109]

1-Linolenoyl-2-*trans*-3-hexadecenoyl-*sn*-glycero-3-phospho-1′-*sn*-glycerol (Ln(3t)PG), 18:3/16:1 PG, $C_{40}H_{71}O_{10}P$. *[R-[R*,S*-(E,Z,Z,Z)]]-9,12,15-Octadecatrienoic Acid, 3-[[(2,3-Dihydroxypropoxy)-hydroxyphosphinyl]oxy]-2-[(1-oxo-3-hexadecenyl)oxy]propyl Ester.*
Ln(3t)PG [R.N. 120444-76-0]: Chromatography: HPLC.[1410,2104]

1,2-Dilinolenoyl-*sn*-glycero-3-phospho-1′-*sn*-glycerol (DLnPG), 18:3/18:3 PG, $C_{42}H_{71}O_{10}P$. *[R-[R*,S*-(all Z)]]-9,12,15-Octadecatrienoic Acid, 1-[[[(2,3-Dihydroxypropoxy) hydroxyphosphinyl]oxy]methyl]-1,2-ethanediyl Ester.*
DLnPG [R.N. 132922-60-2]: Chromatography: HPLC.[1594]

Molecular Species Composition

The molecular species composition of phosphatidylglycerol in plants, determined by any of the methods described on pages 168–171, is given in Table 4-11.

Synthesis

DPPG may be prepared (a) by phosphorylation of 1,2-dipalmitoyl-*sn*-glycerol with $POCl_3$ followed by reaction with 2,3-dibenzyl-*sn*-glycerol,[2217] (b) by condensation of DPPA[2321] or its mono(2,2,2-trichloroethyl) ester[2285] with 2,3-isopropylidene-*sn*-glycerol or 2,3-dibenzyl-*sn*-glycerol, respectively, in the presence of mesitylenesulfonyl chloride, or (c) by enzymatic transesterification of DPPC with glycerol in the presence of phospholipase D.[2330] OPPG may be prepared by the reaction of silver benzyl 2,3-isopropylidene-*sn*-glycero-1-phosphate with 1-oleoyl-2-palmitoyl-*sn*-glycero-3-iodohydrin.[2327] DOPG may be prepared by transesterification of DPPC with glycerol in the presence of phospholipase D.[2260,2329]

Hydrolysis

The hydrolysis of DPPG in acetate buffer (pH 4.0, 30 °C) proceeds at a faster rate than DPPC or DPPE ($k_{obs.} = 17 \times 10^{-2}$, 1.0×10^{-2} and 0.7×10^{-2} days^{-1}, respectively), but the rates are not significantly different if corrections are made for surface charge effects.[2228]

Lysophosphatidylglycerols

Physical Properties

1-Palmitoyl-*sn*-glycero-3-phospho-*sn*-1′-glycerol (PHPG), 16:0 PG, $C_{22}H_{45}O_9P$. *[R-(R*, R*)]-Hexadecanoic Acid, 3-[[(2,3-dihydroxypropoxy)hydroxyphosphinyl]oxy]-2-hydroxypropyl Ester.*

PHPG [R.N. 30227-18-0]: Water soluble[2240]; critical micelle concentration.[2235]
Chromatography: TLC.[2240]

1-Stearoyl-*sn*-glycero-3-phospho-*sn*-1′-glycerol (SHPG), 18:0 PG, $C_{24}H_{49}O_9P$. *[R-(R*,R*)]-Octadecanoic Acid, 3-[[(2,3-dihydroxypropoxy)hydroxyphosphinyl]oxy]-2-hydroxypropyl Ester.*

SHPG: Chromatography: TLC.[2240]

1-Oleoyl-*sn*-glycero-3-phospho-*sn*-1′-glycerol (OHPG), 18:1 PG. $C_{24}H_{47}O_9P$. *[R-[R*,R*-(Z)]]-9-Octadecenoic Acid, 3-[[(2,3-dihydroxypropoxy)hydroxyphosphinyl]oxy]-2-hydroxypropyl Ester.*

OHPG [R.N. 125715-20-0]: Solid, soluble in ethanol, insoluble in ether.
Chromatography: TLC.[2331]

Synthesis

PHPG may be prepared (a) by phosphorylation of 2-benzyl-1-palmitoyl-*sn*-glycerol followed by reaction with 2,3-dibenzyl-*sn*-glycerol[2240] or (b) by enzymatic hydrolysis of DPPG with *Naja naja* snake venom phospholipase A_2.[2235] HPPG may be prepared by enzymatic hydrolysis of DPPG with lipase from the mold *R. arrhizus*.[2232] SHPG may be prepared by phosphorylation of 2-benzyl-1-stearoyl-*sn*-glycerol followed by reaction with 2,3-dibenzyl-*sn*-glycerol[2240] OHPG may be prepared by enzymatic hydrolysis of DOPG with *C. adamanteus* snake venom phospholipase A_2.[2331]

Diphosphatidylglycerol

Physical Properties

1′,3′-bis(1,2-Dipalmitoyl-*sn*-glycero-3-phospho)-*sn*-glycerol (DPDPG), $(16{:}0/16{:}0)_2$, $C_{73}H_{142}O_{17}P_2$. *[R-(R*,R*)]-Hexadecanoic Acid, 5,8,11-Trihydroxy-5,11-dioxido-4,6,10,12-tetraoxa-5,11-diphosphapentadecane-1,2,14,15-tetrayl Ester.*

DPDPG [R.N. 105380-95-8]: Oil, $[\alpha]_D^{20}$ +5.1° ($CHCl_3$).[2326] Spectra: IR.[2326] Chromatography: TLC.[2326]

$Na_2C_{73}H_{140}O_{17}P_2.2H_2O$ [R.N. 64043-42-1]: Mp 199-201 °C, $[\alpha]_D^{21}$ +8.50° ($CHCl_3$/MeOH/H_2O 4:4:1),[2332] DSC.[2333] Spectra: IR, ^{1}H NMR, ^{31}P NMR.[2332] Chromatography: TLC.[2332]

$(NH_4)_2C_{73}H_{140}O_{17}P_2{\cdot}2H_2O$ [R.N. 64043-41-0]: Amorphous solid (X-ray), mp 177-178 °C, $[\alpha]_D^{21}$ +9.25° ($CHCl_3$/MeOH/H_2O 4:4:1), $[\alpha]_D^{24}$ +6.35° ($CHCl_3$)[2332]; DSC.[2333] Spectra: IR, ^{1}H NMR.[2332] Chromatography: TLC.[2332]

1′,3′-bis(1-Stearoyl-2-oleoyl-*sn*-glycero-3-phospho)-*sn*-glycerol (SODPG), $(18{:}0/18{:}1)_2$, $C_{81}H_{154}O_{17}P_2$.

SODPG: Chromatography: PC, TLC.[2334]

1′,3′-bis(1,2-Dioleoyl-*sn*-glycero-3-phospho)-*sn*-glycerol (DODPG), $(18{:}1/18{:}1)_2$, $C_{81}H_{150}O_{17}P_2$ *[R-[R*,R*-(Z)]]-9-Octadecenoic Acid, 5,8,11-Trihydroxy-5,11-dioxido-4,6,10,12-tetraoxa-5,11-diphosphapentadecane-1,2,14,15-tetrayl Ester.*

DODPG [R.N. 115405-83-9]: No data. *R,rac* [R.N. 121545-98-0] (prepared by condensing DOPG with DOPA): Spectra: ^{31}P NMR (fig.) and ^{1}H NMR. Chromatography: TLC.[2335]

$Na_2C_{81}H_{152}O_{17}P_2$: Crystals, mp 202-204 °C, $[\alpha]_D^{20}$ +5.81° ($CHCl_3$).[2336]

$BaC_{81}H_{152}O_{17}P_2$: Colorless crystals, mp 192-194 °C, $[\alpha]_D^{20}$ −6.25° ($CHCl_3$).[2336] A comparable difference in optical activity is found between the sodium and barium salts of ox heart cardiolipin.[2336]

1′,3′-bis(1,2-Dilinoleoyl-*sn*-glycero-3-phospho)-*sn*-glycerol (DLDPG), $(18{:}2/18{:}2)_2$, $C_{81}H_{142}O_{17}P_2$. *[R-[R*,R*-(Z,Z)]]-9,12-Octadecadienoic Acid, 5,8,11-Trihydroxy-5,11-dioxido-4,6,10,12-tetraoxa-5,11-diphosphapentadecane-1,2,14,15-tetrayl Ester.*

DLDPG [R.N. 152595-52-3]: Spectra: ^{2}H and ^{31}P NMR (figs.),[2337] FAB-MS/MS.[2136] Electrophoresis: CE.[2324]

Synthesis

DPDPG may be prepared (a) by the reaction of 1,2-dipalmitoyl-*sn*-glycerol with bis(1,2-dimethylethenylene) pyrophosphate and 2-(*t*-butyldimethylsilyl)glycerol in the presence of triethylamine[2332] or (b) by the reaction of DPPA with 2-benzylglycerol in the presence of mesitylenesulfonyl chloride.[2326] Other methods are available that give DPDPGs that are racemic at either one or more of the three glycerol residues.[2326,2338,2339] DODPG may be prepared by condensation of DOPA with 2-(*t*-butyldimethylsilyl)glycerol in the presence of mesitylenesulfonyl chloride.[2340] SODPG may be prepared by the reaction of silver benzyl SOPA with 1,3-diiodo-2-propanol.[2336]

PHOSPHATIDYLINOSITOLS

This group comprises the phosphatidylinositols (PI), the lysophosphatidylinositols (LPI), and their phosphates (PIP, PIP_2, and LPIP). Note that the stereochemical configuration of the phosphatidylinositols is L-α at the glycerol center like the other phospholipids but D at the *myo*-inositol center like the *myo*-inositol phosphates of Chapters 2 and 3:

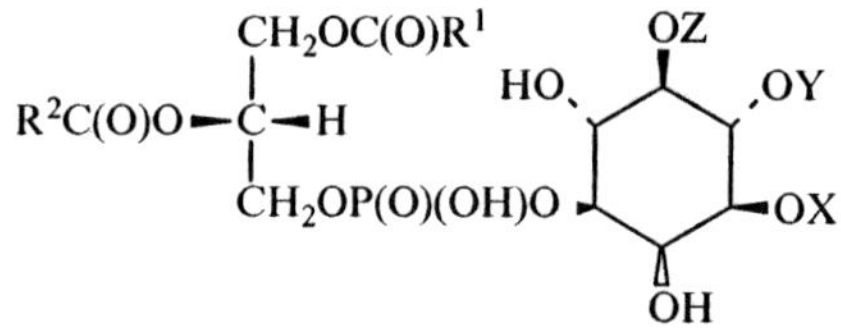

PI, $R^1 = R^2$ = alkyl
1-LPI, R^1 = H, R^2 = alkyl
2-LPI, R^1 = alkyl, R^2 = H
PI(3)P, $R^1 = R^2$ = alkyl, X = $P(O)(OH)_2$, Y = Z = H
PI(4)P, $R^1 = R^2$ = alkyl, Y = $P(O)(OH)_2$, X = Z = H
LPI(4)P, Y = $P(O)(OH)_2$, R^1 = X = Z = H, R^2 = alkyl
PI(3,4)P_2, $R^1 = R^2$ = alkyl, X = Y = $P(O)(OH)_2$, Z = H
PI(4,5)P_2, $R^1 = R^2$ = alkyl, Y = Z = $P(O)(OH)_2$, X = H

Phosphatidylinositols

Physical Properties

1,2-Dipalmitoyl-*sn*-glycero-3-phospho-D-*myo*-inositol (DPPI), 16:0/16:0 PI, $C_{41}H_{79}O_{13}P$. D-myo-*Inositol, (R)-1-[2,3-bis[(1-oxohexadecyl)oxy]propyl Hydrogen Phosphate].*

DPPI [R.N. 28366-80-5]: White solid, mp 181 °C dec., $[\alpha]_D^{20}$ +9.3° ($CHCl_3$).[2341] Spectra: IR,[2342] ^1H NMR,[2341] ^{31}P NMR,[2343] FAB-MS (fig.).[2342] Chromatography: HPLC.[2101]

$NaC_{41}H_{78}O_{13}P$ [R.N. 154459-87-7]: White solid, mp 214-217 °C. Spectra: ^1H and ^{31}P NMR.[2344]

$NH_4C_{41}H_{78}O_{13}P$ [R.N. 34290-57-8]: Precipitate, mp 169-172 °C, $[\alpha]_D^{20}$ +7.48° ($CHCl_3$/ methanol), soluble in chloroform and methanol, insoluble in acetone; ORD (fig.), IR (fig.), TLC.[2345]

1-Palmitoyl-2-oleoyl-*sn*-glycero-3-phospho-D-*myo*-inositol (POPI), 16:0/18:1 PI, $C_{43}H_{81}O_{13}P$. D-myo-*Inositol, [R-(Z)]-1-[3-[(1-oxohexadecyl)oxy]-2-[(1-oxo-9-octadecenyl) oxy]propyl Hydrogen Phosphate].*

POPI [R.N. 51547-49-0]: Spectra: FAB-MS (fig.),[2137] CI-MS.[2137] Chromatography: HPLC,[2101] HPLC-MS.[2106]

$NH_4C_{43}H_{80}O_{13}P$ [R.N. 50730-13-7]: White powder, mp 172-174 °C, $[\alpha]_D$ +6.1° ($CHCl_3$), ORD (fig.).[2346]

1-Palmitoyl-2-linoleoyl-*sn*-glycero-3-phospho-D-*myo*-inositol (PLPI), 16:0/18:2 PI, $C_{43}H_{79}O_{13}P$. D-myo-*Inositol, [R-(Z,Z)]-1-[3-[(1-oxohexadecyl)oxy]-2-[(1-oxo-9,12-octadecadienyl)oxy]propyl Hydrogen Phosphate].*

Table 4-12 Molecular Species Composition: Phosphatidylinositols (mol%)

No.	Plant part	Method	16:0 16:0	16:0 18:1	16:0 18:2	16:0 18:3	18:0 18:1	18:0 18:2	18:0 18:3	18:1 18:1	18:1 18:2	18:1 18:3	18:2 18:2	18:2 18:3	18:3 18:3	References
Camellia sinensis L. (Tea)																
1	Pollen	HPLC				+									+	1558
Dianthus caryophyllus L. (Carnation)																
1	Flower microsomal membr.	GLC	3.8	9.0	56.8	2.4		2.9	(16:0/18:0, 1.9)		3.9	3.0	11.6	1.8		1652
Diptychocarpus strictus (Fisch.) Trautv.																
1	Seed	1R,2R	1.6	5.0	14.0	40.0	0.1	0.7	1.0	1.0	2.0	6.3	1.2	3.3	4.2	959
Glycine max (L.) Merr. (Soybean)																
1	Seed	GLC	0.1	4.3	57.3	6.8	1.0	13.1	1.3	0.3	3.6		6.5	1.0[a]		2114
1	Seed	1R,2R	4.0	4.8	50.7	5.9	1.1	11.9	1.4	0.3	2.8	0.3	5.8	0.7	0.07	2114
1	Seed	HPLC	0.1	17.5[b]	47.1	10.6	1.0		3.4	0.6	5.0	1.1	9.1	3.3	1.1	1693
1	–, hypocotyl	AgTLC			57	24		8			1		3	1	Trace	1690
1	–, cotyledon	AgTLC			61	9		14			1		2	1	Trace	1690
1	Seedling, microsomes	HPLC		12.7	52.7	10.7	Trace	16.3	Trace		3.3		4.3	Trace		1983
Gossypium barbadense L. (Sea Island Cotton)																
1	Seed kernel	1R,2R	8.3	13.1	53.0		1.2	5.0		0.5	2.0		1.8			1709
Gossypium hirsutum L. (Upland Cotton)																
1	Seed kernel	1R,2R	3.2	4.8	39.9		0.9	7.2		1.0	8.4		13.2			1374
Hibiscus cannabinus L. (Kenaf)																
1	Seed	1R,2R	1.7	13.5	52.8	1.0	1.2	4.0	0.1	1.7	6.5	0.1	7.2	0.1		1377
Lilium longiflorum Thunb. (Trumpet Lily)																
1	Pollen	HPLC	0.9		20.3	42.7							4.3	10.3[a]	6.9	1769
Nicotiana tabacum L. (Tobacco)																
1	Cell culture	AgTLC		3.9	79.2	16.9			Trace		Trace	Trace	Trace	Trace		1495
Oryza sativa L. (Rice)																
1	Bran, glutinous	AgTLC		22.3	39.5		(18:2/18:1, 6.0)			9.9	13.9		8.4			1845
1	Bran, glutinous	1R,2R		20.1	36.9		(18:2/18:1, 3.4)			6.4	11.7		6.3			1845
1	Bran, nonglutinous	1R,2R		29.4	37.9		(18:2/18:1, 4.5)			8.1	14.4		5.7			1845

1	Bran, nonglutinous	1R,2R		22.6	33.6		(18:2/18:1, 2.6)			8.8	13.1		3.7			1845
Phaseolus coccineus L. (Scarlet Runner Bean)																
1	Seed	HPLC	0.9	8.8[b]	40.6	34.3	0.5		3.9	0.3	2.1	1.8	2.6	3.2	1.1	1693
Phaseolus vulgaris L. (Kidney Bean)																
1	Seed	HPLC	0.7	8.6[b]	26.2	40.1	0.8		4.1	2.0	3.5	3.7	3.9	4.0	2.4	1693
Pisum sativum L. (Pea)																
1	Seed	AgTLC	6.5	24.4	43.5	3.2	3.7	8.7[c]			6.8		3.2			1904
1	Seed	1R,2R	5.5	21.4	42.2	2.4	2.5	4.9	0.3	2.7	5.3	0.3	2.8	0.2		1904
1	Seed	HPLC	0.7	21.6[b]	47.5	2.6	2.8		0.3	1.8	7.1	5.4	6.5	2.1	1.7	1693
1	Leaf	HPLC			44.9	39.9			6.4		0.9		3.3	4.6		1983
Rauvolfia serpentina Benth. ex Kurz (Indian Snakeroot)																
1	Cell culture	AgTLC			92.3	1.6	0.1	5.4	Trace	0.6			Trace			1495
Secale cereale L. (Rye)																
2	Grain (1)	HPLC		8	44	7				1	9	3	18	7	1	1957
	Grain (2)	HPLC		10	41	3				4	16		22	4	1	1957
Solanum tuberosum L. (Potato)																
1	Tuber, microsomes	HPLC			73.7	22.2		Trace	Trace				4.1			1983
Vicia faba L. (Broad Bean)																
1	Seed	HPLC	0.6	16.7[b]	60.6	2.0	1.2		0.4	2.0	8.8	0.3	6.8	0.1	<0.1	1693
Vigna angularis (Willd.) Ohwi and Ohashi (Adzuki Bean)																
1	Seed	HPLC		8.5[b]	60.9	18.5			2.0		3.1	0.7	3.6	2.3	0.5	2051
1	LEAf	HPLC		4.1[b]	30.1	54.4			3.1		0.3	4.2[d]		0.1	0.6	2051
1	Root	HPLC		7.3[b]	45.9	29.7			2.1		0.7	5.0[d]		7.5	1.9	2051
	Similar data reported for PI molecular species in cotyledon, leaf, hypocotyl, and root tissue of adzuki bean cell culture															2052
Zea mays L. (Corn)																
1	Seedling coleoptile microsome	HPLC		4.2	75.4	4.6		3.4			7.4		5.0			2079
1	–, plasma membrane	HPLC		6.6	82.0	3.6							7.8			2079
Zea mays L. ssp. *saccharata* Sturt (Sweet Corn)																
1	Seed	HPLC		10	41	13				3	11		19	3		2086

[a]Including 18:3/18:2 (LnL).
[b]Including 18:0/18:2 (SL).
[c]Including 18:1/18:1 (OO).
[d]Including 18:2/18:2 (LL).

PLPI [R.N. 126253-19-8]: Spectra: FAB-MS,[2135] FAB-MS/MS (fig.),[2136] MALDI FT-ICR-MS (fig.).[1370] Chromatography: HPLC,[2101] HPLC-MS (fig.)[2106], HPLC-MS/MS.[1415]

1-Palmitoyl-2-linolenoyl-*sn*-glycero-3-phospho-D-*myo*-inositol (PLnPI), 16:0/18:3 PI, $C_{43}H_{77}O_{13}P$. D-myo-*Inositol, [R-(Z,Z,Z)]-1-[3-[(1-oxohexadecyl)oxy]-2-[(1-oxo-9,12,15-octadecatrienyl)-oxy]propyl Hydrogen Phosphate].*

PLnPI [R.N. 124264-81-9]: Spectra: FAB-MS,[2135] FAB-MS/MS.[2136] Chromatography: HPLC,[2101] HPLC-MS,[2106] HPLC-MS/MS.[1415]

1-Stearoyl-2-oleoyl-*sn*-glycero-3-phospho-D-*myo*-inositol (SOPI), 18:0/18:1 PI, $C_{45}H_{85}O_{13}P$. D-myo-*Inositol, [R-(Z)]-1-[2-[(1-oxo-9-octadecenyl)oxy]-3-[(1-oxooctadecyl) oxy]propyl Hydrogen Phosphate].*

SOPI [R.N. 58116-17-9]: FAB-MS,[2135] HPLC.[2100]

$NH_4C_{45}H_{84}O_{13}P$ [R.N. 58165-70-1]: mp 174-176 °C, $[\alpha]_D^{20}$ +6.4° ($CHCl_3$), +7.3° ($CHCl_3$/ methanol 2:2). Spectra: IR.[2347] Chromatography: TLC.[2347]

1-Stearoyl-2-linoleoyl-*sn*-glycero-3-phospho-D-*myo*-inositol (SLPI), 18:0/18:2 PI, $C_{45}H_{83}O_{13}P$. D-myo-*Inositol, [R-(Z,Z)]-1-[2-[(1-oxo-9,12-octadecadienyl)oxy]-3-[(1-oxooctadecyl)oxy]propyl Hydrogen Phosphate].*

SLPI [R.N. 83674-12-8]: Spectra: FAB-MS,[2135] FAB-MS/MS.[2136] Chromatography: HPLC,[2348] HPLC-MS,[2106] HPLC-MS/MS.[1415]

1-Stearoyl-2-linolenoyl-*sn*-glycero-3-phospho-D-*myo*-inositol (SLnPI), 18:0/18:3 PI, $C_{45}H_{81}O_{13}P$. D-myo-*Inositol, [R-(Z,Z,Z)]-1-[2-[(1-oxo-9,12,15-octadecatrienyl)oxy]-3-[(1-oxooctadecyl)oxy]-propyl Hydrogen Phosphate].*

SLnPI [R.N. 182509-37-1]: Spectra: FAB-MS,[2135] FAB-MS/MS.[2136] Chromatography: HPLC.[1983]

1,2-Dioleoyl-*sn*-glycero-3-phospho-D-*myo*-inositol (DOPI), 18:1/18:1 PI, $C_{45}H_{83}O_{13}P$. D-myo-*Inositol, [R-(Z,Z)]-1-[2,3-bis-[(1-oxo-9-octadecenyl)oxy]propyl Hydrogen Phosphate].*

DOPI [R.N. 103065-23-2]: Spectra: IR, ^1H NMR, FAB-MS (fig.),[2342] FAB-MS/MS.[2136]

1-Oleoyl-2-linoleoyl-*sn*-glycero-3-phospho-D-*myo*-inositol (OLPI), 18:1/18:2 PI, $C_{45}H_{81}O_{13}P$. D-myo-*Inositol, [R-(Z,Z,Z)]-1-[2-[(1-oxo-9,12-octadecadienyl)oxy]-3-[(1-oxo-9-octadecenyl)-oxy]propyl Hydrogen Phosphate].*

OLPI [R.N. 182509-35-9]: Spectra: FAB-MS/MS.[2136] Chromatography: HPLC,[2101] HPLC-MS.[2106]

1-Oleoyl-2-linolenoyl-*sn*-glycero-3-phospho-D-*myo*-inositol (OLnPI), 18:1/18:3 PI, $C_{45}H_{79}O_{13}P$. D-myo-*Inositol, [R-(all Z)]-1-[2-[(1-oxo-9,12,15-octadecatrienyl)oxy]-3-[(1-oxo-9-octadecenyl)-oxy]propyl Hydrogen Phosphate].*

OLnPI [R.N. 124211-31-0]: HPLC.[2051]

1,2-Dilinoleoyl-*sn*-glycero-3-phospho-D-*myo*-inositol (DLPI), 18:2/18:2 PI, $C_{45}H_{79}O_{13}P$. D-myo-*Inositol, [R-(all Z)]-1-[2,3-bis-[(1-oxo-9,12-octadecadienyl)oxy]propyl Hydrogen Phosphate].*

DLPI [R.N. 122921-66-8]: Spectra: FAB-MS/MS.[2136] Chromatography: HPLC,[2348] HPLC-MS.[2106]

1-Linoleoyl-2-linolenoyl-*sn*-glycero-3-phospho-D-*myo*-inositol (LLnPI), 18:2/18:3 PI, $C_{45}H_{77}O_{13}P$. D-myo-*Inositol, [R-(all Z)]-1-[2-[(1-oxo-9,12-octadecadienyl)oxy]-3-[(1-oxo-9,12,15-octadecatrienyl)oxy]propyl Hydrogen Phosphate].*

LLnPI [R.N. 126936-49-0]: HPLC.[1983]

1,2-Dilinolenoyl-*sn*-glycero-3-phospho-D-*myo*-inositol (DLnPI), 18:3/18:3 PI, $C_{45}H_{85}O_{13}P$. D-myo-*Inositol, [R-(all Z)]-1-[2,3-bis-[(1-oxo-9,12,15-octadecatrienyl)oxy]propyl Hydrogen Phosphate].*

DLnPI [R.N. 124211-30-9]: HPLC.[1769]

Molecular Species Composition

The molecular species composition of phosphatidylinositol in plants, determined by any of the methods described on pages 168–171, is given in Table 4-12.

Synthesis

DPPI may be prepared (a) by the reaction of 2,3,4,5,6-penta-*O*-benzyl-D-*myo*-inositol with phenyl phosphorodichloridate and 1,2-dipalmitoyl-*sn*-glycerol[2345] or with the phosphoramidite of 1,2-dipalmitoyl-*sn*-glycerol in the presence of tetrazole,[2341] (b) by phosphorylation of 2,3,4,5-di-*O*-cyclohexylidene-6-*O*-(2-trimethylsilylethoxy)methyl-D-*myo*-inositol with methyl *N,N*-diisopropylaminophosphorochloridite followed by reaction with 1,2-dipalmitoyl-*sn*-glycerol in the presence of tetrazole,[2349] (c) by condensation of 1,2-dipalmitoyl-*sn*-glycero-3-(sodium phenyl phosphate) with 2,3,5,6-di-*O*-cyclohexylidene-D-*myo*-inositol [2343] or its 4-(5,6-dihydro-4-methoxy-2*H*-pyran) derivative[2344] in the presence of the triazole MSNT, or (d) by acylation of tetrahydropyran-protected *sn*-glycero-3-phospho-D-*myo*-inositol with palmitic anhydride.[2342] POPI may be prepared by the reaction of silver 2,3,4,5,6-penta-*O*-acetyl-D-*myo*-inositol-1-benzyl phosphate with 1-palmitoyl-2-oleoyl-*sn*-glycero-3-iodohydrin.[2346] PLPI may be prepared by acylation of tetrahydropyran-protected PHPI with linoleoyl chloride.[2350] SOPI may be prepared by the reaction of 2,3,4,5,6-penta-*O*-acetyl-D-*myo*-inositol-1-dihydrogen phosphate with 2-oleoyl-1-stearoyl-*sn*-glycerol in the presence of mesitylenesulfonyl chloride.[2347] Finally, DOPI may be prepared by acylation of tetrahydropyran-protected *sn*-glycero-3-phospho-D-*myo*-inositol with oleic anhydride in the presence of 4-dimethylaminopyridine.[2342]

Lysophosphatidylinositols

Physical Properties

1-Palmitoyl-*sn*-glycero-3-phospho-D-*myo*-inositol (PHPI), 16:0 LPI, $C_{25}H_{49}O_{12}P$. D-myo-*Inositol, (R)-1-[2-Hydroxy-3-[(1-oxohexadecyl)oxy]propyl Hydrogen Phosphate].*

PHPI [R.N. 112573-74-7; see also 106182-20-1]: Spectra: ^{1}H NMR,[2351] FAB-MS/MS (fig.).[2352] Chromatography: TLC,[2353] silica gel column.[2351]

$NH_4C_{25}H_{48}O_{12}P$: Solid, IR.[2351]

1-Linoleoyl-*sn*-glycero-3-phospho-D-*myo*-inositol (LHPI), 18:2 LPI, $C_{27}H_{49}O_{12}P$. D-myo-*Inositol, [R (Z,Z)]-1-[2-Hydroxy-3-[(1-oxo-9,12-octadecadienyl)oxy]propyl Hydrogen Phosphate].*

LHPI [R.N. 124330-66-1]: Spectra: FAB-MS/MS.[2352] L-Isomer [R.N. 142508-96-1]: White solid. Spectra: ^{1}H and ^{13}C NMR, FAB-MS/MS. Chromatography: TLC, HPLC.[2354]

Synthesis

PHPI may be prepared by phosphorylation of 2,3,4,5,6-penta-*O*-acetyl-D-*myo*-inositol with methyl *N,N*-diisopropylaminophosphorodichloridite followed by reaction with 1-palmitoyl-2-benzyl-*sn*-glycerol in the presence of tetrazole.[2351]

LHPI and HLPI: Syntheses of the L-isomers are available.[2354]

Phosphatidylinositol Phosphates

Occurrence

The occurrence of polyphosphoinositides in plant cells was first demonstrated in 1985 by Boss and Massel,[1359] who isolated PI, PIP, and PIP_2 from wild carrot cell cultures in 93:1.7:0.8% yield, based on recovered inositol. This ratio proved to be representative of a variety of plant species (Table 4-13). In contrast to animal cells, the amount of PIP_2 is almost always smaller than that of PIP, and their sum total is usually very low relative to PI. Also, because the amounts of PIP and PIP_2 are so small, the only quantitative method available to date is radiolabeling, which assumes that isotopic equilibrium has been reached prior to analysis. The preferred precursor is D-*myo*-[2-^{3}H]-inositol because it labels PIs exclusively, but [2-^{3}H]-glycerol, [^{32}P]P_i and [γ-^{32}P]-ATP, [^{14}C]-*myo*-inositol, and [1-^{14}C]-acetate have also been employed.

Table 4-13 Phosphatidylinositol Phosphates in Plants*

No.	Plant part	Method	LPI	LPIP	PI	PIP	PIP_2	References
Brassica napus L. ssp. *oleifera* L. (Winter Rape)								
1	Leaf	2-[^{3}H]-Ins			32	9	3	1326
Brassica oleracea L. ssp. *botrytis* L. (Cauliflower)								
1	Cotyledon	[^{32}P]P_i			76.7	16.7	6.6	1244
Catharanthus roseus (L.) G. Don (Madagascar Periwinkle)								
1	Cell culture	2-[^{3}H]-Ins			92	5	2.5	2355
1	Cell culture[a]	[^{32}P]P_i			92	7	1	2356
Daucus carota L. (Wild Carrot)								
1	Cell culture[b]	2-[^{3}H]-Ins	7.4	4.7	76.0	7.0	0.58	2357
1	–, plasma membrane[b]	2-[^{3}H]-Ins	4.5	18.9	54.0	14.2	1.3	2357
1	–, microsomes and mitochondria[b]	2-[^{3}H]-Ins	18.6	2.0	73.5	2.7	0.3	2357
1	–, protoplasts[b]	2-[^{3}H]-Ins	11.10	1.33	80.43	4.87	0.67	2353
Dianthus caryophyllus L. (Carnation)								
1	Flower petal	2-[^{3}H]-Ins, [^{32}P]P_i			91.5	8.3	0.23	1202

Table 4-13 Phosphatidylinositol Phosphates in Plants*—cont'd

No.	Plant part	Method	LPI	LPIP	PI	PIP	PIP_2	References
Glycine max (L.) Merr. (Soybean)								
1	Bean, hypocotyl	2-[^{3}H]-Ins			55	2	2	2358
1	–, microsomal membrane	2-[^{3}H]-Ins			95	0	0	2358
1	Cell culture	2-[^{3}H]-Ins			91	9	Trace	1221
Helianthus annuus L. (Sunflower)								
1	Hypocotyl plasma membrane[a]	[^{32}P]P_i			78.8	8.1	4.0	2359
Hordeum vulgare L. (Barley)								
1	Seed, aleurone layer	2-[^{3}H]-Ins			99.1	0.7	0.2	1204
1	Seed, aleurone layer	[^{32}P]P_i			100	10	1	1204
Lilium longiflorum Thunb. (Trumpet Lily)								
1	Pollen[a]	2-[^{3}H]-Ins,			96	3	1	2360
1	Pollen[a]	[^{32}P]P_i			98.5	1	0.5	2360
Lycopersicon esculentum Mill. (Tomato)								
1	Leaf[c]	[^{32}P]P_i			12.7	12.6	1.6	2361
1	Root	TLC			++	++	+	1231
1	Cell culture	2-[^{3}H]-Glycerol			9	1.1	0.04	2362
Nicotiana tabacum L. (Tobacco)								
1	Cell culture	2-[^{3}H]-Ins			96	3	1	1200
Petroselinum crispum Nym. (Parsley)								
1	Cell culture [d]	2-[^{3}H]-Ins			84	6	8	1221
Petunia hybrida hort. (Petunia)								
1	Pollen[a]	2-[^{3}H]-Ins or [^{32}P]P_i			100	0	0	2363
Peucedanum verticillare (L.) Koch								
1	Callus culture	TLC			35.46	4.88	4.32	2364
Pisum sativum L. (Peas)								
1	Leaf	2-[^{3}H]-Ins				+	0.04	1201
Saccharum spp. (Sugarcane)								
1	Leaf[c]	[^{32}P]P_i			23.1	6.3	2.5	2361
Samanea saman (Jacq.) Merr. (Raintree)								
1	Leaf, pulvini	HPLC				0.3	<0.02	1418
Solanum melongena L. (Brinjal, Eggplant)								
1	Leaf[c]	[^{32}P]P_i			20.3	15.9	1.6	2361
Spirodela polyrrhiza L. (Giant Duckweed)								
1	Whole plant	2-[^{3}H]-Ins			15.2	59.2[e]	4.2[f]	1362
Zea mays L. (Corn, Maize)								
1	Leaf[c]	[^{32}P]P_i			13.8	8.9	3.6	2361
1	Coleoptile	[^{14}C]-Ins			92.3	7.0	0.7	2365
1	Coleoptile	1-[^{14}C]-Acetate			93.4	4.6	2.0	2365

*Molar ratio of PI subclasses.
[a]Percentage of total PI.
[b]Percentage of recovered 2-[^{3}H]-Ins.
[c]Percentage of total PL.
[d]nmol/g fwb.
[e]57.6 PI(4)P and 1.6 PI(3)P.
[f]3.6 PI(4,5)P_2 and 0.6 PI(3,4)P_2.

Separation and Analysis

Extraction of the phosphoinositides from the radiolabeled plant tissue is usually performed by a modified Folch procedure, using, for example, 1:2 chloroform/methanol containing 1% conc. HCl.[2356] The inclusion of acid is critical because at neutral pH, PIP and PIP_2 are not extracted into the organic phase.[1196a] The phosphoinositides are separated by TLC in the order PI > PIP > PIP_2 and quantitated by radiography as follows:

> *Wild carrot cells ... (0.3 g fresh wt) were labeled overnight (19 h) with* myo-*[2-^{3}H]inositol (19.0 Ci/mmol) 16.7 μCi/g fresh wt. The cells were harvested on filter paper and washed with 100 mL water to remove the medium. The polyphosphoinositides were extracted ... with ... 1.5 mL $CHCl_3$:MeOH (1:2), 0.5 mL of 2.4*N *HCl, 0.5 mL of $CHCl_3$, and 0.5 mL of 0.1*N *EDTA. The mixture was centrifuged and the lower phase collected. The upper phase was reextracted twice with $CHCl_3$ and the extracts combined and backwashed twice with MeOH:1*N *HCl (1:1). The $CHCl_3$ layer was collected and dried (in vacuo) .... The dried extracts were reconstituted in $CHCl_3$/MeOH (3:1) and spotted on ... high performance, silica gel LKS plates ... which had been impregnated with 1 % potassium oxalate, dipping the preabsorbent area in the oxalate first, and dried (110 C) for 1 h. The plates were run for 2.5 h in $CHCl_3$:MeOH:4* N *NH_4OH, 90:70:20 and scraped and counted ... in a Beckmann LS 7000 scintillation counter.*
>
> Boss, W. F.; Massel, M. O., "Polyphosphoinositides are Present in Plant Tissue Culture Cells", *Biochem. Biophys. Res. Commun.* **1985**, *132*(3), 1018–1023,[1359] reprinted by permission of Elsevier, © 1985.

Separation of the phosphoinositides has also been carried out by HPLC on an amino-NP silica column with a 0 → 0.8 *M* ammonium acetate gradient. The compounds, monitored by liquid scintillation, were eluted in the order PI > PIP > PIP_2.[1418]

A protocol for separating all known isomers of PIP and PIP_2 by thin-layer chromatography on boric acid-impregnated HPTLC plates has been described.[2366]

Phosphoinositide Isomers

That the PIP and PIP_2 of plant species are in fact PI(4)P and PI(4,5)P_2 as in animal systems has been verified by chemical degradation methods. Removal of the acyl and glyceryl substituents from PIP leaves an inositol bisphosphate identified as Ins(1,4)P_2, and similar treatment of PIP_2 leaves an inositol trisphosphate identified as Ins(1,4,5)P_3.[1201] Brearley and Hanke[1207,1362] later detected PI(3)P and PI(3,4)P_2 (but not PI(3,4,5)P_3) in *Spirodela polyrrhiza* in addition to the PI(4)P and PI(4,5)P_2.

Molecular Species

PIP appears to have the same molecular species distribution as PI. HPLC analysis of the PIP in maize coleoptile shows the principal species to be 16:0:18:2 (59.0%) together with smaller amounts of 16:0/18:1, 16:0/18:3, 18:0/18:2, 18:2/18:2, and traces of 18:1/18:2 species.[2365]

Physical Properties

1,2-Dipalmitoyl-*sn*-glycero-3-phospho-D-*myo*-inositol 3-phosphate (DPPI(3)P), 16:0/16:0 PIP, $C_{41}H_{80}O_{16}P_2$. D-myo-*Inositol, (R)-1-[2,3-bis[(1-oxohexadecyl)oxy]propyl Hydrogen Phosphate] 3-Dihydrogen Phosphate.*

DPPI(3)P [R.N. 165689-81-6]: White solid, mp 108-110 °C, $[\alpha]_D^{22}$ +0.7° ($CHCl_3$/methanol).[2368] Spectra: ^1H and ^{31}P NMR,[2369] FAB-MS.[2370]

$NH_4C_{41}H_{79}O_{16}P_2$: LC-MS.[2369]

1,2-Dipalmitoyl-*sn*-glycero-3-phospho-D-*myo*-inositol 4-phosphate (DPPI(4)P), 16:0/16:0 PIP, $C_{41}H_{80}O_{16}P_2$. D-myo-*Inositol, (R)-1-[2,3-bis[(1-oxohexadecyl)oxy]propyl Hydrogen Phosphate] 4-Dihydrogen Phosphate.*

DPPI(4)P [R.N. 57606-15-2]: White solid. Spectra: ^1H and ^{31}P NMR.[2343]

$(NH_4)_3C_{41}H_{77}O_{16}P_2$ [R.N. 161105-05-1]: mp 151-154 °C, $[\alpha]_D^{20}$ +4.5° ($CHCl_3$/methanol), soluble in $CHCl_3$, insoluble in acetone; ORD (fig.). Spectra: IR (fig.). Chromatography: TLC.[2371]

1,2-Dipalmitoyl-*sn*-glycero-3-phospho-D-*myo*-inositol 3,4-bisphosphate (DPPI(3,4)P), 16:0/16:0 PIP_2, $C_{41}H_{81}O_{19}P_3$. D-myo-*Inositol, (R)-1-[2,3-bis[(1-oxohexadecyl)oxy]propyl Hydrogen Phosphate] 3,4-bis(Dihydrogen Phosphate).*

DPPI(3,4)P_3 [R.N. 165883-85-2]: White solid, mp 62-65 °C, $[\alpha]_D^{23}$ +1.9° ($CHCl_3$).[2370] Spectra: ^1H and ^{31}P NMR,[2370] FAB-MS,[2370] MALDI TOF-MS.[2372]

$Na_5C_{41}H_{76}O_{19}P_3$ [R.N. 195318-65-1]: $[\alpha]_D^{22}$ +6.8° (H_2O).[2368]

$(NH_4)_5C_{41}H_{76}O_{19}P_3$ [R.N. 165883-85-2]: Spectra: ^1H and ^{31}P NMR. Chromatography: LC-MS.[2369]

1,2-Dipalmitoyl-*sn*-glycero-3-phospho-D-*myo*-inositol 4,5-bisphosphate (DPPI(4,5)P_2), 16:0/16:0 PIP_2, $C_{41}H_{81}O_{19}P_3$. D-myo-*Inositol, (R)-1-[2,3-bis[(1-oxohexadecyl)oxy]propyl Hydrogen Phosphate] 4,5-bis(Dihydrogen Phosphate).*

DPPI(4,5)P_2 [R.N. 120595-88-2]: No data.

$Na_5C_{41}H_{76}O_{19}P_3$ [R.N. 183623-41-8]: White solid. Spectra: ^1H and ^{31}P NMR, MALDI MS.[2373]

$(C_6H_{11}NH_3)_5C_{41}H_{76}O_{19}P_3$ [R.N. 120664-89-3]: Spectra: ^1H, ^{13}C, and ^{31}P NMR.[2374]

Synthesis

DPPI(3)P may be prepared (a) by phosphorylation of a 2,4,5,6-protected DPPI with dibenzyl *N,N*-diisopropylphosphoramidite with 1*H*-tetrazole,[2375] or (b) by stepwise reaction of a 2,4,5,6-protected 1D-*myo*-inositol-3-dibenzyl phosphate with methyl *N,N*-diisopropylphosphoramidochloridite followed by 1,2-dipalmitoyl-*sn*-glycerol [2369] or vice versa.[2370] DPPI(4)P may be prepared (a) by phosphorylation of a 2,3,5,6-protected DPPI with diphenyl *N,N*-diethylphosphoramidite in the presence of 1*H*-tetrazole,[2343] (b) by condensation of a 2,3,5,6-protected 1D-*myo*-inositol-4-diphenyl phosphate with DPPA in the presence of mesitylenesulfonyl chloride,[2371] (b) by the reaction of a 2,3,5,6-protected 1D-*myo*-inositol-3-dibenzyl phosphate with benzyl 1,2-dipalmitoyl-*sn*-glycero-*N,N*-diisopropylphosphoramidite in the presence of 1*H*-tetrazole.[2375] DPPI(3,4)P_2 may be prepared by condensation of a 2,5,6-protected 1D-*myo*-inositol-3,4-bis(dibenzyl phosphate) (a) with benzyl 1,2-dipalmitoyl-*sn*-glycero-*N,N*-diisopropylphosphoramidite in the presence of 1*H*-tetrazole,[2368,2370] (b) with methyl *N,N*-diisopropylphosphoramidochloridite in the presence of 1*H*-tetrazole,[2369] or (c) with dibenzyl 1,2-dipalmitoyl-*sn*-glycerophosphite and pyridinium perbromide.[2376] DPPI(4,5)P_2 may be prepared by phosphorylation of a 2,3,6-protected DPPI with dibenzyl *N,N*-diisopropylphosphoramidite in the presence of 1*H*-tetrazole.[2373,2374]

Phosphorylation

Enzymic phosphorylation of PIP with PIP kinase yields PIP_2, as shown in Fig. 4-1, but nonenzymic phosphorylation with ATP in the presence of a bivalent metal ion such as Ca(II) or Mn(II) results in a different PIP_2, the 4-pyrophosphate.[2367]

Lysophosphatidylinositol Phosphates

Physical Properties

1-Palmitoyl-*sn*-glycero-3-phospho-D-*myo*-Inositol 4-Phosphate (16:0 LPI(4)P), $C_{25}H_{50}O_{15}P_2$. *D-*myo*-Inositol, (R)-1-[2-Hydroxy-3-[(1-oxohexadecyl)oxy]propyl Hydrogen Phosphate] 4-Dihydrogen Phosphate.*

PHPI(4)P [R.N. 129485-82-1]: Spectra: FAB-MS/MS.[2352] Chromatography: TLC.[2353]

1-Linoleoyl-*sn*-glycero-3-phospho-D-*myo*-inositol 4-phosphate (18:2 LPI(4)P), $C_{27}H_{50}O_{15}P_2$. *D-*myo*-Inositol, [R-(Z,Z)]-1-[2-Hydroxy-3-[(1-oxo-9,12-octadecadienyl)oxy] propyl Hydrogen Phosphate] 4-Dihydrogen Phosphate.*

LHPI(4)P: Spectra: FAB-MS/MS.[2352] Chromatography: TLC.[2353]

PHOSPHATIDYLSERINES

This group comprises the phosphatidylserines (PS) and lysophosphatidylserines (LPS).

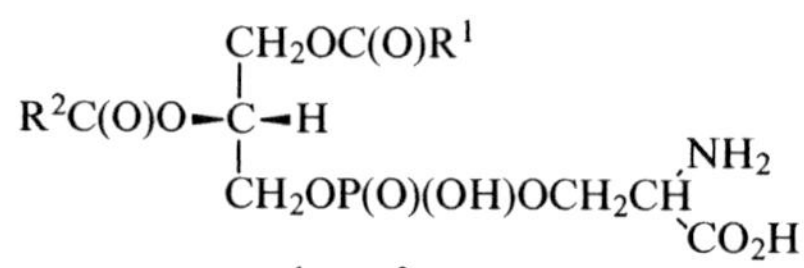

PS, $R^1 = R^2$ = alkyl
2-LPS, R^1 = alkyl, R^2 = H

Phosphatidylserines

Physical Properties

1,2-Dipalmitoyl-*sn*-glycero-3-phospho-L-serine (DPPS), 16:0/16:0 PS, $C_{38}H_{74}NO_{10}P$. *L-Serine, (R)-2,3-bis[(1-Oxohexadecyl)oxy]propyl Hydrogen Phosphate (Ester).*

DPPS [R.N. 40290-42-4; see also 3036-82-6]: Fine white powder, mp 149-150 °C[2377] or 153-155 °C[2378]; $[\alpha]_D^{20}$ +5.3° ($CHCl_3$),[2377] +12.5° ($CHCl_3$)[2211] or +17.7° (benzene),[2378] $[\alpha]_D^{22}$ +5.5° ($CHCl_3/MeOH/H_2O$)[2379]; soluble in chloroform and insoluble in acetone[2377]; pK_a 2.7 (potentiometric)[2380] and 4.6 (FTIR)[2381]; T_C and DSC (fig.).[2256] Spectra: IR,[2377] FTIR,[2382] ^1H NMR,[2274] ^{13}C and ^{31}P NMR,[2211] FAB-MS,[2274] FD-MS (fig.),[2140] CI-MS,[2264] MALDI-MS (fig.),[2178] FAB-MS/MS (fig.),[2383] MALDI FT-ICR-MS/MS (fig.).[1370] Chromatography: TLC,[2211] preparative TLC,[2379] HPLC,[2239] silicic acid column,[2211] DEAE cellulose column.[2379]
$NaC_{38}H_{73}NO_{10}P$: LSI-MS.[2239]

1-Palmitoyl-2-oleoyl-*sn*-glycero-3-phospho-L-serine (POPS), 16:0/18:1 PS, $C_{40}H_{76}NO_{10}P$. *L-Serine, [R-(Z)]-3-[(1-Oxohexadecyl)oxy]-2-[(1-oxo-9-octadecenyl)oxy]propyl Hydrogen Phosphate (Ester).*

POPS [R.N. 40290-44-6; see also 79980-16-8]: mp 127-130 °C, $[\alpha]_D^{20}$ +6.2° ($CHCl_3$).[2377] Spectra: IR (fig.),[2384] Raman,[2385] ESR,[2184] LSIMS (Na salt).[2239] Chromatography: TLC,[2377] HPLC,[2110] LC-MS (fig.),[2106] silicic acid column.[2377] Electrophoresis: CE (fig.).[2181]

$NaC_{40}H_{75}NO_{10}P$: LSI-MS.[2239]

$NH_4C_{40}H_{75}NO_{10}P$: ^{31}P NMR (fig.).[2384]

1-Palmitoyl-2-linoleoyl-*sn*-glycero-3-phospho-L-serine (PLPS), 16:0/18:2 PS, $C_{40}H_{74}NO_{10}P$. *L-Serine, [R-(Z,Z)]-3-[(1-Oxohexadecyl)oxy]-2-[(1-oxo-9,12-octadecadienyl)oxy]propyl Hydrogen Phosphate (Ester).*

PLPS [R.N. 91178-24-4; see also 120574-71-2]: FAB-MS, CI-MS.[2264]

1-Stearoyl-2-oleoyl-*sn*-glycero-3-phospho-L-serine (SOPS), 18:0/18:1 PS, $C_{42}H_{80}NO_{10}P$. *L-Serine, [R-(Z)]-2-[(1-Oxo-9-octadecenyl)oxy]-3-[(1-oxooctadecyl)oxy]propyl Hydrogen Phosphate (Ester).*

SOPS [R.N. 124262-03-7; see also 76091-35-5]: Spectra: FAB-MS,[2135] FAB-MS/MS (fig.),[2386] CID-MS/MS (fig.).[2265] Chromatography: HPLC,[1809] LC-MS.[2106]

1-Stearoyl-2-linoleoyl-*sn*-glycero-3-phospho-L-serine (SLPS), 18:0/18:2 PS, $C_{42}H_{78}NO_{10}P$. *L-Serine, [R-(Z,Z)]-2-[(1-Oxo-9,12-octadecadienyl)oxy]-3-[(1-oxooctadecyl)oxy]propyl Hydrogen Phosphate (Ester).*

SLPS [R.N. 132014-81-4; see also 125874-21-7]: Spectra: FAB-MS.[2135]

1,2-Dioleoyl-*sn*-glycero-3-phospho-L-serine (DOPS), 18:1/18:1 PS, $C_{42}H_{78}NO_{10}P$. *L-Serine, [R-(Z,Z)]-2,3-bis-[(1-Oxo-9-octadecenyl)oxy]propyl Hydrogen Phosphate (Ester).*

DOPS [R.N. 70614-14-1; see also 6811-55-8]: Crystals, mp 165-169 °C, $[\alpha]_D^{22}$ −6.0° ($CHCl_3$), +6.2° ($CHCl_3/MeOH/H_2O$), −13.5° (benzene)[2379]; DSC.[2379] Spectra: IR (fig.),[2384] ^1H NMR (fig.),[2379] ^{31}P NMR (fig.),[2384] LSI-MS.[2239] Chromatography: TLC,[2217] preparative TLC,[2379] HPLC,[2110] LC-MS,[2106] CM-celullose column.[2379]

$BaC_{42}H_{76}NO_{10}P$ [R.N. 71049-19-9]: Softens at 160 °C. Chromatography: TLC, CM-cellulose column.[2379]

1-Oleoyl-2-linoleoyl-*sn*-glycero-3-phospho-L-serine (OLPS), 18:1/18:2 PS, $C_{42}H_{76}NO_{10}P$. *L-Serine, [R-(Z,Z,Z)]-2-[(1-Oxo-9,12-octadecadienyl)oxy]-3-[(1-oxo-9-octadecenyl)oxy]propyl Hydrogen Phosphate (Ester).*

OLPS [R.N. 15045-28-0]: No data available for the *R,L* compound. The *R,DL* isomer (from 1-oleoyl-*sn*-glycero-3-phospho-*N*-phthaloyl-DL-serine and linoleoyl chloride) is an oil, very soluble in benzene, chloroform, ether, and methanol, forming a stable emulsion in water. Spectra: IR (fig.).[2387] Chromatography: TLC, silicic acid column.[2387]

1,2-Dilinoleoyl-*sn*-glycero-3-phospho-L-serine (DLPS), 18:2/18:2 PS, $C_{42}H_{74}NO_{10}P$. *L-Serine, [R-(all Z)]-2,3-bis-[(1-Oxo-9,12-octadecadienyl)oxy]propyl Hydrogen Phosphate (Ester).*

DLPS [R.N. 89616-59-1; see also 14235-22-4]: DSC.[2388] Spectra: ^1H and ^{31}P NMR.[2388] Chromatography: TLC, CM-cellulose column.[2388] *R,DL-DLPS* (from *sn*-glycero-3-phospho-*N*-phthaloyl-DL-serine and linoleoyl chloride): Waxy solid, IR (fig.), silicic acid chromatography[2387]; *rac,L-DLPS* (from *O*-phthalimidomethyl-*N*-(*p*-anisyloxycarbonyl)-*rac*-glycerophospho-L-serine and linoleoyl chloride): IR, PC, TLC, DEAE column chromatography[2389]; *rac,DL-DLPS* (from phosphorylated 1,2-dilinoleoyl-*rac*-glycerol and *N*-phthaloyl-*O*-*t*-butyl-DL-serine): IR (fig.), TLC.[2390]

The only PS phospholipids that contain linolenic acid (16:3) reported to date, according to *Chemical Abstracts*, are the phosphatidylserines OLnPS, LLnPS, and DLnPS from algae.[2391]

Molecular Species Composition

The molecular species composition of phosphatidylserine in plants, determined by any of the methods described on pages 168–171, is given in Table 4-14.

Synthesis

DPPS may be prepared (a) by phosphorylation of 1,2-dipalmitoyl-*sn*-glycerol with $POCl_3$,[2217] phenyl phosphorodichloridite,[2274] phenyl phosphorodichloridate,[2377] or PCl_3/imidazole[2211] followed by reaction with *N*-BOC-L-serine or its *t*-butyl or benzyl ester, (b) by condensation of 1,2-dipalmitoyl-*sn*-glycero-3-iodohydrin with silver phenyl *N*-BOC-*O*-benzyl-L-serine phosphate,[2377] or (c) by acylation of DPPA with *N*-BOC-*O*-benzyl-L-serine in the presence of triisopropylbenzenesulfonyl chloride.[2379] POPS may be prepared by phosphorylation of 1-palmitoyl-2-oleoyl-*sn*-glycerol with phenyl phosphorodichloridate followed by reaction with *N*-BOC-*O*-benzyl-L-serine.[2377] PLPS may be prepared by transesterification of PLPC with L-serine in the presence of phospholipase D.[2264] DOPS may be prepared (a) by phosphorylation of 1,2-dioleoyl-*sn*-glycerol with $POCl_3$ followed by reaction with *N*-BOC-*O*-*t*-butyl-L-serine,[2217] (b) by condensation of DOPA with *N*-BOC-*O*-phthalimidomethyl-L-serine in the presence of triisopropylbenzenesulfonyl chloride,[2379] or (c) by transesterification of DOPC with L-serine in the presence of phospholipase D.[2392] OLPS: None published. The corresponding *R*,DL isomer may be prepared as outlined earlier. DLPS: None published. The corresponding *R*,DL, *rac*,L and *rac*,DL isomers may be prepared as outlined earlier.

Hydrolysis

The rate and extent of hydrolysis of DPPS by phospholipase B from the mold *P. notatum* is the highest of the phospholipids in the series DPPS > DPPI > DPPA > DPPC > DPPE > DPDPG. The products are *sn*-glycero-3-phospho-L-serine and palmitic acid.[2256] Digestion of DOPS with snake venom phospholipase A_2 gives the 2-lyso derivative OHPS, whereas methanolysis with BF_3 removes both acyl groups giving >95% oleic acid by GLC analysis.[2379]

Table 4-14 Molecular Species Composition: Phosphatidylserines (mol%)

No.	Plant part	Method	16:0 18:1	16:0 18:2	18:1 18:2	18:1 18:3	18:2 18:2	18:2 18:3	Reference
Daucus carota L. (Carrot)									
1	Cell suspension culture	AgTLC	3.4	54.7	7.4		24.0	10.5	1648

Lysophosphatidylserines

Physical Properties

1-Palmitoyl-*sn*-glycero-3-phospho-L-serine (PHPS), 16:0 LPS, $C_{22}H_{44}NO_9P$. L-Serine, *(R)-2-Hydroxy-3-[(1-oxohexadecyl)oxy]propyl Hydrogen Phosphate (Ester).*

PHPS [R.N. 116947-34-3; see also 75795-82-3]: Spectra: FAB-MS (fig.).[2393]
Chromatography: TLC.[2240]

1-Stearoyl-*sn*-glycero-3-phospho-L-serine (SHPS), 18:0 LPS, $C_{24}H_{48}NO_9P$. *L-Serine, (R)-2-Hydroxy-3-[(1-oxooctadecyl)oxy]propyl Hydrogen Phosphate (Ester).*

SHPS [R.N. 119786-67-3]: Spectra: FAB-MS,[2135] CID-MS/MS (fig.).[2265]

1-Oleoyl-*sn*-glycero-3-phospho-L-serine (OHPS), 18:1 LPS, $C_{24}H_{46}NO_9P$. *L-Serine, [R-(Z)]-2-Hydroxy-3-[(1-oxo-9-octadecenyl)oxy]propyl Hydrogen Phosphate (Ester).*

OHPS [R.N. 89319-60-8; see also 126575-94-8]: Spectra: FAB-MS (fig.).
Chromatography: TLC.[2394]

Synthesis

PHPS may be prepared by phosphorylation of 2-benzyl-1-palmitoyl-*sn*-glycerol with $POCl_3$ followed by reaction with *O-t*-butyl-*N*-BOC-L-serine.[2240] OHPS may be prepared by enzymatic hydrolysis of DOPS with snake venom phospholipase A_2.[2394]

PHOSPHATIDIC ACIDS

This group comprises the phosphatidic acids (PA) and two kinds of lysophosphatidic acids (LPA):

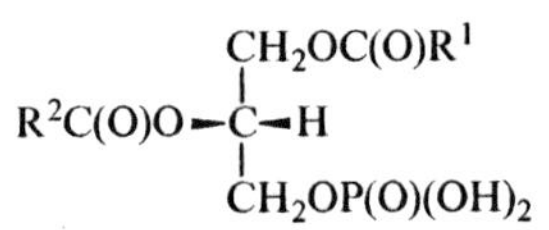

PA, $R^1 = R^2$ = alkyl
1-LPA, R^1 = H, R^2 = alkyl
2-LPA, R^1 = alkyl, R^2 = H

Phosphatidic Acids

Physical Properties

1,2-Dipalmitoyl-*sn*-glycero-3-(dihydrogen phosphate) (DPPA), 16:0/16:0 PA, $C_{35}H_{69}O_8P$. *(R)-Hexadecanoic Acid, 1-[(Phosphonooxy)methyl]-1,2-Ethanediyl Ester.*

DPPA [R.N. 7091-44-3]: White powder,[2395] mp 67-68 °C[2396]; $[\alpha]_D^{26}$ +4.0° ($CHCl_3$)[2397]; solubility in g/100 mL at 21 °C: benzene 4.2, ethanol 2.5, methanol 2.0, acetone 1.6, ether 1.5, acetic acid 1.3, petroleum ether <0.001,[2397] DSC.[2398] Spectra: IR,[2399] FTIR,[2400] Raman,[2401] ^1H and ^{13}C NMR,[118] ^{31}P NMR,[2396] CD,[2402] FAB-MS,[2200] FD-MS,[2403] SI-MS,[2239] FAB-MS-MS.[1415] Chromatography: TLC,[2399] HPTLC,[1405] IEC,[2285] SFC,[2156] FPD-GC,[123] GFC,[2404] HPLC-MS/MS.[1415] *rac-DPPA* [R.N.5129-68-0]: mp 62-63 °C, X-ray,[2405] CE.[2324]

$NaC_{35}H_{68}O_8P$: mp 160 °C.[2406]
$Na_2C_{35}H_{67}O_8P$ [R.N. 71065-87-7]: mp 253 °C.[2406]
$CaC_{35}H_{67}O_8P$ [R.N. 71640-91-0]: mp 220 °C, TLC.[2395]
$(NH_4)_2C_{35}H_{67}O_8P$: Solid.[2285]
$C_6H_{11}NH_3C_{35}H_{68}O_8P$: $[\alpha]_D^{20}$ +8.8° ($CHCl_3$).[2404]
$C_5H_5NHC_{35}H_{68}O_8P$ (pyridinium salt): mp 152-153 °C, $[\alpha]_D^{25}$ +3.8° ($CHCl_3$), soluble in benzene and chloroform, insoluble in ethyl acetate, acetone, CCl_4, and ether.[2397]

1-Palmitoyl-2-oleoyl-*sn*-glycero-3-(dihydrogen phosphate) (POPA), 16:0/18:1 PA, $C_{37}H_{71}O_8P$. *[R-(Z)]-9-Octadecenoic Acid, 1-[[(1-oxohexadecyl)oxy]methyl]-2-(phosphonooxy)ethyl Ester.*
POPA [R.N. 62600-81-1]: Oil,[2407] DSC.[2408] Spectra: IR,[2407] FAB-MS,[1369] CID-MS/MS (fig.).[2265] Chromatography: TLC,[2407] HPLC,[2103] silica gel column.[2407]

1-Palmitoyl-2-linoleoyl-*sn*-glycero-3-(dihydrogen phosphate) (PLPA), 16:0/18:2 PA, $C_{37}H_{69}O_8P$. *[R-(Z,Z)]-9,12-Octadecadienoic Acid, 1-[[(1-oxohexadecyl)oxy]methyl]-2-(phosphonooxy)ethyl Ester.*
PLPA [R.N. 91190-11-3; see also 120661-81-6]: Spectra: FAB-MS.[1369] Chromatography: HPLC.[2103]
$(NH_4)_2C_{37}H_{67}O_8P$: mp 192-194 °C, $[\alpha]_D$ +4.5° ($CHCl_3$).[2298]

1-Palmitoyl-2-linolenoyl-*sn*-glycero-3-(dihydrogen phosphate) (PLnPA), 16:0/18:3 PA, $C_{37}H_{67}O_8P$. *[R-(Z,Z,Z)]-9,12,15-Octadecatrienoic Acid, 1-[[(1-oxohexadecyl)oxy]methyl]-2-(phosphonooxy)ethyl Ester.*
PLnPA [R.N. 221000-71-1]: Chromatography: HPTLC.[1730]

1-Stearoyl-2-oleoyl-*sn*-glycero-3-(dihydrogen phosphate) (SOPA), 18:0/18:1 PA, $C_{37}H_{75}O_8P$. *[R-(Z)]-9-Octadecenoic Acid, 1-[[(1-oxooctadecyl)oxy]methyl]-2-(phosphonooxy)ethyl Ester.*
SOPA [R.N. 10016-26-9; see also 20337-09-1]: Spectra: FAB-MS.[1369] Chromatography: HPLC.[2108]

1-Stearoyl-2-linoleoyl-*sn*-glycero-3-(dihydrogen phosphate) (SLPA), 18:0/18:2 PA, $C_{39}H_{73}O_8P$. *[R-(Z,Z)]-9,12-Octadecadienoic Acid, 1-[[(1-oxooctadecyl)oxy]methyl]-2-(phosphonooxy)ethyl Ester.*
SLPA [R.N. 120661-82-7]: Chromatography: HPLC.[2108]

1,2-Dioleoyl-*sn*-glycero-3-(dihydrogen phosphate) (DOPA), 18:1/18:1 PA, $C_{39}H_{73}O_8P$. *[R-(Z,Z)]-9-Octadecenoic Acid, 1-[(Phosphonooxy)methyl]-1,2-Ethanediyl Ester.*
DOPA [R.N. 61617-08-1]: Viscous oil, highly soluble in chloroform, ethanol, acetone, ether, petroleum ether, and benzene, insoluble in water; $[\alpha]_D$ +3.8° ($CHCl_3$)[2409]; DSC.[2328] Spectra: ^1H NMR,[2379] ^2H and ^{31}P NMR (figs.),[2328] FAB-MS.[1369] Chromatography: TLC,[2379] IEC,[2285] HPLC.[2103]
$NaC_{39}H_{72}O_8P$: Waxlike material, highly soluble in ether and chloroform, slightly soluble (0.6 g/100 mL) in ethanol; $[\alpha]_D^{24}$ +3.3° ($CHCl_3$).[2409]
$(NH_4)_2C_{39}H_{71}O_8P$: Solid.[2285]

1,2-Dilinoleoyl-*sn*-glycero-3-(dihydrogen phosphate) (DLPA), 18:2/18:2 PA, $C_{39}H_{69}O_8P$. *[R-(all Z)]-9,12-Octadecadienoic Acid, 1-[(Phosphonooxy)methyl]-1,2-Ethanediyl Ester.*
DLPA [R.N. 17966-17-5]: Chromatography: HPLC.[2103]

1-Linoleoyl-2-linolenoyl-*sn*-glycero-3-(dihydrogen phosphate) (LLnPA), 18:2/18:3 PA, $C_{39}H_{67}O_8P$. *[R-(all Z)]-9,12,15-Octadecatrienoic Acid, 1-[[(1-oxo-9,12-octadecadienyl)oxy]-methyl]-2-(phosphonooxy)ethyl Ester.*
LLnPA [R.N. 228874-67-3; see also 142010-42-2]: Chromatography: HPLC.[2100]

1,2-Dilinolenoyl-*sn*-glycero-3-(dihydrogen phosphate) (DLnPA), 18:3/18:3 PA, $C_{39}H_{65}O_8P$. *[R-(all Z)]-9,12,15-Octadecatrienoic Acid, 1-[(Phosphonooxy)methyl]-1,2-Ethanediyl Ester.*
DLnPA [R.N. 34961-23-4; see also 94086-57-4]: Oil, $[\alpha]_D^{20}$ +3.1° ($CHCl_3$).[2292] Chromatography: TLC,[2292] HPLC.[2100]

Synthesis

DPPA may be prepared (a) by phosphorylation of 1,2-dipalmitoyl-*sn*-glycerol with diphenyl phosphorochloridate,[2397] *o*-phenylene phosphorochloridate,[2399] $POCl_3$,[2410] bis(2,2,2-trichloroethyl)phosphorochloridate,[2285] or 1,2-dibromo-1-phenylethanephosphonic acid[2396] in the presence of a tertiary amine, (b) by condensation of 1,2-dipalmitoyl-*sn*-glycero-3-iodohydrin with silver dibenzyl phosphate,[2411] (c) by acylation of *sn*-glycero-3-phosphate or one of its derivatives with palmitoyl chloride[2412] or palmitic anhydride,[2404,2413] or (d) by enzymatic hydrolysis of DPPC with phospholipase D.[2395] POPA may be prepared (a) by phosphorylation of 1-palmitoyl-2-oleoyl-*sn*-glycerol with bis(2,2,2-trichloroethyl) phosphorochloridate[2407] or (b) by enzymatic hydrolysis of POPC with phospholipase D.[2414] PLPA may be prepared by condensation of 1-palmitoyl-2-linoleoyl-*sn*-glycero-3-iodohydrin with silver di(*p*-methylbenzyl) phosphate.[2298] DOPA may be prepared (a) by phosphorylation of 1,2-dioleoyl-*sn*-glycerol with bis(2,2,2-trichloroethyl) phosphorochloridate[2379] or $POCl_3$,[2409] (b) by acylation of *sn*-glycero-3-phosphate with oleic anhydride[2415] or (c) by enzymatic hydrolysis of DOPC with phospholipase D.[2379] DLPA may be prepared by phosphorylation of 1,2-dilinoleoyl-*sn*-glycerol with $POCl_3$.[2416] DLnPA may be prepared (a) by phosphorylation of 1,2-dilinolenoyl-*sn*-glycerol with $POCl_3$[2292] or (b) by enzymatic hydrolysis of DLnPC with phospholipase D.[2417]

Hydrolysis

Hydrolysis of DL-DPPA with 0.1*N* NaOH in ethanol (3 h at reflux) gives *rac*-glycerol-3-phosphate with no migration or loss of phosphate.[2418]

Lysophosphatidic Acids

Physical Properties

1-Palmitoyl-*sn*-glycero-3-(dihydrogen phosphate) (PHPA), 16:0 PA, $C_{19}H_{39}O_7P$. *(R)-Hexadecanoic Acid, 2-Hydroxy-3-(phosphonooxy)propyl Ester.*
PHPA [R.N. 7220-34-0]: Colorless solid.[2419] Spectra: EI-MS and CI-MS (figs.).[2420] Chromatography: TLC,[2421] AgTLC,[2422] HPLC,[2422] GFC.[2420]

$NaC_{19}H_{38}O_7P$ [R.N. 17618-08-5]: Colorless powder, $[\alpha]_D^{20}$ +13.2° ($CHCl_3$), IR, PC.[2419]
2-Palmitoyl-*sn*-glycero-3-(dihydrogen phosphate) (HPPA), 16:0 PA, $C_{19}H_{39}O_7P$. *(R)-Hexadecanoic Acid, 1-Hydroxy-3-(phosphonooxy)propyl Ester.*
HPPA: Chromatography: TLC,[2421] AgTLC.[2422]

1-Stearoyl-*sn*-glycero-3-(dihydrogen phosphate) (SHPA), 18:0 PA, $C_{21}H_{43}O_7P$. *(R)-Octadecanoic Acid, 2-Hydroxy-3-(phosphonooxy)propyl Ester.*
SHPA [R.N. 65494-37-3]: Spectra: EI-MS and CI-MS.[2420] Chromatography: GFC,[2420] HPLC,[2422] AgTLC.[2422] *rac-SHPA* [R.N. 19491-32-8]: mp 85-87 °C, TLC.[2423]

2-Stearoyl-*sn*-glycero-3-(dihydrogen phosphate) (HSPA), 18:0 PA, $C_{21}H_{43}O_7P$. *(R)-Octadecanoic Acid, 1-Hydroxy-3-(phosphonooxy)propyl Ester.*
HSPA: No data. For *rac-HSPA*: Colorless crystals, mp 74.5-75.5 °C, IR (fig.), TLC.[2423]

1-Oleoyl-*sn*-glycero-3-(dihydrogen phosphate) (OHPA), 18:1 PA, $C_{21}H_{41}O_7P$. *[R-(Z)]-9-Octadecenoic Acid, 2-Hydroxy-3-(phosphonooxy)propyl Ester.*
OHPA [R.N. 6552-98-5]: Chromatography: TLC,[2421] AgTLC,[2422] HPLC.[2422]

2-Oleoyl-*sn*-glycero-3-(dihydrogen phosphate) (HOPA), 18:1 PA, $C_{21}H_{41}O_7P$. *[R-(Z)]-9-Octadecenoic Acid, 1-Hydroxy-3-(phosphonooxy)propyl Ester.*
HOPA: Chromatography: TLC,[2421] AgTLC.[2422]

1-Linoleoyl-*sn*-glycero-3-(dihydrogen phosphate) (LHPA), 18:2 PA, $C_{21}H_{39}O_7P$. *[R-(Z,Z)]-9,12-Octadecadienoic Acid, 2-Hydroxy-3-(phosphonooxy)propyl Ester.*
LHPA [R.N. 65528-85-0]: Chromatography: AgTLC, HPLC.[2422]

1-Linolenoyl-*sn*-glycero-3-(dihydrogen phosphate) (LnHPA), 18:3 PA, $C_{21}H_{37}O_7P$. *[R-(Z,Z,Z)]-9,12,15-Octadecatrienoic Acid, 2-Hydroxy-3-(phosphonooxy)propyl Ester.*
LnHPA [R.N. 68041-32-7]: Chromatography: AgTLC, HPLC.[2422]

Synthesis

PHPA may be prepared by the phosphorylation of 2-benzyl-1-palmitoyl-*sn*-glycerol with $POCl_3$[2419] or of 1-palmitoyl-*sn*-glycerol with ATP in the presence of DG kinase,[2421] or (b) by hydrolysis of DPPA with snake venom phospholipase A_2.[2424] HPPA may be prepared by phosphorylation of 2-palmitoyl-*sn*-glycerol with ATP in the presence of DG kinase.[2421] SHPA may be prepared (a) by phosphorylation of 1-stearoyl-*sn*-glycerol with ATP in the presence of DG kinase[2422] or (b) by enzymatic hydrolysis of SHPC with phospholipase D.[2420] OHPA may be prepared (a) by phosphorylation of 1-oleoyl-*sn*-glycerol with PCl_3/H_3PO_4[2425] or with ATP in the presence of DG kinase[2421] or (b) by enzymatic hydrolysis of DOPA with snake venom phospholipase A_2.[2424] HOPA may be prepared by phosphorylation of 2-oleoyl-*sn*-glycerol with ATP in the presence of DG kinase.[2421] LHPA may be prepared by phosphorylation of 1-linoleoyl-*sn*-glycerol with ATP in the presence of DG kinase.[2422] LnHPA may be prepared by phosphorylation of 1-linolenoyl-*sn*-glycerol with ATP in the presence of DG kinase (Table 4-15).[2422]

Table 4-15 Phospholipids in Plants*

No.	Plant part	Method	PC	PE	PG	PI	PS	PA	DPG	PL	References
Abelmoschus esculentus (L.) Moench (Okra)											
2	Seed	TLC	+	+		+	+			0.62-0.68	1426
5	Seed	TLC	48.4-52.4	17.7-20.5		24.0-26.2		3.8-5.2		19.3-21.9[a]	1427
1	Cotyledon	TLC				NAPE +					1361
Acacia arabica (Lam.) Willd. (Gum Arabic Tree)											
1	Seed	TLC	+	+		(LPC +)					1428
Acacia catechu (L.f.) Willd. (Catechu)											
1	Seed	TLC	36.0	25.8	7.4	21.8			9.0	0.35[c]	1429
Acer platanoides L. (Norway Maple)											
1	Seed	TLC	53.0	20.8		14.8		11.5		2.3[c,v]	1430
1	Seed[k]	TLC	179	60	22	49		12			1431
Acer pseudoplatanus L. (Sycamore Maple)											
1	Seed[k]	TLC	164	48	19	41	6				1431
1	Cell culture	TLC	47	29	5	16			1.8	160[j]	1432
1	–, mitochondria	TLC	43	35	3	6			13	300[j]	1432
1	–, inner membrane	TLC	41	37	2.5	5			14.5	400[j]	1432
1	–, outer membrane	TLC	54	30	4.5	11			0	830[j]	1432
1	–, amyloplast envelope membrane	TLC	59	1	25	15					1433
1	–, tonoplast	TLC	30.3	46.7	2.3	15.3	2.5	1.4	(LPC 1.6)	44.5[a]	1434
Adansonia digitata L. (Baobab)											
1	Seed, shelled	Frac. pptn.	37	63		0				0.13	1435
Adiantum hispidulum Swarz. (Australian Maidenhair Fern)											
1	Whole plant	TLC	+	+							1436
Aegilops speltoides Tausch. (Goatgrass)											
2	Leaf, chloroplast[m]	TLC	0.07-0.09		0.22-0.27						1437
Aegilops squarrosa L. (Barb Goatgrass)											
2	Leaf, chloroplast[m]	TLC	0.05-0.09		0.27						1437
Aegopodium podagraria L. (Goutweed)											
1	Seed	TLC	47	19	Trace	14		20		0.6[a]	1438
			also LPC trace, LPE trace.								

*Unless stated otherwise; the PL column gives the total phospholipid content of each plant part in percent by weight, and the PC, PE, PG, PI, PS, PA, and DPG columns show the distribution in percent by weight of the total PL; [a]Percent of total lipids; [b]mg lipid P/100 mL juice; [c]Percent dwb; [d]mg PL/100 mg protein; [e]mg PL/g fwb; [f]mg PL/100 g dwb; [g]Percent of polar lipids; [h]μmol PL/mg protein; [i]μmol PL/mg chlorophyll; [j]μg PL/mg protein; [k]μg lipid P/g; [l]μmol PL/g fwb; [m]mol PL/mol chlorophyll; [n]mg PL/g dwb; [o]μmol PL/g dwb; [p]μg PL/100 mg fwb; [q]μg PL/mg chlorophyll; [r]mg PL/cm² leaf area; [s]nmol PL/cm² leaf area; [t]μmol PL/mg protein; [u]IR given; [v]mg lipid P/g lipid; [w]μg PL/mL; [x]μg lipid P/cm² leaf area; [y]Molecules per 10^6 plastids; [z]μg PL/mg chromatin; [aa]Fatty acid composition and *sn*-1/*sn*-2 distribution given for each phospholipid; [bb]μg/g fwb; [cc]Taken from the abstract.

Continued

Table 4-15 Phospholipids in Plants*—cont'd

No.	Plant part	Method	PC	PE	PG	PI	PS	PA	DPG	PL	References
Aesculus hippocastanum L. (European Horse Chestnut)											
1	Leaf, thylakoid[a]	TLC			12.4						1439
Agathis robusta Bailey (Queensland Kauri)											
1	Leaf	TLC	+	+							1436
Agave americana L. (Maguey)											
1	Leaf	TLC			7						1440
Agave miradorensis Jacobi (Dwarf Century Plant)											
1	Leaf	TLC			7						1440
Agave sisalana Perrine (Sisal Hemp)											
1	Leaf	TLC			8						1440
Alchemilla pentaphylla L. (Lady's Mantle)											
1	Leaf	TLC	51	20	17	10			1	0.40	1441
Aleurites montana (Lour.) E. H. Wils. (Chinese Tung)											
1	Seed	TLC	25.5	6.6	34.7	28.6			4.6	1.9[a]	1442
1	Leaf	TLC	50.0	21.2	17.1	7.0	Trace		4.7	17.0[a]	1443
Alhagi persarum Boiss. (Camel's Thorn)											
1	Seed[aa]	TLC	+	+	+	+	(NAPE +, NALPE +)			1.5	1444
Alhagi pseudalhagi (M. Bieb.) Desv.											
1	Seed[aa]	TLC	+	+	+	+		+		1.1	1444
			also NAPE +, LPC +.								
Alisma plantago-aquatica L. (Mad Dog Weed)											
1	Whole plant	TLC	55.8	13.2	16.4	9.7	1.7	2.8		1.49[c]	1445
Allium ascalonicum L. (Scallion)											
1	Leaf	TLC	25.3	59.0	1.2			8.5	6.0	0.254	1446
Allium cepa L. (Onion)											
1	Seed	TLC	50.8	24.2		16.9		8.1		1.24[l]	1447
1	Root	TLC	19.4	21.2	18.8	10.6	8.1	13.1	8.8	16.6[e]	1448
1	Root tip, total particulates	TLC	54.0	21.6	6.1	7.1	2.4	2.5	5.2		1449
			also LPC 0.8, LPE 0.2								
1	–, nuclei	TLC	27.1	20.4	5.6	8.0	3.1	24.6	1.2	1.2	1449
			also LPC 5.2, LPE 4.9								
1	–, nuclear membrane	TLC	52.2	25.8	5.2	7.3	0	4.5	1.8	6.9	1449
			also LPC 1.4, LPE 1.9								

1	–, crude microsome	TLC	19.2	19.3	5.2	7.9	4.3	31.6	1.6	9.2	1449
			also LPC 5.8, LPE 5.3								
1	–, rough microsome	TLC	50.3	26.6	5.2	5.9	2.4	4.6	1.2	13.7	1449
			also LPC 1.5, LPE 2.1								
Allium fistulosum L. (Welsh Onion)											
1	Leaf	TLC					+				1450
Allium porrum L. (Leek)											
2	Leaf (1)	TLC	34.4	7.5	15.6			10.5	32.0	0.818	1446
	Leaf (2)	TLC	43.2	9.5	14.8			11.3	21.2	0.738	1446
1	Leaf[l]	Choline	30								1451
1	7-day seedling, endoplastic reticulum	TLC	75.9	19.2		3.2	1.7				1452
Allium sativum L. (Garlic)											
1	Leaf	TLC	15.2	35.1	16.4			16.6	15.8	0.136	1446
2	Clove *incl. PS	TLC	60.0-61.0*	25.0-25.3	10.1-11.4			3.6		0.112-0.134	1453
Allium turcomanicum Regel											
1	Seed, defatted	TLC	+	+		+				1.1	794
Alnus rubra Bong. Red Alder											
1	Root	HPLC	+	+		+					1454
1	–, nodule	HPLC	+	+		+					1454
Alocasia macrorrhiza (L.) G. Don (Giant Taro)											
1	Tuber, peeled	TLC	41.5	29.3		14.6				0.12[c]	1455
Aloe vera (L.) Burm. f. (Curaçao Aloe)											
1	Overground pieces	TLC	12.0	12.0		2.7	6.5	47.3	(LPI 1.2)	0.92	1456
Althaea rosea (L.) Cav. (Hollyhock)											
1	Seed	TLC	+	+	+	+	+	+	+	1.3	1457
			also LPC +, LPE +, NALPE +								
Amaranthus arthropurpureus											
1	Seed	TLC	+	+		+					1458

*Unless stated otherwise; the PL column gives the total phospholipid content of each plant part in percent by weight, and the PC, PE, PG, PI, PS, PA, and DPG columns show the distribution in percent by weight of the total PL; [a]Percent of total lipids; [b]mg lipid P/100 mL juice; [c]Percent dwb; [d]mg PL/100 mg protein; [e]mg PL/g fwb; [f]mg PL/100 g dwb; [g]Percent of polar lipids; [h]µmol PL/mg protein; [i]µmol PL/mg chlorophyll; [j]µg PL/mg protein; [k]µg lipid P/g; [l]µmol PL/g fwb; [m]mol PL/mol chlorophyll; [n]mg PL/g dwb; [o]µmol PL/g dwb; [p]µg PL/100 mg fwb; [q]µg PL/mg chlorophyll; [r]mg PL/cm^2 leaf area; [s]nmol PL/cm^2 leaf area; [t]µmol PL/mg protein; [u]IR given; [v]mg lipid P/g lipid; [w]µg PL/mL; [x]µg lipid P/cm^2 leaf area; [y]Molecules per 10^6 plastids; [z]µg PL/mg chromatin; [aa]Fatty acid composition and *sn*-1/*sn*-2 distribution given for each phospholipid; [bb]µg/g fwb; [cc]Taken from the abstract.

Continued

Table 4-15 Phospholipids in Plants*—cont'd

No.	Plant part	Method	PC	PE	PG	PI	PS	PA	DPG	PL	References
Amaranthus blitus L. (Wild Blite)											
1	Leaf	CC (silica)	13.1							2.40[l]	1459
Amaranthus cruentus L. (Purple Amaranth)											
1	Seed	TLC	26	16	6	10	0		8	5.0[a]	1460
			also LPC 8, LPE 2								
1	Seed	TLC	+	+			+				1461
1	Leaf[i]	TLC	0.92	0.38	0.49						1462
1	–, chloroplast[i]	TLC	0.05	0	0.30						1462
1	–, bundle-sheath strand[i]	TLC	0.29	0.10	0.42						1462
1	Callus culture[k]	TLC				960					1463
Amaranthus deflexus L.											
1	Leaf[l]	TLC			0.96						1402
Amaranthus hybridus L. (Green Amaranth)											
1	Seed	TLC	50	33		17				0.40	1458
Amaranthus hypochondriacus L. (Prince's Feather)											
1	Seed	TLC	+	+			+				1461
Amaranthus lividus Willd (Wild Blite)											
1	Leaf, chloroplast	TLC			+						1402
Amaranthus powellii S. Wats. (Powell Amaranth)											
1	Leaf[l]	TLC			1.14						1402
Amaranthus retroflexus L. (Redroot Pigweed)											
1	Leaf, thylakoid	TLC	35		56	8				1987[q]	1464
1	Leaf, chloroplast	HPLC	44.8	35.4	11.2	8.5				78.0[a]	1465
Amaranthus spinosus L. (Spiny Amaranth)											
1	Seed	TLC	+	+		+					1458
Amaranthus tricolor L. (Tampala)											
1	Seed	TLC	+	+		+					1458
1	Leaf	TLC	21	25	25	13			16	1.33[c]	1466
Amaryllis spp. (Amaryllis)											
1	Leaf	TLC			10						1440
Amorpha fruticosa L. (Dull Leaf Indigobush)											
1	Seed[s]		+								1467

Anacardium occidentale L. (Cashew)											
1	Seed kernel	TLC	11.6	11.0		(LPC 7.0)		7.5	4.6	0.95	1468
1	Nut kernel	TLC	53.6	14.0	+		6.5	(LPC 5.1)			1469
1	Fruit, juice	TLC	+	+	+			(LPC +)			1469
Androsace helvetica (L.) All. (Swiss Rock Jasmine)											
1	Leaf	TLC	58	20	10	10			3	0.16	1441
Anethum graveolens L. (Dill)											
1	Leaf	TLC	39.4	1.3	55.4	0.9			3.0	1.64[c]	1470
1	Stem	TLC	69.3	3.5	25.4	0			1.8	0.55[c]	1470
1	Fruit	TLC	43.0	17.9		20.5		8.6		0.2	1471
			also LPC 4.2, LPE 1.9								
Angelica sinensis (Olivi) Diels (Danggui)											
3	Leaf?	TLC	35.5-61.0	13.7-28.7		0.1-2.4	0.1-2.0	0.3-4.3	10.9-30.6		1472
Anomodon viticulosus (Hedw.) Hook. and Taylor (Anomodon Moss)											
1	Whole plant *incl. LPE	TLC	76.7	8.4	7.6	4.5*	2.8			0.35[c]	1473
Anthoceros laevis (L.) var. *lorbeer*											
1	Whole plant	TLC	+	+							1436
Anthriscus cerefolium L. (Chervil)											
1	Seed	TLC	28	7	3	42		17	2	1.0[a]	1438
Antirrhinum majus L. (Snapdragon)											
1	Leaf, chloroplast	TLC	5.7		87.7	6.6				2.11[c]	1408
1	–, thylakoid membr.	TLC	5.5		80.4	14.1				1.99[c]	1408
Apium graveolens L. ssp. *dulce* (Celery)											
1	Seed	TLC	Trace	33		33		(LPE 33)		0.78	1474
1	Leaf	TLC	17.4	44.4	32.7			1.6	3.9	0.792	1446
1	Stem	TLC	37.3	35.8	5.4			8.7	12.8	0.795	1446
1	Fruit	TLC	40.6	14.1		29.1		7.1		0.4	1471
			also LPC, 3.0, LPE 3.1								

*Unless stated otherwise; the PL column gives the total phospholipid content of each plant part in percent by weight, and the PC, PE, PG, PI, PS, PA, and DPG columns show the distribution in percent by weight of the total PL; [a]Percent of total lipids; [b]mg lipid P/100 mL juice; [c]Percent dwb; [d]mg PL/100 mg protein; [e]mg PL/g fwb; [f]mg PL/100 g dwb; [g]Percent of polar lipids; [h]μmol PL/mg protein; [i]μmol PL/mg chlorophyll; [j]μg PL/mg protein; [k]μg lipid P/g; [l]μmol PL/g fwb; [m]mol PL/mol chlorophyll; [n]mg PL/g dwb; [o]μmol PL/g dwb; [p]μg PL/100 mg fwb; [q]μg PL/mg chlorophyll; [r]mg PL/cm^2 leaf area; [s]nmol PL/cm^2 leaf area; [t]μmol PL/mg protein; [u]IR given; [v]mg lipid P/g lipid; [w]μg PL/mL; [x]μg lipid P/cm^2 leaf area; [y]Molecules per 10^6 plastids; [z]μg PL/mg chromatin; [aa]Fatty acid composition and *sn*-1/*sn*-2 distribution given for each phospholipid; [bb]μg/g fwb; [cc]Taken from the abstract.

Continued

Table 4-15 Phospholipids in Plants*—cont'd

No.	Plant part	Method	PC	PE	PG	PI	PS	PA	DPG	PL	References
1	Cell culture	TLC	29.9	15.4						24.1[o]	1475
Apodanthera undulata Gray (Melon-Loco)											
1	Seed	TLC	68.5	13.3		13.7		(LPC 4.5)		4.0[a]	1476
Arabidopsis thaliana (L.) Heynh (Mouse-Ear Cress)											
1	Leaf	TLC	41.8	25.1	24.6	8.5					1477
1	Leaf, plasma membrane	TLC	35.5	38.9	9.0	10.3 (incl. PS)		6.4		46.8[a]	1478
1	Leaf, protoplast	TLC	27.9	21.8	23.6				26.7		1479
Arachis hypogaea L. (Peanut)											
1	Nut	TLC	21.0	18.0	6.5	9.0		10.0		1.2[a]	1480
1	Nut[n]	TLC	2.7	0.5		1.5	0	(Phyto GP 0.9)			1481
1	Nut, oil body	TLC	61.6	5.0		8.4	25.0			0.80	1482
1	Seed	HPLC	51.2	16.8	2.0			1.6	17.4	1.3[a]	1483
1	Leaf	TLC	32.8	20.6	16.1	13.7			16.8	6.1[l]	1484
1	–, chloroplast	TLC	33.7		39.1	27.2				191.9[q]	1485
1	–, hypocotyl	TLC	32.4	27.0	17.6	23.0				0.017	1486
Araucaria cunninghamii Ait. (Moreton Bay Pine)											
1	Leaf	TLC	+	+							1436
Araucaria heterophylla (Salisb.) Franco (Norfolk Pine)											
1	Leaf	TLC	+	+							1436
Aronia melanocarpa L. (Chokeberry)											
1	Seed[e]	TLC	19.9	16.6	0.8	29.8	0	15.8	2.0		1487
			also LPC 10.8, LPE 1.4								
Arthrocnemum indicum (Willd.) Moq.											
1	Shoot	TLC	36.3	6.2	18.6	5.8		4.0	26.5	0.226[c]	1488
Arum maculatum L. (Cuckoopint)											
1	Leaf, chloroplast[a]	TLC			9						1489
1	Spadix, mitochondria	TLC	20.0	33.0	6.6	6.9	1.5	2.6	11.0		1490
Asplenium viride Huds. (Green Spleenwort)											
1	Whole plant	TLC	+	+							1436
Astragalus chinensis L. (Chinese Milk Vetch)											
1	Seed[c]	TLC	0.204								1491
Atrichum angustatum (Brid.) Bryol. eur.											
1	Whole plant *incl. LPE	TLC	28.2	21.8	25.2	12.5*	7.7			1.69[c]	1492

Atrichum haussknechtii Jur. et Milde											
1	Whole plant	TLC	39.5	14.8	15.2	10.3	11.0			1.17[c]	1493
Atriplex gmelinic C.A. Mey											
1	Leaf	TLC	52.2	7.5	7.5					2.84[c]	1494
1	Root	TLC	60.4	30.8	Trace					0.38[c]	1494
Atriplex prostrata DC											
1	Leaf[l]	TLC			0.60						1402
Atropa belladonna L. (Belladonna)											
1	Cell culture	TLC	51.1	28.1	7.8	9.6		Trace		0.65[e]	1495
Aulacomnium palustris (Hedw.) Schwägr.											
1	Whole plant	TLC	49.5	20.9	17.4	5.2	7.2			0.26[c]	1493
Avena sativa L. (Oats)											
1	Seed, dehulled *Incl. PA	TLC	29.9	14.8	9.5	3.9*	3.2	(LPC,LPE,LPS 20.4)		1.55[c]	1496
1	Seed	TLC		NAPE 0.41 mg/seed							1497
1	Seed	PC		NAPE +, NALPE +							1498
1	Starch	TLC		LPC 51.6, LPE 5.1, LPI 7.0						1.3	1499
1	Starch	TLC	LPC 67.4, LPE 7.8, LPG 12.0, LPI trace, LPS trace							1.25	1500
1	Coleoptile plasma membrane	TLC	23.0	21.6	4.3	5.5	7.7	35.5	(LPC 2.4)	0.55[h]	1501
1	Leaf	TLC					+				1450
1	Leaf, plasma membrane	TLC	35.1	29.2	6.6	9.4 (incl. PS)		15.3		28.8[a]	1502
1	–, mitochondria	TLC	40.7	33.3	8.4	3.4			14.1		1503
1	–, chloroplast	TLC	30.9	61.4	7.7						1503
1	Stem, internode	TLC	50	31	8	6	5			2.15[c]	1504
1	–, node	TLC	43	35	7	9	5			0.76[c]	1504
1	–, leaf sheath	TLC	31	46	14	3	6			0.35[c]	1504
1	Shoot, plasma membrane[j]	HPLC	77.3	70.3							1505
1	Root, whole[g]	Choline	0.54							1.37	1386
1	–, cell wall[g]	Choline	0.021							0.027	1506
1	–, mitochondria[g]	Choline	0.031							0.045	1506

*Unless stated otherwise; the PL column gives the total phospholipid content of each plant part in percent by weight, and the PC, PE, PG, PI, PS, PA, and DPG columns show the distribution in percent by weight of the total PL; [a]Percent of total lipids; [b]mg lipid P/100 mL juice; [c]Percent dwb; [d]mg PL/100 mg protein; [e]mg PL/g fwb; [f]mg PL/100 g dwb; [g]Percent of polar lipids; [h]µmol PL/mg protein; [i]µmol PL/mg chlorophyll; [j]µg PL/mg protein; [k]µg lipid P/g; [l]µmol PL/g fwb; [m]mol PL/mol chlorophyll; [n]mg PL/g dwb; [o]µmol PL/g dwb; [p]µg PL/100 mg fwb; [q]µg PL/mg chlorophyll; [r]mg PL/cm^2 leaf area; [s]nmol PL/cm^2 leaf area; [t]µmol PL/mg protein; [u]IR given; [v]mg lipid P/g lipid; [w]µg PL/mL; [x]µg lipid P/cm^2 leaf area; [y]Molecules per 10^6 plastids; [z]µg PL/mg chromatin; [aa]Fatty acid composition and *sn*-1/*sn*-2 distribution given for each phospholipid; [bb]µg/g fwb; [cc]Taken from the abstract.

Continued

Table 4-15 Phospholipids in Plants*—cont'd

No.	Plant part	Method	PC	PE	PG	PI	PS	PA	DPG	PL	References
1	–, mictosomes[g]	Choline	0.319							0.555	1506
1	Root, plasma membrane	TLC	18.2	10.5	4.0	3.7		13.9		0.33[c]	1507
			also LPC 4.9, LPE 6.3								
1	–, mitochondria	TLC	9.9	6.5	4.2	3.0		41.3	9.3*	0.19[c]	1507
			also LPC 8.9, LPE 5.4						*Incl. glycolipid		
1	Root, plasmalemma[n]	TLC	18	36		+	+				1508
1	–, microsomes[n]	TLC	29	27		+	+				1508
Avicennia germinans (L.) L. (Black Mangrove)											
1	Root	TLC	35.7	28.9		10.7				32[k]	1509
Avicennia resinifera Forst.											
1	Leaf[l]	TLC			0.46						1402
Azolla filiculoides Lam. (Pacific Mosquitofern)											
1	Whole plant	TLC	+	+							1436
Balanites aegyptiaca (L.) Del. (Desert Date)											
1	Seed, kernel	TLC	+	+		+		+			1510
Barbarea verna (Mill) Aschers (Winter Cress)											
1	Seed oil	TLC	35	25		5	10	10	10		1511
Bartramia halleriana Hedw. (Haller's Bartramia Moss)											
1	Whole plant	TLC	+	+							1436
Benincasa hispida (Thunb.) Cogn. (Wax Gourd)											
1	Seed, ripe	TLC	+	+		+	+			0.10	1512
1	–, unripe	TLC	+	+		+	+			0.26	1512
Beta cicla L. (Swiss Chard)											
1	Leaf	TLC	+	+	+	+					1513
Beta maritima L. (Sea Beet)											
1	Leaf	TLC	+	+	+	+					1513
Beta orientalis Roth.											
1	Leaf	TLC	+	+	+	+					1513
Beta patellaris Moq.											
1	Leaf	TLC	+	+	+	+					1513
Beta procumbens C. Sm.											
1	Leaf	TLC	+	+	+	+					1513

Beta vulgaris L., *crassa* group (Red Beet, Sugar Beet)											
1	Leaf[l]	TLC	0.404								1514
1	Leaf, stem	PC	32.0	15.8	7.7	6.5	2.0	9.9	3.4	0.53	1515
1	–, blade	PC	36.4	12.3	17.0	6.5	2.0	4.0	4.0	1.03	1515
1	–, chloroplast[a]	TLC	2.5		8.0						1516
1	Leaf	PC	52	26	11	11					1517
1	–, chloroplast	PC	38	15	38	9					1517
2	Root (1)	TLC	19.5	3.4	31.2	6.4	4.9	21.6	10.6	0.081	1518
	Root (2)		24.1	2.2	27.5	4.5	5.4	19.1	14.1	0.047	
			also LPC 1.5, LPE 0.9 (1) and (2) LPC 1.9, LPE 1.2 (2)								
1	Root	PC	35.3	16.5	6.4	4.7	2.1	11.6	2.4	0.08	1515
1	Root	Choline	0.190							0.326	1519
1	Root, vacuole	TLC	54	24	4	5	1	12		58[d]	1520
1	Root plasma membrane	TLC	35.4	39.4	1.5			14.6		1.3[h]	1521
Betula pendula Roth (Weeping Birch)											
1	Leaf	TLC	+	+	+	+	+			3.0[a]	1522
1	Bud	TLC	+	+	+	+	+			1.1[a]	1522
Bifora radians M. Bieb. (Wild Bishop)											
1	Seed	TLC	10	5	2	12	7	52	9	0.4[a]	1438
Blechnum fluviatile (R.Br.) Sal. (Ray Water Fern)											
1	Leaf	TLC	35	18	26	6			15	2.28[l]	1523
Blechnum spicant (L.) Roth (Deer Fern)											
1	Whole plant	TLC	+	+							1436
Bougainvillea spp. (Bougainvillea)											
1	Leaf	TLC					+				1450
Bouvardia ternifolia (Cav.) Schl. (Firecracker Bush)											
1	Cell culture	TLC	53.6	26.2	7.3	8.6		0.5		0.83[e]	1495
Brachystegia nigera Hoyle and Jones (Achi)											
1	Seed	TLC	16	26	17	32			8	1.53[c]	1524
Brachythecium spp.											
1	Whole plant *Incl. LPE	TLC	41.8	17.4	24.0	8.1*	4.2			0.64[c]	1492

*Unless stated otherwise; the PL column gives the total phospholipid content of each plant part in percent by weight, and the PC, PE, PG, PI, PS, PA, and DPG columns show the distribution in percent by weight of the total PL; [a]Percent of total lipids; [b]mg lipid P/100 mL juice; [c]Percent dwb; [d]mg PL/100 mg protein; [e]mg PL/g fwb; [f]mg PL/100 g dwb; [g]Percent of polar lipids; [h]μmol PL/mg protein; [i]μmol PL/mg chlorophyll; [j]μg PL/mg protein; [k]μg lipid P/g; [l]μmol PL/g fwb; [m]mol PL/mol chlorophyll; [n]mg PL/g dwb; [o]μmol PL/g dwb; [p]μg PL/100 mg fwb; [q]μg PL/mg chlorophyll; [r]mg PL/cm² leaf area; [s]nmol PL/cm² leaf area; [t]μmol PL/mg protein; [u]IR given; [v]mg lipid P/g lipid; [w]μg PL/mL; [x]μg lipid P/cm² leaf area; [y]Molecules per 10^6 plastids; [z]μg PL/mg chromatin; [aa]Fatty acid composition and *sn*-1/*sn*-2 distribution given for each phospholipid; [bb]μg/g fwb; [cc]Taken from the abstract.

Continued

Table 4-15 Phospholipids in Plants*—cont'd

No.	Plant part	Method	PC	PE	PG	PI	PS	PA	DPG	PL	References
Brassica campestris L. (Turnip Rape)											
2	Leaf, chloroplast	TLC	3.0-4.8	1.6-1.8	7.8-8.0	1.6-3.4 (incl. SL)			1.8-2.0	0.735-0.806[i]	1525
1	–, thylakoid	TLC	35		52	12				1864[q]	1464
Brassica campestris L. ssp. *komatsuma* Mak. (Chinese Cabbage)											
1	Leaf	TLC					+				1450
1	Leaf[l]	Choline	0.2								1451
Brassica campestris L. ssp. *sarson* (Yellow Sarson)											
1	Seed	TLC	+	+	+	+	+	+	+		1526
			also LPC +, LPI +								
Brassica caulorapa Pasq. (Kohlrabi)											
1	Stem	Fr. Prec.	0.033								1527
1	Tuber[l]	Choline	1								1451
Brassica juncea L. (Indian Mustard)											
1	Seed, oil body	TLC	53.1	15.5		13.1	18.3			1.60	1482
Brassica napus L. (Rape)											
3	Seed	TLC	20.0-30.0	10.0-13.0		2.0-10.0	19.0-21.0	10.0-15.0	5.0-11.0		1528
			also LPC 3.0-5.0, LPE 0-3.0, LPS 1.0-4.0, LPA 2.0-4.0								
2	Seed (1)	TLC	37.8	24.9	3.1	26.7		3.7	(LPC 3.8)	630[k]	1529
	Seed (2)	TLC	41.8	27.8	5.2	21.2		1.8	(LPC 2.2)	630[k]	1529
1	Seed	TLC	28.8	7.3	2.2	14.2		41.6			1397
1	Seed	HPLC	25.6	12.8		15.8		44.8			1397
1	Seed, oil body	TLC	59.9	5.9		14.0	20.2			1.97	1482
1	Leaf[c]	TLC	0.52	0.13	0.28	0.06		0.05			1530
1	Leaf	TLC	20.8	26.3	20.1			11.3	20.1	0.431	1446
1	–, thylakoid	TLC	25		44	32				887[q]	1464
1	Root	TLC	55	33		7			5	26[o]	1531
1	Whole floret[a]	HPLC	1.8	1.5		0.5		0.3	(LPC 0.9, LPE 0.4)		1532
1	–, elaioplast[a]	HPLC	3.2	1.0		0.2		0.1	(LPC 2.7, LPE 1.5)		1532
1	–, tapetosome[a]	HPLC	4.2	4.8		1.1		1.3	(LPC 0.4, LPE 0.3)		1532
1	Pollen, interior	HPLC	50.6	35.0	2.6	10.0		1.9			1533
1	Pollen, coat	HPLC	26.7	17.0	0	0		56.3			1533

Brassica napus L. (Rape, low erucic acid (Canola))											
2	Seed (1)	TLC	41.2	23.7	6.8	26.2		1.0	(LPC 1.1)	650[k]	1529
	Seed (2)	TLC	53.8	19.2	4.8	14.8		3.4	(LPC 4.0)	810[k]	1529
Brassica napus L. (Rape, low erucic acid, low glucosinolate (Double Zero Canola))											
2	Seed (1)	TLC	56.5	16.9	4.2	17.8		2.0	(LPC 2.6)	430[k]	1529
	Seed (2)	TLC	50.3	21.0	7.7	18.4		1.5	(LPC 1.1)	740[k]	1529
Brassica napus L. var. *oleifera* (Winter Rape)											
1	Leaf[n]	HPLC	1.8	2.4	2.5			2.7			1534
1	Leaf	TLC	31.5	18.2	26.3	9.6		6.1		1.60[c]	1535
1	Root	TLC	38.2	40.4	4.6	7.9	1.7	3.1		1.13[c]	1535
Brassica nigra (L.) Koch (Black Mustard)											
1	Seed	CC (silica)	20.0	35.3							1536
Brassica oleracea L. ssp. *acephala* DC (Kale)											
1	Leaf[f]	Choline	18								1451
Brassica oleracea L. ssp. *botrytis* L. Cauliflower											
1	Bud, mitochondria	TLC	44	34	4	7			11	356[j]	1537
1	Bud, inner membrane	TLC	41	37	3	5			14	409[j]	1537
1	–, outer membrane	TLC	42	24	10	21			3	634[j]	1537
1	–, microsomes	TLC	50	35	8	6			1	510[j]	1537
1	–, plastid	TLC	54	6	23	14			1.7		1538
1	–, plasma membrane	TLC	38	15	4	8	7		1		1539
			also LPC 4 and LPE 3.								
1	–, plastid envelope membrane	TLC	60	3	26	11					1433
Brassica oleracea L. ssp. *capitata* L. (Cabbage)											
3	Leaf	TLC	12.0-15.9	15.1-31.0	21.5-29.9	3.8-7.8	5.3-11.4	10.8-16.9	1.9-4.0	0.028-0.037	1540
		also LPC 0.9-2.1, LPE 1.4-3.0									
1	Leaf	TLC	20.8	25.0	26.6	7.8	3.9	9.1	(NAPE 6.8)	20.2[a]	1541
1	Leaf	CC	29	27	6		2				1542
1	Stalk	TLC	++	+				+			1396

*Unless stated otherwise; the PL column gives the total phospholipid content of each plant part in percent by weight, and the PC, PE, PG, PI, PS, PA, and DPG columns show the distribution in percent by weight of the total PL; [a]Percent of total lipids; [b]mg lipid P/100 mL juice; [c]Percent dwb; [d]mg PL/100 mg protein; [e]mg PL/g fwb; [f]mg PL/100 g dwb; [g]Percent of polar lipids; [h]μmol PL/mg protein; [i]μmol PL/mg chlorophyll; [j]μg PL/mg protein; [k]μg lipid P/g; [l]μmol PL/g fwb; [m]mol PL/mol chlorophyll; [n]mg PL/g dwb; [o]μmol PL/g dwb; [p]μg PL/100 mg fwb; [q]μg PL/mg chlorophyll; [r]mg PL/cm^2 leaf area; [s]nmol PL/cm^2 leaf area; [t]μmol PL/mg protein; [u]IR given; [v]mg lipid P/g lipid; [w]μg PL/mL; [x]μg lipid P/cm^2 leaf area; [y]Molecules per 10^6 plastids; [z]μg PL/mg chromatin; [aa]Fatty acid composition and *sn*-1/*sn*-2 distribution given for each phospholipid; [bb]μg/g fwb; [cc]Taken from the abstract.

Continued

Table 4-15 Phospholipids in Plants*—cont'd

No.	Plant part	Method	PC	PE	PG	PI	PS	PA	DPG	PL	References
Brasica oleracea ssp. *gemmifera* Zenk. (Brussels Sprouts)											
1	Leaf[l]	Choline	21								1451
Brassica rapa L. ssp. *rapa* (Turnip)											
1	Seed	TLC		$P_{NAPE}/P_{total}=0.9$							1497
2	Root	TLC	+	+	+	+		+	+		1543
Bromus erectus Huds. (Meadow Brome)											
1	Callus	TLC	43	27	25			5			1544
Bryonia alba L. (White Bryony)											
1	Root	TLC	26.6	7.4	66.0						1545
Bryum pseudotriquetrum (Hedw.) Gärtn. *et al.*											
1	Whole plant	TLC	11.2	31.9	20.7	10.6	13.4			1.17[c]	1493
Buchanania lanzan Spreng. (Almondette, Chironji)											
1	Seed, kernel	TLC	23.0	47.0	12.0	10.0	(LPC 3.0, LPE 5.0)			4.06[c]	1546
Buchloe dactyloides Nutt. (Buffalograss)											
1	Root plasma membrane	TLC	19.2	42.6	20.3	6.8	14.3			3.90[a]	1547
Bupleurum rotundifolium L.											
1	Seed	TLC	23	8		40		25	4	1.4[a]	1438
			also LPC trace, LPE trace.								
Butea monosperma (Lam.) Taub. (Palash)											
1	Seed, kernel	TLC	44.6	14.8		27.0		(LPC 4.9)	1.8	0.84	1548
Butomus umbellatus L. (Flowering Rush)											
1	Whole plant	TLC	53.3	25.5	10.4	5.6	1.4	3.7		1.26[c]	1445
Butyrospermum paradoxum (Gaertn.f.) Hep. Shea Butter Tree											
1	Nut, kernel	TLC	50	10			22			1.35[c]	1549
Caesalpinia bonducella (L.) Roxb. Sagargota											
1	Seed	TLC	22.2	25.6		30.6			19.8		1550
Calendula officinalis L. (Pot Marigold)											
1	Seed, defatted	TLC	+	+		+				1.0	794
Calliergon cordifolium (Hedw.) Kindb.											
1	Whole plant	TLC	28.8	26.3	32.7	3.9	4.8			1.41[c]	1551
Calocedrus formosana Florin (Formosa Incense Cedar)											
1	Leaf	TLC	+	+							1436

Caloncoba echinata (Oliv.) Gilg.											
1	Leaf	TLC	++	Trace	+	+		Trace			1552
Calophyllum inophyllum L. (Alexandrian Laurel)											
1	Seed, kernel	TLC	33.8	46.3			6.1	8.1	(LPC 5.7)	0.96[c]	1553
Calotropis gigantea (L.) Ait.f. (Mudar)											
1	Leaf	TLC	32.5	15.7	22.0	14.1	4.7		11.0	2.12[c]	1554
Caltha palustris L. (Cowslip)											
1	Whole plant	TLC	43.8	26.7	9.3	10.0	4.1		3.4	6.97[c]	1555
Camellia japonica L. (Camellia)											
1	Leaf	TLC	43	19	14	4			19	4.65[l]	1523
Camellia sinensis L. (Tea)											
1	Seed kernel[g]	TLC	45	20	3	5					1556
1	Young shoot[g]	TLC	31	14	8	9					1556
1	Coarse leaf[g]	TLC	15	8	9	5					1556
1	Leaf	TLC	59.7	10.9	12.6	8.8			8.0	0.204	1557
1	Pollen	TLC	54	34	1	6				1.30[c]	1558
Canavalia ensiformis (L.) DC (Jack Bean)											
1	Root, nodule	TLC	16	27	16	21	19			0.082	1559
Cannabis sativa L. (Hemp)											
2	Seed (1)	^{31}P NMR	50.5	16.3		25.5	0	3.4		1.0[v]	1560
	Seed (2)	^{31}P NMR	42.4	16.7		27.2	0	7.4		1.2[v]	1560
Capparia spinosa L. (Caper)											
1	Plant, aerial parts	TLC	16.6	20.4	26.1	12.8		2.8		8.0[a]	1561
			also NAPE 14.6, NALPE 6.7								
Capsicum annuum L. (Bell Pepper, Green Pepper)											
1	Leaf[l]	TLC	0.918								1514
1	–, chloroplast	TLC	26.7	9.5	54.9	7.6		1.3		0.32[i]	1562
1	Fruit, whole	TLC	75.8	8.7	2.3	4.1	2.1	Trace		0.008	1563
			also LPC 2.8, LPE trace								
1	–, chloroplast	TLC	33.7	24.8	24.7	12.4		4.4		0.0022[l]	1564
1	–, chromoplast	TLC	58.6	11.4	14.3	14.8		0.9		0.0041[l]	1564
1	–, microsomal membr.	TLC	58.1	28.5	1.8	10.4		1.2		0.036[l]	1564
1	Fruit, pericarp	TLC	48.7	26.3	4.1	9.1		8.8		594[l]	1565
			also LPC (including LPE) 3.0								

*Unless stated otherwise; the PL column gives the total phospholipid content of each plant part in percent by weight, and the PC, PE, PG, PI, PS, PA, and DPG columns show the distribution in percent by weight of the total PL; [a]Percent of total lipids; [b]mg lipid P/100 mL juice; [c]Percent dwb; [d]mg PL/100 mg protein; [e]mg PL/g fwb; [f]mg PL/100 g dwb; [g]Percent of polar lipids; [h]µmol PL/mg protein; [i]µmol PL/mg chlorophyll; [j]µg PL/mg protein; [k]µg lipid P/g; [l]µmol PL/g fwb; [m]mol PL/mol chlorophyll; [n]mg PL/g dwb; [o]µmol PL/g dwb; [p]µg PL/100 mg fwb; [q]µg PL/mg chlorophyll; [r]mg PL/cm^2 leaf area; [s]nmol PL/cm^2 leaf area; [t]µmol PL/mg protein; [u]IR given; [v]mg lipid P/g lipid; [w]µg PL/mL; [x]µg lipid P/cm^2 leaf area; [y]Molecules per 10^6 plastids; [z]µg PL/mg chromatin; [aa]Fatty acid composition and *sn*-1/*sn*-2 distribution given for each phospholipid; [bb]µg/g fwb; [cc]Taken from the abstract.

Continued

Table 4-15 Phospholipids in Plants*—cont'd

No.	Plant part	Method	PC	PE	PG	PI	PS	PA	DPG	PL	References
Capsicum frutescens L. (Chili Pepper)											
1	Fruit	TLC	12.6	26.6	12.7				43.4	0.386	1446
Carica papaya L. (Papaya)											
1	Seed	TLC	28.1	18.7		34.0			(LPC +)	0.15	1548
Carica pubescens Lenné and K. Koch (Mountain Papaya)											
1	Leaf[l]	TLC			0.67						1566
Carthamus oxycantha M. Rieb. (Wild Safflower)											
1	Seed	TLC	36.7	29.4		20.6					1567
Carthamus tinctorius L. (Safflower)											
1	Seed, whole	TLC	32	16		23				0.705	1568
1	Seed	TLC	+	+	+	+	+	+		(LPC +)	1569
1	Cotyledon, oil bodies	TLC	57.6	15.0	0	25.4		1.9			1570
1	–, microsomes	TLC	53.8	20.1	0	22.9		3.1			1570
1	Seed, oil body	TLC					0				1571
Carum carvi L. (Caraway)											
1	Seed (wild form)	TLC	46	20	2	20		5	2	4.5[a]	1438
			also LPC trace, LPE 4.								
1	Fruit	TLC	41.5	18.6	7.5	18.6		2.4		0.5	1471
			also LPC 3.1, LPE 4.3								
Carum copticum (L.) Benth. and Hook. (Bishop's Weed)											
1	Seed	TLC	26	35		26		(LPE 14)		0.84[c]	1572
Carya illinoinensis (Wangenh.) K. Koch (Pecan)											
5	Pollen	TLC	+	+	+	+	+	Trace			1573
Caryocar coriaceum Wittm. (Piqui)											
1	Seed	TLC	Trace	28		72				0.20[c]	1574
Cassia absus L. (Pig's Senna)											
1	Seed[a]	TLC	19.3	15.8				(LPC 0.7, LPE 1.3)		2.41	1575
Cassia fistula L. (Golden Shower)											
1	Seed[a]	TLC	22.6	18.2				(LPC 1.1, LPE 2.4)		2.35	1575
1	Seed	CC (silica)	35.6	64.4							1576
Cassia occidentalis L. (Coffee Senna)											
1	Seed[a]	TLC	23.3	17.2				(LPC 0.9, LPE 1.2)		1.90	1575

Cassia tora L. (Sicklepod)											
1	Seed	TLC	53.1	16.6	1.7	17.2 (incl. PS)		10.7		1.17	1577
Castanea crenata Sieb. and Zucc. (Japanese Chestnut)											
3	Seed, inner, free	TLC	26.4-33.1	7.7-19.5	14.6-26.2	9.8-21.5		(LPC 11.5-25.5)			1578
3	–, inner, bound	TLC	32.0-40.4	6.3-10.4	9.9-20.8	26.7-32.3		(LPC 5.1-15.7)			1578
3	–, outer, free	TLC	18.6-30.7	9.0-22.9	14.3-24.1	21.4-34.8		(LPC 7.0-22.9)			1578
3	Seed, outer, bound	TLC	8.0-36.9	5.4-8.1	6.0-19.9	36.9-43.4		(LPC 5.9-24.0)			1578
Cattleya spp.											
1	Leaf	TLC			10						1440
Cattleya bowringiana Veitch. (Pine Ridge Orchid)											
1	Leaf[l]	TLC			0.14						1402
Celosia argentea L. (Cockscomb)											
1	Seed	TLC	43	35		22				0.47	1458
Ceratodon purpureus (Hedw.) Brid.											
1	Protonemata	TLC	55.8	11.6	20.4	12.2				2.12[m]	1579
Ceratophyllum demersum L.											
1	Whole plant	TLC	36.0	19.1	13.4	9.5	4.9	17.1		2.98[c]	1445
Ceratophyllum submersum L.											
1	Whole plant	TLC	44.0	16.5	14.8	16.4	4.9		2.6	6.28[c]	1555
Ceratopteris thalictroides (L.) Brongn. (Water Sprite)											
1	Whole plant	TLC	47.5	23.7	9.7	7.4	9.1	Trace	4.4	3.00[c]	1555
Chaerophyllum bulbosum L. (Turnip-Rooted Chervil)											
1	Seed	TLC	49 also LPC 1.	8	5	29		6	1	1.2[a]	1438
Chenopodium album L. (Lambsquarters)											
1	Leaf, thylakoid	TLC	46		42	12				983[q]	1464
1	Leaf, chloroplast	HPLC	38.4	46.7	8.7	6.1				82.8[a]	1465
Chenopodium quinoa Willd. (Quinoa)											
1	Seed	TLC	13.2 also LPC 3.9, LPE 46.4	19.8		11.3	4.3	1.2		1.77	1580

*Unless stated otherwise; the PL column gives the total phospholipid content of each plant part in percent by weight, and the PC, PE, PG, PI, PS, PA, and DPG columns show the distribution in percent by weight of the total PL; [a]Percent of total lipids; [b]mg lipid P/100 mL juice; [c]Percent dwb; [d]mg PL/100 mg protein; [e]mg PL/g fwb; [f]mg PL/100 g dwb; [g]Percent of polar lipids; [h]µmol PL/mg protein; [i]µmol PL/mg chlorophyll; [j]µg PL/mg protein; [k]µg lipid P/g; [l]µmol PL/g fwb; [m]mol PL/mol chlorophyll; [n]mg PL/g dwb; [o]µmol PL/g dwb; [p]µg PL/100 mg fwb; [q]µg PL/mg chlorophyll; [r]mg PL/cm^2 leaf area; [s]nmol PL/cm^2 leaf area; [t]µmol PL/mg protein; [u]IR given; [v]mg lipid P/g lipid; [w]µg PL/mL; [x]µg lipid P/cm^2 leaf area; [y]Molecules per 10^6 plastids; [z]µg PL/mg chromatin; [aa]Fatty acid composition and *sn*-1/*sn*-2 distribution given for each phospholipid; [bb]µg/g fwb; [cc]Taken from the abstract.

Continued

Table 4-15 Phospholipids in Plants*—cont'd

No.	Plant part	Method	PC	PE	PG	PI	PS	PA	DPG	PL	References
1	–, flour	TLC	52.6	8.9		13.7	2.9	0.4		0.64	1580
			also LPC 3.6, LPE 17.8								
1	–, bran	TLC	49.1	13.6		5.9	4.0	0.5		1.45	1580
			also LPC 4.3, LPE 22.7								
1	–, hull	TLC	16.4	20.9		10.1	3.3	0.6		2.40	1580
			also LPC 3.1, LPE 45.6								
Chenopodium rubrum L. (Red Goosefoot)											
1	Cell culture	TLC	76	24	Trace					1.65[c]	1581
Cicer arietinum L. (Chickpea)											
1	Seed, immature[g]	TLC	8.1	12.6		18.4		3.2	4.8		1582
1	Seed, mature[g]	TLC	39.3	30.6		0.9		1.4	4.1		1582
1	–, dehulled[g]	TLC	37.4	33.4		1.7		1.1	4.7		1582
1	Root, 6 days old	TLC	34.3	27.3	6.5	6.2	8.2	6.1	5.7	0.83	1583
Cicer soongaricum DC (Channa)											
1	Seed	TLC	50.0	38.0		Trace	11.9			2.15	1584
Cichorium intybus L. (Chicory)											
1	Root[l]	Choline	2								1451
Cistus ladanifer L. (Rockrose)											
1	Seed	TLC	46	32	10	12				0.54[c]	1585
		No significant variation in seeds of 15 other *Cistus* spp.									
Citrofortunella mitis Blanco (Calamondin)											
1	Fruit outer peel, leucoplast[g]	TLC	43.0	4.6	2.5	6.6			Trace		1586
Citrullus colocynthis (L.) Schrad. (Colocynth)											
1	Seed	TLC	41	17	14	11			16	5.66[c]	1524
Citrullus vulgaris Schrad. (Watermelon)											
1	Seed	TLC	36.1	8.6 (incl. PG)		30.4		(LPC 6.6, LPE 6.4)			1587
3	Seed	TLC	24.7-26.1	23.8-27.1*		14.3-19.7	6.7-8.6	(LPC 12.8-19.0)		0.30-0.59	1588
		*Incl. MGDG									
1	Cotyledon[a]	TLC	2.8	2.4		2.8		(LPC 2.4)			1589
1	Root[a]	TLC	3.5	3.1		2.9		(LPC 2.0)			1589

Citrullus vulgaris ssp. *fistulosus* (Squash Melon)											
1	Seed	TLC	34.1	10.9 (incl. PG)		25.4		(LPC 4.7, LPE 6.4)			1587
Citrus aurantifolia (Christm.) Swingle (Key Lime)											
1	Leaf	TLC	73.7	15.8	6.7	3.8				0.495	1590
Citrus aurantium L. (Sour Orange)											
1	Leaf	TLC	+	+	+	+					1591
Citrus iyo hort. ex Tanaka (Iyomikan)											
1	Seed	TLC	+	+		+	+			0.25	1592
Citrus jambhiri Lush. (Rough Lemon)											
1	Leaf	TLC	68.3	19.5	8.3	3.9				0.508	1590
Citrus junos (Sieb.) Tanaka (Yuzu)											
1	Seed, kernel	TLC	55.3	24.9		3.5	9.4	1.6	(LPC 4.0)		1593
1	–, coat	TLC	47.8	31.4		4.5	5.1	1.4	(LPC 7.6)		1593
1	Leaf	TLC	21.0	13.7	29.8	35.5				0.58	1594
1	Leaf	TLC	+	+		+	+				1595
Citrus limon (L.) Burm.f. (Lemon)											
1	Juice	TLC	41	38		15	5	2		1.15[b]	1596
1	Leaf	TLC	+	+		+	+				1595
1	Leaf	TLC	47.8	35.8	8.6	7.8				0.232	1590
Citrus medica L. (Citron)											
1	Leaf	TLC	+	+	+	+					1591
Citrus medioglobosa hort. ex Tanaka (Naruto)											
1	Fruit, flavedo	TLC	52	36		12				1.16[l]	1597
Citrus natsudaidai Hayata (Natsudaidai)											
2	Seed	TLC	+	+		+	+			0.21-0.53	1592
1	Leaf	TLC	54.0	20.6	27.7					0.97	1598
Citrus × *paradisi* Macf. (Grapefruit)											
1	Leaf	TLC	57.2	27.7	8.7	6.3				0.332	1590
1	Fruit, juice	TLC	47	36		12	1	1		0.67[b]	1596

*Unless stated otherwise; the PL column gives the total phospholipid content of each plant part in percent by weight, and the PC, PE, PG, PI, PS, PA, and DPG columns show the distribution in percent by weight of the total PL; [a]Percent of total lipids; [b]mg lipid P/100 mL juice; [c]Percent dwb; [d]mg PL/100 mg protein; [e]mg PL/g fwb; [f]mg PL/100 g dwb; [g]Percent of polar lipids; [h]µmol PL/mg protein; [i]µmol PL/mg chlorophyll; [j]µg PL/mg protein; [k]µg lipid P/g; [l]µmol PL/g fwb; [m]mol PL/mol chlorophyll; [n]mg PL/g dwb; [o]µmol PL/g dwb; [p]µg PL/100 mg fwb; [q]µg PL/mg chlorophyll; [r]mg PL/cm² leaf area; [s]nmol PL/cm² leaf area; [t]µmol PL/mg protein; [u]IR given; [v]mg lipid P/g lipid; [w]µg PL/mL; [x]µg lipid P/cm² leaf area; [y]Molecules per 10^6 plastids; [z]µg PL/mg chromatin; [aa]Fatty acid composition and *sn*-1/*sn*-2 distribution given for each phospholipid; [bb]µg/g fwb; [cc]Taken from the abstract.

Continued

Table 4-15 Phospholipids in Plants*—cont'd

No.	Plant part	Method	PC	PE	PG	PI	PS	PA	DPG	PL	References
C. paradisi Macf. × *C. reticulata* Blanco (Tangelo)											
1	Leaf	TLC	62.1	19.0	5.9	13.1				0.153	1590
Citrus reticulata Blanco (Tangerine)											
3	Leaf	TLC	55.2-65.2	15.1-17.0	6.7-10.3	9.7-13.7				0.135-0.165	1590
Citrus sinensis (L.) Osbeck (Orange)											
1	Seed	TLC	25.2	28.3		30.2			16.3		1550
1	Leaf	TLC	+	+		+	+				1595
1	Fruit juice	TLC	41	38		8	3	7		1.31[b]	1596
1	–, chromoplast	TLC	48	42	0					1.41[l]	1599
1	Fruit flavedo, chloroplast	TLC	Trace	Trace	<90					0.39[l]	1599
1	–, chromoplast	TLC	Trace	Trace	>90					0.52[l]	1599
1	Fruit, essential oil	TLC	+	0					0		1600
(*C. sinensis* × *C. reticulata*) × *C. reticulata* Blanco (Mandarin)											
1	Leaf	TLC	50.8	33.1	4.0	12.1				0.124	1590
Citrus sudachi hort. ex Shirai (Sudachi)											
1	Seed, kernel	TLC	57.0	23.9		4.0	9.3	1.9	(LPC 3.4)		1593
1	–, coat	TLC	45.1	32.3		5.9	7.9	1.1	(LPC 6.0)		1593
Citrus tamurana hort. ex Tanaka (Hyuganatsu)											
1	Fruit juice	TLC	+	+		+	+	+		3.3[b]	1601
			also LPC +, LPE +								
Citrus unshiu Marc. (Mandarin)											
1	Fruit, flavedo	TLC	33.9	37.5	10.6	4.7	1.5	6.9	4.4	0.34	1602
			also LPC 0.3, LPE 0.2								
1	–, albedo	TLC	16.2	37.0	15.1	14.5	6.7	3.6	6.9	0.10	1602
			also LPC 0, LPE 0								
1	–, membrane	TLC	26.1	28.5	19.7	11.8	5.4	1.4	5.7	0.23	1602
			also LPC 0.9, LPE 0.5								
1	–, flesh	TLC	34.9	32.4	12.4	7.5	2.5	4.9	1.9	0.04	1602
			also LPC 2.3, LPE 1.2								
Citrus yuko hort. ex Tanaka (Yuko)											
1	Seed, kernel	TLC	50.4	28.6		2.3	11.2	1.7	(LPC 4.4)		1593
1	–, coat	TLC	42.7	37.7		3.1	6.2	1.0	(LPC 7.4)		1593
Clarkia rubicunda (Lindl.) Lewis											
1	Pollen	TLC	+	+		+					1603
	Of 69 spp. of angiosperm pollens collected in California, 77% contained PC, 54% PE, 6% PG, and 48% PI.										

Climacium dendroides (Hedw.) Web. et Mokr.											
1	Whole plant	TLC	67.0	11.4	10.0	4.0	4.2			0.60[c]	1493
Cochlearia anglica L. (English Scurvy Grass)											
1	Root	TLC	39.0	39.1	4.5	6.2		12.1		0.83[e]	1604
Cocos nucifera L. (Coconut)											
1	Nut, kernel	TLC	34.6	24.6		19.0	4.8	2.5		0.20[a]	1605
			also LPC 4.6 and LPE 3.4								
Coffea arabica L. (Coffee)											
3	Bean[s]		+								1606
1	Leaf	TLC	52	15	22	11					1607
Coix lacryma-jobi L. (Adlay, Job's Tears)											
1	Seed, brown	TLC	43.0	27.5		13.0	2.4	(LPC 3.3)		1.83	1608
1	–, polished	TLC	48.4	24.9		7.4	4.0	(LPC 3.9)		1.78	1608
Colocasia esculenta (L.) Schott (Taro)											
1	Leaf	TLC	46.9	10.7	19.9	16.2	6.6			2.63[c]	1470
1	Stalk	TLC	38.3	18.4	24.7	13.6	5.1			1.11[c]	1470
1	Cormel, peeled	TLC	+	+					(LPC +)		1609
Concephalum conicum (L.) Corda (Scented Liverwort.)											
1	Whole plant	TLC	+	+							1436
Convallaria majalis L. (Lily of the Valley)											
1	Leaf	TLC			14						1440
Conyza canadensis L. (Horseweed)											
1	Leaf, chloroplast[a]	TLC	8.6	5.5	7.0					2.68[i]	1610
Coriandrum sativum L. (Coriander)											
1	Seed	TLC	44.0	29.3		23.1				0.1	1548
1	–, coat endosperm	TLC	49	21		12		19			1611
1	Fruit	TLC	46.1	18.4		22.4		6.6		0.3	1471
			also LPC 3.5, LPE 1.8								
Corylus avellana L. (European Hazel)											
1	Seed, cotyledon	TLC	37.8	47.2	5.5	8.9		0.6			1612
1	–, embryonic axis	TLC	54.6	32.3	8.0	5.1					1612
1	Seed, kernel[a]	HPLC	0.075	0.019		0.015					1613
	and weekly changes during development 134-176 daf										

*Unless stated otherwise; the PL column gives the total phospholipid content of each plant part in percent by weight, and the PC, PE, PG, PI, PS, PA, and DPG columns show the distribution in percent by weight of the total PL; [a]Percent of total lipids; [b]mg lipid P/100 mL juice; [c]Percent dwb; [d]mg PL/100 mg protein; [e]mg PL/g fwb; [f]mg PL/100 g dwb; [g]Percent of polar lipids; [h]μmol PL/mg protein; [i]μmol PL/mg chlorophyll; [j]μg PL/mg protein; [k]μg lipid P/g; [l]μmol PL/g fwb; [m]mol PL/mol chlorophyll; [n]mg PL/g dwb; [o]μmol PL/g dwb; [p]μg PL/100 mg fwb; [q]μg PL/mg chlorophyll; [r]mg PL/cm² leaf area; [s]nmol PL/cm² leaf area; [t]μmol PL/mg protein; [u]IR given; [v]mg lipid P/g lipid; [w]μg PL/mL; [x]μg lipid P/cm² leaf area; [y]Molecules per 10^6 plastids; [z]μg PL/mg chromatin; [aa]Fatty acid composition and *sn*-1/*sn*-2 distribution given for each phospholipid; [bb]μg/g fwb; [cc]Taken from the abstract.

Continued

Table 4-15 Phospholipids in Plants*—cont'd

No.	Plant part	Method	PC	PE	PG	PI	PS	PA	DPG	PL	References
1	Seedling, plumule	TLC	33.4	45.6	3.0	18.0					1612
1	–, root, and hypocotyl	TLC	29.8	39.8	10.5	19.9					1612
1	–, petiole	TLC	36.2	39.6	8.2	16.0					1612
1	–, cotyledon	TLC	38.0	27.0	19.4	10.2		5.4			1612
Crambe abyssinica Hochst (Crambe)											
1	Seed	TLC	34.6	24.4		27.0	6.4				1614
Crambe amabilis Butkov and Majlin											
1	Seed	TLC	56.1	20.6		13.2	3.0			0.85[c]	1615
			also NAPE 2.8, NALPE 2.3, LPC 2.0								
1	Leaf	TLC	23.0	25.2	3.0	33.4		8.3		0.15[c]	1616
			also NAPE 3.5, NALPE 2.6, LPC 1.0								
1	Flower	TLC	37.2	21.0	2.0	20.2	5.4	2.2		0.63[c]	1616
			also NAPE 4.5, NALPE 3.2, LPC 4.3								
1	Stem	TLC	24.0	23.0	5.5	25.2	5.4	6.5		0.10[c]	1616
			also NAPE 4.4, NALPE 4.0, LPC 2.0								
Crambe schugnana Korch.											
1	Seed	TLC	55.0	13.7		19.0	4.6			0.69[c]	1617
			also NAPE 3.4, LPC 4.3								
Crataegus turkestanica Pojark. (Turkestan Hawthorn)											
1	Fruit, kernel	TLC	61.5	8.7		27.9		1.7		0.03	1618
1	Pericarp	TLC	60.1	23.8	Trace	10.3		1.7		0.21	1619
Crotalaria juncea L. (Sunn Hemp)											
1	Pollen	TLC	8.2	23.8	12.9	9.9	13.9	31.3			1620
Crotalaria retusa L. (Rattleweed)											
1	Pollen grain	TLC	58.6	27.3	8.6	8.8 (incl. PS)				5.49[c]	1621
Ctenidium molluscum (Hedw.) Mitt. (Mollusc Ctenidium Moss)											
1	Whole plant	TLC	+	+							1436
Cucumis melo L. (Sweet Melon)											
1	Seed	TLC	+	+						4-10	1622
Cucumis melo L. ssp. *reticulatus* (Muskmelon)											
1	Seed, kernel	TLC	26.0	29.3	7.9	10.2	5.0	3.1		1.04[c]	1623
			also LPC 8.8, LPE 9.7								
Cucumis sativus L. (Cucumber)											
1	Cotyledon	TLC	39	29		29		3		1.80[l]	1447
1	Cotyledon, 7 days[a]	TLC	21	4	2						1403

1	–, protoplast[a]	TLC	8	5	8						1403
1	Leaf	TLC	34.9	21.5	22.6	7.2	4.6	4.6	4.6	1.95[l]	1624
1	Leaf, thylakoid[s]		+		+						1625
1	–, LH (II)[s]		+		35.5						1625
1	Fruit, whole	TLC	45.6	31.3	6.4	5.0	2.2	0.7		0.050	1563
			also LPC 3.2 and LPE 1.0								
1	Fruit, peeled and deseeded	TLC	50.8	36.5	4.4	4.4	0.4	0.9	1.8	45.5[a]	1626
			also LPC 0.4 and LPE 0.8								
2	Root	TLC	25.5-29.6	38.4-40.5	0.9-2.0	14.8-19.0	7.2-12.0	25.8-44.6		15.3-21.6[k]	1627
Cucurbita digitata Gray (Coyote Gourd)											
1	Seed	TLC	74.9	10.5		13.7			(LPC trace)	3.3[a]	1476
Cucurbita ficifolia Bouchè (Figleaf Gourd)											
1	Leaf	TLC			3.60[a]					3.26[l]	1628
1	Root	TLC	32.4	30.0	8.4	15.0	11.3	21.5		21.3[k]	1627
Cucurbita foetidissima HBK (Calabazilla)											
1	Seed	TLC	55.8	18.7		17.2			(LPC 0.8)	2.9[a]	1476
Cucurbita maxima Duch. (Hubbard Squash)											
3	Seed	TLC	22.1-24.6	15.4-23.6		18.6-23.2	18.0-18.9			1.23-1.57	1629
1	Leaf[i]	TLC	3.75	3.75	3.74						1630
Cucurbita moschata Duch. (Warty Squash)											
1	Seed	TLC	21.6	22.0		16.0	16.4			2.12	1629
1	Leaf	TLC	+	+	+	+					1631
1	Leaf thylakoid membrane[a]	TLC			10.9						1632
C. maxima × *C. moschata*											
1	Seed	TLC	25.4	25.9		18.4	18.8			1.64	1629
Cucurbita pepo L. (Pumpkin)											
1	Leaf	TLC	43	27	24	4			2	3.74[l]	1523
1	–, mitochondria	TLC	15.9	35.0	15.1	4.0	7.4	11.9	7.5	0.050[c]	1633
1	Leaf	TLC	36.7	26.6	21.6	12.2	2.9			1651[r]	1634

*Unless stated otherwise; the PL column gives the total phospholipid content of each plant part in percent by weight, and the PC, PE, PG, PI, PS, PA, and DPG columns show the distribution in percent by weight of the total PL; [a]Percent of total lipids; [b]mg lipid P/100 mL juice; [c]Percent dwb; [d]mg PL/100 mg protein; [e]mg PL/g fwb; [f]mg PL/100 g dwb; [g]Percent of polar lipids; [h]μmol PL/mg protein; [i]μmol PL/mg chlorophyll; [j]μg PL/mg protein; [k]μg lipid P/g; [l]μmol PL/g fwb; [m]mol PL/mol chlorophyll; [n]mg PL/g dwb; [o]μmol PL/g dwb; [p]μg PL/100 mg fwb; [q]μg PL/mg chlorophyll; [r]mg PL/cm^2 leaf area; [s]nmol PL/cm^2 leaf area; [t]μmol PL/mg protein; [u]IR given; [v]mg lipid P/g lipid; [w]μg PL/mL; [x]μg lipid P/cm^2 leaf area; [y]Molecules per 10^6 plastids; [z]μg PL/mg chromatin; [aa]Fatty acid composition and *sn*-1/*sn*-2 distribution given for each phospholipid; [bb]μg/g fwb; [cc]Taken from the abstract.

Continued

Table 4-15 Phospholipids in Plants*—cont'd

No.	Plant part	Method	PC	PE	PG	PI	PS	PA	DPG	PL	References
1	Root meristem	TLC	31.5	50.1		6.3	3.5	6.6	2.7		1635
Cucurbita pepo L. ssp. *melopepo* (Zucchini Squash)											
1	Fruit, epidermis	TLC	48.8	33.7	14.9	2.6				0.39[c]	1636
Cuminum cyminum L. (Cumin)											
1	Seed	TLC	35.0	49.0	4.0	7.0			(LPE 5.0)	0.74[c]	1637
1	Fruit	TLC	47.6	11.6		20.1		6.7		0.5	1471
			also LPC 4.1, LPE 3.9								
Cupressus dupreziana A. Camus (Saharan Cypress)											
1	Leaf	TLC	+	+							1436
Cuscuta reflexa Roxb. (Dodder)											
1	Whole vine	TLC	39.3	26.6	11.9	19.7		2.5		0.208	1638
Cycas tharwanara											
1	Leaf	TLC	+	+							1436
Cyclamen persicum Mill. (Cyclamen)											
1	Leaf	TLC					+				1450
Cydonia oblonga Mill. (Quince)											
1	Seed	TLC	30.5	6.0	4.1	28.7		6.7	4.4	0.3	1639
			also LPC 3.5, LPE 5.4								
Cymbidium hybrid Sw.											
1	Leaf[l]	TLC			0.21						1402
1	Leaf	TLC			12						1440
Cynoglossum officinale L. (Hound's Tongue)											
1	Seed (NAPE +)	TLC	+	+		+		+		0.4	1640
Cyperus esculentus L. (Yellow Nutsedge, Chufa)											
1	Seed	TLC	+	+	+	+		+	(LPC +)		1569
1	Tuber	TLC	35.0	22.0		16.0	1.7			5.60[a]	1641
Cypripedium insigne Wall. (Large-Flowered Lady's Slipper)											
1	Leaf	TLC			8						1440
Dactylis glomerata L. (Orchard Grass, Cocksfoot)											
1	Leaf	TLC	33	24	33	3			8	3.38[l]	1523
1	Seedling, plasma membrane	TLC	43.5	37.3	10.6	7.2 (incl. PS)		1.3	0	1.05[t]	1642
1	–, endopl. reticulum and golgi membrane	TLC	45.7	27.1	14.4	10.5 (incl. PS)		0.8	1.4	1.11[t]	1642

1	–, mitochondria	TLC	39.8	34.7	9.9	4.5 (incl. PS)		2.2	8.7	0.44[t]	1642
Datura inoxia Mill. (Downy Thorn Apple)											
1	Leaf[a]	TLC				4					1643
1	Callus cell culture	TLC	33	19	6	28	9	5		1.68[c]	1643
Daucus carota L. (Carrot)											
1	Seed	TLC	29.1	35.4		23.1			8.6	0.12	1548
1	Seed	TLC						P_{NAPE}/P_{total} 1.1			1497
1	Leaf	TLC	30.9	25.2	29.5	14.4	Trace			1128[r]	1634
1	Fruit	TLC	48.2	11.9	0.9	30.1		3.2		0.4	1471
			also LPC 3.1, LPE 0.2								
1	Root[k]	PC	2.36	2.47		2.28		3.40	1.12		1644
1	Root	TLC	39	26	11	8	6	4		0.055	1645
3	Root[l]	Choline	47.83								1451
1	Root	HPLC	54	27		9		9		0.79	1646
1	Cell culture[a]	TLC	30.4	16.8	0.9	9.1	Trace	0.6	1.3		1647
1	Cell suspension culture	TLC	44.7	39.1	3.8		12.4				1648
1	–, microsomal membrane	TLC	50.8	29.8	5.9	7.5		6.0		0.43[l]	1649
Daucus carota L. ssp. *carota* (Wild Carrot)											
1	Pericarp, endosperm	TLC	39	20		12		29			1611
Delonix regia ([Boj.ex] Hook.) Raf. (Royal Poinciana, Flamboyant)											
1	Seed	TLC	+	+				(LPC +)			1428
Dendrobium kingianum Bidw.											
1	Leaf[l]	TLC			0.40						1402
Deschampsia antarctica Desv.											
1	Leaf	TLC	24	34	8	34				1.07[l]	1650
1	Root	TLC	23	25	22	30				1.58[l]	1650
Dianthus caryophyllus L. (Carnation)											
1	Flower, petal	TLC	23	10	45.2	17		4.9			1651
1	–, microsomal membrane	TLC	40	28	18	5					1652
Diapensia lapponica L.											
1	Leafy top[n]	TLC	1.4	1.0	0.4			0.2		0.33[c]	1653

*Unless stated otherwise; the PL column gives the total phospholipid content of each plant part in percent by weight, and the PC, PE, PG, PI, PS, PA, and DPG columns show the distribution in percent by weight of the total PL; [a]Percent of total lipids; [b]mg lipid P/100 mL juice; [c]Percent dwb; [d]mg PL/100 mg protein; [e]mg PL/g fwb; [f]mg PL/100 g dwb; [g]Percent of polar lipids; [h]µmol PL/mg protein; [i]µmol PL/mg chlorophyll; [j]µg PL/mg protein; [k]µg lipid P/g; [l]µmol PL/g fwb; [m]mol PL/mol chlorophyll; [n]mg PL/g dwb; [o]µmol PL/g dwb; [p]µg PL/100 mg fwb; [q]µg PL/mg chlorophyll; [r]mg PL/cm² leaf area; [s]nmol PL/cm² leaf area; [t]µmol PL/mg protein; [u]IR given; [v]mg lipid P/g lipid; [w]µg PL/mL; [x]µg lipid P/cm² leaf area; [y]Molecules per 10^6 plastids; [z]µg PL/mg chromatin; [aa]Fatty acid composition and *sn*-1/*sn*-2 distribution given for each phospholipid; [bb]µg/g fwb; [cc]Taken from the abstract.

Continued

Table 4-15 Phospholipids in Plants*—cont'd

No.	Plant part	Method	PC	PE	PG	PI	PS	PA	DPG	PL	References
Dicranella heteromolla (Hedw.) Schimp. (Dicranella Moss)											
1	Whole plant	TLC	+	+							1436
Dicranum polysetum Sw. (Dicranum Moss)											
1	Whole plant *Incl. LPE	TLC	58.9	17.7	8.8	11.7*	2.9			0.68[c]	1654
Digitaria sanguinalis (L.) Scop. (Hairy Crabgrass)											
1	Leaf[l]	TLC			1.08						1402
Dioscorea aimadoimo (Korean Yam)											
1	Tuber	TLC	45.2	38.1		13.9				0.230	1655
Dioscorea alata L. (Winged Yam)											
1	Tuber	TLC	8	19	15	16	32			0.22[c]	1656
Dioscorea batatas Decne. (Chinese Yam)											
1	Tuber	TLC	26.4	29.3		32.9				0.107	1655
Dioscorea bulbifera L. (Air Potato)											
1	Tuber	TLC	42	20	12	13	6			0.23[c]	1656
Dioscorea cayenensis Lam. (Yellow Guinea Yam)											
1	Tuber	TLC	42	21	11	13	6			0.10[c]	1656
Dioscorea japonica Thunb. (Japanese Yam)											
1	Tuber	TLC	43.4	37.2		15.8				0.081	1655
Dioscorea opposita Thunb. (Chinese Yam)											
1	Rhizome	TLC	39.5	12.7		1.8		(LPC 8.2)		0.161	1657
Dioscorea rotundata Poir. (White Guinea Yam)											
1	Tuber	TLC	42	20	12	13	6			0.03[c]	1656
Diospyros kaki L.f. (Japanese Persimmon, Kaki)											
3	Seed[g]	TLC (MeNH-PE 2.5-4.2, Me_2N-PE 3.8-4.7)				8.3-10.3				0.11-0.18[d]	1658
1	Fruit, flesh	TLC	20.0	8.8	29.8	13.5	7.0	4.2	16.7	0.026	1659
1	Fruit, rind	TLC	26.3	7.0	46.5	13.2	1.8	0	0	0.049	1659
1	Fruit, flesh	TLC	28.2	5.0	37.0	11.0	13.0	1.9	3.9	0.037	1659
1	–, rind	TLC	18.8	6.2	37.5	12.5	25.0	1.8	3.5	0.012	1659
Diptychocarpus strictus (Fisch.) Trautv.											
1	Seed	TLC	50.2	12.2		22.9		(LPC 10.4)		0.45[c]	959

1	Seed	TLC						(NAPE 1.9)			1660
Drepanocladus lycopodioides (Brid.) Warnst.											
1	Whole plant	TLC	32.2	21.3	32.4	4.9	6.6			1.19[c]	1551
Dryopteris dilatata (Hoffm.) A. Gray											
1	Whole plant	TLC	+	+							1436
Dryopteris filix-mas (L.) Schott. (Shield Fern)											
1	Leaf	TLC	+	+	+	+				0.91	1661
Echinochloa colona (L.) Link (Barnyard Millet)											
1	Grain	TLC	30.7	24.0	3.8	6.6	8.3	1.8	(LPC 24.8)	0.44[c]	1662
Eichhornia crassipes (Mart.) Solms (Water Hyacinth)											
1	Leaf	TLC	11.4	25.4	30.7	18.9	3.8		9.8	3.93[c]	1663
1	Leaf stalk	TLC	26.7	21.7	30.0	21.7	0		0	0.05[c]	1663
1	Flower	TLC	31.3	34.2	23.5	0	0		11.0	1.77[c]	1663
1	Root	TLC	40.2	27.7	26.0	0	0		6.1	0.37[c]	1663
1	Whole plant	HPLC	The PC contains a pentacos-17-enoic acid group and the PE, a nonadec-6-enoic acid group								1664
Elaeagnus angustifolia L. (Russian Olive)											
1	Seed	TLC	52.1	16.4		19.2		3.5		(NAPE 1.7, NALPE 3.2)	1665
1	Fruit, pericarp	TLC	45.8	25.0		29.2					1665
Elaeis guineensis Jacq. (Oil Palm)											
1	Kernel, haustorium	TLC	+			+					1666
2	Pollen	TLC	+	+		+					1667
1	Fruit, mesocarp	TLC	36	24	9	22	Trace	3	4	0.1-0.2	1668
			also LPC trace and LPE 2.								
Elettaria cardamomum (L.) Maton. (Cardamom)											
1	Seed	TLC		22.7		4.9	6.5		55.7	0.065	1669
			also LPC 1.4, Me_2N-PE 8.8								
1	Pod (Me_2N-PE 49.5)	TLC				50.5				4.4[a]	1669
Eleusine indica (L.) Gaertn. (Goosegrass)											
1	Leaf[l]	TLC			0.70						1402
Elodea canadensis Michx. (Canadian Waterweed)											
1	Whole plant	TLC	47.6	18.8	11.3	11.4	8.4	2.5		2.71[c]	1445
1	Leaf chloroplast	TLC	+	+	+						1670

*Unless stated otherwise; the PL column gives the total phospholipid content of each plant part in percent by weight, and the PC, PE, PG, PI, PS, PA, and DPG columns show the distribution in percent by weight of the total PL; [a]Percent of total lipids; [b]mg lipid P/100 mL juice; [c]Percent dwb; [d]mg PL/100 mg protein; [e]mg PL/g fwb; [f]mg PL/100 g dwb; [g]Percent of polar lipids; [h]µmol PL/mg protein; [i]µmol PL/mg chlorophyll; [j]µg PL/mg protein; [k]µg lipid P/g; [l]µmol PL/g fwb; [m]mol PL/mol chlorophyll; [n]mg PL/g dwb; [o]µmol PL/g dwb; [p]µg PL/100 mg fwb; [q]µg PL/mg chlorophyll; [r]mg PL/cm^2 leaf area; [s]nmol PL/cm^2 leaf area; [t]µmol PL/mg protein; [u]IR given; [v]mg lipid P/g lipid; [w]µg PL/mL; [x]µg lipid P/cm^2 leaf area; [y]Molecules per 10^6 plastids; [z]µg PL/mg chromatin; [aa]Fatty acid composition and *sn*-1/*sn*-2 distribution given for each phospholipid; [bb]µg/g fwb; [cc]Taken from the abstract.

Continued

Table 4-15 Phospholipids in Plants*—cont'd

No.	Plant part	Method	PC	PE	PG	PI	PS	PA	DPG	PL	References
Elodea nuttallii (Planch.) H. St. John (Western Waterweed)											
1	Whole plant	TLC	46.6	19.0	20.2	11.7	2.5		Trace	7.33[c]	1555
Elyna spicata Schrad.											
1	Leaf	TLC	47	20	22	8			3	0.26	1441
Encyclia fragrans Swartz.											
1	Leaf	TLC			11						1440
Ephedra gerardiana Wall. ex Stapf. (Gerard Joint Fir)											
1	Leaf	TLC	+	+							1436
Epidendron radicans Pav. ex Lindl. (Fiery Reed Orchid)											
1	Leaf[l]	TLC			0.22						1402
Equisetum arvense L. (Field Horsetail)											
1	Whole plant	TLC	+	+							1436
Equisetum scirpoides Michx. (Dwarf Scouring Bush)											
1	Whole plant	TLC	+	+							1436
Equisetum sylvaticum L. (Woodland Horsetail)											
1	Whole plant	TLC	+	+							1436
Eriobotrya japonica (Thunb.) Lindley (Loquat)											
2	Seed[a]	TLC		12.8-18.9 PS, 5.8-12.2 Me_2N-PE							1671
Eruca sativa Mill. (Rocket Salad, Taramira)											
1	Seed	TLC	+	+		+		+	(LPC +, LPE +)		1672
Erysimum diffusum Ehrh. (Gray Wallflower)											
1	Seed	TLC	49.1	19.3		21.3				0.5[c]	1673
			also NAPE 3.2, NALPE 2.1, LPC 5.0								
Erysimum sylvestris (L.) Bess.											
1	Seed	TLC	45.2	19.2		16.4				1.0[c]	1673
			also NAPE 7.0, NALPE 5.2, LPC 7.0								
Euonymus europaeus L. (European Spindle Tree)											
1	Seed, embryo	TLC	49	25		23				262[w]	1674
Euonymus japonica Thunb. (Japanese Spindle Tree)											
1	Leaf	TLC	54	15	18	12					1607
Fagopyrum esculentum Moench (Buckwheat)											
1	Flour	TLC	38.5	21.4		2.2	14.8	7.0		3.4	1675
			also LPC 6.2, LPE 4.5, LPS 5.3								
1	Flour starch	TLC	23.5	28.3	7.8	4.5				0.045	1676
			also LPC 24.8, LPE 4.0								

Fagus sylvatica L. (European Beech)											
1	Seed	TLC	Trace	44.9		55.1				0.28[c]	1677
1	Leaf	TLC	40.0	23.4	27.5	8.8	Trace			80[r]	1634
Feijoa sellowiana O. Berg. (Feijoa)											
1	Fruit, flesh	TLC	16.4	7.2	19.3	10.8		36.7	8.1	0.08	1678
1	–, peel	TLC	12.6	10.1	42.0	19.3		8.4	7.6	0.18	1678
Ferula jaeschkeana Vatke											
1	Seed, defatted	TLC	+	+		+				1.5	794
Ferula karataviensis (Rgl. and Schmalh.) Korov.											
1	Seed, defatted	TLC	+	+		+				0.8	794
Ferula kelleri K.-Pol.											
1	Seed, defatted	TLC	+	+		+				0.5	794
Ferulago silvatica Bess.											
1	Seed	TLC	22	11	8	31	8	10	2	0.5[a]	1438
			also LPC 2, LPE 1.								
Festuca pratensis Huds. (Meadow Fescue)											
2	Leaf	TLC	38	18	19	12			13	0.26	1679
Ficus carica L. (Fig.)											
2	Whole fruit (1)	TLC	18.2	3.6	45.6	4.5	4.8	7.1	1.8	0.046	1680
	Whole fruit (2)	TLC	10.5	1.9	28.3	1.9	21.9	18.4	3.0	0.022	1680
			also LPC 2.1, LPE 8.5 (1) and LPC 1.2, LPE 1.8 (2)								
Fissidens taxifolius Hedw. (Fissidene Moss)											
1	Whole plant	TLC	+	+							1436
Flueggea microcarpa Blume											
1	Seed	TLC	30.2	26.4	1.0	23.6			18.8	0.44	1681
Foeniculum vulgare Mill. (Fennel)											
1	Fruit	TLC	38.2	9.3	2.6	31.2		7.9		0.3	1471
			also LPC 3.9, LPE 2.8								
Fontinalis antipyretica Hedw. (Water Moss)											
1	Whole plant	TLC	32.0	23.7	32.4	6.5	3.1			1.67[c]	1551
Fragaria × *ananassa* Duch.											
10	Root	TLC	+	+		+				0.033-0.051[c]	1682

*Unless stated otherwise; the PL column gives the total phospholipid content of each plant part in percent by weight, and the PC, PE, PG, PI, PS, PA, and DPG columns show the distribution in percent by weight of the total PL; [a]Percent of total lipids; [b]mg lipid P/100 mL juice; [c]Percent dwb; [d]mg PL/100 mg protein; [e]mg PL/g fwb; [f]mg PL/100 g dwb; [g]Percent of polar lipids; [h]μmol PL/mg protein; [i]μmol PL/mg chlorophyll; [j]μg PL/mg protein; [k]μg lipid P/g; [l]μmol PL/g fwb; [m]mol PL/mol chlorophyll; [n]mg PL/g dwb; [o]μmol PL/g dwb; [p]μg PL/100 mg fwb; [q]μg PL/mg chlorophyll; [r]mg PL/cm² leaf area; [s]nmol PL/cm² leaf area; [t]μmol PL/mg protein; [u]IR given; [v]mg lipid P/g lipid; [w]μg PL/mL; [x]μg lipid P/cm² leaf area; [y]Molecules per 10^6 plastids; [z]μg PL/mg chromatin; [aa]Fatty acid composition and *sn*-1/*sn*-2 distribution given for each phospholipid; [bb]μg/g fwb; [cc]Taken from the abstract.

Continued

Table 4-15 Phospholipids in Plants*—cont'd

No.	Plant part	Method	PC	PE	PG	PI	PS	PA	DPG	PL	References
Fragaria spp. (Strawberry)											
1	Fruit	Frac. pptn.	0.062								1527
Fumaria vaillantii Loisel. (Earthsmoke)											
1	Seed	TLC	64.1	4.6		27.5		3.3	(LPC 0.1)	0.25	1683
Geum reptans L. (Creeping Avens)											
1	Leaf	TLC	47	23	21	8			1	0.26	1441
Ginkgo biloba L. (Ginkgo)											
1	Seed	TLC	+	+	+	+					1684
1	Nut, shelled	TLC	20.8	24.5	51.2					0.12	1685
			also LPC 1.6 and LPE 2.4								
1	Leaf[l]	TLC	1.80	0.55	0.85	0.25			0.27		1523
Glaucium fimbrilligerum Boiss. (Sea Poppy)											
1	Seed, defatted	TLC	+	+		+				0.6	794
Gleditsia triacanthos L. (Honey Locust)											
1	Seed oil	TLC	28	24			23			8	1686
Glycine max (L.) Merr. (Soybean)											
1	Seed	TLC	32	20	16	14			18	2.75[c]	1524
1	Seed	PC	59.9	8.3		8.4	4.2	6.4			1687
1	Seed	TLC			NAPE 2.9 mg/seed						1497
1	Seed, meal	HPLC	45.7	25.4	1.0	21.1	4.1		2.6		1688
1	–, mitochondria	HPLC	36.9	39.2		15.0			6.1		1688
1	Seed, 20 daf	Distribution among organelles (*not* within each):									
1	–, oleosome (*incl. MGDG and DGDG)	TLC	15.4	29.2	1.4*	24.5			5.5		1689
1	–, endoplastic reticulum	TLC	6.7	6.7	0.6*	3.3			3.3		1689
1	–, plastid	TLC	41.7	33.7	94.9*	62.1		18.7			1689
1	–, mitochondria	TLC	27.3	22.1	2.6*	9.9			70.3		1689
1	–, peroxisome	TLC	8.9	8.3	0.5*	0.2			2.2		1689
1	Seed, hypocotyl	TLC	41	20	2	16	Trace	2	(LPC 3, LPE 2)		1690
1	–, cotyledon	TLC	47	26	1	15	Trace	2	(LPC 2, LPE trace)		1690
3	Seed, coat	TLC			+			+	+		1691

			45.3-51.7	18.0-23.3		12.5-14.4					
3	–, cotyledon	TLC	45.9-49.3	20.6-27.6	+	17.3-21.5		+	+		1691
3	–, axis	TLC	45.2-47.3	22.8-27.3	+	15.2-17.3		+	+		1691
1	Mixed bean, mature	TLC	39.0	23.3		20.3		4.8		4.87[a]	1692
1	Bean	TLC	37.5	18.8	2.5	13.8			(NAPE 6.2)	1.31	1693
1	–, immature	TLC						(DPA 0.039%)			1694
1	–, lipid bodies	TLC	56.7	13.4	3.4	15.5		10.3	0	0.97	1695
1	–, protein bodies	TLC	+	+	+	+			+ (NAPE +)	2.08[c]	1696
1	Seed, black	TLC	50.5	29.6		10.7	9.2			2.14[c]	1697
1	Leaf	TLC		26	28 (incl. PC)				36	2.60[l]	1698
2	Root (1) *Incl. PS	TLC	43.1	33.1*	7.0	7.8		3.9	5.1		1699
	Root (2)		46.4	33.3*	4.9	5.8		3.0	3.2		1699
1	–, microsomes	TLC	46	31	2	8		2	(LPE 12)		1699
1	–, plasma membrane vesicles	TLC	42.3	32.1	1.7	7.5	6.1	8.9	0.2		1700
1	–, mature (NAPE 0.5)	TLC	46.1	31.0	2.1	23.8		0.6	2.7	7.9[o]	1701
1	–, immature (NAPE 2.7)	TLC	30.4	26.0	5.1	24.5		2.5	8.8	15.2[o]	1701
1	Cotyl, spherosomes	TLC	67.5	20.0		5.9		6.6			1702
Goebelia pachycarpa (C.A. Mey) Bunge											
1	Seed	TLC	41.7	18.0		25.6				1.5[c]	1703
			also NAPE 4.1, NALPE 5.5, LPC 5.1								
Gossypium arboreum L. (Tree Cotton)											
1	Seed	TLC	48.0	8.8		11.6	10.1	(LPC 15.2)			1704

*Unless stated otherwise; the PL column gives the total phospholipid content of each plant part in percent by weight, and the PC, PE, PG, PI, PS, PA, and DPG columns show the distribution in percent by weight of the total PL; [a]Percent of total lipids; [b]mg lipid P/100 mL juice; [c]Percent dwb; [d]mg PL/100 mg protein; [e]mg PL/g fwb; [f]mg PL/100 g dwb; [g]Percent of polar lipids; [h]μmol PL/mg protein; [i]μmol PL/mg chlorophyll; [j]μg PL/mg protein; [k]μg lipid P/g; [l]μmol PL/g fwb; [m]mol PL/mol chlorophyll; [n]mg PL/g dwb; [o]μmol PL/g dwb; [p]μg PL/100 mg fwb; [q]μg PL/mg chlorophyll; [r]mg PL/cm^2 leaf area; [s]nmol PL/cm^2 leaf area; [t]μmol PL/mg protein; [u]IR given; [v]mg lipid P/g lipid; [w]μg PL/mL; [x]μg lipid P/cm^2 leaf area; [y]Molecules per 10^6 plastids; [z]μg PL/mg chromatin; [aa]Fatty acid composition and *sn*-1/*sn*-2 distribution given for each phospholipid; [bb]μg/g fwb; [cc]Taken from the abstract.

Continued

Table 4-15 Phospholipids in Plants*—cont'd

No.	Plant part	Method	PC	PE	PG	PI	PS	PA	DPG	PL	References
Gossypium arboreum L. ssp. *indicum*											
1	Leaf	TLC	48.4	7.0	13.3					0.58[c]	1494
1	Root	TLC	49.0	42.9	Trace					0.27[c]	1494
Gossypium aridum (Rose and Standl.) Skovst. (American Wild Cotton)											
1	Seed	TLC	31.3	12.5		5.2	18.9	(LPC 5.8)			1704
Gossypium armorianum Kearn.											
1	Seed	TLC	36.0	15.5		7.4	18.2	(LPC 6.0)			1704
Gossypium barbadense L. (Sea Island Cotton)											
1	Seed	TLC	38.2	10.9		13.9	11.8	(LPC 12.6)		0.39	1705
			The PC contains traces of malvalic acid in the 2-position								1706
1	Seed, kernel	TLC	28.2	17.9		11.7	14.2	(LPC 13.1)		2.38	1707
1	Seed, kernel	TLC	50.4	14.1		20.4		(LPC 5.6)		1.6[c]	1708,1709
1	Leaf	TLC	+	0	+	+		+	+		1710
Gossypium davidsonii Kellogg											
1	Seed	TLC	32.4	16.0		4.9	16.2	(LPC 4.5)			1704
Gossypium gossypioides (Ulbr.) Stanley											
1	Seed	TLC	23.4	14.3		9.6	13.5	(LPC 9.7)			1704
Gossypium herbaceum L. (Levant Cotton)											
1	Seed	TLC	41.9	7.1		13.5	13.4	(LPC 16.9)			1704
Gossypium hirsutum L. (Upland Cotton)											
1	Seed, glanded	TLC	36.9	21.8		10.6	14.3	(LPC 7.7)			1704
1	Seed	TLC						0.3	(LPI 0.1)		1711
1	Seed	TLC		NAPE 4.1, NALPE 3.3							1712
3	Seed	TLC		NAPE +, NALPE +, LPC +							1713
1	–, kernel	TLC	50.2	13.4		26.6				1.5[c]	1374,1714
1	–, oil body	TLC	58.6	4.6		18.1	18.7			1.18	1482
3	Leaf	TLC	47.8-48.8	0	19.2-23.6	8.4-11.2		2.5-4.4			1710
1	Bud[a]	TLC	30.4	2.6		11.5					1715
1	Bud	TLC	41.3	8.0		9.5		21.2			1716

1	48-h seedling, cotyledon										
1	–, lipid bodies	TLC	35	21	0	34		4	0		1717
1	–, mitochondria	TLC	35	32	1	5		0	20		1717
1	–, glyoxysomes	TLC	49	12	13	15		0	0		1717
1	Flower, 1st day	TLC	42.7	10.9		10.4		6.2			1716
1	Boll	TLC	35.4	8.8		13.1		19.1			1716
1	Bark	TLC	28.5	17.4		10.6		28.9			1716
1	Stem	TLC	17.7	18.1		13.0		37.4			1716
1	Root	TLC	38.7	2.6		10.5		35.2			1716
1	Extrafloral nectar	TLC	44.3	4.4		9.6		(LPC 21.3, LPE 12.7)			1718
1	Fiber, immature	TLC	47.8	30.4	7.0		14.7				1719
1	–, mature	TLC	56.6				43.6				1719
Gossypium hirsutum L., glandless											
3	Leaf	TLC	30.3-39.7	+	21.3-24.4	10.7-12.0		1.9-3.9			1710
1	Leaf	TLC	37.5	20.1		10.9	20.0				1704
1	Seedling	TLC	NAPE in leaf 2.5, stem 3.2, plumule 3.2, cotyledon 1.9, root 3.8								1720
Gossypium klotzschianum Anderss.											
1	Seed	TLC	33.4	17.2		6.1	14.6	(LPC 6.4)			1704
Gossypium raimondii Ulbr.											
1	Seed	TLC	34.5	15.9		5.8	16.9	(LPC 5.8)			1704
Gossypium thurberi Todaro (Thurber's Cotton)											
1	Seed	TLC	34.0	14.0		4.7	22.5	(LPC 5.7)			1704
Grimmia pulvinata (Hedw.) Sm.											
1	Whole plant	TLC	+	+							1436
Halocnemum strobilaceum (Pall.) M. Bieb.											
1	Shoot	TLC	40.6	16.8	11.1	8.0			19.1	0.261[c]	1488
Hancornia speciosa Müll Arg. (Mangaba Tree)											
1	Latex[e,s]	Extn.	0.08								1721
Hedychium gardnerianum Ker-Gawl. (Kahili Ginger)											
1	Leaf[l]	TLC			0.53						1402

*Unless stated otherwise; the PL column gives the total phospholipid content of each plant part in percent by weight, and the PC, PE, PG, PI, PS, PA, and DPG columns show the distribution in percent by weight of the total PL; [a]Percent of total lipids; [b]mg lipid P/100 mL juice; [c]Percent dwb; [d]mg PL/100 mg protein; [e]mg PL/g fwb; [f]mg PL/100 g dwb; [g]Percent of polar lipids; [h]μmol PL/mg protein; [i]μmol PL/mg chlorophyll; [j]μg PL/mg protein; [k]μg lipid P/g; [l]μmol PL/g fwb; [m]mol PL/mol chlorophyll; [n]mg PL/g dwb; [o]μmol PL/g dwb; [p]μg PL/100 mg fwb; [q]μg PL/mg chlorophyll; [r]mg PL/cm² leaf area; [s]nmol PL/cm² leaf area; [t]μmol PL/mg protein; [u]IR given; [v]mg lipid P/g lipid; [w]μg PL/mL; [x]μg lipid P/cm² leaf area; [y]Molecules per 10^6 plastids; [z]μg PL/mg chromatin; [aa]Fatty acid composition and *sn*-1/*sn*-2 distribution given for each phospholipid; [bb]μg/g fwb; [cc]Taken from the abstract.

Continued

Table 4-15 Phospholipids in Plants*—cont'd

No.	Plant part	Method	PC	PE	PG	PI	PS	PA	DPG	PL	References
Helianthus annuus L. (Sunflower)											
1	Seed	TLC	35.2	13.0 (incl. PG)		24.3		13.0	14.6	2.5	1722
1	Seed, dehulled	TLC	56.2	11.7	4.3	7.4	1.8	4.3	(LPC trace)	200[k]	1723
2	Achene, dehulled	TLC		+ (fatty acid distribution given)							1724
1	Seed, oil	HPLC	47.6	20.9		19.7		11.8			1725
1	Seed cotyledon, oleosome[g]	TLC	79	13		8	0				1571
1	–, microsome[g]	TLC	76	12		10	2				1571
1	Hypocotyl, nuclei	TLC	23.0	24.6		10.3		7.4	15.6	58[a]	1726
1	–, nuclear membrane	TLC	25.5	25.3	29.8	7.4				51.3[a]	1726
1	Stems and petioles	TLC	39.9	27.7	12.5	17.0		2.9		0.086	1638
1	Leaf, plasma membrane	TLC	37	33	14	16		0		540[j]	1727
1	–, chloroplast envelope membrane[a,c]	TLC	28.9	1.8	5.3	0.6			0	36.6	1728
1	Shoot, 20 days old	TLC	30.9	23.4	26.2	19.4				0.24[c]	1729
1	Pollen[l]	TLC	9.1	4.3		0.5		(LPC 0.4, LPE 0.5)			1730
1	Root	TLC	51.3	25.9		12.3		10.5			1731
1	–, microsome	TLC	44.4	33.8		12.7		9.1			1731
Helianthus tuberosus L. (Jerusalem Artichoke)											
1	Tuber, plasma membrane	TLC	39.3	43.6	2.6	10.6 (incl. PS)		3.9		1.52[h]	1732
Heliotropium lasiocarpum F and M (Heliotrope)											
1	Seed, defatted	TLC	+	+		+				0.3	794
Helodium blandowi (Web. et Mokr.) Warnst.											
1	Whole plant	TLC	26.8	20.8	27.8	8.4	9.5			0.52[c]	1493
Heracleum angustisectum Recn.											
1	Seed	TLC	37 also LPC 6, LPE 6.	10		29		11		1.5[a]	1438

Heracleum lehmannianum Bunge (Cow Parsnip, Mandan)											
1	Seed	TLC	30.4	19.5	2.8	24.8					1733
			also LPC 2.4, LPI 1.8, NAPE 12.2, NALPE 6.1								
Heracleum sibiricum L.											
1	Seed	TLC	42	14	2	18	1	14	2	4.4[a]	1438
			also LPC 3, LPE 4.								
Heracleum ternatum L.											
1	Seed	TLC	44	11	2	30		6		1.2[a]	1438
			also LPC 3, LPE 4.								
Hevea brasiliensis (Willd.) Mill. (Pará Rubber)											
1	Mature tree, latex	TLC	58.4	21.0		20.6				0.23	1734
1	Seedling, latex[a]	Frac. pptn.	51	3		10.5		0		0.71	1735
Hibiscus cannabinus L. (Kenaf)											
9	Seed	TLC	13.3-38.0	8.5-29.7	4.4-19.7	1.3-5.2	1.0-6.5	1.0-14.1	1.0-6.3	3.91-10.9[a]	1736
			also LPC 1.1-7.5								
2	Seed (1)	TLC	38.4	25.1		22.0					1737
	Seed (2)		39.6	23.7		21.8					
			also NAPE 7.1, NALPE 3.6, LPC 2.3, LPI 1.5 (1), and NAPE 8.0, NALPE 3.1, LPC 2.6, LPI 1.2 (2)								
1	Seed	TLC	38.2	15.6		22.7				1.0[c]	793,1377
1	Bud	TLC	30.0	15.3	9.7	27.2		7.4		0.41[c]	1738
			also NAPE 7.7, NALPE 2.7								
1	Flower	TLC	30.4	15.2	8.5	30.0		5.8		0.32[c]	1738
			also NAPE 7.0, NALPE 3.1								
1	Leaf	TLC	23.1	16.0	11.4	25.0		13.5		0.40[c]	1738
			also NAPE 7.8, NALPE 3.2								
1	Stem	TLC	23.7	15.0	15.0	25.2		12.6		0.20[c]	1738
			also NAPE 6.2, NALPE 2.3								
1	Root	TLC	26.5	16.1	15.7	31.8		0		0.33[c]	1738
			also NAPE 7.5, NALPE 2.4								
Hibiscus sabdariffa L. (Roselle)											
1	Seed	TLC	+	+	+	+	+	+	+	1.7	1457
			also LPC +, LPE +, NALPE +.								

*Unless stated otherwise; the PL column gives the total phospholipid content of each plant part in percent by weight, and the PC, PE, PG, PI, PS, PA, and DPG columns show the distribution in percent by weight of the total PL; [a]Percent of total lipids; [b]mg lipid P/100 mL juice; [c]Percent dwb; [d]mg PL/100 mg protein; [e]mg PL/g fwb; [f]mg PL/100 g dwb; [g]Percent of polar lipids; [h]μmol PL/mg protein; [i]μmol PL/mg chlorophyll; [j]μg PL/mg protein; [k]μg lipid P/g; [l]μmol PL/g fwb; [m]mol PL/mol chlorophyll; [n]mg PL/g dwb; [o]μmol PL/g dwb; [p]μg PL/100 mg fwb; [q]μg PL/mg chlorophyll; [r]mg PL/cm^2 leaf area; [s]nmol PL/cm^2 leaf area; [t]μmol PL/mg protein; [u]IR given; [v]mg lipid P/g lipid; [w]μg PL/mL; [x]μg lipid P/cm^2 leaf area; [y]Molecules per 10^6 plastids; [z]μg PL/mg chromatin; [aa]Fatty acid composition and *sn*-1/*sn*-2 distribution given for each phospholipid; [bb]μg/g fwb; [cc]Taken from the abstract.

Continued

Table 4-15 Phospholipids in Plants*—cont'd

No.	Plant part	Method	PC	PE	PG	PI	PS	PA	DPG	PL	References
Hippophaë rhamnoides L. (Sea Buckthorn)											
1	Seed	TLC	+	+	+	+				0.6	1739
			also NAPE +, NALPE +, LPC +, LPI +								
1	Leaf	TLC	32.8	29.7	21.5	16.0				12.1[a]	1740
1	Fruit pulp	TLC	26.7	13.1		8.9	5.8			0.2-0.5[a]	1741
Hordeum distichon L. (Two-Rowed Barley)											
1	Leaf, plastid[l]	TLC			0.13						1742
Hordeum vulgare L. (Barley)											
2	Whole grain (1)	TLC	44.4	8.8	0.2	1.3	4.8	Trace	1.5	0.56	1743
	Whole grain (2)		44.3	7.6	0.8	1.1	5.0	0.2	1.7	0.58	
			also LPC 37.3 (1) and 36.8 (2)								
3	Grain	TLC	39.8-	9.9-	3.1-	4.1-5.3			6.2-7.0	0.030-	1744
	*Incl. PS		41.8*	10.8	4.1					0.059	
			also LPC 28.5-31.2								
1	Starch, free	TLC	12.1*	4.9	8.3	3.5	(LPC 30.8)		40.4	0.0008[c]	1745
	*Incl. PS										
1	–, bound	TLC	21.2*	3.4	7.1	2.2	(LPC 26.8)		39.3	0.055[c]	1745
	*Incl. PS										
1	Leaf	PC	52	1.4	22.6	14.0					1746
1	Leaf	TLC	36.9	15.4	30.8	7.2		0	1.8	0.76[o]	1747
1	–, thylakoid[a]	TLC	4.3		10.8					3.69[i]	1748
1	–, PS(II) vesicles[a]	TLC	0		5.1					2.36[i]	1748
1	–, PS(I) vesicles[a]	TLC	0		6.2					4.65[i]	1748
1	–, tonoplast	TLC	67.0	12.0	6.2	9.6	5.3				1749
1	–, plasma membrane	TLC	47.1	30.5	3.6			11.8		0.08[a]	1750
1	–, intracellular membrane	TLC	47.0	16.5	17.3			9.0		1.86[a]	1750
1	–, chloroplast[a]	TLC			11						1489
1	Caryopsis, aleurone protoplast	TLC	0	+		+		+	(LPC +, LPE +)		1751
1	–, lipid bodies	TLC	0	0		+		+			1751
			also NAPE +, NALPE +, LPC + and LPI +.								
1	Seedling, 7 days[l]	TLC	1.3	0.83	0.60	0.41	0.35				1752
1	–, mesophyll protoplast	TLC	+	0		+					1751

			also NAPE +, NALPE +, LPC +, LPE + and LPG +.								
1	Root *Incl. PS	TLC	50.6	36.5*	5.2	Trace		2.4	4.7		1699
1	–, endoplastic reticulum †incl. LPE	TLC	59.5	20.6	6.9	9.8†	1.5	1.0	0.1	(LPC 0.7)	1753
1	–, tonoplast and Golgi membrane	TLC	57.6	23.2	6.3	7.4†	2.5	1.8	0.6	(LPC 0.6)	1753
1	–, plasma membrane	TLC	42.7	33.8	3.6	5.1†	6.6	5.6	1.7	(LPC 0.8)	1753
Huperzia selago (L.) Bernh. (Fir Clubmoss)											
1	whole plant	TLC	+	+							1436
Hyacinthus spp. (Hyacynth)											
1	Leaf	TLC			8						1440
Hydnocarpus anthelminticus Pierre (Chaulmoogra Tree)											
1	Leaf	TLC	++	+	+	+		+			1552
Hygrohypnum ochraceum (Turn. ex Wils.) Loeske											
1	Whole plant	TLC	32.5	22.8	21.4	7.0	9.6			0.23	1493
Hylocomium splendens (Lesq. and James) Lindl.											
1	Gametophore	TLC	67	13	7	13				15[a]	1754
Hypnum cupressiforme Hedw. (Hypnum Moss)											
1	Whole plant	TLC	+	+							1436
Ilex aquifolium L. (English Holly)											
1	Leaf	TLC	+	+	+	+			+		1755
Illicium verum Hook. f. (Star Anise)											
1	Seed, dehusked	TLC	19.5		3.3	11.3	12.7		(Me_2N-PE 2.1)	0.067	1669
Impatiens balsamina L. (Garden Balsam)											
1	Leaf	TLC	35.1	25.5	31.5	6.3	1.6			1480[q]	1756
1	–, chloroplast	TLC	25	+	67	8				1220[q]	1756
Indigofera enualphylla (Indigo)											
1	Seed	TLC	37.5	25.7		29.7			8.7	0.16[c]	1757
			also LPC 0.4 and LPE 0.3								

*Unless stated otherwise; the PL column gives the total phospholipid content of each plant part in percent by weight, and the PC, PE, PG, PI, PS, PA, and DPG columns show the distribution in percent by weight of the total PL; [a]Percent of total lipids; [b]mg lipid P/100 mL juice; [c]Percent dwb; [d]mg PL/100 mg protein; [e]mg PL/g fwb; [f]mg PL/100 g dwb; [g]Percent of polar lipids; [h]μmol PL/mg protein; [i]μmol PL/mg chlorophyll; [j]μg PL/mg protein; [k]μg lipid P/g; [l]μmol PL/g fwb; [m]mol PL/mol chlorophyll; [n]mg PL/g dwb; [o]μmol PL/g dwb; [p]μg PL/100 mg fwb; [q]μg PL/mg chlorophyll; [r]mg PL/cm^2 leaf area; [s]nmol PL/cm^2 leaf area; [t]μmol PL/mg protein; [u]IR given; [v]mg lipid P/g lipid; [w]μg PL/mL; [x]μg lipid P/cm^2 leaf area; [y]Molecules per 10^6 plastids; [z]μg PL/mg chromatin; [aa]Fatty acid composition and *sn*-1/*sn*-2 distribution given for each phospholipid; [bb]μg/g fwb; [cc]Taken from the abstract.

Continued

Table 4-15 Phospholipids in Plants*—cont'd

No.	Plant part	Method	PC	PE	PG	PI	PS	PA	DPG	PL	References
Indigofera glandulosa J. C. Wendl.											
1	Seed	CC (silica)	41.2	58.8							1576
Indigofera hirsuta L. (Hairy Indigo)											
1	Seed	CC(silica)	34.4	65.6							1576
Indigofera trifoliata L. (Three-Leaf Indigo)											
1	Seed	CC(silica)	38.4	66.6							1576
Ipomoea aquatica Forssk. (Swamp Cabbage)											
1	Seed	TLC	48.4	27.1		20.9 (incl. PS)		2.8		0.42	1577
Ipomoea batatas (L.) Lam. (Sweet Potato)											
1	Leaf	TLC			0.21[l]						1402
1	Leaf	TLC	+	+	+	+					1631
1	Root, peeled and cured	TLC	26	29	4	19	4	1.5	5.9	0.73[c]	1758
9	Tuber	CC (Florisil)	28.9-39.0	61.0-71.1						0.11-0.31	1759
1	Tuber mitochondrial membrane	TLC	33.2	38.7				(LPC 6.6)			1760
Iris spp. (Iris)											
1	Leaf	TLC			9						1440
Irvingia gabonensis Baill. (Ogbono)											
1	Seed	TLC	47	4	15	15			19	3.49[c]	1524
Ixora parviflora Vahl (Torch Tree)											
1	Seed	TLC	29.1	26.2	3.1	24.5			17.1	0.70	1681
Juglans regia L. (English Walnut)											
1	Nut	TLC	12.5	19.0	0	0		13.0	(LPE 4.5)	1.8[a]	1480
1	Nut[g]	Choline	21	31						1.8	1480
Kalanchoë daigremontiana Ham. and Perr. (Devil's Backbone)											
1	Leaf	TLC					+				1450
1	Leaf, tonoplast	TLC	37.6	36.1	12.9		3.3	10.0			1749
Lablab purpureus (L.) Sweet (Hyacinth Bean)											
1	Seed	TLC	+	+					(LPC +)		1428

Lactuca sativa L. (Lettuce)											
1	Leaf[l]	Choline	2								1451
[l]1	Leaf	TLC	33.2	7.5	42.0	6.4	3.4	3.2	4.3	0.037	1761
1	–, thylakoid	TLC			12.0[a]						1762
1	–, chloroplast	TLC	+	+	++	+		+			1396
1	Stalk	TLC	++	+	+	++		+			1396
Lactuca sativa L. ssp. *asparagina* Bail. (Celtuce)											
1	Leaf	TLC	12.2	61.2	26.6					0.413	1446
1	Shoot	TLC	61.3	4.4	16.1			10.9	(LPC 17.3)	0.204	1446
Lantana camara L. ssp. *aculeata* (Yellow Sage)											
1	Stems and petioles	TLC	40.3	25.1	13.0	17.6		4.0		0.181	1638
Larix sibirica Ledeb. (Siberian Larch)											
1	Cambium	TLC	48	21		0	5	16	(NAPE 9)	0.56	1763
Lathyrus sativus L. (Grass Pea)											
1	Seed	CC (silica)	+	65.8							1764
Lavandula latifolia Medic. (Broadleaved Lavender)											
1	Root *Incl. PS	TLC	52.1	35.5*	5.0	Trace		3.3	3.5		1699
Lavandula vera DC (English Lavender)											
1	Cell Culture	TLC	43	48		1	Trace	3			1765
Lemna minor L. (Lesser Duckweed)											
1	Whole plant	TLC	41.0	24.1	10.2	10.8	3.1	2.6	8.2	3.52[c]	1445
1	Chloroplast	TLC	18		73	9				1372[q]	1766
1	Frond	TLC	30	16	9	6	3 (incl. PA, LPC, LPE)			0.071	1767
Lemna trisulca L. (Star Duckweed)											
1	Whole plant	TLC	35.5	20.2	16.1	7.8	3.5	7.8	9.1	2.42[c]	1445
Lepidium sativum L. (Garden Cress)											
1	Seed	TLC	33.5	29.7	4.0	26.0			6.8	0.80[c]	1429

*Unless stated otherwise; the PL column gives the total phospholipid content of each plant part in percent by weight, and the PC, PE, PG, PI, PS, PA, and DPG columns show the distribution in percent by weight of the total PL; [a]Percent of total lipids; [b]mg lipid P/100 mL juice; [c]Percent dwb; [d]mg PL/100 mg protein; [e]mg PL/g fwb; [f]mg PL/100 g dwb; [g]Percent of polar lipids; [h]μmol PL/mg protein; [i]μmol PL/mg chlorophyll; [j]μg PL/mg protein; [k]μg lipid P/g; [l]μmol PL/g fwb; [m]mol PL/mol chlorophyll; [n]mg PL/g dwb; [o]μmol PL/g dwb; [p]μg PL/100 mg fwb; [q]μg PL/mg chlorophyll; [r]mg PL/cm² leaf area; [s]nmol PL/cm² leaf area; [t]μmol PL/mg protein; [u]IR given; [v]mg lipid P/g lipid; [w]μg PL/mL; [x]μg lipid P/cm² leaf area; [y]Molecules per 10^6 plastids; [z]μg PL/mg chromatin; [aa]Fatty acid composition and *sn*-1/*sn*-2 distribution given for each phospholipid; [bb]μg/g fwb; [cc]Taken from the abstract.

Continued

Table 4-15 Phospholipids in Plants*—cont'd

No.	Plant part	Method	PC	PE	PG	PI	PS	PA	DPG	PL	References
Lespedeza bicolor Turcz. (Bush Clover)											
1	Seed	TLC	36.5	17.3		21.5				4.99[a]	1768
Levisticum officinale Koch (Lovage)											
1	Seed	TLC	17	6		25		47		1.1[a]	1438
			also LPC 3, LPE 2								
1	Leaf?	TLC	37.2	2.8		2.6	0.1	1.2	43.9		1472
Lilium longiflorum Thumb. (Trumpet Lily)											
1	Pollen	TLC	49[H]	27[H]	+	11[H]	+			0.56	1769
Limnanthes alba Hart. ex Benth. (Meadowfoam)											
1	Seed, cotyledon, 10 dap[a]	TLC	35	40				11			1770
v1	–, 60 dap[a]	TLC	3	1				<1			1770
Limnobium spongia (Bosc) Steud. (Frogbit)											
1	Whole plant	TLC	37.5	25.0	12.1	5.6	2.8	13.6	3.4	2.90[c]	1445
Lindelofia macrostyla (Bunge) Pop.											
1	Seed, defatted	TLC	+	+		+				0.3	794
Lindelofia tschimganica (Lipsky) Pop. ex Pazij.											
1	Seed, defatted	TLC	+	+		+				1.2	794
Linum usitatissimum L. (Linseed, Flax)											
1	Seed[p]	TLC	155	220 (incl. PS)		2	(LPC 67)				1771
1	Seed, cotyledon	TLC	58.0	+		+					1772
1	–, oil body	TLC	57.2	2.8		6.9	33.1			0.90	1482
1	–, microsomes	TLC	60.6	10.6	0	21.4		7.5			1570
Lolium cuneatum Nevski (Poison Ryegrass)											
1	Seed, defatted	TLC	+	+		+				0.3	794
Lolium perenne L. (Perennial Ryegrass)											
1	Leaf	TLC	44	18	25	7			7	3.05[l]	1523
Luffa aegyptiaca Mill. (Sponge Gourd)											
1	Leaf	TLC	+	+	+	+					1631
Lupinus albus L. (White Lupine)											
1	Seed[e]	Frac. pptn.	0.411	0.123							718
1	Leaf	TLC	53.1	23.4	17.9			5.5		1.62	1773

2	–, thylakoid	TLC		250-330[q]							1774
1	Root *Incl. PS	TLC	35.2	36.2*	6.7	8.1		8.0	5.1		1699
Lupinus arboreus Sims. (Tree Lupine)											
1	Pollen	TLC	+	+		+					1603
Lupinus luteus L. (Yellow Lupine)											
1	Root	TLC	36.7	23.6	7.1	8.8	9.0	10.8	4.0		1775
1	–, mitochondria *Incl. PS	TLC	32.8	23.7	12.9*	12.7		9.0	9.0	72.5[a]	1776
1	–, microsomes *Incl. PS	TLC	34.9	28.2	9.5*	14.1		5.8	7.5	71.5[a]	1776
Lycopersicon esculentum Mill. (Tomato)											
1	Seed	TLC	+	+		+		Trace	(LPC +)	0.4	1777
1	Leaf	PC	35	1.5	23	21					1746
1	Leaf	TLC	50	21	20	5			5	2.18[l]	1523
1	Leaflet	TLC	8.8	6.4	16.1	12.0		56.6			1778
2	Fruit[l]	Choline	8-19								1451
1	Fruit	TLC	43.2	23.5		10.4	4.5	9.6	8.7	0.152	1779
1	–, microsomes	TLC	53.7 also LPC 1.7 and LPE 2.4	31.0	2.6	6.8		1.8			1780
1	–, crude plastid	TLC	40.2 also LPC 1.4 and LPE 2.8	27.1	12.8	8.2		5.7	1.7		1780
1	–, pericarp	TLC	47.2	27.0	3.8	13.3	3.4	4.0	1.4	0.038	1781
1	–, mitochondria	TLC	34.1	23.7	4.3	14.2	8.0	7.3	8.2		1781
1	–, chloroplast[i]	TLC	2.83	2.68	0.48						1782
1	–, chloroplast[i]	TLC	0.32	0.08	0.29						1782
1	Leaf, thylakoid[m]	TLC	0.13	0.38							1783
1	Leaf, thylakoid	TLC			11.4[a]						1762
1	Root	TLC	48.6	39.1	tr					1.55[c]	1494

*Unless stated otherwise; the PL column gives the total phospholipid content of each plant part in percent by weight, and the PC, PE, PG, PI, PS, PA, and DPG columns show the distribution in percent by weight of the total PL; [a]Percent of total lipids; [b]mg lipid P/100 mL juice; [c]Percent dwb; [d]mg PL/100 mg protein; [e]mg PL/g fwb; [f]mg PL/100 g dwb; [g]Percent of polar lipids; [h]μmol PL/mg protein; [i]μmol PL/mg chlorophyll; [j]μg PL/mg protein; [k]μg lipid P/g; [l]μmol PL/g fwb; [m]mol PL/mol chlorophyll; [n]mg PL/g dwb; [o]μmol PL/g dwb; [p]μg PL/100 mg fwb; [q]μg PL/mg chlorophyll; [r]mg PL/cm² leaf area; [s]nmol PL/cm² leaf area; [t]μmol PL/mg protein; [u]IR given; [v]mg lipid P/g lipid; [w]μg PL/mL; [x]μg lipid P/cm² leaf area; [y]Molecules per 10^6 plastids; [z]μg PL/mg chromatin; [aa]Fatty acid composition and *sn*-1/*sn*-2 distribution given for each phospholipid; [bb]μg/g fwb; [cc]Taken from the abstract.

Continued

Table 4-15 Phospholipids in Plants*—cont'd

No.	Plant part	Method	PC	PE	PG	PI	PS	PA	DPG	PL	References
Lycopersicon esculentum ssp. *cerasiforme* (Dunal) Alef. (Cherry Tomato)											
1	Fruit, outer pericarp	TLC	25.4	27.2	9.2	10.5		27.7		0.15[c]	1784
1	Callus[k]	TLC	80.2	154.5							1785
Lycopersicon esculentum ssp. *eupimpinellifolium* (Mill.) Brezh.											
1	Leaf	TLC	+	+	+	+					1786
Lycopersicon esculentum ssp. *pruniforme* Brezh.											
1	Leaf	TLC	+	+	+	+					1786
Lycopersicon hirsutum Dunal ssp. *glabratum* Mill. (Wild Tomato)											
1	Leaf	TLC	+	+	+	+					1786
Lycopodium annotinum L. (Stiff Clubmoss)											
1	Whole plant	TLC	+	+							4578
Lycoris incarnata Comes											
1	Flower	TLC		1.6						0.009	1787
Lycoris radiata Herb. (Red Spiderlily)											
1	Flower	TLC	30.5*							0.05	1788
			*16:0/16:0 PC and 16:0/18:2 PC isolated from this fraction by preparative TLC								1788
1	Leaf	TLC	+	+	+	+					1631
Lysimachia nummularia L. (Moneywort)											
1	Whole plant	TLC	25.9	16.0	13.8	13.0	4.7	4.9	21.7	21.8	1789
Macrozamia communis L.											
1	Leaf	TLC	+	+							1436
Magnolia denudata Desr. (Yulan)											
1	Leaf	TLC	+	+							1436
Malabajla graveolens L.											
1	Seed	TLC	32	9		42		11	2	2.0[a]	1438
			also LPC 4								
Malus domestica Borkh. (Apple)											
1	Seed	TLC	30.8	16.0	1.8	28.4	2.5	10.7	1.1	0.2	1639
			also LPC 3.6, LPE 2.1								
1	Leaf	TLC	41.6	21.8	19.6	17.0				0.36	1790
1	Stem	TLC	42.0	32.2	7.7	18.2				0.14	1790

1	Shoot, xylem	TLC	54.5	20.1	6.3	19.0				0.027[c]	1791
1	–, phloem	TLC	63.2	18.7	6.6	11.5				0.076[c]	1791
1	Fruit, peeled and cored, preclimacteric[a]	TLC	21.5	14.4	3.1	6.0	0.4	0.3	0.7	46.6	1792
1	–, postclimacteric	TLC	22.7	10.7	0.8	6.2	0.4	0.7	0.5	42.2	1792
2	–, mitochondria	TLC	9.2-16.1	21.8-64.6	4.3-5.6	2.0-2.9	7.6	2.5-3.8	5.0-6.1	52.5-63.1[a]	1793
1	–, parenchyma	TLC	35.2	41.6	Trace	3.2	Trace	16.8	3.2		1794
1	–, epidermis	TLC	33.8	18.2	9.5	18.2	Trace	11.6	8.7		1794
1	Fruit	TLC	34.9	0.6		14.6		19.1	(LPC 20.0, LPE 10.9)		1795
1	–, callus culture	TLC	Trace	44.2	32.4	0.7		12.4	(LPC 8.1, LPE 2.2)		1795
1	Bud, dormant	TLC	40	37	8	15		<1			1796
1	Root	TLC	23.5	47.0	12.2	17.4				0.12	1790
Malus sylvestris Mill. (Crabapple)											
1	Bark	TLC	4.7	45.2	18.4	4.4		(LPC 27.3)		0.429	1797
Mandragora turcomanica Mizger. (Mandrake)											
1	Seed	TLC	+	+		+		+		0.687	1798
1	Fruit, flesh	TLC	+	+		+		+		0.067	1798
Mangifera indica L. (Mango)											
3	Seed	TLC	24.3-25.5	17.8-19.1				(LPC 16.5-18.0)			1799
10	Kernel	TLC	+	+			+ Trace	+	(LPC +)	0.1-0.8[a]	1800
1	Kernel, decorticated	TLC	20.5	13.2	13.7	15.5		(LPE 11.2)		0.12[c]	1801
Manihot esculenta Crantz (Cassava)											
1	Leaf, young	TLC	Trace	35.9	46.0	8.8		1.5	7.9	1.46	1802
1	Root, decorticated	TLC	37.6	21.5	3.1	20.3	1.9	12.7	2.9	0.706[l]	1803

*Unless stated otherwise; the PL column gives the total phospholipid content of each plant part in percent by weight, and the PC, PE, PG, PI, PS, PA, and DPG columns show the distribution in percent by weight of the total PL; [a]Percent of total lipids; [b]mg lipid P/100 mL juice; [c]Percent dwb; [d]mg PL/100 mg protein; [e]mg PL/g fwb; [f]mg PL/100 g dwb; [g]Percent of polar lipids; [h]μmol PL/mg protein; [i]μmol PL/mg chlorophyll; [j]μg PL/mg protein; [k]μg lipid P/g; [l]μmol PL/g fwb; [m]mol PL/mol chlorophyll; [n]mg PL/g dwb; [o]μmol PL/g dwb; [p]μg PL/100 mg fwb; [q]μg PL/mg chlorophyll; [r]mg PL/cm^2 leaf area; [s]nmol PL/cm^2 leaf area; [t]μmol PL/mg protein; [u]IR given; [v]mg lipid P/g lipid; [w]μg PL/mL; [x]μg lipid P/cm^2 leaf area; [y]Molecules per 10^6 plastids; [z]μg PL/mg chromatin; [aa]Fatty acid composition and *sn*-1/*sn*-2 distribution given for each phospholipid; [bb]μg/g fwb; [cc]Taken from the abstract.

Continued

Table 4-15 Phospholipids in Plants*—cont'd

No.	Plant part	Method	PC	PE	PG	PI	PS	PA	DPG	PL	References
Marchantia berteroana Lehm. and Lindenb.											
1	Thallus	TLC	27	14	24	11			24	0.66[l]	1523
Marchantia polymorpha L.											
1	Whole plant	TLC	21.6	21.6	18.3	13*	13.8			0.94[c]	1492
	*Incl. LPE										
Matricaria chamomila L. (Sweet False Camomile)											
1	Seed, defatted	TLC	+	+		+				1.1	794
Medicago sativa L. (Alfalfa, Lucerne)											
1	Leaf	TLC	27	18	24	10			21	2.78[l]	1523
1	Leaf	HPLC	+	+	+	+					1804
3	–, chloroplast	TLC	29-38		55-62	5-12					1805
1	Stems and petioles	TLC	42.3	27.2	9.8	17.4		3.3		0.20	1638
Melilotus alba Desr. (White Sweet Clover)											
1	Leaf	PC	30.0	8.4	24.9	17.5	6.3				1746
Mentha arvensis L. (Field Mint)											
1	Leaf	TLC	61.7	7.0	16.0	15.2			0	1.59[c]	1470
1	Stem	TLC	68.8	2.2	26.1	0			2.9	0.63[c]	1470
Mesembryanthemum crystallinum L. (Ice Plant)											
1	Leaf, tonoplast	TLC	54.4	25.4	9.7		8.0	2.5			1749
Metrosideros excelsus Sol. ex Gaertn. (New Zealand Christmas Tree)											
1	Leaf[l]	TLC			0.25						1402
Mnium cuspidatum Hedw.											
1	Whole plant	TLC	+	+							1436
1	–, gametophore	TLC	58	12	19	12				26[a]	1754
Mnium marginatum (With.) Brid. (Olivegreen Calcareous Moss)											
1	Whole plant	TLC	28.4	22.4	30.2	11.1*	7.9			1.36[c]	1473
	*Incl. LPE										
Mnium medium Bry. Eur.											
1	Gametophore	TLC	64	10	15	10				39[a]	1754
Mnium undulatum Strasb.											
1	Whole plant	TLC	+	+							1436
Mucuna pruriens (L.) DC (Cowitch Bean)											
1	Seed	CC (alumina)	12.5								1806
Mucuna sloanei Fawc. and Rend. (Ukpo)											
1	Seed	TLC	21	25	18	24			12	2.33[c]	1524

Muntingia calabura L. (Calabur)											
1	Seed	TLC	19.3	21.6	3.5	30.8			21.8	0.18	1681
Murraya koenigii (L.) K. Spreng. (Curry-Leaf)											
1	Seed	TLC	28.2	32.1	5.0	7.1		3.0		0.42[c]	1807
			also LPC, 11.1, LPE 14.0								
Musa × *paradisiaca* L. (Banana)											
1	Shoot tip, micropropagated[cc]		+	+	+	+		+			1808
Myrica rubra Sieb. and Zucc. (Bayberry)											
3	Seed	TLC	51.9-55.4	10.1-12.8		1.8-4.7	32.0-33.0			0.14-0.45[c]	1809
3	–, endocarp	TLC	49.0-50.0	11.9-15.9		1.1-4.5	30.4-33.2	(LPC 0.5-0.9)		0.037-0.063[c]	1809
Myriophyllum quitense HBK (Water Milfoil)											
1	Leaf, submerged	TLC	21.1	37.5	15.6	12.5				0.12	1810
1	–, aerial	TLC	26.6	32.8	19.3	13.5				0.33	1810
1	Stem	TLC	18.3	47.2	19.8	8.8				0.13	1810
1	Flower	TLC	34.5	37.2	4.4	19.5				0.24	1810
1	Fruit	TLC	25.8	35.2		28.1				0.54	1810
1	Root	TLC	21.9	34.1		24.3				0.27	1810
Myriophyllum spicatum L. (Eurasian Water Milfoil)											
1	Whole plant	TLC	39.3	14.1	27.2	5.0	7.3	0.6	6.5	5.60[c]	1555
Myriophyllum verticillatum L. (Myriad Leaf)											
1	Whole plant	TLC	38.0	16.7	15.7	18.8	4.5	6.3		4.40[c]	1445
Narcissus spp. (Daffodil)											
1	Leaf	TLC			11						1440
1	Bulb	TLC	+	+	+	+			+		1811
Nasturtium officinale R.Br. (Watercress)											
1	Leaf	TLC	45.6	30.1	7.9	16.3				0.130	1812
Nelumbo nucifera Gaertn. (Sacred Lotus)											
1	Leaf	TLC	17.2	6.9	60.4	9.7			5.8	3.07[c]	1813
Nerium oleander L. (Oleander)											
1	Leaf, thylakoid	TLC	23	23	45	9				0.047	1814
Nicotiana glutinosa L.											
1	Leaf chloroplast	TLC	+	+			+		+		1815

*Unless stated otherwise; the PL column gives the total phospholipid content of each plant part in percent by weight, and the PC, PE, PG, PI, PS, PA, and DPG columns show the distribution in percent by weight of the total PL; [a]Percent of total lipids; [b]mg lipid P/100 mL juice; [c]Percent dwb; [d]mg PL/100 mg protein; [e]mg PL/g fwb; [f]mg PL/100 g dwb; [g]Percent of polar lipids; [h]µmol PL/mg protein; [i]µmol PL/mg chlorophyll; [j]µg PL/mg protein; [k]µg lipid P/g; [l]µmol PL/g fwb; [m]mol PL/mol chlorophyll; [n]mg PL/g dwb; [o]µmol PL/g dwb; [p]µg PL/100 mg fwb; [q]µg PL/mg chlorophyll; [r]mg PL/cm^2 leaf area; [s]nmol PL/cm^2 leaf area; [t]µmol PL/mg protein; [u]IR given; [v]mg lipid P/g lipid; [w]µg PL/mL; [x]µg lipid P/cm^2 leaf area; [y]Molecules per 10^6 plastids; [z]µg PL/mg chromatin; [aa]Fatty acid composition and *sn*-1/*sn*-2 distribution given for each phospholipid; [bb]µg/g fwb; [cc]Taken from the abstract.

Continued

Table 4-15 Phospholipids in Plants*—cont'd

No.	Plant part	Method	PC	PE	PG	PI	PS	PA	DPG	PL	References
Nicotiana quadrivalvis Pursh (Nonchaw)											
1	Leaf	PC	+	+		+	+				1816
			also LPC +								
Nicotiana rustica L. (Aztec Tobacco)											
1	Leaf	TLC	38.2	13.6	27.9						1817
1	Callus	TLC	49.7	26.0	5.3						1817
Nicotiana sylvestris Speg. and Comes (Woodland Tobacco)											
1	Pollen	TLC	50	23	2	19			5	2.5[c]	1818
Nicotiana tabacum L. (Tobacco)											
1	Leaf	TLC	42	12	25	8	8		4	0.092	1819
2	Leaf	TLC	22.0-34.0	11.4-14.3	29.5-30.3						1817
1	Leaf[a]	HPLC	6.3	3.5	3.7	0.9	0	2.3	1.6		1413
1	Leaf	PC	46		28	26					1746
2	Leaf, air-cured	TLC	10.3-11.2	14.5-15.1	6.8-7.0	16.0-19.8		13.9-14.8	7.3-8.3	0.31-0.43[c]	1820
2	Callus	TLC	42.5-53.9	20.2-24.8	11.0-41.1						1817
1	Flower, petal	TLC	46	15	6	21	11		1	0.157	1819
1	–, ovary	TLC	37	15	6	27	13		1	0.171	1819
1	–, pistil	TLC	28	10	8	37	14		1	0.176	1819
1	–, anther	TLC	47	18	5	16	12		1	0.110	1819
1	Cell, culture	TLC	42.5	27.0	3.8	8.4		2.0		0.49[e]	1495
Nigella sativa L. (Black Cumin)											
1	Seed	TLC	32.8	28.5	6.1	25.4			7.2	2.00[c]	1429
1	Seed[n]	HPLC	0.007	0.118		0.0003					1821
Nitraria komarovii Iljin and Lava											
1	Seed, defatted	TLC	+	+		+				0.3	794
Nitraria schoberi L.											
1	Seed, defatted	TLC	+	+		+				0.7	794
Nothofagus dombeyi (Mirb.) Oerst. (Coigue)											
1	Leaf	TLC	+	9	81	6				3.2[i]	1822
1	–, thylakoid	TLC	8	19	52	21					1823
1	–, envelope membrane	TLC	15	12	60	13					1823

Nuphar luteum (L.) Smith (Yellow Water Lily)											
1	Whole plant	TLC	37.6	20.6	12.2	12.6	3.7	6.3	7.1	1.31[c]	1445
Nymphoides alba L.											
1	Whole plant	TLC	46.6	25.1	11.2	9.6	2.3		5.3	9.20[c]	1555
Ocimum basilicum L. (Basil)											
1	Tissue	TLC	+								1824
Oenothera biennis L. (Evening Primrose)											
1	Seed, mature	TLC	31.9	17.6	16.7	27.1		6.7		0.047[c]	1825
1	–, immature	TLC	54	22	4	17					1826
Olea europaea L. (Olive)											
1	Seed	TLC	48.4	6.0		29.3		16.3	(LPC +)	0.02	1827
1	Exocarp	TLC	+			+					1828
1	–, germinated 24 h	TLC	60.2	19.5		10.6		9.7		3781[b]	1829
1	Leaf chloroplast[a]	TLC	6.1		6.1	3.2					1830
1	Fruit, kernel	TLC	47.3	7.5		23.9	Trace			0.41	1831
1	–, flesh	TLC	49.3	5.3		18.0	7.2			0.36	1831
1	Fruit, microsomes	TLC	50.8	33.8		4.5		8.4			1832
1	Pollen	TLC	76.6	13.5		5.8		4.0		5266[b]	1829
1	Callus culture	TLC	42	29	0			16		0.093	1833
Oncidium sphacelatum Lindl. (Golden Shower)											
1	Leaf[l]	TLC			0.42						1402
Onopordon acanthium L. (Scotch Thistle)											
1	Seed, defatted	TLC	+	+		+				0.8	794
Onopordon olgae Bunge and Schmalh.											
1	Seed, defatted	TLC	+	+		+				0.2	794
Ophioglossum vulgatum L. (Southern Adderstongue)											
1	Whole plant	TLC	+	+							1436
Oreodoxa regia HBK (Cuban Royal Palm)											
1	Seed	TLC	27.1	16.5					(LPC 25.1)		1799
Origanum dictamnus L. (Dittany)											
1	Leaf	TLC	35.5	14.4	23.3	9.0	17.8			1.00	1834

*Unless stated otherwise; the PL column gives the total phospholipid content of each plant part in percent by weight, and the PC, PE, PG, PI, PS, PA, and DPG columns show the distribution in percent by weight of the total PL; [a]Percent of total lipids; [b]mg lipid P/100 mL juice; [c]Percent dwb; [d]mg PL/100 mg protein; [e]mg PL/g fwb; [f]mg PL/100 g dwb; [g]Percent of polar lipids; [h]μmol PL/mg protein; [i]μmol PL/mg chlorophyll; [j]μg PL/mg protein; [k]μg lipid P/g; [l]μmol PL/g fwb; [m]mol PL/mol chlorophyll; [n]mg PL/g dwb; [o]μmol PL/g dwb; [p]μg PL/100 mg fwb; [q]μg PL/mg chlorophyll; [r]mg PL/cm² leaf area; [s]nmol PL/cm² leaf area; [t]μmol PL/mg protein; [u]IR given; [v]mg lipid P/g lipid; [w]μg PL/mL; [x]μg lipid P/cm² leaf area; [y]Molecules per 10^6 plastids; [z]μg PL/mg chromatin; [aa]Fatty acid composition and *sn*-1/*sn*-2 distribution given for each phospholipid; [bb]μg/g fwb; [cc]Taken from the abstract.

Continued

Table 4-15 Phospholipids in Plants*—cont'd

No.	Plant part	Method	PC	PE	PG	PI	PS	PA	DPG	PL	References
Orobanche crenata Forsskal. (Broomrape)											
1	Whole plant?	TLC	+	+	0	+	+	0			1835
Oryza sativa L. (Rice)											
1	Grain, glutinous	TLC	49.2	21.6	11.2	4.3		4.9	9.8	0.159	1446
1	–, endosperm	TLC	19.3	10.1				(LPC 24.8, LPE 6.4)		10.9[a]	1836
1	Grain, nonglutinous, raw	TLC	28.3	11.0	+	4.6	Trace	Trace		1.41	1837
1	–, polished	TLC	24.0	9.2	+	3.2				0.65	1837
1	–, polished	Frac. pptn.						(LPC 0.2)			1378
1	–, bran	TLC	38.3	14.3	+	10.9	TRACE	TRACE		5.16	1837
1	–, endosperm	TLC	6.0	3.0				(LPC 39.2, LPE 27.0)	53.3[a]		1836
1	–, brown	^{13}C NMR						(LPC +, LPE +)			1838
1	–, starch	TLC							(LPC 30.0, LPE 6.9)	0.29	1839
1	Seed, defatted	TLC	+	+		+				0.7	794
1	Coleoptile, 7 days old	TLC	50	26	5	7	3	6	1		1840
8	Leaf	TLC	+			+	+		(Me_2N-PE trace)		1841
1	Leaf	TLC	35	19	35	10					1842
1	Anther wall	TLC	41	25	26	23					1842
1	Pollen grain	TLC	36	42	<0.5	22					1842
3	Bran	TLC	35.0-38.4	27.2-29.0	1.4-1.8	21.0-23.3		7.2-9.6		0.98-1.12	1843
			also NAPE 0.7-0.8, LPC 1.0-1.5, LPE 1.0-1.4								
2	Bran (1)	TLC	14.2	34.2	8.9	9.0	7.3		(LPC 1.6)	8.1	1844
	Bran (2)		10.2	40.5	10.3	13.5	5.3		(LPC 1.6)	3.5[a]	

1	Bran, glutinous	TLC	+	+	+	+	+			(NAPE +, LPC +, LPE +)	1845
1	–, nonglutinous	TLC	+	+	+	+	+		(NAPE +, LPC +, LPE +)		1845
1	Bran, spherosomes	TLC	63.8	7.2	1.4	Trace	1.8	2.7	1.8	0.9^{c}	1846
			also NAPE 5.4, NALPE 5.0, LPC 9.5, and LPE 1.4								
2	Starch (1)	TLC	9.5	2.2				(LPC 58.2, LPE 30.0)	42.7		1847
	Starch (2)	TLC					(LPC 81.3, LPE 18.7)			57.7	
2	Protein bodies (1)	TLC	10.7	10.0			(LPC 45.7, LPE 33.6)			14.0	1847
	Protein bodies (2)	TLC	14.1	6.3			(LPC 44.4, LPE 35.2)			14.2	
1	Shoot, plasma membrane	HPLC	39	31		10	20			0.034^{k}	1848
4	Seedling mitochondria	TLC	48.5-51.4	20.2-21.5	3.5-5.5	5.9-6.8	1.1-2.0	8.8-10.4	3.4-4.6	(LPC 1.3-3.3)	1849
1	Cell suspension culture	TLC					(NAPE +)				1361
2	Cell culture, tonoplast	TLC	45.4-50.5	32.9-36.0	5.2-6.1	7.8-9.3 (incl. PS)		2.7-4.2			1850
2	–, plasma membrane	TLC	37.8-40.5	40.5-41.3	7.3-7.8	7.6-9.2 (incl. PS)		3.6-4.5			1850
Paeonia decora Anders. (Peony)											
1	Seed[cc]		+	+							1851
Palisota barteri Hook.f.											
1	Berry, chromoplast		TLC	+	+	+	+				1852
Panax ginseng C.A. Meyer (Ginseng)											
1	Root	TLC	25.0	26.8	28.8	12.0		1.6		0.202	1853
1	–, periderm	TLC	24.8	9.8	11.0	7.3		17.2		0.232	1853
1	–, cortex	TLC	28.5	21.6	32.2	8.1		2.5		0.145	1853
1	–, medulla	TLC	38.2	15.4	33.7	5.2		1.0		0.207	1853

*Unless stated otherwise; the PL column gives the total phospholipid content of each plant part in percent by weight, and the PC, PE, PG, PI, PS, PA, and DPG columns show the distribution in percent by weight of the total PL; [a]Percent of total lipids; [b]mg lipid P/100 mL juice; [c]Percent dwb; [d]mg PL/100 mg protein; [e]mg PL/g fwb; [f]mg PL/100 g dwb; [g]Percent of polar lipids; [h]μmol PL/mg protein; [i]μmol PL/mg chlorophyll; [j]μg PL/mg protein; [k]μg lipid P/g; [l]μmol PL/g fwb; [m]mol PL/mol chlorophyll; [n]mg PL/g dwb; [o]μmol PL/g dwb; [p]μg PL/100 mg fwb; [q]μg PL/mg chlorophyll; [r]mg PL/cm^2 leaf area; [s]nmol PL/cm^2 leaf area; [t]μmol PL/mg protein; [u]IR given; [v]mg lipid P/g lipid; [w]μg PL/mL; [x]μg lipid P/cm^2 leaf area; [y]Molecules per 10^6 plastids; [z]μg PL/mg chromatin; [aa]Fatty acid composition and *sn*-1/*sn*-2 distribution given for each phospholipid; [bb]μg/g fwb; [cc]Taken from the abstract.

Continued

Table 4-15 Phospholipids in Plants*—cont'd

No.	Plant part	Method	PC	PE	PG	PI	PS	PA	DPG	PL	References
Panicum dichotomiflorum Michx. (Fall Panicum)											
1	Leaf[l]	TLC			0.62						1402
Panicum miliaceum L. (Proso Millet)											
1	Seed	TLC	27.5	20.1		22.3	14.0	13.3	(LPC 2.8)	0.41	1854
9	Seed, flour	TLC	+	+			+			0.47-0.89	1855
9	–, bran	TLC	+	+			+			0.30-0.70	1855
Panicum sumatrense Roth (Little Millet)											
1	Grain	TLC	36.5	21.0	8.0	Trace	Trace	5.5	(LPC 28.0)	0.41[c]	1662
Papaver dubium L. (Field Poppy)											
1	Pollen	TLC	54.7	25.4	2.6	11.4 (incl. PS)		1.5	4.5		1856
Papaver somniferum L. (Opium Poppy)											
4	Fruit juice	PC	16.4-70.0	30.0-83.6							1857
Parkinsonia aculeata L. (Mexican Palo Verde)											
1	Seed	TLC	23.3	22.3			29.1		(LPC 25.2)	3.5	1858
Parthenocissus tricuspidata (Sieb. and Zucc.) Planch. (Boston Ivy)											
1	Cell culture, callus	TLC	53	37		8		1			1859
1	–, protoplast	TLC	36	28		9		27			1859
Paspalum dilatatum Poir. (Dallis Grass)											
1	Leaf	TLC	28	13	35	7		(LPC 10)	7	1.38[l]	1523
Paspalum distichum L. (Knotgrass)											
1	Leaf[l]	TLC			0.61						1402
Paspalum scrobiculatum L. (Kodo Millet)											
1	Grain	TLC	21.2	30.2	18.0	Trace	Trace	9.0	(LPC 20.1)	0.20[c]	1662
Pastinaca hirsuta L.											
1	Seed	TLC	36 also LPC 5	9	5	31		11	1	1.7[a]	1438
Pastinaca sativa L. (Parsnip)											
1	Root	TLC	38	20	8	12			22	0.88[l]	1523

Peganum harmala L. (African Rue)											
1	Seed, defatted	TLC	+	+		+				0.2	794
Pellia neesiana (Gott.) Limpr. (Ring Pellia)											
1	Whole plant	TLC	30.1	15.7	23.0	5.8	18.5			2.98[c]	1551
Pennisetum glaucum (L.) R.Br. (Pearl Millet)											
1	Seed	TLC	24.3	6.4	1.1	Trace	Trace	1.2		7.19[c]	1860
			also LPC 41.6, LPE 21.2								
1	Pollen grain	TLC	55.1	31.3	1.3	7.6 (incl. PS)		3.2	1.3	4.99[c]	1861
Penstemon speciosus Dougl.											
1	Pollen	TLC		+	+	+					1603
Pentaclethra macrophylla Bentl. (African Oil Bean)											
1	Seed	TLC	27	18	22	22			11	1.72[c]	1524
Perilla frutescens (L.) Britt. (Perilla Mint)											
2	Seed *Incl.PS	TLC	25-46*	18	20-25	5-17			(LPC 10-14)	4.0-4.5[a]	1862
5	Seed	TLC	17.6-20.6	50.4-57.1		4.8-6.6 (incl. PS)		13.6-19.9		2.0-3.0[c]	1863
2	Seed	TLC	+	+		+					1864
Persea americana Mill. (Avocado)											
1	Fruit, kernel	TLC	44.9	15.0		12.1				4.2	1831
2	–, flesh (1)	TLC	37.0	19.5		18.0				1.63	1831
	–, flesh (2)	TLC	38.5	12.0		16.5				1.43	1831
1	Fruit, mitochondria	TLC	45	21	5	11	4	5	9		1865
1	–, chloroplast	TLC	34	33	6	12	6		9		1865
Petroselinum crispum (Mill.) Nym. (Parsley)											
1	Leaf[a]	TLC	15.5	5.9	5.4					0.18	1866
1	Root[a]	TLC	9.3	6.2	10.8	7.7				0.28	1866

*Unless stated otherwise; the PL column gives the total phospholipid content of each plant part in percent by weight, and the PC, PE, PG, PI, PS, PA, and DPG columns show the distribution in percent by weight of the total PL; [a]Percent of total lipids; [b]mg lipid P/100 mL juice; [c]Percent dwb; [d]mg PL/100 mg protein; [e]mg PL/g fwb; [f]mg PL/100 g dwb; [g]Percent of polar lipids; [h]µmol PL/mg protein; [i]µmol PL/mg chlorophyll; [j]µg PL/mg protein; [k]µg lipid P/g; [l]µmol PL/g fwb; [m]mol PL/mol chlorophyll; [n]mg PL/g dwb; [o]µmol PL/g dwb; [p]µg PL/100 mg fwb; [q]µg PL/mg chlorophyll; [r]mg PL/cm^2 leaf area; [s]nmol PL/cm^2 leaf area; [t]µmol PL/mg protein; [u]IR given; [v]mg lipid P/g lipid; [w]µg PL/mL; [x]µg lipid P/cm^2 leaf area; [y]Molecules per 10^6 plastids; [z]µg PL/mg chromatin; [aa]Fatty acid composition and *sn*-1/*sn*-2 distribution given for each phospholipid; [bb]µg/g fwb; [cc]Taken from the abstract.

Continued

Table 4-15 Phospholipids in Plants*—cont'd

No.	Plant part	Method	PC	PE	PG	PI	PS	PA	DPG	PL	References
1	Cell culture[a]	TLC	32.9	18.0	3.1	Trace				0.08	1866
1	Fruit (also LPC 1.9, LPE 2.9)	TLC	38.5	16.5	0.9	32.0		4.4		0.5	1471
Petunia hybrida hort. (Petunia)											
1	Cell culture, callus	TLC	14	6		4		1	0.5		1859
1	–, protoplast	TLC	41	15		20		24			1859
Pharbitis nil Choisy (Morning Glory)											
1	Leaf	TLC	44	17	27	10		2		37.0[s]	1867
Phaseolus coccineus L. (Scarlet Runner Bean)											
1	Bean	TLC	40.9	23.5	3.5	20.4	(NAPE 9.6)			1.24	1693
1	Leaf	CC (silica)	31	20 (incl. PS)		45					1356
1	Leaf[a]	TLC	12.3	12.5		3.6					1868
1	–, chloroplast[g]	TLC	3.5		5.1						1869
1	–, PS(II) particle[m]	TLC	0.023		0.042						1870
Phaseolus lunatus L. (Lima Bean)											
1	Seed[c]	Frac. pptn.	0.622	0.089							718
1	Bean	TLC	43.7	47.4		8.9				1.391	1446
1	Root tip plasmalemma	TLC	+	+	?	+		(LPC ?)			1871
Phaseolus vulgaris L. (Bean (common))											
3	Seed	TLC	30.2-36.2	30.6-53.6		7.0-15.1	5.6-22.8			0.75-0.98	1872
4	Whole bean	TLC	28.9-68.6	8.9-32.8		3.4-17.5		0.01-18.0		0.28-1.11[c]	1873
1	Seed	TLC	21	28	28	13			9	0.26[c]	1524
1	–, cotyledon vesicle	TLC	60	21		19				0.41[h]	1874
1	–, hypocotyl mitochondria	TLC	30.4	24.0	15.7*	9.9[¶]	*iNcl. DPG	¶Incl. PS		(LPC 2.4)	1875
1	Seed	TLC	37.5	25.7		20.4		0.5		2.0	1876
			also NAPE 7.7, NALPE 3.4, LPC 3.4								
1	Seed, cotyledon	^{31}P NMR	+	+		+		(LPC +)			49
1	–, embryonic axis	^{31}P NMR	+	+							49
1	Starch	TLC						(LPC 5.1, LPE 2.0)		0.15	1499
1	Leaf	TLC	38.9	27.0	23.7	10.4	Trace			1733[r]	1634

1	Leaf	TLC	44	19	28	7		2		71.1[s]	1867
1	Leaf	PC	41	19	30	11					1517
6	Leaf	TLC	34.2-45.0	16.0-20.5	23.5-29.5	5.5-9.0	2.2-3.7	1.7-2.7	1.2-5.0		1877
1	Leaf, chloroplast	TLC			12[a]						1489
1	Leaf, thylakoid membrane	TLC	29.4		67.2					81[n]	1878
1	Leaf, 2 weeks	TLC	41.1	8.5	33.5	6.7				2.24[c]	1879
1	Root, 2 weeks	TLC	32.0	9.0	38.5	6.4				0.78[c]	1879
1	Root[l]	Choline	1								1451
1	Root, mitochondria	TLC	47.3	38.4	4.0				10.3		1880
	– ssp. *humilis*	Bush Bean									
1	Bean[bb]	Choline	1								1451
Philonotis fontana (Hedw.) Brid.											
1	Whole plant	TLC	12.3	35.9	28.9	9.4	13.5			0.62[c]	1493
Phoenix dactylifera L. (Date Palm)											
3	Seed	TLC	20.5-22.5					(LPC 19.5-22.0)			1799
Phoenix sylvestris (L.) Roxb. (Wild Date Palm, Kharjur)											
1	Seed	TLC	38.7	20.7		27.2			12.7	0.45[c]	1757
			also LPC 0.4, LPE 0.3								
Phragmites communis Trin. (Water Reed)											
4	Plasmalemma	TLC	5.4-12.8	5.4-14.0	24.3-61.1	6.1-20.7		11.3-28.0	4.3-18.8		1881
Phyllitis scolopendrium L. (Hart's Tonguefern)											
1	Whole plant	TLC	+	+							1436

*Unless stated otherwise; the PL column gives the total phospholipid content of each plant part in percent by weight, and the PC, PE, PG, PI, PS, PA, and DPG columns show the distribution in percent by weight of the total PL; [a]Percent of total lipids; [b]mg lipid P/100 mL juice; [c]Percent dwb; [d]mg PL/100 mg protein; [e]mg PL/g fwb; [f]mg PL/100 g dwb; [g]Percent of polar lipids; [h]μmol PL/mg protein; [i]μmol PL/mg chlorophyll; [j]μg PL/mg protein; [k]μg lipid P/g; [l]μmol PL/g fwb; [m]mol PL/mol chlorophyll; [n]mg PL/g dwb; [o]μmol PL/g dwb; [p]μg PL/100 mg fwb; [q]μg PL/mg chlorophyll; [r]mg PL/cm^2 leaf area; [s]nmol PL/cm^2 leaf area; [t]μmol PL/mg protein; [u]IR given; [v]mg lipid P/g lipid; [w]μg PL/mL; [x]μg lipid P/cm^2 leaf area; [y]Molecules per 10^6 plastids; [z]μg PL/mg chromatin; [aa]Fatty acid composition and *sn*-1/*sn*-2 distribution given for each phospholipid; [bb]μg/g fwb; [cc]Taken from the abstract.

Continued

Table 4-15 Phospholipids in Plants*—cont'd

No.	Plant part	Method	PC	PE	PG	PI	PS	PA	DPG	PL	References
Phyllospadix iwatensis Makino											
1	Whole plant (NAPE tr)	TLC	42.7	22.4	16.6	13.7	1.5	Trace	3.3	0.14	1882
Phyllostachys pubescens Mazel (Moso Bamboo)											
1	Shoot, peeled	TLC	49.4	21.6	11.0	8.7	5.5	(LPC tr.)	3.7	0.044[k]	1883
			and variation from tip to base of shoot								
Picea abies (L.) Karst. (Norway Spruce)											
1	Needles (leaf)	TLC	48.1	8.4	33.2	10.2				1.21[i]	1884
1	–, thylakoid	TLC	45.8	8.0	32.3	13.9				0.85[i]	1884
1	–, envelope	TLC	32.7	9.2	42.1	16.0				0.067[i]	1884
1	–, chloroplast[q]	TLC			433						1885
Picea mariana (Mill.) BSP (Black Spruce)											
1	Shoot	TLC	37	20	31	12				0.61[c]	1886
Pimpinella acuminata (Edgew.) C. B. Clarke											
1	Seed[a]	TLC	1.2	0.8		1.2	(LPC 0.4)			30.0	1887
Pimpinella anisum L. (Anise)											
1	Fruit	TLC	51.1	11.1	1.9	20.4		4.4		0.3	1471
			also LPC 3.6, LPE 4.1								
Pinus armandii Franch. (Armand Pine)											
1	Seed	TLC	44.9	22.3			12.3			0.38	1888
Pinus banksiana Lamb. (Jack Pine)											
1	Seedling	TLC	31.3		54.8 (incl. PG)	13.9				0.073	1889
Pinus cembra L. (Swiss Stone Pine)											
1	Kernel[n]	TLC	5.4	1.1		3.0	0		(Phyto-GP 0.4)		1481
Pinus halepensis Mill. (Aleppo Pine)											
1	Root *Incl. PS	TLC	34.7	22.2*	10.2	11.0		12.6	9.1		1699
1	Pollen	TLC	44.4	26.5		27.3				1.05[c]	1890
			also LPC 1.5, LPE 0.3								
1	Pollen	HPLC	40.1	25.0	0	22.8	0		0	1.01	1890

Pinus koraiensis Sieb. and Zucc. (Korean Pine)											
1	Nut	TLC	40.2	19.4		7.3	2.6		(LPC 5.3)	0.6	1891
Pinus mugo Turra (Mugo Pine)											
1	Leaf	TLC	+	+							1436
Pinus nigra Arnold (Black Pine)											
1	Needle	HPLC	42	20	23					0.30	1892
Pinus pinaster Ait. (Cluster Pine)											
1	Root *Incl. PS	TLC	39.9	24.2*	6.0	14.0		7.3	8.5		1699
Pinus pinea L. (Italian Stone Pine)											
1	Seed, germ, 7 days	TLC	38.1	15.2	2.9	17.9		4.2	11.3		1893
1	Root *Incl. PS	TLC	33.2	34.6*	10.3	8.2		4.6	9.1		1699
Pinus ponderosa Dougl. (Ponderosa Pine)											
1	Pollen	TLC	60	18	6	13	1		2	2.9[c]	1894
Pinus radiata D. Don (Monterey Pine)											
1	Leaf	TLC	35	12	24	16			13	2.30[l]	1523
1	Root *Incl. PS	TLC	29.5	24.0	13.8	13.8		11.1	7.7		1699
Pinus sibirica Du Tour (Siberian Stone Pine)											
1	Seed, germ	TLC	20.8	17.7		14.8	11.6	17.3	(LPC 10.2)	8.1	1895
1	–, endosperm	TLC	19.3	17.7		12.8	11.1	21.4	(LPC 8.9)	1.7	1895
Pinus sylvestris L. (Scotch Pine, Scots Pine (Brit.))											
1	Needle	TLC	+	+	+	+		+	+		1896
1	Needle (leaf) *Incl. PC,PS	TLC		6.5	41.8	21.9**		29.8		2.17[l]	1897
1	–, thylakoid	TLC		3.9	35.2	30.5*		30.9			1897
1	–, chloroplast	TLC	27.3	0	25.4	23.7	23.5			0.068	1797
1	Seedling[l]	TLC	2.38	1.01	1.10	0.55 (incl. PS)					1898
1	Wood *Incl. PG	TLC	62.1	14.6*		5.8		1.0	14.0	1.71[a]	1899

*Unless stated otherwise; the PL column gives the total phospholipid content of each plant part in percent by weight, and the PC, PE, PG, PI, PS, PA, and DPG columns show the distribution in percent by weight of the total PL; [a]Percent of total lipids; [b]mg lipid P/100 mL juice; [c]Percent dwb; [d]mg PL/100 mg protein; [e]mg PL/g fwb; [f]mg PL/100 g dwb; [g]Percent of polar lipids; [h]µmol PL/mg protein; [i]µmol PL/mg chlorophyll; [j]µg PL/mg protein; [k]µg lipid P/g; [l]µmol PL/g fwb; [m]mol PL/mol chlorophyll; [n]mg PL/g dwb; [o]µmol PL/g dwb; [p]µg PL/100 mg fwb; [q]µg PL/mg chlorophyll; [r]mg PL/cm^2 leaf area; [s]nmol PL/cm^2 leaf area; [t]µmol PL/mg protein; [u]IR given; [v]mg lipid P/g lipid; [w]µg PL/mL; [x]µg lipid P/cm^2 leaf area; [y]Molecules per 10^6 plastids; [z]µg PL/mg chromatin; [aa]Fatty acid composition and *sn*-1/*sn*-2 distribution given for each phospholipid; [bb]µg/g fwb; [cc]Taken from the abstract.

Continued

Table 4-15 Phospholipids in Plants*—cont'd

No.	Plant part	Method	PC	PE	PG	PI	PS	PA	DPG	PL	References
Pistacia vera L. (Pistachio)											
1	Nut	TLC	63	22		+	+	7	7	1.07	1900
			also LPC 5.6								
Pistia stratiotes L. (Water Lettuce)											
1	Whole plant	TLC	46.2	26.2	6.0	8.6	7.7	Trace	5.0	2.87[c]	1555
Pisum sativum L. (Peas)											
3	Whole pea	TLC	54.3	16.1		17.1					1901
3	–, cotyledon	TLC	52.9	14.5		14.1					1901
1	–, mitochondria	TLC	52.2	15.5		13.6			5.1		1901
			also LPC 6.7 LPE 3.8								
3	Cotyledon, starch	TLC	0	+			+		(LPC 0, LPE +, LPS +)		1902
1	Seed	PC				(NAPE +, NALPE +)					1498
1	Seed	TLC	41.6	15.5	6.5	2.6		15.6	10.4	0.38	1903
			also LPC 7.8								
1	Seed	TLC	45.7	16.2	3.2	20.8			(NAPE 12.1)	1.22	1693
1	Seed	TLC	+	+	+	+	+	+	+		1904
			also LPC +, LPE +, NAPE +, and NALPE +								
1	Seed	TLC	54.6	17.0	2.2	21.2	0.6	0.7		1.55[c]	1905
			also LPC 2.5, LPE 1.3								
1	–, starch granule	TLC						(LPC 0, LPE 0)			1905
1	Seed	HPLC	+	+	+	+	+	+			1804
1	Seed	TLC					Alkenyl PC 0.6, PE 0.2				1906
1	3-day seedling	TLC						Alkenyl PC 1.2-1.8, PE 0.4-0.6			1906
		Both plasmalogens are 1-alken-1′-yl-2-acyl PL's where the alkenyl group is mainly 16:0, 18:1, and 18:2									
1	24-h seedling, microsomes[a]	TLC	53.9	11.1		15.8					1907
1	Stems and petioles	TLC	40.5	27.3	12.3	16.2		3.7		0.157	1638
1	Leaf	TLC					+				1450
1	Leaf[m]	TLC	0.55		0.37					1.23[m]	1908
1	–, chloroplast	TLC	22	0	71	7					1909
1	–, mitochondria	TLC	44.7	32.1	1.9	2.7			18.6		1909

1	–, thylakoid[m]	TLC	0.24		0.32					0.75[m]	1908
1	Shoot, chloroplast membrane	TLC	77.2	2.4	12.6	6.9					1910
1	–, inner (heavy)	TLC	49.5	4.4	32.8	10.8					1910
1	–, outer (light)	TLC	78.2	2.7	9.8	9.1					1910
1	Root, microsomes[j]	HPLC	0.54	0.37		0.04					1911
1	–, symbiosomes[j]	HPLC	0.44	0.27		0.01					1911
1	–, plasmalemma[j]	HPLC	0.13	0.34		0.00					1911
1	Root, plastids[g]	TLC	47.5	9.2	16.4	1.7	1.3	7.1		(LPC 4.2, LPA 12.6)	1912
Plagiomnium ellipticum (Brid.) T. Kop.											
1	Whole plant *Incl. LPE	TLC	28.6	21.0	30.6	11.4*	8.4			0.15[c]	1473
Plagiomnium medium (Bruch et Schimp. in B.S.G.) T. Kop.											
1	Whole plant	TLC	35.3	14.0	27.3	12.2	11.4			0.54[c]	1493
Platanus spp. (Plane Tree)											
1	Leaf	TLC	32.3	24.2	22.0	0	Trace			223[r]	1634
Platycerium alcicorne (Willem.) Tardieu											
1	Whole plant	TLC	+	+							1436
Pleurozium schreberi (Brid.) Mitt.											
v1	Whole plant *Incl. LPE	TLC	55.6	20.0	15.5	8.9*				0.34[c]	1654
1	–, gametophore	TLC	67	13	7	13				15[a]	1754
Poa annua L. (Annual Bluegrass)											
1	Leaf, thylakoid[a]	TLC			9.7						1913
Poa pratensis L. (Kentucky Bluegrass)											
1	Leaf	TLC	29.6	19.6	27.1	3.7	1.9	18.1	Trace	0.85[x]	1914
Podocarpus nivalis Hook. (Snow Totara)											
1	Leaf	TLC	+	+							1436
Podocarpus nubigenus Lindl. (Cloud Podocarp)											
1	Needle	TLC	36	21	21	21				1.4[o]	1915

*Unless stated otherwise; the PL column gives the total phospholipid content of each plant part in percent by weight, and the PC, PE, PG, PI, PS, PA, and DPG columns show the distribution in percent by weight of the total PL; [a]Percent of total lipids; [b]mg lipid P/100 mL juice; [c]Percent dwb; [d]mg PL/100 mg protein; [e]mg PL/g fwb; [f]mg PL/100 g dwb; [g]Percent of polar lipids; [h]µmol PL/mg protein; [i]µmol PL/mg chlorophyll; [j]µg PL/mg protein; [k]µg lipid P/g; [l]µmol PL/g fwb; [m]mol PL/mol chlorophyll; [n]mg PL/g dwb; [o]µmol PL/g dwb; [p]µg PL/100 mg fwb; [q]µg PL/mg chlorophyll; [r]mg PL/cm^2 leaf area; [s]nmol PL/cm^2 leaf area; [t]µmol PL/mg protein; [u]IR given; [v]mg lipid P/g lipid; [w]µg PL/mL; [x]µg lipid P/cm^2 leaf area; [y]Molecules per 10^6 plastids; [z]µg PL/mg chromatin; [aa]Fatty acid composition and *sn*-1/*sn*-2 distribution given for each phospholipid; [bb]µg/g fwb; [cc]Taken from the abstract.

Continued

Table 4-15 Phospholipids in Plants*—cont'd

No.	Plant part	Method	PC	PE	PG	PI	PS	PA	DPG	PL	References
Podocarpus salignus D. Don (Willow Podocarp)											
1	Needle	TLC	5	50	15	30				2.0[o]	1915
Polygonum amphibium L. (Water Smartweed)											
1	Whole plant	TLC	46.9	15.9	13.0	12.2	1.3	10.7		0.89[c]	1445
Polypodium vulgare L. (European Polypody)											
1	Whole plant	TLC	+	+							1436
Polytrichum commune Hedw. (Polytrichum Moss)											
1	Whole plant	TLC	+	+							1436
Polytrichum juniperinum Hedw. (Hair Cap Moss)											
1	Whole plant	TLC	24.6	20.4	21.9	9.0	12.0			0.42[c]	1493
Poncirus trifoliata L. × *Citrus paradisi* Macf. (Citrumelo)											
1	Leaf	TLC	64.5	20.9	6.9	7.7				0.363	1590
Pongamia glabra Vent. (Pongam, Karanja)											
1	Seed, kernel	TLC	43.1	18.8		33.3				0.78	1916
Populus euroamericana (Dode) Guin. (Carolina Poplar)											
v1	Bark	CC (silica)	44.3	19.3	10.7	11.4				14.0[o]	1917
Populus nigra L. var. *italica* (Lombardy Poplar)											
1	Leaf	TLC	40	18	24	11			8	4.55[l]	1523
Portulaca grandiflora Hook. (Rose Moss)											
1	Leaf[l]	TLC			0.26						1402
Portulaca oleracea L. (Purslane)											
1	Leaf[l]	TLC			0.26						1402
Posidonia oceanica (L.) Delile (Mediterranean Tapeweed)											
1	Whole plant	TLC	41.9	23.1	17.7	12.8	1.9	Trace	2.9	0.53[c]	1555
Potamogeton filiformis Pers.											
1	Whole plant	TLC	25.6	15.6	28.2	5.2	5.5	2.3	17.6	18.0	1789
Potamogeton natans L. (Floating Leaf Pondweed)											
1	Whole plant	TLC	38.7	16.9	11.4	6.1	4.0	7.4	15.4	29.7	1789
Potamogeton perfoliatus L. (Perfoliate Pondweed)											
1	Whole plant	TLC	41.4	11.0	24.9	9.9	4.1	8.7		1.97[c]	1445
Potentilla asiatica Juz. (Cinquefoil)											
1	Seed, defatted	TLC	+	+		+				0.4	794

Pringlea antiscorbutica R.Br. (Kerguelen Cabbage)											
1	Leaf	TLC	41.5	23.1	22.4	11.6			1.4	1.47[c]	1918
Prunus armeniaca L. (Apricot)											
1	Seed kernel	TLC	43.0	15.2	1.3	31.6	1.1	3.1	0.4	0.5	1919
			also LPC 0.8, LPE 0.5								
1	Root	TLC	41.6	36.0*	8.8	1.1		7.0	5.5		1699
	*Incl. PS										
Prunus avium (L.) L. (Cherry)											
1	Seed kernel	TLC	41.0	14.7	1.0	30.0	2.7	4.7	0.8	0.4	1919
			also LPC 1.0, LPE 1.0								
3	Seed embryo	TLC	40.3-43.1	36.3-39.4		2.4-2.7	12.3-14.4		(LPC 1.8-3.1)	0.34-0.56[c]	1920
3	–, endocarp	TLC	39.3-40.4	39.7-43.9		6.2-7.3	3.8-4.7		(LPC 3.5-4.3)	0.30-0.49[c]	1920
1	Leaf[f]	TLC			1.3-2.2						1921
1	Callus culture	TLC	35	17	39	9 (incl. PS)					1922
Prunus cerasus L. (Morello Cherry)											
1	Seed kernel	TLC	37.4	19.5	2.8	21.5	9.7	4.3	2.8	0.4	1919
			also LPC 0.6, LPE 1.7								
Prunus domestica L. (Plum)											
1	Seed kernel	TLC	59.8	16.0	0.4	13.8	0.0	1.2	1.5	0.2	1919
			also LPC 1.6, LPE 1.9								
Prunus dulcis (Mill.) D.A. Webb (Almond)											
1	seed kernel, mature	TLC	14.3	2.1	8.0*	12.2	0.9[¶]	1.6	58.8		1923
1	–, immature	TLC	23.6	3.0	6.4*	18.8	1.2[¶]	1.6	45.2		1923
			*Incl. sulfolipids								
			¶Incl. lysolipids								
1	Root	TLC	38.2	33.5*	9.0	3.1		10.1			1699
	*Incl. PS										

*Unless stated otherwise; the PL column gives the total phospholipid content of each plant part in percent by weight, and the PC, PE, PG, PI, PS, PA, and DPG columns show the distribution in percent by weight of the total PL; [a]Percent of total lipids; [b]mg lipid P/100 mL juice; [c]Percent dwb; [d]mg PL/100 mg protein; [e]mg PL/g fwb; [f]mg PL/100 g dwb; [g]Percent of polar lipids; [h]μmol PL/mg protein; [i]μmol PL/mg chlorophyll; [j]μg PL/mg protein; [k]μg lipid P/g; [l]μmol PL/g fwb; [m]mol PL/mol chlorophyll; [n]mg PL/g dwb; [o]μmol PL/g dwb; [p]μg PL/100 mg fwb; [q]μg PL/mg chlorophyll; [r]mg PL/cm^2 leaf area; [s]nmol PL/cm^2 leaf area; [t]μmol PL/mg protein; [u]IR given; [v]mg lipid P/g lipid; [w]μg PL/mL; [x]μg lipid P/cm^2 leaf area; [y]Molecules per 10^6 plastids; [z]μg PL/mg chromatin; [aa]Fatty acid composition and *sn*-1/*sn*-2 distribution given for each phospholipid; [bb]μg/g fwb; [cc]Taken from the abstract.

Continued

Table 4-15 Phospholipids in Plants*—cont'd

No.	Plant part	Method	PC	PE	PG	PI	PS	PA	DPG	PL	References
Prunus mume Sieb. and Zucc. (Ume Apricot)											
1	Seed	TLC	+	+		+			(NAPE +)	0.245	1924
Prunus persica (L.) Batsch. (Peach)											
2	Whole seed	TLC	43.7-44.7	35.5-37.4			17.1-19.9			0.09-0.12	1925
2	–, endocarp	TLC	38.5-40.5	50.0-50.3		4.5-6.6	2.4-6.0			0.002-0.006	1925
1	Seed kernel	TLC	38.2	12.9	1.5	29.2	0.9	7.1	1.0	0.2	1919
			also LPC 5.0, LPE 1.8								
1	Leaf	TLC	27.0		73.0						1926
1	Root	TLC	46.6	36.3*	8.7	1.3		4.2	2.9		1699
	*Incl. PS										
Pseudobryum cinclidiodes (Hueb.) T. Kop.											
1	Whole plant	TLC	14.2	20.7	35.0	15.0	15.1			0.27[c]	1493
Psilotum triquetrum Swartz.											
1	Whole plant	TLC	+	+							1436
Psophocarpus tetragonolobus (L.) DC (Winged Bean)											
1	Seed	TLC	37	27	8	18	2		7	0.73	1927
4	Bean	TLC	35-43*	18-22 (incl. PG)		16-27		(NAPE 1-3)		1.18-1.51	1928
	*Incl. LPE										
Psoralea drupacea Bunge (Drupe Scurf Pea)											
1	Seed	TLC	35.3	23.4		21.8				1.1[c]	1929
			Also given: NAPE 7.2, NALPE 3.2, LPC 3.0								
Pteridium aquilinum (L.) Kuhn (Bracken Fern)											
1	Pinnae	TLC	72		9	14			4	1.58	1930
1	Upper rachis	TLC	61		15	15			10	0.61	1930
1	Petiole, upper	TLC	56		20	12			12	0.25	1930
1	–, lower	TLC	86		tr	14			Trace	0.21	1930
1	Rhizome, storage	TLC	58		17	8			17	0.12	1930
1	–, frond-bearing	TLC	50		Trace	Trace			50	0.02	1930

Punica granatum L. (Pomegranate)											
1	Seed	TLC	+	+		+		+	(LPC +, LPI +)	1.0	1931
1	Seed	TLC	+			+	+			2.9	1932
Pyrus communis L. (Pear)											
1	Seed	TLC	30.1	13.8	2.4	29.3	2.7	5.5	3.1	0.2	1639
			also LPC 3.8, LPE 3.1								
2	Fruit, flesh (1)	TLC	11.3	35.4	29.9	2.4	5.3	3.1	7.0	0.111	1933
	Fruit, flesh (2)	TLC	24.5	35.4	32.1	3.4	3.8			0.045	
			also LPC 2.2 (1), 0.5 (2); LPE 2.4 (1), 1.4 (2)								
2	–, peel (1)	TLC	16.7	39.6	29.2	2.8	4.2	2.1	3.5	0.099	1933
	–, peel (2)	TLC	22.3	43.7	27.2	1.9	3.8			0.206	
			also LPC 0.7 (1), 1.0 (2); LPE 1.4 (1), 1.5 (2)								
Pyrus pyrifolia ssp. *culta* (Mak.) Nakai (Nashi)											
4	Seed, kernel	TLC	27.9-30.3	40.3-42.3			24.8-25.6	0.3-0.8		(LPC 0.9-1.3)	1934
4	–, coat	TLC	23.4-24.5	45.3-47.2			25.2-26.8	0.3-0.6		(LPC 0.5-0.9)	1934
Quercus palustris Muenchh. (Pin Oak)											
1	Leaf	TLC	33.5	30.6	3.5	1.3			31.1	0.37	1935
Quercus robur L. (English Oak)											
1	Leaf	TLC	15.4	21.2	25.8	9.1	10.6		18.2	4.16[l]	1936
Ramonda nathaliae Panc. and Petrov											
1	Leaf	TLC	52.2	21.7						0.69[c]	1937
Ramonda serbica Panc.											
1	Leaf	TLC	51.0	21.6						0.51[c]	1937
Ranunculus glacialis L. (Glacier Buttercup)											
1	Leaf	TLC	41	18	34	6			0.9	0.13	1441
Raphanus acanthisformis Becker (Japanese Radish)											
1	Leaf	TLC					+				1450

*Unless stated otherwise; the PL column gives the total phospholipid content of each plant part in percent by weight, and the PC, PE, PG, PI, PS, PA, and DPG columns show the distribution in percent by weight of the total PL; [a]Percent of total lipids; [b]mg lipid P/100 mL juice; [c]Percent dwb; [d]mg PL/100 mg protein; [e]mg PL/g fwb; [f]mg PL/100 g dwb; [g]Percent of polar lipids; [h]µmol PL/mg protein; [i]µmol PL/mg chlorophyll; [j]µg PL/mg protein; [k]µg lipid P/g; [l]µmol PL/g fwb; [m]mol PL/mol chlorophyll; [n]mg PL/g dwb; [o]µmol PL/g dwb; [p]µg PL/100 mg fwb; [q]µg PL/mg chlorophyll; [r]mg PL/cm² leaf area; [s]nmol PL/cm² leaf area; [t]µmol PL/mg protein; [u]IR given; [v]mg lipid P/g lipid; [w]µg PL/mL; [x]µg lipid P/cm² leaf area; [y]Molecules per 10^6 plastids; [z]µg PL/mg chromatin; [aa]Fatty acid composition and *sn*-1/*sn*-2 distribution given for each phospholipid; [bb]µg/g fwb; [cc]Taken from the abstract.

Continued

Table 4-15 Phospholipids in Plants*—cont'd

No.	Plant part	Method	PC	PE	PG	PI	PS	PA	DPG	PL	References
Raphanus sativus L. (Radish)											
1	Seed	TLC						P_{NAPE}/P_{total} 0.5 mg/seed			1497
1	Seed	CC (silica)	15.7	72.5							1536
1	Leaf[a]	TLC	3.6	8.3	5.4	1.7					1938
1	Cotyledon, thylakoid[m]	TLC	118		467						1939
1	–, LHC(II) complex[m]	TLC	11		174						1939
Raphanus sativus L. ssp. *radicola* DC											
1	Leaf	TLC	24	16	36	14			9		1940
Raphia farinifera (Gaertn.) Hyl. (Raffia Palm)											
1	Seed	TLC	+	+		+					1941
1	Fruit, mesocarp	TLC	+			+					1941
1	Pollen	TLC	+	+		+					1941
Raphia hookeri Mann and Wendl. (Ivory Coast Raffia Palm)											
1	Seed	TLC	+	+		+					1941
1	Fruit, mesocarp	TLC	+			+					1941
1	Pollen	TLC	+	+		+					1941
Raphia regalis Becc.											
1	Seed	TLC	+	+		+					1941
1	Fruit, mesocarp	TLC	+			+					1941
1	Pollen	TLC	+	+		+					1941
Raphia sudanica Chev.											
1	Seed	TLC	+	+		+					1941
1	Fruit, mesocarp	TLC	+			+					1941
1	Pollen	TLC	+	+		+					1941
Raphia vinifera Beauv. (Wine palm)											
1	Seed	TLC	+	+		+					1941
1	Fruit, mesocarp	TLC	+			+					1941
1	Pollen	TLC	+	+		+					1941
Rauvolfia serpentina (L.) Benth. ex Kurz (Indian Snakeroot)											
1	Cell culture	TLC	55.0	20.1	2.0	10.2		2.5		1.53[e]	1495
Rhaponticum integrifolium C. Winkl.											
1	Seed, defatted	TLC	+	+		+				1.0	794
Rhododendron spp. (Rhododendron)											
1	Root[c,n]	TLC	0.34	+	+	+		0.45			1942

Ribes nigrum L. (Black Currant)											
1	Seed[e]	TLC	34.7	11.2	6.9	20.6	1.9	7.5	0	(LPC 8.3, LPE 6.4)	1487
Riccia fluitans L. (Crystalwort)											
1	Whole plant	TLC	33.3	16.1	26.3	6.2	15.7			1.73[c]	1551
Ricinus communis L. (Castor Bean)											
1	Seed	TLC	15.1	12.2		19.2	13.8	(LPC 3.1, LPE 3.8)		0.12[a]	1943
1	Seed, endosperm	TLC							(NAPE +)		1361
1	–, glyoxysome	TLC	51.4	27.2	2.7	9.0	1.5		2.3		1944
1	–, endoplastic reticulum	TLC	45.3	28.7	3.6	13.1	2.3		2.5	1.24[h]	1944
1	–, proplastid	TLC	35.6	38.7	8.4	8.0	4.6			0.55[h]	1944
1	–, mitochondria	TLC	33.1	35.7	2.2	3.9	2.2		17.7	0.65[h]	1944
1	–, spherosomes	TLC	30	26	25	5	5		(LPC 0, LPE 0)	0.38[a,c]	1945
1	Leaf[s]	TLC	18.4	10.1	14.9						1946
1	Leaf	TLC					+				1450
Robinia pseudoacacia L. (Black Locust)											
1	Leaf	TLC	40	24	28	8					1607
1	Bark	TLC	+	+				(PC/PE ratio 1.5)			1947
Roemeria refracta DC (Roemer Poppy)											
1	Seed, defatted	TLC	+	+		+				0.5	794
Rorippa amphibia (L.) Bess (Amphibious Marsh Cress)											
1	Whole plant	TLC	12.0	25.1	12.8	12.5	8.1	5.2	13.7	33.5	1789
Rosa canina L. (Dog Rose)											
1	Seed[e]	TLC	46.3	12.2	1.2	20.7	1.7	4.3	0.9		1487
1	Seed, defatted	TLC	+	+		+				0.2	794

*Unless stated otherwise; the PL column gives the total phospholipid content of each plant part in percent by weight, and the PC, PE, PG, PI, PS, PA, and DPG columns show the distribution in percent by weight of the total PL; [a]Percent of total lipids; [b]mg lipid P/100 mL juice; [c]Percent dwb; [d]mg PL/100 mg protein; [e]mg PL/g fwb; [f]mg PL/100 g dwb; [g]Percent of polar lipids; [h]μmol PL/mg protein; [i]μmol PL/mg chlorophyll; [j]μg PL/mg protein; [k]μg lipid P/g; [l]μmol PL/g fwb; [m]mol PL/mol chlorophyll; [n]mg PL/g dwb; [o]μmol PL/g dwb; [p]μg PL/100 mg fwb; [q]μg PL/mg chlorophyll; [r]mg PL/cm² leaf area; [s]nmol PL/cm² leaf area; [t]μmol PL/mg protein; [u]IR given; [v]mg lipid P/g lipid; [w]μg PL/mL; [x]μg lipid P/cm² leaf area; [y]Molecules per 10^6 plastids; [z]μg PL/mg chromatin; [aa]Fatty acid composition and *sn*-1/*sn*-2 distribution given for each phospholipid; [bb]μg/g fwb; [cc]Taken from the abstract.

Continued

Table 4-15 Phospholipids in Plants*—cont'd

No.	Plant part	Method	PC	PE	PG	PI	PS	PA	DPG	PL	References
Rosa rugosa Thunb. (Turkestan Rose)											
1	Hips, chromoplast tubules	TLC	27-40	19-20	6-9	24-28	17-19				1948
1	Leaf	TLC	48	18	19	9			6	3.12^{l}	1523
Rubus spp. (Blackberry)											
2	Callus culture	TLC	+	+	+	+					1949
1	Fruit	Frac. pptn.	0.128								1921
Rumex acetosa L. (Sorrel)											
1	Leaf	TLC	39.6	21.3	20.9	6.8	5.9	1.9	3.6	0.033	1761
Ruta graveolens L. (Rue)											
1	Plant, aerial parts	TLC	22.5	23.1	27.7	19.1	2.3	5.3	Trace	24.7^{a}	1950
1	Shoot, leaf tissue culture	TLC	12.8	30.1	32.9	16.5	2.1	5.5	Trace	17.7^{a}	1950
Saccharum officinarum L. (Sugarcane)											
1	Leaf	TLC	19.4	19.1	41.0	10.2	5.3	2.0		1.89^{c}	1951
1	Juice, raw	TLC	36.7	28.3						7.15^{w}	1952
1	Suspension culture	TLC	3.0	33.2	42.2	11.6	4.8	2.1		1.86^{c}	1951
Sagittaria sagittifolia L. (Swamp Potato)											
1	Whole plant	TLC	42.7	10.9	21.9	7.1	5.3	12.1		2.30^{c}	1445
Salicornia arabica L.											
1	Shoot	TLC	43.0	5.1	11.0	6.8		8.0	19.0	0.235^{c}	1488
Salicornia europaea L. (Samphire)											
1	Leaf	TLC	50.0	10.1	9.1					3.35^{c}	1494
1	Root	TLC	54.0	24.0	Trace					0.10^{c}	1494
Salix herbacea L. (Dwarf Willow)											
1	Leaf	TLC	44	17	31	7			1	0.44	1441
Salix reticulata L. (Netleaf Willow)											
1	Leaf	TLC	51	19	19	8			2	0.25	1441
Salix retusa L. (Notchleaf Willow)											
1	Leaf	TLC	52	21	18	7			1	0.28	1441
Salvadora oleoides Decne (Khakan)											
1	Seed	TLC	17.1	29.9		27.3			22.7		1953

Salvinia nutans (L.) Allioni											
1	Whole plant	TLC	28.0	18.6	13.2	10.6	8.0	7.2	14.4	18.4	1789
Sambucus nigra L. (European Elder)											
1	Leaf	PC	43	11	39	7					1517
Sapium sebiferum (L.) Roxb. (Vegetable Tallow Tree)											
1	Seed	TLC	45.3	12.9	Trace	16.7	7.7			0.7[c]	1954
			also NAPE 6.2, NALPE trace, LPC 5.3, LPE trace								
Saponaria vaccaria L. (Cow Cockle)											
1	Seed, whole	TLC	37	19		27	12	0.4		(LPC 2.3, LPE 1.5)	1955
1	–, dehulled	TLC	41	7		34	11	0.9		(LPC 2.6, LPE 5.0)	1955
1	–, starch	TLC	15	8				13		(LPC 52, LPE 11)	1956
Saussuria depressa (L.) DC											
1	Leaf	TLC	48	19	20	11			2	0.11	1441
Saxegothaea conspicua Lindl. (Prince Albert Yew)											
1	Needle	TLC	14	19	33	33				2.1[o]	1915
Secale cereale L. (Rye)											
4	Seed grain[a]	TLC	4-5	<3		<3			(NAPE +, LPC +)		1957
1	Seed nucleus, diploid	TLC	+	+						61.6[a]	1958
1	–, tetraploid	TLC	+	+						61.8[a]	1958
1	Seed embryo	TLC	34.8	35.3		29.8					1959
1	–, nuclear matrix	TLC	32.1	56.0		11.9					1960
1	Starch	TLC						(LPC 51.9)		0.54[d]	1499
2	Leaf	TLC	25.4-36.5	12.3-12.9	48.6-25.2	10.2-17.0			3.5-8.3	11.4-14.0[a]	1961
1	–, thylakoid	TLC			11						1440
1	–, plasma membrane	TLC	46.7	34.5	5.8	2.2	4.5	5.4		1.07[h]	1962
1	–, LHC (II) complex[a]	TLC	7	2	33						1963
1	Leaf, chloroplast envelope, inner membrane	TLC	8.1	0.2	4.0	0.7 (incl. PS)		1.3			1964

*Unless stated otherwise; the PL column gives the total phospholipid content of each plant part in percent by weight, and the PC, PE, PG, PI, PS, PA, and DPG columns show the distribution in percent by weight of the total PL; [a]Percent of total lipids; [b]mg lipid P/100 mL juice; [c]Percent dwb; [d]mg PL/100 mg protein; [e]mg PL/g fwb; [f]mg PL/100 g dwb; [g]Percent of polar lipids; [h]μmol PL/mg protein; [i]μmol PL/mg chlorophyll; [j]μg PL/mg protein; [k]μg lipid P/g; [l]μmol PL/g fwb; [m]mol PL/mol chlorophyll; [n]mg PL/g dwb; [o]μmol PL/g dwb; [p]μg PL/100 mg fwb; [q]μg PL/mg chlorophyll; [r]mg PL/cm^2 leaf area; [s]nmol PL/cm^2 leaf area; [t]μmol PL/mg protein; [u]IR given; [v]mg lipid P/g lipid; [w]μg PL/mL; [x]μg lipid P/cm^2 leaf area; [y]Molecules per 10^6 plastids; [z]μg PL/mg chromatin; [aa]Fatty acid composition and *sn*-1/*sn*-2 distribution given for each phospholipid; [bb]μg/g fwb; [cc]Taken from the abstract.

Continued

Table 4-15 Phospholipids in Plants*—cont'd

No.	Plant part	Method	PC	PE	PG	PI	PS	PA	DPG	PL	References
1	–, outer membrane	TLC	31.5	0.7	5.0	4.5 (incl. PS)		1.5			1964
2	Shoot, tillering node	TLC	26.7-37.2	15.3-22.5	23.2-15.4	11.6-8.3			23.3-16.6	30.2-23.6[a]	1961
1	Shoot, plasma membrane[j]	HPLC	73.5	109.7							1505
1	Root	TLC	60	23	6	5		3	3	0.24	1965
1	Seedling	PC	56.5	16.7			21.1			0.48	1355
Selaginella kraussiana (Kunze) A. Braun											
1	Whole plant	TLC	+	+							1436
Selaginella pallescens (C. Presl.) Spring (Moss Fern)											
1	whole plant	TLC	+	+							1436
Selaginella selaginoides (L.) P. Beauv. (Club Spikemoss)											
1	Whole plant	TLC	+	+							1436
Selaginella uncinata (Desv. ex Poir.) Spring (Blue Spikemoss)											
1	Whole plant	TLC	+	+							1436
Selaginella willdenowii (Desv. ex Poir.) Baker											
1	''''Whole plant	TLC	+	+							1436
Senecio vulgaris L. (Groundsel)											
1	Leaf, thylakoid[a]	TLC			11.5						1913
1	–, chloroplast	HPLC	36.0	36.8	8.6	7.1		11.5		77.1[a]	1465
Senna septentrionalis (Viv.) Irwin and Barn. (Arsenic Bush)											
1	Leaf[l]	TLC			0.33						1402
Sequoiadendron giganteum (Lindl.) Buchh. (Giant Sequoia)											
1	Leaf	TLC	+	+							1436
Sesamum indicum L. (Sesame)											
1	Seed	TLC	14.0	57.5	7.3	11.9		4.1	5.0	41.54	1446
3	Seed[n]	TLC	13.9-16.1	3.7-5.9		5.7-6.8					1966
1	Seed, oil body	TLC	41.2	15.8		20.9	22.1			0.57	1482
Setaria glauca (L.) Beauv. (Yellow Foxtail)											
1	Leaf[l]	TLC			0.72						1402
Setaria italica (L.) Beauv. (Italian Foxtail Millet)											
1	Seed, glutinous	TLC	+	+	+	+			+ (LPC +)	0.08	1967

Setaria verticillata (L.) Beauv. (Bristly Foxtail)										
1	Leaf[l]	TLC			0.79					1402
Sicyos angulatus L. (Bur Cucumber)										
2	Leaf	TLC			4.17-4.54[a]				3.66-3.75[l]	1628
Simmondsia chinensis (Link) Schneid. (Jojoba)										
1	Seed	TLC	45	38	7	10			0.16	1968
1	Leaf, chloroplast	TLC	29		50	21				1969
1	–, tissue	TLC	58	20	14	2				1970
1	–, protoplast	TLC	24	13	10	13				1970
1	Callus, tissue	TLC	41	25	3	21				1970
1	–, protoplast	TLC		27	5	32				1970
1	Explant, shoot	TLC	19[a]						1.59[e]	1971
1	–, callus	TLC	20[a]						0.82[e]	1971
Sison amomum L. (Stone Parsley)										
1	Seed	TLC	21	4	2	33	22	2	1.1[a]	1438
			also LPC 6, LPE 1							
Smilax spp. (Greenbrier)										
1	Leaf	TLC			7					1440
Solanum acaule Bitt. (Wild Potato)										
1	Leaf[l]	TLC	1.28	1.06 (incl. PG)		0.13				1972
Solanum acaule Bitt. ssp. *depexum* Buk.										
1	Leaf	TLC	+	+	+	+				1786
Solanum acaule Bitt. ssp. *punae* Juz.										
1	Leaf[l]	TLC	1.30	1.09 (incl. PG)		0.10				1972
Solanum acaule Bitt. ssp. *schreiteri* Buk.										
1	Leaf[l]	TLC	1.23	0.92 (incl. PG)		0.15				1972
Solanum argentinum Bitter and Lillo										
1	Seed	TLC	17.4	20.2			(Sphingo-PL 20.5)			1973
Solanum commersonii Dun. (Commerson's Nightshade)										
1	Leaf, plasma membrane	TLC	42.6	36.2	3.7	5 (incl. LPE)	9.5	(LPC 2.6)	48.3[a]	1974

*Unless stated otherwise; the PL column gives the total phospholipid content of each plant part in percent by weight, and the PC, PE, PG, PI, PS, PA, and DPG columns show the distribution in percent by weight of the total PL; [a]Percent of total lipids; [b]mg lipid P/100 mL juice; [c]Percent dwb; [d]mg PL/100 mg protein; [e]mg PL/g fwb; [f]mg PL/100 g dwb; [g]Percent of polar lipids; [h]μmol PL/mg protein; [i]μmol PL/mg chlorophyll; [j]μg PL/mg protein; [k]μg lipid P/g; [l]μmol PL/g fwb; [m]mol PL/mol chlorophyll; [n]mg PL/g dwb; [o]μmol PL/g dwb; [p]μg PL/100 mg fwb; [q]μg PL/mg chlorophyll; [r]mg PL/cm^2 leaf area; [s]nmol PL/cm^2 leaf area; [t]μmol PL/mg protein; [u]IR given; [v]mg lipid P/g lipid; [w]μg PL/mL; [x]μg lipid P/cm^2 leaf area; [y]Molecules per 10^6 plastids; [z]μg PL/mg chromatin; [aa]Fatty acid composition and *sn*-1/*sn*-2 distribution given for each phospholipid; [bb]μg/g fwb; [cc]Taken from the abstract.

Continued

Table 4-15 Phospholipids in Plants*—cont'd

No.	Plant part	Method	PC	PE	PG	PI	PS	PA	DPG	PL	References
Solanum demissum Lindl. (Nightshade)											
1	Leaf[l]	TLC	1.21	1.16 (incl. PG)		0.09					1972
Solanum maglia Schlecht											
1	Leaf	TLC	9.1	10.9	79.9	0.1				0.206	1975
Solanum melongena L. (Eggplant)											
1	Seed	HPLC	29.3	19.1		10.5		41.0		28.8[k]	1976
1	Leaf[l]	TLC	0.635								1514
1	Fruit[l]	TLC	0.092								1514
1	Fruit	TLC	0.3	0.9	6.6		(LPC 0.2)	0.6	90.4	0.205	1446
Solanum pamitanae											
1	Leaf[l]	TLC	0.46	0.77 (incl. PG)		0.11					1972
Solanum rybinii Juz. et Buk.											
1	Leaf[l]	TLC	1.28	1.04 (incl. PG)		0.11					1786,1972
Solanum sanctae-rosae Hawkes											
1	Leaf[l]	TLC	0.54	0.60 (incl. PG)		0.12					1972
Solanum schreiteri Buk. (Wild Potato)											
1	Leaf, chloroplast	TLC	29.7		39.9	9.1		21.3		0.026[l]	1977
Solanum tuberosum L. (Potato)											
1	Leaf	TLC	50.2	13.4	21.7	10.5	1.1	1.4	1.8	2.77[l]	1978
1	Leaf[a]	HPLC	7.9	4.3	2.9	0.6	0.6	0	0		1413
1	–, plasma membrane	TLC	38.4	35.7	3.9		7.5 (incl. LPE)	10.2	(LPC 4.3)	46.4[a]	1974
1	Tuber, peeled[a]	TLC	25.8	12.6	0.9	6.3	0.6	0.3	0.6	47.4	1979
1	Tuber	TLC	41.5	32.1		26.4				0.046	1980
4	Tuber, free	TLC	45.4-46.9	35.4-36.1		17.8-18.5				5.2-8.6[a]	1981
4	–, bound	TLC	23.8-29.3	42.7-44.9	11.1-12.3	14.7-16.3				32.7-34.7[a]	1981
1	Tuber[a]	HPLC	15.9	33.6	0.4	2.5	1.4	0	0		1413
1	Tuber[l]	Choline	14								1451
1	Tuber, mitochondria	TLC	44.3	26.3		12.7			16.8	0.0277	1982
1	–, inner membrane	TLC	27	29		24.5			19.5		1982
1	–, outer membrane	TLC	52.6	25.0		10.3			12.1		1982
1	–, microsomes	TLC	44.3	17.6		19.1			19.0	0.0292	1982,1983

1	Tuber, cell wall	TLC	35.1	33.0		31.9				0.0009	1984
1	–, peroxisome	TLC	52.5	20.9		7.2			2.9	0.0014	1984
1	–, plasmalemma	TLC	60.4	22.9		10.3			6.3	0.0041	1984
1	–, starch	TLC	73.0	15.3		11.6				0.0001	1984
1	Tuber, plasmalemma	TLC	33.1	39.3		14.6	2.7	5.0	2.1	(LPC 3.2)	1985
1	Starch	TLC	+	0			+			(LPC 0, LPE 0, LPS 0)	1902
1	Starch	TLC	+		+					0.006	1986
Sophora japonica L. (Japanese Pagoda Tree)											
1	Seed	TLC	38.5	15.1	5.9	20.1	0	1.2		1.1	1987,1988
			also NAPE 6.8, NALPE 2.1, LPC 2.3								
1	Bud	TLC	20.3	14.0	13.9	17.0	0	16.0		0.28	1987
			also NAPE 9.1, NALPE 6.8, LPC 2.9								
1	Flower	TLC	22.0	19.2	8.2	19.2	0	15.1		0.6	1987
			also NAPE 7.3, NALPE 3.0, LPC 4.0								
Sorbus aucuparia L. (Rowan)											
1	Leaf	TLC	40	31	18	7			5	5.54[l]	1523
Sorghum bicolor (L.) Moench. (Sorghum)											
1	Seed	TLC	36.3	5.5	1.2	Trace	Trace	1.6		0.93[c]	1989
			also LPC 41.8, LPE 13.6								
1	Leaf, chloroplast	TLC	26	4	48	13	4		4		1990
1	Root	TLC	46.8	30.2*	10.0	2.1		5.0	5.2		1699
	*Incl. PS										
Sorghum bicolor L. var. *cernuum* Host. (White Durra)											
1	Grain	TLC		+		+	+			0.59	1991
Spartina patens (Ait.) Muhl. (Saltmeadow Cordgrass)											
1	Root plasma membrane[a]	TLC	8.5	6.1	2.2	1.1	1.0			20.5[c]	1992
Sphagnum nemoreum Scop.											
1	Whole plant	TLC	14.9	26.2	35.4	12.1	11.2			0.22[c]	1493
Sphagnum palustre L. (Prairie Sphagnum)											
1	Whole plant	TLC	+	+							1436

*Unless stated otherwise; the PL column gives the total phospholipid content of each plant part in percent by weight, and the PC, PE, PG, PI, PS, PA, and DPG columns show the distribution in percent by weight of the total PL; [a]Percent of total lipids; [b]mg lipid P/100 mL juice; [c]Percent dwb; [d]mg PL/100 mg protein; [e]mg PL/g fwb; [f]mg PL/100 g dwb; [g]Percent of polar lipids; [h]μmol PL/mg protein; [i]μmol PL/mg chlorophyll; [j]μg PL/mg protein; [k]μg lipid P/g; [l]μmol PL/g fwb; [m]mol PL/mol chlorophyll; [n]mg PL/g dwb; [o]μmol PL/g dwb; [p]μg PL/100 mg fwb; [q]μg PL/mg chlorophyll; [r]mg PL/cm^2 leaf area; [s]nmol PL/cm^2 leaf area; [t]μmol PL/mg protein; [u]IR given; [v]mg lipid P/g lipid; [w]μg PL/mL; [x]μg lipid P/cm^2 leaf area; [y]Molecules per 10^6 plastids; [z]μg PL/mg chromatin; [aa]Fatty acid composition and *sn*-1/*sn*-2 distribution given for each phospholipid; [bb]μg/g fwb; [cc]Taken from the abstract.

Continued

Table 4-15 Phospholipids in Plants*—cont'd

No.	Plant part	Method	PC	PE	PG	PI	PS	PA	DPG	PL	References
Sphagnum squarrosum Crome (Sphagnum)											
1	Whole plant	TLC	11.3	32.7	34.7	8.1	13.2			0.06[c]	1493
Spinacia oleracea L. (Spinach)											
1	Leaf	TLC	28.9	14.2	39.1	3.9	3.1	4.8	6.0	0.105	1761
1	Leaf[a]	HPLC	14.0	8.6	7.4	2.9	0	0	1.7		1413
3	Leaf[l]	Choline	66-141								1451
1	Leaf	PC	56	17	16	11					1517
1	–, chloroplast	PC	39	2	46	13					1517
1	–, chloroplast	TLC	28	67	6						1969
1	–, plasma membrane	TLC	25.4	24.4	3.1			6.1			1993
1	–, mitochondria	TLC	45	39	4	1			11		1994
1	–, lamellae	TLC	24	0	65	11				14,200[q]	1995
1	–, lamellae[f]	CC (DEAE)	55	0	157	28					1379
1	–, PS (II) fraction[g]	TLC	3		9						1996
1	–, thylakoid[d]	TLC	0		7	2				49[f]	1997
1	Leaf chloroplast envelope[g]	TLC	18	0	9	2			0		1998
1	–, inner membrane[d]	TLC	Trace		7	3				107[f]	1997
1	–, outer membrane[g]	TLC	32	0	10	5			0		1998
1	Leaf, inner envelope	HPLC			+						1999
1	–, outer envelope	HPLC			+						1999
1	–, thylakoid	HPLC			+						1999
1	Petiole, mitochondria	TLC	39	42	2	2			12		1994
1	Root	TLC	62.4	32.1	Trace					1.14[c]	1494
1	Root	TLC	33.7	16.3			49.5			9.5[g]	2000
Spirodela oligorrhiza (Kurz) Hegelm. (Duckweed)											
1	Whole tissue	TLC	42.4	20.3	14.9	6.8	1.5	0.7		3.22[l]	1389
1	Leaf, chloroplast	TLC	34.3	5.3	36.9	9.4	1.2	1.6			1389
Spirodela polyrrhiza (L.) Schleid. (Giant Duckweed)											
1	Frond	TLC	56	21	18	6					2001

1	Leaf, chloroplast	TLC	20		80						2002
Sterculia foetida L. (Indian Almond, Jangli Badam)											
1	Nut, kernel	TLC	30.0	23.0		40.6				0.2	1548
Striga asiatica (L.) Kuntze (Witchweed)											
1	Seed	TLC	12	17	29	27					2003
Striga aspera (Willd.) Benth.											
1	Leaf, chloroplast[g]	TLC	10		30						2004
Striga hermonthica (Del.) Benth. (Purple Witchweed)											
1	Leaf, chloroplast[g]	TLC	10		32						2004
Suaeda fruticosa Forsk. (Alkali Seepweed)											
1	Shoot	TLC	54.3	15.0	8.7	3.9		9.4	4.7	0.127[c]	1488
Suaeda maritima (L.) Dum.											
1	Leaf, protoplast and tonoplast	TLC	+	+		+	+				2005
Suaeda maritima ssp. *macrocarpa* (Annual Seablite)											
1	Leaf	TLC	59.0	20.9	16.1	4.1					2006
1	–, chloroplast memb.	TLC	39.9		48.0	12.1				22.3[q]	2007
Swietenia mahogoni Jacq. (West Indian Mahogany)											
1	Seed	TLC	31.5	28.9	5.6	28.2			5.8	3.44[c]	1429
Tamarindus indica L. (Tamarind)											
1	Seed, kernel	TLC	++	++		+	+	+		(LPC +, LPE +)	2008
Tapinanthus dodoneifolius (DC) Danser (African Mistletoe)											
1	Leaf chloroplast[y]	TLC	119		+	+				30	2009
Taxus baccata L. (English Yew)											
1	Leaf	TLC	+	+							1436
Taxus cuspidata Sieb. and Zucc. (Japanese Yew)											
1	Seed	TLC	+	+			(LPC +)				2010
Tetragonia expansa Thunb. (New Zealand Spinach)											
1	Leaf	CC (silica)	28.6							1.16[l]	1459

*Unless stated otherwise; the PL column gives the total phospholipid content of each plant part in percent by weight, and the PC, PE, PG, PI, PS, PA, and DPG columns show the distribution in percent by weight of the total PL; [a]Percent of total lipids; [b]mg lipid P/100 mL juice; [c]Percent dwb; [d]mg PL/100 mg protein; [e]mg PL/g fwb; [f]mg PL/100 g dwb; [g]Percent of polar lipids; [h]μmol PL/mg protein; [i]μmol PL/mg chlorophyll; [j]μg PL/mg protein; [k]μg lipid P/g; [l]μmol PL/g fwb; [m]mol PL/mol chlorophyll; [n]mg PL/g dwb; [o]μmol PL/g dwb; [p]μg PL/100 mg fwb; [q]μg PL/mg chlorophyll; [r]mg PL/cm^2 leaf area; [s]nmol PL/cm^2 leaf area; [t]μmol PL/mg protein; [u]IR given; [v]mg lipid P/g lipid; [w]μg PL/mL; [x]μg lipid P/cm^2 leaf area; [y]Molecules per 10^6 plastids; [z]μg PL/mg chromatin; [aa]Fatty acid composition and *sn*-1/*sn*-2 distribution given for each phospholipid; [bb]μg/g fwb; [cc]Taken from the abstract.

Continued

Table 4-15 Phospholipids in Plants*—cont'd

No.	Plant part	Method	PC	PE	PG	PI	PS	PA	DPG	PL	References
1	Leaf	PC	+	+		+	+				1816
1	Leaf	PC	59		29	12					1746
1	–, chloroplast	PC	37	6	34	21					1746
Theobroma cacao L. (Cocoa)											
3	Bean, shelled	TLC	36.1-40.4	11.3-18.6		25.5-29.0	2.3-3.3	0.6-3.7	(LPC 4.5-9.0)	0.28-0.45	2011
1	Butter *Incl. LPE	TLC	37.7 also LPC 8.0, LPE +	15.6	27.3	10.9*	0.3	0.2		0.34^{c}	2012
Thevetia neriifolia [Juss. ex] Steud. (Yellow Oleander, Pivala Kanher)											
1	Seed	TLC	17.2	27.2		30.1			24.0		1550
Thuidium tamariscinum (Hedw.) Schimp. (Tamarisk Thuidium Moss)											
1	Whole plant	TLC	+	+							1436
Torreya nucifera Sieb. and Zucc. (Kaya Nut)											
1	Seed	TLC	+	+			(LPC +)				2010
Trichodesma incanum (Bunge) A. DC											
1	Seed, defatted	TLC	+	+		+				0.1	794
Trichopilia tortilis Lindl.											
1	Leaf[l]	TLC			0.30						1402
Trifolium alexandrinum L. (Berseem Clover)											
1	Seed	TLC		+	+	+		+	(LPC +, LPE +)		2013
Trifolium pratense L. (Red Clover)											
1	Seed	TLC	20.1	14.6		9.3	7.2	7.0	8.9		2014
1	Leaf	CC (DEAE)	37	15	23	2				0.22	2015
Trifolium repens L. (White Clover)											
1	Leaf	TLC	36	22	29	6			7	3.94^{l}	1523
1	Root	TLC	10.0	44.0	19.2	2.7		24.1		37.0^{e}	1448
Trifolium spp. (Hartshorn Clover)											
1	Seed	TLC	65	13		22				0.46	1481
Trigonella corniculata (L.) L.											
1	Seed	TLC	29.8	24.5	2.5	23.5			19.7	0.13	1681
Trigonella foenum-graecum L. (Fenugreek)											
1	Seed	TLC	54.6	29.5	2.6	3.2	4.1	2.2	(LPC 3.8)	0.79^{c}	2016

1	Leaf	TLC	50.0	24.0	17.4	+	+			21.2[a]	2017
1	Hypocotyl	TLC	+	+	+	+	+				2017
Triticum aegilopoides Bal. (Wild Einkorn)											
2	Leaf, chloroplast[m] (1)	TLC	0.07	0.26							1437
	Leaf, chloroplast[m] (2)	TLC	0.08	0.24							
Triticum aestivum L. (Wheat)											
1	Whole grain	HPLC	26.2	19.7		Trace	Trace	31.3		0.47	2018
			also given: LPC 26.2, LPE 5.6								
14	Seed *Incl. a glycolipid	TLC	10.6-12.6	1.9-3.2			1.7-2.4	20.9-26.8*		0.68-0.80[g]	2019
			also LPC 2.0-3.4								
1	Flour	TLC	25.4	3.5		0.6	0.8			0.032	2020
			also NAPE 21.8, NALPE 12.8, LPC 29.6, LPE 4.0								
1	Flour	CC (silica)	+			(NAPE +, NALPE +, LPC +)					2021
7	Flour	TLC	24.4-55.2	Trace-34.2			Trace-14.7			0.06-0.13[c]	2022
1	Flour	TLC	+	+		(LPC +, LPE +)					2023
1	–, nonstarch	TLC	29	6		+					1392
			also NAPE 31, NALPE 15, LPC 16, LPE 4								
1	Flour, nonstarch	HPLC	6.2	31.2	17.9				(NAPE 16.1)		1417
1	–, starch	TLC	5	1		+			(LPC 86, LPE 8)		1392
2	–, starch	TLC	Variation of LPL content with granule size								2024
1	Starch	HPLC	1.1	0	0						1417
			also NAPE 0, NALPE 0, LPC 87.3, LPE 11.5, LPG 0.1								
6	Starch[cc]	HPLC			(LPC +)						2025
1	Seed	TLC	45.4	11.6 (incl. LPA)		+	3.3	Trace		146[k]	2026
			also NAPE 11.0, NALPE 13.2, LPC 14.0, LPE 1.6								

*Unless stated otherwise; the PL column gives the total phospholipid content of each plant part in percent by weight, and the PC, PE, PG, PI, PS, PA, and DPG columns show the distribution in percent by weight of the total PL; [a]Percent of total lipids; [b]mg lipid P/100 mL juice; [c]Percent dwb; [d]mg PL/100 mg protein; [e]mg PL/g fwb; [f]mg PL/100 g dwb; [g]Percent of polar lipids; [h]µmol PL/mg protein; [i]µmol PL/mg chlorophyll; [j]µg PL/mg protein; [k]µg lipid P/g; [l]µmol PL/g fwb; [m]mol PL/mol chlorophyll; [n]mg PL/g dwb; [o]µmol PL/g dwb; [p]µg PL/100 mg fwb; [q]µg PL/mg chlorophyll; [r]mg PL/cm^2 leaf area; [s]nmol PL/cm^2 leaf area; [t]µmol PL/mg protein; [u]IR given; [v]mg lipid P/g lipid; [w]µg PL/mL; [x]µg lipid P/cm^2 leaf area; [y]Molecules per 10^6 plastids; [z]µg PL/mg chromatin; [aa]Fatty acid composition and *sn*-1/*sn*-2 distribution given for each phospholipid; [bb]µg/g fwb; [cc]Taken from the abstract.

Continued

Table 4-15 Phospholipids in Plants*—cont'd

No.	Plant part	Method	PC	PE	PG	PI	PS	PA	DPG	PL	References
3	–, kernel	TLC	7.3-18.9	2.9-5.2 (incl. PG)		0-7.2		0.9-18.5		0.87-1.17	2027
			also LPC 38.5-71.3, LPE 5.8-8.0, LPG 2.3-19.4, LPI-LPS 0-3.4								
3	–, germ	TLC	44.4-60.4	15.7-16.2 (incl. PG)		15.0-18.9			1.3-4.1	0.11-0.15	2027
			also LPC 2.4-6.2, LPE 2.4-2.8, LPI/LPS 0-8.8								
3	–, aleurone	TLC	50.2-65.4	7.8-15.8 (incl. PG)		13.9-18.5			0.6-5.8	0.11-0.13	2027
			also LPC 6.3-13.2, LPI/LPS 0.5-2.8								
3	–, nonstarch endosperm	TLC	12.5-22.2	1.4-2.6 (incl. PG)						0.04-0.06	2027
			also LPC 19.6-22.4, LPE 2.2-2.5, LPI/LPS 0.9-2.7								
3	–, starch endosperm	TLC	0-0.5					0-2.4		0.43-0.55	2027
			also LPC 66.0-84.6, LPE 10.9-13.2, LPG 3.1-7.1, LPI-LPS 0.6-11.2								
1	–, spherosomes (from aleurone layer)	TLC	67.4	16.8		14.3				8.9[c]	2028
1	Seed, chromatin	TLC	48	18		9	9	(TLC 8)	7	14.6[z]	2029
1	Grain, lipoprotein complex	TLC	41.8	4.2	2.3	7.7		1.8		0.47	2030
			also NAPE 11.6, LPC 15.7, LPE 0.5								
1	Leaf	TLC	48.8	18.4	27.8	5.0				1.54[m]	1579
1	Leaf	TLC	23.6	19.3	43.2	12.1	0.5		1.2	160.3[j]	2031
6	Leaf	TLC			13-16						1440
11	Leaf	TLC	20.5-32.1	21.3-26.7						1.30-2.91	2032
2	–, thylakoid	TLC			18						1440
1	–, chloroplast	TLC	29	5	48	10	5		5		1990
1	–, envelope	TLC	57.2	7.2	25.0	9.2				15.2[a]	2033
1	–, thylakoid	TLC	24.8	0	57.4	14.0				12.9[a]	2033

1	Leaf, stroma lamellae[a]	TLC	2.4		10						2034
1	–, grana[a]	TLC	1.4		8.6						2034
1	Tillering node	TLC	44.9	16.9	17.6	6.2	4.9		9.5	121.1[j]	2031
1	Leaf, protein concentrate[g]	TLC	0	3.5						3.5[g]	2035
1	Bran *Incl. LPE	TLC	28.4	5.2	2.2	11.2	4.6*	4.0	1.7		1393
			also NAPE 5.9, NALPE 13.6, LPC 30.4								
1	Starchy endosperm *Incl. LPE	TLC	2.9	0.5	0	1.0	8.6*	0	0		1393
			also NAPE 7.8, NALPE 6.4, LPC 66.6								
1	Seedling	PC	51.2	15.7			24.1			0.53	1355
1	Root *Incl. PS	TLC	46.9	33.2*	6.4	2.1		5.3	5.6		1699
1	Root	HPLC	51	35	4	6	4 (incl. PA)			0.827[e]	1407
1	–, microsomal membrane[a]	TLC	40.4	14.6	4.1	1.4 (incl. PS)		1.6			2036
1	–, plasma membrane[a]	TLC	42.1	27.2	6.6	10.8	2.4	5.2			2037
1	Root, whole[l]	Choline	0.23							0.98	1506
			Data also given for cell wall, mitochondria, and microsomes								
1	Tillering nodes	TLC	30.9	13.8	5.6	8.1	3.0	11.7	3.2		2038
1	–, mitochondria	TLC	20.9	15.3	1.9	3.6			1.0	7.03	2039
T. aestivum L. × *Secale cereale* L. (Triticale)											
1	Seed	TLC	64.4	16.4		5.2	14.2		(LPC trace)		2040

*Unless stated otherwise; the PL column gives the total phospholipid content of each plant part in percent by weight, and the PC, PE, PG, PI, PS, PA, and DPG columns show the distribution in percent by weight of the total PL; [a]Percent of total lipids; [b]mg lipid P/100 mL juice; [c]Percent dwb; [d]mg PL/100 mg protein; [e]mg PL/g fwb; [f]mg PL/100 g dwb; [g]Percent of polar lipids; [h]μmol PL/mg protein; [i]μmol PL/mg chlorophyll; [j]μg PL/mg protein; [k]μg lipid P/g; [l]μmol PL/g fwb; [m]mol PL/mol chlorophyll; [n]mg PL/g dwb; [o]μmol PL/g dwb; [p]μg PL/100 mg fwb; [q]μg PL/mg chlorophyll; [r]mg PL/cm^2 leaf area; [s]nmol PL/cm^2 leaf area; [t]μmol PL/mg protein; [u]IR given; [v]mg lipid P/g lipid; [w]μg PL/mL; [x]μg lipid P/cm^2 leaf area; [y]Molecules per 10^6 plastids; [z]μg PL/mg chromatin; [aa]Fatty acid composition and *sn*-1/*sn*-2 distribution given for each phospholipid; [bb]μg/g fwb; [cc]Taken from the abstract.

Continued

Table 4-15 Phospholipids in Plants*—cont'd

No.	Plant part	Method	PC	PE	PG	PI	PS	PA	DPG	PL	References
Triticum compactum Host. (Club Wheat)											
1	Seed	TLC	9.5	2.8			2.5	21.2*	(LPC 3.0)	0.75[g]	1864
	*Incl. a glycolipid										
Triticum dicoccoides Körn. (Wild Emmer)											
2	Leaf, chloroplast[m]	TLC	0.08		0.21						1437
Triticum dicoccum Schrank (Emmer)											
2	Leaf, chloroplast[m] (1)	TLC	0.10		0.23						1437
	Leaf, chloroplast[m] (2)	TLC	0.06		0.23						
Triticum durum Desf. (Durum Wheat)											
1	Whole grain	HPLC	21.5	25.2		Trace	Trace	28.0		0.47	2018
			also LPC 21.5, LPE 11.2								
1	Seed, kernel	TLC	17.0	3.4 (incl. PG)		6.1			1.8	1.13	2027
			also LPC 53.4, LPE 6.2, LPG 2.6								
1	–, germ	TLC	55.5	15.0	1.2	13.8		(LPC 11.5)	3.0	0.11	2027
1	–, aleurone	TLC	56.2	12.6 (incl. PG)		16.8	(LPC 3.8)	1.7	2.3	0.16	2027
1	–, nonstarchy endosperm	TLC	6.5	11.4 (incl. PG)				3.2	2.6	0.05	2027
			also LPC 35.1, LPE 0.8								
1	–, starchy endosperm	TLC	0.6			1.0		2.4		0.63	2027
			also LPC 82.5, LPE 7.9, LPG 4.1, LPI-LPS 1.5								
2	Leaf	TLC	23.9-33.0	21.9-24.9						1.55-2.21	2032
2	–, chloroplast[m] (1)	TLC	0.08	0.20							
	–, chloroplast (2)	TLC	0.13	0.22							1437
Triticum macha Dek. and Men. (Macha Wheat)											
2	Leaf, chloroplast[m] (1)	TLC	0.05		0.23						1437

	Leaf, chloroplast[m] (2)	TLC	0.08		0.20						
Triticum monococcum L. (Einkorn)											
2	Leaf, chloroplast[m] (1)	TLC	0.10		0.23						1437
	Leaf, chloroplast[m] (2)	TLC	0.07		0.24						
1	Cell culture	TLC	50.7	35.9	3.3	6.0		4.1			2041
Triticum spelta L. (Spelt)											
2	Leaf, chloroplast[m]	TLC	0.07-0.10		0.22-0.28						1437
Tulipa spp. (Tulip)											
1	Leaf	TLC			13						1440
1	Tuber	TLC	37.1	44.2	2.6			7.0	8.9	0.178	1446
Tussilago farfara L. (Coltsfoot)											
1	Leaf	TLC	48.1	19.5	13.0	9.1	0	10.4		1.54[x]	2042
Typha latifolia L. (Cattail)											
1	Pollen	TLC	+	+	+	+		+	+	4.5[c]	2043
1	Pollen	TLC	63.2	18.3	7.0	7.8 (incl. PS)		1.0	2.7		2044
1	Microsomal membrane	TLC	57.9	17.9	6.6	6.3 (incl. PS)		3.5	5.1		2044
			also LPC 1.8, LPE 1.0								
Urtica dioica L. (Stinging Nettle)											
1	Leaf and stem	HPLC	+	+		+			(LPC +)		2045
1	Root	HPLC	+	+		+			(LPC +)		2045
Utricularia vulgaris L. (Bladderwort)											
1	whole plant	TLC	47.4	20.4	11.0	9.0	3.4		8.8	4.92[c]	1555

*Unless stated otherwise; the PL column gives the total phospholipid content of each plant part in percent by weight, and the PC, PE, PG, PI, PS, PA, and DPG columns show the distribution in percent by weight of the total PL; [a]Percent of total lipids; [b]mg lipid P/100 mL juice; [c]Percent dwb; [d]mg PL/100 mg protein; [e]mg PL/g fwb; [f]mg PL/100 g dwb; [g]Percent of polar lipids; [h]μmol PL/mg protein; [i]μmol PL/mg chlorophyll; [j]μg PL/mg protein; [k]μg lipid P/g; [l]μmol PL/g fwb; [m]mol PL/mol chlorophyll; [n]mg PL/g dwb; [o]μmol PL/g dwb; [p]μg PL/100 mg fwb; [q]μg PL/mg chlorophyll; [r]mg PL/cm^2 leaf area; [s]nmol PL/cm^2 leaf area; [t]μmol PL/mg protein; [u]IR given; [v]mg lipid P/g lipid; [w]μg PL/mL; [x]μg lipid P/cm^2 leaf area; [y]Molecules per 10^6 plastids; [z]μg PL/mg chromatin; [aa]Fatty acid composition and *sn*-1/*sn*-2 distribution given for each phospholipid; [bb]μg/g fwb; [cc]Taken from the abstract.

Continued

Table 4-15 Phospholipids in Plants*—cont'd

No.	Plant part	Method	PC	PE	PG	PI	PS	PA	DPG	PL	References
Vaccinium macrocarpon Ait. (Cranberry)											
1	Seed	TLC	90	5			5		(LPC trace)	0.26[c]	2046
Vaccinium myrtillus L. (Whortleberry)											
1	Fruit		0.086								1527
Vallisneria gigantea Graebn. (Eelgrass)											
1	Whole plant	TLC	42.5	25.9	12.8	6.8	7.0		5.2	5.11[c]	1555
Vallisneria spiralis L.											
1	Whole plant	TLC	46.8	21.1	13.9	7.2	6.6		6.1	4.82[c]	1555
Vanda teres (Roxb.) Lindl.											
1	Leaf[l]	TLC			0.20						1402
Vicia faba L. (Broad (Faba) Bean)											
1	Seed	TLC	40.2	43.1		11.4	5.3			0.92	1872
1	Bean	TLC	50.3	16.4	3.1	15.6			(NAPE 13.0)	1.05	1693
1	Starch	TLC	+	0			+		(LPC +, LPE 0, LPS +)		1902
1	Leaf	TLC	31.9	15.3	25.8	7.9			19.1	0.46	2047
1	Leaf protoplast	TLC	31	17	52					29[d]	2048
1	–, guard cells	TLC	52	23	9	10		3	3		2049
1	–, mesophyll cells	TLC	37	13	35	5		1	9		2049
1	Leaf chloroplast envelope membrane	TLC	77	Trace	22	Trace					2050
1	–, lamella	TLC	34	Trace	66	Trace					2050
1	–, mitochondria	TLC	57	33	10	Trace					2050
1	–, microsomes	TLC	67	20	13	Trace					2050
1	Root	TLC	44.5	30.7	6.7	7.9	4.0	3.7	2.5		1775
1	–, mitochondria *Incl. PS	TLC	47.2	20.3	15.1*	13.2		1	3.2	83.4[a]	1776
1	–, microsomes *Incl. PS	TLC	42.1	26.6	13.6*	10.4		2.1	5.3	82.3[a]	1776
Vicia faba minor L.											
1	Root	TLC	44.5	30.7	6.7	7.9	4.0	3.7	2.5		1775
Vicia villosa Roth. (Hairy Vetch)											
1	Seed[c]	Frac. pptn.	0.304	0.081							718
Vigna angularis (Willd.) Ohwi and Ohashi (Adzuki Bean)											
1	Bean	TLC	50.9	20.9	0.1	21.2		(NAPE 4.2)		1.10	1693

1	Leaf	TLC	32	24	36	4		<4	(NAPE 4)		2051
1	Root	TLC	14	6	<3	3		<1	(NAPE 6)		2051
1	Suspension culture, leaf	TLC	23	23	1	7		(NAPE 4, BPA 7)		0.20	2052
1	–, cotyledon	TLC	22	22	1	5		(NAPE 4, BPA 11)		0.29	2052
1	–, hypocotyl	TLC	24	26	1	6		(NAPE 4, BPA 4)		0.26	2052
1	–, root	TLC	16	20	1	9		(NAPE 4, BPA 13)		0.13	2052
Vigna auriculata (Ronji)											
1	Seed	TLC	40.0	40.0		19.7	Trace			1.38	1872
Vigna mungo (L.) Hepper (Black Gram, Matpe)											
1	Seed	TLC	34.6	30.0	0.5	21.4	+	+	2.1	1.01	1384
			also NAPE 9.0, NALPE 0.8, LPC +, LPE +								
Vigna radiata (L.) Wilcz. (Mung Bean)											
1	Seed	TLC	37.0	31.5	3.6	24.4	1.8	0.4			2053
1	Seed	TLC	61.0	21.1		10.8		1.1[u]		1.4[c]	2054
			also NAPE 2.1, NALPE 1.9, LPC 1.0								
1	–, mitochondrial inner membrane	TLC	33	32	4	5		(LPE 3)	23		2055
1	Seed flour, free	TLC also LPC 2.8	10.8	64.7	0.5	5.3 (incl. PS)			16.7	0.041	2056
1	–, bound	TLC also LPC 1.8	20.5	35.2	0.8	26.7 (incl. PS)			15.0	0.17	2056
1	Hypocotyl, 3 days old	TLC	PC, PE, PI, and PA measured back along 5 mm segments from apex								2057
1	–, plasma membrane	TLC	32.7	38.0	4.5	5.3	3.1	16.3		1.29[h]	2058
1	–, plasma membrane	HPLC	23	53		7		17		0.3-0.6[t]	2059
1	–, tonoplast	TLC	46.5	31.4	4.5	11.2	4.3	2.2		1.93[h]	2058
1	–, mitochondria	TLC	36	46	1	2.5			14	300[j]	1432
1	–, inner membrane	TLC	29	50	1	2			17	380[j]	1432
1	–, outer membrane	TLC	68	24	2	5			0	800[j]	1432
1	Leaf chloroplast	TLC	59.0	0	28.5	11.5				20.0[g]	2060
1	Seedling	TLC	42	16	6	3		16	10 (LPC 8)	0.24	2061
1	Seedling, 2 days old	TLC	43.6	32.7	4.7	15.0	1.9	3.2			2053
1	Seedling, 7 days old, axis	TLC	40.9	39.5	3.1	11.4	1.0	4.1			2053

*Unless stated otherwise; the PL column gives the total phospholipid content of each plant part in percent by weight, and the PC, PE, PG, PI, PS, PA, and DPG columns show the distribution in percent by weight of the total PL; [a]Percent of total lipids; [b]mg lipid P/100 mL juice; [c]Percent dwb; [d]mg PL/100 mg protein; [e]mg PL/g fwb; [f]mg PL/100 g dwb; [g]Percent of polar lipids; [h]µmol PL/mg protein; [i]µmol PL/mg chlorophyll; [j]µg PL/mg protein; [k]µg lipid P/g; [l]µmol PL/g fwb; [m]mol PL/mol chlorophyll; [n]mg PL/g dwb; [o]µmol PL/g dwb; [p]µg PL/100 mg fwb; [q]µg PL/mg chlorophyll; [r]mg PL/cm^2 leaf area; [s]nmol PL/cm^2 leaf area; [t]µmol PL/mg protein; [u]IR given; [v]mg lipid P/g lipid; [w]µg PL/mL; [x]µg lipid P/cm^2 leaf area; [y]Molecules per 10^6 plastids; [z]µg PL/mg chromatin; [aa]Fatty acid composition and *sn*-1/*sn*-2 distribution given for each phospholipid; [bb]µg/g fwb; [cc]Taken from the abstract.

Continued

Table 4-15 Phospholipids in Plants*—cont'd

No.	Plant part	Method	PC	PE	PG	PI	PS	PA	DPG	PL	References
1	–, cotyledon	TLC	35.4	33.3	6.3	12.5	1.9	8.1			2053
1	Root, plasma membrane	TLC	29.2	12.7	2.1	6.1 (incl. PS)		10.5			2062
Vigna unguiculata (L.) Walp. (Cowpea)											
1	Seed	TLC	18	30	24	16			12	0.20[c]	1524
1	Leaf	TLC	38.9	30.4	15.3	8.1		7.4			2063
2	Leaf	HPLC	51-59		19-20					1.20-1.69[a]	2064
1	Internode	TLC	55.4	29.7		10.3		4.6			2063
Vinca minor L. (Periwinkle)											
1	Leaf[a]	TLC	13	5	10						2065
Viscum album L. (Mistletoe)											
1	Leaf chloroplast[y]	TLC	158		+	+				20	2009
Vitis riparia Michx. (Riverbank Grape)											
1	Root *Incl. PS	TLC	42.1	35.8*	9.0	2.8		5.9	4.5		1699
Vitis vinifera L. (Wine Grape)											
10	Seed	TLC	16.5-43.6	22.0-41.7		15.3-28.4	Trace	11.8-17.6		0.81-1.12[a]	2066
1	Leaf	TLC	2.3	15.0	34.2	6.7	16.0		(LPE 7)	1.81[c]	2067
1	–, chloroplast	TLC	9.8	10.5	43.4	12.5	4.4		(LPE 0)	2.94[c]	2067
1	Leaf, plasma membrane	HPLC	60.0	40.0	0				0	88.3[j]	2068
1	–, microsomes	HPLC	20.5	15.7	2.9				60.9	54.7[j]	2068
1	Seed	TLC	+ also LPC +, LPI +	+		+		Trace		0.4	1777
1	Berry	TLC	29.7 also LPC 2.2, LPE 3.0	32.3	10.9	5.6	8.4	3.6	4.3	0.058	2069
1	–, seed	TLC	25.1 also LPC 1.9, LPE 2.2	36.3	6.3	8.1	10.4	6.7	3.0	1.12	2069
1	–, skin	TLC	27.6 also LPC 3.2, LPE 4.0	34.0	10.4	6.1	8.8	3.5	2.4	0.038	2069
1	–, flesh	TLC	31.2 also LPC 2.7, LPE 3.4	27.4	12.3	4.0	7.6	2.2	9.2	0.012	2069

1	Root *Incl. PS	TLC	47.6	34.7*	4.9	1.4		5.0			1699
5	Root	CC (silica)	34.5-68.2	31.8-65.5		Trace				0.024-0.084	2070
Withania somnifera (L.) Dun. (Winter Cherry, Aswagandh)											
1	Seed	TLC	34.0	22.4		24.8			17.8	0.70[c]	1757
			also LPC 0.7, LPE 0.5								
Xanthium orientale L. (Oriental Cocklebur)											
1	Leaf	TLC	50	15	25	5			4	0.99[l]	1523
Xanthium strumarium L. (Cocklebur)											
1	Young fruit	Frac. pptn.	33.2	66.8						0.18	2071
Zantedeschia aethiopica (L.) Spreng. (Calla Lily)											
1	Leaf thylakoid[m]	TLC	0.131		0.191						2072
1	Regreened spathe, thylakoid[m]	TLC	0.215		0.240						2072
Zea mays L. (Corn, Maize)											
10	Seed[cc]	TLC	+							0.065-0.233	2073
1	–, germ	TLC	54.5	11.7	4.3	18.7	0.9	1.2	2.1	2.54[c]	1390
			also NAPE 1.1, NALPE 1.8, LPC 0.8, LPE 0.4								
3	–, germ[f]	TLC	22.9-397	376-748	49-194	19-36	59-333			1224-1453[c]	2074
			also NAPE 73-117, LPC 9-60, LPE 11-18, LPI, 0.28								
1	–, chloroplast envelope membrane[a,c]	TLC	6.7	1.1	2.2	1.4			0.8	11.4	1728
3	–, pericarp[f]	TLC	2-7	5-8	1-4	2-4		2-7		27-42[c]	2074
			also NAPE 5-11, LPC 3-7, LPE 1-3								

*Unless stated otherwise; the PL column gives the total phospholipid content of each plant part in percent by weight, and the PC, PE, PG, PI, PS, PA, and DPG columns show the distribution in percent by weight of the total PL; [a]Percent of total lipids; [b]mg lipid P/100 mL juice; [c]Percent dwb; [d]mg PL/100 mg protein; [e]mg PL/g fwb; [f]mg PL/100 g dwb; [g]Percent of polar lipids; [h]μmol PL/mg protein; [i]μmol PL/mg chlorophyll; [j]μg PL/mg protein; [k]μg lipid P/g; [l]μmol PL/g fwb; [m]mol PL/mol chlorophyll; [n]mg PL/g dwb; [o]μmol PL/g dwb; [p]μg PL/100 mg fwb; [q]μg PL/mg chlorophyll; [r]mg PL/cm² leaf area; [s]nmol PL/cm² leaf area; [t]μmol PL/mg protein; [u]IR given; [v]mg lipid P/g lipid; [w]μg PL/mL; [x]μg lipid P/cm² leaf area; [y]Molecules per 10^6 plastids; [z]μg PL/mg chromatin; [aa]Fatty acid composition and *sn*-1/*sn*-2 distribution given for each phospholipid; [bb]μg/g fwb; [cc]Taken from the abstract.

Continued

Table 4-15 Phospholipids in Plants*—cont'd

No.	Plant part	Method	PC	PE	PG	PI	PS	PA	DPG	PL	References
3	–, tip cap[f]	TLC	15-32	7-41	3-27	3-8		14-24		52-185[c]	2074
			also NAPE 10-15, LPC 10-28, LPE 0-3, LPI 0.18								
1	–, endosperm	TLC	44.6	6.1	1.9	3.6	0	0.7	0.9	0.52[c]	1390
			also NAPE 1.2, NALPE 1.1, LPC 35.1, LPE 3.4								
3	–, nonstarch endosperm[f]	TLC	1-31	1-8	Trace-6	3-9		Trace-12		5-101[c]	2074
			also NAPE trace-3, LPC 1-29, LPE 0-4								
1	–, amyloplast envelope	TLC	37.3	4.5	19.4	23.9			(LPC 4.5)		2075
1	Seed, oil body	TLC	64.1	8.1		7.6	20.2			0.91	1482
1	Seed, oil body	TLC					0				1571
1	–, stele[j]	HPLC	88.9	42.4		+	+	+			2076
1	–, cortex[j]	HPLC	59.0	45.9		+	+	+			2076
1	Seed starch, waxy	TLC							(LPC 0)	0.23[d]	2077
1	–, nonwaxy	TLC							(LPC 17)	0.87[d]	2077
1	–, amylomaize	TLC							(LPC 19)	1.11[d]	2077
1	Leaf	TLC	28	11	44	17					1842
1	Leaf	TLC	15	28	22				18	1.42[l]	1698
1	Leaf	TLC					+				1450
1	Leaf chloroplast	TLC	41.1	0	31.4	22.3				28.7[g]	2060
1	–, bundle sheath[a]	TLC			9						1489
1	–, agranal[a]	TLC	7		12						2078
1	–, mesophyll[a]	TLC			9						1489
1	Leaf thylakoid										
1	–, granum[a]	TLC	4		14						2078
1	–, stroma[a]	TLC	6		12						2078
1	Leaf starch	TLC						LPC 0, LPE +, LPS 0			1902
1	Seedling coleoptile, microsome	TLC				PI[H]					2079
1	–, plasma membrane	TLC				PI[H]					2079
1	Anther wall	TLC	43	31	6	20					1842
1	Pollen grain	TLC	55.3	30.0	1.8	9.5 (incl. PS)		2.2	1.3	6.09[c]	1861
1	Root	TLC	51.2	36.1*	5.7	Trace		3.3	3.5		1699

*Incl. PS

1	Root, 6 days old	TLC	45.6	24.2	6.8		11.5	3.4	6.0	0.88	1583
1	Root[a]	HPLC	30.2	15.9	Trace	Trace	0	13.5	0.3		1413
1	–, endoplastic reticulum	TLC	54	24	8	10 (incl. PS)		3		0.98[j]	2080
1	–, mitochondria	TLC	34	30	10	7 (incl. PS)		20		0.33[j]	2080
2	Root plasma membrane	TLC	47.1-62.2	21.4-23.4	4.9-10.5	8.4-8.9	0-1.5	1.6-9.2	0 (LPC 0-0.9)	668[f]	2081
1	Seedling, root	TLC	48.4	25.3			8.4			25.3[k]	2082
1	Cell suspension culture	^{31}P NMR	+	+			+				2083
Zea mays L. ssp. *indentata* (Sturt.) Bailey (Dent Corn)											
6	Seed	TLC	4.5-22.0			30.9-86.4				0.10-0.31	2084
Zea mays L. ssp. *indurata* (Sturt.) Bailey (Flint Corn)											
	Seed	TLC	27.2-29.5			42.9-79.1	+		+		2085
Zea mays L. ssp. *saccharata* Sturt. (Sweet Corn)											
1	Seed	TLC	65	16		12			(LPC +, LPE +)		2086
Zingiber officinale Rosc. (Ginger)											
14	Rhizome	TLC	+			+		+	(LPC +)		2087
Ziziphus jujuba Mill. (Jujube)											
1	Seed	TLC	42.3	8.4	11.5	11.1		7.2		0.17[c]	2088
			also NAPE 7.4, NALPE 3.8, LPC 4.1, LPE 1.9, LPI 2.3								
1	–, pericarp	TLC	18.8	9.8	28.1	17.5		7.1		0.05[c]	2088
			also NAPE 8.6, NALPE 0, LPC 3.7, LPE 0, LPI 0								
Zostera asiatica Miki											
1	Whole plant (NAPE trace)	TLC	43.8	23.8	14.6	12.6	2.2	Trace	3.0	0.14	1882
Zostera capricorni Asch.											
1	Seed	HPLC	4	21		74				0.56[c]	2089

*Unless stated otherwise; the PL column gives the total phospholipid content of each plant part in percent by weight, and the PC, PE, PG, PI, PS, PA, and DPG columns show the distribution in percent by weight of the total PL; [a]Percent of total lipids; [b]mg lipid P/100 mL juice; [c]Percent dwb; [d]mg PL/100 mg protein; [e]mg PL/g fwb; [f]mg PL/100 g dwb; [g]Percent of polar lipids; [h]μmol PL/mg protein; [i]μmol PL/mg chlorophyll; [j]μg PL/mg protein; [k]μg lipid P/g; [l]μmol PL/g fwb; [m]mol PL/mol chlorophyll; [n]mg PL/g dwb; [o]μmol PL/g dwb; [p]μg PL/100 mg fwb; [q]μg PL/mg chlorophyll; [r]mg PL/cm^2 leaf area; [s]nmol PL/cm^2 leaf area; [t]μmol PL/mg protein; [u]IR given; [v]mg lipid P/g lipid; [w]μg PL/mL; [x]μg lipid P/cm^2 leaf area; [y]Molecules per 10^6 plastids; [z]μg PL/mg chromatin; [aa]Fatty acid composition and *sn*-1/*sn*-2 distribution given for each phospholipid; [bb]μg/g fwb; [cc]Taken from the abstract.

Continued

Table 4-15 Phospholipids in Plants*—cont'd

No.	Plant part	Method	PC	PE	PG	PI	PS	PA	DPG	PL	References
Zostera marina L (Seawrack)											
1	Whole plant (NAPE trace)	TLC	41.8	24.3	15.8	12.1	2.6	Trace	3.4	0.11	1882
Zostera nana Roth											
1	Whole plant	TLC	45.1	24.3	15.5	10.3	4.8			0.11[c]	2090
Zygophyllum album L.f. (White Bean Caper)											
1	Shoot	TLC	29.6	16.9	10.0	5.3		9.0	21.7	0.189[c]	1488

*Unless stated otherwise; the PL column gives the total phospholipid content of each plant part in percentage by weight, and the PC, PE, PG, PI, PS, PA, and DPG columns show the distribution in percent by weight of the total PL.
[a]Percentage of total lipids.
[b]mg lipid P/100 mL juice.
[c]Percentage of dwb.
[d]mg PL/100 mg protein.
[e]mg PL/g fwb.
[f]mg PL/100 g dwb.
[g]Percentage of polar lipids.
[h]µmol PL/mg protein.
[i]µmol PL/mg chlorophyll.
[j]µg PL/mg protein.
[k]µg lipid P/g.
[l]µmol PL/g fwb.
[m]mol PL/mol chlorophyll.
[n]mg PL/g dwb.
[o]µmol PL/g dwb.
[p]µg PL/100 mg fwb.
[q]µg PL/mg chlorophyll.
[r]mg PL/cm^2 leaf area.
[s]nmol PL/cm^2 leaf area.
[t]µmol PL/mg protein.
[u]IR given.
[v]mg lipid P/g lipid.
[w]µg PL/mL.
[x]µg lipid P/cm^2 leaf area.
[y]Molecules per 10^6 plastids.
[z]µg PL/mg chromatin.
[aa]Fatty acid composition and *sn*-1/*sn*-2 distribution given for each phospholipid.
[bb]µg/g fwb.
[cc]Taken from the abstract.

5 Mononucleotides

This group comprises the mono-, di-, and triphosphate derivatives of the ribo- and deoxyribonucleosides of the purine bases adenine and guanine and the pyrimidine bases cytosine, uracil, and thymine. The following is a complete list of the compounds discussed in this chapter, except for the cyclic nucleotides and sugar nucleotides which are discussed later in the chapter:

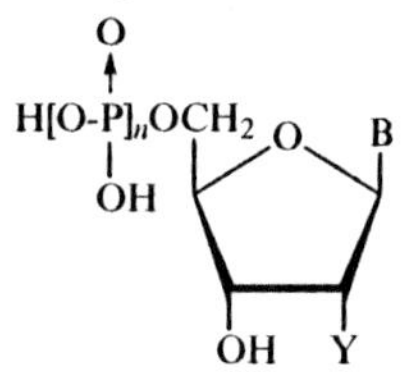

B = A, G, C, U or T
Y = OH or H
n = 1, 2, or 3

B	Base	Nucleoside	Nucleotide
Ribose derivatives, Y=OH			
A	Adenine	Adenosine	AMP, ADP, ATP
G	Guanine	Guanosine	GMP, GDP, GTP
C	Cytosine	Cytidine	CMP, CDP, CTP
U	Uracil	Uridine	UMP, UDP, UTP
Deoxyribose derivatives, Y=H			
A	Adenine	Adenosine	dAMP, dADP, dATP
G	Guanine	Guanosine	dGMP, dGDP, dGTP
C	Cytosine	Cytidine	dCMP, dCDP, dCTP
U	Uracil	Uridine	dUMP, dUDP, dUTP
T	Thymine	Thymidine	TMP, TDP, TTP[a]

[a]Often designated dTMP, dTDP, and dTTP in the literature despite the fact that thymidine already has the 2′-deoxyribose structure.

For general references on this subject, see Feingold,[2426] Pradet and Raymond,[2427] Wasternack,[2428] Brown,[2429] or Townsend.[2430]

Occurrence

Although the structures and properties of the ribonucleotides which comprise the building blocks of ribonucleic acid (RNA) have been known for some time, the occurrence of free nucleotides in higher plants was not established until the mid-1950s. The first nucleotides to be discovered in the free state were ADP and ATP, isolated from mung bean seedlings by Albaum *et al.* in 1950 and shown to be similar if not identical to those found in animal tissues.[2431] The application of ion-exchange chromatography (IEC) in 1956 to separate the acid-soluble ribonucleotides from each other enabled Bergkvist[2432] and

Ginsburg *et al.*[2433] to identify AMP, ADP, ATP, GMP, GDP, GTP, CTP, UMP, UDP, and UTP in wheat and mung beans, respectively. By 1960 all 12 of the ribonucleotide mono-, di-, and triphosphates had been isolated and quantitated. Over the succeeding years, the number of papers devoted to this subject has expanded rapidly, as attested to in the tables that accompany this chapter.

In contrast, very little is known about the occurrence in the free state of the deoxyribonucleotides which comprise the building blocks of deoxyribonucleic acid (DNA). The triphosphates dATP, dGTP, dCTP, and TTP were identified in pine pollen and in sycamore and tobacco callus culture[2434] and the corresponding diphosphates dADP, dGDP, dCDP, and TDP in tomato root tips.[2435] The monophosphates dAMP, dGMP, dCMP, and TMP have yet to be detected in the higher plants.

The metabolic pathways for the biosynthesis of the nucleotides are outlined in Figs. 5-1 and 5-2. The key precursor for the purine nucleotides is inosinic acid (IMP), which is formed from α-D-ribose 5-phosphate (R5P) by a sequence of reactions whose final step is closure of the purine ring. AMP is formed from IMP by ammonolysis of the hydroxyl group at C-6 (Fig. 5-1, step 1), whereas GMP is formed by the introduction of an amino group at C-2 (step 8). Stepwise reversible phosphorylation of AMP or GMP yields ADP, ATP, GDP, and GTP (steps 5-7). Enzymatic reduction of ADP or GDP (step 9) provides the point of entry to the deoxyribonucleotides dADP and dGDP, from which dATP, dGTP, dAMP, and dGMP are formed by reversible phosphorylation (steps 5-7). Steps 2-4 provide a mechanism for purine salvage from AMP, GMP, dAMP, and dGMP. The purine nucleotides needed for the biosynthesis of RNA are furnished by ATP and GTP while those needed for the biosynthesis of DNA are furnished by dATP and dGTP.

The key precursor for the pyrimidine nucleotides is orotidine 5′-phosphate (OMP), which is formed from 5′-phosphoribose-1-pyrophosphate (PRPP) by condensation with

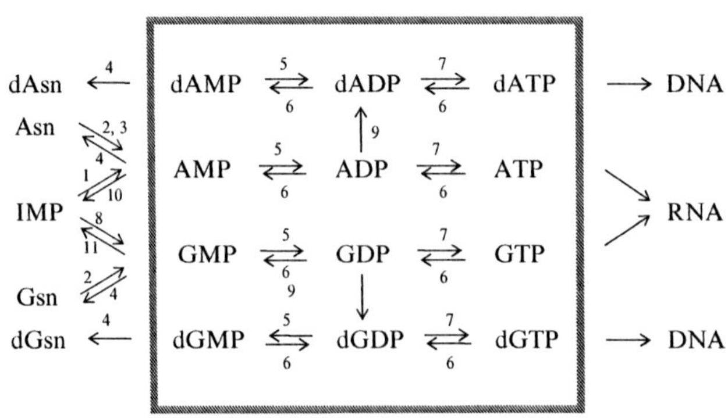

FIGURE 5-1 Purine Nucleotides. (1) Adenylosuccinate synthetase [E.C. 6.3.4.4], aspartate, Mg(II); adenylosuccinate lyase [E.C. 4.3.2.2]. (2) Nucleoside phosphotransferase [E.C. 2.7.1.77]. (3) Adenylate kinase [E.C. 2.7.1.20]. (4) 5′-Nucleotidase [E.C. 3.1.3.5]. (5) Nucleoside monophosphate kinase [E.C. 2.7.4.4]. (6) Nucleoside diphosphatase [E.C. 3.6.1.6]. (7) Nucleoside diphosphate kinase [E.C. 2.7.4.6]. (8) IMP dehydrogenase [E.C. 1.2.1.14], NAD, GMP synthetase [E.C. 6.3.4.1], glutamine. (9) Ribonucleoside diphosphate reductase [E.C. 1.17.4.1], NADPH, Mg(II), thioredoxin. (10) AMP aminohydrolase [E.C. 3.4.5.6]. (11) GMP reductase [E.C. 1.6.6.8].

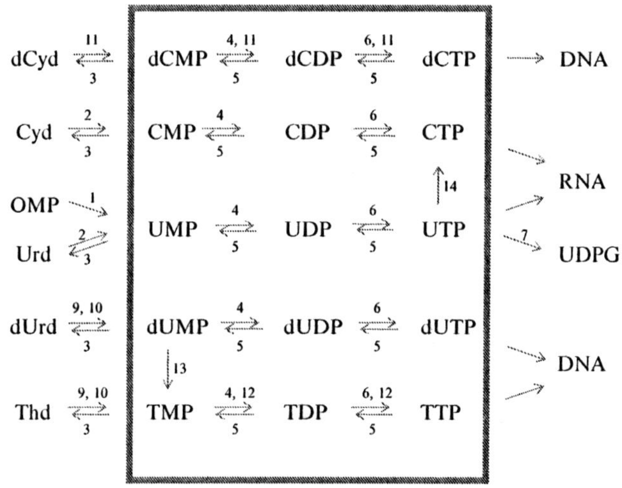

FIGURE 5-2 Pyrimidine Nucleotides. (1) Orotidine 5′-phosphate decarboxylase [E.C. 4.1.1.-23]. (2) Nucleoside phosphotransferase [E.C. 2.7.1.77]. (3) 5′-Nucleotidase [E.C. 3.1.3.5]. (4) Nucleoside monophosphate kinase [E.C. 2.7.4.4]. (5) Nucleoside diphosphatase [E.C. 3.6.1.6]. (6) Nucleoside diphosphate kinase [E.C. 2.7.4.6]. (7) Glucose 1-phosphate uridyltransferase [E.C. 2.7.7.9]. (8) Ribonucleoside diphosphate reductase [E.C. 1.17.4.1], NADPH, Mg(II), thioredoxin. (9) Thymidine kinase [E.C. 2.7.1.75]. (10) Thymidine AMP-phosphotransferase [E.C. 2.7.1.77]. (11) dCMP kinase [E.C. 2.7.4.14]. (12) TMP kinase [E.C. 2.7.4.9]. (13) Thymidylate synthetase [E.C. 2.1.1.6]. (14) CTP synthetase [E.C. 6.3.4.2], NH_3, Mg(II).

orotic acid. Decarboxylation of OMP yields UMP (Fig. 5-2, step 1), from which UDP and UTP are formed by stepwise reversible phosphorylation (steps 4 and 5). CTP is formed from UTP by ammonolysis of the hydroxyl group at C-4 (step 15) and dUDP is formed from UDP by enzymatic reduction (step 8).

Extraction

Many plant tissues contain phosphatases which, if not destroyed or inactivated, are capable of hydrolyzing nucleotides during isolation. The extraction procedures most widely used in current practice employ ice-cold aqueous trichloroacetic acid (TCA) or perchloric acid (PCA), both of which kill the plant tissue and extract the nucleotides without degrading them. After extraction, the excess TCA is removed by extraction with ether and the excess PCA by precipitation as the water-insoluble potassium perchlorate. More concentrated acid should be avoided because it begins to solubilize nucleic acids such as RNA.[2436]

> *Pea seeds were milled in a Glen Creston C580 microhammer mill…, a screen of 0.1 mm. diameter mesh being used. The milled seed (10 g. samples) was homogenized in a MSE homogenizer for 3 min. at 0° with 0.3N-perchloric acid (25-30 ml) and the homogenate filtered under vacuum through a sintered-glass Buchner funnel (porosity 3) into a flask surrounded by crushed ice. Extraction from the almost dry residue was assisted by pressing with an inverted wide-topped glass stopper. The*

homogenizer flask was rinsed out with four successive 10 ml. portions of 0.15N-perchloric acid and these washings were used to effect four further extractions of the seed residue in the funnel. The clear filtrate was adjusted to pH 7.2 with 20% (v/v) potassium hydroxide solution and, after standing at 0° for 45 min., was centrifuged to remove potassium perchlorate.

Bieleski[10] recommended the use of an ice-cold ternary mixture of methanol, chloroform and 1*M* formic acid (12:5:3) as the extractant, whereby any chlorophyll or terpenoid coextracted with the nucleotides are removed in the organic phase. Extraction with boiling 80% ethanol[6] is discouraged because it does not completely inactivate the phosphatases.[10]

The presence of phenolics in such plant extracts interferes with the direct determination of the nucleotide content by UV spectrophotometry. They cannot be separated satisfactorily from the nucleotides by differential extraction, solvent partitioning, gel filtration, ion-exchange chromatography, or sorption,[2438] but they can be removed by passing the extract through a column of polyvinylpyrrolidone (PVP), either alone[2439] or in combination with Amberlite-XAD-2,[2438] or by passing the extract through a disposable cartridge of phenylsilane-bonded silica gel[2440] or Cu(II)-loaded Chelex 100.[2441] Alternatively, the nucleotides themselves may be separated from the extracts by adsorption on charcoal, washing the charcoal to remove the contaminants, and then recovering the nucleotides:

The extract was adjusted to within the range pH 3.5-4.0 with acetic acid and passed slowly through a column, 1.2 cm. diameter, containing a mixture of acid-washed Norit OL charcoal (500 mg.) and Hyflo Supercel (800 mg.) supported on a 2 mm.-thick bed of Hyflo Supercel on a sintered-glass disk (porosity 3). The column was washed with 20 ml. of water and eluted with 15 ml. of 25% (v/v) ethanol containing 0.5% (v/v) of aq. ammonia (sp. gr. 0.88). The charcoal mixture was not allowed to dry out at any stage of the procedure. Percolation of solutions through the column was facilitated by application of a positive air pressure of 1-2 lb./in^2.

For a critical survey of the methods for purifying plant extracts, see Brown.[2429]

Separation and Analysis

(1) Paper Chromatography (PC) Method

In the 1950s and 1960s, two-dimensional PC was the method of choice for the separation of plant nucleotides. The procedures are similar to those used for the sugar phosphates (Chapter 1, pp 2 and 3). Adequate resolution is achieved with the ternary solvents listed in Table 5-2 (Chapter 1) provided that precautions are taken to remove cations such as calcium, magnesium, or iron by the use of acid-washed paper,[2764] by treatment of the

Table 5-1 Travel Constants of the Ribonucleotides

Compound	PC (R_f)[a] 1st	PC (R_f)[a] 2nd	PE (M_P)[a]	TLC (R_f)[b] 1st	TLC (R_f)[b] 2nd	TLE (M_{AMP})[c]
P_i	100	100	100	100	100	
AMP	130	55	19	142	95	100
GMP	59	43	47			
CMP	106	53	9			
UMP	87	62	60	92	86	139
ADP	94	27	63	122	52	178
GDP	44	23	85			
CDP	81	27	59			
UDP	65	28	103	73	51	222
ATP	70	14	93	105	45	219
GTP	35	12	107			
CTP	61	14	92	86	36	
UTP	50	15	127			252
ADPG				113	49	151
UDPG	111[d]	14[d]	88	68	42	185

[a]Bieleski and Young.[9]
[b]Feige *et al.*[33]
[c]Jenner.[2723]
[d]Slight decomposition. Solvents (refer to list in Chapter 1, p 2): PC: B2 and A2; PE, 0.025*M* sodium butyrate buffer; TLC: N3 and A3.

plant extract with an ion-exchange resin,[9] or by the addition of EDTA to the solvent.[28,29] Of the ternary solvents listed, those used most often are B2 and N2 for the first dimension and A2 and N3 for the second dimension, but many variations of these and other ternary solvents are documented in the literature cited in Table 5-2. The R_f values in solvents A2 and B2 for the nucleotides covered in this chapter are given in Table 5-1. Inspection of the data reveals that the mobility of each and every nucleotide decreases with increasing number of phosphate groups, i.e., NMP > NDP > NTP.

(2) Thin-Layer Chromatography (TLC) Method

Bieleski[36] showed that the nucleotides could be separated by two-dimensional TLC in a fifth the time required for PC with no loss of resolution and a 20-fold increase in sensitivity. In most of the cases cited in Table 5-1, the TLC was performed using plates coated with 250-μ-thick layers of Macherey-Nagel unbound cellulose powder using the same solvents as those used for PC. The relative positions of the nucleotides and the patterns obtained by the two methods are almost identical. PEI-impregnated cellulose TLC plates were employed by Nygaard *et al.*[2434,2537,2669] and others.[2704] Silica gel gives inferior results[36] but is adequate for the separation of ATP from the other components of sieve-tube sap.[2693]

(3) Electrophoretic Methods

The nucleotides can also be separated from each other by paper electrophoresis (PE)[9,32] or thin-layer electrophoresis (TLE).[309] These methods are most useful when coupled with PC or TLC, particularly in view of the fact that the order of mobility is reversed. Inspection of the PE data in Table 5-1 reveals that the R_f value of each nucleotide increases with increasing number of phosphate groups (i.e., NMP < NDP < NTP). The TLE values (not shown) follow the same trends.[309] In Bieleski's procedure, electrophoretic separations are carried out at pH 3.6 with 0.25*M* acetate buffer and 600 V for PE[9] or 0.28*M* acetate buffer and 1000 V for TLE.[36] Others have used citrate buffer[12,2723] or butyrate buffer[2578] at pHs ranging from 3.5 to 4.8 and voltages ranging from 300 V[2682] to 1400 V.[331]

(4) Ion Exchange Chromatography (IEC) Method

The use of IEC to separate the acid-soluble nucleotides of plants was introduced in 1956 by Bergkvist,[2432] who used a Dowex 1-X10 (formate) anion exchange column and gradient elution with formic acid/sodium formate. The procedures most often employed are convex[2765] or concave[2766] gradient elution with ammonium formate buffers at pH ~ 5, which separate the nucleotides in the order NMP > NDP > NTP and in the order C > U > A > G. Gradient elution with formic acid/ammonium formate at pH ~ 2 changes the latter to C > A > G > U but not the former.[2765] IEC resins of a lower degree of crosslinking are often used such as Dowex 1-X8,[2437] Dowex 1-X4,[2767] or Dowex 1-X2,[2768] and occasionally different kinds of supports such as DEAE-Sephadex A-25[2654] or PEI-cellulose.[2754] Other eluants have been employed such as HCl/NaCl,[2433] HCl/LiCl,[2689] $HCl/CaCl_2$,[2650] or NH_4Cl/borate.[442] The IEC method has been automated[2769] and adapted for large (kilogram) scale use.[2770]

(5) High-Performance Liquid Chromatography (HPLC) Method

This method, introduced in 1972,[2486] has largely supplanted the foregoing methods because of its vastly improved resolution, despite the highly sophisticated equipment required for its use. The HPLC method is usually performed in the IEC mode with strongly anionic supports similar to those used in IEC, but in some cases better separations are achieved with HPLC in the reversed phase mode or the ion-pairing mode. The following procedure is typical of HPLC in the ion-exchange mode:

> *The HPLC equipment (Waters, Milford, MA, U.S.A.) consisted of two Model 510 pumps, a Model 680 automated gradient controller, a U6K injector and a Model 481 LC spectrophotometer. Nucleotides were detected by monitoring the column effluent at 254 nm with sensitivity fixed at 1V a.u.f.s. Separations were performed on a Waters Radial Pak Partisil SAX10 cartridge (10 cm × 0.8 cm I.D.) using a Waters radial compression Z-module system. The elution buffer system used consisted of eluent A, 7mM ammonium dihydrogenphosphate (pH 3.8), and eluent B, 250 mM ammonium dihydrogenphosphate (pH 4.5) with 500 mM potassium chloride. Nucleotide samples (100 μl), obtained from plant root tips, were injected into the column. A linear gradient of eluant A to eluant B was applied for 20 min., followed by an isocratic period of 10 min with eluent B .... The flow-rate was maintained at 4 ml/min; analyses were performed at ambient temperature. Peaks were integrated ... on a*

Waters 740 data module. Individual components of the acid-soluble nucleotide pools were identified, using retention time relative to known standards and by injecting known internal standards.

Dutta, I; O'Donovan, G. A.; Smith, D. W., "Nucleotide Extraction and Quantitation from Tomato Roots by High-Performance Liquid Chromatography", *J. Chromatogr.* **1988**, *444A*, 183–189,[2618] reprinted by permission of Elsevier, © 1988.

Some nucleotides that are not well resolved by anion exchange HPLC, such as AMP, ADP, and the dinucleotides NAD and NADP, can be separated by reversed phase HPLC using a column packed with Hypersil ODS or LiChrosorb RP18 and gradient elution with 0.2*M* (NH_4)-H_2PO_4 and acetonitrile.[2440] The nucleotides can also be separated by reversed phase HPLC under ion-pairing conditions, substituting tetrabutylammonium hydroxide for acetonitrile as the second component of the gradient.[2623] A further refinement of the latter procedure was derivatization of the nucleotides with chloroacetaldehyde and detection of the etheno derivatives of AMP, ADP, and ATP by fluorescence in order to eliminate any interference by the cytidine derivatives.[2455]

Analysis Without Separation

(1) ^{31}P Nuclear Magnetic Resonance (NMR) Method

This method provides a nondestructive means for examining the nucleotides in their natural environment. It is particularly suited for examining the behavior of the nucleotide di- and triphosphates (including sugar derivatives such as UDPG) because the large differences in chemical shift between the phosphorus atoms of each compound makes them easy to distinguish from other phosphorus-containing compounds that are known to be present, such as G_6P, P_i, or NAD. The resolution, however, is insufficient to distinguish one di- or triphosphate from another; the compounds identified by their authors as ADP, ATP, etc., in Table 5-2 do not necessarily contain adenosine exclusively. The method was originally (1983) applied to tobacco cell cultures[2629] and to maize root tips[544] that were placed in 10 mm NMR tubes under constant aeration and scanned without spinning. In both cases the di- and triphosphates were found to be complexed to Mg(II) and to be present in a cytoplasmic (pH ~ 7.1) rather than a vacuolar (pH ~ 5.5) environment. ^{31}P NMR relaxation measurements showed that 30% of the NTP in maize root tips was also protein-bound.[2771] The method has since been applied to cell suspensions[51,2507,2695] and root tips[44,526,542,543,2589,2771] of other plants, soybean root nodule,[2563] potato tuber mitochondria,[2708] Jerusalem artichoke stolon,[45] lettuce[48] and fiddleneck[2644] sprouts, rice and wheat shoots,[2638] sunflower stem,[2563] spinach leaf and chloroplast,[50] avocado fruit,[515] and tobacco pollen.[2563] Use of the maximum entropy method (MEM) instead of fast Fourier transform (FFT) results in resolution comparable to that obtained with extracts (next paragraph), but so far this technique has only been applied to mung bean tips.[543]

As might be expected, better resolution in obtained with perchloric acid extracts of plant tissues than with living plant tissues, and other signals can now be identified. Using a multinuclear probe at 162.0 MHz, Bligny *et al.*[50] were able to identify AMP, ADP, ATP,

Table 5-2 Ribonucleotides in Plants*

No.	Plant part	Method	AMP	ADP	ATP	GMP	GDP	GTP	CMP	CDP	CTP	UMP	UDP	UTP	References
Abies procera Rehd. (Noble Fir)															
1	Pollen	Lumin.			20-530										2442
Acer mono Max. (Gorose)															
1	Sap[a]		+			+			+			+			2443
Acer opalus															
Mill. (Italian Maple)															
1	Phloem exudate[b]	Enzym.			250										2444
Acer platanoides															
L. (Norway Maple)															
1	Seed, cotyledon[c]	Lumin.			5.71										2445
1	–, embryonic axis[c]	Lumin.			1.58										2445
Acer pseudoplatanus L. (Sycamore Maple)															
1	Leaf bud[b]	Lumin.	10	5	1										2446
1	Callus tissue[b]	TLC/TLC			680			110			50			130	2434
1	Cell suspension culture[b]	IEC	90	136	570	260		64	6			304		44	2447
1	Cell suspension culture	Lumin.			280	1260									2448
1	–, extract	^{31}P NMR			NMP, NDP, NTP, N = mainly A and U										51
1	–, *in vivo*	^{31}P NMR			NDP, NTP, N = mainly A and U										51
1	Cell suspension culture[f]	Enzym.	6	37	62										2449
Acer pseudosieboldianum (Pax) Kom. (Korean Maple)															
1	Sap		+			+			+			+			2443
Acer saccharum Marsh. (Sugar Maple)															
1	Embryonic axis[d]	Lumin.	690.8	180.2	46.9										2450
Actinidia arguta (Sieb. and Zucc.) Planch. ex Miq. (Tara Vine)															
1	Sap[a]	–	+			+			+			+			2451
Actinidia chinensis Planch. ssp. *hispida* G.F. Liang															
1	Seed[a]				+										2452
Actinidia deliciosa Liang and Ferg. (Kiwi)															
1	Fruit	Enzym.		14.6	33.6										222

Agropyron desertorum (Link.) Beauv. (Crested Wheatgrass)															
1	2-day seedling	IEC		+	++									+	831
Agrostemma githago L. (Cow Cockle)															
1	Embryo[g]	Lumin.			0.7										2453
Allium cepa L. (Onion)															
1	Seed[g]	Lumin.			0.30										2454
1	Leaf[d]	HPLC	59.1	155.4	475.9										2455
1	–, minus parenchyma[d]	HPLC	61.3	189.7	504.3										2455
1	–, minus vascular bundles[d]	HPLC	45.0	145.9	575.0										2455
1	Bulb	IEC	24						Trace			15			2456
1	Root[h]	Lumin.	0.38	0.71	1.17										2457
Allium fistulosum L. (Welsh Onion)															
1	Bulb	IEC	27									12			2456
Allium porrum L. (Leek)															
1	Seed embryo[g]	HPLC			0.250			Trace			0.030			0.100	2458
Alnus glutinosa (L.) Gaertn. (Black Alder)															
1	Leaf bud[d]	Lumin.	10	25	17										2446
Alocasia odora (Lodd.) Spach. (Asian Taro)															
1	Leaf (units unstated)	Lumin.	0.15	0.35	0.96										2459
Alyssum bertolonii Desv.															
1	Seed[g]	Enzym.	3.2	1.8	2.1										2460
Amaryllis vittata L'Her. (Amaryllis)															
1	Pollen	Enzym.	4300	2030	330										2461
Apium graveolens (L.) ssp. *dulce* (Celery)															
1	Petiole, vascular bundle	PC/PC		+	+										227
Arabidopsis thaliana (L.) Heynh. (Mouse-Ear Cress)															
1	Leaf			16.2	63.9									17.9	480
Arachis hypogaea L. (Peanut)															
1	Seed[a]	Fluorometric			Low										2462
1	–, embryonic axis[a]	Fluorometric			+										2463
1	Leaf[i]	IEC	27.5	11.5	45.0	4.0	27.0	16.0	25.0	17.0	55.0	2.0	18.0	10.5	2464

*nmol/g fwb unless stated otherwise. [a]Taken from the abstract. [b]µmol/L. [c]µg/plant part. [d]Dry weight basis. [e]nmol/10^7 cells. [f]ppm. [g]nmol/plant part. [h]nmol/mg insoluble polysaccharide. [i]µg/g. [j]µg/mL. [k]nmol/10^6 protoplasts. [l]µmol/m^2. [m]µmol/g. [n]µmol/10^7 cells. [o]nmol/mg protein. [p]nmol/mg chlorophyll. [q]Percentage of total free nucleotides. [r]m*M*. [s]mg/g. [t]Percentage P dwb. [u]Total extinction/callus (plant part). [v]counts per minute × 10^3.

Continued

Table 5-2 Ribonucleotides in Plants*—cont'd

No.	Plant part	Method	AMP	ADP	ATP	GMP	GDP	GTP	CMP	CDP	CTP	UMP	UDP	UTP	References
Arachis hypogaea L. ssp *fastigiata* (Spanish Peanut)															
1	Embryonic axis	Lumin.	16.1	41.0	248.0										2465
Arachis hypogaea L. ssp *hypogaea* (Virginia Peanut)															
1	Embryonic axis	Lumin.	16.8	41.0	138.3										2465
Asparagus officinalis L. (Asparagus)															
1	Green stem	IEC	110			Trace			60			60			2456
1	White stem	IEC	70			Trace			40			80			2456
1	Stem[d]	IEC	780	690	940							2210			2466
1	Stem (top, base)	Lumin.			+										2467
1	Stem/shoot	HPLC	48			6			3			84			2468
Avena fatua L. (Wild Oat)															
1	Shoot[d,i]	Lumin.			826.6										2469
1	3-day seedling[j]	IEC	93.5	12.5	24.0	32.5	49.0	10.0	46.5	7.5	29.0	19.0	23.0	7.5	2470
Avena sativa L. (Oat)															
1	Seed[d]	Enzym.			0										232
1	Seed	IEC	35	47	76	0	7	8	0	4	8	38	30	26	2471
1	Leaf, chloroplast[k]	Lumin.	0.90	1.10	0.80										2472
1	–, mitochondria[k]	Lumin.	0.22	0.56	0.85										2472
1	–, cytosol [k]	Lumin.	0.02	0.90	1.85										2472
1	Seedling	IEC	26.3	45.5	115.9	16.3	20.3	11.2	15.6*	63.4	15.8	229.7	15.3	46.3	2473
	*Incl. NAD														
1	–, coleoptile	Lumin.			121										2474
1	–, mesocotyl	Lumin.			62										2474
1	–, root and stem	IEC	+	+	+	+						+			2475
Avicennia germinans (L.) L. (Black Mangrove)															
1	Root[d]	Lumin.	10	525	1115										2476
Bambusa spinosa Roxb. (Bamboo)															
1	Stalk	IEC	33						15			40			2456
Beta vulgaris L. (Sugar Beet)															
1	Seed[g]	Lumin.			0.04										2477
2	Seed embryo	TLC	+	+	+							+	+	+	2478

2	Seed embryo[g]	Enzym.									0.179-0.202	2478
2	Axis	Lumin.			0.35-0.50			Other NTP 0.40-0.50				481
1	Leaf	PC/PC	8.8	15.3	26.5				18.6	14.1	8.2	2479
1	–, vascular bundle	PC/PC	17.0	7.2	9.7				19.3	30.0	7.2	2479
1	–, petiole parenchyma	PC/PC	2.8	0	2.1				8.0	6.3	2.4	2479
1	Leaf[d]	HPLC	259	419	802							2480
1	–, chloroplast[d]	Enzym.	1010	1230	790							2481
1	Leaf	Lumin.	238	542	583							2482
3	Leaf[l]	Lumin.			1420-2600		and variation among upper, middle, and lower leaves					2483
1	Root	Lumin.			11.8							2484
1	Root[i]	IEC	1570.6	+	2584.7	23.5	+	+	1403.5			2485
1	Taproot [d]	HPLC		<100						7200		2486
1	Fibrous root[d]	HPLC		400			165			1030		2486
Betula costata Trautv.												
1	Sap[a]		+			+		+	+			2451
Betula davurica Pall.												
1	Sap[a]		+			+		+	+			2451
Betula ermanii Cham.												
1	Bark[d]	Lumin.			152							2487
1	Xylem[d]	Lumin.			56							2487
Betula nigra L. (River Birch)												
1	Root[d]	Lumin.	0.1	0.4	1.8							2488
Betula pendula Roth (European White Birch)												
1	Root[d]	Lumin.	0.1	0.3	1.2							2488

*nmol/g fwb unless stated otherwise. [a]Taken from the abstract. [b]μmol/L. [c]μg/plant part. [d]Dry weight basis. [e]nmol/10^7 cells. [f]ppm. [g]nmol/plant part. [h]nmol/mg insoluble polysaccharide. [i]μg/g. [j]μg/mL. [k]nmol/10^6 protoplasts. [l]μmol/m^2. [m]μmol/g. [n]μmol/10^7 cells. [o]nmol/mg protein. [p]nmol/mg chlorophyll. [q]Percentage of total free nucleotides. [r]m*M*. [s]mg/g. [t]Percentage P dwb. [u]Total extinction/callus (plant part). [v]counts per minute $\times 10^3$.

Continued

Table 5-2 Ribonucleotides in Plants*—cont'd

No.	Plant part	Method	AMP	ADP	ATP	GMP	GDP	GTP	CMP	CDP	CTP	UMP	UDP	UTP	References
Betula platyphylla Suk. (Asian White Birch)															
1	Sap[a]		+			+			+			+			2451
Bidens pilosa L. (Bur Marigold)															
1	Cotyledon[g]	Lumin.			0.722										2489
Brassica campestris L. (Turnip Rape)															
1	Leaf[a]				+										2490
Brassica chinensis L. (Pak-Choi)															
1	Seed[a]	Fluorometric			+										2491
Brassica napus L. ssp. *oleifera* (Winter Rape)															
1	Seed[g]	Lumin.	1.47	0.33	0.18										2477
1	Leaf[d]	Lumin.	220	640	660										2492
1	Leaf[i]	Lumin.			107										2493
1	Shoot apex[i]	Lumin.			166										2493
1	Root[d,i]	Lumin.			157.2										2494
Brassica nigra (L.) Koch (Black Mustard)															
1	Leaf petiole cell suspension	Enzym.		4.7	36										485
Brassica oleracea L. ssp. *botrytis* L. (Cauliflower)															
1	Seed[g]	Lumin.			0.02										2477
1	Flowerhead	HPLC	8			99			n.d.			37			2468
Brassica oleracea L. ssp. *capitata* L. (Cabbage)															
1	Seed[d]	Lumin.	513.5 (incl. ADP and ATP)												2495
1	Leaf[i]	IEC	6	–	–	–			5			6			2496
1	Leaf	HPLC	3			n.d.			n.d.			36			2468
Brassica oleracea L. ssp. *caulorapa* (Kohlrabi)															
1	Seed[g]	Lumin.			0.53										2454
Brassica oleracea L. ssp. *italica* Plenck. (Broccoli)															
1	Stems and florets	HPLC	n.d.	131	149	21	29	51	147	152	196	92	71	183	2441
1	Flowerhead	HPLC	9			n.d.			n.d.			32			2468
Brassica pekinensis (Lour.) Rupr. (Chinese Cabbage, Pe-Tsai)															
1	Seed[g]	Lumin.	14400	2500	1000										2497
1	Seed	Fluorometric			+										2491
1	Leaf[i]	IEC	6	2	5	7			4			11			2496
1	Stem	PC/PC	2.3	1.6	0.04		0.2			0.2		0.5			2498
1	–, vascular bundle	PC/PC		+	+		+	+				+	+	+	36

1	–, vascular bundle	TLC/TLC		+	+		+	+				+	+	+	36
1	–, vascular bundle	TLC/TLE		+	+			+				+	+		36
Brassica rapa L. (Turnip Rape)															
1	Seed[d]	Lumin.	312.5 (incl. ADP and ATP)												2495
1	Leaf vascular bundle	TLC/TLC		+	+									+	240
Camellia sinensis L. (Tea)															
3	Green leaf[i]	IEC	83-104	+		+						151-216			2499
Camellia sinensis L. ssp. *assamica* (Mast.) Kitam. (Assam Tea)															
1	Shoot	IEC		41	100						9	11	15	32	2500
1	Shoot	Enzym.	2	51	80										2500
Cannabis sativa L. (Hemp)															
1	Leaf[d,i]	IEC	11.3	15.4	15.5										2501
Capsicum annuum L. (Bell Pepper)															
1	Seed[g]	Lumin.			0.44										2454
1	Leaf[d]	Lumin.			730										2502
1	Leaf, mature	HPLC	11	34	30							13	7	8	2503
1	–, young	HPLC	19	74	160							62	34	82	2503
1	–, petiole	HPLC	5	17	26							31	5	13	2503
1	Fruit	HPLC		++	++	+	+			+		++	+	+	2441
1	Fruit	HPLC	2			n.d.			n.d.			19			2468
Carpinus betulus L. (European Hornbeam)															
1	Phloem exudate[b]	Enzym.			94										2444
Carthamus tinctorius L. (Safflower)															
1	Seed	Fluorometric			+										2491
1	Leaf	HPLC	3	16	11							17	5	10	2503
1	Flower bud	HPLC	4	29	43							28	13	30	2503
Catharanthus roseus (L.) G. Don (Madagascar Periwinkle)															
1	Cell suspension culture[d,e]	Enzym.	49	87	21										2504
1	Cell suspension culture	HPLC	15.9	17.5	100.7	2.7	2.6	26.1	5.6	4.1	7.3	32.7	29.3	55.0	2505

*nmol/g fwb unless stated otherwise. [a]Taken from the abstract. [b]μmol/L. [c]μg/plant part. [d]Dry weight basis. [e]nmol/10^7 cells. [f]ppm. [g]nmol/plant part. [h]nmol/mg insoluble polysaccharide. [i]μg/g. [j]μg/mL. [k]nmol/10^6 protoplasts. [l]μmol/m^2. [m]μmol/g. [n]μmol/10^7 cells. [o]nmol/mg protein. [p]nmol/mg chlorophyll. [q]Percentage of total free nucleotides. [r]m*M*. [s]mg/g. [t]Percentage P dwb. [u]Total extinction/callus (plant part). [v]counts per minute × 10^3.

Continued

Table 5-2 Ribonucleotides in Plants*—cont'd

No.	Plant part	Method	AMP	ADP	ATP	GMP	GDP	GTP	CMP	CDP	CTP	UMP	UDP	UTP	References
1	Cell suspension culture[e]	Lumin.			2.1										2506
1	Cell suspension culture (*in vivo*)	^{31}P NMR		+	+										2507
Chenopodium rubrum L. (Red Goosefoot)															
1	Seedling	Lumin.			+										2508
1	Cell suspension culture	Enzym.	4	15	42							<1	1	5	2509
Chrysanthemum spp. (Chrysanthemum)															
1	Leaf[i]	IEC		286	217	201	88	105	176	52	258	211	250	188	2510
Cicer arietinum L. (Chickpea)															
1	Seedling, cotyledon[g]	Enzym.	105	10	575										247
1	–, embryonic axis[g]	Enzym.	1.5	3.5	11.5										247
Cichorium intybus L. (Chicory)															
1	Explant leaf[d,g]	IEC	32			50			32			45			2511
Citrullus vulgaris Schrad. (Watermelon)															
1	Seed[a]	PC/PE										+			2512
Citrus aurantium L. (Sour Orange)															
1	Fruit juice[b]	Lumin.			7.67										2513
–, var. "Bittersweet" (Bittersweet Orange)															
1	Fruit juice[b]	Lumin.			87.91										2513
Citrus grandis (L.) Osbeck (Pummelo)															
2	Fruit juice[b]	Lumin.			22.67-135.20										2513
Citrus limettioides Tanaka (Sweet Lemon)															
1	Fruit pulp	IEC	7.0	27.0	18.5										2514
Citrus limon L. (Lemon)															
1	Leaf[d]	PC/PE	8600	4200	1800										2515
1	Fruit pulp	IEC	5.5	21.3	15.8										2514
Citrus limonia Osbeck (Lime)															
2	Fruit juice[b]	Lumin.			3.27-4.89										2513
Citrus natsudaidai Hayata (Sour Orange)															
1	Leaf	Lumin.			200										2516
1	Juice vesicle	Lumin.			1910										2516

Citrus paradisi Macf. (Grapefruit)															
2	Fruit juice[b]	Lumin.			9.79-40.20										2513
Citrus paradisi Macf. × *Citrus reticulata* Blanco (Tangelo)															
1	Peeled pulp vesicles[i]	IEC	0.6-7.7	0-3.3		0-0.3				n.d.		0-1.0			2517
1	Peeled pulp vesicles[i]	Lumin.			20.6-25.8										2517
Citrus reticulata × *Fortunella* spp. (Calamondin)															
1	Fruit juice[b]	Lumin.			3.34										2513
Citrus reticulata × *Citrus sinensis* (Murcott)															
1	Fruit juice[b]	Lumin.			91.50										2513
Citrus sinensis (L.) Osbeck (Orange)															
1	Juice	IEC	+	+	+		+	+	+	+	+	+	+	+	248
4	Juice[j]	HPLC	0.5-1.8	6.0-26.0	1.6-5.4	0.2-1.4	0.8-4.3	0.2-0.9	0.8-2.8	2.8-19.0	0.4-0.9	0.3-1.8	2.3-7.6	0.3-2.3	2518
1	Juice[b]	Lumin.			64.90										2513
1	Juice vesicles[i] *Incl. GTP	IEC	1.8	0.2	2.0	2.2			4.6 (incl. NAD)			1.9	0.7	21.9*	2519
2	Peeled pulp vesicles[i]	IEC	0.3-5.0			0.9			0.3-5.6			0.8-1.2			2517
2	Peeled pulp vesicles[i]	Lumin.			5.2-19.9										2517
1	Juice vesicles, mature[b]	Enzym.		+	66										2520
1	–, immature[b]	Enzym.		+	151										2520
Citrus sinensis × *Citrus reticulata* (Orangeuma)															
1	Fruit juice[b]	Lumin.			127.60										2513
Cladrastis lutea (Michx.f.) Koch (Yellowwood)															
1	Phloem exudate[b]	Enzym.			310										2444
Coleus shirensis Gürke															
1	Leaf chloroplast[d]	Enzym.	840	440	480										2481

*nmol/g fwb unless stated otherwise. [a]Taken from the abstract. [b]μmol/L. [c]μg/plant part. [d]Dry weight basis. [e]nmol/10^7 cells. [f]ppm. [g]nmol/plant part. [h]nmol/mg insoluble polysaccharide. [i]μg/g. [j]μg/mL. [k]nmol/10^6 protoplasts. [l]μmol/m^2. [m]μmol/g. [n]μmol/10^7 cells. [o]nmol/mg protein. [p]nmol/mg chlorophyll. [q]Percentage of total free nucleotides. [r]m*M*. [s]mg/g. [t]Percentage P dwb. [u]Total extinction/callus (plant part). [v]counts per minute × 10^3.

Continued

Table 5-2 Ribonucleotides in Plants*—cont'd

No.	Plant part	Method	AMP	ADP	ATP	GMP	GDP	GTP	CMP	CDP	CTP	UMP	UDP	UTP	References
Corchorus capsularis L. (Jute)															
1	Seed	Enzym.	46.5	77.4	30.7										2521
Corylus avellana L. (European Hazel)															
1	Seed[d]	IEC	215	12	3							155			2522
Costus speciosus (J. Kõnig) Sm. (Crepe Ginger)															
1	Cell suspension culture	HPLC	26	48	158			+	+		+				2523
Cucumis melo L. (Melon)															
1	Seed[a]	PC/PE										+			2512
Cucumis sativus L. (Cucumber)															
1	Seed[g]	Lumin.			3.16										2454
1	Cotyledon[d]	Lumin.	65	60	0										2524
1	Leaf	IEC			29.1										250
1	Leaf	Enzym.			530										2525
1	Leaf	Lumin.	108	101	39										2526
1	Fruit	IEC	130						Trace			20			2456
1	Seedling	IEC		+	+										2527
1	Root	Enzym.			440										2525
1	Root	Lumin.	86	35	10										2526
Cucurbita maxima Duch. (Squash)															
1	Seed, cotyledon	Enzym.	19	74	290										2528
1	Leaf, sieve-tube exudate	TLC/TLC		+	+									+	240
Cucurbita pepo L. (Pumpkin)															
1	Young leaf	Lumin.			11800										2529
1	Leaf[d,m]	IEC	2.0	1.5	2.4	1.4	1.8	1.1	3.8			2.4	1.9	0.7	2530
1	–, chloroplast[d,m]	IEC	0.3	0.6	0.6										2530
1	Leaf mitochondria[d]	IEC	406	2779	302										1633
1	Stem exudate[b]	IEC	129	8	549	5	27	207				7	115	101	251
1	Unripe fruit exudate[b]	IEC	34	28	147	3	3	12				4	84	33	251
1	Ripe fruit exudate [b]	IEC	74	8	333	28	51	127				43	293	549	251
1	Sieve tube exudate[b]	Lumin.			17.6-32.9										2531
1	Root	IEC	16.6	20.7	1.4	9.9	7.6		2.7			14.3	0.7		2532
1	Root, meristem[n]	IEC	1.2	4.8	0.5	1.2	3		2.1			2.2	4.4		2533

1	–, growth zone[n]	IEC	1.2	22.6	1.6	2.8	6.2		6.2			12.2	3.1		2533
1	–, hair zone[n]	IEC	2.5	28.3	1.5	4.4	25.2		15			7.6	22		2533
Cyclobalanopsis globosa Lin and Liu															
1	Seed[a,d]	–	+		>20										2534
Datura inoxia Mill. (Downy Thorn Apple)															
1	Tissue culture	HPLC[3465]	22.2	16.2	37.4	31.9	10.8	18.5	n.d.	5.4	7.4	5.9	11.5	25.4	2535
1	Cell suspension culture	HPLC	2.2	7.1	46	0.4	1.8	11.0			2.2	12.7	2.4	13.1	2440
Daucus carota L. (Carrot)															
1	Root	IEC	+			0						+	+		2536
1	Cell suspension culture	TLC/TLC			65			18							2537
1	Cell suspension culture	HPLC		70	300		25	95		n.d.	n.d.		35	95	2538
1	Cell suspension culture (*in vivo*)	^{31}P NMR		+	+										2507
1	Callus suspension culture[l]	Lumin.	0.065	0.10	0.21										2539
Digitalis lanata Ehrh. (Grecian Foxglove)															
1	Leaf[l]	Lumin.	8.0	12.5	33.5										2540
Diospyros kaki Thunb. (Persimmon)															
1	Leaf tea[a]	–	+			+			++			+			2541
Distichlis spicata (L.) Greene (Coastal Saltgrass)															
1	Leaf[d]	Lumin.	27	74	160										2542
Drimys lanceolata (Poir.) Baill. (Pepper Tree)															
1	Leaf	IEC	11.3	12.1		9.4	20.2	1.6	5.2				Trace		2543
Echinochloa cruz-galli Beauv. ssp. *praticola* (Barnyard Grass)															
1	Seed[a]	–			+										2544
Egeria densa Planch. (Brazilian Waterweed)															
1	Leaf	PC/PC			+										2545
1	Leaf	Enzym.		20	73										495
1	Shoot[d]	Enzym.			700										2481
1	–, cytoplasm[d]	Enzym.			3200										2481

*nmol/g fwb unless stated otherwise. [a]Taken from the abstract. [b]μmol/L. [c]μg/plant part. [d]Dry weight basis. [e]nmol/10^7 cells. [f]ppm. [g]nmol/plant part. [h]nmol/mg insoluble polysaccharide. [i]μg/g. [j]μg/mL. [k]nmol/10^6 protoplasts. [l]μmol/m^2. [m]μmol/g. [n]μmol/10^7 cells. [o]nmol/mg protein. [p]nmol/mg chlorophyll. [q]Percentage of total free nucleotides. [r]m*M*. [s]mg/g. [t]Percentage P dwb. [u]Total extinction/callus (plant part). [v]counts per minute × 10^3.

Continued

Table 5-2 Ribonucleotides in Plants*—cont'd

No. Plant part	Method	AMP	ADP	ATP	GMP	GDP	GTP	CMP	CDP	CTP	UMP	UDP	UTP	References
Episcia reptans Mart. (Flame Violet)														
1 Leaf[d]	Lumin.	560	410	200										2546
Evodia rutaecarpa Hook.f. and Thomps. (Evodia)														
1 Phloem exudate[b]	Enzym.			350										2444
Fagopyrum esculentum Moench (Buckwheat)														
1 Seed[i]	Lumin.			12.6										2547
1 Seedling[a]			+	+										2548
Fagus sylvatica L. (European Beech)														
1 Leaf bud[d]	Lumin.	10	8	4										2446
1 Phloem exudate[b]	Enzym.			260										2444
Festuca arundinacea Schreb. (Reed Fescue)														
1 Leaf	Lumin.			0.033										2549
Festuca pratensis L. (Meadow Fescue)														
1 Leaf	Lumin.			0.125										2549
Ficaria verna Huds. (Lesser Celandene)														
1 Tuber[i]	TLC/TLC	16	50	155										2550
Fragaria spp. (Strawberry)														
1 Leaf	IEC		59	98		11	14.4	3	6	18	8.4	28	40	259
1 Leaf	Enzym.		65	92										259
Fraxinus americana L. (White Ash)														
1 Phloem exudate[b]	Enzym.			590										2444
1 Phloem exudate[b]	TLC/TLC			+			0						0	2444
Fraxinus excelsior L. (European Ash)														
1 Leaf bud[d]	Lumin.	0	15	17										2446
1 Apical bud[o]	Lumin.			2.0			(Other NTP 3.3)							2551
1 Phloem exudate[b]	Enzym.			270										2444
Fraxinus mandshurica Rupr. ssp. *japonica* Maxim. (Japanese Ash)														
1 Bark[d]	Lumin.			150										2487
1 Xylem[d]	Lumin.			43										2487

Ginkgo biloba L. (Ginkgo)															
1	Bark[d]	Lumin.			83										2487
1	Xylem[d]	Lumin.			7										2487
Glycine max (L.) Merr. (Soybean)															
1	Seed[d]	IEC	290	40	110	40	0	80*	0		180	90	70		2552
	*Incl. UTP														
1	Seed	IEC	26.0	28.0	34.7	0.3		Trace				+	2.6		2553
1	Seed[c,d]	Lumin.	22	2	1										2554
1	Seed[d]	Lumin.	584.0 (incl. ADP and ATP)												2495
1	Seed[i]	Lumin.			11.4										2555
2	–, axis[d,i]	Lumin.			0.65-2.57										2556
2	–, cotyledon[d,i]	Lumin.			0.20-1.04										2556
7	Embryonic axis[g]	Lumin.			2.98-4.28										2557
1	Leaf[p]	HPLC	8	96	124								8	40	26
1	Leaf[d]	IEC	150	1000	1280										2558
1	Leaf	Lumin.	89	357	464										2482
7	Leaf[o]	Lumin.			16.6-30.9										2559
1	Sprout	IEC	66									21			2456
1	3-day seedling, hypocotyl	IEC	+	+	+	+		+			+	+	+	+	2560
1	Fruit, cotyledon[d]	Lumin.	750 (incl. ADP and ATP)												2561
1	–, seed coat[d]	Lumin.	1300 (incl. ADP and ATP)												2561
1	–, pod wall[d]	Lumin.	300 (incl. ADP and ATP)												2561
1	Root[j]	IEC	60	1130	3600										2558
1	Root	HPLC (lumin.)	7 (18)	2 (3)	45 (67)			+		+					2562
1	Nodule	HPLC	259	140	134			+		+	+	+			2562
1	Nodule (*in vivo*)	^{31}P NMR			~350										2563
1	Nodule	Lumin.	14.4	102.0	148.1										2564
1	–, host cell fraction	Lumin.	8.1	86.7	103.5										2564
1	–, bacteroid fraction	Lumin.	6.3	15.3	44.6										2564
1	Root	Enzym.	4.4	14.1	37.7										2565
1	–, nodule	Enzym.	61.4	121	186	and distribution within the nodule									2565

*nmol/g fwb unless stated otherwise. [a]Taken from the abstract. [b]µmol/L. [c]µg/plant part. [d]Dry weight basis. [e]nmol/10^7 cells. [f]ppm. [g]nmol/plant part. [h]nmol/mg insoluble polysaccharide. [i]µg/g. [j]µg/mL. [k]nmol/10^6 protoplasts. [l]µmol/m^2. [m]µmol/g. [n]µmol/10^7 cells. [o]nmol/mg protein. [p]nmol/mg chlorophyll. [q]Percentage of total free nucleotides. [r]m*M*. [s]mg/g. [t]Percentage P dwb. [u]Total extinction/callus (plant part). [v]counts per minute × 10^3.

Continued

Table 5-2 Ribonucleotides in Plants*—cont'd

No.	Plant part	Method	AMP	ADP	ATP	GMP	GDP	GTP	CMP	CDP	CTP	UMP	UDP	UTP	References
Gossypium arboreum L. (Tree Cotton)															
1	Immature boll, seed, hair[g]	CC (cellulose)	40	78	20							88	57	47	2566
1	Ovule[d]	Enzym.	128	50	353										2567
1	Seed hair (lint)[d]	Enzym.	127	137	413										2567
Gossypium hirsutum L. (Upland Cotton)															
1	Seed	HPLC	+	+	+										2568
1	Seed[a]	Fluorometric			+										2491
1	Leaf[d]	IEC	331	439	444							137	972	166	2569
1	Leaf chloroplast[a]		+	+	0	+		+				+		+	2570
1	Raw fiber	IEC	+	+	+	+	+	+					+	+	2571
1	Raw fiber	HPLC	+	+	+	+	+	+				+	+	+	2571
1	Ovule[d]	Enzym.	136	60	362										2567
1	Seed hair (lint)[d]	Enzym.	217	103	401										2567
1	Root[d]	IEC	162	478	849							80	1387	676	2569
1	Seedling, leaf *Incl. UDPG	IEC	114	139	169	42.4	30.6	84.2	8.4	Trace	53.8	40.5	246*	37.3	2572
1	–, root *Incl. UDPG	IEC	82	53.2	166.5	12.8	15.5	70.5	3.8	Trace	53.2	30.3	438*	17	2572
2	Seedling, root[d]	Lumin.	0.08	0.38-0.63	0.72-0.97										3104
Hedera helix L. (English Ivy)															
1	Leaf	HPLC	29.2	30.1	42.7	15.3	7.8	7.2	6.3	5.0	19.8	25.4	21.5	18.7	2573
Helianthus annuus L. (Sunflower)															
1	Seed[d]	Lumin.	177.1 (incl. ADP and ATP)												2495
1	Leaf[l]	Lumin.	7.4	13.5	19.8										2574
1	Leaf[d]	IEC	10	680	800										2558
1	Stem	^{31}P NMR			+										2563
1	Root[d]	^{31}P NMR	40	220	730										2558
Helianthus tuberosus L. (Jerusalem Artichoke)															
1	Shoot	Lumin.	9.0	4.2	63.0										2575
1	Young stolon (*in vivo*)	^{31}P NMR		+	+										45

1	Tuber	IEC	5	11	11	2	2	3				6	2	4	2576
1	Tuber	Lumin.	22.2	8.2	23.0										2575
Hemerocallis hybrid (Daylily)															
1	Flower petal[q] (*incl. CDP)	TLC/TLC	3.8	8.7	43.4		2.9*	6.8 (incl. CTP)				3.5	1.9	9.3	266
Heracleum mantegazzianum Somm. and Lev. (Giant Hogweed)															
1	Phloem[d]	Enzym.			650										2444
1	Xylem[d]	Enzym.			360										2444
1	Parenchyma[d]	Enzym.			240										2444
Heracleum sosnowskii Mandan (Cow Parsnip)															
1	Vascular bundle, phloem	PC/PC	133	92	81	14	26 (incl. GTP)			16	22	268	199*	64	2577
1	–, xylem (*incl. UDPX)	PC/PC	51	24	50	Trace	20 (incl. GTP)			0	8	156	159*	34	2577
Hevea brasiliensis Müll. (Pará Rubber Tree)															
1	Latex[d]	PE		74	15					39	48		30		2578
1	Latex[b]	Lumin.	26	30	77										2579
Hibiscus cannabinus L. (Kenaf)															
1	Leaf[a]				+										2580
Hippophaë salicifolia D. Don (Willow-Leaved Sea Buckthorn)															
1	Phloem exudate[b]	Enzym.			360										2444
Hordeum distichon Alef. (Two-Rowed Barley)															
1	Seed[a]	PC/PE										+			2512
– ssp. *nutans* (Rode) Alef.															
3	Leaf	PC	+	+	+	+		+	+			+		+	2581
2	Etiolated seedling	PC	+	+	+	+		+	+			+		+	2581
1	1-day seedling, cotyledon[d,g]	Enzym.		45	37.5										267
1	Crown[o]	Lumin.	0.18	1.90	2.19										2582
Hordeum hexastichon Alef. ssp. *parallelum* Körn (Six-Rowed Barley)															
1	Leaf	PC	+	+	+	+		+	+			+		+	2581

*nmol/g fwb unless stated otherwise. [a]Taken from the abstract. [b]µmol/L. [c]µg/plant part. [d]Dry weight basis. [e]nmol/10^7 cells. [f]ppm. [g]nmol/plant part. [h]nmol/mg insoluble polysaccharide. [i]µg/g. [j]µg/mL. [k]nmol/10^6 protoplasts. [l]µmol/m^2. [m]µmol/g. [n]µmol/10^7 cells. [o]nmol/mg protein. [p]nmol/mg chlorophyll. [q]Percentage of total free nucleotides. [r]m*M*. [s]mg/g. [t]Percentage P dwb. [u]Total extinction/callus (plant part). [v]counts per minute × 10^3.

Continued

Table 5-2 Ribonucleotides in Plants*—cont'd

No.	Plant part	Method	AMP	ADP	ATP	GMP	GDP	GTP	CMP	CDP	CTP	UMP	UDP	UTP	References
Hordeum vulgare L. (Barley)															
1	Seed[a]	PC/PE										+			2512
1	Seed	IEC	33	43	40	0	8	7	0	3	5	37	29	18	2471
2	Grain, dehusked[g]	Enzym.											2.5-3.5	4.0-7.5	502
14	Seed, spring[g]	Lumin.			0.057-0.083										2583
13	Seed, winter[g]	Lumin.			0.057-0.087										2583
1	Leaf[i]	Enzym.	6.4	12.1	12.5										2584
1	Leaf	HPLC	20.4	26.3	26.8										2585
1	Leaf	Lumin.			245										524
1	Shoot[d,i]	Lumin.			103.8										2469
1	Root	PC/PC		+	+										2586
1	20-day root	HPLC		81	166		44	19			82		57		2587
1	Root	Lumin.	520	773	1385										2588
1	Root tip (*in vivo*)	^{31}P NMR			+										2589
Hygrohypnum luridum (Hedw.) Jenn. (Hygrohypnum Moss)															
1	Whole moss, hydrated	Lumin.			570										2590
Ilex paraguariensis St. Hil. (Maté)															
1	Leaf	HPLC	71	110	121	9	21	30							2591
Impatiens balsamina L. (Garden Balsam)															
1	Root tip[d,m]	Enzym.			13.3										497
Ipomoea batatas (L.) Lam. (Sweet Potato)															
1	Tuber	HPLC	6			n.d.			100			34			2468
1	Root	IEC	43.6	2.2	35.9							18.5		8.4	2592
1	Root	IEC	12.3	68.9	131										2593
Juglans regia L. ssp. *orientis* Sarg. (English Walnut)															
1	Bark[d]	Lumin.			210										2487
1	Xylem[d]	Lumin.			117										2487
Juncus roemerianus Scheele (Needlegrass Rush)															
1	Stem base[d,i]	Lumin.	+	+	40.0										2594
1	Root[d,i]	Lumin.	+	+	1.3										2594
1	Rhizome[d,i]	Lumin.	+	+	7.5										2594
1	–, less bark[d,i]	Lumin.	+	+	22.4										2594
1	–, bark[d,i]	Lumin.	+	+	6.8										2594

1	–, tip[d,i]	Lumin.	+	+	99.9									2594
Kalanchoë blossfeldiana v. Poelln. (Flaming Katy)														
1	Seed[d,g]	Lumin.	880	310	110									2595
1	Leaf	Lumin.			32									2596
Kalanchoë daigremontiana Ham. and Perr. (Devil's Backbone)														
1	Leaf	Lumin.	14	14	27									2596
Kalanchoë tubiflora (Harv.) Hamet (Chandelier Plant)														
1	Phyllodium[d]	Lumin.			230									2597
Koelreuteria paniculata Laxm. (Golden Rain Tree)														
1	Phloem exudate[b]	Enzym.			220									2444
Lactuca sativa L. (Lettuce)														
1	Seed	HPLC	+	+	+	+	+	+		+			+	2598
2	Seed[d]	Lumin.	164-219	74-76	10									2599
1	Seed[g]	Lumin.			0.24									2454
3	Seed, imbibed	Lumin.			609-698									2600
1	Leaf	IEC	27			Trace					16			2456
1	Leaf[i]	IEC	5	5	6	7			2		6			2496
1	Sprout (extract)	^{31}P NMR	+											48
Larix decidua Mill. (European Larch)														
1	Cambium	IEC	25.3	4.7	2.8	3.9	1.7	0.8			19.3	4.4	2.6	2601
1	Young xylem	IEC	3.3	0.2		0.4					1.7	0.1		2601
Leleba oldhamii (Munro) Nakai (Summer Bamboo)														
1	Desheathed shoot	IEC	67.5	8.8	23.8	7.5					12.6	5.0		2602
Lemna gibba L. (Gibbous Duckweed)														
1	Whole plant	Enzym.	13	47	99 (lumin.)									2603
Lemna paucicostata Hegelm. (Lesser Duckweed)														
1	Frond	Lumin.			156									2604
Lens culinaris Medik. (Lentil)														
1	Sprout	HPLC	+			+					+			2605

*nmol/g fwb unless stated otherwise. [a]Taken from the abstract. [b]μmol/L. [c]μg/plant part. [d]Dry weight basis. [e]nmol/10^7 cells. [f]ppm. [g]nmol/plant part. [h]nmol/mg insoluble polysaccharide. [i]μg/g. [j]μg/mL. [k]nmol/10^6 protoplasts. [l]μmol/m^2. [m]μmol/g. [n]μmol/10^7 cells. [o]nmol/mg protein. [p]nmol/mg chlorophyll. [q]Percentage of total free nucleotides. [r]m*M*. [s]mg/g. [t]Percentage P dwb. [u]Total extinction/callus (plant part). [v]counts per minute × 10^3.

Continued

Table 5-2 Ribonucleotides in Plants*—cont'd

No.	Plant part	Method	AMP	ADP	ATP	GMP	GDP	GTP	CMP	CDP	CTP	UMP	UDP	UTP	References
Lilium henryi Bak. (Lily)															
1	Pollen	PC/PC	+			+			+						2606
Linum usitatissimum L. (Flax, Linseed)															
1	Seed	IEC	+			+			+			+			2607
1	Seed[a]	Lumin.	181.0 (incl. ADP and ATP)												2495
10	Seed	HPLC	71.1-73.4	14.5-25.3	7.6-227.4										2608
Liriodendron tulipifera L. (Tulip Tree)															
1	Leaf	IEC	13.0			16.4			0.5			6.1	4.9	7.4	2543
Lolium perenne L. (Perennial Ryegrass)															
1	Leaf	Lumin.			160										2609
Lotus corniculatus L. (Birdsfoot Trefoil)															
1	Root[b]	HPLC		25	47										508
Lupinus luteus L. (Yellow Lupine)															
1	Seed	IEC	++	++	+	+						+			2437
1	–, axis	Lumin.			100										2610
1	–, cotyledon	Lumin.			10										2610
Lycium chinense Mill. (Box Thorn)															
1	Leaf	IEC							+			+			2611
Lycopersicon esculentum Mill. (Tomato)															
1	Leaf[d]	PC/PC	105	340	615		30	150			250	45	70	400	2612
1	Leaf	IEC	20.7	69.1	143.8		11.7	27.5			8.2	15.9		9.5	2613
2	Fruit, immature, whole	PC/PE	2.7	3.2	1.6-2.2	0.2-0.9	0.5-0.8					0.2	0.2-0.5		2614
1	Fruit	IEC	300						14			68			2456
1	Fruit[m]	IEC	11.5	4.8	38.4							27		75	2615
1	Fruit[i]	Enzym.	0.5	4	16										2616
1	Fruit, pericarp[b]	HPLC	100												2617
1	–, locular	HPLC	40												2617
1	Seedling	IEC		+	+										2527
1	Root tip	HPLC	11.2	3.5	6.5	6.0	1.5	2.9	31.0	2.7	2.7	55.0	3.4	2.9	2618

Machilus thunbergii Sieb. and Zucc.															
1	Seed[a,d]		+		>20										2534
1	Seed[d]	Lumin.	13.3	34.6	55.0										2619
Magnolia grandiflora L. (Southern Magnolia)															
1	Leaf	IEC	27.5	5.1		16.2	10.0	1.5	6.2			12.1	8.1		2543
Magnolia kobus DC ssp. *borealis* Sarg.															
1	Bark[d]	Lumin.			466										2487
1	Xylem[d]	Lumin.			58										2487
Malus baccata (L.) Borkh. (Siberian Crabapple)															
1	Root	Lumin.			17.42										2620
Malus domestica Borkh. (Apple)															
1	Seed, embryonic axis[g]	Lumin.			0.216				(Other NTP 0.115)						2621
1	–, cotyledon[g]	Lumin.			1.254				(Other NTP 0.655)						2621
1	Leaf	IEC	+	+	+		+	+			+				2622
3	Phloem exudate[f]	HPLC	1-2	4-30	2-14	1-2	1-3	2-5	3-7	1-2	1-2	6-24	4-14	1	2623
1	Fruit[i]	Lumin.			2000										2467
1	Bark[d]	Lumin.			154										2487
1	Xylem[d]	Lumin.			42										2487
Malus sylvestris (L.) Mill. (Crabapple)															
1	Stem, phloem tissue	PC/PC		+	+										227
Malus xiaojinensis Cheng and Jiang															
1	Root	Lumin.			9.98										2620
Medicago sativa L. (Alfalfa)															
1	Shoot	PC/PC	+	+	+										280
1	Root	PC/PC	+	+	+										280
1	Root[d]	HPLC		17	31										508
Metasequoia glyptostroboides Hu and Cheng (Dawn Redwood)															
1	Bark[d]	Lumin.			60										2487
1	Xylem[d]	Lumin.			31										2487
Michelia compressa (Maxim.) Sargent															
1	Seed[a,d]		+		<10										2534

*nmol/g fwb unless stated otherwise. [a]Taken from the abstract. [b]μmol/L. [c]μg/plant part. [d]Dry weight basis. [e]nmol/10^7 cells. [f]ppm. [g]nmol/plant part. [h]nmol/mg insoluble polysaccharide. [i]μg/g. [j]μg/mL. [k]nmol/10^6 protoplasts. [l]μmol/m^2. [m]μmol/g. [n]μmol/10^7 cells. [o]nmol/mg protein. [p]nmol/mg chlorophyll. [q]Percentage of total free nucleotides. [r]m*M*. [s]mg/g. [t]Percentage P dwb. [u]Total extinction/callus (plant part). [v]counts per minute × 10^3.

Continued

Table 5-2 Ribonucleotides in Plants*—cont'd

No.	Plant part	Method	AMP	ADP	ATP	GMP	GDP	GTP	CMP	CDP	CTP	UMP	UDP	UTP	References
Morus alba L. (White Mulberry)															
1	Shoot, debarked	Lumin.	5	8	4										2624
1	–, bark	Lumin.	12	25	10										2624
Morus bombycis Koidz.															
1	Bark[d]	Lumin.			213										2487
1	Xylem[d]	Lumin.			86										2487
Mucuna pruriens (L.) DC (Cowitch Bean)															
1	Leaf[f]	Ennzym.	20	20											282
Musa × paradisiaca L. (Banana)															
1	Fruit	Enzym.			113										2625
Neolitsea parvigemma (Hayata) Kaneh. and Sas.															
1	Seed[a,d]		+		>20										2534
Nicotiana glutinosa L. (Tobacco)															
1	Leaf[d]	Lumin.		231	223										2626
Nicotiana tabacum L. (Tobacco)															
1	Leaf	PC/PC		+	+										2627
1	Leaf	HPLC	9.9	15.9	28	0.5	2.4	4.7		3.8	6.0	3.3	4.4	11.6	2440
			and variation from top (apex) leaf to lowest (senescent) leaf of a single plant												2628
1	Leaf	Enzym.	14.4	30.8	71.2										287
1	Pollen grain (*in vivo*)	^{31}P NMR			+										2563
1	Root	PC/PC		+	+										2627
1	Root	HPLC	1.0	7.8	15	0.4	0.5	2.3				4.9	1.0	6.7	2440
1	Cell suspension culture	HPLC	3.9	13.5	85		3.2	15.4		1.2	3.2	19.1	12.2	26	2440
1	Cell suspension culture (extract or *in vivo*)	^{31}P NMR		+	+										2629
1	Callus tissue[d]	TLC/TLC			780			150			270			170	2434
1	Callus culture	Lumin.	20	50	155										2630
1	Seed	Enzym.	20	140	370										2631
1	Seed	Lumin.			0										2632
1	Seed[d]	Lumin.	60.9 (incl. ADP and ATP)												2495

Oryza sativa L. (Rice)															
1	Seed[a]	Fluorometric			+										2491
1	Kernel	IEC	113.04	68.46	35.18		2.96		2.16	1.36	6.64	44.58	11.98	9.88	2633
1	Seed embryo	Enzym.	44	107	76										2634
1	Ripening grain, 10 daf	IEC	30	72	84	31	7	8				67	67	50	2635
5	Leaf[c]	Lumin.			13.8-21.6										2559
1	Leaf and culm	IEC	++									+	+	+	2636
1	Leaf sheath phloem sap[r]	HPLC	Trace	0.31	1.63			0.21			0.08			0.27	2637
1	Upper internode phloem sap[r]	HPLC	Trace	0.27	1.76			0.22			0.04			0.10	2637
1	Shoot	Lumin.			130										2638
1	Shoot (*in vivo*)	^{31}P NMR				NTP									2638
1	Root	HPLC	350	1660	2660										2639
1	Seedling, root[a]	IEC	+	+	+										2640
1	Cell suspension culture	IEC	48.4	66.1	63.1*	14.9	50.0		*Incl. UDP			78.3	32.0	7.4	2641
Panicum hemitomon Schult.															
1	Root[d]	Lumin.	72	38	465										2642
Panicum miliaceum L. (Proso Millet)															
1	Leaf mesophyll chloroplast[p]	Lumin.			4.3										2643
Persea americana Mill. (Avocado)															
1	Fruit[s]	HPLC		80	80										515
1	Fruit (*in vivo*)	^{31}P NMR		+	+										515
Phacelia tanacetifolia Benth. (Fiddleneck)															
1	Seed[g]	Lumin.			0.019										2644
1	Sprout (*in vivo*)	^{31}P NMR					NDP, NTP								2644
Pharbitis nil Choisy (Morning Glory)															
1	Seedling, hypocotyl[d]	Lumin.	190	700	2030										2645
1	–, plumule[d]	Lumin.			4000										2645
1	–, cotyledon	Lumin.			5										2646
1	–, apex	Lumin.			150										2646

*nmol/g fwb unless stated otherwise. [a]Taken from the abstract. [b]μmol/L. [c]μg/plant part. [d]Dry weight basis. [e]nmol/10^7 cells. [f]ppm. [g]nmol/plant part. [h]nmol/mg insoluble polysaccharide. [i]μg/g. [j]μg/mL. [k]nmol/10^6 protoplasts. [l]μmol/m^2. [m]μmol/g. [n]μmol/10^7 cells. [o]nmol/mg protein. [p]nmol/mg chlorophyll. [q]Percentage of total free nucleotides. [r]m*M*. [s]mg/g. [t]Percentage P dwb. [u]Total extinction/callus (plant part). [v]counts per minute × 10^3.

Continued

Table 5-2 Ribonucleotides in Plants*—cont'd

No.	Plant part	Method	AMP	ADP	ATP	GMP	GDP	GTP	CMP	CDP	CTP	UMP	UDP	UTP	References
Phaseolus vulgaris L. (Bush Bean)															
1	Seed	IEC	++	++	+	+						+			2437
1	Seed[d]	HPLC	1630			1040									2647
1	Seed	Enzym.	20	140	370										2631
1	Seed[i]	Lumin.			21.0										2547
1	Embryo[d,i]	IEC	8.4			1.7			n.d.			n.d.			2648
1	Green bean	IEC	53						26			81			2456
1	Leaf[q]	PC/PC								1.5	0.9	8.3	10.5	4.2	12
1	Leaf.	IEC	210	190	140										22
1	Leaf	IEC	19.9	54.2	77.5		8.8	15.8			4.3	8.6		5.4	2649
1	Leaf[d]	Lumin.	890	600	400										2546
1	Etiolated leaf[g]	IEC	3.3		2.7	0.9	0.9		3.4			3.5			2650
1	Leaf	HPLC	22			n.d.			n.d.			70			2468
1	Leaf, mature	HPLC[2651]	59	100	53		14	10			3	17	24	18	2652
1	–, young	HPLC	50	160	162		41	49			20	66	33	55	2652
1	Leaf, petiole	IEC	1	3	33	–		6	1		1	0.5	–	9	2653
1	–, laminae	IEC	12	18	69	7		8				15		14	2653
1	Stem	IEC	22	40	56.5							75	43		2654
1	Fruit, pericarp	IEC	17	11	69							22		54	520
1	Whole seedling	Lumin.	21.5	24.0	81.0										2655
1	Seedling	IEC	25.0	25.0	80.0	10.5		18.0	4.5		6.5	19.5	8.0	21.0	2656
1	–, leaf	IEC	30	60	280	0.0		40				25		60	2657
1	–, leaf	Lumin.			320										2474
1	–, root	IEC	15	22	110	10		140				15		47	2657
1	–, root	Lumin.	5	9	56										2658
Phaseolus vulgaris L. ssp. *napus* (Dwarf Common Bean)															
1	Leaf[d,i]	Lumin.			134										2659
Phellodendron wilsonii Hayata and Kaneh.															
1	Seed[a,d]		+		<10										2534

Phragmites australis (Cav.) Steud. (Reed)														
1	Root		20.8 (incl. ADP and ATP)											2660
1	Rhizome		51.2 (incl ADP and ATP)											2660
Phyllostachys bambusoides Sieb. and Zucc. (Timber Bamboo)														
1	Sap[a]		+			+			+			+				2451
Phyllostachys nigra (Lodd ex Lindl.) Munro (Black Bamboo)														
1	Sap[a]		+			+			+			+				2451
Phyllostachys pubescens Mazel ex Houz. (Moso Bamboo)														
1	Desheathed shoot	IEC	58.4	13.6	4.8	12.4	2.4					15.6	12.8		2661
1	Sap[a]		+			+			+			+			2451
Picea abies (L.) Karst. (Norway Spruce)														
1	Seed, endosperm[d]	Enzym.		+	100										523
1	–, embryo[d]	Enzym.		+	430										523
1	Needle (leaf)[d]	HPLC	5.8	61.0	54.4										2662
1	Needle (leaf)[d]	Lumin.		65	321										2663
Pinus banksiana Lamb. (Jack Pine)														
1	Seed[g]	Lumin.			5.4										2664
1	11-day seedling[q]	IEC	57.0	14.0	11.0	1.0	2.0	1.0	4.0	1.0	0.6	1.0	1.0	0.4	2665
Pinus contorta × *Pinus banksiana* (Lodgepole-Jack Pine)														
1	Needle (leaf)[d]	Lumin.			658										2666
Pinus koraiensis Sieb. and Zucc. (Korean Pine)														
1	Bark[d]	Lumin.			240										2487
1	Xylem[d]	Lumin.			72										2487
Pinus lambertiana Dougl. (Sugar Pine)														
1	Seed[d]	Lumin.			47										2667
1	–, embryo	Lumin.			370										2668
1	–, megagametophyte	Lumin.			200										2668
Pinus mugo Turra (Mugo Pine)														
1	Pollen[d]	TLC/TLC			1350			370			310			700	2434
1	Pollen	Lumin.			300										2669

*nmol/g fwb unless stated otherwise. [a]Taken from the abstract. [b]μmol/L. [c]μg/plant part. [d]Dry weight basis. [e]nmol/10^7 cells. [f]ppm. [g]nmol/plant part. [h]nmol/mg insoluble polysaccharide. [i]μg/g. [j]μg/mL. [k]nmol/10^6 protoplasts. [l]μmol/m^2. [m]μmol/g. [n]μmol/10^7 cells. [o]nmol/mg protein. [p]nmol/mg chlorophyll. [q]Percentage of total free nucleotides. [r]m*M*. [s]mg/g. [t]Percentage P dwb. [u]Total extinction/callus (plant part). [v]counts per minute × 10^3.

Continued

Table 5-2 Ribonucleotides in Plants*—cont'd

No.	Plant part	Method	AMP	ADP	ATP	GMP	GDP	GTP	CMP	CDP	CTP	UMP	UDP	UTP	References
Pinus ponderosa Laws. (Ponderosa Pine)															
1	Mature seed, embryo[g]	Lumin.	0.130	0.350	0.060										2670
1	–, gametophyte[g]	Lumin.	0.600	0.480	0.060										2670
Pinus resinosa Ait. (Red Pine)															
1	Ovule, embryo	Lumin.			0.72										2671
1	–, gametophyte	Lumin.			0.10										2671
Pinus strobus L. (Eastern White Pine)															
1	Bark [d]	Lumin.			101										2487
1	Xylem[d]	Lumin.			29										2487
Pinus sylvestris L. (Scotch Pine)															
1	Seed, endosperm[d]	Enzym.		+	100										523
1	–, embryo[d]	Enzym.		+	250										523
1	Needle (leaf)[d]	Lumin.	37	72	176										2672
1	Root	Lumin.	6	34	101										2673
Pistacia mutica Fisch. and Mey (Mt. Atlas Mastic Tree)															
1	Leaf[i]	Lumin.			6.0										2674
Pisum sativum L. (Pea)															
1	Mature seed	PC/PC		++	+++			+						+	52
1	Mature seed	Enzym.		198	374										52
1	Seed	Enzym.	21	102	286										296
2	Seed embryo	Enzym.												175-351	295
1	Seed[i]	PC/PC	2.7	38.0	8.1										2675
1	Seed[d]	Lumin.	1000.0 (incl. ADP and ATP)												2495
1	Leaf[i]	PC/PC	37.9	149.0	131.4										2675
1	Leaf[i]	Enzym.	5.25	17.8	16.5										2584
1	Stem[i]	PC/PC	64.6	182.0	182.0										2675
1	Seed	IEC	447	218	86	13				Trace		57			2437
1	Seed	HPLC	11	116	133	9	20	27	n.d.	n.d.	39	96	86	289	2441
1	Immature seed with pod	IEC	55						Trace			40			2456
1	Seed coat	Enzym.												272	2676
1	Leaf chloroplast	PC/PC	+	+	+										2677

1	14-day shoot[d]	PC						1708		+			2678
1	2-day cotyledon	Enzym.	70	30	230								2679
1	Seedling, cotyledon	Enzym.									32	67	527
1	Shoot	PC/PC	+	+	+		+				+	+	2680
1	Whole seedling	PC/PC		+	+					+	+	+	2681
1	Whole plant	PE			15								2682
1	Root[i]	PC/PC	42.2	105.5	85.2								2675
1	Root	PC/PC	+	+	+		+				+	+	2680
1	Root	IEC	+	+	+					+	+		2654
1	Root[d]	CC (cellulose)	1490	790	4400		200		570			860	2683
1	Root tip	Enzym.	8.65	23.68	88.1								2684
1	Root tip (*in vivo*)	^{31}P NMR					NDP, NTP						526
Platanus occidentalis L. (Eastern Sycamore)													
1	Bark[d]	Lumin.			157								2487
1	Xylem[d]	Lumin.			81								2487
Poa annua L. (Annual Bluegrass)													
1	Root[t]	HPLC			10^{-4}								696
Podocarpus macrophyllus (Thunb.) Sweet (Yew Plum Pine)													
1	Seed[a,d]		+		>20								2534
Populus gelrica (Houtz.) Houtz.													
1	Twig, living bark[d]	Lumin.	23.5	14.1	84.3								2685
1	–, xylem[d]	Lumin.	25.7	41.2	177.9								2685
Portulaca grandiflora Hook. (Rose Moss)													
1	Callus[u]	IEC	35.46			12.53		32.85		4.5			2686
Prunus padus L. (Bird Cherry)													
1	Phloem exudate[b]	Enzym.			440								2444
Prunus persica (L.) Batsch. (Peach)													
1	Fruit flesh[v]	Lumin.		143.7	26.0								2687
2	Leaf bud[o]	Lumin.			4-20			(Other NTP 7-22)					2688

*nmol/g fwb unless stated otherwise. [a]Taken from the abstract. [b]μmol/L. [c]μg/plant part. [d]Dry weight basis. [e]nmol/10^7 cells. [f]ppm. [g]nmol/plant part. [h]nmol/mg insoluble polysaccharide. [i]μg/g. [j]μg/mL. [k]nmol/10^6 protoplasts. [l]μmol/m^2. [m]μmol/g. [n]μmol/10^7 cells. [o]nmol/mg protein. [p]nmol/mg chlorophyll. [q]Percentage of total free nucleotides. [r]m*M*. [s]mg/g. [t]Percentage P dwb. [u]Total extinction/callus (plant part). [v]counts per minute × 10^3.

Continued

Table 5-2 Ribonucleotides in Plants*—cont'd

No.	Plant part	Method	AMP	ADP	ATP	GMP	GDP	GTP	CMP	CDP	CTP	UMP	UDP	UTP	References
Prunus persica ssp. *nectarina* (Nectarine)															
1	Fruit flesh[v]	Lumin.			75.5										2687
Prunus sargentii Rehd. (Sargent Cherry)															
1	Bark[d]	Lumin.			86										2487
1	Xylem[d]	Lumin.			32										2487
Pseudotsuga menziesii (Mirb.) Franco (Douglas Fir)															
1	Pollen	Lumin.		40-590											2442
Pteridium aquilinum (L.) Kuhn (Bracken Fern)															
1	Rhizome[d]	IEC			1.5	0.4						2.8			2689
Pyrus communis L. (Pear)															
1	Fruit pulp[i]	Enzym.			18.9										2690
1	Bark[d]	Lumin.			89										2487
1	Xylem[d]	Lumin.			69										2487
Quercus borealis Michx. f. (Red Oak)															
1	Phloem exudate[b]	Enzym.				0									2444
Quercus mongolica ssp. *grosseserrata* Rehd. and Wils.															
1	Bark[d]	Lumin.			95										2487
1	Xylem[d]	Lumin.			57										2487
Quercus petraea L. ex Liebl. (Durmast Oak)															
1	Phloem exudate[b]	Enzym.			68										2444
Quercus phellos L. (Willow Oak)															
1	Phloem exudate[b]	Enzym.			68										2444
Quercus robur L. (English Oak)															
1	Leaf bud[d]	Lumin.	4	3	1										2446
Raphanus acanthiformis Backer (Japanese Radish)															
1	Fruit	IEC	38						Trace			44			2456
Raphanus sativus L. (Radish)															
1	Seed[a]	Fluorometric			+										2491
1	Seed[g]	Lumin.	2.2	0.9	<0.1										2691
1	Seed[d]	Lumin.	578.9 (incl. ADP and ATP)													2495

1	Seedling	Lumin.	29	46	211									2692
1	Root	HPLC	4			n.d.			40		35			2468
Ribes sinanense F. Maekawa														
1	Bark[d]	Lumin.			206									2487
1	Xylem[d]	Lumin.			44									2487
Ricinus communis L. (Castor Bean)														
1	Phloem exudate[b]	TLC/TLC		+	400-600									2693
1	Phloem exudate[b]				730								140	2694
1	Phloem exudate[b]	Enzym.	20	80	1300									304
Robinia pseudoacacia L. (Black Locust)														
1	Phloem exudate [b]	Enzym.	6.8	7.4	50									2444
1	Phloem exudate[b]	TLC/TLC			+			0					0	2444
Rosa damascena Mill. (Damask Rose)														
1	Cell susp. culture (*in vivo*)	^{31}P NMR				NDP, NTP								2695
Saccharum officinarum L. (Sugar Cane)														
1	Leaf[q]	PC/PC	19	29	7		–	12			21	–		306
1	Root[q]	PC/PC	11	21	15		5	9			17	12		306
1	Meristem	PC/PC	+						+		+			2696
Salix alba L. (White Willow)														
1	Phloem exudate[b]	Enzym.			360									2444
Salix babylonica L. (Weeping Willow)														
1	Bark[d]	Lumin.			27									2487
1	Xylem[d]	Lumin.			29									2487
Salix daphnoides Vill.														
1	Phloem exudate[b]	Enzym.			250									2444
Salix matsudana Koidz. ssp. *tortuosa* Vilm. (Dragon-Claw Willow)														
1	Bark[d]	Lumin.			118									2487
1	Xylem[d]	Lumin.			51									
Salix triandra L. (Almond Willow)														
1	Phloem exudate	TLC/TLC		+	+					+				309
Salix viminalis L. (Basket Willow)														
1	Sieve-tube sap[i]	Lumin.			850									2697

*nmol/g fwb unless stated otherwise. [a]Taken from the abstract. [b]µmol/L. [c]µg/plant part. [d]Dry weight basis. [e]nmol/10^7 cells. [f]ppm. [g]nmol/plant part. [h]nmol/mg insoluble polysaccharide. [i]µg/g. [j]µg/mL. [k]nmol/10^6 protoplasts. [l]µmol/m^2. [m]µmol/g. [n]µmol/10^7 cells. [o]nmol/mg protein. [p]nmol/mg chlorophyll. [q]Percentage of total free nucleotides. [r]m*M*. [s]mg/g. [t]Percentage P dwb. [u]Total extinction/callus (plant part). [v]counts per minute × 10^3.

Continued

Table 5-2 Ribonucleotides in Plants*—cont'd

No.	Plant part	Method	AMP	ADP	ATP	GMP	GDP	GTP	CMP	CDP	CTP	UMP	UDP	UTP	References
Sambucus sieboldiana ssp. *miquelii* Hara															
1	Bark[d]	Lumin.			478										2487
1	Xylem[d]	Lumin.			117										2487
Sauromatum guttatum Schott (Voodoo Lily)															
1	Appendix	Lumin.		848	909										2698
Secale cereale L. (Rye)															
1	Seed embryo[g]	Lumin.	0.0	0.12	0.69										2699
1	Leaf	HPLC	37.1	60.2	50.1										2585
Sequoiadendron giganteum (Lindl.) Buchholz (Giant Sequoia)															
1	Apex[i]	Lumin.			2.74		1.76	0.71							2700
1	Shoot tip	Lumin.			1200		(Other NTP 600)								2701
Sinapis alba L. (White Mustard)															
1	Seedling[g]	Enzym.	2.01	2.9	12.2										2702
1	Apical bud[g]	Lumin.	0.118	0.008	0.010										2703
Solanum melongena L. (Eggplant)															
1	Seed[g]	Lumin.			0.12										2454
Solanum tuberosum L. (Potato)															
1	Shoot[d]	Enzym.		35,600	69,200										1099
1	Tuber[d]	PC							80			+			2678
1	Tuber	PC/PC	+	+	10		+	+				+	+	+	9
1	Tuber	PC/PE	+	+	10		+	+				+	+	+	9
4	Cooked tuber[i]	TLC/TLC	17.9-30.0			12.9-21.1						16.0-21.4			2704
1	Tuber[q]	IEC	3.0	5.7	17.2							6.0			316
1	Tuber	CC (cellulose)	5.5	26.2	22.8	0						0			2705
1	Tuber	HPLC	78			n.d.			7			15			2468
1	Tuber[t]	HPLC			n.d.										696
5	Tuber, incubated at 52 °C[i]	HPLC	0-400	210-2200	0-1900	0-1900	150-2000	230-2200	190-2400	0-1800	0-920	0-220	0-500	0-2200	2706
1	Tuber[d]	Enzym.		400	800										1099
1	Tuber	Enzym.	5.3	6.4	21.6										287
1	Tuber	Lumin.			3										2707

1	Tuber	Lumin.	and variation in ADP and ATP content along the stolon-apex axis									533
1	–, mitochondria (extract)[o]	^{31}P NMR	0.05	0.19	0.4							2708
1	–, starch[m]	IEC	10.0	22.8	2.2	1.0			1.3			2709
Sonneratia alba J. Smith												
1	Callus culture	HPLC	9.8	6.4	29.7	0.7	0.4	3.1				2710
Sorbus commixta Hedl. (Mountain Ash)												
1	Bark[d]	Lumin.			71							2487
1	Xylem[d]	Lumin.			39							2487
Sorghum bicolor (L.) Moench (Sorghum)												
1	Seed[d]	Lumin.	93.3 (incl. ADP and ATP)									2495
1	Root tip (*in vivo*)	^{31}P NMR					NDP, NTP					44
Sorghum bicolor × *Sorghum sudanense*												
1	Seed[g]	Lumin.	0.67									2454
Spartina alterniflora Loisel. (Smooth Cordgrass)												
1	Leaf[d]	Lumin.	132	802	980							2711
1	–, sheath[d]	Lumin.	93	498	1172							2711
1	Root[d]	Lumin.	408	435	1913							2711
1	Rhizome[d]	Lumin.	219	487	1126							2711
Sparlina Cynosuroides (L.) Roth												
1	Leaf[d]	Lumin.	131	249	522							2542
Spartina patens (Ait.) Muhl. (Saltmeadow Cordgrass)												
1	Leaf[d]	Lumin.	120	347	687							2711
Spinacia oleracea L. (Spinach)												
1	Seed[g]	Lumin.			0.07							2454
1	Leaf	PC/PC			0							2545
1	Leaf (extract)	^{31}P NMR	+	+	+				+	+	+	50
1	Leaf (*in vivo*)	^{31}P NMR					NDP, NTP					50
1	Leaf[p]	Enzym.	8	27	60				4	7	13	322
			and subcellular distribution in chloroplast, mitochondria, cytosol and vacuole									2712
1	Leaf	Lumin.	29	63	133							2713

*nmol/g fwb unless stated otherwise. [a]Taken from the abstract. [b]µmol/L. [c]µg/plant part. [d]Dry weight basis. [e]nmol/10^7 cells. [f]ppm. [g]nmol/plant part. [h]nmol/mg insoluble polysaccharide. [i]µg/g. [j]µg/mL. [k]nmol/10^6 protoplasts. [l]µmol/m^2. [m]µmol/g. [n]µmol/10^7 cells. [o]nmol/mg protein. [p]nmol/mg chlorophyll. [q]Percentage of total free nucleotides. [r]m*M*. [s]mg/g. [t]Percentage P dwb. [u]Total extinction/callus (plant part). [v]counts per minute × 10^3.

Continued

Table 5-2 Ribonucleotides in Plants*—cont'd

No.	Plant part	Method	AMP	ADP	ATP	GMP	GDP	GTP	CMP	CDP	CTP	UMP	UDP	UTP	References
1	–, protoplast[p]	Lumin.	9.2	18.3	26.4										326
1	–, chloroplast[p]	Lumin.	13.9	10.0	14.7										326
1	–, extrachloroplast[p]	Lumin.	4.1	8.5	15.3										326
1	–, chloroplast[p]	IEC		7.80	2.63										20
1	–, chloroplast[p]	Lumin.		6.29	4.79										20
1	–, chloroplast[d]	Enzym.	840	1360	1260										2481
1	–, chloroplast	^{31}P NMR	+	+	+							+			50
Thalassia testudinum Koenig and Sims (Turtle Grass)															
1	Leaf[d,i]	Lumin.			524										2714
Tilia platyphyllos Scop. (Large-leaved Linden)															
1	Phloem exudate[b]	Enzym.	46.6	34.2	140										2444
1	Phloem exudate[b]	TLC/TLC			+			0						0	2444
Tortula ruralis (Hedw.) Gaertn., Meyer and Scherb.															
1	Whole moss, hydrated	Lumin.			2400										2590
Trifolium incarnatum L. (Crimson Clover)															
1	Seed[d,g]	Lumin.	0	0.12	0										2715
Trigonella foenum-graecum L. (Fenugreek)															
1	Seed[d]	IEC	200	96.0	43.2							30			2716
× *Triticosecale* Wittm. (Triticale)															
4	Seed[g]	Lumin.			0.4-0.8										2717
Triticum aestivum L. (Wheat)															
75	Seed[d]	Lumin.			1.7-7.1										2718
1	Seed[d]	Lumin.	120.0 (incl. ADP and ATP)												2495
1	Seed[i]	PC/PC	1.7	7.0	2.0										2675
1	Seed[a]	Fluorometric			+										2491
1	Seed[d]	Enzym.			210										2719
1	Seed caryopsis[g]	Enzym.	0.58	0.94	0.34										2720
1	Leaf[i]	PC/PC	108.2	116.2	112.7										2675
1	Root[i]	PC/PC	66.2	–	80.1										2675
1	Grain	IEC	0.67	8.8	9.5	1.1			12.8			21.3	19.3	1.3	2721
1	Grain, high-yielding[d]	Enzym.	101.0	228.3	120.4										2722

1	Grain, low-yielding[d]	Enzym.	130.2	190.0	121.7										2722
1	Grain	PC/PE	38	77	151							35	135	93	2723
1	–, endosperm	PC/PE	62	119	120							53	186	125	2723
1	–, testa, and pericarp	PC/PE	24	69	99							32	130	109	2723
1	–, aleurone tissue[d]	PC/PC		20	218			34			34		72		2724
1	Kernel	IEC	9.70	39.14	68.36	10.86	3.74	1.86	2.56	7.90	2.64	26.84	27.32	34.32	2633
1	Embryo[g]	HPLC	320	98	15		78	6			n.d.		5	n.d.	2725
1	Embryo[b]	Enzym.										12.2	8.9	11.1	2726
1	Leaf[q]	PE/PC		19	42								38		331
1	Leaf[d,i]	IEC	26			17			13			20			2727
1	Leaf	HPLC	22.4	35.4	39.7										2585
1	Leaf[l]	Enzym.		13	46										2728
1	Leaf	Lumin.	7	20	53										2729
1	–, chloroplast[i]	Lumin.	6.8	13.2	27.9										2730
1	–, mitochondria[i]	Lumin.	−0.2	6.1	3.2										2730
1	–, cytosol[i]	Lumin.	4.7	5.2	39.1										2730
1	Mature leaf	Lumin.	29	54	59										2731
1	Leaf and culm	IEC	2.4	2.3											2721
1	Aerial parts	IEC	35	49	72	0	14	15	0	0	14	44	52	35	2432
1	Shoot	Lumin.			200										2638
1	Shoot (*in vivo*)	^{31}P NMR					NTP								2638
1	Etiolated shoot	IEC	3800	3450	950	350	325	200	3650	325	150	275	1450	125	2732
2	Crown[o] (1)	Lumin.	0.23	2.08	3.67										2582
	Crown[o] (2)		0.12	2.01	2.68										2582
4	Crown[d]	Lumin.			550-820										2733
1	Sprout[g]	Enzym.	1.0	1.48	2.74										2720
1	Embryo	[^{32}P]-PEP			8			46							2734
1	Embryo	RNA									<1			<1	2734
1	Seedling	PC/PC	+	+	+	+			+			+	+	+	2735
1	–, coleoptile	PC/PC	+									+			2735
1	–, root	PC/PC	+									+			2735

*nmol/g fwb unless stated otherwise. [a]Taken from the abstract. [b]μmol/L. [c]μg/plant part. [d]Dry weight basis. [e]nmol/10^7 cells. [f]ppm. [g]nmol/plant part. [h]nmol/mg insoluble polysaccharide. [i]μg/g. [j]μg/mL. [k]nmol/10^6 protoplasts. [l]μmol/m^2. [m]μmol/g. [n]μmol/10^7 cells. [o]nmol/mg protein. [p]nmol/mg chlorophyll. [q]Percentage of total free nucleotides. [r]m*M*. [s]mg/g. [t]Percentage P dwb. [u]Total extinction/callus (plant part). [v]counts per minute × 10^3.

Continued

Table 5-2 Ribonucleotides in Plants*—cont'd

No.	Plant part	Method	AMP	ADP	ATP	GMP	GDP	GTP	CMP	CDP	CTP	UMP	UDP	UTP	References
1	–, root	IEC	+	+	+										2640
1	Root and stem	IEC	++	+		+						+			2475
Triticum compactum Host (Club Wheat)															
1	Leaf	IEC	42	50	26							33	17	38	2736
Triticum durum Desf. (Durum Wheat)															
1	Seed[a]	PC/PE										+			2512
2	Embryo	HPLC	930-943	129-164											2737
Tsuga heterophylla (Raf.) Sarg. (Western Hemlock)															
1	Pollen	Lumin.			160-550										2442
Ulmus davidiana Planch. ssp. *japonica* Nakai (Japanese Elm)															
1	Bark[d]	Lumin.			29										2487
1	Xylem[d]	Lumin.			29										2487
Ulmus glabra Huds. (Wych Elm)															
1	Phloem exudate[b]	Enzym.			310										2444
Vicia faba L. (Faba Bean)															
1	Whole plant	IEC	3.3	1.3	4.8	1.3		1.7	10.0			2.0		6.5	2738
1	Seed	IEC	1.9	1.4	1.5	1.3		0.9	13.3			2.1		0.8	2738
1	Pod	IEC	0.7	0.5	0.5	0.5		1.2	0.5			0.5		1.3	2738
1	Guard cell protoplast[m]	Lumin.	306	168	660										2739
1	Chloroplast[m]	Lumin.	63	9	18										2739
1	Mitochondria[m]	Lumin.	93	54	201										2739
Vicia sativa L. (Spring Vetch)															
1	Seed[g]	Lumin.			6.44										2454
Vigna angularis (Willd.) Ohwi and Ohashi (Adzuki Bean)															
1	Root tip (nmol/mm)	Lumin.			0.028	and variation in ATP content of 1 mm segments from the tip									2740
Vigna mungo (L.) Hepper (Black Gram)															
1	Root nodule	Enzym.			289										2741
1	Root tip (*in vivo*)[r]	^{31}P NMR		0..11	0.41										542,543

1	Pod	IEC	0.7	0.5	0.5	0.5		1.2	0.5			0.5		1.3	2738
1	Guard cell protoplast[m]	Lumin.	306	168	660										2739
1	Chloroplast[m]	Lumin.	63	9	18										2739
1	Mitochondria[m]	Lumin.	93	54	201										2739
Vigna radiata (L.) Wilcz. (Mung Bean)															
1	Seed[i]	Lumin.			63.0										2555
1	Whole sprout	IEC	+	+	+	+	+	+	+	+	+	+		+	2742
1	Sprout	HPLC	16			4			n.d.			30			2468
1	Cotyledon[d]	Lumin.	300	285	240										2524
1	Hypocotyl	Lumin.	12	54	92										2743
1	3-day hypocotyl	IEC	144	28	527				28			21		71	2744
1	5-day seedling[d]	IEC	570	+		210						610	400		2433
1	Seedling	HPLC	2	62	176	16	9	33	n.d.	n.d.	n.d.	26	25	102	2441
1	Root tip	^{31}P NMR	+	+	+										543
Viola tricolor L. (Field Pansy)															
1	Leaf	PC/PC			0										2545
Vitis vinifera L. (Wine Grape)															
1	Leaf[d,m]	IEC	5.4	5.4	5.1	23	4.9	3.4	2.0	1.8		5.6	2.8	6.0	2745
1	Shoot, frost resistant[d,i]	Enzym.	1.84	4.00	7.50										2746
1	Shoot, frost sensitive[d,i]	Enzym.	1.45	3.39	5.13										2746
1	Berry	Enzym.	0.8	9.2	29.2										333
2	Berry	Lumin.	10-13*		35-47			(*Incl. ADP)							2747
1	Berry	Lumin.	Variability in adenine nucleotide content among 50 berries of a single variety (mean ± SD): AMP 3.713 ± 1.079, ADP 2.922 ± 1.207, ATP 20.971 ± 5.836												2748
Wolffia arrhiza (L.) Wimm. (Spotless Water Meal)															
1	Old plant	Enzym.	14	58	97										2749
Xanthium pennsylvanicum Wallr. (Cocklebur)															
1	Seed	Lumin.	456	335	501										2750
1	Root tip	Lumin.			67000										2751

*nmol/g fwb unless stated otherwise. [a]Taken from the abstract. [b]µmol/L. [c]µg/plant part. [d]Dry weight basis. [e]nmol/10^7 cells. [f]ppm. [g]nmol/plant part. [h]nmol/mg insoluble polysaccharide. [i]µg/g. [j]µg/mL. [k]nmol/10^6 protoplasts. [l]µmol/m^2. [m]µmol/g. [n]µmol/10^7 cells. [o]nmol/mg protein. [p]nmol/mg chlorophyll. [q]Percentage of total free nucleotides. [r]m*M*. [s]mg/g. [t]Percentage P dwb. [u]Total extinction/callus (plant part). [v]counts per minute × 10^3.

Continued

Table 5-2 Ribonucleotides in Plants*—cont'd

No.	Plant part	Method	AMP	ADP	ATP	GMP	GDP	GTP	CMP	CDP	CTP	UMP	UDP	UTP	References
Yucca flaccida Haw. (Weak-Leaf Yucca)															
1	Phloem exudate[j]	TLC/TLC (lumin.)		+	+ (615)						+				309
Zea mays L. (Corn)															
1	Seed[d]	Lumin.	97.3 (incl. ADP and ATP)												2495
1	Grain[i]	Lumin.			4200										2467
1	Kernel, embryo[g]	Lumin.	3.9	1.7	0.3										2752
1	Kernel, embryo[d,i]	Enzym.	103	346	291										2753
1	–, endosperm[d,i]	Enzym.	85	226	147										2753
1	Immature grain	IEC	240									532			2754
1	Grain, milky stage	IEC	10.6									10			2755
1	Kernel, endosperm[m]	HPLC	9.10	7.32	10.89							11.48	12.04	22.98	342
1	–, starch granule[m]	HPLC	2.66	1.09	2.43							2.79	1.45	3.83	342
1	Leaf	IEC	25.8	64.3	146.3		17.2	25.0			5.2	1.2		3.2	2613
1	Leaf[l]	Lumin.	9.1	14.5	22.4										2574
1	Shoot apex[d,g]	Lumin.			1300										2756
1	Phloem sap[r]	HPLC	0.26	0.82	1.52	0.00	0.17	0.32	0.00	0.00	0.00	0.06	0.14	0.42	2757
1	Pollen[d,i]	Lumin.		150	2500										2758
1	Culm[g]	HPLC	12.7	30.8	68.7								13.3	21.6	2759
1	Seedling	PC/PC	+	+	+	+	+	+	+		+	+	+	+	2735
1	–, coleoptile	PC/PC	+								–	+			2735
1	–, root	PC/PC	+								+	+			2735
1	Seedling	IEC	25.0	25.5	87.0	15.5		17.5	4.0		6.5	15.5	8.0	18.5	2656
8	Seedling, embryo[g]	HPLC	+	+	16.6-27.9			+		+	+			+	2760
1	Seedling, embryo[g]	Lumin.			11										2752
1	Seedling, root[i]	IEC	+	22.5	36.4	+		362	+		10.7	16.3	1.1	419	2761
1	Root	PC/PC	Trace	107	148						+			118	12
1	Root	TLC/TLC		+	+			+			+			+	12
1	Root tip[g]	HPLC	0.22	0.81	2.13		+	+		+	+	0.49	1.07	1.08	2759
1	Root tip	Lumin.	28	61	64										2762

1	Root tip (*in vivo*)	^{31}P NMR	NDP, NTP			544
Zostera marina L. (Seawrack)						
1	Seed pod[d,g]	Lumin.	108	63	129	2763
1	Leaf[d,g]	Lumin.	47	95	245	2763
1	–, sheath[d,g]	Lumin.	55	49	72	2763
1	Root and rhizome[d,g]	Lumin.	27	13	34	2763

*nmol/g fwb unless stated otherwise.
[a]Taken from the abstract.
[b]μmol/L.
[c]μg/plant part.
[d]Dry weight basis.
[e]nmol/10^7 cells.
[f]ppm.
[g]nmol/plant part.
[h]nmol/mg insoluble polysaccharide.
[i]μg/g.
[j]μg/mL.
[k]nmol/10^6 protoplasts.
[l]μmol/m^2.
[m]μmol/g.
[n]μmol/10^7 cells.
[o]nmol/mg protein.
[p]nmol/mg chlorophyll.
[q]Percentage of total free nucleotides.
[r]m*M*.
[s]mg/g.
[t]Percentage P dwb.
[u]Total extinction/callus (plant part).
[v]counts per minute $\times 10^3$.

UMP, UDP, and UTP in PCA extracts of spinach leaf, together with UDPG, UDPGal, and various other phosphorus-containing compounds. According to Lutz *et al.*,[2772] all of the purine and pyrimidine di- and triphosphates in biological tissue extracts can be quantitated by this method alone without using HPLC or any biochemical method.

(2) Enzyme-Coupled Assay (NAD(P)) Method

This method, first modified for plant use by Rowan[53] in 1957, enables the individual nucleotides in a plant extract to be quantitated rapidly and accurately without prior separation and, with few exceptions, each assay is specific for the compound being analyzed. The basis of the method is the change in absorbance of NAD (or NADP) at 340 nm, which increases as NAD-(P) is reduced to NAD(P)H and decreases as NAD(P)H is oxidized to NAD(P). Since none of the nucleotides interact directly with NAD(P) or NAD(P)H, each nucleotide must first be converted to a reactive substrate. In theory, any of the reactive substrates shown in Table 1-4 could be used, but the only ones that have achieved broad acceptance are G6P (Table 1-4, reaction 6) for the nucleoside triphosphates and PEP (Table 1-4, reaction 5) for the nucleoside mono- and diphosphates. In the following sections we shall see how these enzyme-coupled reactions are used.

(A) ADENOSINE PHOSPHATES

ATP is usually assayed by reaction with glucose in the presence of hexokinase to form G6P, which then reacts with NADP in the presence of G6P dehydrogenase (reaction 6). In a separate analysis, ADP is assayed by reaction with PEP in the presence of pyruvate kinase to form pyruvate, which then reacts with NADH in the presence of lactate dehydrogenase (reaction 5), after which AMP is assayed in the same reaction mixture by addition of myokinase, which converts the AMP to ADP. The amounts of AMP, ADP, and ATP are calculated from the changes in optical density of the NADP or NADH at each stage. The following procedure is typical:

> *Pool sizes of the adenylates, ATP, ADP, and AMP, were measured using assays that coupled adenylate use to NADH or NADPH oxidation/reduction.... The ATP assay mixture contained 100 μL of tissue extract and 500 μL of an assay solution containing 100 mM Tris (pH 8.1), 5 mM $MgCl_2$, 1 mM Glc, 260 mM $NADP^+$, and 0.14 unit of Glc-6-P dehydrogenase. The assay was initiated by the addition of 0.42 units of hexokinase in a volume of 1 μL, with 1 mol of NADPH being produced for each mol of ATP consumed in the reaction. The ADP and AMP assay mixtures contained 100 μL of tissue extract and 500 μL of an assay solution 50 mM Hepes (pH 7.6), 5 mM $MgCl_2$, 24 mM NADH, 1.7 mM PEP, 0.4 unit of lactate dehydrogenase, and 16 mM ATP. Determination of ADP was initiated by the addition of 3 units of pyruvate kinase in 2 μL, with 1 mol of NADH being oxidized for each mol of ADP consumed in the reaction. When the ADP assay was completed, AMP was determined by addition of 5 units of myokinase in a 2-μL volume. One mole of NADH oxidized for each 0.5 mol of AMP present in the sample.*
>
> Oresnik, I. J.; Layzell, D. B., "Composition and Distribution of Adenylates in Soybean (*Glycine max* L.) Nodule Tissue", *Plant Physiol.* **1994**, *104*(1), 217–225,[2565] reprinted by permission of the American Society of Plant Biologists, © 1994.

A few investigators[2481,2846] have assayed ATP by reaction with 3-PGA in the presence of phosphoglycerate kinase to form !,3-DPGA which then reacts with NADH in the presence of G3P dehydrogenase (reaction 4, Table 1-4), but this method does not offer any advantage as the ATP still has to be assayed separately from the AMP and ADP. Many investigators have used the test kits provided by Boehringer Mannheim for this purpose.

(B) URIDINE PHOSPHATES

UMP, UDP, and UTP can be assayed in a single run using a sequence of enzyme-coupled reactions linked to the reaction of G6P with NAD (reaction 6). UDPG is assayed first with UDPG dehydrogenase [E.C. 1.1.1.22], after which UTP, UDP, and UMP are assayed in turn by the addition of UDPG pyrophosphorylase [E.C. 2.7.7.9], NDP kinase [E.C. 2.7.4.6], and NMP kinase [E.C. 2.7.4.4], respectively.[322] The plant tissue should be extracted with TCA rather than PCA because of the acid-sensitivity of the uridine phosphates.[2712] UDP can also be assayed by an enzyme-coupled reaction linked to the reaction of pyruvate with NADH (reaction 5).[527]

(C) GUANOSINE AND CYTIDINE PHOSPHATES

Guanosine and cytidine phosphates should be determinable by this method but there are no examples of these in Table 5-2.

(3) Firefly Luminescence Method

This method, first applied to plant extracts by Pradet[2847] in 1967, takes advantage of the unique specificity of American firefly (*Photinus pyralis*) luciferase for ATP. No special precautions are needed to remove phenolics or pigments from a plant extract prior to the assay. Of the 12 most common naturally occurring ribo- and deoxyribonucleoside triphosphates, only one, dATP, generates more than 0.1% of the light produced by ATP.[2848] The intensity of the light, measured in a luminometer, is proportional to the ATP content of the sample over a wide range—from about 10^{-6} mol/L to a detection limit of about 5×10^{-14} mol/L.[4315] The method can be used for the determination of ATP alone but in most cases the AMP and ADP contents are also needed in order to calculate the "energy charge" of the adenylate system. How this is done is discussed in the following sections.

(A) ADENOSINE PHOSPHATES

The enzymes and other reagents used in this method are the same as those used in method 2 but without the NAD(P) or NAD(P)H inasmuch as the end product in each case is ATP. The assays can be carried out sequentially on the same extract, as done by Pradet, or on separate aliquots of the extract as described below.[2670] In the original procedure,[2847] ATP is assayed first, after which PEP and pyruvate kinase are added, converting the ADP to ATP, and when this assay is finished myokinase is added, converting the AMP to ADP, which is in turn converted to ATP by the excess PEP and pyruvate kinase present in the extract. The amounts of ATP, ADP, and AMP are determined at each stage by the intensity of light released in the luminometer. The following procedure is typical of the method:

> *Four seeds of the designated stage of germination were dissected into gametophyte and embryo or seedling in a moist chamber.... Each part was quickly extracted by 10 ml of boiling, glass-distilled water for 10 min at 100 C. The extract was cooled in an ice bath, and 0.8 ml of the ... extract was incubated at 30 °C for 15 min in each of the following mixtures. A. For ATP determination, 0.1 ml of reaction buffer containing 0.5 M N-2-hydroxyethylpiperazine-N′-2-ethanesulfonic acid (HEPES), pH 7.5, and 0.5 M magnesium acetate and 0.1 ml of glass-distilled water. B. For ADP and ATP determination, 0.1 ml of the reaction buffer and 0.1 ml of a solution containing 20 μg of pyruvate kinase (EC 2.7.1.40) (Sigma, crystalline), and 500 nmoles of trisodium phosphoenolpyruvate (Calbiochem). C. For [total adenosine phosphate], 0.1 ml of the reaction buffer, and 0.1 ml of solution containing 20 μg of pyruvate kinase, 500 nmoles of trisodium phosphoenolpyruvate, and 20 μg of dialyzed (against 1 mM phosphate buffer, pH 7.0) adenylate kinase (EC 2.7.4.3) (Sigma). After incubation, the extracts were either stored at 0 °C overnight or assayed immediately by the luciferin-luciferase method using an Aminco Chem-Glow photometer.... Two replications of 0.4 ml each of the above mixtures were added with 100 μl of firefly lantern extract (Sigma FL-E-50; 50 mg of buffered extract in 5 ml of water), containing 50 mM potassium arsenate and 20 mM magnesium sulfate, pH 7.4. The instant peak height of light emission was used for all three mixtures.... Based on the peak height of light intensity produced from the three reaction mixtures, ATP concentrations were read directly from a standard curve of freshly prepared ATP solution and the particular batch of enzyme-substrate preparation used for the assay. ATP content of the extract was obtained from reaction mixture A, ADP content was the difference of B and A, and AMP was the difference of C and B. The adenylate energy level of the tissue was calculated to molar concentration by the total ATP and water content in each gametophyte or seedling. The adenylate energy charge is calculated according to Atkinson [*Annu. Rev. Microbiol. ***1969***, 23, *329–337]: Energy charge = ([ATP] + 1/2[ADP])/([ATP] + [ADP] + [AMP]).*
>
> Ching, T. M.; Ching, K. K., "Content of Adenosine Phosphates and Adenylate Energy Charge in Germinating Ponderosa Pine Seeds", *Plant Physiol.* **1972**, *50*(5), 536–540,[2670] reprinted by permission of the American Society of Plant Biologists, © 1972.

Though the light emission is in most cases measured by peak height, it can also be measured by integration of the light emitted over a period of 10 s,[2477] or by liquid scintillation counting,[2718] depending on the capability of the instrument. In some cases, the conversion of ADP to ATP has been done by substituting creatine phosphate for the PEP[2687] or of creatine phosphotransferase [E.C. 2.7.3.2] for the phosphokinase.[2687]

(B) OTHER NUCLEOSIDE PHOSPHATES

Nucleoside triphosphates other than ATP, labeled "other NTP" in Table 5-2, can be determined by the bioluminescence method after conversion to ATP by incubation with nucleoside diphosphate kinase [NDPK, E.C. 2.7.4.6] in the presence of excess ATP. The increase in the ATP level corresponds to the NTP level of the extract.[2849] GTP can be determined by incubation of the extract with succinic thiokinase [E.C. 6.2.1.4], which consumes the GTP prior to treatment with NDPK and ATP:

$$[\mathrm{GTP}] = [\mathrm{ATP}]_{\mathrm{endogenous}} + [\mathrm{ATP}]_{\mathrm{NDPK}} - [\mathrm{ATP}]_{\mathrm{thiokinnase}}.$$ [2700]

Similarly, UTP can be determined by incubation of the extract with UDPG pyrophosphorylase [E.C. 2.7.7.9] in the presence of excess G1P, which consumes the UTP prior to treatment with NDPK and ATP:

$$[\mathrm{UTP}] = [\mathrm{ATP}]_{\mathrm{endogenous}} + [\mathrm{ATP}]_{\mathrm{NDPK}} - [\mathrm{ATP}]_{\mathrm{pyrophosphorylase}}.$$ [2478]

(4) Polymerase (RNA) Method

This method, which was used to determine the cellular levels of CTP and UTP in germinating wheat embryos,[2734] is based on the ability of the enzyme RNA polymerase [E.C. 2.7.7.6] to catalyze the formation of an acid-insoluble RNA polymer in the presence of a suitable primer. [H^3]-ATP was used in the assay and the levels of the two nucleotides were determined by collecting the precipitated RNA and measuring its tritium content by scintillation counting.

(5) [^{32}P]-Labeled PEP Method

This method, which was used to determine the cellular levels of ATP and GTP in germinating wheat embryos,[2734] is based on the phosphorylation of ADP or GDP by [^{32}P]-labeled PEP in the presence of pyruvate kinase [E.C. 2.7.1.40]. The extent of labeling provides a direct measure of the ATP (or GTP) content; neither enzyme coupling to NAD(P) nor bioluminescence is required for this assay. The endogenous ATP is converted quantitatively to ADP by hexokinase in the presence of excess G1P prior to [^{32}P] labeling, and this same reaction is used to remove [^{32}P]-labeled ATP when assaying for GTP.[2734]

RIBONUCLEOTIDES

This group comprises the adenosine derivatives AMP, ADP, and ATP, the guanosine derivatives GMP, GDP, and GTP, the cytidine derivatives CMP, CDP, and CTP, and the uridine derivatives UMP, UDP, and UTP (Y=OH in the formula on p 289).

Adenosine Phosphates

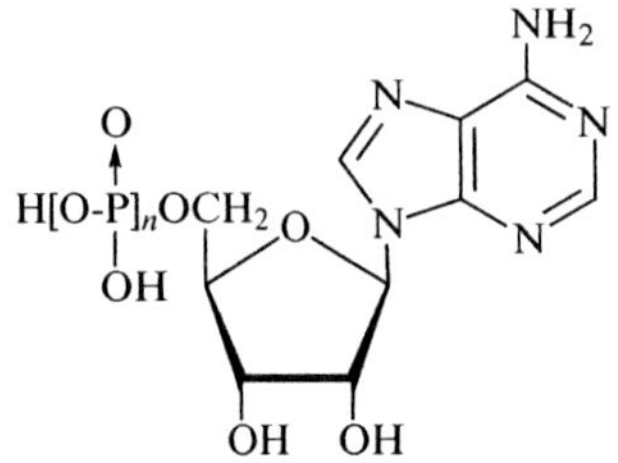

AMP, n = 1
ADP, n = 2
ATP, n = 3

Physical Properties

Adenosine 5′-phosphate (AMP), $C_{10}H_{14}N_5O_7P$. *5′-Adenylic Acid*

AMP [R.N. 61-19-8]: Filamentary crystals,[2850] mp 208 °C dec.,[2851] $[\alpha]_D^{20}$ − 26.0°(10% HCl), −47.5° (2% NaOH),[2852] specific volume 0.53 at 24 °C[2853]; conductivity (calc.)[2854]; polarographic reduction[2855]; diamagnetic susceptibility (χ_m) -171×10^{-6} cm^3/mol at 25 °C[410]; countercurrent distribution[2856]; hydrophobicity[2857]; fluorescence,[2858] phosphorescence,[2859] bioluminescence.[2860] Ionization constants: pK_a 0.89, 3.70, 6.01 (potentiometric, 25 °C),[2861] 13.06 (thermometric, 25 °C),[354] 3.7, 6.1 (UV, 25 °C),[2862] <1.2, 3.9, 6.7 (^{1}H NMR, 32 °C),[2863] 6.11 (^{31}P NMR).[68] Spectra: UV[2862] (fig.),[2864] IR (fig.),[2865] FTIR (fig.),[2866] Raman (fig.)[2867] ORD (fig.),[2868] CD (fig.),[2869] PES,[2870] ESR,[2871] tunneling (fig.),[2872] ultrasound (fig.),[2873] ultrasonic absorption coefficients.[2874] [^{1}H] NMR (fig.),[2875] 2d-[^{1}H] NMR,[2876] [^{13}C] NMR (fig.),[2877] [^{14}N] NMR,[2878] [^{15}N] NMR (fig.),[2879] [^{17}O] NMR (fig.),[2880] [^{31}P] NMR,[68] 2d-[^{13}C,^{15}N] NMR (fig.),[2881] 2d-[^{13}C,-^{31}P] NMR (fig.).[2882] Mass spectra: EI-MS,[2883] FD-MS,[2884] FAB-MS,[2885] SI-MS,[2886] ESI-MS,[2887] EH-MS,[2888] LM-MS,[2889] FAB-MS/MS,[2890] ES-MS/MS.[2814] Chromatography: PC,[9] TLC, [36] HPTLC,[2891] GC (TMS derivative),[2892] GFC,[2893] IEC,[2894] HPLC,[2440] preparative HPLC,[2895] column (activated carbon).[2896] Electrophoresis: PE,[32] TLE,[36] isotachophoresis,[2897] moving boundary (fig.),[2898] CE,[2899] CZE,[2900] MEKC.[2901] Theoretical calculations: LCAO-MO,[116] PCILO,[2902] EHT,[2903] CNDO,[2904] $\Delta G°$,[2905] mol. polarizability and diamagnetic susceptibility.[2906] *AMP·H_2O* [R. N. 18422-05-4]: Needles (photo), dec. 195 °C (sealed tube), $[\alpha]_D^{20}$ − 38.0°(10% HCl), −59.5° (2% NaOH), $[\alpha]_D^{50}$ − 40.5°(H_2O); least water soluble of the four RNA nucleotides[2907]; long, thin monoclinic needles, d_c 1.647, X-ray structure.[2908]

$Na_2C_{10}H_{12}N_5O_7P \cdot nH_2O$ [R.N. 146877-92-1]: FTIR (fig.),[2909] Raman,[2910] FAB-MS,[2911] dielectric properties (TSDC, fig.).[2912] *Dodecahydrate* [R.N. 177980-01-7]: Orthorhombic crystals, X-ray structure.[2913]

$MgC_{10}H_{12}N_5O_7P \cdot 5H_2O$: Cryst. solid, FTIR (fig.).[2866] Hexahydrate, Raman (fig.).[2914]

$CaC_{10}H_{12}N_5O_7P \cdot 6H_2O$: Cryst. solid, FTIR (fig.).[2866]

$BaC_{10}H_{12}N_5O_7P \cdot 7H_2O$ [R.N. 61240-11-7]: Clear, platelike monoclinic crystals, d_m 1.91, d_c 1.886, X-ray.[2915] IR,[2916] FTIR (fig.),[2866] Raman (fig.).[2914]

$(C_{23}H_{26}N_2O_4)_2C_{10}H_{14}N_5O_7P$ (brucine salt): mp 180-182 °C.[2851]

$(C_{13}H_9N)_2C_{10}H_{14}N_5O_7P{\cdot}H_2O$ (acridine salt): Egg-yellow, dense prismatic needles, mp 208 °C,[2917] $[\alpha]_D^{20}$ − 46.4°(H_2O).[1442]

Also given: FTIR (fig.) for the Li, Cs, and NH_4 salts of AMP and data for the K and Rb salts.[2909]

Adenosine 5′-Diphosphate (ADP), $C_{10}H_{15}N_5O_{10}P_2$. *Adenosine 5′-(Trihydrogen Diphosphate).*

ADP [R.N. 58-64-0]: $[\alpha]_D^{25}$ − 25.7°(H_2O),[2918] countercurrent distribution,[2856] liquid-liquid extraction,[2919] fluorescence,[2858] phosphorescence,[2859] bioluminescence,[2860] polarographic reduction,[2855] ODMR,[2920] diamagnetic susceptibility (χ_m) -196×10^{-6} cm^3/mol at 25 °C[410]; ionization constants: pK_a 3.95, 6.26 (potentiometric, 25 °C),[2921] 3.9, 6.3 (UV, 25 °C).[2862] Spectra: UV (fig.),[2864] IR (fig.),[2922] ORD (fig.),[2923] Raman (fig.),[2924] ultrasound (fig.),[2873] ultrasonic absorption coefficients,[2874] EPR (no signal),[2925] [^{1}H] NMR,[461] [^{13}C] NMR (fig.),[2926] [^{17}O] NMR (fig.),[2927] [^{31}P] NMR,[68] 2d-[^{1}H] NMR (fig.),[2928] ESI-MS,[2887] SI-MS.[2929] Chromatography: PC,[9] TLC,[36] HPTLC,[2891] GFC,[2930] IEC,[2894] HPLC.[2440] Electrophoresis: PE,[32] TLE,[36] isotachophoresis,[2897] moving boundary (fig.),[2898] MEKC.[2901] CZE,[2900] Theoretical calculations: LCAO-MO,[116] EHT,[2931] CNDO.[2904] *ADP·3H₂O* [R.N. 72636-41-0]: Needles (photo),[2932] d_m 1.64, d_c 1.64, X-ray structure[2933]; piezoelectric properties.[2934]

$NaC_{10}H_{14}N_5O_{10}P_2{\cdot}H_2O$ [R.N. 1172-42-5]: White powder, ^{1}H, ^{23}Na, ^{31}P NMR.[2935]

$Na_2C_{10}H_{13}N_5O_{10}P_2$: Dielectric properties (TSDC, fig.).[2912]

$KC_{10}H_{14}N_5O_{10}P_2{\cdot}2H_2O$ [R.N. 72696-48-1]: Thin orthorhombic needles, d_c 1.771, X-ray structure[2936]; d_m 1.81, d_c 1.82.[2937]

$(C_{13}H_9)N{\cdot}C_{10}H_{15}N_5O_{10}P_2{\cdot}H_2O$ (acridine salt): Egg-yellow prismatic needles, mp 215 °C dec.,[2917] X-ray powder pattern.[2938]

Adenosine 5′-Triphosphate (ATP), $C_{10}H_{16}N_5O_{13}P_3$. *Adenosine 5′-(Tetrahydrogen Triphosphate).*

ATP [R.N. 56-65-5]: Fluorescence,[2858] phosphorescence,[2859] bioluminescence,[2860] countercurrent distribution,[2856] liquid-liquid extraction,[2919] polarographic reduction,[2855] diamagnetic susceptibility (χ_m) -226×10^{-6} cm^3/mol at 25 °C[410]; dipole moment (*D*) 230 esu cm.[2939] Ionization constants: pK_a 4.00, 6.48 (potentiometric, 25 °C),[2921] 4.1, 6.5 (UV, 25 °C),[2862] other acidic groups.[2940] Spectra: UV (fig.),[2862] far UV,[2941] IR (fig.),[2942] Raman (fig.),[2924] ORD,[2943] CD,[2944] ultrasound,[2945] ultrasonic absorption coefficients,[2874] [^{1}H] NMR (fig.),[2875] [^{13}C] NMR (fig.),[2946] [^{17}O] NMR (fig.),[2880] [^{31}P] NMR,[68] 2d-[^{1}H] NMR (fig.),[2928] PES,[2870] FAB-MS,[2947] ESI-MS,[2887] SI-MS.[2929] Chromatography: PC,[9] TLC,[36] HPTLC,[2891] GFC,[2893] IEC,[2894] HPLC,[2440] preparative HPLC.[2895] Electrophoresis: PE,[32] TLE,[36] isotachophoresis,[2897] moving boundary (fig.),[2898] CE,[2899] CZE,[2900] MEKC.[2901] Theoretical calculations: LCAO-MO,[116] EHT,[2931] CNDO,[2904] $\Delta G°$.[2905]

$Na_2C_{10}H_{14}N_5O_{13}P_3$ [R.N. 987-65-5]: Extremely stable both in the dry state and in neutral solution.[2948] *Trihydrate* [R.N. 51963-61-2]: Crystalline solid (photo), mp 188 °C dec.[2949]; very thin orthorhombic blades, d_m 1.785, d_c 1.787, X-ray structure[2950]; molar conductance (Λ_M).[2951] Spectra: IR,[2951] Raman (fig.),[2952] ^{1}H, ^{23}Na, and ^{31}P NMR,[2935] EPR (fig.),[2925] FAB-MS,[2953] ES-MS.[2954]

$BaC_{10}H_{14}N_5O_{13}P_3$ [R.N. 22887-43-0]: Amorphous solid, unstable in air disproportionating in part into ADP and adenosine 5′-tetraphosphate[2948]; IR, molar conductance (Λ_M).[2951]

$Ba_2C_{10}H_{12}N_5O_{13}P_3$ [R.N. 22887-44-1]: Amorphous solid, stable when dry but slightly degraded in air[2948]; forms a hexahydrate when left in air 2 weeks[2955]; IR, molar conductance (Λ_M)[2951]; X-ray diffraction pattern.[2956]

$(C_{13}H_9N)_3C_{10}H_{16}N_5O_{13}P_3 \cdot H_2O$ (acridine salt): Air-stable, hygroscopic egg-yellow needles, mp 208-209 °C dec.[2957]

Other salts: ATP, unlike other phosphate esters, does not form crystalline salts with morphine, brucine, or strychnine, nor with heavy metal (Ag, Hg, Cu) or alkaline earth (Ca, Ba) bases.[2917]

Conformation

In the solid state, according to X-ray analysis, the plane of the ribose ring in AMP is at a dihedral angle of 76° with respect to the plane of the adenine ring, both being slightly puckered. The relative orientation of the ribose and adenine rings is *anti*. In aqueous solution, according to [^{31}P]-^{1}H NMR analysis, the AMP molecule is flexible with respect to the ribose residue, undergoing interconversion between the two preferred conformations, ^{3}E and ^{2}E (or N and S), where ^{3}E is 3′-*endo* and ^{2}E is 2′-*endo*. The ratio of ^{3}E to ^{2}E is 27:63 at concentrations below 0.01*M*, over 1:1 at concentrations over 2.0*M*, and all ^{3}E in the solid state.[2958] With regard to the exocyclic $CH_2OP(O)(OH)_2$ group, which is flexible in aqueous solution, there are six possible rotamers: three for the C(4′)—C(5′) bond (gg, gt, and tg) and three for the C(5′)—O(5′) bond (g′g′, g′t′, and t′g′). The preferred conformation of AMP in neutral aqueous solution is *anti* gg-g′g′. See Evans and Sarma[2958] for Newman projections and other details.

The conformation of ADP in the solid state, based on X-ray analysis of the free acid[2933] and its potassium salt,[2936,2937] is similar to that of AMP. The orientation is *anti trans* gg in both acid and salt. In neutral aqueous solution, however, the orientation (according to ^{1}H NMR analysis) is only 20% *anti*.[2959] The diphosphate chain is extended rather than folded as in ATP. The ribose unit is predominantly ^{2}E and the base lends only a small contribution to the *syn* conformation of the molecule.[2960] A theoretical study of ADP by the EHT method indicates that the adenosine part has three preferred conformations whereas the diphosphate part is able to adapt its conformation to the environmental conditions.[2931]

The triphosphate chain in the crystalline disodium salt of ATP is in a folded conformation in each of two crystallographically independent molecules, folded to form part of a left-handed and right-handed helix, respectively.[2950] This geometry seems to be maintained in neutral aqueous solution. One of the preferred conformations, determined by the effect of lanthanide shift reagent on its ^{1}H and ^{31}P NMR spectra, corresponds closely to the crystal form having the right-handed helical phosphate chain.[2961] This conclusion is supported by molecular orbital[2962] and classical potential energy[2963] calculations which indicate that the terminal phosphate group may be even closer to the adenine ring in the free molecule than in the crystal.[2962] In aqueous solution, the conformation about the glycosyl bond is *syn* but changes to *anti* if the pH is above 12; the thermodynamics of this isomerization were determined by an ultraviolet relaxation method.[2874]

Synthesis

AMP may be prepared (a) by phosphorylation of 2′,3′-*O*-isopropylideneadenosine with $POCl_3$[2964] or $P_2O_3Cl_4$,[2965] with morpholino[2966] or phenyl[2967] phosphorodichloridate, with *o*-phenylene,[2968] diallyl,[2969] or bis(2,2,2-tribromoethyl)[2970] phosphorochloridate, or with 8-quinolyl,[2971] *p*-nitrophenyl,[2972] or 2-cyanoethyl[2973] dihydrogen phosphate; (b) by phosphorylation of 2′,3′-*O*-benzylideneadenosine with dibenzyl phosphorochloridate[2974] or of 2′,3′-*O*-diacetyladenosine with benzylphosphorous diphenylphosphoric anhydride[2975]; (c) by phosphorylation of unprotected adenosine with $POCl_3$,[2976] $P_2O_3Cl_4$,[2965] di(2-*t*-butylphenyl) phosphorochloridate,[2977] or tris(8-quinolyl) phosphate[2978]; or (d) by fusion of 2,3-di-*O*-benzoyl-5-(diphenylphosphonooxy)-D-ribofura-nosyl bromide with 6-benzamidopurine[2979] or its bis(trimethylsilyl) derivative.[2980]

ADP may be prepared (a) from 5′-*O*-tosyladenosine by reaction with tris(tetra-*n*-butylammonium) pyrophosphate[2981]; (b) from AMP by condensation with phosphoric acid in the presence of dicyclohexylcarbodiimide,[2982] 1,1′-carbonyldiimidazole,[2983] or cyclopentanone oxime *p*-nitrobenzenesulfonate[2984]; (c) from AMP by phosphorylation with benzyl *N*-cyclohexylphos-phoramidate,[2985] dibenzyl phosphorochloridate,[2986] phosphoric triimidazolide,[2987] or 2-cyanoethyl α-pyridyl phosphate[2988]; (d) from an AMP anhydride by reaction with tri-*n*-butylammonium phosphate[2989,2990]; (e) from adenosine 5′-phosphoramidate[2991] or N-substituted phosphoramidate by reaction with anhydrous phosphoric acid[2992–2994]; (f) from other AMP derivatives containing protecting substituents at the 2′-OH and 3′-OH positions,[2995,2996] or (g) from ATP by partial hydrolysis[2997] or methanolysis.[2998]

ATP may be prepared (a) from adenosine by phosphorylation with $POCl_3$ in trimethyl phosphate followed by reaction with tri-*n*-butylammonium phosphate[2999] or bis(tri-*n*-butylammonium) pyrophosphate[3000]; (b) from 5′-*O*-tosyladenosine by reaction with tetrakis(tetra-*n*-butylammonium) triphosphate[2981]; (c) from AMP by condensation with phosphoric acid in the presence of dicyclohexylcarbodiimide,[2981] cyclopentanone oxime *p*-nitrobenzenesulfonate[2984] or 8-(5-chloroquinolyl) *S*-phenylphosphorothioate[3001]; (d) from AMP by phosphorylation with dibenzyl phosphorochloridate,[3002] phosphoric triimidazolide,[2987] or pyrophosphoric tetraimidazolide[3003]; (e) from AMP by enzymatic phosphorylation with acetyl phosphate (preferably the sodium salt[3004]) as the ultimate phosphate donor[3005]; (f) from adenosine 5′-phosphoramidate[3006] or phosphoromorpholidate[3007] by reaction with a bis(trialkylammonium) pyrophosphate; (g) from ADP by phosphorylation with acetyl phosphate in the presence of Fe(III) or [24]-N_6O_2 polyamine catalyst,[3008,3009] with tri(*n*-butylammonium) phosphate in the presence of 1,1′-carbonyldiimidazole[2983] or by reaction with benzyl *N*-cyclohexylphosphoramidate[2985]; or (h) from various other AMP or ADP derivatives containing protected OH groups.[2957,2995,3010–3012]

Hydrolysis

AMP is stable to hydrolysis under alkaline conditions but somewhat less stable under neutral or mildly acid conditions.[205] At intermediate pH (5-9) the hydrolysis follows pseudo-first-order kinetics, giving only the products of P—O bond cleavage, adenosine, and P_i. At pH 5, the first-order rate constant increases steadily from 0.000231 h^{-1} (half-life 3000 h) to 0.816 h^{-1} (half-life 0.85 h) as the temperature is raised from 53 °C to 121 °C.[3013] At pH 3 and 100 °C, in the absence of buffer, the products of N—C bond cleavage, adenine and ribose, also appear.[3013,3014] Under much milder conditions (pH 8.6, 37 °C), the hydrolysis of AMP is catalyzed by Ce(III) and other heavy metal ions[3015] and by lanthanide complexes.[3016] Hydrolysis of AMP with 1*N* H_2SO_4 liberates only 1% of P_i after 30 min at 100 °C.[3017] Hydrolysis with 4% HCl liberates all of the adenine but no P_i.[3018] The probable course of this reaction is discussed below under GMP.

The hydrolysis of ADP to AMP and P_i in aqueous solution at 80-95 °C is first order with respect to ADP and is catalyzed by H^+. The activation energy at pH 8.0 is 24.2 kcal/mol.[3019] The hydrolysis is also catalyzed by bivalent metal ions, notably Cu(II),[3020] by Ce(III) and other heavy metal ions,[3021,3022] and by macrocyclic polyamines such as [24]-N_6O_2.[3023] Tracer studies with ^{18}O-enriched water show that bond fission occurs at both sides of the anhydride oxygen resulting in a partially labeled P_i, but there is no evidence of oxygen isotope exchange with the solvent.[3024]

The hydrolysis of ATP to ADP and P_i in aqueous solution is first order with respect to ATP and is catalyzed by H^+. The rate constant k decreases from about $4\times10^{-6}\ s^{-1}$ at pH 2 to about $10^{-7}\ s^{-1}$ at pH 9.5 (50 °C, $I=0.1$).[3025] The hydrolysis is also catalyzed by bivalent metal ions, notably Cu(II),[3020] by Ce(III) and other heavy metal ions,[3021,3022] and by macrocyclic polyamines such as [24]-N_6O_2.[3023] The effectiveness of the bivalent metal ions decreases in the order Cu(II) > Cd(II) > Zn(II) Ni(II) > Mn(II) > Mg(II) and differs from one nucleotide to another, whereas the first-order rate constants for the uncomplexed nucleotides (ATP, GTP, CTP, and UTP) are virtually identical.[3026] The hydrolysis of ATP is complicated by the liberation under certain conditions of pyrophosphate[3014,3024] or metaphosphate[3027] and by further degradation of the ribose 5-phosphate to ribose and P_i.[3028]

Phosphorylation

Methods for the phosphorylation of AMP to ADP or ATP and of ADP to ADP are discussed under the section "Synthesis" above. These methods, which result in an extension of the phosphate chain, employ either a phosphorylating agent containing a reactive P—Cl, P—N, or P—O bond or a condensing agent to link a P_i to the terminal phosphate group. In the absence of a catalyst, reaction with AMP with sodium trimetaphosphate, best carried out in aqueous solution at pH 12 over a period of 20-30 days at room temperature, yields a 1:1 mixture of 2′,5′-ADP and 3′,5′-ADP, products of phosphorylation of the remaining free OH groups of the ribose ring.[3029] In the presence of Mg(II) catalyst, the reaction takes a different course giving within 1-2 days under ambient conditions a mixture of adenosine

5′-polyphosphates P_nA where A = adenosine and n = 4, 7, and 10.[3030] ADP, under the same conditions, yields a mixture of adenosine polyphosphates P_nA where n = 5, 8, and 11.[3030]

Phosphorylation of AMP with KH_2PO_4 under prebiotic conditions (formamide, 110 °C) for 12 h yields a mixture of 2′-, 3′-, and 5′-AMP, 2′,3′-cyclic AMP, and several diphosphates, like that obtained with adenosine under the same conditions.[3031]

Guanosine Phosphates

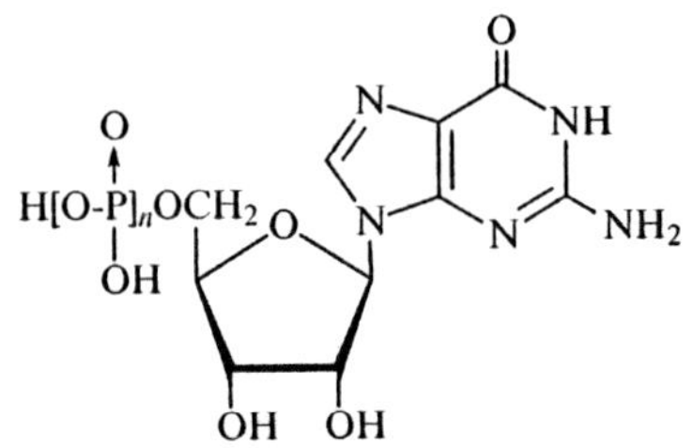

GMP, n = 1
GDP, n = 2
GTP, n = 3

Physical Properties

Guanosine 5′-Phosphate (GMP), $C_{10}H_{14}N_5O_8P$. *5′-Guanylic Acid.*

GMP [R.N. 85-32-5]: Fluorescence,[3032] phosphorescence,[3033] chemiluminescence,[3034] polarography,[3035] lactim-lactam tautomerism,[3036] hydrophobicity,[2857] countercurrent distribution.[2856] Dipole moment (D) 166 esu cm[2939] (calc. μ 11.6, θ 10^{3037}), liquid crystals.[3038] Ionization constants: pK_a 0.7, 2.3, 5.92, 9.36 (potentiometric, 25 °C),[2861] 2.4, 6.1, 9.4 (UV, 25 °C),[2862] <2.1, 6.2 (^{1}H NMR, 32 °C),[2863] 6.15 (^{31}P NMR, 22 °C).[68] Spectra: UV (fig.),[2864] IR (fig.)[3039] FTIR (fig.),[3040] Raman (fig.),[3041] ORD (fig.),[3042] CD (fig.),[3043] PES (fig.),[2870] ESR,[2871] [^{1}H] NMR (fig.),[3044] [^{13}C] NMR (fig.),[3045] [^{15}N] NMR,[2879] [^{31}P] NMR,[68] 2d-[^{1}H] NMR (fig.),[3046] ultrasound,[3047] EI-MS,[2883] FAB-MS,[3048] SI-MS,[2886] FAB-MS/MS.[2890] Chromatography: PC,[9] TLC,[3049] HPTLC,[2891] IEC,[2894] HPLC,[2440] preparative HPLC,[2895] GC (TMS derivative),[2892] GFC,[2893] column (activated carbon.).[2896] Electrophoresis: PE,[9] TLE,[3050] CZE,[2900] isotachophoresis,[2897] MEKC.[2901] Theoretical calculations: PCILO,[3051] CNDO.[3052] *GMP·2H₂O*: Long tapering needles (photo),[2907,3053] mp 180 °C (sealed capillary),[2907] $[\alpha]_D^{20}$ − 13.5°(H_2O), −48.5° (5% NH_4OH),[3053] $[\alpha]_D^{25}$ − 7.5°(H_2O), +1.5° (10% HCl), −57.0° (2% NaOH).[2907] *GMP·3H₂O* [R.N. 26166-34-7]: Orthorhombic needle-shaped crystals, d_m 1.644, d_c 1.650, X-ray structure.[3054,3055] Conformation similar to AMP but unlike IMP.[3054]

$NaC_{10}H_{13}N_5O_8P$: IR (fig.),[3039] FAB-MS/MS.[2890]

$Na_2C_{10}H_{12}N_5O_8P$ [R.N. 5550-12-9]: IR (fig.).[3039] *Tetrahydrate* [R.N. 20813-77-8]: Monoclinic crystals, d_m 1.705, TGA, dielectric relaxation, X-ray structure.[3056] *Heptahydrate* [R.N. 20813-78-9]: Orthorhombic crystals, d_m 1.69, d_c 1.65, X-ray structure,[3057] far IR (fig.),[2769] FTIR (fig.),[2909] Raman (fig.),[3058] [^{23}Na] NMR,[3059] FAB-MS,[2911] TGA,[3056] dielectric relaxation.[3056]

$BaC_{10}H_{12}N_5O_8P{\cdot}2H_2O$ [R.N. 3387-75-5]: IR (fig.).[3039] *Hexahydrate* [R.N. 143256-05-7]: Rectangular orthorhombic crystals, d_c 1.970, X-ray structure.[3060]

$(C_{23}H_{26}N_2O_4)_2C_{10}H_{14}N_5O_8P{\cdot}7H_2O$ (brucine salt): mp 233 °C, $[\alpha]_D^{20}$ – 26.0° (35% ethanol).[3061]

Other salts: FTIR (figs.) for the Li, Cs, and NH_3 salts of GMP and data for the K and Rb salts.[2909]

Guanosine 5′-Diphosphate (GDP), $C_{10}H_{15}N_5O_{11}P_2$. *Guanosine 5′-(Trihydrogen Diphosphate).*

GDP [R.N. 146-91-8]: Liquid-liquid extraction,[2919] countercurrent distribution,[2856] bioluminescence,[2860] chemiluminescence.[3034] Ionization constants: pK_a 7.19 (potentiometric, 25 °C),[3062] 2.9, 6.3, 9.6 (UV, 25 °C).[2862] Spectra: UV (fig.),[2864] PES (fig.),[2870] CD (fig.),[3063] [^{1}H] NMR,[3044] [^{31}P] NMR.[68] Chromatography: PC,[9] TLC,[36] HPTLC,[2891] IEC,[2894] GFC,[2930] HPLC,[2440] preparative HPLC.[2895] Electrophoresis: PE,[9] TLE,[36] CZE,[2900] isotachophoresis,[2897] MEKC.[2901]

$Na_2C_{10}H_{13}N_5O_{11}P_2$ [R.N. 7415-69-2]: FAB-MS.[2911]

Guanosine 5′-Triphosphate (GTP), $C_{10}H_{16}N_5O_{14}P_3$. *Guanosine 5′-(Tetrahydrogen Triphosphate).*

GTP [R.N. 86-01-1]: Bioluminescence,[2860] chemiluminescence,[3034] countercurrent distribution,[2856] liquid-liquid extraction,[2919] polarographic reduction,[3035] lactim-lactam tautomerism.[3036] Ionization constants: pK_a 7.65, 9.6 (potentiometric, 25 °C),[3062,3064] 3.3, 6.5, 9.3 (UV, 25 °C).[2862] Spectra: UV (fig.),[2862] PES (fig.),[2870] [^{1}H] NMR,[3044] [^{31}P] NMR,[68] FAB-MS.[2911] Chromatography: PC,[9] TLC,[36] HPTLC,[2891] GFC,[2930] IEC,[2894] HPLC,[2440] preparative HPLC.[2895] Electrophoresis: PE,[9] TLE,[36] CZE,[2900] isotachophoresis,[2897] MEKC.[2901]

$Na_3C_{10}H_{13}N_5O_{14}P_3$ [R.N. 36051-31-7]: ES-MS.[2954]

Conformation

The conformation of GMP in the crystalline disodium[3057] and barium[3060] salts is identical to that of AMP, and likewise shows a preference for the *anti* ($^2E \leftrightarrow {}^3E$) gg-g′g′ conformation in neutral aqueous solution.[3065] Temperature has little effect on this conformation, but if the pH is reduced to 4 or less (the pK at which the base is protonated at N^7) the ribose conformation changes from S to N, the exocyclic rotamer distribution changes, but the *anti* conformation remains the same.[3044] In contrast, the tetrabutylammonium salt of GMP dissolves in dimethylformamide forming aggregates in which the GMP has the *syn* conformation.[3046]

The conformation about the glycosidic bond in both GDP and GTP is predominantly *syn* (>90%) in aqueous solution, according to ^{1}H NMR. The β-phosphate group is oriented toward the pyrimidine base and close to H(3′).[2959]

Synthesis

GMP may be prepared (a) from 2′,3′-*O*-isopropylideneguanosine by phosphorylation with $POCl_3$,[2964] $P_2O_3Cl_4$,[2965] phenyl phosphorodichloridate,[2967] or bis(2,2,2-tribromoethyl) phosphorochloridate,[2970] with *p*-nitrophenyl,[2972] 2-cyanoethyl,[3066] or 8-quinolyl[2971] dihydrogen phosphate, or with tetrakis(*p*-nitrophenyl)pyrophosphate[3067]; (b) from 2′,3′-*O*-di(*t*-butyldimethylsilyl)guanosine by phosphorylation with diallyl phosphorochloridate,[2969] or (c) from unprotected guanosine by phosphorylation with $POCl_3$,[2976] $P_2O_3Cl_4$,[3068] or bis(2-*t*-butylphenyl) phosphorochloridate.[2977]

GDP may be prepared (a) from 5′-*O*-tosylguanosine by reaction with tris(tetra-*n*-butylammonium) pyrophosphate,[2981] (b) from GMP by condensation with phosphoric acid in the presence of DCC[3069] or 1,1′-carbonyldiimidazole,[2983] (c) from GMP by reaction with benzyl hydrogen phosphoramidate[3070] or phosphoric triimidazolide,[2987] (d) from guanosine 5′-phosphoramidate by reaction with anhydrous phosphoric acid,[3071] or (e) from a GMP anhydride by displacement with tri-*n*-butylammonium phosphate.[2989,3010]

GTP may be prepared (a) from GMP by condensation with phosphoric acid in the presence of DCC[3069] or 1,1′-carbonyldi(2-methylimidazole)[1756] or with tri-*n*-butylammonium pyrophosphate in the presence of 1,1′-carbonyldiimidazole,[3072] (b) from GMP by phosphorylation with phosphoric triimidazolide,[2987] (c) from GMP by enzymatic phosphorylation PEP, pyruvate kinase, and guanylate kinase in the presence of a catalytic amount of ATP,[3072] or (d) from a GMP anhydride,[2989] ester,[3012] or morpholide[3073] by displacement with bis(tri-*n*-butylammonium) pyrophosphate.

Hydrolysis

Hydrolysis of GMP with 5% H_2SO_4 is essentially complete within 30 min at 100 °C, yielding a mixture of guanine (96.2%), guanosine (1.2%), ribose (65.6%), ribose phosphate (32.7%), and P_i (68.0%). Based on these results and a separate determination of the behavior of each product under identical conditions, Ishikawa concludes that the principal reaction is glycolytic cleavage of the GMP to guanine and ribose phosphate.[3018] At intermediate pHs (5-9), GMP is hydrolyzed to guanosine and P_i in the same manner as AMP but at a somewhat faster rate,[3013] and is also subject to catalysis under mild conditions by Ce (III) and other heavy metal ions[3015] and by lanthanide complexes.[3016]

The behavior of GDP toward hydrolysis with $NaClO_4$ ($10^{-3}M$, $I=0.1$, 50 °C) parallels that of ADP, the rate decreasing with increasing pH and independent of the kind of nucleic base. The hydrolysis is significantly accelerated by Cu(II), passing through a pH optimum at 6.0 where the Cu(II) interaction is strongest.[3074]

Hydrolysis of GTP over the pH range 2-10 parallels that of ATP, and is likewise catalyzed by Cu(II) though weaker at its maximum (pH 6.2) by a factor of 2.[3074] The hydrolysis is also catalyzed by Ce(III) and Th(IV).[3022]

Phosphorylation

Methods for the phosphorylation of GMP to GDP or GTP are discussed under the section "Synthesis" above. These methods, which result in an extension of the phosphate chain, employ either a phosphorylating agent containing a reactive P—Cl, P—N, or P—O bond or a condensing agent to link a P_i to the terminal phosphate group. The reaction of GMP with sodium trimetaphosphate, however, takes a different course, giving a 1:1 mixture of 2′,5′-GDP and 3′,5′-GDP, products of phosphorylation of the two remaining ribose ring OH groups.[3029]

Cytidine Phosphates

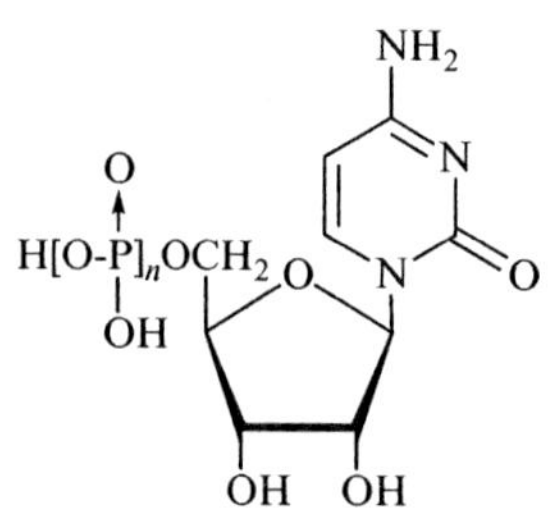

CMP, n = 1
CDP, n = 2
CTP, n = 3

Physical Properties

Cytidine 5′-Phosphate (CMP), $C_9H_{14}N_3O_8P$. *5′-Cytidylic Acid.*

CMP [63-37-6]: Dense white monoclinic wedge-shaped crystals[3075] or elongated plates (photo),[2907] mp 227 °C,[3075] 230 °C[3076] or dec. 230-233 °C (sealed tube),[2907] $[\alpha]_D^{20}$ + 49.1°(H_2O)[3076] or +23.43° (0.5*N* NaOH),[3075] $[\alpha]_D^{25}$ + 26.0°(10% HCl) or +25.5° (2% NaOH),[2907] $[\alpha]_D^{50}$ + 47.5°(H_2O)[2907]; fluorescence,[3077] phosphorescence,[3033] hydrophobicity,[2857] countercurrent distribution,[2856] polarographic reduction.[3078] Ionization constants: pK_a 0.80, 4.24, 5.97, ~13.2 (potentiometric, 25 °C),[2861] 4.5, 6.3 (UV, 25 °C).[2862] Spectra: UV (fig.),[2864] IR,[3079] far IR (fig.),[3080] FTIR (fig.),[3081] Raman (fig.),[3082] ORD (fig.),[3042] CD (fig.),[3043] [^{1}H] NMR,[3083] [^{13}C] NMR,[2946] [^{31}P] NMR,[68] 2d-[^{1}H] NMR (fig.),[3084] FD-MS,[3085] EI-MS (TMS derivative),[2883] FAB-MS/MS,[2890] ultrasound.[3047] piezoelectric resonance (fig.).[3086] Chromatography: PC,[9] TLC,[3049] HPTLC,[2891] IEC,[2894] HPLC,[2440] GC (etheno derivative),[3087] column (activated carbon),[2896] GFC.[2893] Electrophoresis: PE,[9] TLE,[3050] isotachophoresis,[2897] CZE,[2900] MEKC.[2901]

$Li_2C_9H_{12}N_3O_8P$ [R.N. 19326-12-6]: $[\alpha]_D$ +6.9° (H_2O).[3088]

$NaC_9H_{13}N_3O_8P$ [R.N. 3106-19-2]: FAB-MS/MS.[2890]

$Na_2C_9H_{12}N_3O_8P$ [R.N. 6757-06-8]: Long orthorhombic needles, d_m 1.622, d_c 1.581, X-ray structure[3089]; IR (fig.),[3090] FAB-MS.[2911] *Tetrahydrate*: Raman (fig.).[3041]

$BaC_9H_{12}N_3O_8P{\cdot}8.5H_2O$: Long, thin orthorhombic needles, d_m 1.845, d_c 1.844, X-ray structure.[3091] *Anhydrous* [R.N. 13435-44-4]: $[\alpha]_D^{20}$ + 14.0°(H_2O).[3092]

$(C_{23}H_{26}N_2O_4)_2C_9H_{14}N_3O_8P$ (brucine salt): Crystalline solid, mp 180-182 °C, soluble in hot ethanol, insoluble in water.[3075]

Cytidine 5′-Diphosphate (CDP), $C_9H_{15}N_3O_{11}P_2$. *Cytidine 5′-(Trihydrogen Diphosphate).*

CDP [63-38-7]: Needles (photo)[2932]; bioluminescence,[2860] liquid-liquid extraction,[2919] countercurrent distribution,[2856] polarographic reduction.[3035] Ionization constants: pK_a 7.18 (potentiometric, 25 °C),[3062] 4.6, 6.4 (UV, 25 °C).[2862] Spectra: UV (fig.),[2864] [^{1}H] NMR,[461] [^{31}P] NMR,[68] tunneling (fig.),[2872] ultrasound (fig.).[3093] Chromatography: PC,[9] HPTLC,[2891] IEC,[2770] HPLC,[2440] GFC.[2930] Electrophoresis: PE,[9] CZE,[2900] isotachophoresis,[2897] MEKC.[2901] *Monohydrate*: Air-stable, d_m 1.745, d_c 1.749, X-ray structure.[3094]

Cytidine 5′-Triphosphate (CTP), $C_9H_{16}N_3O_{14}P_3$. *Cytidine 5′-(Tetrahydrogen Triphosphate).*

CTP [65-47-4]: Bioluminescence,[2860] liquid-liquid extraction,[2919] countercurrent distribution,[2856] polarographic reduction.[3035] Ionization constants: pK_a 7.65 (potentiometric, 25 °C),[3062] 4.8, 6.6 (UV, 25 °C).[2862] Spectra: UV (fig.),[2862] [^{1}H] NMR (fig.),[3072] [^{31}P] NMR,[68] ultrasound (fig.),[3093] FD-MS.[3095] Chromatography: PC,[9] HPTLC,[2891] IEC,[2894] HPLC,[2440] GFC.[2930] Electrophoresis: PE,[9] CZE,[2900] isotachophoresis,[2897] MEKC.[2901]

$Na_2C_9H_{14}N_3O_{14}P_3$ [R.N. 36051-68-0]: ES-MS.[2954]

$Na_4C_9H_{12}N_3O_{14}P_3$: White nonhygroscopic water-soluble powder.[3096]

Conformation

The conformation of CMP in the solid state and in aqueous solution is similar to that of AMP, based on X-ray analysis of the barium salt[3091] and on ^{1}H and ^{13}C NMR spectroscopic analysis of the sodium salt in neutral aqueous solution.[3097,3098] The C(4′)—C(5′) bond and *anti* conformation of CMP are preserved in CDP. The diphosphate group is oriented toward the ribose hydroxyl groups in CDP but away from them in CDP-choline.[3094]

Synthesis

CMP may be prepared (a) by phosphorylation of 2′,3′-*O*-isopropylidenecytidine with $POCl_3$[2964] or $P_2O_3Cl_4$,[2965] with diallyl[2969] or dibenzyl[376] phosphorochloridate, with *p*-nitrophenyl[2972] or 8-quinolyl[2971] dihydrogen phosphate, or with H_3PO_4/P_2O_5[3071]; (b) by phosphorylation of unprotected cytidine with $POCl_3$,[2976] $P_2O_3Cl_4$,[3068] or di(2-*t*-butylphenyl) phosphorochloridate,[2977] or (c) by fusion of 2,3-di-*O*-benzoyl-5-(diphenylphosphonooxy)-D-ribofuranosyl bromide with bis(trimethylsilyl)-*N*-acetylcytosine[3088] or trimethylsilyl 4-ethoxy-2(1*H*)-pyrimidinone.[2980]

CDP may be prepared (a) from 5′-*O*-tosylcytidine by reaction with tris(tetra-*n*-butylammonium)pyrophosphate[2981]; (b) from CMP by phosphorylation with phosphoric triimidazolide,[2987] 2-cyanoethyl-α-pyridyl phosphate,[2988] or benzyl dihydrogen phosphate,[3099] (c) from cytidine 5′-phosphoramidate[3071] or 5′-phosphoromorpholidate[3007] and anhydrous phosphoric acid, (d) by displacement of diphenyl phosphate[3010] or di-*n*-butylphosphinothioic acid[2989] from a CMP anhydride or of α-pyridone or 8-hydroxyquinoline from a CMP ester by reaction with tri-*n*-butylammonium phosphate.

CTP may be prepared (a) from CMP by phosphorylation with phosphoric triimidazolide,[2987] (b) from CMP by enzymatic phosphorylation with PEP in the presence of pyruvate kinase and adenylate kinase,[3072] (c) from CMP by condensation with tri-*n*-butylammonium pyrophosphate in the presence of 1,1′-carbonyldiimidazole,[3072] or (d) from a CMP anhydride,[2989] ester,[3012] amidate,[3006] or morpholidate[3096] by displacement with bis(triethylammonium) or bis(tri-*n*-butylammonium) pyrophosphate.

Hydrolysis

Hydrolysis of CMP with 10% sulfuric acid at 125 °C gives cytidine, isolated as the picrate.[3092] Almost no effect is observed when aqueous solutions of CMP are heated at 100 °C under various pH conditions.[3100]

Phosphorylation

Methods for the phosphorylation of CMP to CDP or CTP are discussed under the section "Synthesis" above. These methods, which result in an extension of the phosphate chain, employ either a phosphorylating agent containing a reactive P—Cl, P—N, or P—O bond or a condensing agent to link P_i to the terminal phosphate group. The reaction of CMP with sodium trimetaphosphate, however, takes a different course giving a 1:1 mixture of 2′,5′-CDP and 3′,5′-CDP, products of phosphorylation of the two remaining ribose ring OH groups, after 48 days at room temperature.[3029]

Uridine Phosphates

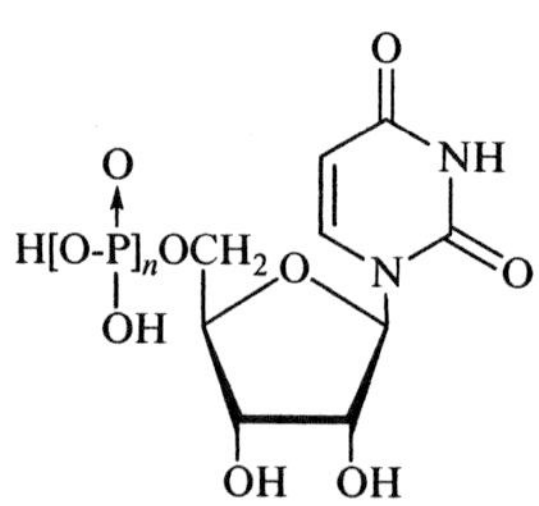

UMP, n = 1
UDP, n = 2
UTP, n = 3

Physical Properties

Uridine 5′-Phosphate (UMP), $C_9H_{13}N_2O_9P$. *5′-Uridylic Acid*

UMP [R.N. 58-97-9]: Elongated prisms with pointed ends,[2907] mp 202 °C (sealed capillary),[3101] $[\alpha]_D^{20}$ + 10.5°(H_2O),[3101] +11.1° (H_2O),[3076] or +6.5° (2% NaOH)[2907]; $[\alpha]_D^{28}$ + 3.44(10% HCl)[3102]; fluorescence,[3077] phosphorescence,[3033] polarographic reduction.[3035] Ionization constants: pK_a 1.02, 5.88, 9.43, ~13.9 (potentiometric, 25 °C),[2861] 6.4, 9.5 (UV, 25 °C).[2862] Spectra: UV (fig.),[2864] IR (fig.),[3103] ORD (fig.),[3042] CD (fig.),[3043] [^{1}H] NMR (fig.),[3097] [^{13}C] NMR (fig.),[3105] [^{31}P] NMR,[68] 2d-[^{13}C,^{31}P] NMR (fig.),[2882] EI-MS (TMS derivative),[2883] FD-MS,[3085] EH-MS,[2888] tunneling (fig.),[2872] ultrasound.[2873] Chromatography: PC,[9] TLC,[36] HPTLC,[2891] IEC,[2894] HPLC,[2440] GC (TMS derivative),[2892]

GFC,[2893] column (activated carbon).[2896] Electrophoresis: PE,[32] TLE,[36] CZE,[3106] isotachophoresis,[2897] MEKC.[2901] Theoretical calculations: PCILO,[2902] CNDO.[3052]

$Na_2C_9H_{11}N_2O_9P{\cdot}7H_2O$ [R.N. 64426-16-0]: Orthorhombic crystals, d_m 1.64, d_c 1.64, X-ray structure.[3107] Spectra: IR (fig.),[3103] Raman,[2910] FT Raman (fig.),[3108] laser Raman (fig.),[3109] FAB-MS.[2911]

$BaC_9H_{11}N_2O_9P$ [3486-23-5]: Long needles (photo).[3110] *Dihydrate*: Amorphous solid, IR (fig.).[3103] *Nonahydrate*: Platelike crystals, d_m 2.05, X-ray structure[3111]; laser Raman (fig.).[3109]

$NH_3C_9H_{13}N_2O_9P$: Prismatic needles (photo), s. 200 °C, dec. 242 °C (sealed tube), $[\alpha]_D^{50}$ + 10.5°(H_2O),[2907] $[\alpha]_D^{20}$ + 13.0°(H_2O),[3112] +1.5° (2% NaOH),[2907] $[\alpha]_D^{25}$ + 2.5°(10% HCl).[2907]

$(NH_3)_2C_9H_{13}N_2O_9P$: Heavy elongated prisms, dec. 185 °C, $[\alpha]_D^{20}$ + 21.0°(H_2O),[3112] +21.9° (H_2O).[3076]

$(C_{23}H_{26}N_2O_4)_2C_9H_{13}N_2O_9P{\cdot}3.5H_2O$ (brucine salt) [R.N. 21145-59-5]: Colorless needles, s. 170 °C, dec. 229-232 °C,[2980] $[\alpha]_D^{18}$ – 70.4° (pyridine).[376]

Uridine 5′-Diphosphate (UDP), $C_9H_{14}N_2O_{12}P_2$. *Uridine 5-(Trihydrogen Diphosphate)*

UDP [R.N. 58-98-0]: Crystalline solid, very stable in alkaline solution[2770]; bioluminescence,[2860] liquid-liquid extraction,[2919] countercurrent distribution.[2856] Ionization constants: pK_a 7.16 (potentiometric, 25 °C),[3062] 6.5, 9.4 (UV, 25 °C).[2862] Spectra: UV (fig.),[2864] [^{1}H] NMR,[461] [^{31}P] NMR.[68] Chromatography: PC,[9] TLC,[36] HPTLC,[2891] IEC,[2770] HPLC,[2440] GFC.[2930] Electrophoresis: PE,[9] TLE,[36] CZE,[3106] isotachophoresis,[2897] MEKC.[2901]

$K_2C_9H_{12}N_2O_{12}P_2{\cdot}3H_2O$ [R.N. 71229-83-9]: Flat orthorhombic plates, d_x 1.84, X-ray structure.[3113]

Uridine 5′-Triphosphate (UTP), $C_9H_{15}N_2O_{15}P_3$. *Uridine 5′-(Tetrahydrogen Triphosphate)*

UTP [R.N. 63-39-8]: Bioluminescence,[2860] liquid-liquid extraction,[2919] countercurrent distribution,[2856] polarographic reduction.[3035] Ionization constants: pK_a 7.58, 9.5 (potentiometric, 25 °C),[3062,3064] 2.86, 4.05, 6.72 (potentiometric, 35 °C),[3114] 6.6, 9.5 (UV, 25 °C).[2862] Spectra: UV (fig.),[2862] Raman (fig.),[3115] [^{31}P] NMR.[68] Chromatography: PC,[9] TLC,[36] HPTLC,[2891] IEC,[2894] HPLC,[2440] GFC.[2930] Electrophoresis: PE,[32] TLE,[36] CZE,[2900] isotachophoresis,[2897] MEKC.[2901]

$Na_3C_9H_{12}N_2O_{15}P_3$ [R.N. 19817-92-6]: ES-MS,[2954] SI-MS,[3116] thermospray MS.[3117]

Conformation

The conformation of UMP in the solid state, based on X-ray analysis of the disodium[3107] and barium[3111] salts, is similar to that of AMP; the preferred conformation is *anti* C(2′)-endo gg.[3107] The *anti* and gg conformations are both stabilized by the electrostatic attractive interactions between the 5′-phosphate group and the pyrimidine base.[3118] The conformation of UMP in aqueous solution, based on ^{1}H and ^{13}C NMR analysis, is not significantly different from that of the crystalline salt.[3097,3098]

Synthesis

UMP may be prepared (a) by phosphorylation of 2′,3′-*O*-isopropylideneuridine with $POCl_3$[2964] or $P_2O_3Cl_4$,[2965] with diallyl[2969] or dibenzyl[376] phosphorochloridate, with *p*-nitrophenyl[2972] or 8-quinolyl[2971] dihydrogen phosphate, with tris(8-quinolyl)phosphate,[3119] or with H_3PO_4-P_2O_5,[2986] (b) by phosphorylation of 2′,3′-*O*-benzylideneuridine with diphenyl[3120] or dibenzyl[3121] phosphorochloridate, (c) by phosphorylation of unprotected uridine with $POCl_3$,[2976] $P_2O_3Cl_4$,[3068] or tris(8-quinolyl)phosphate,[2978] (d) by fusion of 2,3-di-*O*-benzoyl-5-(diphenylphosphonooxy)-D-ribofuranosyl bromide with bis(trimethylsilyl)uracil,[2980] or (e) by deamination of CMP.[3122]

UDP may be prepared (a) from 5′-*O*-tosyluridine by reaction with tris(tetra-*n*-butylammonium)pyrophosphate[2981]; (b) from UMP by phosphorylation with dibenzyl phosphorochloridate,[2986] phosphoric triimidazolide,[2987] 2-cyanoethyl α-pyridyl phosphate,[2988] or ethoxyvinyl dibenzyl phosphate[3123]; (c) from uridine 5′-phosphoramidate[3124] or 5′-phosphoromorpholidate[3007] by condensation with anhydrous phosphoric acid or tri-*n*-butylammonium phosphate; (d) from a UMP anhydride by reaction with tri-*n*-butylammonium phosphate[2989]; or (e) from other UMP derivatives containing protecting substituents at the 2′-OH and 3′-OH positions.[2995,3125]

UTP may be prepared (a) from UMP by phosphorylation with phosphoric triimidazolide,[2987] (b) from UMP by enzymatic phosphorylation with PEP in the presence of pyruvate kinase and adenylate kinase,[3072] (c) from UMP by condensation with phosphoric acid in the presence of DCC[3126] or with tri-*n*-butylammonium pyrophosphate in the presence of 1,1′-carbonyldiimidazole,[3072] or (d) from a UMP anhydride,[2989,3010] ester,[3012] or amidate[3006] by displacement with bis-(triethylammonium) or bis(tri-*n*-butylammonium)pyrophosphate, (e) from various other UMP derivatives containing protecting substituents at the 2′-OH and 3′-OH positions,[2995,3127] or (f) from CTP by deamination with sodium nitrite.[3072]

Hydrolysis

The hydrolysis of UTP over the pH range 2-10 parallels that of ATP, GTP, and CTP. Its behavior toward Cu(II) and other metal ion catalysts differs from that of ATP and GTP owing to the absence of metal ion-pyrimidine interaction, but the rate of hydrolysis is significantly enhanced if the Cu(II) to UTP ratio is raised from 1:1 to 2:1. The reactivity of the metal ion decreases in the order $Cu > Ni > Zn$.[3128]

Phosphorylation

Methods for the phosphorylation of UMP to UDP or UTP are discussed under the section "Synthesis" above. These methods, which result in an extension of the phosphate chain, employ either a phosphorylating agent containing a reactive P—Cl, P—N, or P—O bond or a condensing agent to link a P_i to the terminal phosphate group. The reaction of UMP with sodium trimetaphosphate, however, can take a different course, depending on the reaction conditions. It gives either a mixture of uridine 5′-polyphosphates P_nU, where U = uridine and $n = 4$, 7, and 10,[3030] or a 1:1 mixture of 2′,5′-ADP and 3′,5′-ADP, products of phosphorylation of the remaining free OH groups of the ribose ring.[3029]

Table 5-3 Deoxyribonucleotides in Plants

No.	Plant part	Method	dADP	dATP	dGDP	dGTP	dCDP	dCTP	TDP	TTP	References
Acer pseudoplatanus L. (Sycamore Maple)											
1	Callus culture[a]	TLC		3.3		2.3		0.6		2.0	2434
Lycopersicon esculentum L. (Tomato)											
1	Root tip	HPLC	2.67	5.40	0.68	8.41	0.97	4.31	3.12	2.81	2435
Nicotiana tabacum L. (Tobacco)											
1	Callus culture[a]	TLC		51.3		4.3		5.8		6.8	2434
Pinus mugo Turra (Mugo Pine)											
1	Pollen[a]	TLC		2.0		2.0		3.3		2.8	2434
Vicia faba L. (Faba Bean)											
1	Root tip[b]	FFL		51							3129

nmol/g fwb unless stated otherwise.
[a]Dry weight basis.
[b]nmol/20 root tips.

DEOXYRIBONUCLEOTIDES

This group comprises the adenosine derivatives dAMP, dADP, and dATP; the guanosine derivatives dGMP, dGDP, and dGTP; the cytidine derivatives dCMP, dCDP, and dCTP; the uridine derivatives dUMP, dUDP, and dUTP; and the thymidine derivatives TMP, TDP, and TTP (Y=H in the formula on p 289).

Extraction, Separation, and Analysis

The deoxyribonucleotides are extracted together with the ribonucleotides by ice-cold 0.3*M* perchloric acid[2434] or 6% trichloroacetic acid.[2435] Because of the disparity in concentration levels between the two classes of compounds, it is advisable to destroy the latter by treatment of the extract with periodate prior to analysis.[2435,3129] The deoxyribonucleotides may be separated by two-dimensional TLC or HPLC and analyzed by UV,[2435] by ^{32}P labeling,[2434] or by the firefly luminescence (FFL) method.[3130] The results are shown in Table 5-3.

Deoxyadenosine Phosphates

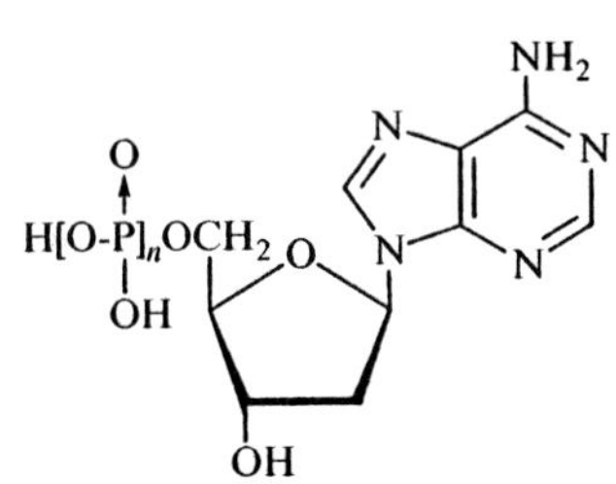

dAMP, $n = 1$
dADP, $n = 2$
dATP, $n = 3$

Physical Properties

2′-Deoxyadenosine 5′-Phosphate (dAMP), $C_{10}H_{14}N_5O_6P$. Beilstein **26**, E_{IV} 3591-2. *2′-Deoxy-5′-Adenylic Acid*

dAMP [R.N. 653-63-4]: Plates (from water),[3131] mp 142.5 °C,[3132] hydrophobicity,[2857] phosphorescence,[3133] liquid-liquid distribution coefficient[3134]; DC[3135] and AC[3136] polarographic reduction; conformation in solution.[3097] Ionization constants: pK_a ~4.4, 6.4 (potentiometric),[3137] 3.79 (UV, 20 °C),[3138] 6.65 (thermometric, 25 °C).[3139] Themodynamics: N–H ionization,[3140] PO–H ionization,[3139] theoretical calculation.[3141] Spectra: UV (fig.),[3043] far UV,[365] Raman (fig.),[3142] UV Raman (fig.),[3143] ORD (fig.),[3042] CD (fig.),[3043] [^{1}H] NMR (fig.),[2875] [^{13}C] NMR (fig.),[2946] [^{31}P] NMR,[3144] ultrasound.[3047] Mass spectra: EI,[3145] FAB,[3146] ESI,[3147] PD,[3148] CID.[3149] Chromatography: PC,[3150] TLC,[3049] HPTLC,[3151] IEC,[3151] HPLC (ion exchange,[3152] ion pairing,[3153] ion suppression,[3154] reversed phase[3155]), GC,[3156] GFC,[3157] CC (activated carbon),[2896] MEKC.[3158] Electrophoresis: PE,[3159,3160] TLE,[3161] CZE,[3162] CE.[3163] Theoretical calculations: Extended Hückel,[3164] Hartree-Fock,[3165,3166] CNDO,[3052] *ab initio* SCF,[3167] PCILO.[2902]

$NaC_{10}H_{13}N_5O_6P{\cdot}6H_2O$ [R.N. 39925-66-1]: Orthorhombic crystals, d_m 1.53, d_c 1.511, X-ray structure.[3168] [^{23}Na] NMR,[3169] HF-CO,[3165] XANES (fig.).[3170]

$CaC_{10}H_{12}N_5O_6P$: Crystals (fig.),[3171] $[\alpha]_D^{19}$ − 38°(H_2O),[3131] Raman,[3172] HF-CO.[3166]

Pentahydrate: [R.N. 90550-55-3]: Colorless orthorhombic plates, d_c 1.697, X-ray structure.[3173]

2′-Deoxyadenosine 5′-Diphosphate (dADP), $C_{10}H_{15}N_5O_9P_2$. *2′-Deoxyadenosine 5′-(Trihydrogen Diphosphate)*

dADP [R.N. 2793-06-8]: AC Polarography.[3136] Chromatography: PC,[3025] TLC,[3174] IEC,[3175] HPLC (anion exchange,[2435] reversed phase,[3176] ion pairing[3177]). Electrophoresis: PE,[3178] CE.[3163]

2′-Deoxyadenosine 5′-Triphosphate (dATP), $C_{10}H_{16}N_5O_{12}P_3$. *2′-Deoxyadenosine 5′-(Tetrahydrogen Triphosphate)*

dATP [R.N. 1927-31-7]: Countercurrent distribution,[2856] bioluminescence,[2860] AC polarography,[3136] pK_a 3.9, 7.0 (potentiometric, 70 °C).[3179] Spectra: UV,[3180] Raman,[3181] ^{1}H NMR,[3180] ^{31}P NMR,[3182] EI-MS (TMS derivative),[3183] FAB-MS,[3184] MALDI-MS,[3184] PD-MS.[3185] Chromatography: PC,[3025] TLC,[3049] IEC,[3175] HPLC (anion exchange,[2435] reversed phase,[3176] ion pairing [3177]), CC.[3180] Electrophoresis: PE,[3006] CE,[3163] CZE.[3186]

$Na_4C_{10}H_{12}N_5O_{12}P_3{\cdot}3H_2O$: White nonhygroscopic water-soluble powder.[3096]

Conformation

The conformation of dADP, like that of ADP, is extended rather than folded in aqueous solution. The ribose is predominantly ^{2}E and the base has only a small *syn* contribution, as determined by NMR spectroscopy.[2960]

Synthesis

dAMP may be prepared by phosphorylation (a) of 3′-acetyl-2′-deoxyadenosine with dibenzyl phosphorochloridate,[3187] (b) of *N*-benzoyl-3′-DMT-2′-deoxyadenosine with phosphoryl tris(triazolide),[3188] or (c) of unprotected 2′-deoxyadenosine with $POCl_3$ in trimethyl phosphate.[3189] Treatment of 2′-deoxyadenosine with monoammonium phosphate in formamide under simulated prebiotic conditions (several weeks at 37 °C) yields predominantly 5′-dAMP, together with 3′-dAMP and other by-products. The yield of 5′-dAMP was 21%, compared to 19%, 26%, and 10% for 5′-dGMP, dCMP, and TMP, respectively.[3190] For the effect of reaction temperatures varying from 25 °C to 140 °C on these reactions, see Schoffstall *et al.*[3031]

dADP may be prepared (a) by condensation of a dAMP amidate,[3191] imidazolidate,[3188] or anhydride[3178] with tri-*n*-butylammonium phosphate or (b) by enzymatic phosphorylation of dAMP with ATP.[3192]

dATP may be prepared (a) by phosphorylation of 2′-deoxyadenosine with $POCl_3$ in trimethyl phosphate followed by reaction with bis(tri-*n*-butylammonium)pyrophosphate,[3000] (b) by condensation of a dAMP amidate,[3006,3191] imidazolidate,[3180,3193] or morpholidate[3096] with bis(tri-*n*-butylammonium)pyrophosphate, or (c) by enzymatic phosphorylation of dAMP or dADP with ATP.[3192,3194] Phosphorylation of 2′-deoxyadenosine with sodium trimetaphosphate yields predominantly 3′-dATP.[3195]

Hydrolysis

dAMP is resistant to hydrolysis under neutral or basic conditions but is readily split by acid, giving adenine and 2-deoxyribose-5-phosphate.[3196,3197] The rate constant k is $15.4 \times 10^{-4}\ s^{-1}$ in 0.001*M* HCl at 363.2°K, less than half the $32.0 \times 10^{-4}\ s^{-1}$ rate constant for 2′-deoxyadenosine.[3198] In contrast, hydrolysis of dAMP in the presence of a Ce(IV) catalyst cleaves the phosphate ester bond under neutral conditions (pH 7.2, 50 °C) giving 2′-deoxyadenosine.[3199] The pseudo-first-order rate constant k is $6.7 \times 10^{-4}\ min^{-1}$. This hydrolysis can also be performed under basic conditions if cyclodextrin is added to suppress the formation of cerium hydroxide gels.[3200]

The hydrolysis of dADP to dAMP in aqueous solution (70 °C, $I=0.2$) is first order with respect to dADP and is acid catalyzed, the rate constant k decreasing from $10.7 \times 10^{-6}\ s^{-1}$ at pH 5 to $0.4 \times 10^{-6}\ s^{-1}$ at pH 8.5,[3179] paralleling that of ADP in this pH range.[3027] Likewise, the acid-catalyzed hydrolysis of dATP to dADP under the same conditions is first order with respect to dATP, the rate constant k decreasing from $10.4 \times 10^{-6}\ s^{-1}$ at pH 5.1 to $0.9 \times 10^{-6}\ s^{-1}$ at pH 8.5,[3179] paralleling that of ATP in this pH range.[3027] At pH 4 and below, however, rapid cleavage of the *N*-glycosidic bond occurs in both cases owing to the inherent sensitivity of 2′-deoxyadenosine to acid; 2′,3′-dideoxyadenosine is even more acid-sensitive.[3179]

Phosphorylation

Methods for the phosphorylation of dAMP to dADP or dATP are described under the section "Synthesis" above. dAMP is unaffected by sodium trimetaphosphate, a reagent that readily phosphorylates AMP.[3029]

Deoxyguanosine Phosphates

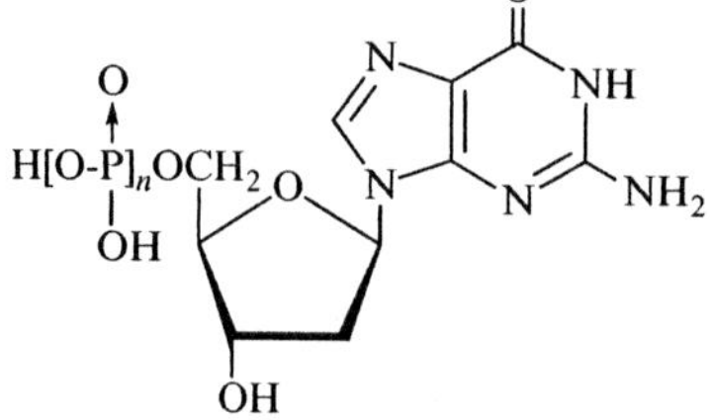

dGMP, n = 1
dGDP, n = 2
dGTP, n = 3

Physical Properties

2′-Deoxyguanosine 5′-Phosphate (dGMP), $C_{10}H_{14}N_5O_7P$. Beilstein **26**, E_{IV} 3899-900. *2′-Deoxy-5′-Guanylic Acid*

dGMP [R.N. 902-04-5]: Long needles (fig.),[3171,3201] mp 180-182 °C[3137]; phosphorescence,[3133] chemiluminescence,[3034] liquid-liquid distribution coefficient,[3134] lactim-lactam tautomerism,[3036] conformation in solution.[3097] Ionization constants: pK_a 2.9, 6.4, 9.7 (potentiometric),[3137] 9.33 (UV, 20 °C),[3138] 6.87 (^{31}P NMR, 27 °C)[3202]; thermodynamics of N–H ionization.[3140] Spectra: UV (fig.),[3043] far UV,[365] FTIR (fig.),[2909] Raman (fig.),[3142] UV Raman (fig.),[3143] ORD (fig.),[3042] CD (fig.),[3043] [^{1}H] NMR,[3203] [^{13}C] NMR (fig.),[2946] EI-MS,[3145] ESI-MS,[3147] FAB-MS,[3146] PD-MS,[3148] SI-MS,[3204] CID-MS.[3149] Chromatography: PC,[3150] TLC,[3049] HPTLC,[3151] IEC,[3205] HPLC (ion exchange,[1924] reversed phase,[3155] ion suppression[3154]), CC (activated carbon),[2896] MEKC.[3158] Electrophoresis: PE,[3159] TLE,[3161] CE,[3163] CZE.[3162] Theoretical calculations: Extended Hückel,[3164] Hartree-Fock,[3165] CNDO,[3052] thermodynamics.[3141]
$Na_2C_{10}H_{12}N_5O_7P$: FTIR (fig.),[2909] Raman (fig.),[3041] HF-CO,[3165] XANES (fig.).[3170] *Tetrahydrate* [R.N. 52558-16-4]: Thin monoclinic plates, d_c 1.646,[3206] d_m 1.64,[3207] X-ray structure.[3206]

$BaC_{10}H_{12}N_5O_7P$: Needles, $[\alpha]_D^{19}$ – 31.1°(H_2O).[3201] *Heptahydrate* [R.N. 133342-19-5]: FTIR, X-ray powder pattern.[3208]

2′-Deoxyguanosine 5′-Diphosphate (dGDP), $C_{10}H_{15}N_5O_{10}P_2$. *2′-Deoxyguanosine 5′-(Trihydrogen Diphosphate)*

dGDP [R.N. 3493-09-2]: Spectra: SI-MS (fig.).[3209] Chromatography: PC,[3025] TLC,[3174] HPLC (anion exchange,[2435] reversed phase,[3176] ion pairing[3177]). Electrophoresis: PE,[3178] CE.[3163]

2′-Deoxyguanosine 5′-Triphosphate (dGTP), $C_{10}H_{16}N_5O_{13}P_3$. *2′-Deoxyguanosine 5′-(Tetrahydrogen Triphosphate)*

dGTP [R.N. 2564-35-4]: Bioluminescence,[2860] countercurrent distribution.[2856] Spectra: UV,[3180] Raman,[3181] [^{1}H] NMR and [^{31}P] NMR,[3180] FAB-MS and MALDI-MS.[3184] Chromatography: PC,[3025] TLC,[3049] IEC,[3175] HPLC (anion exchange,[2435] reversed phase,[3176] ion pairing[3177]), CC.[3180] Electrophoresis: CE,[3163] CZE.[3186]
$Na_4C_{10}H_{12}N_5O_{13}P_3{\cdot}3H_2O$: White nonhygroscopic water-soluble powder.[3096]

Synthesis

dGMP may be prepared by phosphorylation (a) of 3′-acetyl-2′-deoxyguanosine with benzylphosphorous diphenylphosphoric anhydride[3187] or (b) of *N*-isobutyryl-3′-DMT-2′-deoxyguanosine with phosphoryl tris(triazolide).[3188] dGDP may be prepared (a) by condensation of a dGMP imidazolide[3188] or anhydride[3178] with tri-*n*-butylammonium phosphate or (b) by enzymatic phosphorylation of dGMP with UTP.[3192] dGTP may be prepared (a) by condensation of a dGMP imidazolidate[3180,3193] or morpholidate[3096] with bis(tri-*n*-butylammonium)pyrophosphate or (b) by enzymatic phosphorylation of dGDP with UTP.[3192] Phosphorylation of 2′-deoxyguanosine with sodium trimetaphosphate yields predominantly 3′-dGTP.[3195]

Hydrolysis

dGMP is resistant to hydrolysis under neutral or basic conditions but is readily split by acid, giving guanosine and 2-deoxyribose-5-phosphate.[3196,3197] Under neutral conditions (pH 7.2, 50 °C), however, hydrolysis in the presence of $CeCl_3$ cleaves the P—O bond giving 2′-deoxyguanosine. The rate constant k is 2.8×10^{-2} min^{-1}.[3199]

Deoxycytidine Phosphates

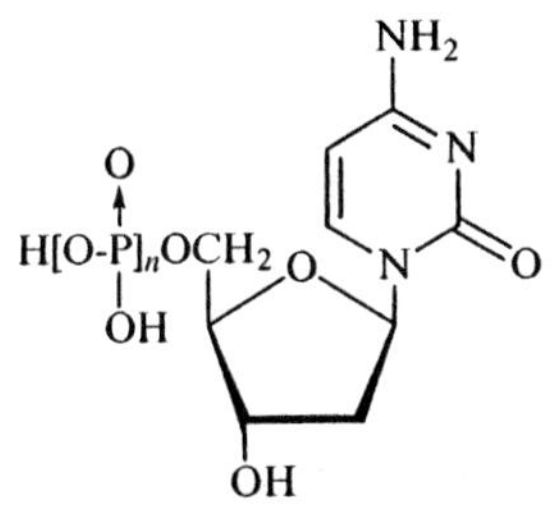

dCMP, $n = 1$
dCDP, $n = 2$
dCTP, $n = 3$

Physical Properties

2′-Deoxycytidine 5′-Phosphate (dCMP), $C_9H_{14}N_3O_7P$. Beilstein **25**, E_{IV} 3664.
2′-Deoxy-5′-Cytidylic Acid

dCMP [R.N. 1032-65-1]: Long, thin prisms (fig.), mp 183-187 °C dec.[3171] or 185-187 °C,[3132] $[\alpha]_D^{21}$ + 35°(H_2O), soluble in water and difficultly soluble in ethanol[3171]; hydrophobicity,[2857] liquid-liquid distribution coefficient,[3134] phosphorescence,[3133] polarographic reduction,[3078] conformation in solution.[3097] Ionization constants: pK_a 4,6, 6.6 (potentiometric),[3137] 4.52 (UV, 25 °C)[3210] thermodynamics of N–H ionization.[3140] Spectra: UV (fig.),[3043] far UV,[365] IR (fig.),[3211] FTIR,[3212] UV Raman (fig.),[3143] ORD (fig.),[3042] CD (fig.),[3043] [^{1}H] NMR,[3213] [^{13}C] NMR,[2946] ultrasound,[3047] EI-MS,[3145] EH-MS,[2888] ESI-MS,[3147] FAB-MS,[3146] PD-MS,[3148] SI-MS,[3204] CID-MS.[3149] Chromatography: PC,[3150] TLC,[3049] HPTLC,[3151] IEC,[3211] HPLC (ion exchange,[3152] ion pairing,[3153] ion suppression,[3154] reversed phase[3155]), GFC,[3157] GC-MS,[3214] CC (activated carbon),[2896] MEKC.[3158] Electrophoresis: PE,[3159,3160] TLE,[3161] CZE,[3162] CE.[3163] Theoretical calculations: Atomic charge,[3215] extended Hückel,[3216] Hartree-Fock,[3165] CNDO,[3052] PCILO,[2902] SCF-MO,[3217]

thermodynamics.[3141] *Monohydrate*: Small needles, mp 183-184 °C dec., $[\alpha]_D^{17}$ + 38.5°(H_2O)[3211] or colorless monoclininc prisms, d_m 1.687, d_c 1.675, X-ray structure[3218]; dipole moment (calc.) μ 14.4, θ 160.[3037] Spectra: Raman (fig.),[3082] ESR.[3219]
$Na_2C_9H_{12}N_3O_7P\cdot7H_2O$ [R.N. 85003-63-0]: Orthorhombic crystals, d_c 1.67, X-ray structure[3220]; Raman (fig.)[3041] HF-CO,[3165] XANES (fig.).[3170]
$BaC_9H_{12}N_3O_7P$: $[\alpha]_D^{19}$ + 14.4°(H_2O).[3171]
$CdC_9H_{12}N_3O_7P.2H_2O$ [R.N. 90468-86-3]: Orthorhombic crystals, d_c 2.02, X-ray structure[3221]; Raman (fig.).[3041]

2′-Deoxycytidine 5′-Diphosphate (dCDP), $C_9H_{15}N_3O_{10}P_2$. *2′-Deoxycytidine 5′-(Trihydrogen Diphosphate)*
dCDP [R.N. 800-73-7]: Chromatography: PC,[3025] TLC,[3222] IEC,[3223] HPLC (anion exchange,[2435] reversed phase,[3176] ion pairing[3177]). Electrophoresis: CE.[3163]

2′-Deoxycytidine 5′-Triphosphate (dCTP), $C_9H_{16}N_3O_{13}P_3$. *2′-Deoxycytidine 5′-(Tetrahydrogen Triphosphate)*
dCTP [R.N. 2056-98-6]: Countercurrent distribution.[2856] Spectra: UV,[3180] Raman,[3181] [^{1}H] NMR,[3180] [^{31}P] NMR,[3180] FAB-MS,[3184] MALDI-MS.[3184] Chromatography: PC,[3025] TLC,[3049] IEC,[3223] HPLC (anion exchange,[2435] reversed phase,[3176] ion pairing[3177]), CC.[3180] Electrophoresis: CE,[3163] CZE.[3186]
$Na_4C_9H_{12}N_3O_{13}P_3\cdot H_2O$: White nonhygroscopic water-soluble powder.[3096]

Synthesis

dCMP may be prepared by phosphorylation (a) of *N*,3′-diacetyl-2′-deoxycytidine with dibenzyl phosphorochloridate,[3211] (b) of *N*-benzoyl-3′-DMT-2′-deoxycytidine with phosphoryl tris(triazolide),[3188] or (c) of unprotected 2′-deoxycytidine with $POCl_3$ in trimethyl phosphate.[3189]

dCDP may be prepared (a) by condensation of a dCMP imidazolidate[3188] or anhydride[3178] with tri-*n*-butylammonium phosphate or (b) by enzymatic phosphorylation of dCMP with ATP.[3192]

dCTP may be prepared (a) by condensation of a dCMP imidazolidate[3180,3193] or morpholidate[3096] with bis(tri-*n*-butylammonium) pyrophosphate or (b) by enzymatic phosphorylation of dCDP with ATP.[3192] Phosphorylation of 2′-deoxycytidine with sodium trimetaphosphate yields predominantly 3′-dCTP.[3195]

Hydrolysis

dCMP is resistant to hydrolysis under neutral or basic conditions but is hydrolyzed upon heating in acid, giving cytosine and 2-deoxyribose-5-phosphate.[3196] The rate increases rapidly in the vicinity of the p*K* of cytosine and then levels off to a plateau; the rate constant k at pH 2.1 and 95 °C is 0.610×10^{-4} s^{-1} (ΔH, ΔS given).[3224] In contrast, hydrolysis of dCMP under neutral (pH 7.2, 50 °C) or acid (pH 4.0, 70 °C) conditions in the presence of Ce (IV) or Th(IV) catalyst cleaves the P–O bond giving 2′-deoxycytidine.[3199,3225]

Phosphorylation

Methods for the phosphorylation of dCMP to dCDP or dCTP are described under the section "Synthesis" above. dCMP is unaffected by sodium trimetaphosphate, a reagent that readily phosphorylates CMP.[3029]

Deoxyuridine Phosphates

Physical Properties

2′-Deoxyuridine 5′-Phosphate (dUMP), $C_9H_{13}N_2O_8P$. Beilstein **24**, E_{IV} 1200-01. *2′-Deoxy-5′-uridylic Acid*

dUMP [R.N. 964-26-1]: Finely crystalline hygroscopic solid[3226]; conformation in solution.[3097] Spectra: UV (fig.),[3043] UV Raman (fig.),[3143] CD (fig.),[3043] [^{13}C] NMR,[3227] [^{31}P] NMR,[3144] EI-MS.[3228] Chromatography: PC,[3150] TLC,[3222] IEC,[3223] HPLC (ion exchange,[3229] ion pairing,[3230] ion suppression,[3154] reversed phase[3231]), GFC.[3157] Electrophoresis: PE,[3226] CE.[3163] Theoretical calculations: CNDO.[3052]

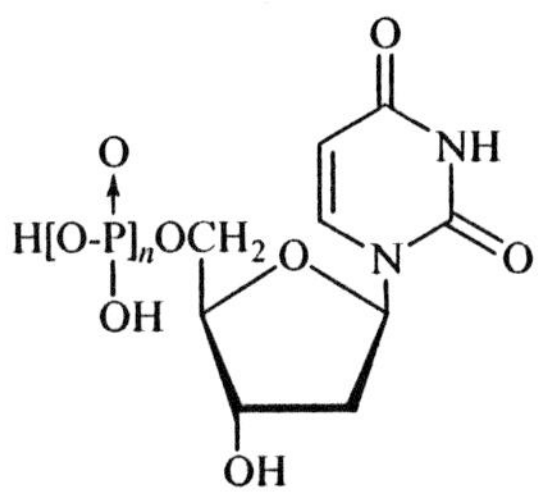

dUMP, n = 1
dUDP, n = 2
dUTP, n = 3

$Na_2C_9H_{11}N_2O_8P{\cdot}5H_2O$ [R.N. 75322-08-6]: Monoclinic crystals, d_m 1.68, d_c 1.64, X-ray structure[3232]; FT Raman (fig.).[3108]

$CdC_9H_{11}N_2O_8P{\cdot}3H_2O$ [R.N. 92484-16-7]: Colorless crystals, d_c 2.14, X-ray structure.[3233]

2′-Deoxyuridine 5′-Diphosphate (dUDP), $C_9H_{14}N_2O_{11}P_2$. *2′-Deoxyuridine 5′-(Trihydrogen Diphosphate)*

dUDP [R.N. 4208-67-7]: Chromatography: PC,[3150] HPLC (reversed phase,[3176] ion pairing).[3177] Electrophoresis: CE.[3163]

2′-Deoxyuridine 5′-Triphosphate (dUTP), $C_9H_{15}N_2O_{14}P_3$. *2′-Deoxyuridine 5′-(Tetrahydrogen Triphosphate)*

dUTP [R.N. 1173-82-6]: Bioluminescence.[2860] Chromatography: PC,[3150] TLC,[3222] HPLC (reversed phase),[3176] IC.[3234] Electrophoresis: CE.[3163]

Synthesis

dUMP may be prepared (a) by phosphorylation of 3′-acetyl-2′-deoxyuridine with dibenzyl phosphorochloridate or tetra-*p*-nitrophenyl pyrophosphate,[3226] (b) by phosphorylation of unprotected 2′-deoxyuridine with $POCl_3$ in trimethyl phosphate,[3189] (c) by hydrolysis of 2′-deoxyuridine 5′-phosphorodiamidate,[3235] or (d) by deamination of dCMP with nitrous acid.[3236]

Thymidine Phosphates

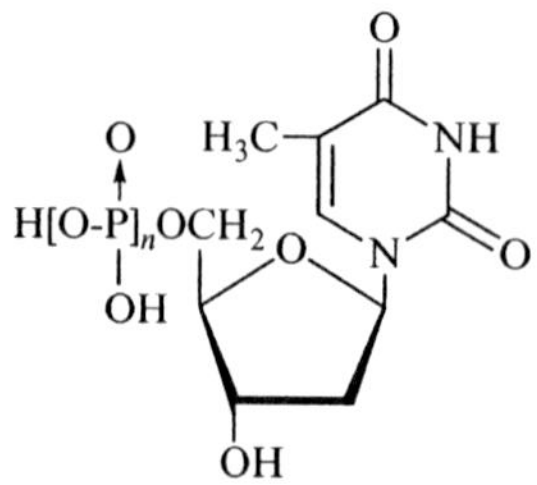

TMP, n = 1
TDP, n = 2
TTP, n = 3

Physical Properties

Thymidine 5′-Phosphate (TMP), $C_{10}H_{15}N_2O_8P$. Beilstein **24**, E_{IV} 1301-02.
5′-Thymidylic Acid

TMP [R.N. 365-07-1]: Microcrystalline solid, soluble in water and dilute ethanol,[3171] $[\alpha]_D$ −12.3° (HCl)[3088] hydrophobicity,[2857] fluorescence,[3077] phosphorescence,[3033,3133] liquid-liquid distribution coefficient[3134] countercurrent distribution,[2856] polarographic reduction,[3035] conformation in solution.[3097] Ionization constants: pK_a 6.36, 9.90 (potentiometric, 25 °C),[353] 9.65 (UV, 20 °C),[3138] <2.3, 6.4, 10.1, >13.4 (NMR, 23 °C).[3083] Spectra: UV (fig.),[3043] far UV,[365] Raman (fig.),[3237] UV Raman (fig.),[3143] ORD (fig.),[3042] CD (fig.),[3043] [^{1}H] NMR (fig.),[3097] [^{13}C] NMR (fig.),[2946] [^{31}P] NMR,[68] 2d-[^{13}C-^{31}P] NMR,[2882] ESR (fig.),[2871] triplet-triplet absorption,[3238] ultrasound.[3047] Mass Spectra: EI-MS,[3145] FAB-MS,[3146] ESI-MS,[3147] PD-MS,[3148] EH-MS,[2888] SI-MS,[3204] CID-MS.[3149] Chromatography: PC,[3150] TLC,[3049] HPTLC,[3151] IEC,[3211] HPLC (ion exchange,[3152] ion pairing,[3153] ion suppression,[3154] reversed phase [3155]), GC,[2892] GC-MS,[3214] GFC,[2930] CC (activated carbon),[2896] MEKC.[3158] Electrophoresis: PE,[3159] TLE,[3161] CZE,[3163] CE,[2899] isotachophoresis.[3239] Theoretical calculations: extended Hückel,[3216] Hartree-Fock,[3166] thermodynamics,[3141] conformational energy.[3118]

$CaC_{10}H_{13}N_2O_8P{\cdot}2H_2O$ [R.N. 105280-62-4]: Monoclinic crystals, d_c 1.697, X-ray structure[3240]; Raman,[3172] HF-CO.[3166] *Hexahydrate* [R.N. 175884-31-8]: Thin monoclinic plates, d_m 1.58-1.61, X-ray structure[3241]; polarized crystal absorption spectrum (fig.).[3242]

$BaC_{10}H_{13}N_2O_8P$: Long, thin prisms (fig.), $[\alpha]_D^{21}$ − 4.4°(H_2O),[3171] IR.[3079]

$(C_{23}H_{26}O_4N_2)_2C_{10}H_{15}N_2O_8P{\cdot}7H_2O$ (dibrucine salt): Needles, soften at 140 °C, mp ~175 °C.[3243]

Thymidine 5′-Diphosphate (TDP), $C_{10}H_{16}N_2O_{11}P_2$. *Thymidine 5′-(Trihydrogen Diphosphate)*

TDP [491-97-4]: Countercurrent distribution.[2856] Spectra: ^{1}H, ^{13}C, and ^{31}P NMR.[2981] Chromatography: PC,[3025] IEC,[2894] HPLC (anion exchange,[2435] reversed phase,[3176] ion pairing [3177]), GFC.[2930] Electrophoresis: PE,[3244] CE,[3163] CZE,[2900] capillary isotachophoresis.[3239]

Ammonium salt (lyophilized): White powder.[2981]

$Ba_2(C_{10}H_{13}N_2O_{11}P)_2{\cdot}6H_2O$: White powder[3244]; IR.[3079]

Thymidine 5′-Triphosphate (TTP), $C_{10}H_{17}N_2O_{14}P_3$. *Thymidine 5′-(Tetrahydrogen Triphosphate)*

TTP [R.N. 365-08-2]: Countercurrent distribution.[2856] Spectra: UV,[3180] Raman,[3181] [^{1}H] and [^{31}P] NMR,[3180] FAB-MS,[3184] MALDI-MS,[3184] PD-MS.[3185] Chromatography: PC,[3025] TLC,[3049] IEC,[2894] HPLC (anion exchange,[2435] reversed phase,[3176] ion pairing[3177]), GFC,[2930] CC.[3180] Electrophoresis: PE,[3244] CE,[3163] CZE,[3186] capillary isotachophoresis.[3239]
$Na_4C_{10}H_{13}N_2O_{14}P_3{\cdot}2H_2O$: White nonhygroscopic water-soluble powder.[3096]
$Ba_2C_{10}H_{13}N_2O_{14}P_3{\cdot}5H_2O$. White solid.[3244]

Synthesis

TMP may be prepared (a) by phosphorylation 3′-acetylthymidine with dibenzyl phosphorochloridate[3243] or 2-cyanoethyl phosphate[3066] or of 3′-DMT-thymidine with phosphoryl tris(triazolide),[3188] (b) by phosphorylation of unprotected thymidine with dibenzyl phosphorochloridate,[3245] di(2-*t*-butylphenyl)phosphorochloridate,[2977] diethylazodicarboxylate and triphenylphosphine,[3246] or $(CCl_3CMe_2O)_2P(O)Cl$,[3247] (c) by oxidation of thymidine 5′-phosphite with 2,2′-dipyridyl disulfide,[3248] (d) by hydrolysis of thymidine 5′-phosphorodiamidate,[3235] or (e) by fusion of 2,3-di-*O*-benzoyl-5-diphenylphosphoryl-D-ribofuranosyl bromide with trimethylsilylthymine.[3088]

TDP may be prepared (a) by the reaction of 5′-*O*-tosylthymidine with tris(tetra-*n*-butylammonium)pyrophosphate,[2981] (b) by successive reactions of 3′-*O*-benzylthymidine with phosphorylating agents containing protective substituents,[3244] (c) by phosphorylation of TMP with dibenzyl phosphorochloridate[2986] or ATP,[3192] or (d) by condensation of a TMP ester,[3012] thioester,[3249,3250] anhydride,[2989,3178,3180] or imidazolidate[3188] with tri-*n*-butylammonium phosphate.

TTP may be prepared (a) by successive reactions of 3′-*O*-acetylthymidine[3251] or 3′-*O*-benzylthymidine[3244] with phosphorylating agents containing protective substituents, (b) by condensation of a TMP ester,[3012] thioester,[3249] anhydride,[2989] imidazolidate,[3180,3193] or morpholidate[3096] with bis(tri-*n*-butylammonium)pyrophosphate, or (c) by enzymatic phosphorylation of TDP with ATP.[3192]

Hydrolysis

TMP is resistant to hydrolysis under neutral or basic conditions but is hydrolyzed upon heating in acid, giving thymine and 2-deoxyribose-5-phosphate.[3196] A plot of log rate vs. H_O at 75 °C is linear with slope −0.66 (ΔH, ΔS given).[3224] In contrast, hydrolysis of TMP under neutral (pH 7.2, 50 °C) conditions in the presence of $CeCl_3$ catalyst cleaves the phosphate ester bond, giving thymidine. The rate constant k is 4.5×0^{-2} min^{-1}.[3199]

The hydrolysis of TTP over the pH range 2-10 parallels that of the ribonucleotides ATP, GTP, CTP, and UTP; the first-order rate constants are identical within experimental error. Its behavior toward Cu(II) and other metal ion catalysts is identical to that of UTP.[3128]

Phosphorylation

Methods for the phosphorylation of TMP to TDP or TTP are described under the section "Synthesis" above. TMP is unaffected by sodium trimetaphosphate.[3029] The phosphorylation of thymidine with sodium dihydrogen phosphate under possibly primordial conditions (2 h at 160 °C) produces 6.3% of a mixture of 3′- and 5′-TMP [3252] (see also Ref. [3253]). The same reaction carried out under abiogenic conditions (1-9 h in vacuum UV) produces a 4.59% mixture of 3′- and 5′-TMP.[3254]

Cyclic Ribonucleotides

Each of the foregoing ribo- and deoxyribonucleotides is capable of forming a cyclic monophosphate in which the phosphate group is linked to the 3′ and 5′ positions of the ribose ring, as illustrated:

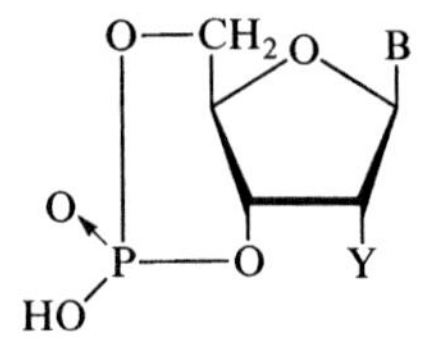

B = A, G, C, U, or T
Y = H (T) or OH (others)

Occurrence

Reports of the occurrence of cAMP in barley[3255] and lettuce[3256] in 1970 and others that followed were met with considerable skepticism. The levels were thought to be too low to play a significant role in plant metabolism and the methodology was inadequate. Several new methods that were specific for cAMP had to be developed before the existence of cAMP in higher plants was accepted; these methods are described in detail below. cGMP was identified in tobacco in 1973[2816] and all five of the cyclic nucleotides shown above were proven to be present in peas in 1989,[2829] though not in all tissues at the same time.

The metabolic pathway for the biosynthesis and degradation of cAMP is shown in Fig. 5-3. The pathways for the other cyclic nucleotides are unknown.

Extraction and Separation

The cyclic nucleotides are extracted from plant tissue in the same manner as noncyclic nucleotides, but because the levels are so low (much lower than in animal or microbial tissues) no simple method of separation is adequate. In the procedure developed by

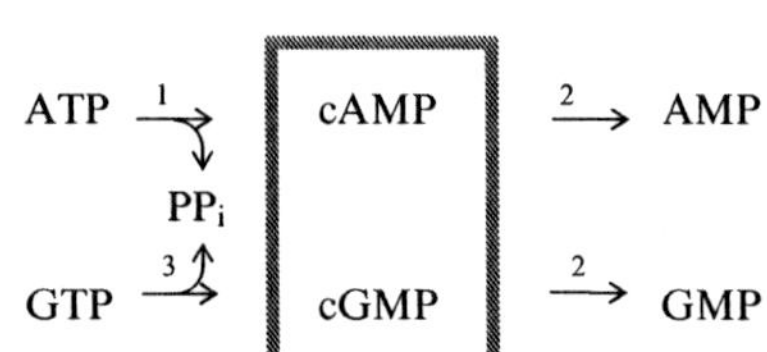

FIGURE 5-3 Cyclic NMPs. (1) Adenylate cyclase [E.C. 4.6.1.1], Mg(II) or Mn(II). (2) cNMP phosphodiesterase [E.C. 3.1.4.17], Mg(II). (3) Guanylate cyclase [E.C. 4.6.1.2], Mg(II), or Mn(II).

Brown and Newton,[2823] 6-day-old seedlings of *Phaseolus vulgaris* were extracted with methanol, chloroform, and 1*M* formic acid (12:5:3). The nucleotides were separated from other components by adsorption on charcoal and then separated from each other by IEC on Dowex 2-X8 (Cl^-), followed by IEC on Dowex 50-X4 (H^+), TLC on silica gel G, and finally high-voltage PE. The final product produced a single spot when cochromatographed with authentic cAMP by PC with five different eluants, TLC with three different eluants, and PE with three different buffers. The amounts determined by UV-visible spectrophotometry were between 2.6 and 9.2 nmol of cAMP per gram of seedling (dry wt.).[2823] In a later paper, Brown *et al.*[2824] confirmed these results by determining the cAMP level at each stage of the separation using the protein binding assay (PBA) described below, which is specific for cAMP. Most of the assay methods described on the following pages require prior purification of the cAMP (or cGMP), though not to the extent just described, in order to remove unknown substances that interfere with the assays. cCMP, cUMP, and cTMP have only been detected in plants by the FAB-MS/MS method; efforts to quantitate this method (as of this writing) have not been successful.[2829]

Pollard[3255] showed that 8-[^{14}C]-labeled adenine fed to germinating barley seeds produced a substance that appeared to be cAMP based on the usual criteria—chromatographic behavior, hydrolysis to 5′-AMP by phosphodiesterase, and deamination to cIMP—but similar experiments with other plants have been negative[2791] or inconclusive.[3257,3258]

Special Methods for Cyclic Nucleotides

(1) Protein Binding Assay (PBA) Method

The Gilman competitive protein binding assay, first applied to plant extracts independently by Bachofen,[2841] Becker and Ziegler,[2830] Brewin and Northcote,[2777] and Wellburn *et al.*[2780] in 1973, takes advantage of the specificity of cAMP-dependent protein kinase for cAMP. The method relies on the displacement of [^{3}H]-labeled cAMP from the protein by the cAMP in the extract. The following procedure is typical:

> *The reaction mixture in 200 μl of 0.05 M sodium acetate buffer, pH 4.0, contained: 3 pmoles ^{3}H-cAMP (about 40000 cpm), varying concentration of unlabeled cAMP (ranging from 3 to 50 pmoles), or 50 μl of plant extract, respectively, and 20 μg of binding protein. The assay tubes were incubated in ice bath for 90 min. After that, the content of each reaction tube was diluted to 1 ml with 0.8 ml of cold 0.02 M potassium phosphate buffer, pH 6.0, and this mixture was filtered through a separate cellulose nitrate filter disc (pore size 0.45 μm). Then the filters were washed with 10 ml of cold buffer, dried, and placed in toluene scintillator, and finally counted in a liquid scintillation spectrometer.*
>
> Stroiński, A.; Floryszak-Wieczorek, J., "Adenosine 3':5'-Cyclic Monophosphate Changes during Germination of *Hordeum vulgare*", *Acta Physiol. Plant.* **1985**, *7*(2), 35–43,[2800] reprinted by permission of Elsevier, © 1985.

To avoid interference by substances that mimic the action of cAMP, it is essential that the cAMP, even after separation from other the other nucleotides, be further purified before assay by chromatography on alumina followed by Dowex 50.[2775,2781] Many investigators have used the kits available from Boeringer Mannheim, Amersham, and other suppliers for this assay. cGMP may be assayed in the same manner using ^{3}H-cGMP and a cGMP binding protein from silkworm pupae.[2827]

(2) Protein Kinase Activation (PKA) Method

This assay was first applied to plant extracts in 1973 by Raymond *et al.*,[2774] based on a procedure developed 3 years earlier by Kuo and Greengard. The amount of cAMP in the extract is determined by the amount of ^{32}P transferred by the kinase from [γ-^{32}P]ATP to the substrate (histone).[2778] The following procedure is typical:

> *Under our standard assay conditions, each tube contained 0.05 M sodium acetate buffer (pH 6), 2 μmol magnesium acetate, 40 μg histone mixture, cAMP or plant extract in 100 μl, and 25 μl of cAMP-dependent protein kinase [from rabbit skeletal muscle]. Tubes were kept in ice during the addition of the enzyme. The reaction was commenced by the addition of 1000 pmol γ-^{32}P-ATP and tubes were incubated for 5 min at 30 C. The reactions were terminated by adding 2 ml of ice-cold trichloroacetic acid-tungstate-sulfuric acid mixture and processed [Kuo and Greengard, 1972]. The protein was finally collected by centrifugation, dissolved in 0.1 ml of 1 N NaOH, and suspended in 10 ml of Bray's scintillation fluid for counting. A standard curve with authentic cAMP was prepared for each experiment....*
>
> Handa, A. K.; Johri, M. M., "Cyclic Adenosine 3':5'-Monophosphate in Moss Protonema", *Plant Physiol.* **1977**, 59(3), 490–496,[2793] reprinted by permission of the American Society of Plant Biologists, © 1977.

This method is much more specific for cAMP than the protein binding method, but it too requires prior purification of the crude plant extract because of the presence of protein inhibitors.[2781] To avoid the need for ^{32}P-labeled ATP, Handa and Bressan[2817] developed a modified protein kinase activation assay for use with plant extracts in which the consumption of ATP is measured instead by the firefly luminescence method.

(3) Radioimmunoassay (RIA) Method

This method, developed by Steiner in 1969, utilizes antisera that are specific for cAMP or cGMP. Like the protein binding method, it is based on competitive ligand binding between labeled and unlabeled cyclic nucleotide but is less subject to interference.[2429] The method was first applied to plant extracts independently by Lundeen *et al.*[2816] and Mascarenhas and Salomon[2779] in 1973. The following procedure is typical:

> *The dry residue after [chromatography on neutral alumina and] freeze-drying was taken up in 1 M perchloric acid; 0.1 mL of a 9 M KOH and 0.25 M EDTA solution was added to 0.5 mL of the perchloric extract; EDTA is needed to chelate Mg^{2+}. The precipitate (mainly $KClO_4$) was pelleted, and 0.15 mL of the supernatant was added to 6.75 mg of succinic anhydride. The succinylated sample was diluted six-fold*

with H_2O. Final pH was between 5 and 6. Radioactive tracer (2'-O-succinyl cAMP [^{125}I]iodotyrosine methyl ester in 0.1 M citrate buffer, pH 6.2) was added, and the mixture was dialyzed against anti-cAMP antibodies dissolved in the same buffer. The standard curve and determinations were performed as in [T.T. Volker et al.[3259]*].*

Spiteri, A.; Viratelle, O. M.; Raymond, P.; Rancillac, M.; Labouesse, J.; Pradet, A., "Artefactual Origins of Cyclic AMP in Higher Plant Tissues", *Plant Physiol.* **1989**, *91*(2), 624–628,[2795] reprinted by permission of the American Society of Plant Biologists, © 1989.

The analysis of cGMP requires an antiserum specific for cGMP. Many investigators have used the kits available from Amersham (UK), Yamasa (Japan), or Boehringer Mannheim (FRG) for this assay.

(4) Mass Spectrometry (MS) Methods

In 1980 Newton *et al.*[2820] provided the first mass spectrometric evidence confirming the existence of cAMP in a higher plant tissue, *P. vulgaris* leaf. The EI-MS spectrum of the trimethylsilylated cAMP exhibited ion fragments of the five- and six-membered rings that could only be derived from 3′:5′-cAMP and not from 2′:3′-cAMP nor any deoxy derivative. Janistyn[2838,2839] and Johnson *et al.*[2788] used GC-MS of the trimethylsilylated derivatives to quantify the content of cAMP[2788,2839] and cGMP[2838] in various plant tissues.

Using FAB-MS in combination with MIKES analysis of some of the resulting ions by collision-induced dissociation (CID), Newton *et al.*[2829] were able to identify all five of the common cyclic 3′,5′ nucleotides (cAMP, cGMP, cCMP, cUMP, and cTMP) as well as cIMP in purified but underivatized extracts of pea root. A closer examination of the cGMP extracted from bean leaf showed none of the 2′,3′ isomer.[2821] Witters *et al.*[2814] developed a procedure for measuring levels of these compounds in the 10^{-6} to 5×10^{-9} M range in plant extracts by means of negative electrospray tandem MS in combination with ion suppression HPLC. Miniaturization of the HPLC setup by the use of capillary HPLC columns and precolumn switching lowered the detection limit to 25 fmol, enabling the analysis to be performed on mg amounts of plant tissue.[2815]

Physical Properties

Cyclic Adenosine 3′,5′-Phosphate (cAMP), $C_{10}H_{12}N_5O_6P$. *Adenosine, Cyclic 3′,5′-(Hydrogen Phosphate)*

cAMP [R.N. 60-92-4]: Fine orthorhombic needles, anhydrous, d_m 1.75 (X-ray structure),[3260] DSC (fig.),[3261] countercurrent distribution[2776]; solubility in water 0.0236*M*, diffusion coefficient 4.44×10^{-6} cm^2/s, E_{max} 1.38×10^{-4} (all at pH 7.0 and 20 °C)[3262]; polarographic reduction.[3135] Ionization constants: pK_a 9.364 (potentiometric, 25 °C),[3263] ~13 (thermometric, 25 °C),[354] 3.6 (^{1}H NMR, 32 °C),[2863] 3.62 (electrophoretic, 20-25 °C).[3264] Spectra: UV,[3265] IR (fig.),[3266] Raman (fig.),[3142] UV Raman,[3267] ORD (fig.),[3268] CD (fig.),[3269] [^{1}H] NMR (fig.),[3270] [^{1}H] FT-NMR (fig.),[3271] [^{13}C] NMR (fig.),[3130] 2d-[^{13}C-^{1}H] NMR,[3272] [^{31}P] NMR[68]; EI-MS (TMS derivative),[2883] FAB-MS,[2911] FD-MS,[3095,3273] CIMS (fig.),[3274] ES-MS,[3275] LM-MS,[2889] thermospray MS,[3117]

MIKES (fig.),[2829] FAB-MS/MS,[2821] ES-MS/MS.[2814] Chromatography: PC,[3265] TLC, [3276] HPTLC,[3277] HPLC,[2825] GFC,[3278] GC-MS,[2838] alumina CC,[3279] Amberlite-XAD CC.[3280] Electrophoresis: PE,[3281] isotachophoresis,[3239] CZE,[3106] MEKC.[2901] Theoretical calculations: *ab initio* SCF,[3282] EHT,[3283,3284] PCILO.[3272] *cAMP·H$_2$O*: Crystalline solid.[3265] *cAMP·1.75H$_2$O*: Platelets with a pearly lustre, mp 219-220 °C dec. (corr.).[3266] Specific volume 0.60 at 24 °C; MW (ultracentrifugation) 320 (calc. 329).[2853]
$NaC_{10}H_{11}N_5O_6P$ [R.N. 33116-15-3]: DSC (fig.).[3261] *Tetrahydrate*: Monoclinic crystals, d_c 1.558 (X-ray).[3285]

Cyclic Guanosine 3′,5′-Phosphate (cGMP), $C_{10}H_{12}N_5O_7P$. *Guanosine, Cyclic 3′,5′-(Hydrogen Phosphate)*

cGMP [R.N. 7665-99-8]: Orthorhombic crystals (X-ray structure),[3286] chemiluminescence,[3034] liquid crystals.[3038] Ionization constants: pK_a −6.3 (*NH_3^+*), 18.6 (NH_2) (^{1}H NMR, 0 °C).[3287] Spectra: UV,[3265] IR,[3288] FT Raman (fig.),[3289] UV Raman,[3267] CD (fig.),[3290] [^{1}H] NMR (fig.),[3291] [^{13}C] NMR,[3130] [^{31}P] NMR,[3292] FAB-MS,[2821] MIKES,[2829] ES-MS/MS.[2814] Chromatography: PC,[3265] TLC,[2787] HPLC,[2825] GFC,[3293] GC-MS,[2838] alumina CC,[3279] Amberlite-XAD CC.[3280] Electrophoresis: PE,[3294] isotachophoresis,[3239] CZE.[3106]
$NaC_{10}H_{11}N_5O_7P$ [R.N. 40732-48-7]: DSC (fig.),[3261] Raman (fig.),[3041] FT Raman (fig.),[3289] ESR (fig.),[3295] ENDOR (fig.).[3295] *Tetrahydrate* [R.N. 52922-48-2]: Orthorhombic crystals, d_m 1.66, d_c 1.665 (X-ray structure).[3296]

Cyclic Cytidine 3′,5′-Phosphate (cCMP), $C_9H_{12}N_3O_7P$. *Cytidine, Cyclic 3′,5′-(Hydrogen Phosphate)*

cCMP [R.N. 3616-08-8]: Crystalline solid; DSC (fig.).[3261] Ionization constants: pK_a 3.89 (potentiometric, 37 °C), ~4.0 (UV, 37 °C).[3297] Spectra: UV,[3265] IR (fig.),[3298] [^{1}H] NMR (fig.),[3291] [^{13}C] NMR,[3130] CD (fig.),[3299] EI-MS,[2821] FAB-MS,[2821] FD-MS,[3273] MIKES,[2829] ES-MS/MS.[2814] Chromatography: PC,[3265] TLC,[3300] HPTLC,[3277] HPLC,[3301] GC (3,*N*-etheno, TMS derivative),[3087] alumina CC,[3279] Amberlite-XAD CC.[3280] Electrophoresis: PE,[3294] gel. [3298] *cCMP·3H$_2$O*: Transparent hexagonal platelike crystals, d_c 1.40, d_m 1.42 (X-ray structure).[3302]
$NaC_9H_{11}N_3O_7P \cdot 4H_2O$: Raman (fig.).[3041]

Cyclic Uridine 3′,5′-Phosphate (cUMP), $C_9H_{11}N_2O_6P$. *Uridine, Cyclic 3′,5′-(Hydrogen Phosphate)*

cUMP [R.N. 4004-57-3]: Transparent colorless monoclinic crystals, d_m 1.530, d_c 1.375 (X-ray structure),[3303] pK_a 8.524 (potentiometric, 37 °C).[3297] Spectra: UV,[3265] CD (fig.),[3304] [^{1}H] NMR (fig.),[3305] [^{13}C] NMR,[3130] [^{31}P] NMR,[3144] FD-MS,[3273] MIKES.[2829] Chromatography: PC,[3265] TLC,[3300] HPTLC,[3277] HPLC,[3301] GFC,[3293] alumina CC,[3279] Amberlite-XAD CC.[3280] Electrophoresis: PE,[3294] CZE.[3106]
$NaC_9H_{10}N_2O_6P \cdot 4H_2O$ [R.N. 41092-63-1]: DSC (fig.).[3261]
$Et_3N \cdot C_9H_{11}N_2O_6P$ [R.N. 20212-92-4]: Colorless crystalline solid, mp 220-223 °C dec.[3306]; tabular monoclinic plates, d_m 1.38, d_c1.38 (X-ray structure).[3307]

Cyclic Thymidine 3′,5′-Phosphate (cTMP), $C_{10}H_{13}N_2O_7P$. *Thymidine, Cyclic 3′,5′-(Hydrogen Phosphate)*

cTMP [R.N. 6453-60-7]: Spectra: UV,[3308] IR,[3288] ^{1}H NMR (fig.),[3309] ^{13}C NMR,[3130] ^{31}P NMR,[3144] MIKES.[2829] Chromatography: PC,[3308] TLC,[3310] HPTLC,[3277] HPLC.[3311] Electrophoresis: PE.[3308]

$NH_3 \cdot C_{10}H_{13}N_2O_7P$ [R.N. 119999-00-7]: White solid. Spectra: ^{1}H NMR, ^{31}P NMR.[3312]

Conformation

Crystal X-ray analysis of cAMP shows that the asymmetric unit cell contains two molecules of cAMP which differ in their orientation about the glycosidic bond, one having a torsional angle of +102° and the other a torsional angle of −50°.[3260] In aqueous solution, according to ^{1}H NMR analysis, the fused ribose and phosphate rings provide a rigid framework with the five-membered rings in the ^{3}E conformation (3′-endo, 4′-exo) and the six-membered phosphate ring in the chair conformation.[3291,3309] The conformation of the adenine ring relative to the ribose phosphate framework, however, remains unresolved despite numerous efforts to determine its *syn-anti* distribution by ORD,[3268] ultrasonic relaxation,[3313] proton NMR lanthanide complexing,[3271,3314] [^{13}C] NMR analysis,[3130] potential energy[3315] or PCILO[3272] calculations, or supermolecular modeling.[3316] The most recent NMR studies suggest a 1:1 ratio of *syn* to *anti* at pH 2.0[3317] or 8.0,[3313] similar to that of the crystalline solid. In any event, the energy difference between the two conformers is only 1 kcal/mol[3283] and the barrier to rotation 6.6 kcal/mol.[3284]

Single-crystal X-ray analysis of cGMP, either as the acid[3286] or the sodium salt,[3296] shows that it has but one molecule per unit cell in which the base displays the *syn* conformation with a glycosyl torsional angle of −102°.[3296] In aqueous solution, according to ^{1}H NMR analysis, the conformation is similar to that of cAMP with respect to the ribose and phosphate rings,[3291] but is predominantly *syn* about the glycosyl bond owing to hydrogen bonding between the guanine NH_2 and the axial ribose OH. This conformation is supported by AMI,[3318] Hartree-Fock,[3318] PCILO,[3272] potential energy,[3315] and EHMO[3319] calculations.

The conformation of cCMP in aqueous solution[3268] is similar to that of cAMP, but the ribose ring of cCMP.$3H_2O$ in the solid state is 3′-endo, 2′-exo,[3302] unlike any of the other cyclic NMPs. Theoretical calculations by the potential energy[3315] or PCILO[3272] methods suggest that the conformation about the glycosidic linkage is predominantly *anti*.

Single-crystal X-ray analysis of cUMP shows that it has an asymmetric unit cell that contains two molecules, one having a torsional angle of −77° and the other of −58° between the base and the ribose.[3307] In aqueous solution, the conformation is similar to that of cAMP with respect to the ribose and phosphate rings[3291] but probably *anti* with respect to the glycosyl bond.[3320] The conformation is *anti* in both molecules in the crystal,[3307] and theoretical calculations based on EHMO,[3319] PCILO,[3272] or potential energy[3315] methods suggest that it is predominantly *anti* in solution.

The conformation of cTMP in aqueous solution, based on ^{1}H NMR analysis, is similar to that of cAMP with the five-membered deoxyribose ring in the ^{3}E conformation (4′-exo) and the six-membered phosphate ring in the chair conformation,[3309] but the ribose ring is flatter owing to the absence of the 2′-OH group.[3291] The torsional angle around the glycosidic bond is in the *anti* range in the preferred conformation.[3321] There are no X-ray data available for cTMP so its conformation in the solid state is unknown.

Synthesis

cAMP may be prepared (a) from adenosine by phosphorylation with $POCl_3$ in trimethyl phosphate followed by cyclization with aqueous KOH[3322]; (b) from AMP by cyclization with DCC[3265] or TPP/PDS[3323] in pyridine; (c) from a 5′-AMP *p*-nitrophenyl or 8-quinolyl ester by cyclization with acetic acid/pyridine,[3324] CuCl/pyridine,[3325] or *t*-BuOK/DMSO[3326]; (d) from a 5′-AMP thioester by cyclization with I_2/pyridine[3294]; (e) from 5′-AMP diphenylphosphinothioic anhydride by cyclization with Bu_3N/pyridine[2989]; (f) from a 2′-protected 3′-AMP diphenyl ester by cyclization with *t*-BuOK/DMSO[3327]; (g) from the *p*-trichloromethyl derivative of AMP by cyclization with *t*-BuOK/DMSO[3328]; or (h) from ATP by hydrolysis with aqueous $Ba(OH)_2$.[3266]

cGMP may be prepared from guanosine by method (a),[3322] from GMP or its *N*-benzoyl derivative by method (b),[3265,3323] from a 5′-GMP ester by method (c),[3324–3326] from a 5′-GMP thioester by method (d),[3294] or from the P-CCl_3 derivative of GMP by method (g).[3328]

cCMP may be prepared from CMP or its *N*-benzoyl derivative by method (b),[3265,3323] from a 5′-CMP ester by method (c),[3324–3326] from a 5′-CMP thioester by method (d),[3294] or from the P-CCl_3 derivative of CMP by method (g).[3328]

cUMP may be prepared from UMP by method (b),[3265,3323] from a 5'-UMP ester by method (c),[3324–3326] from a 5′-UMP thioester by method (d),[3294] from a 5-UMP anhydride by method (e),[2989] from a 2′-protected 3′-UMP diphenyl ester by method (f),[3327] or from the P-CCl_3 derivative of UMP by method (g).[3328]

cTMP may be prepared from TMP by method (b),[3308,3323] from a 3′- or 5′-TMP *p*-nitrophenyl, 8-quinolyl or diphenyl ester by method (c),[3324,3325,3327] from a 5′-TMP anhydride by method (e),[2989] or from the methyl ester of cTMP by refluxing with *t*-butylamine.[3312]

Phosphorylation of adenosine with phosphorus pentoxide under prebiotic conditions yields only traces of 3′,5′-cAMP. The principal products are 2′-AMP, 3′-AMP, and 2′,3′-cAMP.[3329]

Hydrolysis

Hydrolysis of cAMP with 1*N* HCl at 100 °C yields adenine and P_i with only trace amounts of ribose phosphates.[3265] The rate constant at 90 °C is $0.184 \times 10^{-3}\ s^{-1}$ compared to $2.5 \times 10^{-3}\ s^{-1}$ for AMP; the difference is ascribed to the rigidity of the fused ring which retards the formation of a planar cyclic oxocarbenium ion.[3330] Hydrolysis of cAMP with 0.2*M* $Ba(OH)_2$ for 30 min at 100 °C results in a 5:1 mixture of 3′-AMP and 5′-AMP with no trace of 2′-AMP.[3265] Under much milder conditions (pH 8, 37 °C), the hydrolysis of cAMP is catalyzed by Co^{III} complexes such as $Co(trien)(H_2O)^{3+}$[3331] and by lanthanides[3332] or their germanium complexes.[3333] By far the most efficient lanthanides are Ce^{III} and its cluster $[Ce_3(OH)_5]^{4+}$, which decrease the half-life of cAMP from 600,000 years to 1-2 min and 35 s, respectively, an effect attributed to the oxidation of the Ce^{III} to Ce^{IV} in air; the P–O(5′) bond is cleaved first.[3332]

The behavior of cGMP toward acid- or base-catalyzed hydrolysis parallels that of cAMP, its half-life in 1*N* HCl at 100 °C being 28 min vs. 30 min for cAMP, 26 min for cCMP, 8 min for cUMP,[3265] and 13 min (at 50 °C) for cTMP.[3334] Under much milder conditions (pH 7, 30 °C), the hydrolysis of cGMP is catalyzed by Ce(III)[3332] or its germanium complex[3333] and even more strongly by Ce(IV).[3332,3335]

The second-order rate constant for the hydrolysis of cCMP with 1*N* HCl at 90 °C is $0.200 \times 10^{-3}\ s^{-1}$, the product being a mixture of 2′-, 3′-, and 5′-CMP owing to rapid phosphate migration. Under less acidic conditions (pH > 2) cleavage of the *N*-glycosidic bond predominates, liberating cytidine.[3330] Hydrolysis of cCMP with 0.2*M* $Ba(OH)_2$ at 100 °C results in a 5:1 mixture of 3′- and 5′-monophosphates but the reaction is complicated by extensive deamination of cytidine to uracil.[3265]

The glycosyl bond in cUMP is strikingly acid-labile; the hydrolysis evidently proceeds by a different mechanism, involving opening of the ribose ring via a Schiff base intermediate in contrast to the cyclic oxycarbenium ion mechanism proposed for the other cNMPs. The rate constants for the hydrolysis of cUMP with 1*N* HCl at 90 °C are $0.251 \times 10^{-3}\ s^{-1}$ for the glycosyl bond and $0.272 \times 10^{-3}\ s^{-1}$ for the competing hydrolysis of the phosphate group to 5′-UMP and 3′(2′)-UMP.[3330]

The first-order rate constant for the hydrolysis of cTMP in 1*N* HCl at 90 °C is $4.07 \times 10^{-3}\ s^{-1}$ compared to $0.0053 \times 10^{-3}\ s^{-1}$ for TMP.[3330] The only chromophoric product is thymine.[3334] Hydrolysis of cTMP with 1*N* NaOH at 100 °C ($t_{1/2}$ 2 h) yields a 5:1 mixture of 3′-TMP and 5′-TMP.[3308] Under much milder conditions (pH 8.0, 30 °C) the pseudo-first-order rate constant for the hydrolysis of cTMP by Ce(IV) is 1.2 min^{-1}.[3335]

Energetics of Hydrolysis

In a study of the adenylate cyclase reaction, the standard free energy of hydrolysis ($\Delta G°$) of cAMP to 5′-AMP at pH 7.3 was found to be −11.9 kcal/mol, compared to −8.9 kcal/mol for the hydrolysis of ATP to ADP and P_i.[3336] This value is unexpectedly high and places cAMP in a class with "energy-rich" phosphates like PEP or creatine phosphate. Several studies were undertaken in the 1970s to explain this phenomenon.

The free energy of a reaction (ΔG) comprises an enthalpy component (ΔH) and an entropy component (ΔS). The enthalpy of hydrolysis of cAMP to 5′-AMP at pH 7.3 and 25 °C was found to be −14.1 kcal/mol in a microcalorimetric study of the phosphodiesterase reaction.[3337] In a later study this figure was revised to −11.1 kcal/mol.[3338] Nevertheless, these studies indicate that the large exergonic change that takes place during the hydrolysis of cAMP to AMP is associated primarily with a change in enthalpy rather than entropy. From the relation $\Delta G' = \Delta H + T\Delta S$, the value of $\Delta S'$ calculated from the data above is only −7.4 or +2.7 cal/deg/mol, respectively.

The enthalpy of hydrolysis of cAMP is about 8 kcal/mol more exothermic than that of trimethylene phosphate, a model compound with a similar six-membered phosphate ring.[3282] Theoretical calculations using molecular mechanics and *ab initio* methods suggest that about 4-5 kcal/mol arise from the *trans* ring fusion, which results in a puckering of the ribose ring, and about 1-2 kcal/mol arise from an unfavorable O–C–C–O interaction which is relieved upon cleavage of the phosphate ring.[3282] The remaining 1-2 kcal/mol appear to arise from solvation of 5′-AMP (but not of cAMP) by water.[3339]

The enthalpy of hydrolysis of cGMP to 5′-GMP by bovine phosphodiesterase at 25 °C was found to be −10.5 kcal/mol,[3337] compared to −14.1 kcal/mol for cAMP and −12.0 kcal/mol for the hydrolysis of cUMP to 5′-UMP by canine phosphodiesterase at 25 °C.[3340]

Table 5-4 Cyclic Ribonucleotides in Plants

No.	Plant part	Method	cAMP	cGMP	cCMP	cUMP	cTMP	References
Acer pseudoplatanus L. (Sycamore Maple)								
1	Seed embryo and axis	PBA	0.018					2773
1	–, embryo, and axis	RIA		0.083				2773
1	–, cotyledon	PBA	0.007					2773
1	–, cotyledon	RIA		0.009				2773
1	Cell suspension culture	FFL	0.140					2774
1	Cell suspension culture	PKA	0.123					2774
Agave americana ssp. *marginata* Trel (Maguey)								
1	Leaf	PBA	0.001					2775
1	Leaf	PKA	0.0009					2775
Apium graveolens L. (Celery)								
1	Leaf	PBA	<0.0028					2776
Avena sativa L. (Oats)								
1	Coleoptile	PBA	0.120					2777
1	Coleoptile	PKA	<0.0025					2778
1	Coleoptile	RIA	0.520					2779
1	Etioplast[a]	PBA	0.067					2780
1	Seedling, leaf, and coleoptile	PBA	0.0464					2781
1	Seedling, leaf, and coleoptile	PKA	0.0019					2781
Brassica chinensis L. (Chinese Cabbage)								
1	Leaf	FFL	0.890					2774
Brassica oleracea L. ssp. *capitata* L. (Cabbage)								
1	Leaf	FFL	2.00					2782
Campanula spp. (Chinese Bellflower)								
1	Root	PBA	0.00288					2783
Catharanthus roseus (L.) G. Don (Madagascar Periwinkle)								
1	Cell tissue culture	PKA	<0.0005					2778
1	Callus culture[b]	RIA	0.143					2784

Table 5-4 Cyclic Ribonucleotides in Plants—cont'd

No.	Plant part	Method	cAMP	cGMP	cCMP	cUMP	cTMP	References
Ceratophyllum demersum L. (Hornwort)								
1	Whole plant	FFL	0.030					2785
1	Whole plant	PBA	0.026					2785
Cicer arietinum L. (Chickpea)								
1	Seed	^{14}C-adenine	0.036					2786
Citrus limettiodes Tanaka (Sweet Lime)								
1	Leaf	PBA	0.230					2787
Citrus limon L. (Lemon)								
1	Young leaf	GC-MS	0					2788
1	Young leaf	RIA	0.028					2788
Citrus sinensis (L.) Osbeck Shamouti Orange								
1	7-day growth flush	PBA	0.380					2787
Citrus unshiu Marc. (Mandarin)								
1	Leaf	PBA	0.340					2787
1	Root	PBA	0.350					2787
Cocos nucifera L. (Coconut)								
1	Nut, endosperm	RIA	4.4-4.6					2789
Coleus blumei Benth. (Painted Nettle)								
1	Leaf	PBA	<0.0015					2776
1	Stem	PBA	<0.0037					2776
Cucumis sativus L. (Cucumber)								
1	Leaf	PBA	<0.0061					2776
1	Stem	PBA	<0.0080					2776
1	Seedling, cotyledon	PBA	0.252					2787
1	Root	PBA	0.0533					2776
1	Root	PBA	0.216					2787
Datura spp. (Thorn Apple)								
1	Leaf	PBA	0.132					2787
Daucus carota L. (Carrot)								
1	Root	PBA	0.00114					2783
1	Pith tissue culture	FFL	0.153					2774
1	Pith tissue culture	PKA	0.134					2774
1	Cell culture[b]	RIA	0.050					2790
Dicranella coarctata (C. Müll) Bosch and Sande Lac.								
1	Protonemal tissue	^{14}C-Adenine	<0.001					2791
Evodia officinalis Dode								
1	Fruit[c]	RIA		36.8				2792
1	–, peduncle[c]	RIA		11.0				2792
Evodia rutaecarpa (Juss.) Benth.								
1	Fruit[c]	RIA		31.2				2792
1	–, peduncle[c]	RIA		27.0				2792

[a]nmol/mg protein. [b]nmol/L. [c]Dry weight basis. [d]nmol/10^6 protoplasts. [e]nmol/mg chlorophyll.

Continued

Table 5-4 Cyclic Ribonucleotides in Plants—cont'd

No.	Plant part	Method	cAMP	cGMP	cCMP	cUMP	cTMP	References
Fumaria hygrometrica (L.) Sibth								
1	Cell suspension culture	PBA	0.063					2793
1	Cell suspension culture	PKA	0.0315					2793
1	Protonemal filaments	^{14}C-Adenine	<0.001					2791
1	Gametophyte tissue	^{14}C-Adenine	<0.001					2791
Glycine max (L.) Merr. (Soybean)								
1	Seed, hypocotyl	RIA	0.0006					2794
1	Cell suspension culture	RIA	0					2794
1	Callus culture	PBA	0.027					2777
Helianthus annuus L. (Sunflower)								
1	Seed	RIA	<0.0005					2795
1	Embryo	RIA	<0.0005					2795
1	Seedling	RIA	<0.0005					2795
1	Callus culture	PBA	0.125					2796
1	Callus culture	PKA	0.0096					2796
1	Callus culture[b]	RIA	0.131					2784
Helianthus tuberosus L. (Jerusalem Artichoke)								
1	Tuber, dormant	RIA	0.060					2797
1	–, sprouting	RIA	0.025					2797
Hordeum vulgare L. (Barley)								
1	Seed[c]	FFL	0.420					2782
1	Seed	PBA	n.d.					2798
1	Seed	PKA	n.d.					2798
1	Seed[c]	RIA	0.2-2					2799
1	Seed	PBA	0.0002					2800
1	Seed	RIA		0.000032				2801
1	–, embryo	PBA	0.160					2787
1	Seedling, coleoptile	PBA	0.0037					2800
1	–, coleoptile	RIA		0.000200				2801
1	–, root	PBA	0.0112					2800
1	–, root	RIA		0.000147				2801
Impatiens sultanii Hook. f. (Sultana)								
1	Young leaf	PBA	0.92					2802
1	Old leaf	PBA	0.22					2802
Kalanchoë daigremontiana Ham. and Per. (Devil's Backbone)								
1	Leaf	PBA	0.006					2775
1	Leaf	PKA	0.005					2775

Table 5-4 Cyclic Ribonucleotides in Plants—cont'd

No.	Plant part	Method	cAMP	cGMP	cCMP	cUMP	cTMP	References
Lactuca sativa L. (Lettuce)								
1	Seed	FFL	0.241					2774
1	Seed	PKA	0.269					2774
1	Seed	RIA	0.0085					2795
1	–, without external tegument	RIA	<0.0005					2795
1	–, germinated	PBA	0.022					2781
1	–, germinated	PKA	0.0026					2781
1	Seedling leaf and root	PBA	0.265					2787
Lemna gibba L. (Duckweed)								
1	Whole plant	PBA	1.040					2787
Lemna paucicostata Hegelm. (Lesser Duckweed)								
1	Whole plant	HPLC	0.075					2803
1	6-day plant[c]	RIA	0.014					2804
1	10-day plant[c]	RIA	0.023	0.17				2805
Lilium longiflorum Thunb. (Trumpet Lily)								
1	Flower, anther	RIA	0.0089					2806
1	–, filament	RIA	0.0009					2806
1	–, ovary	RIA	0.0059					2806
1	–, upper style	RIA	0.0151					2806
1	–, lower style	RIA	0.0043					2806
1	–, tepal	RIA	0.0008					2806
Lolium multiflorum Lam. (Italian Ryegrass)								
1	Endosperm cell suspension culture	PBA	0.014					2807
1	Endosperm cell suspension culture	PKA	0.012					2807
Lupinus arboreus Sims (Tree Lupine)								
1	Root	PBA	<0.0012					2776
1	–, nodule	PBA	0.0412					2776
Lycopersicon esculentum Mill. (Tomato)								
1	Seedling	PKA	<0.0015					2778
1	Cell suspension culture	PBA	0.00164					2808
Mangifera indica L. (Mango)								
1	Young (apical) leaf	PBA	4.80					2787
1	Old (basal) leaf	PBA	2.80					2787
1	Root	PBA	1.56					2787
Medicago sativa L. (Alfalfa)								
1	Callus suspension culture FAB-MS		+					2809
1	Callus suspension culture FAB-MS	RIA	0.0004					2809

[a]nmol/mg protein. [b]nmol/L. [c]Dry weight basis. [d]nmol/10^6 protoplasts. [e]nmol/mg chlorophyll.

Continued

Table 5-4 Cyclic Ribonucleotides in Plants—cont'd

No.	Plant part	Method	cAMP	cGMP	cCMP	cUMP	cTMP	References
Najas flexilis (Willd.) Rostk. and Schmidt (Water Nymph)								
1	Whole plant	FFL	0.050					2785
1	Whole plant	PBA and diurnal variation	0.053					2785 2810
Nicotiana glutinosa L. (Tobacco)								
1	Leaf	RIA		0.000343				
	2811							
Nicotiana suaveolens Lehm. × *N. langsdorffii*								
1	Stem	RIA	0.0236					2812
Nicotiana tabacum L (Tobacco)								
1	Leaf	HPLC	0.0028					2813
1	–, chloroplast	HPLC-MS	0.0029					2814
1	Seedling, leaf	PBA	0.055					2787
1	–, apical bud	PBA	0.216					2787
1	–, stem	PBA	0.056					2787
1	–, root	PBA	0.070					2787
1	Shoot, meristem apex	HPLC-MS	0.0178					2815
1	Cell tissue culture	PKA	<0.0005					2778
1	Callus culture	PKA	0.0061					2781
1	Callus culture	GC-MS	0.137					2788
1	Callus culture	RIA	0.378					2788
1	Pith tissue culture	FFL	0.095					2774
1	Pith tissue culture	PKA	0.084					2774
1	Pith parenchyma	PKA	<0.0009					2778
1	Pith parenchyma[c]	RIA	0.40	0.050				2816
1	Axenic cell culture	PKA 0.0005-0.003						2817
Nuphar advena Ait. (Spatterdock)								
1	Whole plant	FFL	0.026					2785
1	Whole plant	PBA	0.029					2785
Oryza sativa L. (Rice)								
1	Embryo	RIA	<0.0005					2795
2	Shoot (1)	PBA	0.0042					2818
1	Shoot (2)	PBA	0.0020					2818
1	Seedling	RIA	<0.0005					2795
Panax ginseng C. Meyer (Ginseng)								
1	Root	PBA	0.00113					2783
Persea americana Mill. (Avocado)								
1	Young (apical) leaf	PBA	9.60					2787
1	Old (basal) leaf	PBA	4.90					2787
1	White root	PBA	1.30					2787
Petroselinum crispum Mill. (Parsley)								
1	Cell suspension culture	RIA	0.006					2819
1	Protoplast[d]	RIA	0.003					2819

Table 5-4 Cyclic Ribonucleotides in Plants—cont'd

No.	Plant part	Method	cAMP	cGMP	cCMP	cUMP	cTMP	References
Phaseolus vulgaris L. (Beans)								
1	Leaf	GC-MS	+					2820
1	Leaf	FAB-MS		+				2821
1	Leaf, protoplast[a]	PBA	0.400					2822
1	Seedling[c]	UV-Vis	2.6-9.2					2823
1	Seedling[c]	PBA	2.2					2824
1	Seedling[c]	PKA	0.0136					2825
1	Seedling[c]	HPLC	0.0197					2825
1	Shoot[c]	PBA	0.0046					2826
1	–, chloroplast[a]	PBA	0.0383					2826
1	–, chloroplast[a]	RIA		0.020				2826
1	Root, meristem	RIA		0.020				2805
		and variation in concentration along the root						2805
Phaseolus vulgaris L. ssp. *napus* (Dwarf Bean)								
1	Seedling	GC-MS	0.100					2788
1	Seedling	RIA	0.281					2788
Pinus densiflora Sieb. and Zucc. (Japanese Red Pine)								
1	Pollen	PBA	0.23	0.06				2827
Pinus pinaster Ait. (Cluster Pine)								
1	Embryo	RIA	<0.0005					2795
1	Needle (leaf)	RIA	<0.0005					2795
Pinus radiata D. Don (Monterey Pine)								
1	Seed	PBA	0.0168					2776
1	Old needle (leaf)	PBA	0.0384					2776
1	Young needle	PBA	0.0096					2776
1	Spring shoot	PBA	<0.0017					2776
1	Male flower	PBA	0.0627					2776
1	Female flower	PBA	<0.0024					2776
Pinus taeda L. (Loblolly Pine)								
1	Callus culture	PBA	0.125					2828
Pisum sativum L. (Pea)								
1	Seed, embryo	PBA	0.176					2787
1	Leaf[a]	PBA	0.179					2822
1	Root tip	FAB-MS	+	+	+	+	0	2829
1	Nonmeristematic root segment	FAB-MS	+	+	+	+	+	2829
Potamogeton zosteriformis Fern. (Flatstem Pondweed)								
1	Whole plant	FFL	0.028					2785
1	Whole plant	PBA	0.025					2785
Raphanus sativus L. (Radish)								
1	Root	PBA	0.00120					2783
Robinia pseudoacacia L. (Black Locust)								
1	Sieve-tube sap[b]	PBA	9.0					2830

[a]nmol/mg protein. [b]nmol/L. [c]Dry weight basis. [d]nmol/10^6 protoplasts. [e]nmol/mg chlorophyll.

Continued

Table 5-4 Cyclic Ribonucleotides in Plants—cont'd

No.	Plant part	Method	cAMP	cGMP	cCMP	cUMP	cTMP	References
Rubus spp. (Bramble)								
1	Pith tissue culture	FFL	0.142					2774
Scirpus subterminalis Torr.								
1	Whole plant	FFL	0.048					2785
1	Whole plant	PBA	0.053					2785
		and diurnal variation						2810
Secale cereale L. (Rye)								
1	Seedling	HPLC	0.700					2831
Sinapis alba L. (White Mustard)								
1	Seedling	PBA	0.114					2832
Solanum tuberosum L (Potato)								
1	Tuber	PBA	0.028					2833
1	Tuber	RIA	30.2-40.1	14.0-15.2				2834
Spirodela oligorrhiza (Kurz.) Hegelm. (Duckweed)								
1	Whole plant	PBA	0.0110					2776
Trifolium repens L. (White Clover)								
1	Leaf[b]	PBA	0.0132	and diurnal variation				2835
Triticum aestivum L. (Wheat)								
1	Embryo (germ)	PBA	0.0111					2776
1	Leaf	PBA	0.180					2787
1	Root	PBA	0.170					2787
Vicia faba L. (Faba Bean)								
1	Seedling	UV-Vis	0					2823
1	Internode	PBA	0					2836
Vigna mungo (L.) Hepper (Black Gram)								
1	Seedling, shoot	PBA	0.0174					2837
1	–, root	PBA	0.0085					2837
Zea mays L. (Corn)								
1	Kernel	GC-MS	0.079					2788
1	Kernel	RIA	0.274					2788
1	Seedling	GC-MS		0.072				2838
1	Seedling[c]	GC-MS	6.808					2839
1	Seedling	PBA	0.035					2840
1	–, chloroplast[e]	PBA	0.594					2840
1	–, coleoptile tip[c]	PBA	1.3					2841
1	–, shoot[c]	HPLC	0.1014					2842
1	–, root[c]	HPLC	0.0294					2842
1	–, root tip[c]	PBA	4.2					2841
		and variation in cAMP content from top to bottom						2841
1	Root	PBA	0.309					2843

Table 5-4 Cyclic Ribonucleotides in Plants—cont'd

No.	Plant part	Method	cAMP	cGMP	cCMP	cUMP	cTMP	References
Ziziphus jujuba Mill (Jujube)								
1	Mature fruit	RIA	100-150					2844
1	–, flesh[c]	RIA		30-50				2845
1	Mature fruit	PBA	100-150					2844
1	Immature fruit	PBA	<0.0025					2844
Ziziphus vulgaris, ssp *inermis*								
1	Fruit, flesh	RIA		19.3				2845
1	–, stone	RIA		0.14				2845
Ziziphus vulgaris, ssp. *spinosus*								
1	Fruit, flesh	RIA		5.4				2845
1	–, stone	RIA		0.41				2845
		Of some 180 medicinal herbs from China, only the fruits of *Ziziphus jujuba* and *Evodia rutaecarpa* contain cGMP[2845]						

nmol/g fwb unless stated otherwise.
[a]nmol/mg protein.
[b]nmol/L.
[c]Dry weight basis.
[d]nmol/10^6 protoplasts.
[e]nmol/mg chlorophyll.

Sugar Nucleotides

This group comprises the sugar derivatives of three of the four ribonucleoside diphosphates (ADP, GDP, and UDP) and one deoxyribonucleoside diphosphate (TDP), and related compounds such as their hexosamine and hexuronic acid derivatives. The following is a complete list of the sugar nucleotides (NDPX, where X is the sugar component) mentioned in this section:

Nucleotide	Sugar, XH
ADP	D-Glucose (Glc),[a] D-galactose (Gal), D-fructose (Fru), D-mannose (Man), D-ribose (Rib), L-arabinose (Ara), D-xylose (Xyl), *N*-acetyl-D-glucosamine (GlcNAc)
GDP	D-Glucose (Glc), D-galactose (Gal), D-mannose (Man), D-rhamnose (Rha), D-ribose (Rib), D-xylose (Xyl), L-arabinose (Ara), D-fucose (Fuc)
CDP	None
UDP	D-Glucose (Glc), D-galactose (Gal), D-fructose (Fru), D-rhamnose (Rha), D-mannose (Man), D-digitoxose, D-ribose (Rib), L-arabinose (Ara), D-xylose (Xyl), D-fucose (Fuc), D-apiose (Api), D-glucuronic acid (GlcA), D-galacturonic acid (GalA), *N*-acetyl-D-glucosamine (GlcNAc), *N*-acetyl-D-galactosamine (GalNAc)
TDP	D-Glucose (Glc), D-galacturonic acid (GalA)

[a]Frequently abbreviated to G, but given throughout this section as Glc to avoid any uncertainty.

Since this class of compounds is so large, the only compound whose chemistry is described in any detail is its predominant member, UDPGlc. The coverage in Table 5-6, however, is intended to be comprehensive:

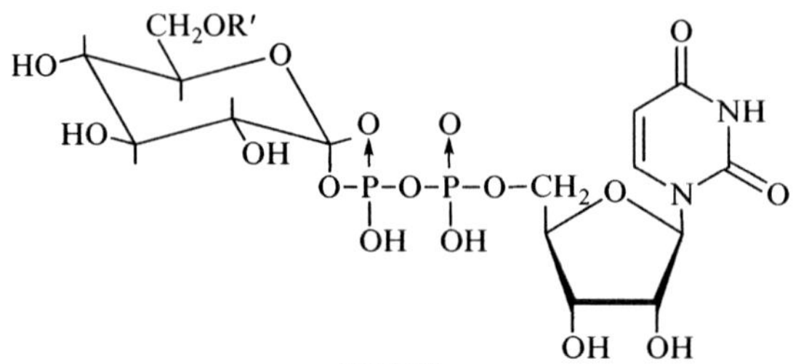

UDPGlc

Occurrence

UDPGlc and UDPGal were first isolated from sugar beet by Buchanan *et al.* in 1953[3344] and shown to be identical with the compounds isolated by Leloir from yeast. In 1956 Ginsburg *et al.*,[2433] who were mentioned at the beginning of this chapter in connection with the parent nucleotides, were able to identify UDPGlc, UDPGal, UDPXyl, and UDPAra (66:28:4:2) in one fraction of an extract of mung beans. In 1957 Solms and Hassid identified UDPGlc-NAc and UDPGlcA in mung beans.[3377] Many other sugar nucleotides were discovered in plants over the next two decades, the majority being derivatives of UDP. The metabolic pathways for the biosynthesis of the uridine sugar nucleotides are outlined in Fig. 5-4. For a more complete review, see Feingold.[2426]

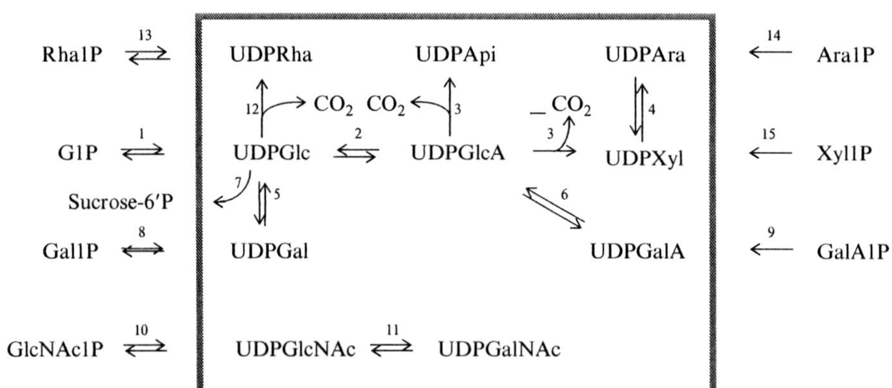

FIGURE 5-4 Uridine Sugar Nucleotides. (1) UDPGlc pyrophosphorylase [E.C. 2.7.7.9], UTP. (2) UDPGlc dehydrogenase [E.C. 1.1.1.22]. (3) UDPGlcA decarboxylase [E.C. 4.1.1.35]. (4) UDPXyl 4-epimerase [E.C. 5.1.3.5]. (5) UDPGlc 4-epimerase [E.C. 5.1.3.2]. (6) UDPGlcA 4-epimerase [E.C. 5.3.3.6]. (7) Sucrose synthase [E.C. 2.4.1.13] (8) UDPGal pyrophosphorylase [E.C. 2.7.7.10], UTP. (9) UDPGalA pyrophosphorylase, UTP. (10) UDPGlcNAc pyrophosphorylase, UTP. (11) UDPGlcNAc 4-epimerase.

Separation and Analysis

(1) Paper Chromatography (PC) Method

Two-dimensional PC using the solvents and conditions outlined in Chapter 1, pp 2-3, can be used to separate the sugar nucleotides from the other nucleotides, but UDPGlc is slightly degraded by solvents A2 and B2.[9,327] An alternative procedure, which can also be used for the separation of ADPGlc from the other nucleotides,[3369] is that of Paladini and Leloir[3382] in which ethanol/1*M* ammonium acetate, pH 7.5 (70:30) (solvent N2), is used in one direction followed by ethanol/1*M* ammonium acetate buffer, pH 3.8 (70:30), in the opposite direction. Dankert *et al.*[2755] published a list of PC solvent systems that can be used to separate intact UDPGlc, UDPGal, UDPGlcNAc, and UDPGalNAc from each other and of intact ADPGlc, ADPGal, ADPMan, and ADPGlcNAc from each other; all of these were found in sweet corn grains after separation of the UDP-sugar and ADP-sugar fractions by IEC.[2755] Even better separations are achieved with solvents containing organic borates.[3383] In most cases, however, it is more convenient to split the sugars from the sugar nucleotides by brief (15 min) hydrolysis at pH 2, after which the individual sugars can be separated and quantitated by PC, PE, TLC, etc., and identified by the usual tests.

(2) Thin-Layer Chromatography (TLC) Method

Two-dimensional TLC separations of the sugar nucleotides are often carried out under the conditions described by Bieleski,[36] using the same solvents (B2 and A2) as those used for PC. In general, the TLC results parallel those of the PC method.

(3) Electrophoretic Methods

Paper electrophoresis (PE) and thin-layer electrophoresis (TLE) have been employed occasionally in the first or second dimension of a two-dimensional PC/PE[331,337,2723] or TLC/TLE[36] chromatographic separation. The mobility M_{AMP} relative to AMP = 100 is 185 for UDPGlc and 151 for ADPGlc in citrate buffer (pH 4.8).[2723]

(4) Ion Exchange Chromatography (IEC) Method

This method, developed by Bergkvist[2432] and Ginsburg *et al.*[2433] in 1956, is often used to separate the sugar nucleotides from other nucleotides prior to their separation from each other. Upon IEC on an anion exchange column such as Dowex 1-X8 (formate) and a formic acid/ammonium formate gradient, the ADP sugars (ADPX) elute as a band just ahead of ADP and the UDP sugars (UDPX) elute as another band just ahead of UDP. The addition of boric acid to the gradient eluent produces a dramatic change in the elution profile allowing the ADPX and UDPX bands to elute before any of the other nucleotides,[2754] yet this simple modification of the procedure has seldom been adopted.

(5) High-Performance Liquid Chromatography (HPLC) Method

This method, introduced by Liu and Shannon[342] and Carpita and Delmer[2571] in 1981, provides much higher resolution than the IEC method, but it is still not sufficient to ensure complete separation of the sugar nucleotides. With rare exceptions, the separation is carried out in the ion-exchange mode with strongly anionic supports such as Partisil 10 SAX or Hypersil APS. In general, the sugar nucleotides are well separated according to their nucleobases, eluting in the order CDPG > UDPG > ADPG > GDPG between the nucleoside monophosphates and the nucleoside diphosphates, but the separation of compounds having similar or epimeric sugar moieties is difficult. UDPGlc and UDPGal are poorly separated but the corresponding uronic acids (UDPGlcA and UDPGalA) are well separated, appearing slightly before GDP. Another well-resolved peak eluting just before UDPGlc was identified as UDP-*N*-acetylglucosamine (UDPGlcNAc) based on its retention time, coelution with an authentic sample and UV analysis, but the presence of small amounts of UDP derivatives containing other *N*-acetylated amino sugars could not be excluded.[2440] The separation of UDPGlc and ADPGlc from each other and from other nucleotides has also been carried out by HPLC in the ion-pair mode with reversed phase supports such as Lichrosorb RP-18[2623] or Adsorbosphere.[3379]

Analysis Without Separation

(1) Enzymatic Methods

(A) URIDINE DIPHOSPHATE SUGARS

UDPGlc may be assayed directly without separation from other nucleotides by either of the enzymatic reactions listed in Table 5-5, the only requirement being the removal of substances that interfere with the spectrophotometric analysis. The following is a typical procedure for reaction (1):

> *For the analysis of UDP-glucose, a 25 μl sample [of a 0.5 ml barley leaf extract containing 100 mM Tris-HCl (pH 8.0) and 10 mM $MgCl_2$] was added to 0.95 ml assay solution containing 350 mM glycine (pH 8.7), 6 mM EDTA, and 0.12 IU UDP-glucose dehydrogenase. Assays were initiated by the addition of 2.1 mM NAD^+ and increased absorbance was measured at 340 nm.*
>
> Sicher, R. C.; Kremer, D. F.; Harris, W. G., "Control of Photosynthetic Sucrose Synthesis in Barley Primary Leaves. Role of Fructose 2,6-Bisphosphate", *Plant Physiol.* **1986**, *82*(1), 15–18,[3358] reprinted by permission of the American Institute of Plant Biologists, © 1986.

This procedure can be used as the first step in a sequence of enzyme-coupled reactions in which UDPGlc, UTP, UDP, and UMP are assayed in turn in a single run (see p 296).[3027] Alternatively, UDPGlc may be assayed as the next to last step in a sequence of enzyme-coupled reactions in which G6P, F6P, G1P, ATP, UDPGlc, and ADPGlc are assayed in turn

Table 5-5 NAD(P) Coupled Reactions

Method	Reaction	For analysis of
1	UDPGlc + NAD → UDPGlcA + NADH	UDPGlc
2	UDPGlc + NADP → G1P + PP_i	UDPGlc

Enzymes: (1) UDPGlc dehydrogenase [E.C. 1.1.1.22]. (2) UDPGlc pyrophosphorylase [E.C. 2.7.7.9].

in a single run (see Chapter 1, p 39). This sequence uses reaction (2) to assay the UDPGlc, as in the following procedure:

The assay mixture (1.5 mL) in the cuvette contained the following: buffered extract [of wheat grain], 0.5 mL; Tris-HCl buffer (pH 7.8), 60 μmol; $MgSO_4$, 10 μmol; NADP, 1 μmol; sodium pyrophosphate, 3 μmol. The absorbance (E_1) at 340 nm was recorded. Two units (5 μL) of G-6-P dehydrogenase was added and the increased absorbance (E_2) noted. The extinctions E_3, E_4, E_5, E_6, and E_7 were likewise recorded after successive additions of 2 units (5 μL) each of phosphoglucoisomerase, phosphoglucomutase, hexokinase, UDPG pyrophosphorylase, and 5 μL of ADPG pyrophosphorylase. From these values, concentrations were calculated as ΔOD(G-6-P) = $E_2 - E_1$, ΔOD(F-6-P) = $E_3 - E_2$, ΔOD(G-1-P) = $E_4 - E_3$, ΔOD(ATP) = $E_5 - E_4$, ΔOD (UDPG) = $E_6 - E_5$, and ΔOD(ADPG) = $E_7 - E_6$.

(B) ADENOSINE DIPHOSPHATE SUGARS

ADPGlc may be assayed by the method just described.

(2) Firefly Luminescence Method

This method is similar to the bioluminescence method for the determination of UTP (see p 297) except that the plant extract is incubated with UDPGlc pyrophosphorylase and excess PP_i (method 2, Table 5-5) instead of excess G1P prior to treatment with NDPK and ATP[2849]:

$$[\mathrm{UDPGlc}] = [\mathrm{ATP}]_{\mathrm{endogenous}} + [\mathrm{ATP}]_{\mathrm{NDPK}} - [\mathrm{ATP}]_{\mathrm{pyrophosphorylase}}$$

If the assay is repeated with a mixture of UDPGlc pyrophosphorylase and UDPGal epimerase in the first incubation, the amount of UDPGal in the extract can be calculated by the expression[481]:

$$[\mathrm{UDPGal}] = [\mathrm{ATP}]_{\mathrm{endogenous}} + [\mathrm{ATP}]_{\mathrm{NDPK}} + [\mathrm{ATP}]_{\mathrm{pyrophosphorylase}} - [\mathrm{ATP}]_{\mathrm{epimeraase}}$$

(3) ^{31}P Nuclear Magnetic Resonance (NMR) Method

Because of molecular asymmetry, the sugar diphosphate nucleotides display separate resonances for the two phosphorus atoms in their ^{31}P NMR spectra, further split into a pair of doublets owing to P—O—P coupling. The chemical shifts (relative to 85% phosphoric acid) are −10.38 and −11.94 ppm for UDPGlc, −10.34 and −11.83 ppm for UDPGal, −10.74 and −13.05 ppm for UDPMan, and −10.57 and −12.22 ppm for UDPGlcNAc.[68] The principal sugar nucleotides, UDPGlc and UDPGal, are readily identifiable in the ^{31}P NMR spectra of perchloric acid extracts of maize root tips,[587] spinach leaves and intact chloroplasts,[50] and sycamore cell suspension cultures,[3341] but in the *in vivo* studies which comprise the majority of those listed in Table 5-6 the resolution is insufficient to

Table 5-6 UDPGlc and Other Sugar Nucleotides in Plants

No.	Plant part	Method	UDPGlc	Other NDPX	Total	References
Acer pseudoplatanus L. (Sycamore Maple)						
1	Cell suspension culture[a]	IEC	90	UDPGal trace, UDPGalA trace		2447
1	Cell suspension culture[a]	^{31}P NMR	+	UDPGal +		3341
Actinidia deliciosa Liang and Ferg. (Kiwi)						
1	Fruit	Enzym.	40			222
1	Callus culture	^{31}P NMR	+			475
Allium porrum L. (Leek)						
1	Seed embryo[b]	HPLC	0.23			2458
Amorphophallus konjac C. Koch (Konjak)						
1	Corm[c]	IEC	57	ADPGlc 19, GDPMan 24		3342
Apium graveolens L. ssp. *dulce* (Celery)						
1	Petiole, vascular bundle	PC/PC	+			227
Arum maculatum L. (Cuckoopint)						
1	Spadix	HPLC	292	ADPGlc 28.8		3343
Avena sativa L. (Oats)						
1	Seedling	IEC	73.4			2510
1	–, root, and stem	IEC	+	ADPGlc +		2475
1	–, aerial parts	IEC	38	UDPGlcNAc 12		2471
Beta vulgaris L. (Sugar Beet)						
2	Seed embryo[b]	Lumin.	0.206-0.245	UDPGal 0.81-0.93		2478
1	Leaf	PC/PC	+	UDPGal +		3344
1	Leaf	IEC	13.9			2479
1	–, vascular bundle	IEC	20.7			2479
1	–, petiole, and parenchyma	IEC	8.0			2479
2	Axis[b]	Lumin.	0.15-0.20	UDPGal 0.15		481
1	Root[d]	IEC	3386.8			2485
1	Root	IEC		UDPX (X=Glc, Gal, Xyl) 18, TDPX (X=Glc, GalA) +, GDPGlc +		3345
Brassica campestris L. (Turnip Rape)						
1	Pod, deseeded[a]	Enzym.	136	ADPGlc 60		238

Table 5-6 UDPGlc and Other Sugar Nucleotides in Plants—cont'd

No.	Plant part	Method	UDPGlc	Other NDPX	Total	References
Brassica oleracea L. ssp. *italica* Plenck. (Broccoli)						
1	Stems and florets	HPLC	216	ADPRib 0		2441
Brassica pekinensis (Lour.) Rupr. (Chinese Cabbage)						
1	Petiole, vascular bundle	PC/PC	+			227
1	Petiole, vascular bundle	TLC/TLC	+			36
1	Petiole, vascular bundle	TLC/TLE	+			36
Brassica rapa L. (Turnip)						
1	Leaf, vascular bundle	TLC/TLC	+	UDPX +		240
Bryophyllum tubiflorum Harv. (Chandelier Plant)						
1	Phyllodia	TLC/TLC	+			3346
Camellia sinensis L. ssp. *assamica* (Mast.) Kitam. (Assam Tea)						
1	Leaf	IEC	50			2500
Capsicum annuum L. (Bell Pepper)						
1	Leaf, mature	HPLC	29			2503
1	–, young	HPLC	183			2503
1	–, petiole	HPLC	19			2503
Carthamus tinctorius L.(Safflower)						
1	Leaf	HPLC	13			2503
1	Flower bud	HPLC	51			2503
Catharanthus roseus (L.) G. Don (Madagascar Periwinkle)						
1	Cell suspension culture	HPLC	95	UDPGal 4, UDPXyl 0.3, UDPAra 3, UDPGlcA 18		3347
1	Cell suspension culture	^{31}P NMR	+			2507
Chenopodium rubrum L. (Red Goosefoot)						
1	Cell suspension culture	Enzym.	40			2509
Cichorium intybus L. (Chicory)						
1	Root		+	No others present		3348
Citrus sinensis (L.) *Osbeck* (Orange)						
1	Fruit, juice	IEC	+			248

[a]Dry weight basis. [b]nmol/plant part. [c]Percent of total sugar nucleotide. [d]μg/g. [e]nmol/mL. [f]μmol/mg chlorophyll. [g]Percentage of total free nucleotide. [h]mol/mol CABP binding site. [i]μmol/mg protein. [j]nmol/mg starch.

Continued

Table 5-6 UDPGlc and Other Sugar Nucleotides in Plants—cont'd

No.	Plant part	Method	UDPGlc	Other NDPX	Total	References
Cucumis sativus L. (Cucumber)						
1	Cotyledon[c]	HPLC	73.15	UDPGal 6.40, UDPXyl 0.89, UDPRha 0.58, UDPAra 1.53, UDPMan 1.27, UDPGlcA 16.18	0.68	3349
1	Leaf	IEC	66.0	UDPX 81.5, ADPGlc 0		250
Cucurbita maxima Duch. (Squash)						
1	Leaf, sieve-tube exudate	TLC/TLC	+			240
Cucurbita pepo L. (Pumpkin)						
1	Stem, phloem exudate[e]	IEC	758			251
1	Ripe fruit, phloem exudate[e]	IEC	15			251
1	Unripe fruit, phloem exudate[e]	IEC	343			251
Cuscuta reflexa Roxb. (Dodder)						
1	?	TLC/TLC	+			33
Dahlia pinnata (Dahlia)						
1	Tuber	IEC	+	UDPGal +, UDPFru +, UDPGalNAc +		3350
Datura inoxia Mill. (Downy Thorn Apple)						
1	Cell suspension culture	HPLC	102*	UDPGlcNAc 4.1, UDPGlcA 4.8, UDPGalA 0.6		2440
	*Incl. UDPGal			GDPGlc (previous day) 1.7		3351
1	Cell suspension culture	HPLC	199.1	ADPGlc 6.6		2535
Daucus carota L. (Carrot)						
1	Cell suspension culture	HPLC	220		450	2538
1	Cell suspension culture	^{31}P NMR	+			2507
Digitalis purpurea L. (Digitalis)						
1	Leaf	PC	+	UDPDigitoxose 2.8, UDPGal +, UDPAra +, UDPXyl +	360	3352
Drimys lanceolata (Poir.) Baill.						
1	Leaf	IEC		ADPX 17.65		2543
Egeria densa Planch. (Brazilian Waterweed)						
1	?	TLC/TLC	+			33

Table 5-6 UDPGlc and Other Sugar Nucleotides in Plants—cont'd

No.	Plant part	Method	UDPGlc	Other NDPX	Total	References
Fragaria spp. (Strawberry)						
1	Leaf	IEC	90.5	UDPGal 20.9, UDPXyl 8.4, UDPAra 10.9, UDPFru 24.3, UDPGlcA 4.2, GDPGlc 0.03, GDPGal 0.03, GDPMan 1.15, GDPXyl 0.29		259
1	Leaf	Enzym.	106			2500
Glycine max (L.) Merr. (Soybean)						
1	Seed	IEC	2.5			2553
1	Leaf[f]	HPLC	79	ADPGlc 15		26
1	Root nodule	^{31}P NMR	+			2563
1	Sprout	PC	+	UDPGal +		3353
1	Cell suspension culture	IEC	102	UDPGal 34, UDPXyl 2.4, UDPAra 2.4, UDPRha 1.9, ADPGlc +, ADPRib +, ADPAra +, GDPGlc +, GDPMan +, GDPAra +, GDPFuc +	740	3354
1	Cell suspension culture[e]	HPLC	6.1	ADPGlc 0.43		3343
Gossypium arboreum L. (Tree Cotton)						
1	Boll, primary wall fiber[b]	IEC	145	UDPGal 32, UDPXyl 10, UDPAra 12, UDPGalA 18	218	2566
1	–, secondary wall fiber[b]	IEC	250	UDPGal 28	278	2566
Gossypium hirsutum L. (Upland Cotton)						
1	Boll[b]	Enzym.	200			2571
1	–, fiber[g]	HPLC	>75	UDPGal 13, UDPMan trace, UDPRha trace, UDPFuc trace, UDPXyl +, UDPAra +		2571
1	Seedling, leaf	IEC	246 (incl. UDP)			2572
1	–, chloroplast		+	UDPGal +, ADPMan +, GDPGlc +, GDPAra +		2570
1	–, root	IEC	438 (incl. UDP)			2572
Helianthus annuus L (Sunflower)						
1	Leaf	IEC	+			3355
1	Stem	^{31}P NMR	+			2563
Helianthus tuberosus L. (Jerusalem Artichoke)						
1	Tuber	IEC	7.75	UDPGal 0.78, UDPFru 0.13, UDPHexNAc 0.6, GDPMan 0.9		2576,3356
1	Young stolon	IEC	+			45

[a]Dry weight basis. [b]nmol/plant part. [c]Percent of total sugar nucleotide. [d]μg/g. [e]nmol/mL. [f]μmol/mg chlorophyll. [g]Percentage of total free nucleotide. [h]mol/mol CABP binding site. [i]μmol/mg protein. [j]nmol/mg starch.

Continued

Table 5-6 UDPGlc and Other Sugar Nucleotides in Plants—cont'd

No.	Plant part	Method	UDPGlc	Other NDPX	Total	References
Hemerocallis spp. (Daylily)						
1	Flower petal[g]	TLC/TLC	16.7	ADPGlc 2.9		266
Heracleum sosnowski Manden (Cow Parsnip)						
1	Vascular bundle, phloem	PC/PC	40	ADPGlc 16, UDPGal +, UDPPentose +	104	2577
1	–, xylem	PC/PC	+	ADPGlc 0, UDPGal +, UDPPentose +	88	2577
Hordeum vulgare L. (Barley)						
1	Seed	IEC	+			3357
2	Immature grain, dehusked	Enzym.	9-12			502
1	Leaf[f]	Enzym.	50	and variation during the daily cycle		3358
1	Root	HPLC	1177			2587
1	Root	^{31}P NMR	+			2589
1	Seedling, aerial parts	IEC	44	UDPGlcNAc 14		2471
Impatiens balsamina L. (Garden Balsam)						
1	Leaf node	IEC		UDPFru +		3359
Ipomoea batatas (L.) Lam. (Sweet Potato)						
1	Root	IEC	68.1	ADPGlc 5.9		2592
Isoetes duriei Bory						
1	?	TLC/TLC	+			33
Lactuca sativa L. (Lettuce)						
1	Seed	HPLC	+			2598
Larix decidua Mill. (European Larch)						
1	Cambial tissue	IEC	9.1	UDPGal 1.0, UDPXyl 1.2, UDPAra 0.7, UDPFru 0.3, ADPGlc 4.3, UDPGalA 0.3, ADPRib 0.1, ADPGal trace, ADPFru trace, GDPG 1.4, GDPGal 0.2, GDPMan 1.2	22.1	2601
1	Xylem tissue	IEC	8.4	UDPGal 1.1, UDPXyl 0.3, UDPAra 0.2, UDPFru 1.5, ADPG 1.3, ADPRib 0.7, ADPGal 0.1, ADPFru 0.2, GDPG 0.3, GDPGal 0.3, GDPMan 0.1	14.6	2601
Leleba oldhami (Munro) Nakai (Summer Bamboo)						
1	Desheathed shoot	IEC	72.5	UDPXyl 13.8, UDPGlcNAc 2.5 (incl. UDPGalNAc) 2.5		2602
Lemna gibba L. (Gibbous Duckweed)						
1	?	TLC/TLC	+			33
Lemna major L.						
1	Whole plant	TLC/TLC	+			3360
Liriodendron tulipifera L. (Tulip Tree)						
1	Leaf	IEC		UDPX 1.64, ADPX 10.92, GDPX 4.75		2543

Table 5-6 UDPGlc and Other Sugar Nucleotides in Plants—cont'd

No.	Plant part	Method	UDPGlc	Other NDPX	Total	References
Lycopersicon esculentum Mill. (Tomato)						
1	Leaf[a]	IEC	540			473
1	Leaf	IEC		GDPGlc 2.48		2613
1	Fruit[h]	IEC	225			2615
Magnolia grandiflora L. (Southern Magnolia)						
1	Leaf	IEC		UDPX 6.72, ADPX 25.95, GDPX 35.68		2543
Malus domestica Borkh. (Apple)						
3	Phloem exudate[g]	HPLC	17-70			2623
Malus sylvestris (L.) Mill. (Crabapple)						
1	Stem, phloem tissue	PC/PC	+			227
Medicago sativa L. (Alfalfa)						
1	Root	PC/PC	+			280
Musa cavendishii Lamb. ex Paxton (Dwarf Banana)						
1	Peeled fruit	PC/PC	+	UDPGal trace		3361
Nicotiana alata Link and Otto (Jasmine Tobacco)						
1	Pollen tube	HPLC	1600			3362
1	Cell suspension culture	HPLC	30			3362
Nicotiana tabacum L. (Tobacco)						
1	Leaf *Incl. UDPGal	HPLC	23*	ADPGlc 1.1, GDPGlc 2.8, UDPGlcNAc 10.4		2440
1	Leaf, old[b]	HPLC	67 (incl. UDPGal and UDPGlcNAc)			3363
1	Leaf, young[b]	HPLC	132 (incl. UDPGal and UDPGlcNAc)			3363
1	Terminal bud[b]	HPLC	73 (incl. UDPGal and UDPGlcNAc)			3363
1	Root *Incl. UDPGal	HPLC	9.0*	UDPGlcNAc 3.3		2440
1	Apex *Incl. UDPGal	HPLC	147*	UDPGlcA 7.1, UDPGalA 2.0, GDPGlc 5.3		2628
1	Pollen grain	^{31}P NMR	+			2563

[a]Dry weight basis. [b]nmol/plant part. [c]Percent of total sugar nucleotide. [d]μg/g. [e]nmol/mL. [f]μmol/mg chlorophyll. [g]Percentage of total free nucleotide. [h]mol/mol CABP binding site. [i]μmol/mg protein. [j]nmol/mg starch.

Continued

Table 5-6 UDPGlc and Other Sugar Nucleotides in Plants—cont'd

No.	Plant part	Method	UDPGlc	Other NDPX	Total	References
1	Cell suspension culture	HPLC	213*	GDPGlc 3.6, UDPGlcNAc 13.8, UDPGalA 0.6	*incl. UDPGal	2440
1	Callus culture	IEC	38.0	ADPGlc 0		3364
Oenanthe javanica (Blume) DC (Chinese Celery)						
1	Stem	PC/PC	7.5*	UDP(D-GalA) 2.0 *Incl. UDPGal and UDPXyl		2498
Ophioglossum lusitanicum L. (Least Adderstongue)						
1	Leaf	TLC/TLC	+			33
Ophioglossum vulgatum L. (Southern Adderstongue)						
1	Leaf	TLC/TLC	+			33
Oryza sativa L. (Rice)						
1	Grain	IEC	+	ADPGlc +		2636
1	–, 10 daf	IEC	56			2635
1	–, mid-milky stage	IEC	100	ADPGlc 33		3365
1	–, ripening kernel	IEC	88.0	ADPGlc 45.6		2633
1	Leaf and culm	IEC	+	ADPGlc trace		2636
1	Shoot	^{31}P NMR	+			2638
1	Cell suspension culture	IEC	32.0	UDPGal 10.0, UDPXyl 3.1, UDPAra 4.0, UDPFuc 2.8, UDPRha 2.9, UDPRib 3.0, UDPMan 2.2, UDPGlcA 0.6, UDPGalA 0.2, ADPGlc 6.2, ADPRib 8.5, ADPXyl 0.3, ADPAra 0.4, GDPGlc 13.2, GDPXyl 14.3, GDPFuc 1.0, GDPRha 2.0, GDPRib 4.3	111.3	2641
Petroselinum crispum (Mill.) Nym. (Parsley)						
1	Whole plant[a]	IEC	19.1	UDPGal 6.1, UDPXyl 0.9, UDPAra 0.5, UDPFru 0.1, UDPRha 0.5, UDPApi 0.04	27.3	3366
Phacelia tanacetifolia Benth. (Fiddleneck)						
1	Seedling	^{31}P NMR	+			2644
Phaseolus vulgaris L. (Bean)						
1	Leaf[g]	PC/PC	58.2			12
1	Leaf[h]	Enzym.	3.8			519
1	Mature leaf	IEC	4.2	GDPGlc 1.7		2649
1	Mature leaf	HPLC	79	(UDPGlcNAc +)[2651]		2652
1	Young leaf	HPLC	363			2652
1	Etiolated leaf[b]	IEC	0.5	UDPXyl 2.1		2650
1	Leaf, petiole	IEC	24	ADPRib 0.5		2653
1	–, laminae	IEC	24	ADPRib 3		2653
1	Seedling	IEC	51.0	ADPGlc 5.0, ADPRib 3.5		2656

Table 5-6 UDPGlc and Other Sugar Nucleotides in Plants—cont'd

No.	Plant part	Method	UDPGlc	Other NDPX	Total	References
1	Seedling, leaf	IEC	230			2657
1	–, root	IEC	160			2657
1	Fruit, pericarp	IEC	117			520
Phyllostachys pubescens [Mazel ex] Houz. (Moso Bamboo)						
1	Desheathed shoot	IEC	24.0*	UDPXyl 7.6, UDPGalA 3.6, UDPGlcNAc 3.6 (incl. UDPGalNAc)		2661
		*Incl. UDPGal				
Pinus banksiana Lamb. (Jack Pine)						
1	Seedling[g]	IEC	1.0			2665
Pisum sativum L. (Pea)						
1	Seed	IEC	216			24
1	Seed	IEC	+	UDPGal +, UDPXyl +, UDPAra +, UDPRha +, UDPGlcNAc +, UDPGalNAc +, ADPGlc +	217	3367
1	Mature seed	IEC	79	ADPRib 39, UDPGal trace		2437
1	Seed	HPLC	309	ADPRib trace		2441
2	Seed embryo	HPLC	324-421			295
1	Seed coat	Enzym.	496			2676
1	Leaf[f]	HPLC	4.2			3368
1	Root	IEC	3600			2683
1	Root tip	^{31}P NMR	+			526
1	Seedling	PC/PC	0			2681
1	–, cotyledon	Enzym.	238			527
Populus deltoides Bartr. ex Marsh (Eastern Cottonwood)						
1	Leaf[f]	Enzym.	72			297
Pteridium aquilinum (L.) Kuhn (Bracken Fern)						
1	Rhizome[a]	IEC	9.1	UDPGal 2.3, UDPXyl 1.3, UDPAra 0.4, UDPGalA 0.6, GDPMan 0.9, GDPFuc 0.4		2689
Ricinus communis L. (Castor Bean)						
1	Seedling cotyledon	Enzym.	259			303
1	Phloem exudate[e]	HPLC	400			304
1	Phloem exudate[e]	Enzym.	450			2694
Saccharum officinarum L. (Sugar Cane)						
1	Leaf[g]	PC/PC		UDPX 11.9		306
1	Stalk, third internode[i]	HPLC	8.6	and variation from third to ninth internode		307
1	Root[g]	PC/PC	6.1	UDPX 4.5		306

[a]Dry weight basis. [b]nmol/plant part. [c]Percent of total sugar nucleotide. [d]μg/g. [e]nmol/mL. [f]μmol/mg chlorophyll. [g]Percentage of total free nucleotide. [h]mol/mol CABP binding site. [i]μmol/mg protein. [j]nmol/mg starch.

Continued

Table 5-6 UDPGlc and Other Sugar Nucleotides in Plants—cont'd

No.	Plant part	Method	UDPGlc	Other NDPX	Total	References
Salix triandra L. (Almond Willow)						
1	Phloem exudate	TLC/TLC	+			309
Selaginella denticulata (L.) Link (Tooth-Leaved Clubmoss)						
1	?	TLC/TLC	+			33
Sinapis alba L. (White Mustard)						
1	?	TLC/TLC	+			33
Solanum tuberosum L. (Potato)						
1	Tuber	PC/PC	+			9
1	Tuber[g]	IEC	33.2			316
1	Tuber	Enzym.	60	Variation in UDPGlc content along the stolon-apex axis		533
1	Tuber, mature	HPLC	29			532
1	–, immature	HPLC	105			532
1	Starch grain	IEC	0	ADPGlc 2.0, ADPRib 0		2709
Sorghum bicolor (L.) Moench (Sorghum)						
1	Seed [a]	PC/PC	1750*	*Incl. ADPGlc	2500	3369
1	Root tip	^{31}P NMR	+			44
Spinacia oleracea L.						
1	Leaf[f]	Enzym.	90	and subcellular distribution		322
1	Leaf	Enzym.		Diurnal variation in UDPGlc		534
1	Leaf	^{31}P NMR	+	UDPGal + (both absent in chloroplast)		50
Spirodela oligorrhiza (Kurz.) Hegelm. (Duckweed)						
1	Whole plant	TLC/TLC	+			3360
Tilia platyphyllos Scop. (Large-Leaved Linden)						
1	Phloem exudate	TLC/TLC	+			328
Tillandsia usneoides (L.) L. (Spanish Moss)						
1	?	TLC/TLC	+			33
Trifolium subterraneum L. (Sub Clover)						
1	Whole plant[a]	IEC	250			329
Trigonella foenum-graecum L. (Fenugreek)						
1	Seed[a] *Incl. UDPGal	IEC	39.2*			2716
1	Sprout[a] *incl. UDPGal	IEC	123.8*	GDPMan 32.0		2716
×*Triticosecale* Wittm. (Triticale)						
1	Grain	IEC	3.4*	ADPGlc 0.4, GDPGlc 0.2 *Incl. UDPGal and UDPMan		3370
Triticum aestivum L. (Wheat)						
1	Grain	IEC	57.7	ADPGlc 22.0		2721

Table 5-6 UDPGlc and Other Sugar Nucleotides in Plants—cont'd

No.	Plant part	Method	UDPGlc	Other NDPX	Total	References
1	–, ripening kernel	IEC	72.1	ADPGlc 18.5		2633
1	Grain[b]	Enzym.	16.3	ADPGlc 8.6		537
1	Grain	PC/PE	346	ADPGlc 98		2723
1	–, endosperm	PC/PE	381	ADPGlc 162		2723
1	–, testa, pericarp	PC/PE	230	ADPGlc 32		2723
1	Seed, aleurone tissue	PC/PC	90.3	UDPFru 39.7, UDPGal 39.7, UDPXyl 34.3		2724
1	Grain, peeled	HPLC	700	ADPGlc 160		3371
2	Seed, embryo[a]	Enzym.	24-36			3372
1	Seed, embryo	HPLC	+			2598
1	Leaf[g]	PC/PC	5.80			331
1	Leaf[g]	PC/PE	6.58			331
1	Leaf[f]	Enzym.	40			3373
1	Flag leaf[a]	HPLC	360	UDPGlcNAc 150		3374
1	Shoot	^{31}P NMR	+			2638
1	Seedling, embryo	IEC	+			2768
1	–, leaf	PC/PC	+			3375
1	–, shoot	PC/PC	+			3376
1	Seedling, root	PC/PC	+			3376
1	–, scutellum	PC/PC	+			3376
1	–, root, and stem	IEC	+	ADPGlc ?		2475
1	–, aerial parts	IEC	27	UDPX 4		2432
Triticum compactum Host (Club Wheat)						
1	Leaf	IEC	+			2736
Vicia faba L. (Faba Bean)						
1	?	TLC/TLC	+			33
Vigna mungo (L.) Hepper (Black Gram)						
1	Root tip[e]	^{31}P NMR	620			542,543
Vigna radiata (L.) Wilcz. (Mung Bean)						
1	Sprout	IEC	+	UDPGal +, UDPXyl +, UDPAra +, UDPGlcNAc +, GDPMan +, UDPGalNAc +, UDPGlcA +, UDPGalA +		2742

[a]Dry weight basis. [b]nmol/plant part. [c]Percent of total sugar nucleotide. [d]μg/g. [e]nmol/mL. [f]μmol/mg chlorophyll. [g]Percentage of total free nucleotide. [h]mol/mol CABP binding site. [i]μmol/mg protein. [j]nmol/mg starch.

Continued

Table 5-6 UDPGlc and Other Sugar Nucleotides in Plants—cont'd

No.	Plant part	Method	UDPGlc	Other NDPX	Total	References
1	Seedling	IEC	33	UDPGal 14, UDPXyl 2, UDPAra 1, ADPRib 10	50	2433
1	Seedling	IEC		UDPGlcNAc 1.4, UDPGlcA 0.4		3377
1	Seedling	HPLC	175	ADPRib 0		2441
1	–, hypocotyl	IEC	204	UDPGal 28, UDPXyl trace, UDPAra trace, UDPRha 6, UDPGlcNAc 38, GDPMan trace		2744
Yucca flaccida Haw. (Weak-Leaf Yucca)						
1	Phloem exudate	TLC/TLC	+			309
Zea mays L. (Corn)						
1	Grain, milky stage	PC/PC		ADPGlc 8		3378
1	Grain, milky stage[g]	IEC/PC *Incl. UDPGal	6.8*	ADPGlc 4.1, ADPGal 1.7, ADPMan 2.1, ADPGlcNAc 0.7 UDPGlcNAc 1.5, UDPGalNAc 2.4		2755
1	Immature grain	IEC		UDPX 220, GDPX 472		2754
1	Immature kernel, endosperm[j]	HPLC		UDPX 17.32, ADPX 7.48		342
1	–, starch granule[j]	HPLC		UDPX 1.91, ADPX 0.68		342
1	Grain, endosperm[a]	HPLC	1416	ADPGlc 596		3379
1	Leaf	PC/PE	57			337
1	Leaf	IEC	1.76	GDPGlc 1.38		2613
1	Ovary[b]	Enzym.	0.4			343
1	Root	PC/PC	168			12
1	Root	TLC/TLC	+			12
1	Root tip[b]	HPLC	4.54			2759
1	Root tip	HPLC	+	ADPGlc 0, UDPGlcNAc +, GDPGlc +		3380
1	Root tip	^{31}P NMR	+			544,3381
1	Root tip	^{31}P NMR	+	UDPGal +		587
1	Culm[b]	HPLC	97.5			2759
1	Seedling	IEC	54.0	UDPGlcNAc 1.4, UDPGlcA 0.4		2656

nmol/g fwb unless stated otherwise.
[a]Dry weight basis.
[b]nmol/plant part.
[c]Percent of total sugar nucleotide.
[d]μg/g.
[e]nmol/mL.
[f]μmol/mg chlorophyll.
[g]Percentage of total free nucleotide.
[h]mol/mol CABP binding site.
[i]μmol/mg protein.
[j]nmol/mg starch.

distinguish between them. The sugar nucleotide under the UDPG peak could well contain UDPGal or other UDPX, though UDPGlc is usually predominant in plants. Uchiyama and Sakata,[543] however, were able to positively identify UDPGlc in mung bean root tips *in vivo* by applying the maximum entropy method (MEM) to the FT-NMR signals to obtain super-resolved ^{31}P NMR spectra.

Physical Properties

Uridine Diphosphate Glucose (UDPG or UDPGlc), $C_{15}H_{24}N_2O_{17}P_2$. *Uridine 5′-(Trihydrogen Diphosphate), Mono-α-D-glucopyranosyl Ester*

UDPG [R.N. 133-89-1]: White powder, $[\alpha]_D^{21}$ + 45°(H_2O), +35° (8*M* urea).[3384] Spectra: UV,[3239] ^{1}H NMR (fig.),[461] ^{13}C NMR,[405] ^{31}P NMR,[68] ESI-MS (3:1 α:β),[422] ESI-MS/MS,[403] FAB-MS/MS.[3385] Chromatography: PC,[9] TLC,[36] HPTLC,[2891] IEC,[3365] HPLC,[2440] GFC.[3386] Electrophoresis: PE,[9] TLE,[36] capillary isotachophoresis,[3239] CZE,[3387] MECC.[3387]

$Na_2C_{15}H_{22}N_2O_{17}P_2$ [R.N.117756-22-6]: FAB-MS,[2911] SI-MS.[3388] *Dihydrate* [R.N. 89705-69-1]: Monoclinic platelike crystals, d_m 1.80, d_c 1.796, X-ray structure[419]; laser Raman spectrum.[3389]

$Li_2C_{15}H_{22}N_2O_{17}P_2 \cdot 6H_2O$: White powder. PC, PE.[3390]

Conformation

UDPG takes the extended form both in the crystal[419] and in solution.[3391] In the crystalline state the conformation about the *N*-glycosidic bond is *anti*, the ribose is C(2′)-*endo* (^{2}E), and the pyrophosphate linkage takes a staggered form.[419] Kim, however, suggests that UDPG has a folded conformation with the glucosyl 2- and 3-OH groups located inside the folded structure, based on the relative reactivity of the OH groups toward acylation and supported by semiempirical and *ab initio* energy calculations.[3392]

Synthesis

UDPG may be prepared (a) from UMP in poor yield by condensation with glucose 1-phosphate (G1P) in the presence of dicyclohexyl carbodiimide,[3393] (b) from a UMP ester,[3394] thioester,[3395] morpholide,[3390,3396] triazolide,[3397] anhydride,[2989] or UMP derivative containing 2′,3-OH protective substituents[3398,3399] by condensation with G1P in pyridine, (c) from UDP by condensation with 2-(β-D-glucosyloxy)-3-methoxypyridine in DMF,[422] (d) from the *O,O*-diphenyl ester of UDP by displacement of the terminal phosphate group with G1P,[3010] or (e) from UTP and G1P by the action of coupled enzymes.[3072]

Hydrolysis

UDPG is stable in buffered aqueous solution at neutral pH but decomposes at high pH[3382] or in the presence of Mn(II) and other metal ions[3400] giving UMP and glucose 1,2-cyclic phosphate. The rate of hydrolysis at pH 2.70 and 20 °C is 2.5×10^{-6} min^{-1}, which corresponds to a half-life of 20 days.[3401] The standard free energy change $\Delta G'^{\circ}$ for the hydrolysis of UDPG to UMP and G1P, based on the −7.3 kcal/mol value reported by Avigad[262] for the hydrolysis of UDPG to UDP and glucose, was calculated to be −10.3 kcal/mol at pH 7.0 and 25 °C.[3402]

6 Dinucleotides

This group comprises the pyridine adenine dinucleotides NAD and NADP, the riboflavin adenine dinucleotide FAD, and their respective dihydro derivatives. The pyridine dinucleotides are considered first.

Pyridine Adenine Dinucleotides

This group comprises the dinucleotides NAD, NADP and their dihydro derivatives NADH and NADPH.

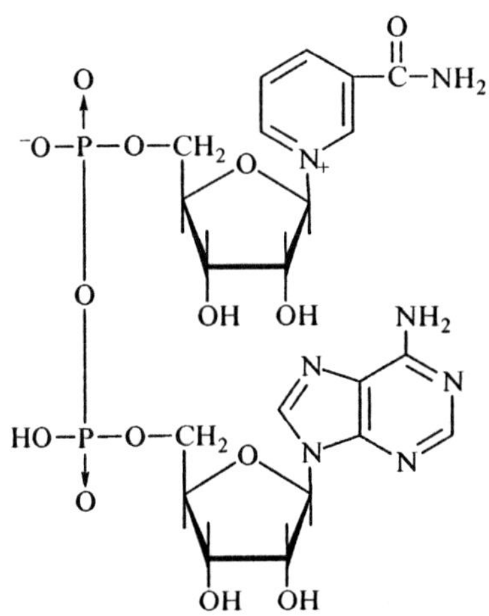

NAD, X = H
NADP, X = P(O)(OH)$_2$

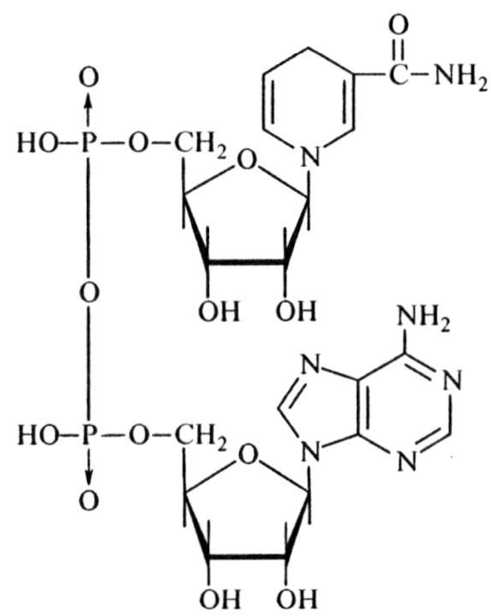

NADH, X = H
NADPH, X = P(O)(OH)$_2$

Nomenclature

The coenzyme NAD was originally (1923) called cozymase, then codehydrogenase I, coenzyme I (CoI), or diphosphopyridine nucleotide (DPN, 1936). Its current name, nicotinamide adenine dinucleotide (NAD), was adopted by the IUB in 1960.[3403] The charged symbol NAD^+ currently in vogue is not sanctioned by the IUB and should be discouraged because, as Alberty pointed out,[3404] NAD actually has a negative charge in the neutral pH range in which it normally exists. Similar comments apply to the coenzyme NADP, previously called codehydrogenase II, coenzyme II (CoII), or triphosphopyridine nucleotide (TPN).

Occurrence

The first reports of the occurrence of coenzymes I and II in the higher plants were those of von Euler and Nordenfelt,[3405] who isolated coenzyme I from barley root tips in 1926 and Whatley,[3406] who isolated coenzyme II from the leaves and other tissues of clover, chickweed, oats, potato, elder, and runner beans in 1951. The chemical structures of the two coenzymes were both known by 1950. Although it had been long known that the reduced coenzymes NADH and NADPH could be extracted without degradation from yeast or animal tissue by dilute alkali, they were not isolated from plants until the early 1960s. NADPH was isolated in 1961 from peas by Marré and Bianchetti,[3407] and NADH and NADPH were isolated in 1963 from corn, tobacco, and various other plants by Fritz *et al.*,[3408] Sunderland and Merrett,[3409,3410] and Yamamoto,[3411] respectively.

The function of these coenzymes in plants, as in animals and bacteria, is to act as redox reagents for a broad variety of metabolic reactions. As a general rule, the NADP/NADPH couple functions in synthetic (anabolic) reactions and the NAD/NADH couple functions in degradative (catabolic) reactions.[3412] The two couples are usually found in separate compartments within the plant cell. In the leaf, for example, the NADH/NADPH couple resides within the chloroplast and the NAD/NADH couple in the cytosol.[3413]

The metabolic pathways for the biosynthesis of the dinucleotides are outlined in Fig. 6-1. NAD and its precursors nicotinic acid mononucleotide (NAMN) and nicotinic acid adenine dinucleotide (NAAD) are components of the pyridine nucleotide cycle.[3414] NAAD, synthesized from NAMN by reaction with ATP, is converted to NAD by amidation with glutamine (step 1). Cleavage of the pyridine-ribose bond (step 2) yields ADP-Ribose and nicotinamide. Deamidation of the nicotinamide to nicotinic acid and reaction of the latter with R5P to form NAMN completes the cycle. In step 3, NAD is reversibly phosphorylated by ATP giving NADP, this being the only known mechanism for the biosynthesis of NADP.[3411] An ATP-dependent NADH kinase [E.C. 2.7.1.86] that converts NADH to NADPH has been reported for yeasts[3415] but not yet for plants. In steps 4 and 5, NAD and NADP are reversibly reduced to NADH and NADPH, respectively, by reaction with a dehydrogenase and its substrate. About 200 dehydrogenases have been identified in plants,[3416] some (such as alcohol dehydrogenase) specific for NAD, others (such as G6P dehydrogenase) specific for NADP, and

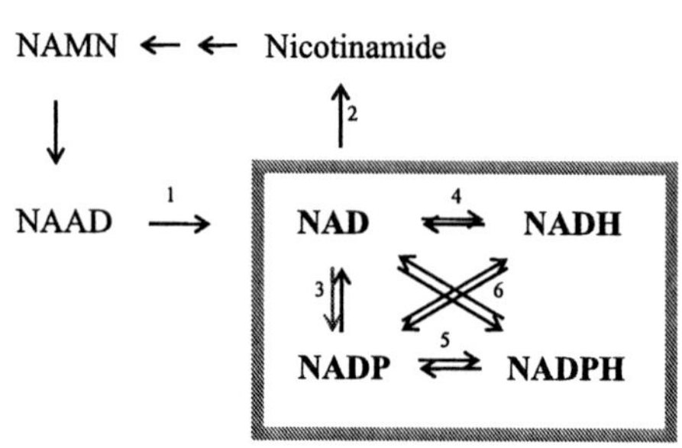

FIGURE 6-1 Pyridine Dinucleotides. (1) NAD synthase [E.C. 6.3.5.1], L-glutamine and ATP. (2) NAD nucleosidase [E.C. 3.2.2.5]. (3) NAD kinase [E.C. 2.7.1.23] and ATP. (4) An NAD-specific dehydrogenase and its substrate, for example, alcohol dehydrogenase [E.C. 1.1.1.1] and ethanol. (5) An NADP-specific dehydrogenase and its substrate, for example, G6P dehydrogenase [E.C. 1.1.1.49] and G6P. (6) Ferredoxin-NADP reductase [E.C. 1.18.1.2] and reduced ferredoxin.

still others (such as glutamate dehydrogenase) specific for both. Spinach ferredoxin-NADP reductase [E.C. 1.6.7.1] has been shown to catalyze transhydrogenation reactions among the four dinucleotides (step 6).[3417] There appears to be no enzymatic mechanism for salvage of the reduced dinucleotides other than return to their oxidized forms.

Extraction

The conditions under which the dinucleotides can be extracted without degradation are governed by their unusually high sensitivity to pH. The oxidized compounds NAD and NADP are stable to acid and unstable to base, whereas the reduced compounds NADH and NADPH are stable to base and unstable to acid.[3418] Extraction with aqueous solutions of acids such as perchloric acid or trichloroacetic acid under the conditions normally employed for the extraction of mononucleotides can therefore only be used for the analysis of NAD and NADP although, as Brown demonstrated, the degradation products of NADH and NADPH can be used as a measure of their initial concentration.[2437] In general, however, it is preferable to extract the oxidized and reduced dinucleotides separately, as illustrated by the following example:

> *Leaves (2 g) and globe "roots" (2 g) of 25-day old radish plants were chopped up separately with a razor blade and ground in a homogenizer at 90-95 °C for 2 min with 10 ml of 0.1N HCl to extract oxidized coenzymes (NAD and NADP). Additional tissue (2 g each) was similarly treated with 0.1N NaOH to extract reduced coenzymes (NADH and NADPH). Each of the four homogenates was then rapidly chilled in ice bath and adjusted to pH 6.5 with NaOH or pH 7.5 with HCl and then 0.5 ml of 0.2 M glycylglycine buffer at pH 6.5 or pH 7.5 was added, in the case of oxidized and reduced coenzymes respectively. Total volumes were measured, and then each homogenate was centrifuged at 10,000 g for 20 min at 4 °C. Supernatants were stored frozen at −80 °C before assay of coenzymes.*
>
> Tezuka, T.; Yamaguchi, F.; Ando, Y., "Physiological Activation in Radish Plants by UV-A Radiation", *Photochem. Photobiol.* **1994**, *24B*(1), 33–40,[3419] reprinted by permission of Elsevier, © 1994.

Separation and Analysis

(1) Paper (PC) or Thin Layer (TLC) Chromatography Methods

NAD and NADP can be separated from other nucleotides and from each other by two-dimensional PC[9] or TLC[33] on methanol/chloroform/7*M* formic acid extracts, but the only quantitative data reported to date are for 2d-PC/PE separations performed on TCA extracts of developing wheat grain.[2723]

(2) Ion Exchange Chromatography (IEC) Method

Under the conditions employed by Bergqvist[2471] for the separation of acid soluble nucleotides referred to in the previous chapter, NAD and NADP are well separated from each other but tend to coelute with other nucleotides. NAD elutes early under the CMP peak and NADP later under the CDP peak.[2471] The reduced dinucleotides NADH and NADPH cannot be separated by this method because they are degraded by acidic solvents, but it has been found that their main scission products—ADP-Ribose and ADP-Ribose-2′-phosphate, respectively—are well resolved and can be taken as a measure of the initial concentration of each compound.[2437] All four dinucleotides can be separated without degradation by IEC on a DEAE-Sephadex A-25 column with LiCl gradient elution.[2894] This procedure has been used in a few instances to separate NAD and NADP from plant extracts,[2566,2654,2683] but has yet to be used for the separation of intact NADH or NADPH from plants.

(3) High-Performance Liquid Chromatography (HPLC) Method

This method has been used since the mid-1980s for the separation of NAD and NADP from other nucleotides in plant extracts. It can be used in the IEC mode with strongly anionic column packings such as Partisil-10 SAX[2651] or μ-Bondapak NH_2[2441] or in the reverse-phase (RP) mode with packings such as Hypersil ODS.[2440] NAD is eluted before NADP in the IEC mode and in the reverse order in the RP mode. Curiously, none of the studies in Table 6-2 mentions the presence of ADP-Ribose-2′-phosphate, the degradation product of NADPH, and only two of them mention the presence of ADP-Ribose, the degradation product of NADH.[2441,3494] In their studies on *Costus speciosus* cell suspension cultures, Li *et al.*[2523] circumvented this problem by using a 0.5*M* perchloric acid extract for the NAD and NADP analysis and a 0.5*M* NaOH extract for the NADH and NADPH analysis. The extracts were freeze dried, redissolved in pH 6.0 and pH 8.0 buffer, respectively, and subjected separately to HPLC on a Supelcosil LC-18 RP column and a K_2HPO_4/citrate (pH 4.5) mobile phase. Aso *et al.*[3417] developed a similar procedure for separating the four dinucleotides simultaneously and used it to measure the effect of each dinucleotide on the transhydrogenase activity of ferredoxin-NADP reductase [E.C. 1.18.1.2] isolated from spinach leaves, but did not determine their endogenous levels.

(4) Fluorimetric Methods

Bevilacqua and Scuncia[3420] used the fluorimetric method of Robinson *et al.*[3501] to measure the NAD(P) content of a wide variety of plants. In this method, a TCA plant extract containing a few milligrams of $Ce(SO_4)_2$ to oxidize any reduced NAD(P), was treated with acetone and 6N NaOH, resulting in the formation of a highly fluorescent compound ($\lambda = 360$ nm), which was measured fluorimetrically. Green leaves require an additional oxidation with permanganate prior to measurement to destroy their intrinsic fluorescence.[3421] The method was unable to distinguish between NAD and NADP and was soon superseded by the fluorimetric method of Ciotti and Kaplan,[3502] discussed later in method 7C.

(5) Cyanide Addition Method

Colowick *et al.*[3503] developed a spectrophotometric method suitable for the analysis of NAD and NADP based on the adduct formed when each compound is treated with 1.0*M* KCN in aqueous solution (pH 11). The UV spectra of the adducts show reduced absorption at 260 nm and a new peak at 325 nm, much like those of NADH and NADPH, both of which show reduced absorption at 260 nm and a new peak at 340 nm. This method was applied in its original form by Brown[2437,3471] and Mangat[2655,2656] to perchloric acid extracts of various plant seeds or seedlings after fractionation by IEC, and in modified form by Pao *et al.*[3487] who extracted wheat leaves with water, precipitated the dinucleotides with acetone, and measured the cyanide adducts fluorometrically at an excitation wavelength of 325 nm and an emission wavelength of 430 nm.

(6) Luminometric Method

This method utilizes the bacterial luminescence produced from NAD(P)H by the following coupled reactions:

$$NAD(P)H + FMN \leftrightarrow NAD(P) + FMNH_2 \quad \text{Enzyme : FMN reductase[EC 1.6.8.1]}$$
$$FMNH_2 + RCHO + O_2 \leftrightarrow FMN + RCO_2H + H_2O + h\nu \quad \text{Enzyme : Luciferase[EC 1.14.14.3]}$$

Stuhlfauth *et al.*[2540] and Krishnaraj and Thorpe[3488] used the commercial monitoring kits available for this purpose to measure the NADH and NADPH contents of *Digitalis lanata* and wheat leaves, respectively.

(7) Enzymatic Methods

In this method, each of the four dinucleotides is analyzed separately using one of the enzymatic reactions listed in Table 6-1. Several procedures are available that differ primarily in the manner in which the products are detected and measured. The introduction of enzyme recycling by Lowry in 1961 brought about such a dramatic improvement in sensitivity and accuracy that almost all the analyses done since then have been carried out in the recycling mode regardless of the procedure adopted.

(A) MANOMETRIC METHODS

In their early studies on cozymase (NAD), von Euler *et al.*[3405,3428] measured the NAD content of various plant tissues by the amount of CO_2 evolved when an aliquot of the aqueous extract was added to a solution of glucose and NAD-free brewer's yeast (apozymase) in pH 4.6 phosphate buffer. The results were compared to the values obtained with authentic NAD assayed under the same conditions. Okunuki[3430] and von Euler *et al.*[3428] used this

Table 6-1 Some NAD(P)-Linked Reactions

No.	Reaction	Enzyme	
1	NAD + ethanol ↔ NADH + acetaldehyde	Alcohol dehydrogenase	[E.C. 1.1.1.1]
2	NAD + lactate ↔ NADH + pyruvate	Lactate dehydrogenase	[E.C. 1.1.1.27]
3	NAD + PP_i ↔ NMN + ATP	NAD pyrophosphorylase	[E.C. 2.7.7.1]
4	NADP + glutamate + H_2O ↔ NADPH	Glutamate dehydrogenase + α-ketoglutarate + NH_3	[E.C. 1.4.1.4]
5	NADP + G6P ↔ NADPH + 6-PG	G6P dehydrogenase	[E.C. 1.1.1.49]
6	NADP + isocitrate ↔ NADPH	Isocitrate dehydrogenase + α-ketoglutarate + CO_2	[E.C. 1.1.1.42]

method to determine the NAD content of a wide variety of pollens. More recently, Greenbaum *et al.*[3467] developed a modified enzyme recycling method (see below) that uses an oxygen electrode inserted into the sealed reaction vessel to measure the consumption of oxygen corresponding to the rate of enzymatic reduction. All four dinucleotides can be measured at levels as low as 0.01-0.02 μg. This method has the advantage that it can be used with extracts that contain reducing, fluorescent, or colored compounds, for example, green leaves, but so far it has been used only with extracts of pea leaf,[3467] tobacco callus culture,[2630] and pineapple orange fruit juice vesicles.[3435]

(B) SPECTROPHOTOMETRIC METHODS

All four dinucleotides have a maximum absorption at 260 nm in their UV spectra, but the reduced dinucleotides NADH and NADPH have an additional maximum at 340 nm. Anderson and Vennesland[3458] used the increase in light absorption at 340 nm to measure the NAD and NADP contents of TCA extracts of various plant leaves using reactions 1 and 5 of Table 6-1 to reduce the dinucleotides to NADH and NADPH, respectively. Whatley[3406] developed a method for the analysis of NADP in plant tissues based on reactions 5 or 6 of Table 6-1 together with Warburg's "old yellow enzyme" and methylene blue, measurement being made by comparing the rate of disappearance of the blue color with that of authentic NADP assayed under the same conditions.

These methods soon gave way to the enzymatic cycling method introduced by Lowry in 1961. In this method, the NADH or NADPH produced in the initial reaction is reoxidized to NAD or NADP via a carrier such as phenazine methosulfate (PMS) and an acceptor such as 2,6-dichlorophenolindophenol (DCPIP) present in the reaction mixture, the reaction cycling back and forth until all of the initial dinucleotide is consumed. The method is best illustrated by an example of one of the most widely used procedures, that of Slater and Sawyer (1962).

> *The NAD^+ assay contained leaf extract, 32 μmoles tris-HCl (pH 7.4), 780 μmoles ethanol, 0.75 μmoles phenazine methosulphate (PMS) and 108 mμmoles DCIP in a final volume of 1.50 ml. Enzyme (50 μg Boehringer alcohol dehydrogenase) was added after temperature equilibration at 23°; water replaced enzyme in the blank. The reaction was measured by continuously recording the rate of dye reduction at 610 mμ using a Cary recording spectrophotometer. To calculate the amounts of NAD^+ in the leaf extracts, these rates were compared with those obtained using known amounts of NAD^+... NADH was estimated in a reaction mixture containing leaf extract, 25 μmoles tris–HCl (pH 7.4), 620 μmoles ethanol, 0.72 μmole PMS, and 86 mμmoles DCIP in a final volume of 1.20 ml. Complete NADH reoxidation, requiring 15 minutes, was attained before the addition of 150 μg of enzyme. Rates were compared with those obtained using [known amounts of] NADH. The $NADP^+$ assay contained leaf extract, 30 μmoles tris-HCl (pH 7.4), 3 μmoles glucose-6-P, 0.60 μmole PMS, 86 mμmoles DCIP, and 2 μmoles EDTA. The total volume was 1.20 ml. After a period of temperature equilibration, 5 μg of enzyme (Boehringer glucose-6-P dehydrogenase) was added. These rates were compared with those obtained using [known amounts of] $NADP^+$... NADPH was estimated in a reaction mixture containing leaf*

extract, 34 μmoles tris–HCl (pH 7.4), [3.4 μmoles G6P], 0.72 μmole PMS, and 142 mμmoles DCIP in a total volume of 1.30 ml. To increase reaction rates, the concentrations of PMS and DCIP were higher than those used in $NADP^+$ assays; EDTA was excluded. Enzyme (6.2 μg) was added after a reoxidation period of 25 minutes. Rates were compared with those obtained using [known amounts of] NADPH.

Ryrie, I. J., and Scott, K. J., "Changes in Nicotinamide Nucleotide Coenzymes in Barley Leaves Infected with Powdery Mildew", *Plant Physiol.* **1968**, *43*(5), 687–692,[3447] reprinted by permission of the American Society of Plant Biologists, © 1968.

Several modifications of this procedure are available. In the modification described by Yamamoto,[3411] phenazine methosulfate is replaced as the carrier by NAD diaphorase (for NADH) or NADP diaphorase (for NADPH), other conditions remaining the same. In the modification of Lechevallier *et al.*,[3485] the carrier is phenazine ethosulfate (PES) and DCPIP is replaced as the acceptor by thiazolyl blue, monitored at 570 nm, other conditions remaining the same. Both of these modified procedures have been widely used.

(C) FLUORIMETRIC METHODS

In the fluorimetric method of Ciotti and Kaplan,[3502] the oxidized forms of NAD and NADP react with methyl ethyl ketone in 3.5*N* NaOH solution followed by acidification to give a fluorescent product, which is monitored in a fluorimeter at an excitation wavelength of 360 nm and an emission wavelength of 450 nm. NADH and NADPH are destroyed by this treatment. Since this reaction does not distinguish between the dinucleotides, it is necessary to carry out additional assays. NAD and NADP are measured by the *decrease* in fluorescence following incubation of a portion of the extract with alcohol dehydrogenase (reaction 1, Table 6-1) or isocitrate dehydrogenase (reaction 6, Table 6-1).[3454,3456] Conversely, NADH and NADPH are measured by the *increase* in fluorescence following incubation of a portion of the extract with alcohol dehydrogenase (reaction 1, Table 6-1) or glutathione reductase.[3456] To avoid the manipulations required in the Ciotti and Kaplan method, Wigge *et al.*[3413] made use of enzymatic indicator solutions that permit all four of the dinucleotides to be measured directly. Rhodes and Wooltorton[3455] used the indicator resazurin, which is highly fluorescent in its reduced form resorufin, in procedures analogous to those of the spectrophotometric method described in method 7B. Asali *et al.*[3432] described the use of a bioreactor to monitor the fluorescence of NAD(P)H online in *Catharanthus roseus* callus suspension cultures while being subjected *in vivo* to various metabolic perturbations.

(D) ^{32}P RADIOACTIVITY METHOD

A novel method was developed by Cheung and Suhadolnik[3504] in which the [^{32}P]-labeled PP_i-NAD pyrophosphorylase reaction (reaction 3, Table 6-1) is coupled with the glucose-hexokinase reaction (Chapter 1, Figure 1-5, reaction 14) to convert NAD to β-[^{32}P]ADP, which is then selectively adsorbed on activated charcoal and counted by liquid scintillation. This procedure, used to quantitate NAD in crude extracts of germinating wheat embryos, is capable of measuring NAD in amounts as low as 1 pmol.[3504]

Analysis Without Separation

^{31}P Nuclear Magnetic Resonance (NMR) Method

The proton-decoupled ^{31}P NMR spectra of NAD and NADP in neutral aqueous solutions exhibit overlapping AB quartets for the two pyrophosphate P atoms in the −9 to −12 ppm region (relative to $P_i = 0$) and NADP shows an additional singlet for the 2′-P atom in the +4 ppm region. In contrast, the reduced dinucleotides NADH and NADPH show only singlets for the diphosphate P atoms despite their obviously different chemical environments.[68,3505,3506] Bligny *et al.* found such patterns in the ^{31}P spectra of perchloric acid extracts of spinach leaves and chloroplasts,[50] sycamore cell suspension cultures[3341] and potato tuber mitochondria,[2708] but assigned the two AB quartets to (NAD + NADP) and (NADH + NADPH). This is inconsistent with the known behavior of the reduced dinucleotides in acid solution discussed earlier. It appears more likely that the downfield AB quartet represents NAD alone and the upfield AB quartet and associated singlet represent NADP alone. The results are so listed in Table 6-2.

The ^{31}P method has also been applied to living tissues kept under constant aeration as described previously for other phosphorus-containing metabolites. A small peak in the −10 ppm region has been ascribed to NAD in the ^{31}P NMR spectra of sycamore,[3341] *Rosa damascena*,[2695] carrot,[2507] and Madagascar periwinkle[2507] cell suspension cultures; in pea[526] and corn[544] root tips; and in fiddleneck sprouts,[2644] spinach leaves,[50] Jerusalem artichoke stolons,[45] tobacco pollen grains,[2563] soybean nodules,[2563] and sunflower stems,[2563] but other phosphorus-containing compounds (notably UDPG) also have signals in this region and the resolution is insufficient for positive identification.

Physical Properties

Nicotine Adenine Dinucleotide (NAD), $C_{21}H_{27}N_7O_{14}P_2$, also called Diphosphopyridine Nucleotide (DPN). *Adenosine 5′-(Trihydrogen Diphosphate), P′ → 5′ Ester with 3-(Aminocarbonyl)-1-β-D-ribofuranosylpyridinium, Inner Salt.*

NAD [R.N. 53-84-9]: Luminescence,[3507] phosphorescence (fig.),[2920] molecular self-association,[3508] molar absorptivity (ε) 17.4×10^3 L/mol/cm (260 nm, 25 °C),[3509] diamagnetic susceptibility (χ_M) -342×10^{-6} cm^3/mol at 25 °C,[3510] dipole moment (D) 250 esu cm,[2939] countercurrent distribution,[3511,3512] heat capacity,[3513] ODMR,[2920] cyclic voltammetry (fig.),[3514] polarography,[3515] NOE.[3516] Ionization constants: pK_a 3.88 (potentiometric, 25 °C),[3517] 12.13, 12.17, 12.75, and 12.78 for the four ribosyl OH groups (^{13}C NMR, 20 °C).[3518] Spectra: UV (fig.),[3519] FTIR (fig.),[3520] Raman (fig.),[3521] UV Raman (fig.),[3522] [^{1}H] NMR (fig.),[3523,3524] [^{13}C] NMR (fig.),[359] [^{15}N] NMR (fig.),[3525] [^{31}P] NMR (fig.),[3505] 2d[^{13}C-^{31}P] NMR (fig.),[2882] ORD (fig.),[2923] CD (fig.),[3526] SERS (fig.),[3527] FAB-MS (fig.),[3528] FD-MS (fig.),[3528] ESI-MS,[2887] LID-MS (fig.).[3529] Chromatography: PC,[3530] TLC,[3531] HPTLC,[2891] IEC,[2894] HPLC,[2440] preparative HPLC,[3532] GC,[3156] GFC,[2893] DEAE cellulose column.[3533] Electrophoresis: PE,[9] CZE,[3534] capillary isotachophoresis,[3239] MECC.[3535] Theoretical calculations: LCAO-MO,[3536,3537]

Table 6-2 Nicotinamide Adenine Dinucleotides in Plants*

No.	Plant part	Method	NAD	NADH	NADP	NADPH	References
Abutilon molle Sweet (Indian Mallow)							
1	Seed[a]	Fluor.	11.2[b]				3420
Acacia diethriciana F. Müll.							
1	Seed[a]	Fluor.	18.8[b]				3420
Acanthus mollis L. (Bear's Breech)							
1	Leaf[a]	Fluor.	34[b]				3421
Acer pseudoplatanus L. (Sycamore Maple)							
1	Cell suspension culture[c]	IEC	6				2447
1	Cell suspension culture[d]	Enzym.	18	5	1.9	1.0	2449
1	Cell suspension culture	^{31}P NMR	+		+		3341
Aegilops speltoides Tausch (Goatgrass)							
1	Leaf	Enzym.	14.9	1.4	28.1	10.3	3416
Aegilops squarrosa L. (Barb Goatgrass)							
1	Leaf	Enzym.	7.3	3.2	3.3	25.6	3416
Albizia lophantha (Willd.) Benth. (Plume Albizzia)							
1	Seed	Fluor.	12[e]				3420
Allium neapolitanum Cirillo (White Garlic)							
1	Seed[a]	Fluor.	2.8[b]				3420
Angelica silvestris L. (Woodland Angelica)							
1	Seed[a]	Fluor.	4.1[b]				3420
Aquilegia vulgaris L. (European Columbine)							
1	Seed[a]	Fluor.	2[b]				3420
Asplenium nidus L. (Hawaii Birdnest Fern)							
1	Adult leaf[c]	Enzym.	41	23	28	53	3422
Avena sativa L. (Oats)							
1	Seed[a]	Fluor.	5[b]				3423
1	Sprout, embryo	Enzym.	40.5	6.5	9.2	20.0	3424
1	–, coleoptile node[f]	Enzym.	4.0	6.0	8.9	5.2	3425
1	Leaf, chloroplast[g]	Enzym.	4.1	4.0	2.4	0.9	231
1	–, mitochondria[g]	Enzym.	3.5	1.9	<0.5	<0.5	231
1	–, cytosol and vacuole[g]	Enzym.	12.0	1.5	1.8	6.5	231
1	Seedling[a]	Fluor.	2.58[b]				3423
1	Seedling	IEC	15.6 (incl. CMP)				2473
1	Mature, aerial parts	IEC			12		2471
Beta vulgaris L. (Sugar Beet)							
1	Leaf[h]	Enzym.	8.87	2.23	3.28	2.24	3426
1	–, chloroplast[c,i]	Enzym.	1.1		28	7	3427
1	–, cytoplasm[c,i]	Enzym.	4.3		0.5	3	3427
1	Root	IEC	0		8		3345
Beta vulgaris L. ssp. *rapacea* (Koch) Döll (Fodder Beet)							
1	Leaf[c]	Enzym.				31	237
Betula pubescens Ehrh. (Downy Birch)							
1	Pollen[j]	von Euler	572				3428

*nmol/g fresh weight basis unless stated otherwise; [a]μg/g; [b]Incl. NADP and NMN; [c]Dry weight basis; [d]nmol/10^7 cells; [e]Including NADH; [f]nmol/plant part; [g]nmol/mg chlorophyll; [h]μmol/m^2; [i]nmol/mg protein; [j]Relative to brewer's yeast cozymase = 100; [k]pmol/seedling; [l]μmol/mg chlorophyll; [m]Including NADPH; [n]Based on its degradation product, ADPRib 2′ = phosphate; [o]Cozymase units; [p]nmol/–10^9 plastids; [q]μmol/mmol chlorophyll; [r]m*M*; [s]μg/seedling; [t]ng/g.

Continued

Table 6-2 Nicotinamide Adenine Dinucleotides in Plants*—cont'd

No.	Plant part	Method	NAD	NADH	NADP	NADPH	References
Borago officinalis L. (Common Borage)							
1	Seed[a]	Fluor.	4.3[b]				3420
Brassica napus L. ssp. *oleifera* (Winter Rape)							
1	Leaf	Enzym.	6	2	9	3	3429
Brassica nigra (L.) Koch. (Black Mustard)							
1	Seed[a]	Fluor.	11.1[b]				3420
Brassica oleracea L. ssp. *italica* Plenck. (Broccoli)							
1	Stems and florets	HPLC	35				2441
Bryophyllum tubiflorum Harv. (Chandelier Plant)							
1	Phyllodia	TLC/TLC	+		+		3346
Camellia japonica L. (Camellia)							
1	Pollen[j]	Okunuki	111				3430
Camellia sasanqua Thunb. (Sasanqua)							
1	Pollen[j]	Okunuki	50				3430
Cannabis sativa L. (Hemp)							
1	Seed[a]	Fluor.	6.6[b]				3420
Capsicum annuum L. (Bell Pepper)							
1	Seed[a]	Fluor.	3[b]				3420
1	Fruit	HPLC	+				2441
Carduus crispus L. (Curly Plume less Thistle)							
1	Seed[a]	Fluor.	2.6[b]				3420
Carica papaya L. (Papaya)							
1	Seed[a]	Fluor.	2[b]				3420
Carthamus tinctorius L. (Safflower)							
1	Seed[a]	Fluor.	6[b]				3420
Castanea sativa Mill. (Chestnut)							
1	Seed[a]	Fluor.	17.8[b]				3420
Catharanthus roseus (L.) G. Don (Madagascar Periwinkle)							
1	Cell suspension culture	Enzym.			4.4		3431
1	Cell suspension culture	Fluor.		+		+	3432
1	Cell suspension culture	^{31}P NMR	+				2507
Chelidonium majus L. (Celandine)							
1	Seed[a]	Fluor.	12[b]				3420
Chenopodium rubrum L. (Red Goosefoot)							
1	Seedling[k]	Enzym.	114	10	19	35	2508
Chrysanthemum leucanthemum L. (Oxeye Daisy)							
1	Leaf[a]	Fluor.	29[b]				3421
Chrysanthemum segetum L. (Corn Marigold)							
1	Seed[a]	Fluor.	7[b]				3420
Cicer arietinum L. (Chickpea)							
1	Seed[a]	Fluor.	23.8[b]				3420
Cichorium intybus L. (Chicory)							
1	Root[c]	Enzym.	74	36	6	25	3433
Cistus monspelliensis L. (Montpelier Cistus)							
1	Seed[a]	Fluor.	5.5[b]				3420
Citrullus vulgaris Schrad. (Watermelon)							
1	Seedling, cotyledon[f]	Enzym.	7.60	0.02	0.01	1.30	3411
1	–, hypocotyl[f]	Enzym.	0.16	0.29	2.09	3.83	3411

Table 6-2 Nicotinamide Adenine Dinucleotides in Plants*—cont'd

No.	Plant part	Method	NAD	NADH	NADP	NADPH	References
Citrus aurantium L. (Sour Orange)							
1	Seed[a]	Fluor.	20[b]				3420
Citrus paradisi Macf. (Grapefruit)							
1	Mature peeled pulp vesicles	Enzym.	6.0	0.77	0.93	1.2	3434
Citrus sinensis (L.) Osbeck (Orange)							
1	Mature peeled pulp vesicles	Enzym.	11.8	2.5	1.2	1.6	3434
1	Juice	IEC	+				248
Citrus spp. (Pineapple Orange)							
1	Fruit juice vesicles	Enzym.	3430		2700		3435
Corylus avellana L. (European Hazel)							
1	Seed[a]	Fluor.	9[b]				3420
Costus speciosus (J. König) Sm. (Crepe Ginger)							
1	Cell suspension culture	HPLC	56	296	22	33	2523
Cucumis sativus L. (Cucumber)							
1	Leaf	Enzym.	27.0	4.8	12.9	5.8	3436
1	–, plastid	Enzym.	NADH/NAD and NADPH/NADP ratios given				3437
1	–, chloroplast[l]	Fluor.	40.8[b]				3438
1	Seedling, cotyledon	Enzym.	80.4	19.4	13.0	27.9	3439
1	–, hypocotyl[i]	Enzym.	2.4	0	0.8	0	3440
1	–, cell wall[i]	Enzym.	5.1	0	0	0	3440
1	–, mitochondria[i]	Enzym.	0.5				3441
1	–, root	Enzym.	0.323	0.438			3442
Cucurbita maxima Duch. (Squash)							
1	Seedling, cotyledon	Enzym.	11.6	1.62	16.8	12.4	3443
1	–, leaf	Enzym.	36.8	4.20	32.5	11.6	3443
Cucurbita pepo L. (Pumpkin)							
1	Leaf[c]	IEC	2300	900	1400	1900	2530
1	–, chloroplast[c]	IEC	150	160	140	140	2530
1	–, mitochondria[c]	Enzym.	173	176	149	164	1633
1	Stem, phloem exudate	IEC	+		+		251
1	Fruit, unripe	IEC	+		+		251
1	–, ripe	IEC	+		+		251
Cuscuta reflexa Roxb. (Dodder)							
1	Leaf	TLC/TLC	+		+		33
Datura inoxia Mill. (Downy Thorn Apple)							
1	Cell suspension culture	HPLC	26		7.2		2440
Daucus carota L. (Carrot)							
1	Cell suspension culture	^{31}P NMR	+				2507
Digitalis lanata Ehrh. (Grecian Foxglove)							
1	Leaf[h]	Lumin.				23.5	2540
Digitalis purpurea L. (Digitalis)							
1	Seed[a]	Fluor.	9.5[b]				3420

*nmol/g fresh weight basis unless stated otherwise; [a]µg/g; [b]Incl. NADP and NMN; [c]Dry weight basis; [d]nmol/10^7 cells; [e]Including NADH; [f]nmol/plant part; [g]nmol/mg chlorophyll; [h]µmol/m^2; [i]nmol/mg protein; [j]Relative to brewer's yeast cozymase = 100; [k]pmol/seedling; [l]µmol/mg chlorophyll; [m]Including NADPH; [n]Based on its degradation product, ADPRib 2′ = phosphate; [o]Cozymase units; [p]nmol/–10^9 plastids; [q]µmol/mmol chlorophyll; [r]m*M*; [s]µg/seedling; [t]ng/g.

Continued

Table 6-2 Nicotinamide Adenine Dinucleotides in Plants*—cont'd

No.	Plant part	Method	NAD	NADH	NADP	NADPH	References
Diospyros kaki L.f. (Japanese Persimmon)							
1	Seed[a]	Fluor.	5[b]				3420
Echinochloa oryzicola Vasin. (Rice Barnyardgrass)							
1	Sprout	–		0.090[m]			3444
Egeria densa Planch. (Brazilian Waterweed)							
1	Leaf	TLC/TLC	+		+		33
Elodea canadensis Michx. (Canadian Waterweed)							
1	Leaf[a]	Fluor.	20[b]				3421
Escholtzia californica Cham. (California Poppy)							
1	Seed[a]	Fluor.	12.3[b]				3420
Euonymus europaeus L. (European Spindletree)							
1	Seed[a]	Fluor.	20[b]				3420
Euphorbia lathyris L. (Caper Spurge)							
1	Seed[a]	Fluor.	4[b]				3420
Flaveria brownii A. Powell (Brown's Yellowtops)							
1	Leaf[c]	Enzym.	19.1	17.7	15.7	20.8	3422
Flaveria cronquistii A. Powell (Yellowtops)							
1	leaf[c]	Enzym.	39.8	27.4	36.0	39.2	3422
Flaveria pubescens Rudb. (Yellowtops)							
1	Leaf[c]	Enzym.	35.1	16.0	29.8	26.2	3422
Flaveria trinervia (Spreng) C. Mohr (Clustered Yellowtops)							
1	Leaf[c]	Enzym.	84.5	55.8	67.4	70.2	3422
Glycine max (L.) Merr. (Soybean)							
1	Seed, immature[c]	IEC	3900 (incl. CMP)				2552
1	Seed, mature[c]	IEC	0				2552
1	Seed	IEC			2.5		2553
1	Leaf	Enzym.	20.5	4.8	19.6	15.4	2558
1	Leaf[g]	HPLC			18	22	26
1	Root	Enzym.	7.7	7.0	1.0	21.5	2558
10	Nodule	Enzym.	15.8-118	7.0-33.1	1.9-11.8	9.7-26.6	3445
1	Nodule	HPLC	143 (incl. UMP)		37.1 (incl. UDP)		3446
1	Nodule	^{31}P NMR	+				2563
1	Seedling, cotyledon[f]	Enzym.	36.90	21.20	1.80	4.03	3411
1	–, hypocotyl[f]	Enzym.	0.58	0.29	1.02	1.89	3411
Glycine soja (L.) Sieb. and Zucc. (Soybean)							
1	Seed[a]	Fluor.	70.9[b]				3420
Gossypium arboreum L. (Tree Cotton)							
1	Seed hair, primary wall[f]	IEC	62		8		2566
1	–, secondary wall[f]	IEC	87		7		2566
Gossypium hirsutum L. (Upland Cotton)							
1	Leaf[c]	IEC	+		70	53[n]	2569
1	Root[c]	IEC	+		78	96[n]	2569
Hedera helix L. (English Ivy)							
1	Juvenile leaf[c]	Enzym.	36	23	47	53	3422
Helianthus annuus L. (Sunflower)							
1	Leaf	IEC	+				3355
1	Leaf	Enzym.	24.4	2.4	6.5	18.5	2558
1	Root	Enzym.	5.9	11.1	2.5	16.4	2558

Table 6-2 Nicotinamide Adenine Dinucleotides in Plants*—cont'd

No.	Plant part	Method	NAD	NADH	NADP	NADPH	References
1	Stem	^{31}P NMR	+				2563
1	Seedling, cotyledon[f]	Enzym.	11.10	0.79	0.70	2.58	3411
1	–, hypocotyl[f]	Enzym.	0.38	0.01	2.63	9.50	3411
Helianthus tuberosus L. (Jerusalem Artichoke)							
1	Tuber	IEC	0		2		2576
1	Young stolon	^{31}P NMR	+				45
Hordeum vulgare L. (Barley)							
1	Seed[a]	Fluor.	10[b]				3420
1	Leaf	Enzym.	39	2.0	12	3.0	3447
1	Leaf	HPLC	15.2		16.0		2585
1	–, protoplast[g]	Enzym.	20.1[e]		24.4[m]		3413
1	–, chloroplast[g]	Enzym.	5.2	2.1	9.8	9.7	3413
1	–, mitochondria[g]	Enzym.	2.0	0.1	<0.5	<0.5	3413
1	–, cytosol[g]	Enzym.	8.6	2.0	1.8	3.4	3413
1	Primary leaf[c]	HPLC	100-200		200-300		3448
1	Root[c]	HPLC	~25		~25		3448
1	Root tip[o]	von Euler	3.5				3405
1	Mature, aerial parts	IEC			9		2471
Hyphaenae thebaica Mart. (Doum Palm)							
1	Seed[a]	Fluor.	0.6[b]				3420
Isoetes duriei (Bory)							
1	Leaf	TLC/TLC	+		+		33
Juglans regia L. (English Walnut)							
1	Seed[a]	Fluor.	33[b]				3420
Kalanchoë daigremontiana Ham. and Perr. (Devil's Backbone)							
1	Leaf[c]	Enzym.	28	15	67	76	3422
Lactuca sativa L. (Lettuce)							
1	Sprout	HPLC	+				2598
Larix decidua Mill. (European Larch)							
1	Primary wall (cambium)	IEC	0		0		2601
1	Secondary wall (xylem)	IEC	0.2		0.04		2601
Leleba oldhami (Summer Bamboo)							
1	Desheathed shoot	IEC			2.5		2661
Lemna gibba L. (Gibbous Duckweed)							
1	Whole plant	TLC/TLC	+		+		33
1	Whole plant	Enzym.	17.5	1.4	2.0	3.4	3449
		and diurnal variation					3449
Lemna paucicostata Hegelm. (Lesser Duckweed)							
1	Frond, at floral induction[a]	Enzym.	31.7				3450
Lens culinaris Medik. (Lentil)							
1	Seed[a]	Fluor.	43[b]				3420

*nmol/g fresh weight basis unless stated otherwise; [a]μg/g; [b]Incl. NADP and NMN; [c]Dry weight basis; [d]nmol/10^7 cells; [e]Including NADH; [f]nmol/plant part; [g]nmol/mg chlorophyll; [h]μmol/m^2; [i]nmol/mg protein; [j]Relative to brewer's yeast cozymase = 100; [k]pmol/seedling; [l]μmol/mg chlorophyll; [m]Including NADPH; [n]Based on its degradation product, ADPRib 2′ = phosphate; [o]Cozymase units; [p]nmol/–10^9 plastids; [q]μmol/mmol chlorophyll; [r]m*M*; [s]μg/seedling; [t]ng/g.

Continued

Table 6-2 Nicotinamide Adenine Dinucleotides in Plants*—cont'd

No.	Plant part	Method	NAD	NADH	NADP	NADPH	References
Lilium auratum Lindl. (Golden-Rayed Lily)							
1	Pollen[j]	Okunuki	118				3430
Lilium longiflorum Thunb. (Trumpet Lily)							
1	Pollen[j]	Okunuki	109				3430
Lilium philippinense Bak. (Philippine Lily)							
1	Pollen[j]	Okunuki	103				3430
Luffa aegyptiaca Mill. (Sponge Gourd)							
1	Seed[a]	Fluor.	12.2				3420
Lupinus albus L. (White Lupine)							
1	Seed[a]	Fluor.	27[b]				3420
Lupinus luteus L. (Yellow Lupine)							
1	Seed	IEC	+	+[n]	+	+[n]	2437
1	Root cortex	Enzym.	18.8	21.4	5.5	11	3451
1	Whole nodule	Enzym.	66	68	16	46	3451
1	Nodule cortex	Enzym.	47	47	18	24	3451
Lycopersicon esculentum Mill. (Tomato)							
1	Leaf	Fluor.		+			3452
1	Leaf[c]	IEC	122	16[n]	137	48[n]	3453
1	Leaf	Enzym.	32.7		29.3		3454
1	Stem	Fluor.		+			3452
1	Seedling, root	Enzym.	0.057	0.320			3442
Magnolia grandiflora L. (Southern Magnolia)							
1	Leaf	IEC	+		+		2543
Malus domestica Borkh. (Apple)							
1	Fruit, peel	Enzym.	22.2	2.1	2.2	6.1	3455
1	–, pulp	Enzym.	10.0	1.0	1.0	1.1	3455
Malva rotundifolia L. (Dwarf Mallow)							
1	Seed[a]	Fluor.	7.7[b]				3420
Matthiola incana (L.) R.Br. (Tenweeks Stock)							
1	Leaf[p]	Fluor.	35[b]				3421
Medicago sativa L. (Alfalfa)							
1	Seed[a]	Fluor.	11.9[b]				3420
2	Root nodule (1)	Enzym.	58.2	37.1	2.2	7.5	3445
	Root nodule (2)		17.9	16.1	0.7	4.9	3445
Melianthus majus L. (Honey Flower)							
1	Seed[a]	Fluor.	18.8[b]				3420
Milium multiflorum Cav. (Millet Grass)							
1	Seed[a]	Fluor.	3.1[b]				3420
Mimosa pudica L. (Shameplant)							
1	Seed[a]	Fluor.	6[b]				3420
Mirabilis jalapa L. (Marvel of Peru)							
1	Seed[a]	Fluor.	10[b]				3420
Molinia coerulea (L.) Moench. (Purple Moorgrass)							
1	Seed[a]	Fluor.	1.6[b]				3420
Momordica charantia L. (Balsam Pear)							
1	Seed[a]	Fluor.	5.5[b]				3420
Muscari comosum (L.) Mill. (Tassel Grape Hyacinth)							
1	Seed[a]	Fluor.	7[b]				3420

Table 6-2 Nicotinamide Adenine Dinucleotides in Plants*—cont'd

No.	Plant part	Method	NAD	NADH	NADP	NADPH	References
Nelumbo nucifera Gaertn. (Sacred Lotus)							
1	Seed[a]	Fluor.	12.7[b]				3420
Nicotiana glutinosa L. (Tobacco)							
1	Leaf[a]	Enzym.	5.5	0.8	3.0	0.2	3409
Nicotiana longiflora Cav. (Longflower Tobacco)							
1	Leaf[a]	Fluor.	28[b]				3421
Nicotiana tabacum L. (Tobacco)							
1	Seed[a]	Fluor.	5.6[b]				3420
1	Leaf	HPLC	22.0		15.2		2440
1	Root	HPLC	2.7		0.7		2440
1	Seedling, young leaf[f]	Enzym.	0.38	0	3.96	1.12	3411
1	–, old leaf[f]	Enzym.	11.46	0	49.5	3.84	3411
1	Seedling, young leaf[f]	HPLC	30		33		3363
1	–, old leaf[f]	HPLC	10		12		3363
1	–, terminal bud[f]	HPLC	13		10		3363
1	Apex	HPLC	15.7		10.4		2628
1	Callus culture[d]	Enzym.	38	9	16	5	3410
1	Cell suspension culture	HPLC	17.6		6.3		2440
1	Pollen grain	^{31}P NMR	+				2563
Ophioglossum lusitanicum L. (Least Adderstongue)							
1	Leaf	TLC/TLC	+		+		33
Ophioglossum vulgatum L. (Southern Adderstongue)							
1	Leaf	TLC/TLC	+		+		33
Oryza sativa L. (Rice)							
1	Grain, 10 daf	IEC	42		~18		2635
1	Ripening kernel	IEC			1.6		2540
1	Seedling	Enzym.	4600	500	3000	8500	3456
1	–, endosperm[f]	Enzym.	1.48	0.16	0.15	0	3411
1	–, coleoptile[f]	Enzym.	0.04	0.02	0.13	0.28	3411
Paeonia albiflora Pall. (Chinese Peony)							
1	Pollen[q]	Okunuki	95				3430
Papaver orientale L. (Oriental Poppy)							
1	Pollen[j]	Okunuki	63				3430
Papaver somniferum L. (Poppy)							
1	Pollen[j]	Okunuki	130				3430
Persea americana Mill. (Avocado)							
1	Mature fruit[r]	Enzym.	0.064	0.002			3457
Petroselinum crispum (Mill.) Nym. (Parsley)							
1	Leaf[a]	Enzym.	8.6		9.1		3458
Peumus boldus Molina (Boldo)							
1	Seed[a]	Fluor.	2.4[b]				3420
Phacelia tanacetifolia Benth. (Fiddleneck)							
1	Seed	^{31}P NMR	+				2644

*nmol/g fresh weight basis unless stated otherwise; [a]µg/g; [b]Incl. NADP and NMN; [c]Dry weight basis; [d]nmol/10^7 cells; [e]Including NADH; [f]nmol/plant part; [g]nmol/mg chlorophyll; [h]µmol/m^2; [i]nmol/mg protein; [j]Relative to brewer's yeast cozymase = 100; [k]pmol/seedling; [l]µmol/mg chlorophyll; [m]Including NADPH; [n]Based on its degradation product, ADPRib 2′ = phosphate; [o]Cozymase units; [p]nmol/–10^9 plastids; [q]µmol/mmol chlorophyll; [r]m*M*; [s]µg/seedling; [t]ng/g.

Continued

Table 6-2 Nicotinamide Adenine Dinucleotides in Plants*—cont'd

No.	Plant part	Method	NAD	NADH	NADP	NADPH	References
Pharbitis nil Choisy (Morning Glory)							
1	Seedling, cotyledon[f]	Enzym.			2.0	1.2	3459
Phaseolus coccineus L. (Scarlet Runner Bean)							
1	Etiolated leaf[f]	Enzym.	1.00	0.20	0.94	0.82	3460
Phaseolus vulgaris L. (Beans)							
1	Seed[a]	Fluor.	53[b]				3423
1	Leaf	Enzym.	15.9	7.4	28.0	14.2	3461
1	Leaf[f]	IEC	<0.2	<0.2[n]	0.9	0.7[n]	2650
1	–, petiole	IEC	–		–	2[n]	2653
1	–, laminae	IEC	8		1.8	7[n]	2653
1	Leaf, young	HPLC	73		54		2652
1	–, mature	HPLC	40		56		2652
1	–, mature	IEC	11.4	8.3[n]			2649
1	Leaf, midribs removed[a]	Enzym.	13.7		9.7		3458
1	Whole leaf[q]	Enzym.	38	1	11	5	3462
1	–, chloroplast[q]	Enzym.	26	1	9	4	3462
1	Root	Enzym.	9.1	5.4	17.6	6.0	3461
1	Lateral root	Enzym.	19	1	11	12	3463
1	Seedling[s]	Fluor.	63.40[b]				3423
1	Seedling	Cyanide	11.5		5.5		2656
Phaseolus vulgaris L. ssp. *napus* (Dwarf Kidney Bean)							
1	Seed	IEC	+	+[n]	+	+[n]	2437
Phlomis fruticosa L. (Shrubby Jerusalem Sage)							
1	Seed[a]	Fluor.	8[b]				3420
Phoenix dactylifera L. (Date Palm)							
1	Seed[a]	Fluor.	3.1[b]				3420
Phyllocactus spp.							
1	Pollen[a]	von Euler	685				3428
Phyllostachys pubescens [Mazel ex] Houz. (Moso Bamboo)							
1	Desheathed shoot	IEC			0		2661
Picea abies (L.) Karst. (Norway Spruce)							
1	Needle (leaf)[c]	Enzym.	135	48	82	42	2663
Pinus densiflora Sieb. and Zucc. (Japanese Red Pine)							
1	Pollen[j]	Okunuki	52				3430
Pinus mugo Turra (Mugo Pine)							
1	Pollen[a]	von Euler	302				3428
Pinus sylvestris L. (Scotch Pine)							
1	Seedling (1 year old), root	Enzym.	58.1	24.9	9.5	22.2	3464
1	Pollen	Enzym.	9.6[b]	9.2[m]			3465
1	Pollen[a]	von Euler	299				3428
Pisum sativum L. (Peas)							
1	Seed[a]	Fluor.	96[b]				3423
1	Seed	Enzym.	135				296
1	Seed	IEC	123				24
1	Seed	HPLC	73				2441
1	Seed[o]	von Euler	6.0				3466
1	Mature seed	Cyanide	1		3		2437
1	Mature seed	IEC		40[n]		14[n]	2437

Table 6-2 Nicotinamide Adenine Dinucleotides in Plants*—cont'd

No.	Plant part	Method	NAD	NADH	NADP	NADPH	References
1	Leaf[a]	Enzym.	11.7	1.9	7.0	0.7	3467
1	Leaf'	Enzym.	46.3	5.0	38.9	24.1	3455
1	–, chloroplast[q]	Enzym.	2.52	0.75	1.23	0.41	3468
1	Leaf and stem[a]	Enzym.	8.5		7.5		3458
1	Root[c]	Enzym.	18	8	3	16	2683
1	Root tip	Enzym.	45.70	1.97	3.16	10.60	3469
1	Root tip	^{31}P NMR	+				526
1	Root nodule	Enzym.	53	17	7.0	4.5	3470
1	Sprout	Cyanide	2		8		3471
1	Sprout	IEC		25[n]		16[n]	3471
1	Seedling[c]	Enzym.	399.6	38.4	130.8	20.4	3472
1	Seedling[s]	Fluor.	27.20[b]				3423
1	–, internode[a]	Enzym.			8.1	6.4	3407
Plantago lanceolata L. (Narrowleaf Plantain)							
1	Root	Enzym.	4.09	1.29	4.81	4.22	3473
Platycerium alcicorne (Willem.) (Tardieu)							
1	Adult leaf[c]	Enzym.	13	5	7	11.5	3422
Polytrichum commune L.							
1	Gametophyte without rhizoides[c]	Enzym.	18	13	27	34	3422
Populus deltoides × *Populus* cv. *caudina*							
1	Leaf[g]	Enzym.	9.7	14.4	3.7	4.5	3474
Populus nigra L. (Black Poplar)							
1	Pollen[a]	von Euler	698				3428
Populus tremula L. (European Aspen)							
1	Pollen[a]	von Euler	688				3428
Prunus armeniaca L. (Apricot)							
1	Seed[a]	Fluor.	23[b]				3420
Prunus cerasus L. (Morello Cherry)							
1	Seed[a]	Fluor.	18[b]				3420
Prunus dulcis Webb (Almond)							
1	Seed[a]	Fluor.	7.7[b]				3420
Prunus persica (L.) Batsch (Peach)							
1	Seed[a]	Fluor.	17[b]				3420
Pyrus communis L. (Pear)							
1	Fruit, peel	Enzym.	43.4	7.4	5.0	13.0	3455
1	–, pulp	Enzym.	12.0	1.5	4.1	6.1	3455
Pyrus serotina (Japanese Pear)							
1	Fruit, flesh[t]	Enzym.	5.9	2.2	1.2	<0.1	3475
Raphanus sativus L. (Radish)							
1	Leaf	Enzym.	7.2	2.1	3.8	1.5	3419
1	Globe root	Enzym.	2.5	2.0	0.9	1.0	3419
Reseda lutea L. (Yellow Mignonette)							
1	Seed[a]	Fluor.	1[b]				3420

*nmol/g fresh weight basis unless stated otherwise; [a]μg/g; [b]Incl. NADP and NMN; [c]Dry weight basis; [d]nmol/10^7 cells; [e]Including NADH; [f]nmol/plant part; [g]nmol/mg chlorophyll; [h]μmol/m^2; [i]nmol/mg protein; [j]Relative to brewer's yeast cozymase = 100; [k]pmol/seedling; [l]μmol/mg chlorophyll; [m]Including NADPH; [n]Based on its degradation product, ADPRib 2′ = phosphate; [o]Cozymase units; [p]nmol/–10^9 plastids; [q]μmol/mmol chlorophyll; [r]m*M*; [s]μg/seedling; [t]ng/g.

Continued

Table 6-2 Nicotinamide Adenine Dinucleotides in Plants*—cont'd

No.	Plant part	Method	NAD	NADH	NADP	NADPH	References
Ricinus communis L. (Castor Bean)							
1	Seed	HPLC	+		0		3476
1	Seedling, endosperm[f]	Enzym.	25.30	14.75	8.57	4.73	3411
1	–, hypocotyl.[f]	Enzym	2.24	0.01	0.05	8.24	3411
1	–, cytosol[i]	Enzym.	0.33[e]		47.6[m]		3477
1	–, mitochondria[i]	Enzym.	4.05	0	46.9	0	3477
1	–, glyoxysome[i]	Enzym.	0.48	0.12	3.5[m]		3477
1	–, plastid[i]	Enzym.	0.41[e]		2.0	0	3477
Robinia pseudoacacia L. (Black Locust)							
1	Seed[a]	Fluor.	20[b]				3420
Rosa damascena (Damask Rose)							
1	Cell suspension culture	^{31}P NMR	+				2695
Saccharum officinarum L. (Sugar Cane)							
1	Leaf[c]	Enzym.	20	13	51	50	3422
Salix alba L. (White Willow)							
1	Pollen[a]	von Euler	730				3428
Salix cinerea L. (Large Gray Willow)							
1	Pollen[a]	von Euler	740				3428
Salix fragilis L. (Crack Willow)							
1	Pollen[a]	von Euler	718				3428
Salix nigricans Sm. (Dark-Leaved Willow)							
1	Pollen[a]	von Euler	1023				3428
Salix viminalis L. (Basket Willow)							
1	Pollen[a]	von Euler	722				3428
Scilla hispanica Mill. (Spanish Bluebell)							
1	Seed[a]	Fluor.	6.5[b]				3420
Scirpus holoschoenus L. (Roundhead Bulrush)							
1	Seed[a]	Fluor.	6.3[b]				3420
Secale cereale L. (Rye)							
1	Seed[a]	Fluor.	5[b]				3420
1	Leaf	HPLC	28.6		61.3		2585
Selaginella denticulata (L.) Link. (Tooth-Leaved Clubmoss)							
1	Leaf	TLC/TLC	+		+		33
Sesamum indicum L. (Sesame)							
1	Seed[a]	Fluor.	52.3[b]				3420
Setaria verticillata (L.) Beauv. (Bristly Foxtail)							
1	Seed[a]	Fluor.	3.7[b]				3420
Sinapis alba L. (White Mustard)							
1	Leaf	TLC/TLC	+		+		33
1	Sprout, cotyledon[c,k]	Enzym.	500	35	40	170	3478
1	–, hypocotyl[c,k]	Enzym.		23		68	3478
1	–, root[c,k]	Enzym.		10		51	3478
Solanum tuberosum L. (Potato)							
1	Tuber	PC/PC	+	+	+		9
1	Tuber	PC/PE	+	+	+		9
1	Tuber	Enzym.	7.2	1.2	0.4	1.1	3479
1	Tuber, mitochondria	Enzym.	0.68	0.06	0.11	0.02	3479
1	–, mitochondria[i]	^{31}P NMR	0.25		0.72		2708
1	–, starch grain	IEC	+				3480

Table 6-2 Nicotinamide Adenine Dinucleotides in Plants*—cont'd

No.	Plant part	Method	NAD	NADH	NADP	NADPH	References
1	Tuber explant	Enzym.	12		7		3481
1	Cell suspension culture[c]	Enzym.	570	100	110	120	3482
Sorghum bicolor (L.) Moench. (Sorghum)							
1	Seed[a]	Fluor.	2[b]				3420
Spinacia oleracea L. (Spinach)							
1	Leaf, midribs removed[a]	Enzym.	13.2		8.6		3458
1	Leaf, stroma[g]	Enzym.	4.8	0	14.8	7.2	3483
1	–, cytosol and mitochondria[g]	Enzym.	3.0	0	0.9	12.0	3483
1	Whole leaf[q]	Enzym.	30	1	10	17	3462
1	–, chloroplast[q]	Enzym.	14	0	12	8	3462
1	–, chloroplast[g]	HPLC	3.7		+		3484
1	Leaf	^{31}P NMR	+			766	
1	–, chloroplast	^{31}P NMR	+		+		50
Spirodela polyrrhiza (L.) Schleid. (Giant Duckweed)							
1	Frond[c]	Enzym.	325	59	585	197	3485
1	Plastid[c,p]	Enzym.	40	1	97	17	3485
Stachys officinalis (L.) Trev. (Common Hedgenettle)							
1	Seed[a]	Fluor.	5[b]				3420
Stellaria media (L.) Vill. (Chickweed)							
1	Leaf	Enzym.	Trace		2.7		3406
Strophanthus hispidus D.C.							
1	Seed[a]	Fluor.	11[b]				3420
Thea sinensis L. (Tea)							
1	Pollen[j]	Okunuki	92				3430
Tillandsia usneoides (L.) L. (Spanish Moss)							
1	Whole plant	TLC/TLC	+		+		33
1	Whole plant[c]	Enzym.	7	5	6	11	3422
Trifolium pratense L. (Red Clover)							
1	Seed[a]	Fluor.	7.7[b]				3420
Trigonella foenum-graecum L. (Fenugreek)							
1	Seed	IEC	+				2716
Triticum aestivum L. (Wheat)							
1	Seed[a]	Fluor.	6[b]				3423
1	Seed, embryo	Enzym.	16.3	9.8	0	0	3486
1	Seed, embryo	HPLC	+				2598
1	Whole grain	PC/PE	40				2723
1	–, endosperm	PC/PE	50				2723
1	–, testa and pericarp	PC/PE	11				2723
1	Ripening kernel	IEC			2.7		2540
1	Leaf	Enzym.	10.5	1.9	3.8	5.6	3416
1	Leaf	HPLC	20.3		15.2		2585
1	Leaf[a]	Cyanide	27[b]				3487
1	Leaf	Lumin.		1.55		0.27	3488

*nmol/g fresh weight basis unless stated otherwise; [a]μg/g; [b]Incl. NADP and NMN; [c]Dry weight basis; [d]nmol/10^7 cells; [e]Including NADH; [f]nmol/plant part; [g]nmol/mg chlorophyll; [h]μmol/m^2; [i]nmol/mg protein; [j]Relative to brewer's yeast cozymase = 100; [k]pmol/seedling; [l]μmol/mg chlorophyll; [m]Including NADPH; [n]Based on its degradation product, ADPRib 2′ = phosphate; [o]Cozymase units; [p]nmol/–10^9 plastids; [q]μmol/mmol chlorophyll; [r]m*M*; [s]μg/seedling; [t]ng/g.

Continued

Table 6-2 Nicotinamide Adenine Dinucleotides in Plants*—cont'd

No.	Plant part	Method	NAD	NADH	NADP	NADPH	References
1	Flag leaf[c]	HPLC	210		290		3374
1	Leaf, plastid[p]	Enzym.	13	1.5	16	11	3485
1	Leaf, protoplast[c,g]	Enzym.	23	2	16	3	3472
1	–, chloroplast[c,g]	Enzym.	13	7	2	0.5	3472
1	Seedling[s]	Fluor.	2.92[b]				3423
1	Seedling	Enzym.	6000	2500	16,000	39,500	3456
1	Seedling, endosperm[f]	Enzym.	1.88	0.04	0.44	0.06	3411
1	–, coleoptile[f]	Enzym.	0.10	0.05	0.74	0.91	3411
1	–, shoot[c]	Enzym.	78.3	33.1	86.5	75.7	3422
1	–, root tip	Enzym.	7.45	3.73	1.24	1.38	3489
Triticum boeoticum Boiss. (Wild Einkorn)							
1	Leaf	Enzym.	14.3	5.1	50.3	25.0	3416
Triticum compactum Host (Club Wheat)							
1	Leaf[a]	Cyanide	31.0[b]				3487
Triticum dicoccoides ([Körn. ex] Asch. and Grabner) Aarons (Wild Emmer)							
1	Leaf	Enzym.	12.9	4.5	10.6	13.1	3416
Triticum dicoccum Schrank. (Emmer)							
1	Leaf	Enzym.	10.1	4.9	17.6	18.7	3416
1	–, chloroplast[c]	Enzym.	399	22.7	113	17.6	3490
Triticum durum Desf. (Durum Wheat)							
1	Leaf	Enzym.	8.6	4.6	32.8	34.4	3416
Triticum monococcum L. (Einkorn)							
1	Leaf	Enzym.	23.1	7.0	21.4	27.4	3416
Triticum spelta L. (Spelt)							
1	Leaf	Enzym.	5.3	3.7	9.6	12.8	3416
Verbascum thapsus L. (Mullein)							
1	Seed[a]	Fluor.	3[b]				3420
Vicia faba L. (Faba Bean)							
1	Seed	IEC	2.0				2738
1	Pod	IEC	4.7				2738
1	Leaf	TLC/TLC	+		+		33
1	Guard cell protoplast[f]	Enzym.	0.540	0.071	0.058	0.220	3491
1	Whole seedling	IEC	4.0				2738
Vicia sativa L. (Spring Vetch)							
1	Seed[a]	Fluor.	13.3[b]				3420
Vigna angularis (Willd.) Ohwi and Ohashi (Adzuki Bean)							
1	Seedling, cotyledon[f]	Enzym.	34.70	6.56	4.99	1.89	3411
1	–, epicotyl[f]	Enzym.	3.00	2.81	4.66	5.77	3411
Vigna radiata (L.) Wilcz. (Mung Bean)							
1	Sprout	IEC	+		+		2742
1	–, hypocotyl	IEC			29		2744
1	Seedling[c]	IEC	428				2433
1	Seedling	HPLC	32				2441
1	Seedling, leaf	Enzym.	49.0	1.4	9.7	4.1	3492
Vigna sesquipedalis (L.) Fruh. (Yard-Long Bean)							
1	Seedling, cotyledon[f]	Enzym.	3.86	0	0.29	0.27	3411
1	–, hypocotyl[f]	Enzym.	7.20	0	26.80	2.12	3411
1	–, root tip[f]	Enzym.	0.064	0.034	0.122	0.085	3411
		and variation from root base to tip					3411

Table 6-2 Nicotinamide Adenine Dinucleotides in Plants*—cont'd

No.	Plant part	Method	NAD	NADH	NADP	NADPH	References
Vigna unguiculata (L.) Walp. (Cowpea)							
1	Seed[a]	Fluor.	46.6[b]				3420
1	Sprout, root mitochondria[i]	Enzym.	7.2				3493
Wolffia arrhiza (L.) Wimm. (Spotless Water Meal)							
1	Whole plant[c]	Enzym.	225	92.7	474	427	3422
Yucca aloifolia L. (Aloe Yucca)							
1	Seed[a]	Fluor.	2[b]				3420
Zea mays L. (Corn)							
1	Seed[a]	Fluor.	4[b]				3420
2	Mature grain[c]	HPLC	23-27		15-17		3494
1	–, endosperm[c]	HPLC	391				3379
1	Leaf	IEC	23.0				2613
1	Leaf[h]	Enzym.	19.7	16.8	7.3	8.9	2574
1	Mature leaf	Enzym.	36.0	6.0	18.0	33.0	3495
1	Leaf, midribs removed[a]	Enzym.	7.5		8.8		3458
1	Leaf, mesophyll chloroplast[g]	HPLC			9.3	2.8	3496
1	Sprout	IEC	+		+		3497
1	Sprout[a]	Enzym.	74	0	38	1020	3408
1	Sprout, embryo	HPLC	+				2760
1	Seedling	Cyanide	9.0		5.5		2656
1	Seedling, endosperm[f]	Enzym.	5.72	1.20	1.05	0.64	3411
1	–, coleoptile[f]	Enzym.	0.62	0.77	4.62	1.89	3411
1	–, shoot[c]	Enzym.	112	44.6	245	282	3422
1	–, root	Enzym.	129.5	38.4	6.1	150.0	3498
1	Root tip	Enzym.	123.5	12.9	7.6	10.7	3499
1	Root tip[f]	HPLC	0.63				2759
1	Root tip	^{31}P NMR	+				544
1	Apical root[a]	Fluor.	27		30		3500
1	–, meristem[a]	Fluor.	45		20		3500

*nmol/g fresh weight basis unless stated otherwise.
[a]μg/g.
[b]Incl. NADP and NMN.
[c]Dry weight basis.
[d]nmol/10^7 cells.
[e]Including NADH.
[f]nmol/plant part.
[g]nmol/mg chlorophyll.
[h]μmol/m^2.
[i]nmol/mg protein.
[j]Relative to brewer's yeast cozymase = 100.
[k]pmol/seedling.
[l]μmol/mg chlorophyll.
[m]Including NADPH.
[n]Based on its degradation product, ADPRib 2′ = phosphate.
[o]Cozymase units.
[p]nmol/–10^9 plastids.
[q]μmol/mmol chlorophyll.
[r]m*M*.
[s]μg/seedling.
[t]ng/g.

Gibbs energy of formation ($\Delta_f G$) −418.40 kJ/mol.[3538] *Trihydrate*: Needles or flat prisms (fig.), mp 140-142 °C dec.[3539] *Tetrahydrate*: Triclinic crystals, X-ray structure.[3540]
$LiC_{21}H_{26}N_7O_{14}P_2{\cdot}2H_2O$: Orthorhombic crystals (fig.), fp 203-204 °C, X-ray structure[3541]; d_c 1.648.[3542]
$(C_{20}H_{24}N_2O_3)_3(C_{21}H_{27}N_7O_{14}P_2)_2$ (quinine salt): Needles (fig.), mp 162-170 °C dec.; UV (260 nm, ε 9.43), PE.[3543]

Nicotine adenine dinucleotide, reduced form (NADH), $C_{21}H_{29}N_7O_{14}P_2$, also called reduced Diphosphopyridine Nucleotide (DPNH). *Adenosine 5′-(Trihydrogen Diphosphate), P′→5′ Ester with 1,4-Dihydro-1-β-D-ribofuranosyl-3-pyridinecarboxamide.*

NADH [R.N. 58-68-4]: Colorless but turns yellow on exposure to moist air; solubility in 1,2-propanediol 15 mg/mL.[3544] Solutions in phosphate buffer at pH 7.4 are more stable at 4 °C than at 25 °C or frozen,[3545] but tris buffer is preferred to phosphate buffer.[3546] Fluorescence,[3547] phosphorescence (fig.),[2920] molecular self-association,[3508] molar absorptivity (ε) 6.18×10^3 L/mol/cm (334 nm, 25 °C),[3548] diamagnetic susceptibility (χ_M) -341×10^{-6} cm^3/mol at 25 °C,[3510] ODMR.[2920] Ionization constant: pK_a 3.88 (potentiometric, 25 °C).[3549] Spectra: UV (fig.),[3519] Raman (fig.),[3521] UV Raman (fig.),[3522] [^{1}H] NMR (fig.),[3523,3524] [^{13}C] NMR (fig.),[359] [^{15}N] NMR (fig.),[3525] [^{31}P] NMR (fig.),[3505] ORD (fig.),[2923] CD (fig.),[3550] LID-MS (fig.).[3529] Chromatography: PC,[3530] TLC,[3531] HPTLC,[2891] IEC,[3544] HPLC,[3544] DEAE cellulose column.[3533] Electrophoresis: PE,[9] CZE,[3534] capillary isotachophoresis.[3239] Theoretical calculations: LCAO-MO,[3536,3537] Gibbs energy of formation ($\Delta_f G$) −397.51 kJ/mol.[3538]
$Na_2C_{21}H_{27}N_7O_{14}P_2$: Fluorescence kinetics.[3551] Spectra: FAB-MS (fig.),[3528] FD-MS (fig.).[3528]
$BaC_{21}H_{27}N_7O_{14}P_2{\cdot}4H_2O$: Faintly yellow solid, UV_{max} 340 nm.[3552]

Nicotine adenine dinucleotide phosphate (NADP), $C_{21}H_{28}N_7O_{17}P_3$, also called triphosphopyridine nucleotide (TPN). *Adenosine 5′-(Trihydrogen Diphosphate), 2′-(Dihydrogen Phosphate), P′→5′ Ester with 3-(Aminocarbonyl)-1-β-D-ribofuranosylpyridinium, Inner Salt.*

NADP [53-59-8]: Countercurrent distribution,[3511] diamagnetic susceptibility (χ_M) -370×10^{-6} cm^3/mol at 25 °C,[3510] polarography,[3515] cyclic voltammetry (fig.).[3514] Ionization constants: pK_a 3.874. 6.255 (potentiometric, 25 °C),[3553] 6.40 (2′-phosphate, ^{31}P NMR, 25 °C).[3554] Spectra: ^{1}H NMR,[3523,3524] [^{13}C] NMR (fig.),[359] [^{31}P] NMR (fig.),[3505] CD (fig.),[3526] ESI-MS (fig.).[2887] Chromatography: PC,[3530] TLC,[3531] HPTLC,[2891] IEC,[2894] HPLC,[2440] DEAE cellulose column,[3533] GFC.[3555] Electrophoresis: PE,[9] CZE,[3534] capillary isotachophoresis,[3239] MECC.[3535] Theoretical calculations: Gibbs energy of formation ($\Delta_f G$) −418.40 kJ/mol.[3538]
$Na_2C_{21}H_{26}N_7O_{17}P_3$: FAB-MS (fig.).[3528]

Nicotine adenine dinucleotide phosphate, reduced form (NADPH), $C_{21}H_{30}N_7O_{17}P_3$. Also called reduced triphosphopyridine nucleotide (TPNH). *Adenosine 5′-(Trihydrogen Diphosphate), 2′-Dihydrogen Phosphate, P′→5′ Ester with 1,4-Dihydro-1-β-D-ribofuranosyl-3-pyridinecarboxamide*

NADPH [R.N. 53-57-6]: Somewhat stable in aqueous solution at pH 8 and above but sensitive to changes in temperature, ionic strength, and buffer salts. It has a shorter half-life than NADH under most conditions. Phosphorescence,[3556] molar absorptivity $\varepsilon = 6.18 \times 10^3$ L/mol/cm (334 nm, 25 °C);[3548] Gibbs energy of formation ($\Delta_f G$) −396.53 kJ/mol.[3538] Ionization constant: pK_a 6.52 (2′-P, ^{31}P NMR, 25 °C).[3554] Spectra: ^{1}H NMR (fig.),[3523,3524] [^{13}C] NMR (fig.),[359] [^{31}P] NMR (fig.).[3505] Chromatography: PC,[3530] TLC,[3531] IEC,[2894] HPTLC,[2891] DEAE cellulose column.[3533] Electrophoresis: CZE,[3534] capillary isotachophoresis.[3239]

Conformation

The conformation of NAD is folded in the free acid form, both in the crystalline state and in solution, but extended in its complexes with enzymes. In the free acid, each phosphate carries a negative charge, nicotinamide is positively charged at N-1, and adenine is protonated at N-1. The conformation of the free acid is similar to that of AMP with respect to its components—*anti* across the glycosidic bonds, *gg* across the C_4'—C_5' bonds of the two bases, *endo* pucker at C_2' in both riboses, and *trans* to C_4' for both phosphates. The crystal structure exhibits a novel intermolecular stacking of adenine and water molecules on opposite sides of the nicotinamide resulting in a nicotinamide sandwich.[3540] In contrast, the lithium salt $Li^+NAD^+ \cdot 2H_2O$ has an extended conformation, similar to, but distinctly different from, that of NAD in its enzyme complexes. The lithium ion forces the molecule to adopt a conformation in which the NAD molecules are stacked head to tail.[3541,3542] These conformations are supported by detailed studies of the UV,[3557] [^{1}H] NMR,[3558] [^{13}C] NMR,[3559] [^{31}P] NMR,[3505] and CD[3526] spectra and by theoretical calculations using semiempirical[3560] and *ab initio* MO[3561] methods and by modular dynamics simulations.[3562] The existence of right- and left-handed conformers of the folded form has been discounted.[3558]

The conformation of NADH in aqueous solution has one folded structure similar to that of NAD in rapid equilibrium with an open (extended) form. The 1,4-dihydropyridine ring in the folded form is folded against the adenine in such a manner that the ring is puckered, H_A is axial, and H_B is equatorial. The bases move apart if the temperature is raised, the pH is lowered, or the polarity of the solvent is reduced.[3558] These results are supported by ^{31}P NMR studies.[3505]

The conformations of NADP and NADPH in aqueous solution are primarily folded.[3523,3558] The ribose rings in NADP, according to ^{1}H, ^{13}C, and ^{31}P NMR analysis, are almost equally distributed between 2′-*endo* and 3′-*endo* modes in the AMP component but are predominantly in the 2′-*endo* mode in the NMN component. The conformations about the bridging C(5′)—O(5′) and C(4′)—C(5′) bonds resemble those of the corresponding mononucleotides.[3563]

Synthesis

NAD may be prepared (a) by condensation of NMN with AMP in the presence of dicyclohexylcarbodiimide,[3564] di-*p*-tolylcarbodiimide,[3565] or trifluoroacetic anhydride,[3566] (b) by reaction of NMN with an AMP morpholide[3567] or di-*n*-butylphosphinothioic anhydride[2989] but not from $AMPNH_2$[3568] or (c) from NMN and ATP by the combined action of the enzymes NAD pyrophosphorylase and PP_i pyrophosphatase.[3569]

NADH may be prepared from NAD (a) by chemical reduction with sodium hydrosulfite[3552] or sodium cyanoborohydride,[3570] (b) by electrolytic reduction with mercury,[3571] or (c) by enzymic reduction with yeast alcohol dehydrogenase and ethanol.

NADP may be prepared (a) by condensation of NMN with 2′,5′-ADP in the presence of dicyclohexylcarbodiimide[3564] or (b) by enzymic phosphorylation of NAD with ATP in the presence of NAD kinase.[3569]

NADPH may be prepared from NADP (a) by chemical reduction with sodium cyanoborohydride[3570] or (b) by electrolytic reduction with mercury.[3571] Sodium borohydride cannot be used for the preparation of NADH or NADPH because it produces a mixture of 1,2-, 1,4-, and 1,6- isomers.[3572]

Hydrolysis

NAD is stable to acid but unstable to base. Dilute alkali (0.1-1.0*N* NaOH) splits the nicotinamide-ribose linkage, which is followed by splitting of the pyrophosphate group, releasing nicotinamide and adenylic acid, respectively.[3573] Stronger alkali (5*N* NaOH) results in the formation of a bluish fluorescent product, λ 360 nm, subsequently identified as 2-hydroxynicotinaldehyde.[3574] The hydrolysis is second-order (pH-dependent) between pH 8.5 and 11 but becomes first-order (pH-independent) below pH 6.5 and above pH 12.5.[3518] The pK_{eq} for the pH-dependent reaction is 11.7 at 37 °C, close to the pK_a of the ribose diols (12.17 for C-2′, 12.13 for C-3′ at 20 °C), which suggests that cleavage of the nicotinamide-glycosyl bond is facilitated by the ionization of the adjacent ribose diol. A mechanism for the hydrolysis involving an oxocarbenium intermediate[3518] is supported by recent studies.[3575] For further details see Oppenheimer.[3576] Hydrolysis of NAD in the presence of Zr^{4+} catalyst results in selective cleavage of the pyrophosphate bond giving NMN and AMP, a reaction previously possible only with the enzyme NAD pyrophosphorylase.[3577]

NADH, in contrast to NAD, is stable to base and unstable to acid. The degradation of NADH in unbuffered acid solution is initiated by a general acid-catalyzed hydration of the 1,4-dihydronicotinamide ring, followed by β to α anomerization of the nicotinamide-ribose bond and cyclization, forming a new oxazolidine ring.[3578] The hydrolysis is pH-dependent at moderately low pH but becomes pH-independent at very low pH and at high pH (9-12). The rate changes if a buffer is present owing to further interaction between the buffer and the 1,2-dihydronicotinamide ring. The reaction of NADH with potassium orthophosphate in aqueous solution, for example, results in a series of three consecutive second-order reactions, each of which gets completed before the next step begins.[3579] Prolonged hydrolysis of NADH with 6N HCl at reflux results in a 36.8% conversion to glycine owing to incomplete degradation of the adenosine ring.[3580]

Hydrolysis of NADP under mild conditions (0.02*M* NaOH, 100 °C) splits the nicotinamide-ribose bond giving nicotinamide and ADPR 2′-phosphate,[3581] together with small amounts of ADP 2′-phosphate and AMP 2′-phosphate.[3582]

Like NADH, hydrolysis of NADPH in weakly acidic buffers follows first-order kinetics, but the rate is considerably greater at a given pH. The rate lines converge at very acidic pH.[3583]

For the effects of time, temperature, pH, and salts on the stability of NAD, NADH, NADP, and NADPH in aqueous solution, see Lowry *et al.*[3418]

Riboflavin Nucleotides

This group comprises the riboflavin derivatives flavin mononucleotide (FMN), flavin adenine dinucleotide (FAD), and the corresponding dihydro derivatives.

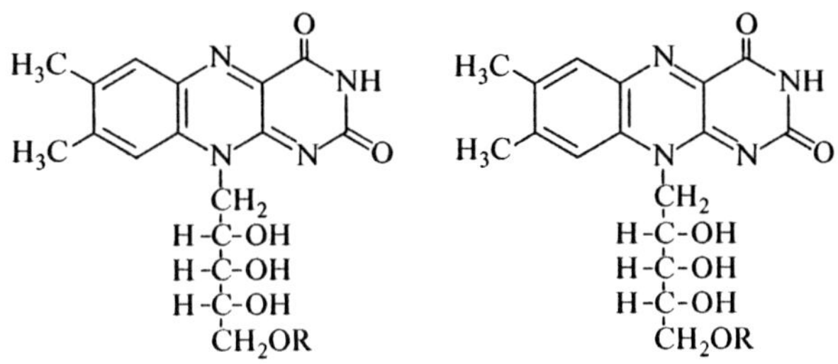

RF, R = H
FMN, R = P(O)(OH)$_2$
FAD, R = see below

RFH$_2$, R = H
FMNH$_2$, R = P(O)(OH)$_2$
FADH$_2$, R = see below

FAD, FADH$_2$: R =

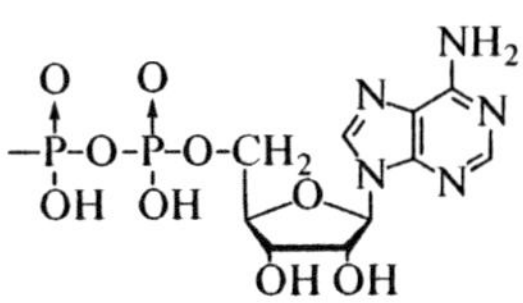

Occurrence

Since the late 1930s, it has been known that Riboflavin (vitamin B_1) is widespread in plants, especially in leafy vegetables, but its true nature was not revealed until the 1950s when Whatley[3406] isolated FAD from the leaves of clover, chickweed, potato, elder, and runner beans and Bevilacqua[3584] showed that the acid soluble "riboflavin" extracted from seeds or seedlings of wheat, oats, and other plants comprises FMN and FAD exclusively and contains no free riboflavin (RF) at all. During germination most of the FAD is found in the radicle and most of the FMN in the plumule.[3585] On the basis of this and other evidence, Appaji Rao and his group, who isolated and characterized all the enzymes listed in Fig. 6-2, concluded that the site of synthesis of the flavins was the radicle.[3585]

The metabolic pathways for the biosynthesis of the riboflavin nucleotides are outlined in Fig. 6-2. In step 1, riboflavin is phosphorylated by ATP giving FMN and ADP.[3586] In step 2, FMN reacts with ATP in the presence of FAD pyrophosphorylase giving FAD and PP_i.[3587] This pathway was confirmed by Koop and Barz,[3588] who found that [2-^{14}C]riboflavin is incorporated stepwise into FMN and FAD in mung bean, chickpea, and parsley cell

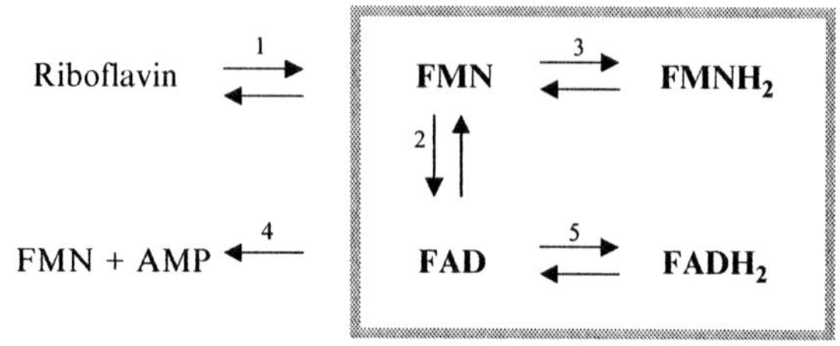

FIGURE 6-2 Riboflavin Nucleotides. (1) Riboflavin kinase [E.C. 2.7.1.26], ATP. (2) FAD pyrophosphorylase [E.C. 2.7.7.2], ATP. (3, 5) FMN reductase [E.C. 1.6.8.1], NAD(P)H. (4) FAD pyrophosphatase [E.C. 3.6.1.18], H_2O.

suspension cultures. In steps 3 and 5, FMN and FAD interact with NADH or NADPH giving $FMNH_2$ and $FADH_2$, respectively. In step 4, FAD is hydrolyzed by FAD pyrophosphatase giving FMN and AMP.[3589] There appears to be no enzymatic mechanism for salvage of the reduced nucleotides other than by way of returning to their oxidized forms.

The source of the riboflavin in plants is unknown. In bacteria, the building blocks for its biosynthesis are known to be guanosine 5′-triphosphate (GTP) and ribulose 5-phosphate (Ru5P), assembled through a series of seven enzyme-linked reactions.[3590]

Extraction

Methods recommended by the AOAC, AACC, and other groups for the extraction of riboflavin from foodstuffs, clinical samples, and pharmaceuticals are too harsh for use in the determination of FMN and FAD.[3591] Extraction with ice-cold trichloroacetic acid under the conditions normally used for other nucleotides can be used for the simultaneous assay of RF, FMN, and FAD. An example of this procedure is described later in the fluorometric method. Several investigators have obtained good results with aqueous extraction using pH 7.4-7.7 phosphate buffer.[3452,3592,3593] Treadwell and Metzler[3594] used saturated ammonium sulfate as both extractant and protein precipitant.

Separation

(a) From Acid Extracts

Whatley[3406] used the procedure of Warburg and Christian[3595] in which ammonium sulfate is added to precipitate the protein and the FAD is separated from the other nucleotides by extraction of the supernatant with liquid phenol and precipitation with acetone. Sanwal and Waygood[3596] separated FMN and FAD from the other nucleotides by adsorption on Florisil and elution with aqueous pyridine.

(b) From Neutral Extracts

Koziolowa[3597] and Koziol[3593] separated RF, FMN, and FAD in phosphate buffer extracts of whole wheat, flour, bran, and gluten by ion exchange chromatography on a sulfonic acid cation exchange resin (K-28, H^+ form). Van Gestelen *et al.* separated the flavins in phosphate buffer extracts of bean[3598] and corn[3599] plasma membranes by high-performance liquid chromatography using a reversed-phase column (Bio-Sil C_{18} HL 90-5s).

Analysis

(1) Enzymatic Method

This method is based on reactivation of the enzyme D-amino acid oxidase [E.C. 1.4.3.3] when its prosthetic group (FAD) is restored after being removed by treatment with acid. The uptake of oxygen catalyzed by the reactivated oxidase upon the addition of the plant extract serves as a measure of its FAD content.[3406]

(2) Fluorometric Method

Most of the analyses reported in Table 6-3 were carried out by the method developed by Bessey *et al.*[3612] for the determination of RF, FMN, and FAD in animal tissue. The following is a typical procedure applied to a plant tissue:

> *Tissue from tobacco plants was ground in an Omni mixer (Ivan Sorvall, Inc., Norwalk, Conn.) with 20 times its volume of ice water and mixed immediately with an equal volume of 20 per cent trichloroacetic acid. After centrifuging, an aliquot of the supernatant was neutralized with 4M K_2HPO_4. A second aliquot of the solution was stored in the dark at 38 °C. overnight to produce hydrolysis of FAD to FMN. After hydrolysis this sample was neutralized with K_2HPO_4. FAD was determined by differences in the fluorescence between the hydrolyzed and unhydrolyzed sample as outlined by Bessey* et al.[3612] *Free riboflavin was determine by shaking an aliquot of the unhydrolyzed sample with equal volumes of water-saturated benzyl alcohol. An aliquot of the benzyl alcohol layer was mixed with an equal volume of water, which was 0.05M with respect to sodium acetate and acetic acid, and 15 volumes of water-saturated toluene. An aliquot of the aqueous layer was then assayed fluorometrically for riboflavin.*[3612] *FMN was calculated by subtracting FAD and free riboflavin from total riboflavin.*
>
> Welkie, G. W.; Miller, G. W., "Iron Nutrition of *Nicotiana tabacum* L. in Relation to Riboflavin, Riboflavin-5-Phosphate and Flavin Adenine Dinucleotide Content", *Plant Physiol.* **1960**, *35*(4), 516–520,[3608]
> reprinted by permission of the American Society of Plant Biologists, © 1960.

Swaraj *et al.*[3592] used a modification of this method in which FMN and FAD are determined on the same sample at different pH values.[3613]

Some Comments on the Reduced Flavins

The reader may have noticed the absence of data in Table 6-3 for $FMNH_2$ or $FADH_2$. The reasons for this are not hard to find. Both substances are extraordinarily air-sensitive, being readily reoxidized to FMN and FAD, respectively. All of the flavins are light-sensitive and need to be handled in subdued light, preferably in the dark. On the other hand, none of them exhibit the pH-sensitivity that is characteristic of the pyridine nucleotides discussed earlier in this chapter. The FMN/$FMNH_2$ couple is stable over the 0.85-12.95 pH range[3614,3615] and the FAD/$FADH_2$ couple over the 2.4-12.4 pH range.[3615] Decomposition products such as lumichrome and lumiflavin, which develop in acid and alkaline solutions, respectively, are the result of photodegradation and not of any inherent instability of the flavins.

Table 6-3 Riboflavin (RF) Nucleotides in Plants*

No.	Plant part	Method	Free RF	FMN	FAD	References
Avena sativa L. (Oats)						
2	Seed[a]	Fluor.		0.48[b]	0.41	3600
1	–, endosperm[a]	Fluor.	0	0	0.55	3600
1	Seed[a]	Fluor.		0.41	0.09	3584
1	Sprout[a]	Fluor.		1.56	0.56	3584
1	Seedling, leaf[a]	Fluor.	1.2	6.5	15.3	3601
1	–, endosperm[a]	Fluor.		0.8[b]	3.7	3600
1	–, root[a]	Fluor.		6.6[b]	5.0	3600
1	, coleoptile	Fluor.		24.0[c]		3602
		and variation from tip to base				
1	Shoot, etiolated	TLC	+	0	0	3594
Beta vulgaris L. (Beet)						
1	Root	TLC	+	0	0	3594
Brassica chinensis L. (Chinese Colza)						
1	Pollen, mature		0	10.98	3.92	3603
1	–, immature		0	2.90	10.60	3603
1	–, senescent		11.96	1.74	0	3603
Camellia japonica L. (Camellia)						
1	Pollen, mature		+	8.34	0.96	3603
1	–, immature		0	3.26	4.94	3603
1	–, senescent		5.92	3.08	0	3603
Camellia sinensis L. (Tea)						
1	Pollen, mature		+	11.34	1.36	3603
1	–, immature		0	4.48	8.42	3603
1	–, senescent		5.76	6.24	+	3603
Chaenomeles japonica (Thunb.) [Lindl. ex] Spach (Japanese Quince)						
1	Pollen, mature		0	6.07	2.53	3603
1	–, immature		0	0.95	5.95	3603
1	–, senescent		4.16	2.74	0	3603
Chlorophytum comosum (Thunb.) Jacq. (Spider Plant)						
1	Leaf[d]	Fluor.		50[b]	94	3604
Cicer arietinum L. (Bengal Gram)						
1	Seed[a]	Fluor.	n.d.[e]	0.41	0.57	3585
Coix lacryma-jobi L. (Job's Tears)						
1	Shoot, etiolated	TLC	+	0	0	3594
Cosmos spp. (Aster)						
1	Pollen, mature		0	13.06	3.54	3603
1	–, immature		0	3.68	8.72	3603
1	–, senescent		8.92	5.08	0	3603
Daucus carota L. (Carrot)						
1	Root	TLC	+	0	0	3594
1	Root[f]	Enzym.	0.10	0.04	0.42	3605
Glycine max (L.) Merr. (Soybean)						
1	Whole bean[f]	Enzym.	0.15	0.90	1.62	3605
1	Hypocotyl, tonoplast[g]	Fluor.			2.0	3606

Table 6-3 Riboflavin (RF) Nucleotides in Plants*—cont'd

No.	Plant part	Method	Free RF	FMN	FAD	References
1	–, plasma membrane[g]	Fluor.			0.5	3606
1	Root nodule[g]	Fluor.		1.2	3.8	3592
Gossypium herbaceum L. (Levant Cotton)						
1	Seed	TLC	+	0	0	3594
Helianthus annuus L. (Sunflower)						
1	Pollen, mature		0	14.31	4.49	3603
1	–, immature		0	2.44	14.36	3603
1	–, senescent		4.16	2.74	0	3603
Hizikia fusiformis Harvey (Hijiki)						
1	Air-dried plant	Fluor.			6.0	3607
Hordeum vulgare L. (Barley)						
1	Seedling, leaf [a]	Fluor.	7.3	7.7	7.5	3601
2	Seedling, coleoptile (1)	Fluor.		10.6[c]		3602
	Seedling, coleoptile (2)	Fluor.		30.0[c]		3602
		and variation from tip to base				
Ipomoea batatas (L.) Lam. (Sweet Potato)						
1	Tuber	TLC	+	0	0	3594
1	Tuber[f]	Enzym.	0.90	0.03	0.87	3605
Lablab purpureus (L.) Sweet (Hyacinth Bean)						
1	Seed[a]	Fluor.	n.d.[e]	0.58	1.43	3585
Laminaria japonica Aresch. (Brown Seaweed)						
1	Air-dried plant	Fluor.			2.2	3607
Lilium longiflorum Thunb. (Trumpet Lily)						
1	Pollen, mature		0	9.80	4.50	3603
1	–, immature		0	1.35	11.25	3603
1	–, senescent		4.67	3.84	0.99	3603
Lycopersicon esculentum Mill. (Tomato)						
1	Stem	Fluor.			+	3452
1	Leaf	Fluor.			+	3452
1	Leaf	Fluor.		0.073[b]	0.051	3596
Macrotyloma uniflorum (Lam.) Verdc. (Horse Gram)						
1	Seed[a]	Fluor.	n.d.[e]	0.51	0.76	3585
Magnolia × soulangiana Soul.-Bod. (Chinese Magnolia)						
1	Leaf	TLC	+	0	0	3594
Nicotiana tabacum L. (Tobacco)						
1	Leaf[a]	Fluor.	1.4	0.1	0.2	3608
Oenothera biennis ssp. *grandiflora* (Large-Flowered Evening Primrose)						
1	Pollen, mature		+	13.13	2.37	3603
1	–, immature		0	4.34	9.56	3603
1	–, senescent		6.67	6.23	0	3603
Oryza sativa L. (Rice)						
1	Flour[f]	Enzym.	0.32	0.08	0.12	3605
1	Flour[f]	Enzym.	0.37	0.69	0.93	3605

*μg/g fresh weight basis, unless stated otherwise; [a]Dry weight basis; [b]Including RF; [c]Including FAD; [d]μg/g protein nitrogen; [e]Not detected; [f]μg/g; [g]nmol/mg protein; [h]pmol/mg protein.

Continued

Table 6-3 Riboflavin (RF) Nucleotides in Plants*—cont'd

No.	Plant part	Method	Free RF	FMN	FAD	References
Phaseolus coccineus L. (Scarlet Runner Bean)						
1	Leaf	Enzym.			0.11	3406
Phaseolus vulgaris L. (Common Bean)						
1	Seed[a]	Fluor.	n.d.[e]	1.11	2.01	3585
1	Seed[a]	Fluor.		0.65	0.29	3584
1	Apical hook plasma membrane[i]	HPLC		470	n.d.[e]	3598
1	Sprout[a]	Fluor.		1.11	0.82	3584
1	Seedling	Fluor.		9.7[b]	9.7	3609
Pisum sativum L. (Peas)						
1	Seed[a]	Fluor.	n.d.[e]	1.20	0.90	3585
1	Seed[a]	Fluor.		0.40	0.28	3584
1	Sprout[a]	Fluor.		1.25	1.35	3584
Porphyra tenera Kjell. (Nori)						
1	Air-dried plant	Fluor.			17.8	3607
Portulaca grandiflora Hook. (Rose Moss)						
1	Pollen, mature		+	12.76	1.84	3603
1	–, immature		0	4.83	6.67	3603
1	–, senescent		5.61	3.19	0	3603
Raphanus sativus L. (Radish)						
1	Leaf[f]	Enzym.	1.72	0.65	1.58	3352
1	Tuber[f]	Enzym.	0.06	0.02	0.24	3352
Robinia pseudoacacia L. (Black Locust)						
1	Pollen, mature		0	11.33	2.57	3603
1	–, immature		0	4.08	9.12	3603
1	, senescent		7.09	2.81	0	3603
Sambucus nigra L. (European Elder)						
1	Leaf	Enzym.			0.28	3406
Secale cereale L. (Rye)						
1	Seed[a]	Fluor.		0.76[b]	0.31	3600
1	–, endosperm[a]	Fluor.		0.38[b]	0.37	3600
1	Seedling, leaf[a]	Fluor.	1.2	6.1	9.1	3601
1	–, endosperm[a]	Fluor.		2.6[b]	1.8	3600
1	–, root[a]	Fluor.		8.7[b]	6.2	3600
Solanum tuberosum L. (Potato)						
1	Tuber	TLC	+	0	0	3594
1	Tuber[f]	Enzym.	0.06	0.66	0.42	3605
1	Leaf	Enzym.			0.27	3406
Spinacia oleracea L. (Spinach)						
1	Leaf	PE	+	+	+	3594
1	Leaf[f]	Enzym.	2.28	0.68	1.01	3605
1	–, chloroplast and mitochondria[f]	Enzym.	0.26	0.33	0.73	3605
1	–, cytosol[f]	Enzym.	1.39	0.39	0.17	3605
Stellaria media (L.) Vill. (Chickweed)						
1	Leaf	Enzym.			0.20	3406

Table 6-3 Riboflavin (RF) Nucleotides in Plants*—cont'd

No.	Plant part	Method	Free RF	FMN	FAD	References
Trifolium repens L. (White Clover)						
1	Leaf	Enzym.			0.42	3406
Triticum aestivum L. (Wheat)						
1	Seed[a]	Fluor.		0.25	0.17	3584
1	Grain	IEC	1.2	0.7	0.5	3597
1	Flour	IEC	0.38	0.73[c]		3593
1	Flour [f]	Enzym.	0.64	0.16	0.32	3605
1	Bran[f]	Enzym.	0.28	0.51	0.64	3605
1	Bran	IEC	0.73	1.03[c]		3593
1	Gluten	IEC	0.32	0.28[c]		3593
1	Seedling, leaf[a]	Fluor.	1.0	5.8	9.2	3601
1	Seedling, leaf[f]	Enzym.	1.06	0.13	0.92	3605
1	Leaf[a]	Fluor.	2	93	46	3610
1	–, etiolated[a]	Fluor.	5	53	108	3610
1	Sprout[a]	Fluor.		1.55	0.20	3584
1	Shoot, etiolated	TLC	+	0	0	3594
Undaria pinnatifida (Harvey) Sur. (Wakame)						
1	Air-dried plant	Fluor.			6.8	3607
Vicia faba L. (Horse Bean)						
1	Leaf[f]	Enzym.	2.14	0.61	2.01	3605
Vigna angularis (Willd.) Ohwi and Ohashi (Adzuki Bean)						
1	Whole bean[f]	Enzym.	0.01	0.45	1.06	3605
Vigna mungo (L.) Hepper (Black Gram)						
1	Seed[a]	Fluor.	n.d.[e]	0.12	1.90	3585
Vigna radiata (L.) Wilcz. (Green Gram)						
1	Seed[a]	Fluor.	n.d.[e]	0.10	3.30	3585
Vigna unguiculata (L.) Walp. (Cowpea)						
1	Seed[a]	Fluor.	n.d.[c]	0.20	2.40	3585
Vitis spp. (Grape)						
4	Vine	Fluor.		0.32-0.36	0-0.16	3611
Zea mays L. (Corn)						
1	Shoot or seedling	TLC	+	0	0	3594
1	Root plasma membrane [h]	HPLC	52	121	89.6	3599

*µg/g fresh weight basis, unless stated otherwise.
[a]Dry weight basis.
[b]Including RF.
[c]Including FAD.
[d]µg/g protein nitrogen.
[e]Not detected.
[f]µg/g.
[g]nmol/mg protein.
[h]pmol/mg protein.

Physical Properties

Flavin mononucleotide (FMN), $C_{17}H_{21}N_4O_9P$. *Riboflavin 5′-(Dihydrogen Phosphate).*

FMN [R.N. 146-17-8]: Slightly hygroscopic, greenish-yellow substance, soluble in water, glacial acetic acid, phenol and *p*-cresol, insoluble in acetone and ether,[3616] and slightly soluble in ethanol.[3617] Countercurrent distribution,[3618] phosphorescence,[3619] dipole moment (D) 150 esu cm,[2939] molecular weight (sedimentation, 25 °C),[3620] diamagnetic susceptibility (χ_m) -226×10^{-6} cm^3/mol at 25 °C,[410] redox potential (E_M) $-$ 0.205 V at pH 7.0 and 20 °C,[3614] cyclic voltammetry (fig.),[3621] polarography.[3622] Ionization constants, potentiometric: pK_a 6.60, 10.12 (18-20 °C)[3623] or 10.35 (20 °C).[3614] Spectra: UV (fig.),[3623] IR (fig.),[3623] Raman (fig.),[3624] ORD (fig.),[3625] CD (fig.),[3626] magnetic CD (fig.),[3627] ^{1}H NMR (fig.),[3628] ^{13}C NMR (fig.),[2926] ^{15}N NMR (fig.),[3629] ^{31}P NMR,[68] ESR (fig.),[3630] ENDOR (fig.),[3631] SERS (fig.),[3632] photoacoustic spectrum (fig.),[3633] thermospray MS,[3634] ES-MS,[3634] LC-MS.[3634] Chromatography: PC,[3635] circular PC,[3636] TLC,[3637] IEC,[3638] silica gel column,[3639] GFC,[3555] MECC.[3535] Electrophoresis: PE,[3640] CZE.[3641] Thermodynamics: $\Delta_f G°$, E′°.[447]

$NaC_{17}H_{20}N_4O_9P{\cdot}2H_2O$: Yellow crystals. Solubility in water (mg/mL): 112 at pH 6.9, 68 at pH 5.6, and 43 at pH 3.8.[3642a]

$C_{13}H_9N{\cdot}C_{17}H_{21}N_4O_9P{\cdot}0.5H_2O$ (acridine salt): Brownish yellow needles (photo), mp 217 °C dec., soluble in hot water and insoluble in ethanol and ether.[3643]

Dihydroflavin mononucleotide ($FMNH_2$), $C_{17}H_{23}N_4O_9P$. *Riboflavin 5′-(Dihydrogen Phosphate), 1,5-Dihydro.*

$FMNH_2$ [R.N. 5666-16-0]: Molecular weight (sedimentation, 25 °C),[3620] polarography.[3622] Ionization constant: pK_a 6.72 (potentiometric, 20 °C).[3614] Spectra: UV (fig.),[3644] ^{13}C NMR,[3645] ^{14}N NMR,[3645] ^{31}P NMR,[3646] laser-excitation transient absorption (fig.).[3647] Thermodynamics: $\Delta_f G°$.[447]

Flavin Adenine Dinucleotide (FAD), $C_{27}H_{33}N_9O_{15}P_2$. *Riboflavin 5′-(Trihydrogen Diphosphate), 5′→5′-Ester with Adenosine.*

FAD [R.N. 146-14-5]: Bright yellow hexagonal crystals (photo),[3648] soluble in water, pyridine, phenol, and *p*-cresol and insoluble in acetone, ether, ethanol, and chloroform.[3616] Phosphorescence,[3619] diamagnetic susceptibility (χ_m) -416×10^{-6} cm^3/mol at 25 °C,[3510] countercurrent distribution,[3612] redox potential (E_M) $-$0.219 V at pH 7.0 and 30 °C,[3615] INDO charge density calc.,[3649] polarography,[3622] CCSA polarography (fig.),[3650] cyclic voltammetry (fig.).[3651] Spectra: UV-Vis. (fig.),[3649] Raman (fig.),[3624] FTIR (fig.),[3651] ORD (fig.),[3625] CD (fig.),[3626] magnetic CD (fig.),[3652] ^{1}H NMR (fig.),[3653] 2d-^{1}H NMR (fig.),[3654] ^{13}C NMR (fig.),[2926] ^{15}N NMR (fig.),[3629] ^{31}P NMR,[68] ESR,[3630] ENDOR (fig.),[3631] SERS (fig.),[3632] CIDNP (fig.),[3655] thermospray MS,[3634] ES-MS.[3634] Chromatography: PC,[3635] circular PC,[3636] IEC,[3638] HPLC,[3656] GFC,[3157] LC-MS,[3634] column (active carbon).[3657] Electrophoresis: PE,[3640] cellulose column,[3648] isotachophoresis.[3239] Thermodynamics: $\Delta_f G°$, E′°.[447] Theoretical calculations: LCAO-MO,[3536] SCF-MO.[3658]

Ba $C_{27}H_{31}N_9O_{15}P_2$: Small yellow spheres in grapelike clusters, UV$_{max}$ 366, 445 nm.[3642b] Solubility in water, 2.5 g/L at 25 °C.[3595]

Dihydroflavin adenine nucleotide ($FADH_2$), $C_{27}H_{35}N_9O_{15}P_2$. *Riboflavin 5′-(Trihydrogen Diphosphate), 1,5-Dihydro, 5′→5′-Ester with Adenosine.*

$FADH_2$ [R.N. 1910-41-4]: Polarography,[3622] PSA polarography (fig.).[3650] Spectra: FTIR (fig.),[3651] laser-excitation transient absorption.[3647] Thermodynamics: $\Delta_f G°$.[447] Theoretical calculations: MO (INDO method).[3659]

Conformation

That strong interaction exists between the adenine and isoalloxazine rings of FAD is evident from the self-quenching of fluorescence compared to that of riboflavin or FMN.[3612] Proton NMR data indicate that the rings are hydrogen bonded to each other. A similar association occurs in aqueous solution between FMN and AMP (the components of FAD) but not between FMN and GMP or CMP.[3660]

In aqueous solution, FAD exists as an equilibrium mixture of open and folded conformers. This conclusion was reached by NMR analysis and confirmed by ORD measurements[3626] and MO calculations.[3658] In the folded conformation the two rings are vertically stacked, but the precise nature of the stacking is unclear; for a discussion of several proposed models see Kainosho and Kyogoku.[3628] The closed conformer also fluoresces but its lifetime (0.3 ns at room temperature) is very short.[3661]

Recent studies based on ^{13}C and ^{15}N NMR shift data suggest that the conformation of $FMNH_2$ is slightly bent about the N_5-N_{10} axis in nonpolar solvents but almost planar in aqueous solution.[3645]

Synthesis

FMN may be prepared (a) by phosphorylation of 2′,3′,4′-triacetylriboflavin with $POCl_3$ in pyridine,[3662] (b) by phosphorylation of unprotected riboflavin with metaphosphoric acid,[3663] monochlorophosphoric acid,[3664] *N*-morpholinophosphorodichloridate,[3665] or an alkyl phosphorodichloridate,[3666] (c) by oxidation of riboflavin 5′-phosphite with bromine or *N*-bromosuccinimide,[3667] or (d) by acid hydrolysis of riboflavin 4′,5′-phosphate.[3668]

$FMNH_2$ may be prepared (a) by catalytic hydrogenation of FMN in the presence of platinized asbestos,[3669] (b) by the reaction of FMN with dithionite,[3615] or (c) by phosphorylation of dihydroriboflavin with partially hydrolyzed $POCl_3$.[3670]

FAD may be prepared (a) by condensation of FMN with AMP in the presence of di-*p*-tolylcarbodiimide [3671] or trifluoroacetic anhydride,[3672] (b) by reaction of FMN with an AMP ester,[3123] amide,[3673] morpholide,[3674] imidazolide,[2993] or anhydride,[2989,3010] or (c) by reaction of FMN with an AMP derivative that contains 2′,3′-protecting substituents.[3399,3675]

$FADH_2$ may be prepared (a) by catalytic hydrogenation of FAD in the presence of platinized asbestos,[3622] or (b) by the reaction of FAD with dithionite.[3615]

Hydrolysis

Hydrolysis of FMN to riboflavin and P_i is the predominant reaction in the 2-7 pH range, reaching a maximum at pH 4 in common with other phosphate esters. At pH 4 and 80 °C, the rate constant is $6.48 \times 10^{-7}\ s^{-1}$. Below pH 2, the predominant reaction is acid-catalyzed rearrangement; at pH 1 and 50 °C, the reaction mixture at thermodynamic equilibrium contains riboflavin and 5′-, 4′-, 3′-, and 2′-phosphates in the ratio 66:11:8:15. The rearrangement is pH-dependent, increasing linearly with the acid concentration.[3676] Prolonged hydrolysis of FMN in 1*N* HCl at 100 °C results in the formation of 2′,5′-anhydroriboflavin.[3677] Under much milder conditions (pH 8.6, 37 °C), the hydrolysis is catalyzed by Ce(III) and other heavy metal ions.[3678] Heating FMN in NaOH solution ruptures the isoalloxazine ring giving a keto acid.[3679]

Phosphorylation

Condensation of FMN with phosphoric acid in the presence of *p*-tolylcarbodiimide results in a 17% conversion to riboflavin 5′-diphosphate. Use of the more reactive DCC produces only cyclic FMN.[3680]

7 Vitamins

This chapter is concerned with the vitamins thiamine (B_1) and pyridoxine (B_6) and their phosphates. For the vitamin riboflavin (B_7) and its phosphates, see Chapter 6. The Vitamin B_1 complex is considered first.

Vitamin B_1

The Vitamin B_1 complex comprises thiamine and its monophosphate (TMP), thiamine di- or pyrophosphate (TDP, also known as cocarboxylase), and thiamine triphosphate (TDP).

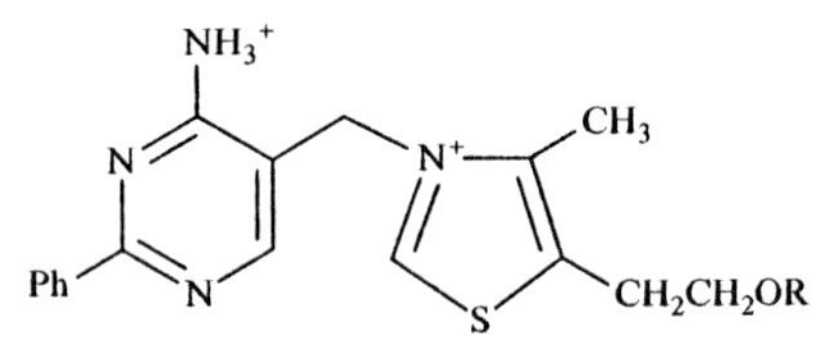

TMP, R = $P(O)(O^-)_2$
TDP, R = $P(O)(O^-)OP(O)(OH)O^-$
TTP, R = $P(O)(O^-)OP((O)(OH)OP(O)(OH)O^-$

Thiamine Monophosphate (TMP), $C_{12}H_{17}N_4O_4PS$ [R.N. 495-23-8]; Beilstein **27** E IV 1776. *3-[(4-Amino-2-methyl-5-pyrimidinyl)methyl]-4-methyl-5-[(2-phosphonooxy)ethyl]thiazolium Betaine.*

Thiamine Di- or Pyrophosphate (TDP, Cocarboxylase), $C_{12}H_{18}N_4O_7P_2S$ [R.N. 136-09-4]; Beilstein **27** E IV 1777. *3-[(4-Amino-2-methyl-5-pyrimidinyl)-methyl]-4-methyl-5-(4,6,6-trihydroxy-3,5-dioxa-4,6-diphosphahex-1-yl)thiazolium Betaine P,P′-Dioxide.*

Thiamine Triphosphate (TTP), $C_{12}H_{19}N_4O_{10}P_3S$ [R.N. 15666-52-1]; Beilstein **27** E IV 1778. *3-[(4-Amino-2-methyl-5-pyrimidinyl)methyl]-4-methyl-5-(4,6,8,8-tetrahydroxy-3,5,7-trioxa-4,6,8-triphosphaoct-1-yl)thiazolium Betaine P,P′,P″-Trioxide.*

Occurrence

Thiamine (vitamin B_1) is widely distributed in the higher plants as in other forms of life.[3681] The most abundant form in plant tissue is free thiamine, but it also occurs in varying degrees in the form of mono-, di-, and triphosphate esters. The diphosphate ester TDP, cocarboxylase, is an important coenzyme for pyruvate decarboxylase and pyruvate dehydrogenase in the Krebs cycle and for transketolase in the pentose shunt pathway, but TMP and TTP (like thiamine itself) exhibit no cocarboxylase activity.[3682] For a review of the catalytic functions of the thiamine phosphates, see Krampitz.[3683]

The discovery of cocarboxylase in plants dates to 1937, when Tauber[3684] found it to be present in a variety of vegetables, fruits, and nuts in amounts ranging from 0.4% to 30% of

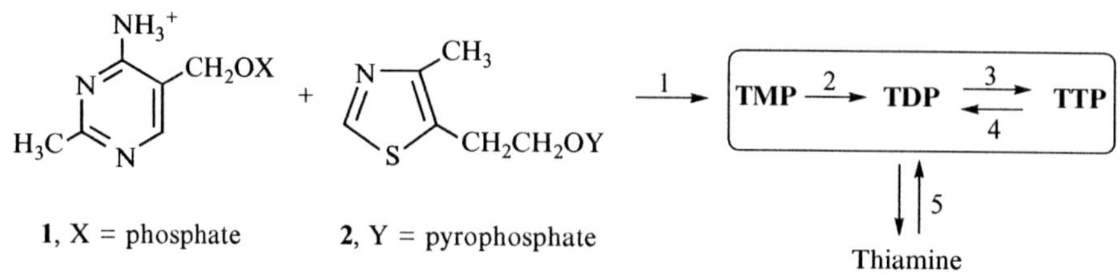

FIGURE 7-1 Thiamine Phosphates. (1) TMP pyrophosphorylase [E.C. 2.5.1.3]. (2) TMP kinase [E.C. 2.7.4.16], ATP. (3) TDP kinase [E.C. 2.7.4.15], ATP. (4) Thiamine triphosphatase [E.C. 3.6.1.28]. (5) Thiamine pyrophosphokinase [E.C. 2.7.6.2], ATP.

the total thiamine content (Table 7-1). The presence of the other phosphorylated thiamine esters in plants was not established until 1961.[3685]

The metabolic pathways for the biosynthesis of the thiamine phosphates are outlined in Fig. 7-1. In step 1, TMP is formed by the coupling of (4-amino-2-methyl-5-pyrimidinyl) methyl pyrophosphate (**1**) and (4-methyl-5-thiazolyl)ethyl phosphate (**2**). In steps 2 and 3, TMP is phosphorylated to TDP and thence to TTP. Cell-free biosynthesis of **1** and **2** has not been established, but mechanistic sequences have been proposed for the biosynthesis of **1** from 5-aminoimidazole riboside, an intermediate on the purine biosynthetic pathway, and of **2** from 1-deoxyxylulose 5-phosphate (dXu5P), cysteine, and tyrosine.[3590]

Physical Properties

Each of the phosphorylated thiamine derivatives can exist either as a betaine in which the two positive charges are balanced by negative charges on the phosphate residue or as a salt in which one or both positive charges are balanced by a stronger acid, usually HCl.

Thiamine Monophosphate

TMP [R.N. 495-23-8]: Colorless rods, mp 228-230 °C dec., soluble in water (pH 5.5 for a 5% solution at 20 °C), insoluble in lower alcohols or acetone.[3686]
TMP·HCl: mp 193 °C dec., PE.[3687]
TMP·HCl·2H_2O [R.N. 16028-14-1]: Large white plates, mp 200 °C, soluble in water, insoluble in ethanol[3688]; isoelectric pH 8.3[3689]; pK_a.[3690,3691] Spectra: UV,[3688] IR (fig.),[3692] fluorescence,[3693,3694] phosphorescence,[3693] ^{1}H NMR,[3695] ^{13}C NMR,[3696] ^{31}P NMR (fig.).[3695] Chromatography: PC,[3697,3698] TLC,[3699,3700] IEC (Sephadex C-25,[3701] cellulose,[3697] starch[3697]), HPLC,[3702] Sephadex gel.[3703] Electrophoresis: Paper,[3697,3704] starch column,[3697] high-voltage zonal.[3705] Polarographic reduction: AC,[3691] DC.[3690,3706]
TMP·2HCl·2H_2O: mp 215-217 °C dec.,[3688] pK_a,[3707] TLC.[3700]

Thiamine Diphosphate

TDP·4H_2O [R.N. 136-09-4]: Triclinic prisms, mp 207 °C,[3708] d 1.542,[3708] 1.56,[3709] stable on storage.[3710,3711] Spectra: IR,[3712] ^{1}H and ^{31}P NMR.[3695] PC,[3710] IEC,[3710] X-ray structure,[3708,3709] high-voltage zonal electrophoresis.[3705]

TDP·4.5H_2O [R.N. 71222-11-2]: Triclinic colorless tabular crystals, mp 232 °C, d 1.567, X-ray structure.[3713]
TDP·HCl: Rods, mp 239-240 °C dec.[3686]; monoclinic plates, d 1.656, X-ray structure.[3714] The cell dimensions do not allow for the 1/2 mol water of crystallization usually reported for this salt (see next entry).
TDP·HCl·1.5H_2O [R.N. 154-87-0]: Nonhygroscopic white needles (photo[3692]), mp 242-244 °C,[3688] readily soluble in cold water, insoluble in alcohols and acetone, colorless in acid or neutral solution, yellow in alkaline solution,[3688] isoelectric pH 3.6,[3689] pK_a (potentiometric),[3690,3715,3716] phase diagrams for H_2O solutions exhibit eutectics for stable and metastable modifications at −0.8 and −2.3 °C, respectively.[3717] Spectra: UV,[3690,3718] IR (fig.),[3692,3719] fluorescence,[3693,3694] phosphorescence,[3693] ^1H NMR,[3720] ^{13}C NMR,[3696] ^{31}P NMR,[3721,3722] DTA.[3723] Electronic structure calculations: MO,[3724] point charge.[3725] Chromatography: PC,[3697,3726] TLC,[3699] IEC (resin,[3697] Sephadex C-25[3701]), column (starch,[3697] cellulose,[3697] Norite[3727]), HPLC,[3702,3728] Sephadex gel,[3703,3729] GC.[3730] Electrophoresis: PE,[32,3697,3704] starch column.[3697] Polarography,[3690,3731,3732] oscillopolarography,[3733] fast cyclic voltammetry.[3734] Electron microscopy (photo).[3735]

Thiamine Triphosphate

TTP·0.5H_2O [R.N. 3475-65-8]: Crystals, mp 194 °C dec.,[3736] water soluble. Spectra: ^1H NMR,[3737] luminescence.[3694] Chromatography: PC,[3697,3738] TLC,[3699] IEC (cellulose, starch),[3697] HPLC.[3702,3728] Electrophoresis: PE,[3697,3704] starch.[3697] Polarography.[3739]
TTP·2H_2O: Crystals, mp 194-197 °C dec.[3740]
TTP·HCl·H_2O [R.N. 15666-52-1]: Short hygroscopic rods, mp 237-239 °C (photo), UV (fig.); PC,[3741] TLC,[3700] PE, cellulose column electrophoresis.[3741]

Synthesis

TMP may be prepared (a) by phosphorylation of thiamine with pyrophosphoric acid,[3742] metaphosphoric acid,[3711] or polyphosphoric acid[3743]; (b) by hydrolysis of TDP with dilute HCl or H_2SO_4 at 100 °C[3688,3744,3745]; (c) by hydrolysis of TTP with dilute $Ba(OH)_2$ or piperazine at 0 °C[3746,3747]; or (d) by the reaction of 4-methyl-5-thiazoleethyl phosphate with 4-amino-5-bromomethyl-2-methylpyrimidine.[3687]

TDP may be prepared (a) by phosphorylation of thiamine with metaphosphoric acid,[3742,3748] polyphosphoric acid,[3743] or (in lower yield) pyrophosphoric acid[3749,3750]; (b) by the reaction of 4-methyl-5-thiazoleethyl diphosphate with 4-amino-5-bromomethyl-2-methylpyrimidine[3687,3742] or 4-amino-5-chloromethyl-2-methylpyrimidine[3751]; or (c) by metathesis between 5-bromothiamine and silver pyrophosphate.[3752]

TTP may be prepared by phosphorylation of thiamine with metaphosphoric acid[3748,3753] or phosphoric acid-phosphorus pentoxide.[3736]

The phosphorylation of thiamine invariably produces mixtures of mono-, di-, tri-, and polyphosphates in proportions that vary with the phosphorus content of the reagent.[3743]

The solubility of the products in ethanol decreases in the order thiamine > TMP > TDP > TTP.[3738] Separation of the mixture into its components is usually accomplished by ion exchange chromatography.[3697,3743]

Conformation

The structures of TMP·H_3PO_4·$3H_2O$,[3754] TMP·HPF_6·$2H_2O$,[3755] and TDP·HCl[3714] have been determined by single crystal X-ray analysis. The geometry is similar to that of thiamine, with the thiazole and pyrimidine rings (both positively charged) almost at right angles to each other. The side chain is folded back over the thiazole ring in such a manner that the O(5γ) lone pair is directed toward the positively charged S(1). The inner phosphate group in TDP·HCl carries a negative charge. TDP itself crystallizes in two polymorphic forms: prismatic TDP·$4H_2O$ and tabular TDP·$4.5H_2O$.[3708,3713] Both show the same geometry as thiamine, but the pyrophosphate moiety extends away from the thiazole ring and each phosphate group in each polymorph carries a negative charge.

NMR studies on TDP in solution provide further evidence that the molecule can exist in both folded and extended conformations, depending on the metal ion used as the probe. Molecular orbital calculations on TDP in various states of ionization have been used to define the regions of preferred conformation.[3724] Quantum chemical point charge calculations have also been performed on TDP, its ylide, and some of its metal complexes.[3712]

Hydrolysis

Aqueous solutions of the phosphorylated thiamines undergo the opening of the thiazole ring above pH 7 that is a characteristic of thiamine. The reaction is reversible but if the solution is allowed to stand in contact with air, the thiol may be oxidized irreversibly to the disulfide.[3690] In aqueous solution, the phosphorylated thiamines, like thiamine itself, are sensitive to irradiation by UV[3756] or X-rays.[3757,3758]

TMP is resistant to acid hydrolysis even at 100 °C, but becomes progressively less stable as the pH increases. Hydrolysis at pH > 5 is complicated by breakdown of the thiamine to degradation products that give a negative thiochrome reaction.[3690,3759] Under mildly alkaline conditions, the hydrolysis is catalyzed by La(III) and Ce(III).[3760]

Hydrolysis of TDP yields thiamine, TMP, or both, depending on the reaction conditions. Hydrolysis with dilute HCl or H_2SO_4 at 100 °C is a useful method for the preparation of TMP (p 443). Under alkaline conditions, the products are thiamine and pyrophosphoric acid.[3745] The hydrolysis is catalyzed by La(III) and Ce(III)[3760]; by Th(III), Zr(IV), Hf(IV), and Ti(IV)[3015]; and by Cu(II).[3761] For measurements of the kinetics under a variety of conditions, see refs. 3690,3762–3764. At all temperatures, TDP is most stable to hydrolysis at pH 6.0.[3763]

Hydrolysis of TTP at 100 °C gives mixtures of thiamine, TMP, and TDP, the ratio of pyrophosphate to phosphate increasing with pH.[3745] Pure TTP is most shelf-stable as the dilithium salt.[3765]

The first-order rate constants *k* for the hydrolysis of TMP, TDP, and TTP at 100 °C are given in Table 7-1.[3745]

Table 7-1 First-Order Rate Constants *k* (min^{-1})

Compound	pH 1.0	pH 7.2	pH 8.7
TMP	0.002	0.009	0.081
TDP	0.033	0.014	0.025
TTP	0.130	0.012	0.008

Phosphorylation

Phosphorylation of TMP or TDP with phosphoric acid in the presence of acetic anhydride or DCC invariably gave mixtures of TMP, TDP, and TTP and in some cases, thiamine polyphosphate.[3710,3766] With acetic anhydride alone at 0 °C, the product (not characterized) was the mixed acetic-phosphoric or acetic-pyrophosphoric anhydride.[3767] The reaction of TMP with adenosine (or inosine)-5′-phosphate and DCC in pyridine/DMF gave the mixed thiamine-adenosine (or inosine) diphosphate.[3768] For the reactions of thiamine, TMP, and TDP with other phosphorylating agents, see ref. 3766.

Complex Formation

Divalent metal ions such as Mg(II) or Ni(II) react with TMP and TDP to form 1:1 complexes. The structures of these complexes are important in relation to the mechanisms of the interaction of cocarboxylase with enzymes, several of which require Mg(II) as a cofactor. ^{1}H NMR data suggest that the metal ions in the TMP complexes are coordinated solely to the phosphate group,[3769,3770] but a single crystal X-ray analysis of the Rh(II) complex TMP·$Rh(OAc)_2$·3/4H_2O showed that the TMP ligand is coordinated to the two axial positions of the $[Rh(OAc)_2]_2$ cage through the pyrimidine nitrogen N(1′).[3771] The conformation of the TMP ligand in the Rh(II) complex is folded like that of TMP·H_3PO_4 (p 444).

In most of the complexes with TDP, the metal is bonded exclusively to the pyrophosphate group. The conformation may be folded or unfolded, depending apparently on the extent of further interaction between the metal ion and the pyrimidine nitrogen N(1′). Thus, in the series Ni(II) ≫ Mn(II) > Mg(II), NMR studies indicate that in solution, the TDP is folded, partly folded, and unfolded, respectively.[3737,3769,3772,3773, but see ref. 3725] A single crystal X-ray analysis on the ternary complex TDP·phen·$Cu(NO_3)_2$·2H_2O revealed a partly folded but on the whole rather extended conformation more akin to that of neutral TDP than to that of TDP·HCl.[3774]

The thiamine-binding proteins in seeds increase in amount during seed maturation and decrease during germination. For this form of complexing, which is beyond the scope of this book, see Kozik and Rapala-Kozik.[3775]

Extraction, Separation, and Analysis

The vitamin B_1 content of plant tissues is usually measured by methods developed for animal tissues, for which a preliminary hydrolysis is necessary to liberate the thiamine from its protein- or phosphate-bound forms. Several methods have been developed in recent years for the separation and quantitative analysis of mixtures of thiamine and its mono-, di-, tri-, and polyphosphate esters. The best of these are paper chromatography (PC), ion exchange chromatography (IEC), and high-performance liquid chromatography (HPLC). The other methods described later are suitable if only TDP and total thiamine are required. For periodic reviews of these methods and their applications, see the sections on thiamine in *Methods in Enzymology*.[3776–3778]

In most cases, the final analysis is carried out by the thiochrome method after enzymatic hydrolysis to convert the thiamine phosphates in each fraction to free thiamine. The thiamine is then oxidized with potassium ferricyanide in NaOH solution and the fluorescence of the resulting thiochrome is measured in a fluorimeter set for excitation at 339 nm and emission at 432 nm. Two other rarely used methods are the Roussin salt fluorometric method described by Teysseire in method 2[3779] and the azo colorimetric method employed by Yusa in method 3b.[3780]

(1) Chromatographic Methods

(A) ION EXCHANGE CHROMATOGRAPHY (IEC)

This was the first method that enabled all four compounds—thiamine, TMP, TDP, and TTP—to be determined simultaneously on a single plant sample.

> *A weighed sample containing about 15-30 μg of total T [thiamine] was extracted with cold 5% trichloroacetic acid solution, by homogenization, the extraction being twice repeated and the extracts centrifuged. The combined extracts, adjusted to pH 6.7-6.9, were passed through a column made up of a mixture of cholesteryl stearate-treated charcoal and cellulose powder, which retained T and its esters. After a washing with H_2O, T compounds were quantitatively eluted with aqueous propanol (9:1). After concentration to a small volume in a rotating evaporator at 37°C, the eluate, acidified to pH 4-4.5, was passed through a column of Dowex 1 resin, X8, 200-400 mesh (acetate form), which retained TDP and TTP and allowed T and TMP to pass through. T and TMP were then determined fluorometrically (thiochrome method) in the solution before and after incubation with acid phosphatase. TDP and TTP were subsequently eluted from the resin column with 0.02*M *and 1.0*M *acetate buffer solution, respectively, and determined in the same manner after incubation with acid phosphatase.*
>
> Rindi, G.; de Giuseppe, L., "Content of Thiamine and its Mono-, Di-, and Triphosphoric Esters in Different Biological Materials", *Int. Z. Vitaminforsch.* **1961**, *31*, 321–326,[3685] reprinted by permission of Verlag Hans Huber, © 1961.

(B) PAPER CHROMATOGRAPHY (PC)

The thiamine compounds were extracted with trichloroacetic acid and adsorbed on charcoal as described by Rindi and de Giuseppe.[3685] The eluate from the charcoal was treated with 1 mL of 1*M* acetate buffer (pH 5.0), concentrated *in vacuo* to about 1 mL, separated by ascending paper chromatography using *n*-propanol/water/0.5*M* acetate buffer (pH 5.0) (7:2:1), and detected by the thiochrome method.[3780]

(C) HIGH-PERFORMANCE LIQUID CHROMATOGRAPHY (HPLC)

The thiamine compounds were extracted with 5% sulfosalicylic acid in the presence of amprolium as an internal standard. The eluate was passed through a column packed with Bio-Rad AG2-X8 anion-exchange resin to remove the sulfosalicylic acid, which fluoresces strongly and would interfere with the thiochrome detection. The solution was then injected into a Perkin-Elmer C_{18} reversed-phase column and separated into its components by isocratic elution with 0.1*M* sodium phosphate buffer, using postcolumn conversion of the thiamines to their thiochrome derivatives. The thiamines eluted in the order TDP > TMP > thiamine. TTP was not found in any of the samples that were analyzed.[3781] The HPLC method was also employed in a study of TDP binding by seed proteins.[3775]

(D) GEL FILTRATION CHROMATOGRAPHY (GFC)

A 0.2-mL aliquot of the sample (wheat flour extract or orange juice) was transferred to a 10 × 600-mm Sephadex G-25 column and eluted with 0.1% $NaHSO_4$. The fraction (12 mL) containing the thiamines was analyzed for thiamine and TDP by the thiochrome method before and after extraction with amyl alcohol.[3729] This method does not distinguish between TDP and other amyl alcohol-insoluble thiamine phosphates.

(2) Teysseire Method

The tissue, softened if necessary by soaking with three times its weight of water, was ground in a mortar with 40-50 g of anhydrous sodium sulfate, filtered, and rinsed three times with saturated ammonium sulfate. The filtrate was treated with 120-150 mg of $Na[Fe_4(NO)_7S_3]$ (Roussin's salt) and a few milliliters of 20% sodium benzoate solution and allowed to stand until the precipitation of salts was complete (15 h). The precipitate was collected on a filter, washed with water, and dried at 50 °C. The Roussin salt of thiamine was extracted from the precipitate with methylal and the Roussin salt of TDP with 95% ethanol. The salts were then analyzed separately for thiamine and TDP by a method developed by the authors[3782] based on the salts' absorption at 340 nm in the presence of strong alkali.[3779]

(3) Enzymatic Methods

These methods make use of the enzymatic reactions shown in Table 7-2.

Table 7-2 Enzymatic Reactions

Method	Reaction	Enzyme
3a	Pyruvate → acetaldehyde + CO_2	Pyruvate decarboxylase [E.C. 4.1.1.1]
3b	TDP + $2H_2O$ → thiamine + 2 H_3PO_4	Acid phosphatase [E.C. 3.1.3.2]

(A) MANOMETRIC METHOD (MANOM.)

This method is based on the restoration of the activity of pyruvate decarboxylase (Table 7-2, reaction 3a) when TDP is reunited with the enzyme in acid-washed Baker's yeast. Thiamine itself has a stimulating effect on this reaction and can also be measured by this method, but this was not done for any of the studies cited in Table 7-3. The following procedure is typical:

> *The well ground plant tissue was extracted by boiling for 5 minutes with 3 to 5 times their weight of pH 6.239 Sõrensen buffer. A 2 mL aliquot of the clear filtrate was treated with 1 mL of dry yeast that had been washed practically free of cocarboxylase, and incubated at 35°C in a Warburg apparatus with 0.5 mL of pyruvate/$MgCl_2$ solution as the substrate. The amount of CO_2 evolved in 60 min. was measured. Under these conditions, synthetic cocarboxylase liberated 56 mm^3 CO_2, equivalent to 5 μg vitamin B_1.*
>
> Tauber, H., "Co-Carboxylase (Vitamin B_1 Pyrophosphate) Content of Plants", *Proc. Soc. Exp. Biol. Med.* **1937**, *37*(3), 541–543,[3684] reprinted by permission of the Royal Society of Medicine Press, UK, © 1937.

(B) PHOSPHATASE METHOD

Incubation of a plant extract with an acid phosphatase results in the quantitative conversion of all the thiamine present to the free form (Table 7-2, reaction 3b). Some phosphatases that have been employed for this purpose are Clarase,[3801] takadiastase,[3780,3796] and extracts of beef kidney,[3797] yeast,[3792] or wheat.[3790,3801] The concentration of phosphorylated thiamine is estimated by the difference between thiamine assays before and after hydrolysis.

Table 7-3 Vitamin B1 in Plants*

No.	Plant Part	Method	Thiamine	TMP	TDP	TTP	References
Aesculus turbinata Blume (Japanese Horse Chestnut)							
1	Bud[a]	PC	+	−	±	+	3783
Allium cepa L. (Onion)							
1	Bulb[b]	Manom.			120		3684
1	Bulb	Manom.	0.40		+		3784
Arachis hypogaea L. (Peanut)							
1	Seed cotyledon	PC	+	+	+ (ratio 33:10:1)		3785
Avena sativa L. (Oats)							
2	Seed	IEC	3.34-4.48	0	0		3786
2	Seed	Phosphatase	3.43-4.48		2.08-2.11		3786
1	Seed embryo[a]	PC	+	−	±	+	3783
1	Seedling[c]	Manom.			0.40		3787
Brassica oleracea L., ssp. *capitata* L. (Cabbage)							
1	Leaf, bleached[b]	Manom.			80		3684
Brassica rapa L. (Turnip Rape)							
1	Leaf[b]	Manom.			93		3684
Camellia japonica L. (Camellia)							
1	Pollen[d]	Manom.			57		3430

Table 7-3 Vitamin B1 in Plants*—cont'd

No.	Plant Part	Method	Thiamine	TMP	TDP	TTP	References
Camellia sasanqua Thunb. (Sasanqua)							
1	Pollen[d]	Manom.			51		3430
Camellia sinensis L. (Tea)							
1	Pollen[d]	Manom.			41		3430
Capsicum annuum L. (Green Pepper)							
1	Fruit[b]	Manom.			86		3684
Carya illinoinensis (Wang.) K. Koch (Pecan)							
1	Nut[b]	Manom.			170		3684
Castanea spp. (Chestnut)							
1	Nut	Manom.	2.83		0.18		3784
Citrus natsudaidai Hayata (Sour Orange)							
1	Seed embryo[a]	PC	–	–	±	±	3783
Citrus sinensis (L.) Osbeck (Orange)							
1	Seed[b]	Manom.			250		3684
1	Fruit, juice[b]	Manom.			80		3684
1	–, juice	GFC	0.630		0.764		3729
1	–, peel[b]	Manom.			125		3684
Coffea arabica L. (Coffee)							
1	Seed	Phosphatase	11.68		Low		3788
Colocasia esculenta (L.) Schott. (Taro)							
1	Leaf	Manom.	0.90		0.11		3784
Cucurbita pepo L. (Pumpkin)							
1	Germ, axis[a]	PC	–	±	+	+	3783
1	–, cotyledon[a]	PC	+	–	–	–	3783
Cucurbita spp. (Squash)							
1	Fruit[b]	Manom.			107		3684
Cycas revoluta Thunb. (Sago Palm)							
1	Seed	HPLC			+[e]		3775
Fagopyrum esculentum Moench (Buckwheat)							
1	Seed	Manom.	2.04		1.89		3784
1	Whole seedling	PC	–	–	+	+	3783
Glycine max (L.) Merr. (Soybean)							
1	Seed	Phosphatase			19.1		3789
1	Germ, axis	PC	–	–	±	+	3783
1	–, cotyledon	PC	+	–	+	+	3783
1	Seedling[a]	PC			+	–	3780
1	Sprout	Phosphatase			31.8		3789
1	Meal	Phosphatase	4.4		6.6		3790
Helianthus tuberosus L. (Jerusalem Artichokie)							
1	Excised sprout[a]	PC	+	–	±	+	3783
Hordeum vulgare L. (Barley)							
2	Seed	IEC	2.07-2.85	0	0		3786
2	Seed	Phosphatase	2.16-2.54	1.72-1.91	563		
1	Seedling[f]	Manom.			0.24		3791

*μg/g fresh weight basis, unless stated otherwise; [a]Levels: +, clearly detectable; ±, ambiguous; –, trace or not detectable; [b]μL CO_2/g; [c]μg/plant part; [d]Relative to yeast cocarboxylase = 100; [e]Protein binding; [f]mg CO_2/g; [g]nmol/g; [h]Dry weight basis.

Continued

Table 7-3 Vitamin B1 in Plants*—cont'd

No.	Plant Part	Method	Thiamine	TMP	TDP	TTP	References
Larix leptolepis Gord. (Japanese Larch)							
1	Bud[a]	PC	–	±	±	±	3783
Lilium auratum Lindl. (Golden-Rayed Lily)							
1	Pollen[d]	Manom.			39		3430
Lilium longiflorum Thunb. (Trumpet Lily)							
1	Pollen[d]	Manom.			45		3430
Lilium philippinense Bak. (Philippine Lily)							
1	Pollen[d]	Manom.			45		3430
Lilium spp. (Lily)							
1	Bulb	Manom.	2.34		0.28		3784
Lycopersicon esculentum Mill. (Tomato)							
1	Fruit	Manom.	0.70		+		3784
Oryza sativa L. (Rice)							
1	Seed, rough	IEC	1.53	0.0	0.053	0.017	3685
1	Seed, embryo[a]	PC	+	–	±	+	3783
1	–, endosperm[a]	PC	+	–	–	–	3783
1	Seed, unmilled	Manom.	2.70		1.15		3784
1	Polish	Phosphatase	203		42		3792
1	Germ	IEC	36.00	0.0	0		3685
1	Bran	IEC	2.440	0.0	0.0	0.0	3685
Paeonia albiflora Pall. (Chinese Peony)							
1	Pollen[d]	Manom.			33		3430
Papaver orientale L. (Oriental Poppy)							
1	Pollen[d]	Manom.			15		3430
Papaver somniferum L. (Opium Poppy)							
1	Pollen[d]	Manom.			53		3430
Persea americana Mill. (Avocado)							
1	Fruit	Phosphatase			1.046		3793
Pharbitis nil (L). Choisy (Morning Glory)							
1	Seed embryo[a]	PC	–	–	±	±	3783
Phaseolus vulgaris L. (String Beans)							
1	Bean	Manom.	2.08		0.11		3784
1	Bean, cotyledon	Phosphatase	3.45		0.72		3794
1	–, seed coat	Phosphatase	0.53		0.20		3794
1	–, decotylized embryo	Phosphatase	12.52		0.96		3794
1	–, first leaf	Phosphatase	6.17		2.56		3794
1	–, hypocotyl	Phosphatase	9.51		0.94		3794
1	–, transitional region	Phosphatase	26.15		4.01		3794
1	–, radicle	Phosphatase	36.13		9.93		3794
Picea abies (L.) Karst. (Spruce)							
1	Seed	HPLC			+[e]		3775
Pinus densiflora Sieb. and Zucc. (Japanese Red Pine)							
1	Whole seedling[a]	PC	+	–	±	+	3783
1	Pollen[d]	Manom.			6		3430
Pisum sativum L. (Peas)							
1	Seed	Phosphatase	1.0		1.0		3780

Table 7-3 Vitamin B1 in Plants*—cont'd

No.	Plant Part	Method	Thiamine	TMP	TDP	TTP	References
1	Germ, axis	Phosphatase	0		3.25		3780
1	Germ, axis[a]	PC	–	–	±	+	3783
1	–, cotyledon[a]	PC	+	–	+	+	3783
1	Seedling[a]	PC			–	+	3780
1	Seedling	Phosphatase	57.3		23.4		3795
1	Stem and leaf	Phosphatase	25.4		11.4		3795
Platanus occidentalis L. (Eastern Sycamore)							
1	Bud[a]	PC	–	–	±	±	3783
Prunus armeniaca L. (Apricot)							
1	Fruit	Teysseire	0.60		0		3779
Prunus dulcis (Mill.) D. A. Webb (Almond)							
1	Seed, shelled[g]	HPLC	5.1	0.35	0.29		3781
Pseudosasa japonica Mak. (Arrow Bamboo)							
1	Bud[a]	PC	+	+	+	+	3783
Pyrus communis L. (Pear)							
1	Fruit	Phosphatase			0.703		3793
Raphanus sativus L. (Radish)							
1	Whole seedling[a]	PC	+	–	+	+	3783
Sagittaria sagittifolia L. (Arrowhead)							
1	Seed	Phosphatase	0.84		0.40		3784
Solanum tuberosum L. (Potato)							
1	Tuber	Phosphatase	0.55		+		3784
Spinacia oleracea L. (Spinach)							
1	Seed	HPLC			+[c]		3775
1	Leaf[b]	Manom.			10		3684
1	Leaf	Manom.	1.30		+		3784
Triticum aestivum L. (Wheat)							
2	Grain	Phosphatase	1.28-1.65		1.61-2.42		3786
2	Grain	IEC	1.16-1.68	0	0		3786
1	Grain	Teysseire	1.00		0		3779
75	Flour	Phosphatase	1.6-7.5		0		3796
1	Flour	GFC	0.222		0		3729
1	Seed, embryo[a]	PC	+	–	+	+	3783
1	–, endosperm[a]	PC	+	–	–	–	3783
1	Germ	Phosphatase	20-25		0.5		3797
1	Germ	IEC	4.56	0.0	0.33	0.0	3685
1	Bran	Teysseire	4.50		0		3779
1	Seedling	Teysseire	1.00		0.50		3779
1	Seedling	Phosphatase	3.7		3.7		3798
Vitis coignetiae Pull. (Crimson Glory Vine)							
1	Bud[a]	PC	–	–	–	±	3783
Xanthium strumarium L. (Cocklebur)							
1	Seed embryo[a]	PC	+	+	+	+	3783

*μg/g fresh weight basis, unless stated otherwise; [a]Levels: +, clearly detectable; ±, ambiguous; –, trace or not detectable; [b]μL CO_2/g; [c]μg/plant part; [d]Relative to yeast cocarboxylase = 100; [e]Protein binding; [f]mg CO_2/g; [g]nmol/g; [h]Dry weight basis.

Continued

Table 7-3 Vitamin B1 in Plants*—cont'd

No.	Plant Part	Method	Thiamine	TMP	TDP	TTP	References
Zea mays L. (Corn)							
1	Seed	PC			–	–	3780
1	Seed	HPLC			+[e]	+	3775
1	–, embryo[a]	PC	+	–	±	+	3783
1	Seedling[a]	PC			–	+	3780
Vegetables							
1	Edible parts[b]	–		+	+	+	3799
Collected by honey bees from various floral sources							
1	Pollen[g,h]	HPLC	11-40	1-6	1-13		3800

*µg/g fresh weight basis, unless stated otherwise.
[a]Levels: +, clearly detectable; ±, ambiguous; –, trace or not detectable.
[b]µL CO_2/g.
[c]µg/plant part.
[d]Relative to yeast cocarboxylase = 100.
[e]Protein binding.
[f]mg CO_2/g.
[g]nmol/g.
[h]Dry weight basis.

Vitamin B_6

The vitamin B_6 complex comprises pyridoxine (PN, also called pyridoxol), pyridoxal (PL), and pyridoxamine (PM), together with the monophosphate esters pyridoxal 5′-phosphate (PLP, codecarboxylase) and pyridoxamine 5′-phosphate (PMP). The missing member of this group, pyridoxine 5′-phosphate (PNP), has no vitamin B_6 activity.

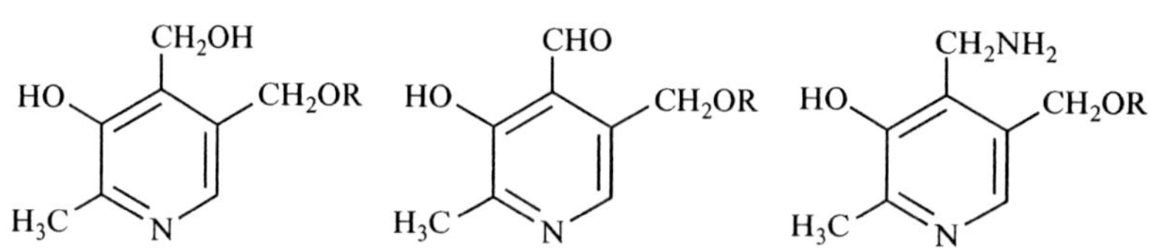

PN, R = H
PNP, R = P(O)(OH)$_2$

PL, R = H
PLP, R = P(O)(OH)$_2$

PM, R = H
PMP, R = P(O)(OH)$_2$

Pyridoxine 5′-Phosphate (PNP), $C_8H_{12}NO_6P$ [R.N. 447-05-2]; Beilstein **21** III/IV 2516. *3,4-Pyridinedimethanol, 5-Hydroxy-6-methyl-, α³-(Dihydrogen Phosphate).*

Pyridoxal 5′-Phosphate (PLP), $C_8H_{10}NO_6P$ [R.N. 54-47-7]; Beilstein **21** IV 6420. *4-Pyridinecarboxaldehyde, 3-Hydroxy-2-methyl-5-[(phosphonooxy)methyl]-.*

Pyridoxamine 5′-Phosphate (PMP), $C_8H_{13}N_2O_5P$ [R.N. 529-96-4]; Beilstein **22** IV 6065. *3-Pyridinemethanol, 4-(Aminomethyl)-5-hydroxy-6-methyl-, α-(Dihydrogen Phosphate).*

Occurrence

Pyridoxal 5′-phosphate (PLP) was first identified in peas and cabbage by Gale and Epps[3802] in 1944, pyridoxamine 5′-phosphate (PMP) in Cerophyl, a grass product, by Rabinowitz and Snell[3803] in 1947, and pyridoxine 5′-phosphate (PNP) in rice bran by Yasumoto *et al.*[3804] in 1976. PLP is a component of potato phosphorylase [E.C. 2.4.1.1], which contains one PLP group per subunit.[3805] Removal of the PLP by incubation with acidic hydroxylamine leaves the apoenzyme catalytically inactive.[3806] It is also a component of other plant enzymes such as those of peas (starch phosphorylase, E.C. 2.4.1.1),[3807] blue lupine (β-cyanoalanine synthetase, E.C. 4.4.1.9, one mol PLP/mol),[3808] rape leaf (cysteine synthetase, E.C. 4.2.99.8, two mols PLP per mol),[3809] *Acacia farnesiana* seedling (*S*-alkyl-L-cysteine lyase, E.C. 4.4.1.6, one mol PLP per mol),[3810] and grass pea seedling (arginine decarboxylase, E.C. 4.1.1.19).[3811] More recent studies show that PNP occurs to some extent in the form of glucosides.[3804,3812]

The metabolic pathways for the biosynthesis of the B_6 vitamins in plants are unknown, but they may be identical to those for microbial organisms such as *Escherichia coli*, which are outlined in Fig. 7-2. In step 1, PNP is formed by the coupling of 1-deoxy-D-xylulose 5-phosphate (**3**, dXu5P) and 4-hydroxy-1-threonine 4-phosphate (**4**)[3813]; dXu5P is also an intermediate in the biosynthesis of TDP (Fig. 7-1). In step 2, PNP is oxidized to PLP, and in step 3, PLP undergoes reversible transamination giving PMP. All three vitamin B_6 phosphates are hydrolyzed to the phosphate-free forms by phosphatase (step 4), and all three of the latter are phosphorylated by ATP in the presence of the same enzyme, pyridoxal kinase (step 5).

Physical Properties

Pyridoxine 5′-Phosphate

PNP [R.N. 447-05-2]: White needles, mp 208-210 °C,[3814] soluble in water, insoluble in alcohol and ether; pK_a (potentiometric and UV).[3815] Spectra: UV,[3816] luminescence,[3817] ^{1}H NMR (fig.).[3778] Chromatography: PC,[3816] TLC,[41] IEC,[3816,3818] HPLC,[3819] GC,[3820] Sephadex gel.[3821] Electrophoresis: paper,[3822] thin layer,[41,3823] HPCE.[3824]

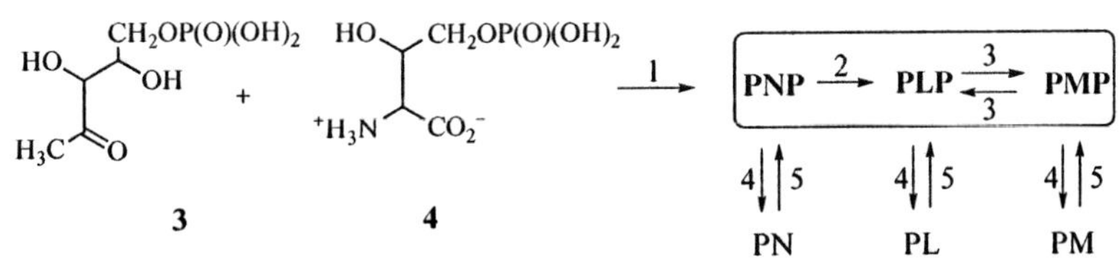

3 **4**

FIGURE 7-2 Vitamin B_6 Phosphates. (1) PNP synthase. (2) Pyridoxine 4-dehydrogenase [E.C. 1.1.1.65], NADP. (3) PMP oxidase [E.C. 1.4.3.5], FMN, O_2, H_2O. (4) Phosphatase [E.C. 3.1.3.2]. (5) Pyridoxal kinase [E.C. 2.7.1.35], ATP.

Pyridoxal 5′-Phosphate

PLP·H_2O [R.N. 54-47-7]: Microscopic lemon-yellow needles[3825] (photo[3826]); mp 139-143 °C dec. or 143 °C reported for the commercial product[3827,3828]; hardly soluble in water, methanol, ethanol, and acetone; insoluble in ether, chloroform, and benzene, but easily soluble in formic acid and DMF[3826]; sensitive to light.[3814] pK_a: potentiometric,[3829] NMR.[3830,3831] Spectra: IR[3829] (fig.[3832]), UV,[3833,3834] fluorescence,[3835] phosphorescence,[3836] ^{1}H NMR (fig.),[3837] ^{13}C NMR,[3831,3838] ^{31}P NMR,[3830] MS (field desorption).[3839] X-Ray structure.[3840] Chromatography: PC,[3816] TLC[41,3841] (two-dimensional[3842]), IEC (resin,[3818] cellulose[3841]), HPLC,[3819,3843] GC,[3820] Sephadex gel.[3157] Electrophoresis: paper,[32,3844] starch column,[3844] thin layer,[41,3823] HPCE.[3824] Polarographic reduction.[3828,3845]

$CaC_8H_8NO_6P \cdot 3H_2O$ [R.N. 5913-70-2]: Lemon-yellow powder, slightly soluble in water.[3846]

$BaC_8H_8NO_6P \cdot 3H_2O$: Yellow solid, soluble in water and ethanol, insoluble in ether.[3847]

$C_{13}H_9N$·PLP·H_2O (acridine salt): Fine lemon-yellow needles, readily soluble in water and ethanol, insoluble in ether.[3847]

Pyridoxamine 5′-Phosphate

PMP [R.N. 529-96-4]: Fine, very hygroscopic needles.[3848]

PMP·$2H_2O$: White rods (photo)[3826] or rhombic plates[3848]; hardly soluble in water, alcohol, and pyridine and insoluble in ether, chloroform, and benzene[3826]; photosensitive.[3849] pK_a: potentiometric,[3850] NMR.[3830,3848] Spectra: UV,[3816,3850] fluorescence,[3849,3851] phosphorescence,[3852] ^{1}H NMR (fig.),[3837] ^{13}C NMR,[3831,3838] ^{31}P NMR,[3830] MS (field desorption).[3839] Chromatography: PC,[3816] TLC[41,3841] (two-dimensional[3842]), IEC (resin,[3818] cellulose[3841]), HPLC,[3819,3843] Sephadex gel.[3157] Electrophoresis: paper,[32,3844] thin layer,[41,3823] starch column,[3844] HPCE.[3824] Polarographic reduction.[3828]

PMP·HCl·H_2O: Colorless crystals, mp 222-224 °C dec.[3853]; IR (fig.)[3832]; X-ray structure.[3854]

$CaC_8H_{11}N_2O_5P \cdot 4H_2O$: Solid. UV.[3853]

Synthesis

Pyridoxine 5′-phosphate (PNP) may be prepared (a) by phosphorylation of pyridoxine with phosphorus oxychloride[3833] or phosphoric acid-phosphorus pentoxide,[3822] (b) by phosphorylation of isopropylidenepyridoxine with phosphoric acid-phosphorus pentoxide[3826] or 2-cyanoethyl phosphate,[3778] (c) by reduction of PLP with sodium borohydride,[3814,3815] (d) by deamination of PMP with sodium nitrite,[3816] or (e) by oxidation of pyridoxine 5′-phosphite with potassium permanganate.[3855]

Pyridoxal 5′-phosphate (PLP) may be prepared (a) by oxidation of PNP[3833,3856] or PMP[3825] with manganese dioxide, (b) by transamination of PMP with pyruvic acid in the presence of a Cu(II) salt,[3857] or (c) by phosphorylation of an aldehyde-protected PLP with phosphoric acid-phosphorus pentoxide, polyphosphoric acid, or metaphosphoric acid, followed by removal of the protecting group (hydrazone,[3846] oxime, Schiff base,[3858,3859] enamine,[3860] acetal, and the like). Phosphorylation of pyridoxal with phosphorus oxychloride gives poor yields of PLP even when the aldehyde group is protected.[3833]

Pyridoxamine 5′-phosphate (PMP) may be prepared (a) by transamination of PLP with glutamic acid[3803] or (b) by phosphorylation of pyridoxamine with phosphoric acid-phosphorus pentoxide,[3848] metaphosphoric acid,[3853] or (in low yield) phosphorus oxychloride.[3861] The initial product of phosphorylation with metaphosphoric acid is the 5′-triphosphate, which yields PMP upon hydrolysis at 100 °C with water (2 h) or 1*N* HCl (10 min).[3853]

Hydrolysis

PMP is more resistant to hydrolysis than PNP or PLP. Less than 0.4% of its phosphorus was detected as P_i after 54 days' storage in aqueous solution at 7 °C, compared to 2.6% for PNP and 2.1% for PLP. Hydrolysis with 1*N* HCl at 100 °C for 15.3 h gave 23% P_i compared to 35% for PNP and 73% for PLP; hydrolysis with 1*N* NaOH under the same conditions, however, gave 10% P_i versus 11% (0%[3815]) for PNP and 6% for PLP. Complete hydrolysis was achieved with 6*N* H_2SO_4 after 55 h (30 h for PLP) at 100 °C.[3816]

The hydrolysis of PLP is catalyzed by acids[3816] and metal ions.[3678] Unlike most other phosphate esters, PLP is most stable at pH 2; the first-order rate constant K_D is 1.9×10^{-4} (pH 1.92, 87 °C)[3862] or 3.96×10^{-4} (pH 2.05, 98 °C).[3827]

The phosphate groups were cleaved during the preparation of 5′-deoxypyridoxine and 5′-deoxypyridoxamine from PLP and PMP by catalytic hydrogenolysis with Pd/C[3863,3864] but not from PLP with sodium borohydride.[3814,3815]

Complex Formation

Bivalent and trivalent metal ions such as Cu(II), Zn(II), or Dy(III) react with PLP and PMP forming 1:1 complexes that are insoluble, intractable, and probably polymeric.[3832,3865] Such complexes are important in relation to the metal ion catalysis of transamination reactions between the vitamin B_6 compounds and amino acids *in vitro* and *in vivo*.

Extraction, Separation, and Analysis

(1) Ion Exchange chromatography (IEC) Method

de Obieta and Munoz[3866] used this method to separate the vitamin B_6 components of germinating *Pinus pinea* seeds, using the azine method of Soda *et al.*[3867] to assay for PLP.

(2) High-Performance Liquid Chromatography (HPLC) Method

This method is now the method of choice for the analysis of the B_s vitamins because they can be separated and quantitated in a single run, usually carried out in the reversed-phase mode with HPLC column packings such as μ-Bondapak C_{18}[3868] or TSK gel

ODS-120A,[3869,3870] or, in one instance, in the ion-pairing mode with Beckman Ultrasphere IP.[3871] The following procedure is typical:

*Solid samples (2 g) were mixed with 9 mL of 5 % (w/v) SSA [sulfosalicylic acid] and 1 mL of a solution of 4-dPN [4′-deoxypyridine] (typically 120 nmol/mL in 5 % SSA) and then homogenized with a Polytron homogenizer . . . for 45 s . . . Ten milliliters of methylene chloride was added, and the mixture was blended for 20 s . . . The homogenates were centrifuged at 9000 g for 15 min. at 2°C, after which an aliquot of the aqueous layer was removed . . . . Preparative anion-exchange chromatography was performed [on the aliquot] with a column packed with Bio-Rad AG2-X8 . . . to remove SSA and purify the sample extracts. The biologically active B_6 vitamers, 4-dPN, and PN-glucoside eluted as a single fluorescing peak near the void volume (retention time approximately 4 min). The eluate either was analyzed directly or was subjected to enzymatic hydrolysis prior to HPLC analysis . . . . The HPLC analytical system . . . consisted of two solvent metering pumps, [a gradient controller, injection valve, and a fluorometric detector]. Fluorometric detection was enhanced by using the postcolumn phosphate-bisulfite reagent . . . . The separation was performed at ambient temperature by using a linear gradient . . . from 100 % solvent A1 [0.033*M *K phosphate and 8 m*M *octanesulfonic acid, pH 2.2] to 100 % solvent A2 [solvent A1 + 2.5 % (v/v) 2-propanol, pH 2.2] in 12 min., followed by a programmed switch to 100 % solvent B [0.033*M *K phosphate and 6.5 % 2-propanol, pH 2.2] 15 min after injection . . . . Quantitation was done relative to the response of the internal standard 4′-deoxypyridine.*

Reprinted with permission from Gregory, J. F., III; Ink, S. L. J., "Identification and Quantification of Pyridoxine β-Glucoside as a Major Form of Vitamin B_6 in Plant-Derived Foods", *J. Agric. Food Chem.* **1987**, *35*(1), 76–82,[3872] © 1987 American Chemical Society.

(3) *Enzymatic Methods*

(A) TRYPTOPHANASE METHOD

In this method, PLP is assayed with the apoenzyme of *E. coli* tryptophanase [E.C. 4.1.99.1], the amount of indole produced from added tryptophan being measured colorimetrically by Ehrlich's aldehyde reaction. The apoenzyme of Baker's yeast or pig heart aspartate aminotransferase [E.C. 2.6.1.1] proved to be unsuitable for this purpose.[3873]

A weighed sample of plant tissue (100 mg) was dissolved in the cold in a porcelain mortar or glass blender with 0.3 ml of freshly prepared 5 % HPO_3; 2.6 ml of distilled water was added to the mixture, and after supplementary grinding of the tissue, the mixture was heated for 5 min at 80°. Then the mixture was chilled, centrifuged, and the supernatant liquid used to determine PALP. No more than 0.3 ml of the HPO_3 extract of the plant tissue was taken for analysis . . . . All the work associated with the production of HPO_3 extracts of plant tissue, with the exception of the stage of

heating, was conducted at 0-4°, the tissue shielded from the action of bright light. The PALP content was determined in the extracts . . . 1 h after their preparation.

Zavenyagina, T. N.; Bukin, Yu. V., "Content of Pyridoxal 5-Phosphate in Wheat and Pea Seed and Seedlings", *Biokhimiya* **1971**, *36*(1), 205–215 [Engl. 172–180],[3873] reprinted by permission of the Editor-in-Chief, Biochemistry (Moscow)/Biokhimiya, © 1971.

(B) PHOSPHATASE METHOD

Yasumoto *et al.*[3804] determined the vitamin B_6 content of cereals and other seeds by microbial assay with *Saccharomyces carlsbergensis* following hydrolysis with either takadiastase or acid. In most cases, digestion with takadiastase was sufficient, but part of the pyridoxine in rice bran, kaoliang, and trefoil was bound in a takadiastase-resistant form. In rice bran, about 4 μg/g of the bound pyridoxine was identified as pyridoxine β-glucoside and about 1 μg/g as PNP.

(4) Fluorometric Method

Zavenyagina and Bukin[3873] described a method in which the PLP content of peas and wheat was determined after oxidation to pyridoxa-4,5-lactone, a highly fluorescent substance, in the presence of oxygen in slightly alkaline solution. The protein-free extracts (pH 7.4) were incubated with KCN for 2.5 h at 25 °C, acidified to pH 3.5 with glacial acetic acid, and the fluorescence intensity measured at 420 nm with extinction at 313 nm. The results were comparable to those of the tryptophanase method.[3873]

Table 7-4 Vitamin B_6 Content of Plants*

No.	Plant part	Method	PNP	PLP	PMP	Total B_6	References
Asparagus officinalis L. (Asparagus)							
1	Stem	HPLC			0.144	0.183	3869
Avena sativa L. (Oats)							
1	Bran	HPLC		0.707	0	1.15	3871
Brassica oleracea L. ssp. *capitata* L. (Cabbage)							
1	Juice[a]	HPLC	0.03	0.27	0.05	1.21	3870
Brassica oleracea L. ssp. *italica* Plenck. (Broccoli)							
1	Raw head	HPLC		1.44	0.41	2.71	3872
1	Head	HPLC	0.11	1.15	0.38	2.73	3870
Citrus sinensis (L.) Osbeck (Orange)							
1	Fruit juice[a]	HPLC		n.d.	0.26	1.47	3872
1	Fruit juice[a]	HPLC	0.021	0.051	0.028	0.199	3869
Daucus carota L. (Carrot)							
1	Root	HPLC		0.53	0.18	4.42	3872
1	Root	HPLC	+	+		+	3812
Glycine max (L.) Merr.							
1	Flour	HPLC		0	0	5.13	3871

Continued

Table 7-4 Vitamin B_6 Content of Plants*—cont'd

No.	Plant part	Method	PNP	PLP	PMP	Total B_6	References
Helianthus tuberosus L. (Sunflower)							
1	Seed	HPLC	+	+		+	3812
1	Sprout	HPLC	+	+		+	3812
Malus domestica Borkh. (Apple)							
1	Fruit juice[a]	HPLC		0.018	0.006	0.074	3869
Medicago sativa L. (Alfalfa)							
1	Sprout	HPLC		0.48	0.45	8.10	3872
Musa paradisiaca L. (Banana)							
1	Fruit	HPLC		0.79	0.07	3.65	3872
1	Fruit	HPLC	0.022	1.29	0.040	3.273	3869
Oryza sativa L. (Rice)							
1	Bran	HPLC	0.097			2.107	3869
1	Bran	HPLC		1.80	0	12.10	3871
1	Bran	Enzym.	16.0	<0.5	<0.5	19.5	3804
Pennisetum spp.							
1	Seed	HPLC	+	+		+	3812
Phaseolus lunatus L. (Lima Bean)							
1	Bean	HPLC		5.61	0.22	6.99	3868
Phaseolus vulgaris L. (Bean)							
1	Bean, raw green	HPLC		1.03	0.31	2.69	3872
Pinus pinea L. (Italian Stone Pine)							
1	Seed, embryo	IEC		42.0		11.9	3866
1	–, endosperm	IEC		107.6		1.1	3866
Pisum sativum L. (Peas)							
1	Whole seed[b]	Enzym.		0.17			3873
1	Seedling, leaf[b]	Enzym.		1.39			3873
1	–, root[b]	Enzym.		0.25			3873
1	Leaf	Fluor.		+			3873
Seaweed							
1	–[c]	HPLC	+	+		+	3874
Secale cereale L. (Rye)							
1	Seed	HPLC	+	+		+	3812
Solanum tuberosum L. (Potato)							
1	Tuber	HPLC		1.09		12.05	3875
Triticum aestivum L. (Wheat)							
1	Whole seed[b]	Enzym.		0.19			3873
1	Whole seed	Fluor.		+			3873
1	Sprout, leaf[b]	Fluor. and variation in PLP content from base to tip		+			3873
1	–, leaf	Fluor.		+			3873
1	–, root	Fluor.		+			3873
1	–, root[b]	Enzym.		0.32			3873
1	Seed	HPLC	+	+		+	3812
1	Whole flour	IEC		0.26	0.27	4.30	3876
1	Flour	HPLC	0.031	0.18	0.022	0.289	3869
1	Bran	HPLC			0	7.93	3871

*μg/g fresh weight basis, unless stated otherwise.

[a]μg/mL.

[b]Dry weight basis.

[c]Not given in the abstract.

8

Distribution of Phosphorus Compounds in Plants

This chapter describes early efforts to develop a comprehensive fractionation scheme for determining the relative amounts of the major classes of phosphorus-containing compounds—lipids, sugars, phytic acid, nucleotides, nucleic acids, proteins, and P_i—in plants that would hopefully enable a balance sheet to be drawn up. Many such schemes have been proposed but few have achieved even limited acceptance. The earliest efforts were those of Ivanoff,[3877] Bernardini,[3878] and Zlataroff,[3879] who determined the distribution of P_i, P_{lipid}, P_{nucl}, and $P_{protein}$ in germinating vetch,[3877] rice,[3878] and chickpea,[3879] respectively. The culmination of these efforts appears in the last table of this chapter in which a balance sheet is drawn up for the distribution of individual phosphorus compounds in a single plant species, duckweed (*Spirodela oligorrhiza*), using modern methods of analysis.[327]

The principal methods can be broadly separated into four categories: selective precipitation methods, selective extraction methods, chromatographic methods, and ^{31}P NMR methods.

Selective Precipitation Methods

In the LePage and Umbreit method (Scheme 8-1), which focuses on the acid-soluble components, the plant tissue is extracted exhaustively with 10% trichloroacetic acid (TCA), and the extract is separated by fractional precipitation into three fractions: a

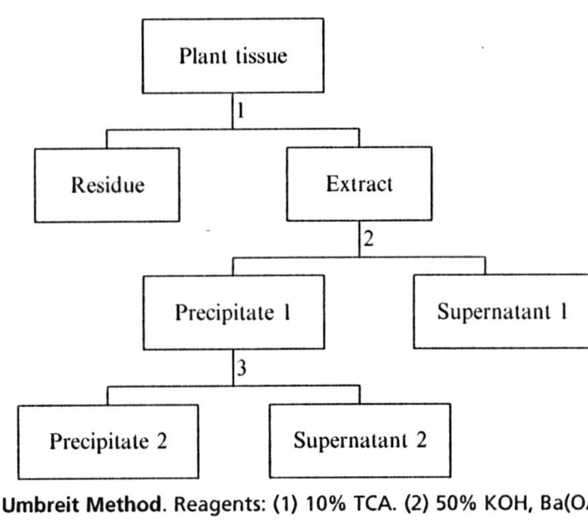

SCHEME 8-1 LePage and Umbreit Method. Reagents: (1) 10% TCA. (2) 50% KOH, $Ba(OAc)_2$, EtOH. (3) 2*N* HCl, $Ba(OAc)_2$, H_2SO_4.

Table 8-1 Selective Precipitation Methods*

No.	Plant part	P_i	P_{lipid}	P_{sugar}	$P_{phytate}$	$P_{nucleotide.}$	$P_{protein}$	P_{total}	References
		LePage and Umbreit Method							
Avena sativa L. (Oats)									
1	Seed embryo	34.2		8.7	0.69	10.2		5.240[a]	14
Beta vulgaris L. (Sugar Beet)									
1	Leaf		7.6		5.3			0.091	15
1	Root		5.9		5.1			0.057	15
Lycopersicum esculentum Mill. (Tomato)									
1	Flower ovary	30.2		10.0		49.8		1.060	18
Oryza sativa L. (Rice)									
1	Grain	Change during maturation 3-46 daf							3882
Solanum tuberosum L. (Potato)									
2	Tuber			31	68			0.063	315
		Klein Method							
Gossypium hirsutum L. (Cotton)									
1	Leaf	50.8	0	18.5		18.5	12.3	1.30	3883
1	Ovule, 10 days	48.1	10.0	23.8		14.0	4.0	0.420	3883
1	Seed kernel	8.6	8.9		74.9	5.0	2.7	0.935	3883
Lycopersicon esculentum Mill. (Tomato)									
1	Seedling hypocotyl	19.5	8.5			23.8	6.8	0.411	3881
		Other Methods							
Cucurbita pepo L. (Pumpkin)									
1	Leaf mitochondria	49.5	18.2	9.0		23.2		0.276	1633
Hordeum vulgare L. (Barley)									
1	Seed	5.8	11.1	10.3	60.0	12.8	0.2	0.253	996
1	–, flour		3.8			89.1	2.9	0.37	3884
	and change on germination 0-6 days[3885]								
	and ripening 19-48 dap[996]								
Linum usitatissimum L. (Flax)									
1	Sprout, 7 days	49.4		19.1	7.7			0.994	3886
1	Root	61.8		17.7	0			0.902	3886
Lupinus luteus L. (Yellow Lupine)									
1	Whole plant	88.3	4.0			4.8		0.516	791

Oryza sativa L. (Rice)									
1	Leaf (straw)	54.0	8.1		2.7	35.1		0.74	1052
1	Seed	5.9	8.3		2.8	83.0		5.84	1052
1	–, aleurone particle	1.7	0.1		83.6	0.3		11.61	1047
	and data for growth from germination to maturity[1052]								
Petroselinum crispum (Mill.) A.W. Hill (Parsley)									
1	Leaf	59.2	16.1			24.7		1.571	792
1	Root	64.9	13.0			22.1		1.043	792
	and change during growth, August 4 to November 11[792]								
Secale cereale L. (Rye)									
1	Seed	6.3	9.6	10.2	60.5	12.0	0.2	0.342	996
	and change on germination 0-6 days[3885] and ripening 19-48 dap[996]								
Trifolium pratense L. (Red Clover)									
1		76.7	7.2		0	16.0		0.223	791
Triticum aestivum L. (Wheat)									
1	Seed	4.6	9.1	8.8	67.2	11.3	0.1	0.331	996
1	Seed, ripe	10.1	5.0	6.8	57.9	20.2		0.73	3887
8	–, flour	5.3-15.1	11.6-13.1		26.3-42.6	12.4-17.4		0.092-0.140	759
	and change on germination 0-6 days[3885] and change on ripening 19-48 dap[996] or 0-36 daf[3887]								
Triticum durum Desf. (Durum Wheat)									
2	Seed flour	5.1-5.5	11.9-12.9		31.7-38.1	17.6-22.3		0.112-0.125	759
Zea mays L. (Corn)									
1	Seed flour	12.9	4.1		75.5	11.5		0.349	791
1	Root	52.5	0.44	15.7		31.3			3888
	and change during growth, August 2 to October 30[791]								

*P_{total} column: Percent P by weight of plant tissue, fresh weight basis. Other columns: Percent of P_{total}.

[a]mg/100 embryos.

barium-insoluble fraction (precipitate 2) containing PGA, F1,6P, ATP, phytic acid, and P_i; a barium-soluble fraction (supernatant 2) containing the hexose monophosphates (G1P, G6P, F6P), PEP, and TP; and an ethanol-soluble fraction (supernatant 1) containing unidentified phosphorus compounds. The components of each fraction are identified by their characteristic properties.[3] When this method is applied to animal or bacterial tissue, it is possible to draw up a balance sheet that accounts for over 90% of the compounds present, but the method is not as successful when applied to plant tissue.[14]

In the earlier Kursanov method, phytic acid and P_i are removed from the TCA extract by precipitation with magnesium chloride and ammonia prior to the addition of barium acetate.[3880]

In the Klein method (Scheme 8-2), which focuses on the acid-insoluble components, the residue from the TCA extraction is separated by fractional precipitation into an ether-soluble fraction (supernatant 1) containing the phospholipids, a magnesium-insoluble fraction (precipitate 2) containing P_i, a hot TCA-soluble fraction (supernatant 3) containing the nucleic acids (RNA, DNA), and a hot TCA-insoluble fraction (precipitate 3) containing the phosphoproteins.[3881]

The rest of the entries in Table 8-1 are data from other fractionation schemes that contain one or more precipitation steps.

Selective Extraction Methods

In the Pons method (Scheme 8-3), the plant tissue is defatted with petroleum ether and then extracted with 0.75*N* trichloroacetic acid (TCA) to remove the acid-soluble compounds (filtrate 3). Another portion of the plant tissue is extracted with benzene-ethanol azeotrope, bp 68.6 °C, to remove the phosphatides (filtrate 2) and then with 2% HCl to remove acid-soluble compounds (filtrate 4). P_i is determined on an aliquot of filtrate 3 by the isobutyl alcohol/molybdate method,[3889] and phytic acid on an aliquot of filtrate 4 by the McCance-Widdowson method.[735] The P_{sugar} content was calculated by subtracting P_i and $P_{phytic\ acid}$

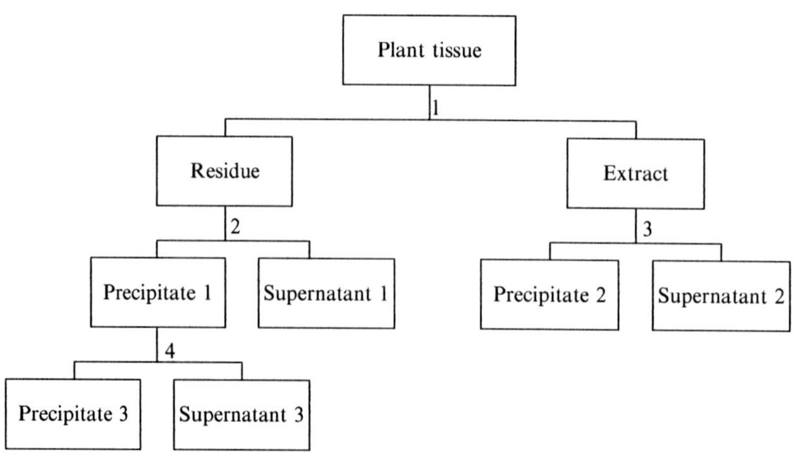

SCHEME 8-2 Klein Method. Reagents: (1) 10% TCA. (2) Et_2O. (3) $MgCl_2$, NH_4OH. (4) 5% TCA (90 °C).

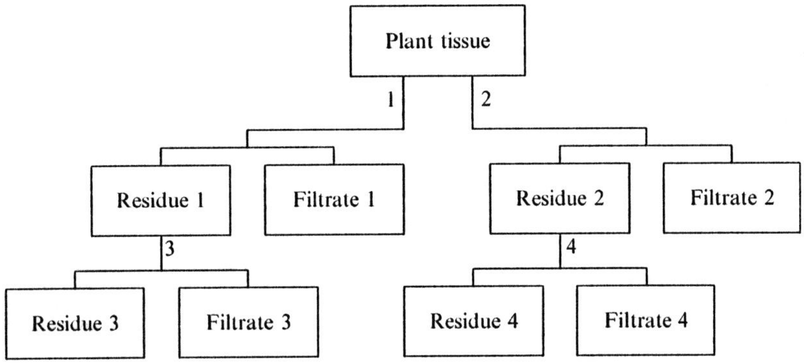

SCHEME 8-3 Pons Method. Reagents: (1) Petroleum ether. (2) Benzene-ethanol azeotrope. (3) 0.75*N* TCA. (4) 2% HCl.

from $P_{acid\text{-}soluble}$ and the $P_{nucleic\ acid}$ content by subtracting $P_{acid\text{-}soluble}$ and $P_{phosphatide}$ from P_{total} (Table 8-2).[1034]

In the Martin and Morton method (Scheme 8-4), the plant tissue is defatted with 1:1 chloroform-methanol and then extracted successively with 0.2*N* perchloric acid, 1*N* perchloric acid, and 1*N* sodium hydroxide, giving fractions $P_{acid\ soluble}$ (supernatant 1), P_{nucl} (supernatant 2), and $P_{protein}$ (supernatant 3), respectively.[3892] P_i is determined independently by the Weil-Malherbe and Green method.[204]

Kedrowski[3917] has described an automated procedure for the simultaneous analysis of P and N in P_{lipid}, P_{sugar}, phytic acid, P_{nucl}, and P_i fractions of plant extracts based on the Technicon® autoanalyzer. The method, which has been applied to nearly 1200 tree samples, is estimated to place 90% of the P_i and 85-90% of the P_{nucl} in their proper place in the scheme.

In the Bieleski method (Scheme 8-5), the plant tissue is extracted successively with MCF, formic acid/methanol, and 0.3*M* lithium hydroxide, giving fractions P_{lipid} (supernatant 2), P_{ester} (supernatant 1), and P_{nucl} (supernatant 3), respectively.[327] P_i is determined separately on supernatant 1 by the triethylamine/molybdate method.[3918]

Two other methods for which few details are available are those of Sokolov[3919] and Nilova.[3920]

Chromatographic Methods

Two-dimensional paper chromatography (PC) and thin-layer chromatography (TLC) have been found to be useful for the separation of individual P_{lipid}, P_{sugar}, P_i, and P_{nucl} compounds from their mixtures in plant extracts (Table 8-3). The solvent systems selected for extraction and separation should be capable of handling all four classes simultaneously, unlike those described in earlier chapters devoted to the individual classes. The following solvent systems are suitable for extraction:

- Methanol/$CHCl_3$/H_2O (12:5:3), then formic acid/ethanol/H_2O (2:5:20)[9]
- Methanol/$CHCl_3$/formic acid/H_2O (12:5:1:2)[327]
- Formic acid (80%), then ethanol (80%)[12]
- Formamide[331]

Table 8-2 Selective Extraction Methods*

No.	Plant part	P_i	P_{lipid}	P_{sugar}	$P_{phytate}$	$P_{nucleotide}$	$P_{protein}$	P_{total}	References
	Pons Method								
Arachis hypogaea L. (Peanut)									
1	Seed, kernel	7.2	9.0	11.1	70.0	2.7		0.334	1034
1	–, defatted cotyledon				53.3	3.6		0.90	847
1	–, aleurone grain				85.6	1.6		1.87	847
Cicer arietinum L. (Bengal Gram)									
1	Seed, mature	11.2	15.7		45.2	10.0		0.350	3890
1	Seed pod, immature and change in seeds and pods from day 20 to day 43[3890] for seed distribution among cotyledon, coat, and embryo, see Lal[3891]	65.2	11.3		9.9			0.435	3890
Gossypium hirsutum L. (Cotton)									
8	Seed kernel	1.8-3.1	6.3-7.0	7.0-11.0	74.5-78.7	4.1-5.9		0.965-1.113	983
3	6-day seedling, cotyledon	42.9-54.0	11.0-15.2	15.2-16.8	5.9-8.5	8.8-14.5		0.716-1.175	984
3	–, hypocotyl and root	59.3-70.2	10.5-14.3	5.5-12.2	0	6.9-9.5		0.568-0.753	984
Glandless cotton									
1	Seed, defatted kernel	3.2	5.5	1.1	85.9	4.2		1.85	985
1	–, globoid	2.5	0.5		95.8			14.21	986
Medicago sativa L. (Alfalfa)									
1	Seed meal	44.5	16.6	21.2	2.1	15.5		0.283	1034
Oryza sativa L. (Rice)									
8	White grain	2.0-3.8	0.7-1.1	0-13.0	34.6-55.0	44.0-52.5		0.096-0.121	1043
8	–, bran	1.9-2.9	0.9-1.2	1.3-3.0	84.0-93.3	1.0-10.4		2.476-2.868	1043
3	Brown grain	3.03-3.30			45.4-62.4			0.320-0.348	1044
3	–, bran	2.15-2.57			32.9-38.7			1.955-2.264	1044
Sesamum indicum L. (Sesame)									
1	Seed	3.5	4.7	1.6	85.9	4.4		0.637	1034
Triticum aestivum L. (Wheat)									
5	Seed flour	2.6-3.3	3.3-5.4	27.0-32.3	1.7-6.6	55.9-59.0		0.123-0.137	16

Vicia faba L. (Broad Bean)									
1	Seed, aleurone grain				64	10		1.4	812
Zea mays L. (Corn)									
1	Seed	11.4	1.4	6.4	74.4	6.4		0.281	1034
	Martin and Morton Method								
Beta vulgaris L. (Silver Beet)									
1	Leaf petiole		17.8	60.7		10.0	8.0	0.280	3892
1	–, mitochondria	18.7	23.8	32.8		5.2	18.8	1.12	3892
1	–, microsome	11.5	50.9	23.6		5.2	8.2	1.35	3892
Triticum aestivum L. (Wheat)									
1	Seed, protein body and changes in endosperm, testa-pericarp, and embryo 5-40 daf[3893]		2.8		53.9	30.9	12.4	1.78	805
Triticum durum Desf. (Durum Wheat)									
1	Seed, protein body and changes in endosperm, testa-pericarp, and embryo 5-40 daf[3893]	4.8	51.1		15.6	10.5	17.0	3.52	805
Vicia faba L. (Broad Bean)									
1	Leaf		22.1	38.9		34.6	4.9	1.22	3894
1	Stem		18.5	59.0		19.7	4.3	0.96	3894
1	Shoot		21.9	28.3		39.3	8.3	1.39	3894
	Bieleski Method								
Alnus crispa (Ait.) Pursh. (Green Alder)									
1	Leaf and seasonal changes in leaves and stems	28.8	16.9	8.5 negl.		28.8		0.059	833
Betula papyrifera Marsh. (Paper Birch)									
1	Leaf and seasonal changes in leaves and stems	41.2	18.1	8.0 negl.		26.6		0.199	833
Chionochloa crassiuscula									
1	Shoot	66.8	7.6	12.1		13.4		0.835	3895
1	Root	75.4	6.4	6.0		12.2		1.069	3895
Chionochloa pallens									
1	Shoot	52.4	12.0	13.0		22.5		0.381	3895
1	Root	52.0	13.4	7.1		27.4		0.349	3895
Larix laricina (Du Roi) K. Koch (American Larch)									
1	Leaf and seasonal changes in leaves and stems	63.2	4.7	9.4 negl.		13.2		0.106	833

*P_{total} column: Percent P by weight of plant tissue, fresh weight basis. Other columns: Percent of P_{total}.

Continued

Table 8-2 Selective Extraction Methods*—cont'd

No.	Plant part	P_i	P_{lipid}	P_{sugar}	$P_{phytate}$	$P_{nucleotide}$	$P_{protein}$	P_{total}	References
Picea mariana (Mill.) B.S.P. (Black Spruce)									
1	Leaf	31.1	21.4	6.8 negl.		21.4		0.103	833
	and seasonal changes in leaves and stems								
Solanum tuberosum L. (Potato)									
5	Tuber	11.7-31.0	5.0-5.9	32.5-46.3[a]	23.4-31.7	12.0-15.1		0.205-0.266	667
Spirodela oligorrhiza (Kurz.) Hegelm. (Duckweed)									
1	Frond tissue culture	70.7	8.2	4.7	0.14	15.5		1.445	327
			Other Methods						
Arachis hypogaea L. (Peanut)									
1	Seed	4.2	4.2		77.9	9.8		0.35	3896
Artemisia turanica H. Krasch. (Wormwood)									
1	Leaf	17.4		14.7	6.5	17.9		0.281	3897
Avena sativa L. (Oats)									
1	Seed	12	7		53	14	6	0.55	875
	and change in root and shoot on germination 2-8 days[875]								
Brassica juncea L. (Mustard)									
1	Leaf	32.0	27.3			16.8	0.4	0.057	3898
Brassica oleracea L. var. *botrytis* L. (Cauliflower)									
2	Leaf	20.2-27.3	16.8-20.6			25.6-33.0	1.0-1.7	0.071-0.158	3898
Cicer arietinum L. (Chickpea)									
1	Seed	11.8	14.2		24.4		48.9	0.436	3879
1	Seedling (25 days)	82.4	8.5		2.7		11.2	0.659	3879
	and data for interval between 0 and 25 days								
Climacoptera crassa (M.B.) Botsch.									
1	Leaf	14.5		8.0	15.6	37.3		0.228	3897
Cucurbita pepo L. (Vegetable Marrow)									
1	Leaf	54.0	8.6			19.0	0.4	0.116	3898
Cuscuta campestris Yunck. (Field Dodder)									
1	Seed	13.3			15.9	29.6		3.9[b]	3899
	and change during seed development (0-35 daf)[3899] and germination (1-9 days)[3900]								

Glycine max (L.) Merr. (Soybean)									
1	Whole bean	4.5	12.6		74.9	5.3		0.602	3901
1	–, protein body		9.4		69.2	21.6		0.43	804
1	Seed	3.1	9.7		66.9	17.2		0.54	3896
Gossypium hirsutum L. (Cotton)									
	Data given for leaf, stem, root, flower, and shoot apex over a 3-month growth period[3902] and for seeds of several spp. during ripening[3903]								
Helianthus annuus L. (Sunflower)									
1	Leaf	12.7	17.5			48.0	3.1	0.052	3898
1	Seed	2.9	2.9		81.9	9.2		0.65	3896
	and data for seed, leaf, stalk, and calathide during ripening[3904]								
Hordeum vulgare L. (Barley)									
1	Seed	9.9	7.2		63.8		19.1	0.379	3905
1	Grain	6.9	3.2	39.5	36.8	3.0	7.1		3906
1	Root	60.5		12.3		8.5			2586
	and rate of uptake of P_i during first 24 h[2586]								
Lens culinaris Medik. (Lentil)									
1	Seed	7.1	8.6		62.6	14.6		0.40	3896
1	–, whole flour	26.2	12.1		50.4	11.0		0.381	3907
Linum usitatissimum L. (Flax)									
1	Whole plant	18.6	5.4	3.7	31.2	41.1		0.602	3908
	and P distribution in stem, leaf, and boll during early growth, flowering, and early yellow ripeness phases								
Lycopersicon esculentum Mill. (Tomato)									
1	Leaf	27.9	13.8			26.0	1.8	0.078	3898
Nicotiana tabacum L. (Tobacco)									
1	Leaf								3909
1	–, fiber	9.3	35.1			52.3		0.290	3910
1	–, chloroplast		30.7			56.8		0.303	3910
Oryza sativa L. (Rice)									
1	Seed	1.8	4.4		77.8	9.2		0.29	3896
1	Seed	tr	2.2		45.9	52.6		0.41	3878
1	–, embryo	0.6	4.2		82.9	12.3		2.70	3878
Phaseolus vulgaris L. (Bean)									
1	Seed	6.1	5.6		73.4	12.0		0.50	3896

*P_{total} column: Percent P by weight of plant tissue, fresh weight basis. Other columns: Percent of P_{total}.

Continued

Table 8-2 Selective Extraction Methods*—cont'd

No.	Plant part	P_i	P_{lipid}	P_{sugar}	$P_{phytate}$	$P_{nucleotide}$	$P_{protein}$	P_{total}	References
Pisum sativum L. (Peas)									
1	Seed	3.9	9.2		65.5	16.2		0.52	3896
1	Seed	6.1		2.8	52.0	2.0		0.167	52
	and change during growth 12-40 daf[52]								
1	Cotyledon	37.3		15.7	31.4	8.3	7.6	0.738[c]	3911
Ricinus communis L. (Castor Bean)									
1	Seed	0.9	1.8		85.3	8.3		0.76	3896
1	–, embryo	8	4		78			1.00	3912
	and data for whole embryo, cotyledon, and root-shoot axis over 10 development stages from 10 to 60 dap[3912]								
1	–, endosperm	3	3		93			0.72	3912
	and data for whole endosperm and nucellus-integument over 10 developmental stages from 10 to 60 dap[3912]								
Salsola gemmascens Pall.									
1	Leaf	14.5		16.8	18.6	22.5		0.190	3897
Salsola sclerantha C.A.M.									
1	Leaf	31.2		6.3	13.8	18.4		0.259	3897
Secale cereale L. (Rye)									
1	Seed	3.6	6.3		77.2	7.6		0.39	3896
Solanum tuberosum L. (Potato)									
1	Leaf	45.8	7.9		22.2		24.3	0.428	798
Sorghum bicolor (L.) Moench (Sorghum)									
6	Seed	5.0-7.0	3.9-4.9		75.7-79.3	6.4-8.1		0.387-0.456	3913
	and change on germination 3-6 days[3913]								
Triticum aestivum L. (Wheat)									
1	Whole plant	39.4	3.1		57.5			0.192	3914
	and data for 18 weeks growth at biweekly intervals								
3	Grain								3906
1	Seed	3.7	5.9		79.0	8.5		0.36	3896
1	Seed	6.6	5.0		68.7		19.7	0.309	3905
1	–, whole germ	21.3	8.8		41.8	28.0		1.356	3907

Vicia faba L. (Broad Bean)									
1	Seed	7.2	7.1		65.7	12.5		0.46	3896
Vigna radiata (L.) R. Wilcz. (Mung Bean)									
1	Seed	4.5	7.6		64.4			0.530	3915
Vigna unguiculata (L.) Walp. (Cowpea)									
1	Seed	6.0	9.7		30.4			0.474	3915
Zea mays L. (Corn)									
1	Seed (yellow)	3.6	10.1		82.6			0.347	3915
1	Seed (white)	4.9	15.2		76.5			0.319	3915
1	Root	53.4	3.9	28.3		16.5		0.032	3916
Darso									
1	Seed	5.1	7.3		87.1			0.312	3915
Zea mays L. var. *indentata* Sturtev (Dent Corn)									
2	Aerial portion and change during growth, July 6 to August 22	36-38	12-15	29		20-21		0.232-0.270	1182
Pasture Grass									
1	Leaf	80.3	1.9		0		17.8	0.564	3905

*P_{total} column: Percent P by weight of plant tissue, fresh weight basis. Other columns: Percent of P_{total}.

[a] Starch.

[b] μg P/seed.

[c] Mg P/pair of cotyledons.

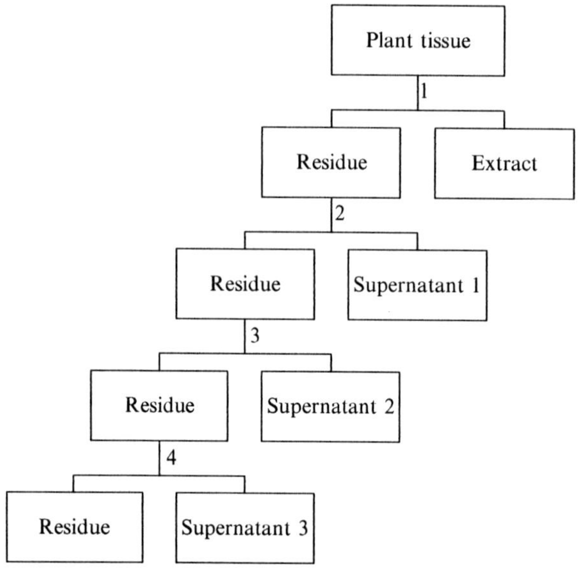

SCHEME 8-4 Martin and Morton Method. Reagents: (1) $CHCl_3$/methanol (1:1). (2) 0.2*N* perchloric acid. (3) 1*N* perchloric acid. (4) 1*N* NaOH.

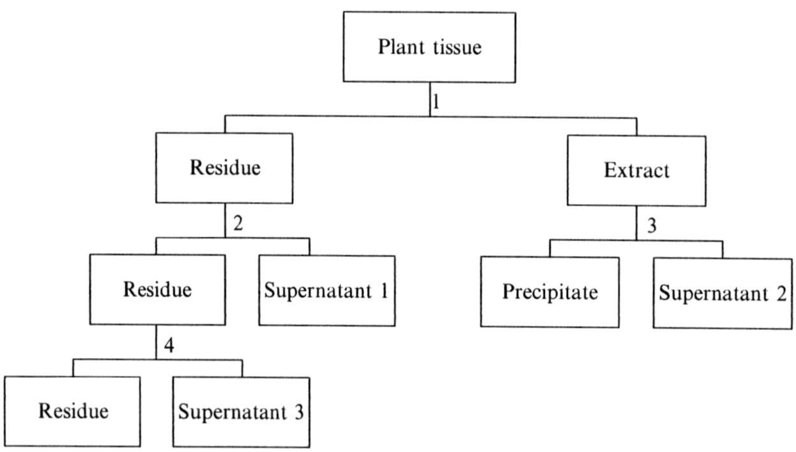

SCHEME 8-5 Bieleski Method. Reagents: (1) Methanol/$CHCl_3$/formic acid/H_2O (MCF, 12:5:1:2). (2) 1% Formic acid in 20% methanol. (3) $CHCl_3$, H_2O. (4) 0.3*M* LiOH.

The following pairs of solvent systems are suitable for two-dimensional chromatography:

- *n*-Propanol/NH_4OH/H_2O (6:3:1) and *n*-propyl acetate/formic acid/H_2O (11:5:3)[9]
- Isobutyric acid/NH_4OH/H_2O (57:4:39) and methanol/1*M* ammonium acetate (7:3)[12]

The method of Wade and Morgan[32] yields good resolution but leaves residual salts on the paper; the method of Wawszkiewicz[29] separates sugar phosphates well but does not separate the nucleotides.[9]

Table 8-3 Chromatographic Methods*

No.	Plant part	P_i	P_{lipid}	P_{sugar}	$P_{phytate}$	$P_{nucleotide}$	P_{total}	References
	PC Method							
Apium graveolens L. var. *dulce* (Celery)								
1	Petiole, vascular bundle	+	+	+		+		227
Brassica pekinensis (Lour.) Rupr. (Pe-tsai)								
1	Petiole, vascular bundle	+	+	+		+		227
Brassica rapa L. (Turnip)								
1	Petiole, vascular bundle	+	+	+		+		240
Malus sylvestris (L.) Mill. (Apple)								
1	Stem, phloem tissue	+	+	+		+		227
Nicotiana tabacum L. (Tobacco)								
1	Leaf	+	+	+		+		62
1	Leaf	+		+		+		2627
1	Root	+		+		+		2627
Persea americana Mill. (Avocado)								
1	Fruit	+	+	+		+		9
Saccharum officinarum L. (Sugarcane)								
2	Leaf	+	+	+		+		306
2	Root	+	+	+		+		306
Solanum tuberosum L. (Potato)								
1	Tuber	+	+	+		+		9
Spirodela oligorrhiza (Kurz.) Hegelm. (Duckweed)								
1	Frond tissue culture	+	+	+		+		327
Triticum aestivum L. (Wheat)								
1	Leaf	+		+		+		331
Zea mays L. (Corn)								
1	Root	+		+		+		12
	PC/PE Method							
Triticum aestivum L. (Wheat)								
1	Leaf	+		+		+		331
	TLC Method							
Brassica pekinensis (Lour.) Rupr. (Pe-tsai)								
1	Petiole, vascular bundle	+	+	+		+		36
Bryophyllum tubiflorum Harv. (Chandelier Plant)								
1	Phylloides	+	+	+		+		3346
Cuscuta reflexa Roxb. (Dodder)								
1	Callus culture	+	+	+		+		33
Helodea densa								
1	Callus culture	+	+	+		+		33
Isoetes duriei								
1	Callus culture	+	+	+		+		33
Lemna gibba L. (Gibbous Duckweed)								
1	Callus culture	+	+	+		+		33
Ophioglossum lusitanicum								
1	Callus culture	+	+	+		+		33

*P_{total} column: Percent P by weight of plant tissue, fresh weight basis. Other columns: Percent of P_{total}.

Continued

Table 8-3 Chromatographic Methods*—cont'd

No.	Plant part	P_i	P_{lipid}	P_{sugar}	$P_{phytate}$	$P_{nucleotide}$	P_{total}	References
Ophioglossum vulgatum								
1	Callus culture	+	+	+		+		33
Salix triandra L.								
1	Phloem exudate	+	+	+		+		309
Selaginella denticulata								
1	Callus culture	+	+	+		+		33
Sinapis alba L. (White Mustard)								
1	Callus culture	+	+	+		+		33
Tillandsia usneoides								
1	Callus culture	+	+	+		+		33
Vicia faba L. (Broad Bean)								
1	Callus culture	+	+	+		+		33
Yucca flaccida H.A.W.								
1	Phloem exudate	+	+	+		+		309
Zea mays L. (Corn)								
1	Root	+		+		+		12
				TLC/TLE Method				
Brassica pekinensis (Lour.) Rupr. (Pe-tsai)								
1	Petiole, vascular bundle	+	+	+		+		36
				Cellulose Column Method				
Fragaria spp. (Strawberry)								
1	Leaf	78.4		21.2[a]			0.025	8
Pisum sativum L. (Peas)								
1	Seed	13.4		5.9[a]	80.7		0.187	8
Solanum tuberosum L. (Potato)								
1	Tuber	85.3		5.3[a]	18.8		0.032	8
				IEC Method				
Agropyron desertorum (Fisch.) Schult. (Crested Wheatgrass)								
1	Seed and change on germination 6-48 h	+	+	+		+		831
Citrus sinensis (L.) Osbeck (Orange)								
1	Juice	+	+	+		+		248
Lycopersicon esculentum Mill. (Cherry Tomato)								
1	Seed	+	+	+		+		24
Phaseolus vulgaris L. (Kidney Bean)								
1	leaf	+		+		+		22
Pisum sativum L. (Snap Peas)								
1	Seed	+	+	+		+		24
Spinacia oleracea L. (Spinach)								
1	Leaf chloroplast	+	+	+		+		20
				HPLC Method				
Glycine max (L.) Merr. (Soybean)								
1	Leaf			+		+		26
Poa annua L. (Annual Bluegrass)								
1	Root	+			+	+		696
Solanum tuberosum L. (Potato)								
1	tuber	+			+	+		696

*P_{total} column: Percent P by weight of plant tissue, fresh weight basis. Other columns: Percent of P_{total}.

[a]Including $P_{nucleotide}$.

A few two-dimensional separations have been performed in which paper electrophoresis (PE) or thin-layer electrophoresis (TLE) was substituted for PC or TLC in either the first or second dimension. The following buffers are suitable for PE or TLE:

- Acetic acid/NH_4OH/H_2O (15:1:61:984, pH 3.6)[36]
- Acetic acid/pyridine/H_2O (12.5:7.5:980, pH 4.5)[331]

The preferred coating for the plates used in TLC or TLE is an unbound cellulose powder like Macherey-Nagel MN 300. Silica gel gives inferior results.[36]

Isherwood and Barrett[8] developed a method for separating P_i, $P_{phytate}$, and P_{nucl} by two-stage chromatography on a cellulose powder column. Phytic acid, sugars, and amino acids are removed in the first stage by elution with *n*-propanol/NH_4OH/H_2O (90.4:0.1:9.5) and P_i, P_{sugar}, P_{nucl}, and organic acids in the second stage by elution with diisopropyl ether/formic acid/*t*-butanol/H_2O (18:6:5:5).

A more useful method is ion exchange chromatography (IEC) on a column containing a strongly basic anion exchange resin such as Dowex 1-X8. Wilson and Harris[831] and Heldt *et al.*[20] used gradient elution with formic acid/ammonium formate, whereas Mori *et al.*[316] and others[22,24,248] used the borate gradient of Khym and Cohn.[21] In some cases, the IEC column was coupled to a Bessman phosphorus autoanalyzer[442] for ease of quantitation.[24,248]

High-performance liquid chromatography (HPLC) is possible on a column containing a strongly basic anion exchange resin,[26,696] but this method is clearly unsuitable for use with crude plant extracts without extensive prior cleanup.

^{31}P Nuclear Magnetic Resonance (NMR) Method

The advent of high field ^{31}P FT NMR spectroscopy in the early 1980s provided a unique opportunity for study of the behavior of phosphorus compounds in living tissue (Table 8-4). These experiments must be done under aeration to prevent anoxia.[544] The strongest signals are usually those of P_i, which shows two signals owing to the difference in pH between the cytoplasmic (pH~7.1) and vascular (pH~5.5) environments.[42] G6P, UDPG, ATP, and ADP are readily identifiable in the ^{31}P NMR spectra of root tips, leaves, and suspension cells. Phytic acid is identifiable in potato tubers and germinated lettuce seeds, but dry seeds show only a broad featureless peak despite their high content of phytic acid and other organic phosphates.[47–49]

In contrast, the ^{31}P NMR spectra of extracts of plant tissues show much higher resolution, especially when EDTA is added to eliminate the line broadening caused by Mg(II) and other paramagnetic ions. Many more sugar phosphates and nucleotides are identifiable in extracts with perchloric acid[48,50] and other acids,[49] and several phospholipids are identifiable in chloroform-methanol extracts.[1366] These studies are beyond the scope of this chapter and are discussed in detail in other chapters.

Table 8-4 ^{31}P NMR Method (*In Vivo*)

No.	Plant part	Compounds identified	References
Acer pseudoplatanus L. (Sycamore Maple)			
1	Cell suspension	P_i, G6P, F6P	46
1	Cell suspension	P_i, G6P, UDPG, ATP (or UTP), NAD	51
Catharanthus roseus (L.) G. Don (Madagascar Periwinkle)			
1	Cell suspension	P_i, G6P	46
Glycine max (L.) Merr. (Soybean)			
1	Cell suspension	P_i, G6P	46
1	Seedling cotyledon	P_i, P_{sugar}	47
Helianthus tuberosus L. (Jerusalem Artichoke)			
1	Stolon	P_i, G6P, ATP, ADP, NAD (or UDPG)	45
Lactuca sativa L. (Lettuce)			
1	Sprout	P_i, phytic acid	48
Nicotiana tabacum L. (Tobacco)			
1	Cell suspension	P_i, G6P, ATP, ADP	2629
Pisum sativum L. (Peas)			
1	Root tip	P_i, G6P, UDPG, NTP, NDP	526
Solanum tuberosum L. (Potato)			
1	Tuber	P_i, phytic acid	43
Sorghum bicolor (L.) Moench (Sorghum)			
1	Root tip	P_i, G6P, UDPG, ATP (or ADP)	44
Spinacia oleracea L. (Spinach)			
1	Leaf	P_i, G6P, UDPG, ATP (or UTP), NAD	51
Zea mays L. (Corn)			
1	Root tip	P_i, G6P	42
1	Root tip	P_i, G6P, UDPG, ATP (or UTP)	43

Distribution of Individual Phosphorus Compounds in Plants

The culmination of these efforts was Bieleski's 1968 paper on duckweed (*S. oligorrhiza*), the first to provide a complete balance sheet for phosphorus in a plant tissue.[327] Using PC and TLC methods in combination with ^{32}P labeling, he was able to identify 98% of the phosphorus compounds in the tissue. *Spirodela*, a minute floating water plant, was ideally suited for this purpose as it grows vegetatively in axenic culture on a completely defined medium, doubling its weight every 2 days. It was grown for several generations in $^{32}P_i$ to provide a uniformly labeled plant tissue. The principal components, in percent of total phosphorus, were P_i (70.7%), RNA (11.1%), PC (3.7%), PE (1.7%), DNA (1.4%), PG (1.1%), and G6P (1.1%). Complete results are given in Tables 8-5. In columns 3 and 4, the data for *S. oligorrhiza* are compared with Bieleski's model P_{ester} pattern for young photosynthetic tissue drawn from a variety of published sources.[1]

Table 8-5 Distribution of Phosphorus Compounds in *S. oiigorrhiza*

Phosphorus compound	Spirodela oligorrhiza[a] % of P_{total}	nmol/g	Model tissue [b] nmol/g
Inorganic phosphates			
P_i	70.7	32,980	
PP_i	0.170	85	
Sugar phosphates			
G6P	1.076	515	200
PGA (s)	0.522	250	85
α-GP	0.405	195	10
Phosphorylethanolamine	0.282	135	20
M6P	0.263	125	40
F6P	0.207	100	60
SDP?	0.207	100	
G1P	0.113	55	15
Phosphorylcholine	0.113	55	20
Sucrose P?	0.094	45	
F1,6P	0.085	40	20
Gal6P	0.085	~40	
PEP	0.080	38	15
6-PG	0.038	19	5
TP	0.033	15	12[c]
Pentose P	0.019	10	13
Phosphorylserine			5
Se7P			2
Inositol phosphates			
Phytic acid	0.139	70	15-300
$InsP_2$?	0.113	55	
$InsP_3$	0.078	40	
Phospholipids			
PC	3.67	1735	630
PE	1.74	820	350
PG	1.08	510	300
PI	0.55	260	130
PS	0.16	75	50
PA	0.06	35	20
DPG			20
Nucleotides			
ATP	0.638	300	100
UDPG	0.357	170	80
ADP	0.275	130	30
UTP	0.264	130	40
UMP	0.227	110	5-40
UDP	0.175	85	50
GTP	0.086	42	10
UDPX	0.085	40	
GDP	0.045	22	5
NAD?	0.042	20	45

Continued

Table 8-5 Distribution of Phosphorus Compounds in *S. oiigorrhiza*—cont'd

Phosphorus compound	Spirodela oligorrhiza[a] % of P_{total}	nmol/g	Model tissue [b] nmol/g
AMP	0.030	12	8
CDP	0.014	7	3
Nucl-3′P	0.014	7	
CMP	0.005	2	1
GMP	0.005	2	1
CTP			10
ADPG			25
GDPG			2
NADP			7
Nucleic acids			
RNA	11.10	5230	
DNA	1.442	680	
RNA-P?	0.186	90	
RNA-P_2?	0.124	60	

[a]Bieleski.[327]
[b]Bieleski.[1]
[c]DHAP.

Appendices

Appendix A: Changes Occurring During Germination

All references to studies that are concerned primarily with changes that occur during germination are gathered together in this appendix, regardless of the nature of the phosphorus compound. No data are given and no attempt has been made to avoid duplication. For further details see the cited references:

No.	Plant species	Compounds analyzed	References
Acer platanoides L. (Norway Maple)			
1	Seed (axis, cotyledon)	PC, PE, PG, PI	3921
Actinidia deliciosa Liang and Ferg. (Kiwi)			
1	Seed	ATP	2452
Agropyron desertorum (Link.) Schult. (Crested Wheatgrass)			
1	Seed	P_{hexose}, phytic acid, $InsP_2$, $InsP_3$, $InsP_4$, $InsP_5$, ADP, ATP, UTP, UDPX, P_i	831
Allium porrum L. (Leek)			
1	Seed	ATP, GTP, CTP, UTP, UDPG	2458
Amaranthus caudatus L. (Love-Lies-Bleeding)			
1	Seed	Phytic acid	3922
Amaranthus cruentus L. (Purple Amaranth)			
1	Seed	Phytic acid	3922
Amaranthus spp.			
2	Seed	Phytic acid	3923
Amaryllis vittata Ait. (Amaryllis)			
1	Pollen	G3P, DHAP, PEP, F1,6P	226
1	Pollen	AMP, ADP, ATP	2461
Arachis hypogaea L. (Peanut)			
6	Seed	Phytic acid	846
1	Seed	PC, PE, PG, PA, DPG	1483
1	Seed	ATP	2462,2463
1	Seed (axis, cotyledon)	ATP	3924

Continued

No.	Plant species	Compounds analyzed	References
1	Seed (cotyledon, embryonic axis)	Total soluble nucleotides	3925
–, ssp. *fastigiata* (Spanish Peanut)			
1	Seed (embryonic axis)	AMP, ADP, ATP	2465
–, ssp. *hypogaea* Virginia (Peanut)			
1	Seed (embryonic axis)	AMP, ADP, ATP	2465
Avena nuda L. Naked (Oats)			
1	Seed	Phytic acid	3923
Avena sativa L. (Oats)			
2	Seed	Phytic acid	861
1	Seed	Phytic acid	748,3923,3926
1	Seed (embryo)	PGA, G1P, F6P, HDP, phytic acid	14
1	Seed	3-PGA, PEP, G6P, F2,6P, ATP	232
1	Seed	NAPE	1497
1	Seed	NAD(P)	3927
1	Seed (leaf, root, endosperm)	FMN, FAD	3600
1	Seed	TDP	3787
1	Seed (shoot, root, endosperm)	P_i, P_{sugar}, $P_{phytate}$, P_{lipid}, $P_{nucleotide}$, $P_{protein}$	875
Avena strigosa Schreb. (Bristle Oats)			
1	Seed	Phytic acid	861
Avena strigosa × *Avena brevis*			
1	Seed	Phytic acid	861
Bauhinia purpurea L. (Purple Bauhinia)			
1	Seed (cotyledon, root, hypocotyl)	PEP	234
Benincasa hispida (Thunb.) Cogn. (Wax Gourd)			
1	Seed	PC, PE, PI, PS	1512
Beta vulgaris L.(Sugar Beet)			
1	Seed	AMP, ADP, ATP	2763
2	Seed	ATP, UDPG	3928
2	Seed	UDPG, UDPGal	481
Betula palyrifera Marsh (White Birch)			
1	Seed	ATP	3929
Brassica campestris L. (Turnip Rape)			
2	Seed	Phytic acid	886
1	Seed	ATP	2490
Brassica chinensis L. (Pak-Choi)			
1	Seed	ATP	2491

No.	Plant species	Compounds analyzed	References
Brassica juncea L. Czern. (Indian Mustard)			
1	Seed	Phytic acid, P_i	3930
Brassica napus L. (Rape)			
2	Seed	Phytic acid	3931
1	Seed	Phytic acid	3932
4	Seed (Canola)	Phytic acid	886
1	Seed (embryo)	PC, PE, PG, PA, LPC, LPA	3933,3934
1	Seed (hull)	PC, PE	3935
2	Seed (pod)	AMP, ADP, ATP	3936
–, ssp. *oleifera* (Winter Rape)			
2	Seed	AMP, ADP, ATP	2477
Brassica oleracea L. ssp. *botrytis* L. (Cauliflower)			
1	Seed	AMP, ADP, ATP	2477
Brassica pekinensis (Lour.) Rupr. (Chinese Cabbage)			
1	Seed	ATP	2491
Cajanus cajan (L.) Huth (Pigeon Pea)			
1	Seed	Phytic acid	3937
6	Seed	Phytic acid	898
Camellia sinensis L. (Tea)			
1	Pollen	PC, PE, PI	1558
Canavalia ensiformis (L.) DC (Jackbean)			
1	Seed	Phytic acid	3938
Cannabis sativa L. (Hemp)			
1	Seed	Phytic acid	3939
Carthamus tinctorius L. (Safflower)			
1	Seed	PC, PE, PG, PI	3940
1	Seed (cotyledon)	PC, PE, PG, PI, PA	1570
1	Seed	ATP	2491
Carum copticum (L.) Benth. and Hook. (Bishop's Weed)			
1	Seed (cotyledon, root)	PC, PE, PI, LPE	3941
Cassia absus L. (Pig's Senna)			
1	Seed	PC, PE, LPC, LPE	1575
Cassia fistula L. (Golden Shower)			
1	Seed	PC, PE, LPC, LPE	1575
Cassia occidentalis L. (Coffee Senna)			
1	Seed	PC, PE, LPC, LPE	1575
Chenopodium quinoa Willd. (Quinoa)			
2	Seed	Phytic acid, $InsP_3$, $InsP_4$, $InsP_5$	916
Chenopodium rubrum L. (Red Goosefoot)			
1	Seed	AMP, ADP, ATP, NAD, NADH, NADP, NADPH	2508

Continued

No.	Plant species	Compounds analyzed	References
Cicer arietinum L. (Chickpea, Bengal Gram)			
1	Seed	G3P, DHAP, PEP, G6P, F6P, F1,6P, AMP, ADP, ATP	247
2	Seed	Phytic acid	3942
1	Seed	Phytic acid	924,925,1065, 3879,3890, 3943,3944
1	Seed	PC	927,3879
1	Seed	PC, PE	3945
1	Seed	cAMP	2786
1	Seed (pod)	$P_{phytate}$, P_{lipid}, $P_{nucleotide}$, P_i, P_{total}	3890
1	Seed (cotyledon, embryo, coat)	$P_{phytate}$, P_{lipid}, $P_{nucleotide}$, $P_{protein}$, P_i, P_{totao}	3891
1	Seed	$P_{phytate}$, P_{lipid}, $P_{nucleotide}$, $P_{protein}$, P_i, P_{total}	3879
Cichorium intybus L. (Chicory)			
1	Seed	Phytic acid	2767
Citrullus vulgaris Schrad. (Watermelon)			
1	Seed (cotyledon)	F2,6P	492
1	Seed (cotyledon, root)	PC, PE, PI, LPE	1589
Cocos nucifera L. (Coconut)			
1	Seed (endosperm)	cAMP	2789
Corchorus capsularis L. (Jute)			
1	Seed	AMP, ADP, ATP	2521
Coriandrum sativum L. (Coriander)			
2	Seed	Phytic acid	3946
1	Seed (endosperm)	PC	1611
Corylus avellana L. (European Hazel)			
1	Seed	PC, PE, PG, PI, PA, NAPE	1612
1	Seed (kernel)	PC, PE, PI	1613
1	Seed	AMP, ADP, ATP, UMP	2522
Crotolaria juncea. L. (Sunn Hemp)			
1	Pollen	PC, PE, PG, PI, PS, PA	1620
Cucumis sativus L. (Cucumber)			
1	Seed	Phytic acid	941
1	Seed (cotyledon)	AMP, ADP, ATP	2524
1	Seed	NAD, NADH	3442
Cucurbita moschata Duch. (Warty Squash)			
1	Seed (leaf thylakoid membrane)	PG	1632
Cuphea lanceolata Ait. (Cigar Flower)			
1	Seed (embryo)	PC	3947

No.	Plant species	Compounds analyzed	References
Cuscuta campestris Yunck (Field Dodder)			
1	Seed	$P_{phytate}$, P_{DNA}, P_{RNA}, P_i, P_{total}	3899,3900
Cyclobalanopsis globosa Lin and Liu			
1	Seed	AMP, ADP	2534
Echinochloa cruz-galli ssp. *praticola* (Barnyard Grass)			
1	Seed	ATP	2544
Echinochloa oryzicola Yasing. Rice (Barnyard Grass)			
1	Seed	NADP(H)	3444
Elaeis guineensis Jacq (Oil Palm)			
1	Seed (kernel)	PC	3948
Eleusine coracana (L.) Gaertn. (Ragi)			
15	Seed	Phytic acid	960
Foeniculum vulgare Mill. Fennel			
1	Seed	Phytic acid	964
Glycine max (L.) Merr. (Soybean)			
1	Seed	Phytic acid	1329,3930, 3949–3952
1	Seed (embryonic axis)	Phytic acid	974
1	Seed	PC, PE, PG, PI, PS, PA, DPG, NAPE	3953
1	Seed	PC, PE, PG, PI, PA, DPG	1801
1	Seed	PC, PE, PI, PS, PA, DPG, phytoglycolipid	3954
4	Seed	PC, PE, PG, PI, PA, DPG, NAPE	3955
1	Seed (cotyledon)	PC, PE, PG, PI, PA, DPG	3956,3957
1	Seed (cotyledon, axis)	PC, PE, PG, PI, LPE	3958
1	Seed	PC, PE molecular species	2093
1	Seed	NAPE	1497
1	Seed	AMP, ADP, ATP, NAD, NADH, NADP, NADPH	2554
1	Seed	AMP, ADP, ATP, GMP, GDP, GTP, CMP, CTP, UMP, UDP, UTP, NAD	2552
1	Seed (cotyledon)	AMP, ADP, ATP, GMP, GTP, UDP, UDPG, NADP	2553
1	Seed	TMP, TDP, TTP	3780
1	Seed	TDP	3789
Gossypium barbadense L. (Sea Island Cotton)			
1	Seed	PC, PE, PI, PS, LPC	1705
1	Seed (kernel)	PC, PE, PI, PS, LPC	1707
Gossypium hirsutum L. (Upland Cotton)			
1	Seed	α-GP, 3-PGA, G1P, F6P, F1,6P	3959
1	Seed	3-PGA, F6P, F1,6P, phytic acid	3960
1	Seed	Phytic acid	984

Continued

No.	Plant species	Compounds analyzed	References
1	Seed (aleurone grain)	Phytic acid	3961
1	Seed	PC, PE, PI, PA	3962
1	Seed	PC, PE, PG, PI, PA, DPG	3963
1	Seed	PC, PI	3964
1	Seed	ATP	2491
1	Seed	AMP, ADP, ATP, GMP, GTP, UMP, UDP, UTP, TTP	3965
2	Seed	AMP, ADP, ATP, GMP, GDP, GTP, CMP, CDP, CTP, UMP, UDP, UTP	2572
1	Seed	TDP	3789
1	Seed	$P_{phytate}$, P_{lipid}, $P_{nucleotide}$, P_i, P_{total}	3903
Gossypium spp. (Cotton)			
1	Seed	PC	3966
Helianthus annuus L. (Sunflower)			
1	Seed	Phytic acid	810,3939
1	Seed	Phytic acid, $InsP_3$, $InsP_4$	3967
1	Seed	PC, PE, PG, PI, PS, PA	3968
1	Seed (cotyledon, oil bodies)	PC, PE, PI, PS	3969
1	Seed	PA	3970
1	Seed	P_{sugar}, $P_{phytate}$, P_{lipid}, $P_{nucleotide}$, $P_{protein}$, $P_{\alpha\text{-}GP}$, P_i, P_{total}	3904
Hibiscus cannabinus L. (Kenaf)			
1	Seed	PC, PE, PG, PI, PA, NAPE, NALPE	1738
Hordeum distichon Alef. (Two-Rowed Barley)			
1	Seed	LPC (starch-bound)	3971
–, ssp. *nutans* (Rode) Alef.			
1	Seed	G3P, DHAP, G1P, G6P, F6P, F1,6P, ADP, ATP	267
Hordeum hexastichon Alef. ssp. *parallelum* Kŏrn (Six-Rowed Barley)			
1	Seed	AMP, ADP, ATP, GMP, GTP, CMP, UMP, UTP	2581
Hordeum vulgare L. (Barley)			
2	Seed	G1P, G6P, F6P, UDPG	502
1	Seed (starch)	LPC, LPE, LPI	3972
1	Seed	Phytic acid	993,3926
1	Seed	Phytic acid, no lower InsPs	3973
1	Seed (embryo)	AMP, ADP, ATP, UMP, UDP	3357
1	Seed (endosperm)	Total soluble nucleotides	3974
1	Seed	cAMP	2800
1	Seed	cGMP	2801
1	Seed	TDP	3791

No.	Plant species	Compounds analyzed	References
1	Seed (embryo, endosperm)	P_{sugar}, $P_{phytate}$, P_{lipid}, $P_{nucleotide}$, $P_{protein}$, P_{RNA}, P_{DNA}, P_i, P_{total}	996,3885,3906
Iva imbricata Walter (Seacoast Marsh Elder)			
1	Seed	TDP	3789
Kalanchoë blossfeldiana v. Poelln. (Flaming Katy)			
1	Seed	AMP, ADP, ATP	2595
Lablab purpureus (L.) (Sweet Hyacinth Bean)			
1	Seed	Phytic acid	3942
Lactuca sativa L. (Lettuce)			
1	Seed	PEP, AMP, ADP, ATP	276
1	Seed	Phytic acid	1005,3975
1	Seed	Phytic acid, InsPs	48
1	Seed	AMP, ADP, ATP	3976
Lens culinaris Medik. (Lentil)			
2	Seed	Phytic acid	1008
1	Seed	Phytic acid	1065,3937
Lilium henryi Bak. (Lily)			
1	Pollen	AMP, ADP, ATP	2606
Lilium longiflorum Thunb. (Trumpet Lily)			
1	Pollen	Phytic acid	1012
1	Pollen	PC, PE, PG, PI, PA, PIP	2360
1	Pollen	PC, PE	1769
Limnanthes alba Hart. ex Benth. (Meadowfoam)			
1	Seed (cotyledon)	PC, PE, PA	1770
Linum usitatissimum L. (Flax)			
1	Seed	Phytic acid	941,3886
1	Seed	PC, PE, PI, PS, LPC, LPE	1771
1	Seed (cotyledon)	PC, PE, PG, PI, PA	1570
Lupinus albus L. (White Lupine)			
1	Seed	Phytic acid, no lower InsPs	1020
Lupinus angustifolius L. (Blue Lupine)			
1	Seed	Phytic acid	1021
Lupinus luteus L. (Yellow Lupine)			
1	Seed	Phytic acid, $InsP_3$, $InsP_4$, $InsP_5$	1020
1	Seed	Phytic acid, $InsP_4$, $InsP_5$	1333
1	Seed (axis, cotyleon)	ATP	2610
Lycopersicon esculentum Mill. (Tomato)			
1	Seed	Phytic acid	1027
1	Seed	NAD, NADH	3442
Machilus thunbergii Sieb. and Zucc.			
1	Seed	AMP, ADP, ATP	2619
1	Seed	AMP, ADP	2534

Continued

No.	Plant species	Compounds analyzed	References
Macrotyloma uniflorum (Lam.) Verdc. (Horse Gram)			
1	Seed	Phytic acid	1029
Medicago sativa L. (Alfalfa)			
1	Seed	PC, PE, PG, PI, DPG	3977,3978
1	Seed	PLP, PMP	3872
Michelia compressa (Maxim.) Sargent			
1	Seed	AMP, ADP	2534
Neolitsea parvigemma (Hayata) Kanak. and Sas.			
1	Seed	AMP, ADP	2534
Nicotiana sylvestris Speg. and Comes			
1	Pollen	PC, PE, PG, PI, PA, DPG	1818
Nicotiana tabacum L. (Tobacco)			
1	Seed	PC, PE, PG, PI, PA	3979,3980
1	Seed	TDP	3789
Oenothera biennis L. (Evening Primrose)			
1	Seed	PC, PE, PG, PI	1826
Olea europaea L. (Olive)			
1	Pollen	PC, PE, PI, PA	1829
Oryza sativa L. (Rice)			
1	Seed	F2,6P	514
1	Seed, dehulled	F2,6P	3981
1	Seed	Phytic acid	694,783,1187, 1218,2632,3982
1	Seed (aleurone particle)	Phytic acid	1048,1185
1	Seed	PC, LPC	3983
1	Seed (coleoptile)	PC, PE, PG, PI, PS, PA, DPG	1840
1	Seed	ATP	2491,2632
1	Seed (embryonic axis)	AMP, ADP, ATP	2634
1	Seed (embryo, endosperm, bran)	ATP	3984
1	Seed (kernel, hull)	ATP	3985
1	Seed	NAD, NADH, NADP, NADPH	3456
3	Seed	P_{sugar}, $P_{phytate}$, P_{lipid}, P_{RNA}, P_{DNA}, P_i, P_{total}	3986
1	Seed (embryo, pericarp, endosperm)	P_{sugar}, $P_{phytate}$, P_{lipid}, $P_{nucleotide}$, $P_{protein}$, P_i, P_{total}	3882
1	Seed	$P_{phytate}$, P_{lipid}, $P_{nucleotide}$, P_i, P_{total}	1052
Papaver dubium L. (Field Poppy)			
1	Pollen	PC, PE, PG, PI, PS, PA, DPG	3987
Papaver somniferum L. (Opium Poppy)			
1	Seed	Phytic acid	810

No.	Plant species	Compounds analyzed	References
Pennisetum glaucum (L.) R. Br. (Pearl Millet)			
1	Seed	Phytic acid	3988
Perilla frutescens (L.) Britt. (Perilla Mint)			
1	Seed	Phytic acid	3952
–, ssp. *japonica* (Egoma)			
2	Seed	PC, PE, PG, PI, PS, LPC	1862
Petroselinum crispum (Mill.) A. W. Hill (Parsley)			
1	Seed	P_1, P_{lipid}, $P_{nucleotide}$, P_i	792
Petunia hybrida L. (Petunia)			
1	Pollen	Phytic acid	3989
1	Pollen	PC, PE, PI, PA (no PG, PS, PIP or PIP_2)	2363
Phacelia tanacetifolia Benth. (Fiddleneck)			
1	Seed	NAD(P)(H)	2644
1	Seed	Phytic acid	1065
Phaseolus coccineus L. (Scarlet Runner Bean)			
1	Seed	F2,6P	518
1	Seed	Phytic acid	1065
Phaseolus lunatus L. (Lima Bean)			
1	Seed	PC, PE, PG, PI	3990
Phaseolus vulgaris L. (Bean)			
1	Seed	F6P, F2,6P, ATP	521
1	Seed (cotyledon)	Phytic acid	3991,3992
2	Seed	Phytic acid	3993,3994
1	Seed (radicle, plumule)	PC	3966
1	Seed	AMP, ADP, ATP	3995
1	Seed	NAD(P)	3996
1	Seed	FMN,FAD	3997
1	Seed (cotyledon, hypocotyl, root)	Free and total thiamine	3998
1	Seed	TDP, vitamin B_1	3999
Phellodendron wilsonii Hayata and Kaneh. (Cork Tree)			
1	Seed	AMP, ADP	3929
Picea abies (L.) Karst. (Norway Spruce)			
1	Seed (embryo, endosperm)	G6P, ADP, ATP	523
Picea pungens Engelm. (Colorado Spruce)			
1	Pollen	Phytic acid	818
Pinus banksiana Lamb. Jack Pine			
1	Seed	ATP	3929
Pinus densiflora Sieb. and Zucc. (Japanese Red Pine)			
1	Pollen	cAMP, cGMP	2827

Continued

No.	Plant species	Compounds analyzed	References
Pinus mugo Turra (Mugo Pine)			
1	Pollen	ATP, GTP, CTP, UTP, UDPGlc, UDPGal	2669
Pinus pinea L. Italian (Stone Pine)			
1	Seed	PLP	3866
Pinus ponderosa Laws. (Ponderosa Pine)			
1	Seed (embryo, endosperm)	AMP, ADP, ATP	2670
1	Pollen	Phytic acid	818
Pinus sylvestris L. (Scotch Pine)			
1	Seed (embryo, endosperm)	G6P, ADP, ATP	523
1	Pollen	NAD(P), NAD(P)H	3465
Pisum sativum L. (Peas)			
1	Seed (cotyledon)	G6P, F6P	2679
2	Seed (embryo)	F2,6P	295
3	Seed	Phytic acid	1083
1	Seed	Phytic acid	1084,3911, 3937,4000
1	Seed	Phytic acid, $InsP_3$, $InsP_4$, $InsP_5$,	1333
1	Seed	PC	4001
1	Seed	PC, PE	4002
1	Seed	PC, PE, PG, PI	3990
1	Seed	PC, PE, PA, LPC	4003
1	Seed	PC, PE, PI, PS, DPG, LPC, LPE	4004
1	Seed	NAPE	1497
1	Seed	ATP	2682
1	Seed	AMP, ADP, ATP	2846
1	Seed (cotyledon)	AMP, ADP, ATP, GMP, UMP, IMP, NAD, NADH, NADP, NADPH	2679,3471
1	Seed	UDPG	4005
1	Seed	UDPFru	3359
1	Seed	NAD	3466
1	Seed	NAD, FAD	4006
1	Seed	TMP, TDP, TTP	3780
1	Seed	TDP	3789,3795
1	Seed (coleoptile, leaf, root)	PLP	3873
Podocarpus macrophyllus (Thunb.) Sweet (Yew Plum Pine)			
1	Seed	AMP, ADP	2534
Prunus dulcis (Mill.) D. A. Webb (Cherry)			
1	Seed (kernel)	PC, PE, PG, DPG	4007

No.	Plant species	Compounds analyzed	References
Prunus persica (L.) Batsch (Peach)			
1	Seed (endocarp)	PC, PE, PI, PS	3086
Pseudotsuga menziesii (Mirb.) Franco (Douglas Fir)			
1	Seed	Total soluble nucleotides	4008
Psophocarpus tetragonolobus (L.) DC (Winged Bean)			
1	Seed	Phytic acid	1101
1	Seed	PC, PE, PG, PI, PS, DPG	1927
Raphanus sativus L. (Radish)			
1	Seed	ATP	2491
1	Seed	AMP, ADP, ATP	2691
Ricinus communis L. (Castor Bean)			
1	Seed	3-PGA, PEP, G6P, F6P, F1,6P	302
1	Seed (endosperm)	F2,6P	528
1	Seed	Phytic acid	1106
1	Seed	Phytic acid (no lower InsPs)	684
1	Seed (cotyledon)	Phytic acid	813
1	Seed (endosperm)	Phytic acid	1105,4009
1	Seed (aleurone grain)	Phytic acid	4009
1	Seed (endosperm)	PC, PE, PG, PS, PA, DPG	4010
1	Seed	NAD, NADP, NMN	3476
1	Seed (embryo, cotyledon, endosperm)	P_{sugar}, $P_{phytate}$, P_{lipid}, P_i, P_{total}	3912
Secale cereale L. (Rye)			
2	Seed	Phytic acid	3926
4	Seed	Phytic acid	1110
1	Seed	Phytic acid	3923,4011
1	Seed (chromatin)	PC, PE, PI	1959
1	Seed (nuclear matrix)	PC, PE, PI	1960
1	Seed	AMP, ADP, ATP	2699
1	Seed (leaf, root, endosperm)	FMN, FAD	3600
1	Seed (embryo, endosperm)	P_{sugar}, $P_{phytate}$, P_{lipid}, $P_{nucleotide}$, $P_{protein}$, P_{RNA}, P_{DNA}, P_i, P_{total}	996
Sesamum indicum L. (Sesame)			
1	Seed	Phytic acid	3952
Sinapis alba L. (White Mustard)			
1	Seed	PC, PI, PA	4012
1	Seed	AMP, ADP, ATP	2702
Solanum tuberosum L. (Potato)			
1	Seed (core)	PC, PE	4013
Sophora japonica L. (Japanese Pagoda Tree)			
1	Seed	PC, PE, PI, PS, PA, NAPE, LPC, NALPE	1987

Continued

No.	Plant species	Compounds analyzed	References
Sorghum bicolor (L.) Moench. (Sorghum)			
1	Seed	Phytic acid	4014
4	Seed	Phytic acid	4015
2	Seed	UMP, UDPG, ADPG	3369
6	Seed	P_i, $P_{phytate}$, P_{lipid}, $P_{nucleotide}$	3913
Striga asiatica (L.) Kuntze (Witchweed)			
1	Seed	PC, PE, PG, PI	2003
Trifolium incarnatum L. (Crimson Clover)			
1	Seed	AMP, ADP, ATP	4016
Trifolium repens L. (White Clover)			
1	Seed	Phytic acid	1128
Trifolium subterraneum L. (Sub Clover)			
1	Seed	Phytic acid	693
Trigonella foenumgraecum L. (Fenugreek)			
1	Seed	Phytic acid	1131,4017,4018
1	Seed	PC, PE, PG, PI, PS, PA, LPC	4019
1	Seed	AMP, ADP, ATP, UMP	2716
Triticum aestivum L. (Wheat)			
1	Seed (root, shoot)	G3P, DHAP, F1,6P	330
1	Seed (endosperm)	G3P, DHAP, G1P, G6P, F6P, F1,6P, F2,6P	537
1	Seed (endosperm)	G1P, G6P, G1,6P, F6P, F1,6P, F2,6P	538
1	Grain	HMP, HDP, phytic acid	3887
3	Seed	Phytic acid	3926
1	Seed	Phytic acid	693,925,1137, 3923,3930, 4020–4022
1	Seed (bran)	Phytic acid	4023
2	Seed (aleurone particle)	Phytic acid	1185,4024
1	Seed (testa, embryo, aleurone, endosperm)	Phytic acid, ATP, P_i	789
2	Seed	Phytic acid, InsP, $InsP_2$, $InsP_3$, $InsP_4$, $InsP_5$	1226
1	Seed	Phytic acid, $InsP_4$	3967
1	Seed	No lower InsPs	1198
1	Grain (embryo removed)	PC	4025
1	Seed	PC, PE, PI, PS, PA, NAPE, LPC, LPE, LPA, NALPE	2026
1	Seed (endosperm)	PC, PE, PG, PI, PS, PA, DPG, NAPE, LPC. LPE, LPA, NALPE	4026
2	Seed	PC, PE, PG, PI, PS, LPC	4027

No.	Plant species	Compounds analyzed	References
1	Seed (kernel)	PC, PE	4028
1	Seed (coleoptile, mitochondria)	PC, PE, PS, PA, DPG, LPE	4029
1	Seed (chromatin)	PC, PE, PI, PS, DPG, LPC	2029
1	Seed	ATP	2491,2695
1	Seed	AMP, ADP, ATP, GDP, GTP, CTP, UDP, UTP	2725
1	Seed	AMP, ADP, ATP	2720,2722
1	Seed (embryo)	AMP, ADP, ATP	4030,4031
1	Seed (embryo)	ATP, GTP, CTP, UTP	2734
1	Seed (endosperm, testa, pericarp)	AMP, ADP, ATP, ADPG, UMP, UDP, UTP, UDPG, NAD	2723
2	Seed	UDPG	3372
2	Grain	AMP, ADP, ATP, NAD, NADH, DADP, NADPH	4032
1	Seed	NAD, NADH, NADP, NADPH	3416,3456
1	Seed (embryo)	NAD, NADH, NADP, NADPH	3486
1	Seed	Total soluble nucleotides	4033
2	Seed (embryo, endosperm)	Total soluble nucleotides	2768
1	Seed	TDP	3798,4034
1	Seed (leaf, root)	PLP	3873
4	Seed	P_{sugar}, $P_{phytate}$, P_{lipid}, $P_{nucleotide}$, $P_{protein}$, P_{RNA}, P_{DNA}, P_i, P_{total}	996,3906
1	Seed	P_{sugar}, $P_{phytate}$, P_i, P_{total}	4035
1	Seed (embryo, endosperm)	P_{sugar}, $P_{phytate}$, P_{lipid}, $P_{nucleotide}$, $P_{protein}$, P_{RNA}, P_{DNA}, P_i, P_{total}	3885
1	Grain (testa/pericarp, embryo, endosperm)	$P_{phytate}$, P_{lipid}, $P_{nucleotide}$, $P_{protein}$, P_{RNA}, P_{DNA}, P_i, P_{total}	3893
T. aestivum × *Secale cereale* (× *Triticosecale* Wittm.) Triticale			
5	Seed	Phytic acid	1146,4036
4	Seed	ATP	2717
Triticum durum Desf. (Durum Wheat)			
1	Seed	Phytic acid	4037
2	Seed (embryo)	AMP, ADP, ATP	2737
1	Grain (testa/pericarp, embryo, endosperm)	$P_{phytate}$, P_{lipid}, $P_{nucleotide}$, $P_{protein}$, P_{RNA}, P_{DNA}, P_i, P_{total}	3893
Triticum spelta L. (Spelt)			
1	Seed	PC	4038
Uniola paniculata Walter (Sea Oats)			
1	Seed	TDP	3789

Continued

No.	Plant species	Compounds analyzed	References
Vicia faba L. (Faba Bean)			
2	Seed	Phytic acid	4039,4040
1	Seed	Phytic acid	810,1065
8	Seed	Phytic acid	4041
1	Seed	PC, PE, PG, PI	3990
1	Seed	NAPE	1497
1	Seed	AMP, ADP, ATP, GMP, GTP, CMP, UMP, CTP, UTP	4042
1	Seed	AMP, ADP, ATP, GMP, GTP, CMP, UMP, UTP, UDPX, NAD	2738
1	Seed	TDP	4043
–, ssp. *minor* L. (Small Horse Bean)			
1	Seed	Phytic acid, $InsP_3$, $InsP_4$, $InsP_5$	1333
Vigna aconitifolia (Jacq.) Mar. (Moth Bean)			
1	Seed	Phytic acid	1029,1158
Vigna angularis (Willd.) Ohwi and Ohashi (Adzuki Bean)			
1	Seed	Phytic acid	3951
1	Seed	PC, PE, PI	4044
Vigna mungo (L.) Hepper (Black Gram)			
1	Seed (cotyledon, axis)	G6P, 6-PG, F6P, ATP	540
1	Seed	Phytic acid	3942,3993,4045
6	Seed	Phytic acid	898
1	Seed	PC, PE, PG, PI, PS, PA, DPG, NAPE, LPC, LPE, NALPE	1384
Vigna mungo × *Vigna radiata*			
1	Seed	Phytic acid	1165
Vigna radiata (L.) Wilcz. (Mung Bean, Green Gram)			
1	Seed	G3P	4046
1	Seed	Phytic acid	925,1051,3943, 3951,3952, 4047,4048
5	Seed	Phytic acid	4049
1	Seed	Phytic acid, lower InsPs	693
1	Seed	Phytic acid, $InsP_1$, $InsP_2$, $InsP_3$, $InsP_4$, $InsP_5$,	4050
1	Seed	PC, PE, PG, PI, PS, PA	2053
1	Seed	PC, PE, PI, PA	2059
1	Seed	AMP, ADP, ATP	2524
1	Seed (hypocotyl)	AMP, CMP	4051
1	Seed	AMP, ADP, ATP, GMP, GDP, GTP, CMP, UMP	4016
1	Seed	AMP, ADP, ATP, UDP, UDPGlc, UDPGlcNAc	4052

No.	Plant species	Compounds analyzed	References
1	Seed	AMP, ADP, ATP, CMP, UMP, UTP, UDPX, X=Glc, Gal, Xyl, Ara, Rha, GlcNAc; GDPMan, NADP	2744
1	Seed	AMP, ADP, ATP, GMP, GDP, GTP, CMP, CDP, CTP, UMP, UDP, UTP, UDPX, X=Glc, Gal, Ara, Xyl, GlcA, GalA, GalNAc; GDPMan	2742
1	Seed (plumule, radicle, cotyledon)	FMN, FAD	3585
1	Seed	TDP	3789
Vigna sesquipedalis (L.) Fruh. Yard-Long Bean			
1	Seed (cotyledon, hypocotyl, root)	NAD, NADH, NADP, NADPH	3411
Vigna umbellata (Thunb.) Ohwi and Ohashi (Rice Bean)			
5	Seed	Phytic acid	4049
Vigna unguiculata (L.) Walp. (Cowpea)			
1	Seed	Phytic acid	824,1069, 3993,4053
1	Seed (plumule, radicle, cotyledon)	FMN, FAD	3585
Xanthium pennsylvanicum Wallr. (Cocklebur)			
1	Seed (cotyledon, axis)	AMP, ADP, ATP	4054
Zea mays L. (Corn)			
3	Grain	Phytic acid	4055
1	Seed	Phytic acid	1180,1182
1	Seed	PC	3966
1	Seed	PC, PE, PA, LPC	4003
1	Seed (kernel)	PC, PE, PG, PI, PA, DPG, LPC	4056
1	Seed	PC, PG, PI	4057
1	Seed (germ, endosperm, pericarp)	PC, PE, PI, PA, LPC, LPE	4058
1	Seed	AMP, ADP, ATP, GMP, GDP, GTP, CMP, CTP, UMP, UDP, UTP, NAD, NADP	3497
1	Seed	ATP	4059
2	Seed (embryo)	ATP	2760
1	Seed (embryo)	ATP	4060
1	Seed (embryo, endosperm)	AMP, ADP, ATP	2753
1	Seed	cAMP	4061
2	Grain	NAD, NADP, NAAD, NMN	3494
1	Seed	NAD, NADH, NADP, NADPH	3408
1	Seed (root, shoot)	NAD, NADH, NADP, NADPH	3498

Continued

No.	Plant species	Compounds analyzed	References
1	Seed (axis, scutellum, endosperm)	Total soluble nucleotides	4062
1	Seed	TMP, TDP, TTP	3780
1	Seed	TDP	3789,4034,4063
1	Seed	P_{sugar}, $P_{phytate}$, P_{lipid}, $P_{nucleotide}$, P_i, P_{total}	791

Appendix B: Changes Occurring During Growth and Senescence

All references to studies that are concerned primarily with changes that occur during growth and senescence are gathered together in this appendix, regardless of the nature of the phosphorus compound. No data are given and no attempt has been made to avoid duplication. For further details see the cited references:

No.	Plant species	Compounds analyzed	References
Acanthus mollis L. (Boar's Breech)			
1	Leaf	NAD(P)	4064
Acer pseudoplatanus L. (Sycamore maple)			
1	Cell culture	PC, PE, PI, PA, LPC, LPE, NALPE	4065
1	Cell culture	AMP, ADP, ATP, NAD, NADH, NADP, NADPH	2449
Actinidia deliciosa Liang and Ferg. (Kiwi)			
1	Callus culture	G6P	475
Ailanthus glandulosa Desf. (Chinese Sumac)			
1	Cataphylls	PC, PE, PG, PI, PS, PA, DPG	4066
Albizia lebbeck (L.) Benth. (Lebbek Tree)			
1	Leaf	PEP	223
Alnus crispa (Ait.) Pursh. (Green Alder)			
1	Leaf, bud, terminal stem	P_{sugar}, $P_{phytate}$, P_{lipid}, $P_{nucleotide}$, P_i, P_{total}	833
Arabidopsis thaliana (L.) Heynh (Mouse-Ear Cress)			
1	Leaf	PC, PE, PG, PI	4067
Arachis hypogaea L. (Peanut)			
1	Leaf	PEP	228
Artemisia turanica H. Krasch. (Wormwood)			
1	Leaf	3-PGA, phytic acid, $P_{nucleotide}$, P_i	3897
Arum maculatum L. (Cuckoopint)			
1	Spadix	2-PGA, 3-PGA, PEP, G1P, G6P, F6P, F1,6P	230
1	–, mitochondria	PC, PE, PG, PI, PS, PA, DPG	1490

No.	Plant species	Compounds analyzed	References
Avena sativa L. (Oats)			
1	Leaf	FMN, FAD	3601
Bauhinia purpurea L. (Purple Bauhinia)			
1	Fruit	PEP	233
Bauhinia variegata L. (Orchid Tree)			
1	Flower bud	PEP	235
1	Leaf, flower	PEP	233
Beta vulgaris L. (Sugar Beet)			
1	Root	PC, PE, PG, PI, PS, PA, DPG	1515
1	Root	AMP, ADP, ATP, GMP, GDP, CMP, UMP	2485
1	Leaf, root	ADP, ATP	4068
Betula papyrifera Marsh. (Paper Birch)			
1	Leaf, bud, terminal stem	P_{sugar}, $P_{phytate}$, P_{lipid}, $P_{nucleotide}$, P_i, P_{total}	833
Betula pendula Roth (European White Birch)			
1	Leaf	F2,6P, ADP, ATP	4069
1	Leaf, bud	PC, PE, PG, PI, PS	4070
Brassica campestris L. (Turnip Rape)			
1	Pod, deseeded	G3P, DHAP, 3-PGA, RuBP, F6P, F1,6P, F2,6P	238
1	Flower	PC, PE, PG, PI, DPG	4071
Brassica napus L. ssp. *oleifera* (Winter Rape)			
2	Stem, apex	PC	4072
2	Leaf, shoot apex	ATP	2493
Brassica oleracea L. ssp. *capitata* L. (Cabbage)			
6	Leaf	AMP, ADP, ATP	4073
Cajanus cajan (L.) Huth (Pigeon Pea)			
1	Leaf	PEP	4074
1	Flower	PEP	241
Canavalia ensiformis (L.) DC (Jack Bean)			
1	Root nodule	PC, PE, PG, PI, PS	1559
Cannabis sativa L. (Hemp)			
1	Leaf	AMP, ADP, ATP	2501
Capsicum annuum L. (Sweet Pepper)			
1	Fruit, pericarp	F2,6P	4075
1	Fruit (chloroplast, chromoplast, microsomal membrane)	PC, PE, PG, PI, PA	1564
Catharanthus roseus (L.) G. Don (Madagascar Periwinkle)			
1	Cell suspension culture	G3P, DHAP, 2-PGA, 3-PGA, PEP, G1P, G6P, F6P, F1,6P	4076
1	Cell suspension culture	UDPX, X=Glc, Gal, Xyl, Ara, GlcA	3347
Centaurea moschata L. (Sweet Sultan)			
1	Leaf	PEP	244

Continued

No.	Plant species	Compounds analyzed	References
Chenopodium rubrum L. (Red Goosefoot)			
1	Cell suspension culture	UDPG	2509
1	Stem	PC, PE, PG, PI, PA, DPG	4077
Cicer arietinum L. (Chickpea, Bengal Gram)			
1	Root	PC, PE, PG, PS, PA, DPG	4078
Cichorium intybus L. (Chicory)			
1	Explant leaf	AMP, GMP, CMP, UMP	2511
1	Root	NAD, NADH, NADP, NADPH	3433
Citrus hassaku hort. ex Tanaka (Hassaku)			
1	Fruit juice	PC, PE	1601
Citrus junos (Sieb.) Tanaka (Yuzu)			
1	Leaf	PC, PE, PG	1598
Citrus limettioides Tanaka (Sweet Lemon)			
1	Fruit	AMP, ADP, ATP	2514
Citrus limon L. (Sour Lemon)			
1	Fruit	AMP, ADP, ATP	2514
1	Flower	$P_{acid\text{-}soluble}$, P_{lipid}, $P_{protein}$, P_{RNA}, P_{DNA}	4079
1	Leaf	$P_{acid\text{-}soluble}$, P_{lipid}, $P_{protein}$, P_{RNA}, P_{DNA}	4080
Citrus medioglobosa hort. ex Tanaka (Naruto)			
1	Fruit (flavedo)	PC, PE, PI	4081
Citrus natsudaidai Hayata (Sour Orange)			
1	Leaf	PC, PE, PG	1598
1	Fruit juice	PC, PE	1601
Citrus sinensis (L.) Osbeck (Orange)			
1	Leaf	PC, PE, PG, PI, DPG	1557
1	Leaf	ATP	4082
1	Juice vesicles	ADP, ATP, NAD, NADH, NADP, NADPH	2520
Citrus tamurana hort. ex Tanaka (Hyuganatsu)			
1	Fruit juice	PC, PE	1601
Climacoptera crassa (M.B.) Botsch.			
1	Leaf	3-PGA, phytic acid, $P_{nucleotide}$, P_i	3897
Crambe amabilis But. and Maj. (Crambe)			
1	Various plant parts	PC, PE, PG, PI, PS, PA, NAPE, NALPE, LPC	1616
Cucumis sativus L. (Cucumber)			
1	Leaf	PC, PE, PG, PI	4083
Cucurbita pepo L. (Pumpkin)			
1	Leaf	PC, PE, PG, PI, PS	1634
1	Root	AMP, ADP, ATP, GMP, GDP, CMP, UMP, UDP	2532
Datura inoxia Mill. (Downy Thorn Apple)			
1	Cell suspension culture	AMP, ADP, ATP, GMP, GDP, GTP, CTP, UMP, UDP, UTP, NAD, NADP	3351
Daucus carota L. (Carrot)			
1	Leaf	PC, PE, PG, PI, PS	1634

No.	Plant species	Compounds analyzed	References
Diapensia lapponica L. (Pincushion Plant)			
1	Leafy top	PC, PE, PG	1653
Distichlis spicata (L.) Greene (Coastal Saltgrass)			
1	Leaf	AMP, ADP, ATP	2542
Elaeeis guineensis Jacq. (Oil Palm)			
1	Fruit (mesocarp)	PC, PE, PG, PI, PS, PA, DPG, LPC	4084
Erythronium japonicum Decne (Katakuri)			
1	Leaf	3-PGA, Ru1,5P	256
Evodia rutaecarpa Hook.f. and Thomson			
1	Fruit	cGMP	2792
Fagopyrum esculentum Moench (Buckwheat)			
1	Leaf	ADP, ATP	2548
Fagus sylvatica L. (European Beech)			
1	Leaf	PC, PE, PG, PI, PS	1634
Festuca pratense Huds. (Meadow Fescue)			
2	Leaf	PC PE, PG, PI, DPG	1679
Ficaria verna Huds. (Lesser Celandene)			
1	Tuber	AMP, ADP, ATP	2550
Fraxinus excelsior L. (European Ash)			
1	Apical and axillary buds	ATP, NTP	2551
Glycine max (L.) Merr. (Soybean)			
1	Leaf	GAP, hexose P	56
1	Leaf	F2,6P	4085
1	Leaf	Phytic acid	4086
1	Root	PC, PE, PG, PI, PA, DPG, NAPE	1701
Gossypium arboreum L. (Tree Cotton)			
1	Fiber	AMP, ADP, ATP	2567
1	Boll (fiber)	NAD, NADP	2566
Gossypium barbadense L. (Sea Island Cotton)			
1	Leaf	3-PGA, G1P, F1P	263
1	Fiber	AMP, ADP, ATP	2567
Gossypium hirsutum L. (Upland Cotton)			
1	Leaf	TP, 3-PGA, HMP, HDP	4087
1	Boll, fiber	G1P, G6P, UDPX, X=Glc, Gal, Xyl, Ara, Man, Rha, Fuc	2571
1	Whole plant	PC, PE, PI, PA	4088
1	Fiber	PC, PE, PG, PI, PS	1719
1	Leaf (chloroplast)	AMP, ADP, ATP	4089
1	Leaf	$P_{phytate}$, P_{lipid}, $P_{nucleotide}$, P_i, P_{total}	4090
1	Leaf, stem, root, flower, shoot apex	P_{sugar}, $P_{phytate}$, P_{lipid}, $P_{nucleotide}$, P_i, P_{total}	3902

Continued

No.	Plant species	Compounds analyzed	References
Helianthus annuus L. (Sunflower)			
1	Leaf, stalk, calathide	P_{sugar}, $P_{phytate}$, P_{lipid}, $P_{nucleotide}$, $P_{protein}$, $P_{\alpha\text{-GP}}$, P_i, P_{total}	3904
1	Callus culture	cAMP	2796
Helianthus tuberosus L. (Jerusalem Artichoke)			
1	Tuber (plasmalemma)	PC, PE, PG, PI, PS, PA	4091
Hibiscus cannabinus L. (Kenaf)			
1	Leaf, bud, stem, flower, root	PC, PE, PG, PI, PA, NAPE, NALPE	1738
Hibiscus rosa-sinensis L. (Hibiscus)			
1	Leaf	PC, PG	1566
Hordeum vulgare L. (Barley)			
1	Leaf	G6P, F2,6P	3358
2	Leaf	PC, PE, PG, PI, PA; no PS	4092
2	Leaf segments	PC, PE, PG, PI	4093
1	Leaf (chloroplast)	PC, PG	4094
1	Leaf	AMP, ADP, ATP, NAD, NADP	4095
1	Leaf	FMN, FAD	3601
2	Seedling	UDPG	502
Impatiens balsamina L. (Garden Balsam)			
1	Leaf	PC, PE, PG, PI, PS	1634
Ipomoea aquatica Forssk. (Swamp Morning Glory)			
1	Cell suspension culture	PC, PE, PG	4096
Ipomoea batatas (L.) Lam. (Sweet Potato)			
1	Root	AMP, ADP, ATP, UMP, UTP, UDPG, ADPG	2592
Larix laricina (Du Roi) K. Koch. (American Larch)			
1	Leaf, bud, terminal stem	P_{sugar}, $P_{phytate}$, P_{lipid}, $P_{nucleotide}$, P_i, P_{total}	833
Larix sibirica (Siberian Larch)			
1	Cambium	PC, PE, PI, PS, PA, NAPE	1763,4097
Lemna gibba L. (Gibbous Duckweed)			
1	Frond (axenic culture)	Phytic acid	695
Lemna paucicostata Hegelm. (Lesser Duckweed)			
1	Whole plant	cAMP, cGMP	2805
Linum usitatissimum L. (Flax)			
1	Leaf, stalk, boll	P_{sugar}, $P_{phytate}$, P_{lipid}, $P_{nucleotide}$, P_i, P_{total}	3908
Lolium multiflorum Lam. (Italian Ryegrass)			
1	Endosperm cell suspension culture	cAMP	2807
Lotus corniculatus L. (Birdsfoot Trefoil)			
1	Root	Phytic acid	1016
Lupinus luteus L. (Yellow Lupine)			
1	Root nodule	NAD, NADH, NADP, NADPH	4098

No.	Plant species	Compounds analyzed	References
Lycopersicon esculentum Mill. (Tomato)			
1	Fruit	TP, PEP, G6P, F6P, F1,6P	278
1	Fruit	PC, PE, PS, PA, DPG	1779
1	Fruit (pericarp)	PC, PE, PG, PI, PS, PA, DPG	4099
1	Fruit	AMP, ADP, ATP, UMP, UTP, UDPG	2615
Malus domestica Borkh. (Apple)			
1	Fruit	G6P, F6P, 6-PG	4100
1	Stem (cortex)	Phytic acid	1030
1	Shoot (phloem, xylem)	PC, PE, PG, PI	1791
1	Bud (axial, terminal)	PC, PE, PG, PI	4101
Mangifera indica L. (Mango)			
1	Leaf	cAMP	2787
Medicago sativa L. (Alfalfa)			
1	Root	Phytic acid	1016
Morus alba L. (Mulberry)			
1	Shoot	AMP, ADP, ATP	2624
1	Stem, bark	AMP, ADP, ATP	4102,4103
Myriophyllum quitense H.B.K. (Water Milfoil)			
1	Leaf, stem, root	PE, PI	1810
Nicotiana suaveolens Lehm. × *Nicotiana langsdorffii*			
1	Stem	cAMP	2812
Nicotiana tabacum L. (Tobacco)			
1	Leaf	3-PGA, PEP, G6P	285
1	Leaf	3-PGA	4104
1	Shoot	3-PGA, G1P, G6P, F6P, F2,6P	288
1	Root	3-PGA, G6P, F2,6P	288
1	Cell suspension culture	InsP, $InsP_2$, $InsP_3$	1200
1	Leaf	PC, PE, PG, PI, PS	4105
1	Leaf, flower (petal, pistil, ovary, anther)	PC, PE, PG, PI, PS, DPG	1819
1	Leaf	AMP, ADP, ATP, GMP, GDP, GTP, CDP, CTP, UMP, UDP, UTP, UDPX (X=Glc, Gal, GlcA, GalA, GlcNAc), GDPG, NAD, NADP	2628
1	Leaf, stem, root, apical bud	cAMP	2787
1	Callus culture	UDPG, ADPG	3364
1	Cell suspension culture	AMP, ADP, ATP, GDP, GTP, CDP, CTP, UMP, UDP, UTP, UDPX (X=Glc, Gal, GalA, GlcNAc), GDPG, NAD, NADP	4106
1	Leaf	FMN, FAD	3608
1	Leaf	P_{sugar}, $P_{phytate}$, P_{lipid}, $P_{nucleotide}$, $P_{protein}$, P_i, P_{total}	3909

Continued

No.	Plant species	Compounds analyzed	References
Oryza sativa L. (Rice)			
1	Seedling	AMP, ADP, ATP, UMP, UDP, UTP, UDPG	4107
3	Leaf	$P_{phytate}$, P_{lipid}, $P_{nucleotide}$, P_i, P_{total}	1052
Padus racemosa Lam. (Bird Cherry)			
1	Stem (cortex)	Phytic acid	1030
Persea americana Mill. (Avocado)			
1	Leaf	cAMP	2787
Petunia hybrida hort. (Petunia)			
1	Flower (bud, anther)	Phytic acid	2363
Phaseolus vulgaris L. (Kidney Bean)			
1	Leaf	2-PGA, G1P, G6P, F6P, F1,6P, AMP, ADP, ATP	22
1	Leaf	PC, PE, PG, PI	4108
1	Leaf	PC, PE, PG, PI, PS	1634
1	Leaf	PC, PG	1566
1	Leaf	PC	4109
1	Leaf (chloroplast envelope membrane)	PC, PG	4110
1	Leaf	AMP, ADP, ATP, GDP, GTP, CTP, UMP, UTP, UDPG, GDPG, NAD, ADPR	2649
1	Leaf	ATP	2756
1	Leaf (petiole, laminae)	AMP, ADP, ATP, GMP, GTP, CMP, CTP, UMP, UTP	2653
1	Root	ATP	4111
Picea abies (L.) Karst. (Norway Spruce)			
1	Needle	G6P, F6P	4112
1	Needle	F6P, F2,6P	4113
1	Needle	P_1, G6P, F6P, F2,6P	530
1	Shoot	PC, PE, PI	2115
Picea mariana (Mill.) B.S.P. (Black Spruce)			
1	Leaf, bud, terminal stem	P_{sugar}, $P_{phytate}$, P_{lipid}, $P_{nucleotide}$, P_i, P_{total}	833
Pinus nigra Arnold (Black Pine)			
1	Needle	PC, PE, PG	1892
Pinus pinea L. (Italian Stone Pine)			
1	Leaf	PLP	3866
Pinus resinosa Ait. (Red Pine)			
1	Ovule	ATP	2671
Pinus tadea L. (Loblolly Pine)			
1	Callus culture	cAMP	2828
Pisum sativum L. (Peas)			
1	Development	P_i, HMP, HDP, ADP, ATP, GTP, UTP	52
1	Leaf	PC, PA	4114
1	Leaf	PC	4109

No.	Plant species	Compounds analyzed	References
Platanus spp. (Plane Tree)			
1	Leaf	PC, PE, PG, PI, PS	1634
Populus gelrica (Houtz.) Houtz.			
1	Bark	G3P, 3-PGA, E4P, R5P, G1P, G6P, 6-PG, F6P, F1,6P, Se7P	4115
1	Bark (xylem)	AMP, ADP, ATP	2685
Populus nigra L. (Black Poplar)			
1	Bark	G3P, E4P, R5P, G1P, G6P, 6-PG, F6P, F1,6P, Se7P	298
Portulaca grandiflora Hook. (Rose Moss)			
1	Leaf	PEP	244
Portulaca oleracea L. (Purslane)			
1	Leaf	3-PGA, PEP, P_{sugar}	4116
Prunus armeniaca L. (Apricot)			
1	Flower	$P_{acid\text{-}soluble}$, P_{lipid}, $P_{protein}$, P_{RNA}, P_{DNA}	4117
Prunus avium (L.) L. (Cherry)			
1	Leaf	PG	1921
Prunus mume Sieb. and Zucc. (Ume Apricot)			
1	Fruit	PC, PE	4118
Prunus persica (L.) Batsch (Peach)			
1	Leaf	$P_{acid\text{-}soluble}$, P_{lipid}, $P_{protein}$, P_{RNA}, P_{DNA}	4119
1	Flower	$P_{acid\text{-}soluble}$, P_{lipid}, $P_{protein}$, P_{RNA}, P_{DNA}	4120
Pyrus communis L. (Pear)			
1	Fruit skin	R5P, F1,6P, Se7P	19
Raphanus sativus L. (Radish)			
1	Leaf	PC, PE, PG, PI	1938
Ricinus communis L. (Castor Bean)			
1	Root-shoot axis, nucellus-integument	P_i, P_{sugar}, $P_{phytate}$, P_{lipid}, P_i, P_{total}	3912
Robinia pseudoacacia L. (Black Locust)			
1	Bark	PC, PE, PG, PI, PS, DPG	4121
Saccharum hybrid (Sugarcane)			
1	Leaf	3-PGA, HMP	4122
Salsola gemmascens Pall.			
1	Leaf	3-PGA, phytic acid, P_i, $P_{nucleotide}$	3897
Salsola sclerantha C.A.M.			
1	Leaf	3-PGA, phytic acid, P_i, $P_{nucleotide}$	3897
Secale cereale L. (Rye)			
1	Leaf	FMN, FAD	3601
Sesamum indicum L. (Sesame)			
1	Oil bodies, microsomes	PC, PE, PG	4123
Solanum acaule Bitt. (Wild Potato)			
1	Leaf	PC, PE, PG, PI	1972

Continued

No.	Plant species	Compounds analyzed	References
–, ssp. *punae* Juz.			
1	Leaf	PC, PE, PG, PI	1972
Solanum acaule Bitt. ssp. *schreiteri* Buk.			
1	Leaf	PC, PE, PG, PI	1972
Solanum demissum Lindl. (Nightshade)			
1	Leaf	PC, PE, PG, PI	1972
Solanum maglia Schlecht.			
1	Leaf	PC, PE, PG, PI	1975
Solanum pamitanae			
1	Leaf	PC, PE, PG, PI	1972
Solanum rybinii Juz. and Buk.			
1	Leaf	PC, PE, PG, PI	1972
Solanum sanctae-rosae Hawkes			
1	Leaf	PC, PE, PG, PI	1972
Solanum tuberosum L. (Potato)			
3	Tuber	F2,6P, UDPG	532
1	Tuber	TP, 3-PGA, PEP, G1P, G6P, F6P, F1,6P, F2,6P	533
2	Tuber	Starch P, phytic acid	4124,4125
6	Tuber	PC, PE, PI	4126
1	Root, shoot	ADP, ATP	4127
1	Tuber (starch grain)	ADP, ADPG, NAD	3480
Sorghum bicolor (L.) Moench (Sorghum)			
1	Leaf	3-PGA	4128
1	Leaf (chloroplast)	PC, PG, PI	2060
Spartina alterniflora Loisel. (Smooth Cordgrass)			
1	Leaf	AMP, ADP, ATP	2542
Spartina cynosuroides (L.) Roth			
1	Leaf	AMP, ADP, ATP	2542
Spartina patens (Ait.) Muhl. (Saltmeadow Cordgrass)			
1	Leaf	AMP, ADP, ATP	2542
Spinacea oleracea L. (Spinach)			
1	Leaf	TP, PGA, PEP, G1P, G6P, F6P, F1,6P	4129
1	Leaf	PC, PG	1566
Trifolium pratense L. (Red Clover)			
1	Whole plant	Phytic acid	1016
Trifolium repens L. (White Clover)			
1	Leaf	6-PG	474
Triticum aestivum L. (Wheat)			
1	Leaf	TP, 3-PGA, RuBP	4130
2	Leaf	PC, PE, PG, PI, PS, PA	4092
1	Leaf	PC, PG	4131
1	Leaf (chloroplast)	PC, PG, PI	2060

No.	Plant species	Compounds analyzed	References
1	Leaf (chloroplast)	PG	4132
1	Stem (apex)	PC	4072
1	Tillering nodes	PC, PE, PG, PI, PS, PA, DPG	2038
1	Stem, leaf, flower	LPC	4133
1	Leaf	AMP, ADP, ATP, NAD, NADP	4095
1	Leaf	AMP, ADP, ATP, UDPG, UDPGlcNAc, NAD, NADP	3374,4134
1	Seedling	UDPG, UDPGal	537
1	Leaf	NAD, NADH, NADP, NADPH	3460
1	Leaf	FMN, FAD	3601
1	Flag leaf	P_{sugar}, P_{lipid}, P_i, P_{total}	4035
1	Leaf, root, tillering node	P_{sugar}, P_{lipid}, $P_{nucleotide}$, P_{RNA}, P_{DNA}, P_i, P_{total}	4135
1	Whole plant, straw, ears	$P_{phytate}$, P_{lipid}, P_i, P_{total}	3914
Vicia faba L. (Faba Bean)			
1	Leaf	PG	2047
1	Seedling, pod	AMP, ADP, ATP, GMP, GTP, UMP, UTP, UDPG, NAD	2738
1	Leaf, stem, cotyledon, pericarp, testa and hilum	FMN	4136
Vigna mungo (L.) Hepper (Black Gram)			
1	Root (nodule)	ATP	2741
Vigna ratiata (L.) Wilcz. (Mung Bean)			
1	Leaf (chloroplast)	PC, PG, PI	2060
Vigna unguiculata (L.) Walp. (Cowpea)			
1	Leaf	PC, PE, PG, PI	4137,4138
Vinca minor L. (Periwinkle)			
1	Leaf	PC, PE, PG	2065
Vitis vinifera L. (Wine Grape)			
1	Seedless berry	TP, PEP, G6P, F6P, F1,6P, AMP, ADP, ATP	333
1	Berry	PC, PE, PG, PI, PS, PA, LPC, LPE	4139
4	Vine	FMN, FAD	3611
Wolffiella floridana (Sm.) Thomps. (Water Meal)			
1	Frond (axenic culture)	Phytic acid	695
Zea mays L. (Corn)			
1	Ear	Phytic acid	810,1181
1	Leaf	PC, PG	1566
1	Leaf	PC, PE, PG	4140
1	Leaf (chloroplast)	PC, PG, PI	2060,4057
1	Leaf (chloroplast, thylakoid membrane)	PC, PG, PI, PS	4141
1	Root	PC, PE, PG, PS, PA, DPG	4078
1	Root	PC, PE, PG, DPG	4142
1	Root	ATP	4143

Continued

No.	Plant species	Compounds analyzed	References
2	Shoot, apex, leaf	ATP	2756
1	Leaf	AMP, ADP, ATP	4144
Zea mays L. ssp. *indentata* Sturt. (Dent Corn)			
2	Aerial portion	P_{sugar}, P_{lipid}, $P_{nucleotide}$, P_i, P_{total}	1182
Ziziphus vulgaris ssp. *inermis*			
1	Fruit (flesh, stone)	cGMP	2845

References

1. Bieleski, R. L. *Annu. Rev. Plant Physiol.* **1973**, *24*, 225–252.
2. Donaldson, N.; Powell, W. H.; Rowlett, R. J., Jr.; White, R. W.; Yorke, K. V. *J. Chem. Doc.* **1974**, *14*(1), 3–15.
3. LePage, G. A.; Umbreit, W. W. In: *Manometric Techniques and Related Methods for the Study of Tissue Metabolism;* Umbreit, W. W.; Burris, R. H.; Stauffer, J. F., Eds.; Burgess Publishing Co: Minneapolis, MN, 1945; pp 160–198.
4. MacDonald, D L., 2nd ed. Pigman, W.; Horton, D., Eds.; Academic Press: New York, 1972; Vol. 1A, pp 253–277, xxiii + 642 pp.
5. Bergmeyer, J.; Grassl, M., Eds.; *Methods of Enzymatic Analysis, Metabolites: Carbohydrates*; Vol. 6; Verlag Chemie: Weinheim, 1984; xxix + 701 pp. (1984). (a) p 335. (b) pp 416–427.
6. Benson, A. A. Paech, K.; Tracey, M. V., Eds.; Vol. 2; Springer-Verlag: Berlin, 1955 pp 113–144.
7. Murata, S.; Sekiya, J. *Plant Cell Physiol.* **1992**, *33*(8), 1239–1242.
8. Isherwood, F. A.; Barrett, F. C. *Biochem. J.* **1967**, *104*(3), 922–933.
9. Bieleski, R. L.; Young, R. E. *Anal. Biochem.* **1963**, *6*(1), 54–68.
10. Bieleski, R. L. *Anal. Biochem.* **1964**, *9*(4), 431–442.
11. von Caemmerer, S.; Edmondson, D. L. *Aust. J. Plant Physiol.* **1986**, *13*(5), 669–688.
12. Cole, C. V.; Ross, C. *Anal. Biochem.* **1966**, *17*(3), 526–539.
13. Arreguin-Lozano, B.; Bonner, J. *Plant Physiol.* **1949**, *24*(4), 720–738.
14. Albaum, H. G.; Umbreit, W. W. *Am. J. Bot.* **1943**, *30*(8), 553–558.
15. Pavlinova, O. A. *Dokl. Akad. Nauk SSSR* **1952**, *83*(4), 597–600.
16. Watson, C. A.; Miller, B. S.; Johnson, J. A.; Bequette, R. K. *Agron. J.* **1963**, *55*(6), 526–528.
17. Kursanov, A. L.; Pavlinova, O. A.; Afanas'eva, T. P. *Fiziol. Rast.* **1959**, *6*(3), 286–295 (Engl. 299–306).
18. Marré, E.; Teubner, F. G.; Murneek, A. E. *Am. J. Bot.* **1954**, *41*(9), 722–726.
19. Meynhardt, J. T.; Romani, R. J.; Maxie, E. C. *S. Afr. J. Agric. Sci.* **1965**, *8*(3), 691–701.
20. Heldt, H. W.; Portis, A. R.; Lilley, R. Mc.C.; Mosbach, A.; Chon, C. J. *Anal. Biochem.* **1980**, *101*(2), 278–287.
21. Khym, J. X.; Cohn, W. E. *J. Am. Chem. Soc.* **1953**, *75*(5), 1153–1156.
22. Adedipe, N. O.; Fletcher, R. A. *Plant Physiol.* **1970**, *46*(4), 614–617.
23. Alabran, D. M.; Mabrouk, A. F. *J. Agric. Food Chem.* **1973**, *21*(2), 205–208.
24. Sayler, D. F.; Geiger, P. J. *J. Food Sci.* **1993**, *58*(4), 890–892.
25. Srere, P.; Cooper, J. R.; Tabachnik, M.; Racker, E. *Arch. Biochem. Biophys.* **1958**, *74*(2), 295–305.
26. Sawada, S.; Kasai, M.; Yamada, T.; Takahashi, T. *J. Chromatogr.* **1987**, *402*, 283–292.
27. Smits, H. P.; Cohen, A.; Buttler, T.; Nielsen, J.; Olsson, L. *Anal. Biochem.* **1998**, *261*(1), 36–42.
28. Tyszkiewicz, E. *Anal. Biochem.* **1962**, *3*(2), 164–172.
29. Wawszkiewicz, E. J. *Anal. Chem.* **1961**, *33*(2), 252–254.

30. Ballou, C. E. *Methods Enzymol.* **1963**, *6*, 479–484.
31. Mortimer, D. C. *Can. J. Chem.* **1952**, *30*(9), 653–660.
32. Wade, H. E.; Morgan, D. M. *Biochem. J.* **1955**, *60*(2), 264–270.
33. Feige, B.; Gimmler, H.; Jaschke, W. D.; Simonis, W. *J. Chromatogr.* **1969**, *41*(1), 80–90.
34. Hanes, C. S.; Isherwood, F. A. *Nature* **1949**, *164*, 1107.
35. Bandurski, R. S.; Axelrod, B. *J. Biol. Chem.* **1951**, *193*(1), 405–410.
36. Bieleski, R. L. *Anal. Biochem.* **1965**, *12*(2), 230–234.
37. Bieleski, R. L.; Redgwell, R. J. *Aust. J. Plant Physiol.* **1977**, *4*(1), 1–10.
38. Vill, J.; Laish, A.; Oja, V.; Pärnik, T. *Photosynthetica* **1977**, *11*(3), 251–259.
39. Schwimmer, S.; Bevenue, A.; Weston, W. J. *Arch. Biochem. Biophys.* **1956**, *60*(2), 279–283.
40. Schürmann, P. *J. Chromatogr.* **1969**, *39*(4), 507–509.
41. Ahrens, H.; Korytnyk, W. *Anal. Biochem.* **1969**, *30*(3), 413–420.
42. Roberts, J. K. M.; Ray, P. M.; Wade-Jardetzky, R.; Jardetzky, O. *Nature* **1980**, *283*, 870–872.
43. Kime, M. J.; Ratcliffe, R. G.; Williams, R. J. P.; Loughman, B. C. *J. Exp. Bot.* **1982**, *33*(135), 656–669.
44. Colmer, T.; Fan, T. W.-M.; Higashi, R. M.; Läuchli, A. *J. Exp. Bot.* **1994**, *45*, 1037–1044.
45. Candelier, P.; Dauphin, G.; Gendraud, M. *Plant Physiol. Biochem. (Paris)* **1989**, *27*(2), 281–287.
46. Martin, J.-B.; Bligny, R.; Rebeille, F.; Douce, R.; Leguay, J.-J.; Mathieu, Y.; Guern, J. *Plant Physiol.* **1982**, *70*(4), 1156–1161.
47. Ishida, N.; Kobayashi, T.; Masuda, R.; Kano, H.; Yoshida, T.; Ogawa, H. *Agric. Biol. Chem.* **1990**, *54*(6), 1359–1365.
48. Delfini, M.; Angelini, R.; Bruno, F.; Conti, F.; Di Cocco, M. E.; Giuliani, A. M.; Manes, F. *Cell. Mol. Biol.* **1985**, *31*(5), 385–389.
49. Crans, D. C.; Mikuš, M.; Marshman, R. W. *Anal. Biochem.* **1993**, *209*(1), 85–94.
50. Bligny, R.; Gardestrom, P.; Roby, C.; Douce, R. *J. Biol. Chem.* **1990**, *265*(3), 1319–1326.
51. Bligny, R.; Foray, M.-F.; Roby, C.; Douce, R. *J. Biol. Chem.* **1989**, *264*(9), 4888–4895.
52. Rowan, K. S.; Turner, D. H. *Aust. J. Biol. Sci.* **1957**, *10*(4), 414–425.
53. Rowan, K. S. *J. Exp. Bot.* **1958**, *9*(27), 436–443.
54. Passoneau, J. V.; Lowry, O. H. *Enzymatic Analysis: A Practical Guide.* Humana Press: Totowa, NJ, 1993; x + 403 pp, p 313.
55. Norris, L.; Norris, R. E.; Calvin, M. *J. Exp. Bot.* **1955**, *6*(16), 64–74.
56. Grma-Pettenati, J.; Piquemal, M.; Latché, J.-C. *Physiol. Vég.* **1986**, *24*(5), 569–580.
57. Lee, S. H.; Ikeda, M.; Kang, Y. H.; Yamada, Y. *J. Fac. Agric. Kyushu Univ.* **1999**, *24*(1), 1–9.
58. Gupta, R. K.; Rathore, V. S. *Photosynthetica* **1982**, *17*(4), 532–536.
59. Zelitch, I. *J. Biol. Chem.* **1965**, *240*(5), 1869–1876.
60. Goyal, A.; Gupta, R. K. *J. Nucl. Agric. Biol.* **1985**, *14*(1), 4–7.
61. Goyal, A.; Gupta, R. K. *J. Nucl. Agric. Biol.* **1985**, *14*(3), 118–119.
62. Ta, T. C.; Joy, K. W.; Ireland, R. J. *Plant Physiol.* **1985**, *78*(2), 334–337.
63. Kumar, S.; Gupta, R. K. *Photosynthetica* **1986**, *20*(4), 390–396.
64. Larsson, C.; Albertsson, P.-A. *Biochim. Biophys. Acta* **1974**, *357*(3), 412–419.
65. Huh, D. H.; Kursell, A. V.; Husic, H. D. *Phytochemistry* **1989**, *28*(3), 719–720.

66. Kiessling, W. *Berichte* **1935**, *68B*, 243–248.
67. Kempf, I.; Zundel, G. *J. Mol. Struct.* **1992**, *269*(1/2), 65–74.
68. Robitaille, P.-M. L.; Robitaille, P. A.; Brown, G. G., Jr.; Brown, G. G. *J. Magn. Reson.* **1991**, *92*(1), 73–84.
69. Lis, T. *Acta Crystallogr.* **1993**, *C49*(4), 696–705.
70. Kent, S. S.; Hemming, B. C. *J. Chromatogr.* **1979**, *177*(2), 372–375.
71. Szewczuk, Y.; Wolny, E.; Wolny, M.; Baranowski, T. *Acta Biochim. Polon.* **1961**, *8*(2), 201–208.
72. Fleury, P.; Courtois, J. *Bull. Soc. Chim. Fr.* **1942**, *9*(3/4), 570–580.
73. Fleury, P.; Courtois, J. *Bull. Soc. Chim. Fr.* **1941**, *8*(1/2), 69–75.
74. Lis, T. *Acta Crystallogr.* **1996**, *C52*(1), 2326–2332.
75. Wang, P.; Murugaiah, V.; Yeung, B.; Vouros, P.; Giese, R. W. *J. Chromatogr.* **1996**, *721A*(2), 289–296.
76. Lis, T. *Acta Crystallogr.* **1994**, *C50*(2), 181–185.
77. Lis, T. *Acta Crystallogr.* **1995**, *C51*(5), 997–1001.
78. Cherbuliez, E.; Rabinowitz, J. *Helv. Chim. Acta* **1956**, *39*(5), 1461–1467.
79. Courtois, J.; Ramet, M. *Bull. Soc. Chim. Fr.* **1944**, *11*(11/12), 539–540.
80. Courtois, J. *Bull. Soc. Chim. Fr.* **1942**, *9*(1/2), 136–146.
81. Rose, Z. B.; Pizer, L. I. *J. Biol. Chem.* **1968**, *243*(18), 4806–4809.
82. Cherbuliez, E.; Rabinowitz, J. *Helv. Chim. Acta* **1956**, *39*(6), 1844–1855.
83. Ballou, C. E.; Fischer, H. O. L. *J. Am. Chem. Soc.* **1955**, *77*(12), 3329–3331.
84. Mavrovouniotis, M. J. *J. Biol. Chem.* **1991**, *266*(22), 14440–14445.
85. Serianni, A. S.; Pierce, J.; Barker, R. *Biochemistry* **1979**, *18*(7), 1192–1199.
86. Bedetti, G.; D'Agnolo, G.; Pocchiari, F. *J. Chromatogr.* **1970**, *49*(1), 53–56.
87. Rendina, A. R.; Cleland, W. W. *Anal. Biochem.* **1981**, *117*, 213–218.
88. Vestermark, A.; Sjödin, B. *J. Chromatogr.* **1972**, *73*(1), 211–218.
89. Shen, P.; Hauri, D.; Ross, J.; Oefner, P. J. *J. Capill. Electrophor.* **1996**, *3*(3), 155–163.
90. Schaeper, J. P.; Shamsi, S. A.; Danielson, N. D. *J. Capill. Electrophor.* **1996**, *3*(4), 215–221.
91. Klybas, V.; Schramm, M.; Racker, E. *Arch. Biochem. Biophys.* **1959**, *80*(2), 229–235.
92. Fischer, H. O. L.; Baer, E. *Berichte* **1932**, *65B*, 337–345.
93. Kiessling, W. *Biochem. Z.* **1934**, *273*, 103–108, from *Chem. Abstr.* **1935**, *29*, 25.
94. Swenson, C. A.; Barker, R. *Biochemistry* **1971**, *10*(16), 3151–3154.
95. Belasco, J. G.; Knowles, J. R. *Biochemistry* **1983**, *22*(1), 122–129.
96. Harvey, D. J.; Horning, M. G. *J. Chromatogr.* **1973**, *76*(1), 51–62.
97. Duke, C. C.; MacLeod, J. K.; Williams, J. F. *Carbohydr. Res.* **1981**, *95*(1), 1–26.
98. Lim, W. A.; Raines, R. T.; Knowles, J. R. *Biochemistry* **1988**, *27*(4), 1165–1167.
99. Vogt, A. M.; Ackermann, C.; Noe, T.; Jensen, D.; Kübler, W. *Biochem. Biophys. Res. Commun.* **1998**, *248*(3), 527–532.
100. Feurle, J.; Jomaa, J.; Wilhelm, M.; Gutsche, B.; Herderich, M. *J. Chromatogr.* **1998**, *803A*(1/2), 111–119.
101. Wong, C.-H.; Whitesides, G. M. *J. Org. Chem.* **1983**, *48*(19), 3199–3205.
102. Avigad, G.; England, S.; Listowsky, I. *Carbohydr. Res.* **1970**, *14*(3), 365–373.
103. Gray, G. R.; Barker, R. *Biochemistry* **1970**, *9*(12), 2454–2462.

104. Gray, G. R. *Biochemistry* **1971**, *10*(25), 4705–4711.

105. Schürmann, P.; Buchanan, B. B.; Arnon, D. I. *Biochim. Biophys. Acta* **1972**, *267*(1), 111–124.

106. Giersch, C. *J. Chromatogr.* **1979**, *172*, 153–161.

107. Smith, R. E.; MacQuarrie, R. A. *Anal. Biochem.* **1988**, *170*(2), 308–315.

108. Kiessling, W. *Berichte* **1934**, *67B*, 869–874.

109. Baer, E. *Biochem. Prep.* **1952**, *2*, 31–38.

110. Arrondo, J. L. R.; Goñi, F. M.; Macarulla, J. M. *Biochim. Biophys. Acta* **1984**, *794*(1), 165–168.

111. Hippmann, H.; Heinz, E. *Z. Pflanzenphysiol.* **1976**, *79*(5), 408–418.

112. Karrer, P.; Benz, P. *Helv. Chim. Acta* **1926**, *9*(4), 598–602.

113. Kiessling, W.; Schuster, P. *Berichte* **1938**, *71B*(1), 123–128.

114. Kalckar, H. *Biochem. J.* **1939**, *33*(1), 631–641.

115. Datta, S. P.; Grzybowski, A. K. *Biochem. J.* **1958**, *69*(2), 218–223.

116. Fukui, K.; Morokuma, K.; Nagata, C. *Bull. Chem. Soc. Jpn.* **1960**, *33*(9), 1214–1219.

117. Greenwald, J.; Halmann, M. *J. Chem. Soc., Perkin Trans. 2* **1972**, (9), 1095–1101.

118. Birdsall, N. J. M.; Feeney, J.; Lee, A. B.; Levine, Y. K.; Metcalfe, J. C. *J. Chem. Soc., Perkin Trans. 2* **1972**, (10), 1441–1445.

119. Wober, W.; Thiele, O. W. *Biochim. Biophys. Acta* **1966**, *116*(1), 163–166.

120. Duncan, H. J. *J. Chromatogr.* **1971**, *62*(3), 391–397.

121. Lairon, D.; Amic, J.; Lafont, H.; Nalbone, G.; Domingo, N.; Hauton, J. *J. Chromatogr.* **1974**, *88*(1), 183–186.

122. Urakami, C.; Murahashi, H.; Kakutani, Y. *Bull. Chem. Soc. Jpn.* **1957**, *30*(4), 379–381.

123. Handa, Y.; Matano, K.; Takeuchi, K.; Tsukatani, H. *J. Chromatogr.* **1981**, *206*(2), 387–391.

124. Charpentier, P.; Bocquet, M. *Bull. Soc. Chim. Fr.* **1932**, *51*(3), 413–419.

125. Fenn, R. H.; Marshall, G. E. *Biochem. J.* **1972**, *130*(1), 1–10.

126. Collatz, H. *Cellul. Chem.* **1936**, *17*, 128–129, from *Chem. Abstr.* **1937**, *31*, 2622g.

127. Inoue, M.; Inoue, A.; Mino, Y.; Ishida, T.; Nakagaki, M. *Chem. Pharm. Bull.* **1980**, *28*(5), 1491–1496.

128. King, H.; Pyman, F. L. *J. Chem. Soc.* **1914**, *105*, 1238–1259.

129. Ballou, C. E.; Fischer, H. O. L. *J. Am. Chem. Soc.* **1954**, *76*(12), 3188–3193.

130. Achilles, W.; Cumme, G. A.; Hoppe, H. *Acta Biol. Med. Germ.* **1973**, *31*(6), 763–770.

131. Rabiller, C.; Eymard, P.; Mesbahi, M. *Can. J. Chem.* **1991**, *69*(8), 1281–1287.

132. Curzon, E. H.; Hawkes, G. E.; Randall, E. W.; Britton, G. H.; Fazakerley, B. V. *J. Chem. Soc., Perkin Trans. 2* **1981**, (3), 494–499.

133. Conyers, R. A. J.; Newsholme, E. A.; Brand, K. *Biochem. Soc. Trans.* **1976**, *4*, 1040–1042.

134. Hofer, H. W. *Anal. Biochem.* **1974**, *61*(1), 54–61.

135. van Alebeek, G.-J. W. M.; Hermans, J. M. H.; Keltjens, J. T.; Vogels, G. D. *J. Chromatogr.* **1992**, *606*(1), 65–71.

136. Ciringh, Y.; Lindsey, J. S. *J. Chromatogr.* **1998**, *816A*(2), 251–259.

137. Lis, T. *Acta Crystallogr.* **1985**, *C41*(11), 1878–1880.

138. Neuberg, C. *Arch. Biochem.* **1943**, *3*(1), 105–112.

139. Meyerhof, O.; Kiessling, W. *Biochim. Biophys. Acta* **1947**, *1*, 292–314.

140. Neuberg, C.; Lustig, H. *Arch. Biochem.* **1942**, *1*(2), 311–318.

141. Plaut, G. W. E.; Kuby, S. A.; Lardy, H. A. *J. Biol. Chem.* **1950**, *184*(1), 243–249.

142. Wurmser, R. *Traite Biochim. Gen.* **1967**, *3*, 28–35.

143. Hashizume, T.; Sasaki, Y. *Anal. Biochem.* **1966**, *15*(2), 346–350.

144. Smrcka, A. V.; Jensen, R. G. *Plant Physiol.* **1988**, *86*(2), 615–618.

145. Embden, G.; Deuticke, H. J.; Kraft, G. *Z. Physiol. Chem.* **1934**, *230*, 12–28.

146. Fewster, P. F.; Fenn, R. H. *Acta Crystallogr.* **1982**, *B38*(1), 282–284.

147. Lis, T.; Jerzykiewicz, B. *Acta Crystallogr.* **1995**, *C51*(5), 1001–1005.

148. Kamiya, K.; Cruse, W. B. T.; Kennard, O. *Biochem. J.* **1983**, *213*(1), 217–223.

149. Pelavin, M.; Hendrickson, D. N.; Hollander, J. M.; Jolly, W. L. *J. Phys. Chem.* **1970**, *74*(5), 1116–1121.

150. Mostad, A.; Rosenqvist, E. *Acta Chem. Scand.* **1971**, *25*(1), 147–157.

151. Lis, T.; Jerzykiewicz, L. B. *Acta Crystallogr.* **1996**, *C52*(1), 129–134.

152. Zerzykiewicz, L. B.; Lis, T. *Acta Crystallogr.* **1997**, *C53*(9), 1249–1251.

153. Negelein, E.; Brömel, H. *Biochem. Z.* **1939**, *303*, 132–144.

154. Goux, W. J. *Carbohydr. Res.* **1983**, *122*(2), 332–339.

155. Jost, H. *Z. Physiol. Chem.* **1927**, *165*(4–6), 171–214.

156. Hobish, M. K.; Powers, D. A. *Biophys. Chem.* **1983**, *18*(4), 407–411.

157. Damiani, A.; Lapiccirella, A.; Kovacs, A. L. *J. Theor. Biol.* **1975**, *52*(2), 383–397.

158. Dellaripa, L.; Vickers, L. P.; Hopkins, H. P., Jr. *J. Solution Chem.* **1983**, *12*(11), 791–800.

159. Baer, E.; Robinson, R. *Can. J. Biochem.* **1971**, *49*(3), 300–305.

160. Dietrich, C. P.; Dietrich, S. M. C.; Pontis, H. G. *J. Chromatogr.* **1964**, *15*(2), 277–278.

161. Mayr, G. W. *Biochem. J.* **1988**, *254*(2), 585–591.

162. Baer, E. *J. Biol. Chem.* **1950**, *185*(2), 763–767.

163. Weichsel, A.; Lis, T.; Kuczek, M. *Carbohydr. Res.* **1989**, *194*, 63–70.

164. Benkovic, S. J.; Schray, K. J. *Biochemistry* **1968**, *7*(11), 4090–4096.

165. George, P.; Witonsky, R. J.; Trachtman, M.; Wu, C.; Dorwart, W.; Richman, L.; Rich-man, W.; Shurayh, F.; Lentz, B. *Biochim. Biophys. Acta* **1970**, *223*(1), 1–15.

166. Gáspár, R., Jr. *Acta Biochim. Biophys. Acad. Sci. Hung.* **1972**, *7*(3), 285–287.

167. Hayes, D. M.; Kenyon, G. L.; Kollman, P. A. *J. Am. Chem. Soc.* **1978**, *100*(14), 4331–4340.

168. Wold, F.; Ballou, C. E. *J. Biol. Chem.* **1957**, *227*(1), 301–312.

169. Cohn, M.; Pearson, J. E.; O'Connell, E. L.; Rose, I. A. *J. Am. Chem. Soc.* **1970**, *92*(13), 4095–4098.

170. Hirschbein, B. L.; Mazenod, F. P.; Whitesides, G. M. *J. Org. Chem.* **1982**, *47*(19), 3765–3766.

171. Katti, S. K.; Hosur, M. V.; Viswamitra, M. A. *Acta Crystasllogr.* **1981**, *B37*(4), 534–538.

172. Cramer, F.; Voges, D. *Chem. Ber.* **1959**, *92*(4), 952–958.

173. Kiessling, W. *Berichte* **1935**, *68B*, 597–601.

174. Weichsel, A.; Lis, T. *Acta Crystallogr.* **1989**, *C45*(8), 1177–1180.

175. Souhassou, M.; Schaber, P. M.; Blessing, R. H. *Acta Crystallogr.* **1996**, *B52*, 865–875.

176. Hosur, M. V.; Viswamitra, M. A. *Acta Crystallogr.* **1981**, *B37*(4), 839–843.

177. Lis, T.; Kuczek, M. *Acta Crystallogr.* **1991**, *C47*(8), 1598–1600.

178. Weichsel, A.; Lis, T. *Pol. J. Chem.* **1994**, *68*(10), 2079–2096.
179. Lis, T. *Acta Crystallogr.* **1996**, *C52*(11), 2720–2724.
180. Blessing, R. H. *J. Mol. Struct.* **1998**, *470*(1/2), 167–182.
181. Pederson, R. L.; Liu, K. K.-C.; Rutan, J. F.; Chen, L.; Wong, C.-H. *J. Org. Chem.* **1990**, *55*(16), 4897–4901.
182. Meyerhof, O. *Bull. Soc. Chim. Biol.* **1938**, *20*(9/10), 1033–1042.
183. Langheld, K. *Berichte* **1912**, *45*, 1125–1127.
184. Ballou, C. E.; Fischer, H. O. L. *J. Am. Chem. Soc.* **1956**, *78*(8), 1659–1661.
185. Jung, S.-H.; Jeong, J.-H.; Miller, P.; Wong, C.-H. *J. Org. Chem.* **1994**, *59*(23), 7182–7184.
186. Valentin, M.-L.; Bolte, J. *Bull. Soc. Chim. Fr.* **1995**, *132*(11), 1167–1171.
187. Gefflaut, T.; Lemaire, M.; Valentin, M.-L.; Bolte, J. *J. Org. Chem.* **1997**, *62*(17), 5920–5922.
188. Meyerhof, O.; Lohmann, K. *Biochem. Z.* **1934**, *273*, 413–418.
189. Fessner, W.-D.; Sinerius, G. *Angew. Chem. Int. Ed. Engl.* **1994**, *33*(2), 209–212.
190. Rios-Mercadillo, V. M.; Whitesides, G. M. *J. Am. Chem. Soc.* **1979**, *101*(19), 5828–5829.
191. Baer, E.; Kates, M. *J. Biol. Chem.* **1950**, *185*(2), 615–623.
192. Tankó, B. *Biochem. J.* **1963**, *89*(3), 40–41P.
193. Hartman, F. C.; Wold, F. *Biochim. Biophys. Acta* **1967**, *141*(3), 445–453.
194. Katagiri, J.; Murakami, S. *Biochem. J.* **1939**, *33*(2), 1257–1261.
195. Furfine, C. S.; Velick, S. F. *J. Biol. Chem.* **1965**, *240*(2), 844–855.
196. Baer, E.; Fischer, H. O. L. *J. Biol. Chem.* **1949**, *180*(1), 145–154.
197. Clark, V. M.; Kirby, A. J. *Biochem. Prep.* **1966**, *11*, 101–104.
198. Sekine, M.; Futatsugi, T.; Yamada, K.; Hata, T. *J. Chem. Soc., Perkin Trans. I* **1982**, (11), 2509–2513.
199. Trentham, D. R.; McMurray, C. H.; Pogson, C. I. *Biochem. J.* **1969**, *114*(1), 19–24.
200. Reynolds, S. J.; Yates, D. W.; Pogson, C. I. *Biochem. J.* **1971**, *122*(3), 285–297.
201. Beck, W. S. *Methods Enzymol.* **1957**, *3*, 201–207.
202. Kozlova, N. Ya.; Mel'nichenko, I. V.; Yasnikov, A. A. *Ukr. Khim. Zh.* **1968**, *34*(10), 1041–1046 (Engl. 56–60).
203. Mironova, D. F. *Ukr. Khim. Zh.* **1964**, *30*(3), 257–261.
204. Weil-Malherbe, H.; Green, R. H. *Biochem. J.* **1951**, *49*(3), 286–292.
205. Dimroth, K.; Witzel, H.; Hülsen, W.; Mirbach, H. *Ann. Chem.* **1959**, *620*, 94–108.
206. Bamann, E.; Nowotny, E. *Chem. Ber.* **1948**, *81*(5), 563–566.
207. Bamann, E.; Trapmann, H.; Oechsner, B. *Arch. Pharm.* **1962**, *295*(9), 663–670.
208. Kennedy, J. F.; Barker, S. A.; Humphreys, J. D. *J. Chem. Soc., Perkin Trans. 1* **1977**, (7), 753–757.
209. Bailly, M. C. *Bull. Soc. Chim. Fr.* **1942**, *9*(3/4), 340–350.
210. Kugel, L.; Halmann, M. *J. Am. Chem. Soc.* **1966**, *88*(15), 3566–3572.
211. Verkade, P. E.; Stoppelenburg, J. C.; Cohen, W. D. *Rec. Trav. Chim.* **1940**, *59*, 586–592.
212. Tsuboi, K. K.; Hudson, P. B. *Arch. Biochem. Biophys.* **1956**, *61*(1), 197–210.
213. Baer, E.; Fischer, H. O. L. *J. Biol. Chem.* **1940**, *135*(1), 321–328.
214. Wold, F.; Ballou, C. E. *J. Am. Chem. Soc.* **1959**, *81*(10), 3368–3371.
215. Trapmann, H.; Devani, M. *Enzymologia* **1964**, *28*(2), 65–84.
216. Bamann, E.; Fischler, F.; Trapmann, H. *Biochem. Z.* **1954**, *325*(6), 413–428.

217. Trapmann, H.; Devani, N. *Naturwissenschaften* **1965**, *52*(9), 208.

218. Cohn, M. *Biochim. Biophys. Acta* **1956**, *20*(1), 92–99.

219. Greenwald, I. *J. Biol. Chem.* **1925**, *63*(2), 339–349.

220. Benkovic, S. J.; Schray, K. J. *Biochemistry* **1968**, *7*(11), 4097–4102.

221. Schwarzenbach, G.; Anderegg, G. *Helv. Chim. Acta* **1957**, *40*(5), 1229–1231.

222. MacRae, E.; Quick, W. P.; Benker, C.; Stitt, M. *Planta* **1992**, *188*(3), 314–323.

223. Mukherjee, D. *Plant Biochem. J.* **1977**, *4*(1), 34–38.

224. Leegood, R. C.; Von Caemmerer, S. *Planta* **1988**, *174*(2), 253–262.

225. Höhler, T.; Schaub, H. *Biochem. Physiol. Pflanz.* **1979**, *174*(1), 58–67.

226. Sharma, S.; Malik, C. P.; Singh, M. B. *J. Palynol.* **1980**, *16*(1/2), 155–156.

227. Bieleski, R. L. *Plant Physiol.* **1966**, *41*(3), 447–454.

228. Rao, G. G.; Rao, G. R. *Indian J. Plant Physiol.* **1979**, *22*(2), 121–126.

229. Loske, D.; Raschke, K. *Planta* **1988**, *173*(2), 275–281.

230. ap Rees, T.; Fuller, W. A.; Wright, B. W. *Biochim. Biophys. Acta* **1977**, *461*(2), 274–282.

231. Hampp, R.; Goller, M.; Füllgraf, H. *Plant Physiol.* **1984**, *75*(4), 1017–1021.

232. Larondelle, Y.; Corbineau, F.; Dethier, M.; Come, D.; Hers, H.-G. *Eur. J. Biochem.* **1987**, *166*(3), 605–610.

233. Mukherjee, D.; Laloraya, M. M. *Plant Biochem J.* **1975**, *2*(1), 1–11.

234. Mukherjee, D.; Laloraya, M. M. *Plant Biochem J.* **1980**, *7*(2), 120–125.

235. Mukherjee, D.; Laloraya, M. M. *J. Indian Bot. Soc.* **1975**, *54*(3/4), 207–210.

236. Rao, I. M.; Arulanatham, A. R.; Terry, M. *Plant Physiol.* **1989**, *90*(3), 820–826.

237. Santarius, K. A. *Planta* **1967**, *73*(3), 228–242.

238. Singal, H. R.; Laura, J. S.; Singh, R. *Indian J. Biochem. Biophys.* **1993**, *30*(5), 270–276.

239. Paul, M. J.; Lawlor, D. W.; Driscoll, S. P. *J. Exp. Bot.* **1990**, *41*(226), 547–555.

240. Bieleski, R. L. *Plant Physiol.* **1969**, *44*(4), 497–502.

241. Mukherjee, D. *J. Indian Bot. Soc.* **1974**, *53*(1/2), 115–118.

242. Steer, B. T. *Plant Physiol.* **1974**, *54*(5), 758–761.

243. Kubota, K.; Ashihara, H. *Biochim. Biophys. Acta* **1990**, *1036*(2), 138–142.

244. Mukherjee, D. *Plant Biochem. J.* **1976**, *3*(1), 24–32.

245. Hatch, M. D. *Arch. Biochem. Biophys.* **1979**, *194*(1), 117–127.

246. Grisolia, S.; Joyce, B. K. *J. Biol. Chem.* **1959**, *234*(6), 1335–1337.

247. Aldasoro, J. J.; Nicolas, G. *Int. J. Biochem.* **1979**, *10*(11), 947–950.

248. Lifshitz, A.; Geiger, P. J. *Lebensm. Wiss. Technol.* **1985**, *18*(1), 43–46.

249. Reddy, A. R.; Das, V. S. R. *Proc. Indian Natl. Sci. Acad.* **1987**, *53B*(5/6), 493–498.

250. Matsumoto, H.; Wakiushi, N.; Takahashi, E. *Physiol. Plant.* **1969**, *22*(3), 537–545.

251. Lehmann, J. *Z. Pflanzenphysiol.* **1981**, *102*(5), 415–424.

252. Leegood, R. C.; ap Rees, T. *Biochim. Biophys. Acta* **1978**, *542*(1), 1–11.

253. Stitt, M.; Cséke, C.; Buchanan, B. *Plant Physiol.* **1986**, *80*(1), 246–248.

254. Kowalczyk, S. *Biochem. Physiol. Pflanz.* **1989**, *184*(5/6), 371–376.

255. Fan, T. W.-M.; Colmer, T. D.; Lane, A. N.; Higashi, R. M. *Anal. Biochem.* **1993**, *214*(1), 260–271.

256. Sawada, S.; Harada, A.; Aseri, Y.; Asano, S.; Kuninake, M.; Kawamura, H.; Kasai, M. *Ecol. Res.* **1999**, *14*(2), 119–130.

257. Effer, W. R.; Ranson, S. L. *Plant Physiol.* **1967**, *42*(8), 1042–1052.

258. Barker, J.; Jakes, R.; Solomos, T.; Younis, M. E.; Isherwood, F. A. *J. Exp. Bot.* **1964**, *15*(44), 284–296.

259. Isherwood, F. A.; Selvendran, R. R. *Phytochemistry* **1970**, *9*(11), 2265–2269.

260. Adams, C. A.; Rinne, R. W. *Plant Cell Physiol.* **1981**, *22*(6), 1011–1021.

261. Fredeen, A. L.; Raab, T. K.; Rao, I. M.; Terry, N. *Planta* **1990**, *181*(3), 399–405.

262. Avigad, G. *Anal. Lett.* **1984**, *17*(B5), 371–384.

263. El-Hinnawy, S. I.; Rachenisky, V. V. *Ann. Agric. Sci.* **1967**, *10*(1), 97–112.

264. Law, R. D.; Crafts-Brandner, S. J. *Plant Physiol.* **1999**, *120*(1), 173–181.

265. Mertens, E.; Van Schaftingen, E.; Hers, H.-G. *FEBS Lett.* **1987**, *221*(1), 124–128.

266. Bieleski, R. L.; Reid, M. S. *Plant Physiol.* **1992**, *98*(3), 1042–1049.

267. Latzko, E.; Kotzé, J. P. *Z. Pflanzenphysiol.* **1965**, *53*(5), 377–387.

268. Labate, C. A.; Leegood, R. C. *Plant Physiol.* **1989**, *91*(3), 905–910.

269. Stitt, M.; Schreiber, U. *J. Plant Physiol.* **1988**, *133*(3), 263–271.

270. Leegood, R. C.; Labate, C. A.; Huber, S. C.; Neuhaus, H. E.; Stitt, M. *Planta* **1988**, *176*(1), 117–126.

271. Winter, H.; Lohaus, G.; Heldt, H. W. *Plant Physiol.* **1992**, *99*(3), 996–1004.

272. Kendall, A. C.; Keys, A. J.; Turner, J. C.; Lea, P. J.; Miflin, B. J. *Planta* **1983**, *159*(6), 505–511.

273. Pierre, J. N.; Queiroz, O. *Planta* **1979**, *144*(2), 143–151.

274. Cockburn, W.; McAulay, A. *Plant Physiol.* **1977**, *59*(3), 455–458.

275. Kenyon, W. H.; Holaday, A. S.; Black, C. C. *Plant Physiol.* **1981**, *68*(5), 1002–1007.

276. Perl, M.; Globerson, D. *Phytochemistry* **1981**, *20*(9), 2289–2290.

277. Galmiche, J. M. *Plant Physiol.* **1973**, *51*(3), 512–519.

278. Chalmers, D. J.; Rowan, K. S. *Plant Physiol.* **1971**, *48*(3), 235–240.

279. Barker, J.; Khan, M. A. A. *New Phytol.* **1968**, *67*(1), 205–212.

280. Hanker, I.; Kůdelová, A. *Biol. Plant.* **1986**, *28*(4), 288–296.

281. Curioni, P. M. G.; Hartwig, U. E.; Nösberger, J.; Schuller, K. A. *Plant Physiol.* **1999**, *119*(2), 445–453.

282. Loreto, F.; Sharkey, T. D. *Planta* **1993**, *189*(3), 420–424.

283. Salminen, S. O.; Young, R. E. *Plant Physiol.* **1975**, *55*(1), 45–50.

284. Sicher, R. C.; Kremer, D. F. *Environ. Exp. Bot.* **1996**, *36*(4), 385–391.

285. Macnicol, P. K.; Young, R. E.; Biala, J. B. *Plant Physiol.* **1973**, *51*(4), 793–797.

286. Macnicol, P. K. *Plant Physiol.* **1973**, *51*(4), 798–801.

287. Macnicol, P. K. *Anal. Biochem.* **1972**, *45*(2), 624–633.

288. Paul, M. J.; Stitt, M. *Plant Cell Environ.* **1993**, *16*(9), 1047–1057.

289. Nakamura, Y.; Yuki, K.; Park, S.-Y.; Ohya, T. *Plant Cell Physiol.* **1989**, *30*(6), 833–839.

290. Yahia, E. M.; Carrillo-Lopez, A. *Food Sci. Technol. (Lond.)* **1993**, *26*(4), 307–311.

291. Badger, M. R.; Sharkey, T. D.; von Caemmerer, S. *Planta* **1984**, *160*(4), 305–313.

292. Farineau, J. In: *Photosynthesis and Photorespiration;* Hatch, M. D.; Osmond, C. B.; Slatyer, R. O., Eds.; Wiley-Interscience: New York, 1971; pp 202–210.

293. Einig, W.; Hampp, R. *Trees* **1990**, *4*(1), 9–15.

294. Day, D. A.; Lambers, H. *Physiol. Plant.* **1983**, *58*(2), 155–160.

295. Edwards, J.; ap Rees, T. *Phytochemistry* **1986**, *25*(9), 2033–2039.

296. Barker, J.; Khan, M. A. A.; Solomos, T. *New Phytol.* **1967**, *66*(4), 577–596.

297. Lorenc-Plucinska, G. *J. Exp. Bot.* **1998**, *49*, 213–219.

298. Sagisaka, S. *Low Temp. Sci., Ser. B Biol. Sci.* **1971**, *29*, 19–28.

299. Kerbel, E. L.; Kader, A. A.; Romani, R. J. *Plant Physiol.* **1988**, *86*(4), 1205–1209.

300. Bukhov, N. G.; Bondar, V. V.; Drozdova, I. S.; Kara, A. N.; Kotov, A. A.; Maevskaya, S. N.; Vasil'ev, A. A.; Voevudskaya, S. Yu.; Voronin, P. Yu.; Mokronosov, A. T. *J. Plant Physiol.* **1996**, *149*(3/4), 405–412.

301. Bourne, D. T.; Ranson, S. L. *Plant Physiol.* **1965**, *40*(6), 1178–1190.

302. Kobr, M. J.; Beevers, H. *Plant Physiol.* **1971**, *47*(1), 48–52.

303. Geigenberger, P.; Stitt, M. *Planta* **1991**, *185*(1), 81–90.

304. Geigenberger, P.; Langenberger, S.; Wilke, I.; Heineke, D.; Heldt, H. W.; Stitt, M. *Planta* **1993**, *190*(4), 446–453.

305. Du, Y.-C.; Nose, A.; Wasano, K. *Plant Cell Environ.* **1999**, *22*(3), 317–324.

306. Cheong, Y. W. Y.; Chan, P. Y. *Plant Soil* **1973**, *38*(1), 113–123.

307. Whittaker, A.; Botha, F. C. *Plant Physiol.* **1997**, *115*(4), 1651–1659.

308. Du, Y.-C.; Nose, A.; Wasano, K.; Uchida, Y. *Aust. J. Plant Physiol.* **1998**, *25*(2), 253–260.

309. Kluge, M.; Becker, D.; Ziegler, H. *Planta* **1970**, *91*(1), 68–79.

310. Reddy, A. R.; Das, V. S. R. *Acta Physiol. Plant.* **1995**, *17*(3), 249–254.

311. Haake, V.; Geiger, M.; Walch-Liu, P.; Engels, C.; Zrenner, R.; Stitt, M. *Plant J.* **1999**, *17*(5), 479–489.

312. Heineke, D.; Kruse, A.; Flügge, U.-I.; Frommer, W. B.; Riesmeier, J. W.; Willmitzer, L.; Heldt, H. W. *Planta* **1994**, *193*, 174–180.

313. Vecher, A. S.; Masny, M. M. *Vestsi Akad. Navuk BSSR, Ser. Biyal. Navuk* **1966**, (1), 62–68.

314. Bieleski, R. L.; Laties, G. G. *Plant Physiol.* **1963**, *38*(5), 586–594.

315. Schwimmer, S.; Bevenue, A.; Weston, W. J. *J. Agric. Food Chem.* **1955**, *3*(3), 257–260.

316. Mori, K.; Nakamura, M.; Funahashi, S. *Bull. Agric. Chem. Soc. Jpn.* **1960**, *24*(4), 344–350.

317. Hajirezaei, M.; Stitt, M. *Plant Sci.* **1991**, *77*(2), 177–183.

318. Isherwood, F. A. *Phytochemistry* **1976**, *15*(1), 33–41.

319. Nielsen, T. H.; Wischmann, B.; Enevoldsen, K.; Møller, B. L. *Plant Physiol.* **1994**, *105*(1), 111–117.

320. Reddy, A. R.; Das, V. S. R. *Biochem. Int.* **1990**, *20*(1), 25–35.

321. Heber, U.; Willenbrink, J. *Biochim. Biophys. Acta* **1964**, *82*(2), 313–324.

322. Quick, P.; Neuhaus, E.; Feil, R.; Stitt, M. *Biochim. Biophys. Acta* **1989**, *973*(2), 263–273.

323. Champigny, M. L.; Brauer, M.; Bismuth, E.; Manh, C. T.; Siegl, G.; Quy, L. V.; Stitt, M. *J. Plant Physiol.* **1992**, *139*(3), 361–368.

324. Dietz, K.-J.; Heber, U. *Biochim. Biophys. Acta* **1984**, *767*(3), 432–443.

325. Gerhardt, R.; Stitt, M.; Hecht, H. W. *Plant Physiol.* **1987**, *83*(2), 399–407.

326. Wirtz, W.; Stitt, M.; Heldt, H. W. *Plant Physiol.* **1980**, *66*(1), 187–193.

327. Bieleski, R. L. *Plant Physiol.* **1968**, *43*(8), 1297–1308.
328. Becker, D.; Kluge, M.; Ziegler, H. *Planta* **1971**, *99*(2), 154–162.
329. Wilson, A. M.; Huffaker, R. C. *Plant Physiol.* **1964**, *39*(4), 555–560.
330. Washüttl, J.; Bancher, E. *Angew. Bot.* **1968**, *42*(3/4), 155–158.
331. Bouillenne-Comhaire, F. *Bull. Soc. R. Sci. Liege* **1967**, *36*(1/2), 56–63.
332. Ashihara, H.; Komamine, A. *Plant Sci. Lett.* **1974**, *2*(6), 331–337.
333. Ruffner, H. P.; Hawker, J. S. *Phytochemistry* **1977**, *16*(8), 1171–1175.
334. Lafon-Lafourcade, S. *Connaiss. Vigne Vin* **1973**, *7*(3), 179–187.
335. Chaumont, M.; Osório, M. L.; Chaves, M. M.; Vanacker, H.; Morot-Gaudry, J.-F.; Foyer, C. H. *J. Plant Physiol.* **1997**, *150*(6), 743–751.
336. Fischer, E.; Raschke, K.; Stitt, M. *Planta* **1986**, *169*(4), 536–545.
337. Farineau, J. *Planta* **1969**, *85*(2), 135–156.
338. Usuda, H. *Plant Physiol.* **1985**, *78*(4), 859–864.
339. Stitt, M.; Heldt, H. W. *Biochim. Biophys. Acta* **1985**, *808*(3), 400–414.
340. Leegood, R. C. *Planta* **1985**, *164*(2), 163–171.
341. Weiner, H.; Heldt, H. W. *Planta* **1992**, *187*(2), 242–246.
342. Liu, T.-T. Y.; Shannon, J. C. *Plant Physiol.* **1981**, *67*(3), 525–529.
343. Zinselmeier, C.; Jeong, B. R.; Boyer, J. S. *Plant Physiol.* **1999**, *121*(1), 25–35.
344. Michalik, I.; Ivanko, S. *Biologia (Bratislava)* **1976**, *31*(4), 221–232.
345. Arora, K. K.; Collins, J. G.; MacLeod, J. K.; Williams, J. F. *Biol. Chem. Hoppe-Seyler* **1988**, *369*(7), 549–557.
346. Wood, T. *J. Chromatogr.* **1968**, *35*(3), 352–361.
347. Lobarzewski, J.; Dénes, G. *Acta Biochim. Biophys.* **1966**, *1*(4), 403–405.
348. Pierce, J.; Serianni, A. S.; Barker, R. *J. Am. Chem. Soc.* **1985**, *107*(8), 2448–2456.
349. Sieben, A. S.; Perlin, A. S.; Simpson, F. J. *Can. J. Biochem.* **1966**, *44*(6), 663–669.
350. Ballou, C. E.; Fischer, H. O. L.; McDonald, D. L. *J. Am. Chem. Soc.* **1955**, *77*(2), 5967–5970.
351. Baxter, J. N.; Perlin, A. S.; Simpson, P. J. *Can. J. Biochem. Physiol.* **1959**, *37*(2), 199–209.
352. Horecker, B. L.; Smyrniotis, P. Z.; Seegmiller, J. E. *J. Biol. Chem.* **1951**, *193*(1), 383–396.
353. Massoud, S. S.; Sigel, H. *Inorg. Chem.* **1988**, *27*(8), 1447–1453.
354. Izatt, R. M.; Rytting, J. H.; Hansen, L. D.; Christensen, J. J. *J. Am. Chem. Soc.* **1966**, *88*(12), 2641–2645.
355. Govil, G.; Saran, A. *J. Theor. Biol.* **1971**, *30*(3), 621–630.
356. Ito, A.; Ito, T. *Photochem. Photobiol.* **1986**, *44*(3), 355–358.
357. Laylayan, V. A.; Ismail, A. A. *Carbohydr. Res.* **1995**, *276*(2), 253–266.
358. Carmona, P.; Molina, M. *J. Raman Spectrosc.* **1990**, *21*(7), 395–400.
359. Birdsall, B.; Feeney, J. *J. Chem. Soc., Perkin Trans. 2* **1972**, (11), 1643–1649.
360. Gross, A.; Abril, O.; Lewis, J. M.; Geresh, S.; Whitesides, G. M. *J. Am. Chem. Soc.* **1983**, *105*(25), 7428–7435.
361. Gascon, A.; Wood, T.; Chitemerere, L. *Anal. Biochem.* **1981**, *118*(1), 4–9.
362. Levene, P. A.; Harris, S. A.; Stiller, E. T. *J. Biol. Chem.* **1934**, *105*(1), 153–156.
363. Furberg, S.; Mostad, A. *Acta Chem. Scand.* **1962**, *16*(7), 1627–1636.

364. Levene, P. A.; Stiller, E. T. *J. Biol. Chem.* **1934**, *104*(2), 299–306.

365. Voet, D.; Gratzer, W. B.; Cox, R. A.; Doty, P. *Biopolymers* **1963**, *1*(3), 193–208.

366. Wood, T. *Carbohydr. Res.* **1975**, *40*, 407–413.

367. Runeckles, V. C.; Krotkov, G. *Arch. Biochem. Biophys.* **1957**, *70*(2), 442–453.

368. Jakob, R.; Saenger, W. *FEBS Lett.* **1985**, *183*(1), 111–114.

369. Pontremoli, S.; Mangiarotti, G. *J. Biol. Chem.* **1962**, *237*(3), 643–645.

370. Rice, S. C.; Pon, N. G. *Anal. Biochem.* **1978**, *87*(1), 39–48.

371. Horecker, B. L.; Hurwitz, J.; Weissbach, A. *J. Biol. Chem.* **1956**, *218*(2), 785–794.

372. Wong, C.-H.; Pollak, A.; McCurry, S. D.; Sue, J. M.; Knowles, J. R.; Whitesides, G. M. *Methods Enzymol.* **1982**, *89D*, 108–121.

373. Zimmermann, F. T.; Schneider, A.; Schörken, U.; Sprenger, G. A.; Fessner, W.-D. *Tetrahedron: Asymmetry* **1999**, *10*(9), 1643–1646.

374. Franke, F. P.; Kapuscinski, M.; Lyndon, P. *Carbohydr. Res.* **1985**, *143*, 69–76.

375. Mocali, A.; Aldinucci, D.; Paoletti, F. *Carbohydr. Res.* **1985**, *143*, 288–293.

376. Michelson, A. M.; Todd, A. R. *J. Chem. Soc.* **1949**, (10), 2476–2486.

377. Khym, J. X.; Doherty, D. G.; Cohn, W. E. *J. Am. Chem. Soc.* **1954**, *76*(21), 5523–5530.

378. Fromm, H. J. *J. Biol. Chem.* **1959**, *234*(12), 3097–3101.

379. Simpson, F. J.; Bhuyan, B. K. *Can. J. Microbiol.* **1962**, *8*(5), 663–672.

380. Hurwitz, J.; Horecker, B. L. *J. Biol. Chem.* **1956**, *223*(2), 993–1998.

381. Borenfreund, E.; Dische, Z. *Biochim. Biophys. Acta* **1957**, *25*(1), 215–216.

382. Horecker, B. L.; Smyrniotis, P. Z.; Hurwitz, J. *J. Biol. Chem.* **1956**, *223*(2), 1009–1019.

383. Tewari, Y. B.; Steckler, D. K.; Goldberg, R. N.; Gitomer, W. L. *J. Biol. Chem.* **1988**, *263*(8), 3670–3675.

384. Haga, M.; Mikelsaar, P.; Laane, A.; Aaviksaar, A.; Peenema, E.; Arukaevu, H. *Prikl. Biokhim. Mikrobiol.* **1979**, *15*(5), 686–692, from *Chem. Abstr.* **1980**, *92*, 2424q.

385. Paech, C.; Pierce, J.; McCurry, S. D.; Tolbert, N. E. *Biochem. Biophys. Res. Commun.* **1978**, *83*(3), 1084–1092.

386. Horecker, B. L.; Hurwitz, J.; Heppel, L. A. *J. Am. Chem. Soc.* **1957**, *79*(3), 701–702.

387. Kobza, J.; Edwards, G. E. *Plant Physiol.* **1987**, *83*(1), 69–74.

388. Latzko, E.; Gibbs, M. *Plant Physiol.* **1969**, *44*(3), 396–402.

389. Sassenrath-Cole, G. F.; Pearcy, R. W. *Plant Physiol.* **1992**, *99*(1), 227–234.

390. Byrd, G. T.; Ort, D. R.; Ogren, W. L. *Plant Physiol.* **1995**, *107*(2), 585–591.

391. Masnyi, M. N. *Fiziol. Biokhim. Usloviya Povysh. Prod. S-kh. Rast.* **1978**, 53–58.

392. Herrmann, H.; Vogel, J. *Nahrung* **1962**, *6*(7/8), 693–700.

393. Robinson, J. M. *Plant Physiol.* **1984**, *75*(2), 397–409.

394. Perchorowicz, J. T.; Jensen, R. O. *Plant Physiol.* **1983**, *71*(4), 955–960.

395. Mott, K. A.; Jensen, R. G.; O'Leary, J. W.; Berry, J. A. *Plant Physiol.* **1984**, *76*(4), 968–971.

396. Sicher, R. C.; Kremer, D. F. *Plant Sci.* **1990**, *67*(1), 47–56.

397. Redgwell, R. J.; Bieleski, R. L. *Phytochemistry* **1978**, *17*(3), 407–409.

398. Cori, C. F.; Colowick, S. P.; Cori, G. T. *J. Biol. Chem.* **1937**, *121*(2), 465–477.

399. Ashby, J. H.; Clarke, H. B.; Crook, E. M.; Datta, S. P. *Biochem. J.* **1955**, *59*(2), 203–208.

400. Gáspár, R., Jr. *Acta Biochim. Biophys. Acad. Sci. Hung.* **1975**, *10*(1/2), 79–82.

401. Trachtman, M.; Halmann, M. *Carbohydr. Res.* **1971**, *19*(2), 245–248.

402. Pace, N.; Tanford, C.; Davidson, E. A. *J. Am. Chem. Soc.* **1964**, *86*(15), 3160–3162.

403. Wolucka, B. A.; Rush, J. S.; Waechter, C. J.; Shibaev, V. N.; de Hoffmann, E. *Anal. Biochem.* **1998**, *255*(2), 244–251.

404. Dallinga, J. W.; Rinkema, F. D.; Heerma, W. *Biomed. Environ. Mass Spectrom.* **1989**, *18*(4), 241–246.

405. O'Connor, J. V.; Nunez, H. A.; Barker, R. *Biochemistry* **1979**, *18*(3), 500–507.

406. Fitchett, M.; Gilbert, B. C.; Willson, R. L. *J. Chem. Soc., Perkin Trans. 2* **1988**, (5), 763–789.

407. Hatzack, F.; Rasmussen, S. K. *J. Chromatogr.* **1999**, *B736*(1/2), 221–229.

408. Leblanc, D. J.; Ball, A. J. S. *Anal. Biochem.* **1978**, *84*(2), 574–578.

409. Narendra, N.; Seshadri, T. P.; Viswamitra, M. A. *Acta Crystallogr.* **1984**, *C40*(8), 1338–1340.

410. Takahashi, F.; Sakai, Y.; Yoshida, T. *Bull. Chem. Soc. Jpn.* **1993**, *66*(2), 371–373.

411. Reynhardt, E. C.; Froneman, S. *Mol. Phys.* **1989**, *66*(5), 921–927.

412. Stoll, R.; Röllgen, F. W. *Org. Mass Spectrom.* **1979**, *14*(12), 542–545.

413. Lis, T. *Carbohydr. Res.* **1992**, *229*(1), 33–39.

414. Posternak, T. *J. Am. Chem. Soc.* **1950**, *72*(10), 4824–4825.

415. Beevers, C. A.; Maconochie, G. H. *Acta Crystallogr.* **1965**, *18*(2), 232–236.

416. Clarke, H. B.; Datta, S. P. *Biochem. J.* **1957**, *66*(3), 451–456.

417. Parker, F. S. *Biochim. Biophys. Acta* **1960**, *42*(3), 513–519.

418. McDowell, C. A.; Naito, A.; Sastry, D. L.; Takegoshi, K. *J. Magn. Reson.* **1988**, *78*(3), 498–510.

419. Sugawara, Y.; Iwasaki, H. *Acta Crystallogr.* **1984**, *C40*(3), 389–393.

420. Clarke, H. B.; Datta, S. P. *Biochem. J.* **1956**, *64*(4), 604–606.

421. Putman, R. W.; Hassid, W. Z. *J. Am. Chem. Soc.* **1957**, *79*(18), 5057–5060.

422. Hanessian, S.; Lu, P. P.; Ishida, H. *J. Am. Chem. Soc.* **1998**, *120*(51), 13296–13300.

423. Wolfrom, M. L.; Smith, C. S.; Pletcher, D. E.; Brown, A. E. *J. Am. Chem. Soc.* **1942**, *64*(1), 23–26.

424. Viscontini, M.; Olivier, C. *Helv. Chim. Acta* **1953**, *36*(2), 466–470.

425. Robison, R.; King, E. J. *Biochem. J.* **1931**, *25*(1), 323–338.

426. Degani, C.; Halmann, M. *J. Am. Chem. Soc.* **1966**, *88*(17), 4075–4082.

427. Hirade, J. *J. Biochem.* **1934**, *20*(1), 161–191.

428. Henderson, S. K.; Henderson, D. E. *J. Chromatogr. Sci.* **1986**, *24*(5), 198–203.

429. Schützner, K. *Cesk. Farm.* **1974**, *23*(1), 34–35.

430. Lis, T. *Carbohydr. Res.* **1985**, *135*(2), 187–194.

431. Schulten, H. R.; Beckey, H. D.; Bessell, E. M.; Foster, A. B.; Jarman, M.; West-wood, J. H. *J. Chem. Soc., Chem. Commun.* **1973**, (13), 416–418.

432. Saito, T.; Noguchi, J. *Nippon Kagaku Zasshi* **1961**, *82*(4), 469–471.

433. Lardy, H. A.; Fischer, H. O. L. *J. Biol. Chem.* **1946**, *164*(2), 513–519.

434. Levene, P. A.; Raymond, A. L. *J. Biol. Chem.* **1931**, *92*(3), 757–763.

435. Katti, S. K.; Seshadri, T. P.; Viswamitra, M. A. *Acta Crystallogr.* **1982**, *B38*(4), 1136–1140.

436. Wood, W. A.; Horecker, B. L. *Biochem. Prep.* **1953**, *3*, 71–73.

437. Cohen, J. S.; Chen, C.-W.; Bax, A. *J. Magn. Reson.* **1984**, *59*(1), 181–187.

438. Hanna, R.; Mendicino, J. *J. Biol. Chem.* **1970**, *245*(16), 4031–4037.

439. Buck, K. W. *Carbohydr. Res.* **1968**, *6*(2), 247–249.

440. Posternak, T. *J. Biol. Chem.* **1949**, *180*(3), 1269–1278.

441. London, R. E.; Matwiyoff, N. A.; Mueller, D. D. *J. Chem. Phys.* **1975**, *63*(10), 4442–4449.

442. Bessman, S. P.; Geiger, P. J.; Lu, T.-C.; McCabe, E. R. B. *Anal. Biochem.* **1967**, *59*(2), 533–546.

443. Levene, P. A.; Raymond, A. L. *J. Biol. Chem.* **1931**, *91*(2), 751–760.

444. Smith, G. D.; Fitzgerald, A.; Caughlan, C. N.; Kerr, K. A.; Ashmore, J. P. *Acta Crystallogr.* **1974**, *B30*(7), 1760–1766.

445. Jarori, G. K.; Maitra, P. K. *FEBS Lett.* **1991**, *278*(2), 247–251.

446. Wells, W. W.; Katugi, T.; Bentley, R.; Sweeley, C. C. *Biochim. Biophys. Acta* **1964**, *82*(2), 408–411.

447. Alberty, R. A. *Arch. Biochem. Biophys.* **1998**, *358*(1), 25–39.

448. Wolff, J. B.; Kaplan, N. O. *J. Biol. Chem.* **1956**, *218*(2), 849–869.

449. Leavitt, A. L.; Sherman, W. R. *Methods Enzymol.* **1982**, *89D*, 9–18.

450. Faulkner, P. *Biochem. J.* **1956**, *64*(3), 436–441.

451. Sim, M. M.; Kondo, H.; Wong, C.-H. *J. Am. Chem. Soc.* **1993**, *115*(6), 2260–2267.

452. MacDonald, D. L. *J. Org. Chem.* **1962**, *27*(3), 1107–1109.

453. McCready, R. M.; Hassid, W. Z. *Biochem. Prep.* **1955**, *4*, 63–70.

454. Seegmiller, J. E.; Horecker, B. L. *J. Biol. Chem.* **1951**, *192*(1), 175–180.

455. Tanaka, T. *Yukagaku Zasshi* **1961**, *81*(6), 797–800.

456. Atherton, F. R.; Howard, H. T.; Todd, A. R. *J. Chem. Soc.* **1948**, (7), 1106–1111.

457. Orlowski, M. *J. Biol. Chem.* **1959**, *234*(7), 1651–1652.

458. Colowick, S. P.; Sutherland, E. W. *J. Biol. Chem.* **1942**, *144*(2), 423–437.

459. Leloir, L. F.; Paladini, A. C. *Methods Enzymol.* **1957**, *3*, 143–147.

460. Moir, R. D.; Stokes, G. B. *Biochem. J.* **1988**, *256*(1), 69–73.

461. Lee, C.-H.; Sarma, R. H. *Biochemistry* **1976**, *15*(3), 697–704.

462. Bailey, J. M.; Fishman, P. H.; Pentchev, P. G. *J. Biol. Chem.* **1968**, *243*(18), 4827–4831.

463. Chance, E. M.; Hess, B.; Plesser, T.; Wurster, B. *Eur. J. Biochem.* **1975**, *50*(2), 419–424.

464. Cori, O.; Lipmann, F. *J. Biol. Chem.* **1952**, *194*(1), 417–425.

465. Bauer, H. P.; Srihari, T.; Jochims, J. C.; Hofer, H. W. *Eur. J. Biochem.* **1983**, *133*(1), 163–168.

466. Desjobert, A. *Bull. Soc. Chim. Biol.* **1951**, *33*(1/2), 42–49.

467. Bunton, C. A.; Llewellyn, D. R.; Oldham, K. G.; Vernon, C. A. *J. Chem. Soc.* **1958**, (10), 3588–3594.

468. Cohn, M. *J. Biol. Chem.* **1949**, *180*(2), 771–781.

469. Bunton, C. A.; Humeres, E. *J. Org. Chem.* **1969**, *34*(3), 572–576.

470. O'Connor, J. V.; Barker, R. *Carbohydr. Res.* **1979**, *73*, 227–234.

471. Bunton, C. A.; Chaimovich, H. *J. Am. Chem. Soc.* **1966**, *88*(17), 4082–4089.

472. Patwardhan, V. N. *Biochem. J.* **1934**, *28*, 1854–1862.

473. Viola, R.; Davies, H. V.; Sommerville, L. B. *Phytochem. Anal.* **1994**, *5*(1), 10–14.

474. Johnson, E. L.; Heck, W. W.; Fites, R. C.; Blum, U. *Environ. Exp. Bot.* **1987**, *27*(4), 441–448.

475. Sacchi, G. A.; Morgutti, S.; Abruzzese, A.; Alisi, C.; Cocucci, M.; Espen, L.; Leva, A. R.; Muleo, R.; Negrini, N.; Cocucci, S. M. *Plant Sci.* **1995**, *106*, 107–113.

476. Sicher, R. C.; Baysdorfer, C.; Kremer, D. F. *Plant Physiol.* **1987**, *83*(4), 768–771.

477. Stitt, M.; Huber, S.; Kerr, P. *Biochem. Plants* **1987**, *10*, 327–409.

478. Fahrendorf, T.; Holtum, J. A. M.; Mukherjee, U.; Latzko, E. *Plant Physiol.* **1987**, *84*(1), 182–187.

479. Sola, M. M.; Gutiérrez, M.; Vargas, A. M. *J. Plant Physiol.* **1994**, *144*(4/5), 569–575.

480. Sicher, R. C.; Kremer, D. F. *Physiol. Plant.* **1992**, *85*(3), 446–452.

481. Gendraud, M.; Lafleuriel, J. *Physiol. Vég.* **1978**, *16*(4), 679–691.

482. Servaites, J. C.; Geiger, D. R.; Tucci, M. A.; Fondy, B. R. *Plant Physiol.* **1989**, *89*(2), 403–408.

483. Buchanan, J. G. *Arch. Biochem. Biophys.* **1953**, *44*(1), 140–149.

484. Doman, N. G.; Kagan, Z. S. *Biokhimiya* **1952**, *17*(6), 719–724.

485. Duff, S. M. G.; Moorhead, G. B. G.; Lefebvre, D. D.; Plaxton, W. C. *Plant Physiol.* **1989**, *90*(4), 1275–1278.

486. Graham, D.; ap Rees, T. *Nature* **1965**, *208*, 88–89.

487. Nakamura, N.; Suzuki, Y.; Suzuki, H. *Physiol. Plant.* **1992**, *86*(4), 616–622.

488. Selvendran, R. R.; Isherwood, F. A. *Phytochemistry* **1970**, *9*(3), 533–536.

489. Phelps, D. C.; McDonald, R. E. *Plant Physiol.* **1989**, *90*(2), 455–462.

490. Nielsen, T. H.; Veierskov, B. *Plant Physiol.* **1990**, *93*(2), 637–641.

491. Rembold, H.; Weigner, C. *Z. Naturforsch.* **1990**, *45C*(7/8), 922–923.

492. Botha, A.-M.; Botha, F. C. *Plant Physiol.* **1993**, *101*(4), 1385–1390.

493. Parekh, L. J.; Sakariah, K. K.; Shah, V. J. *Enzymologia* **1970**, *38*(1), 23–28.

494. Van Praag, E.; Gonzalez, P.; Brune, A.; Goren, R.; Zehavi, U.; Echeverria, E. *Biochem. Mol. Biol. Int.* **1998**, *44*(1), 117–125.

495. Bellando, M.; Sacco, S.; Albergoni, F.; Rocco, P.; Marre, M. T. *Bot. Acta* **1997**, *110*(5), 388–394.

496. de Veau, E. J. I.; Robinson, J. M.; Warmbrodt, R. D.; Kremer, D. F. *Plant Physiol.* **1992**, *99*(3), 1105–1117.

497. Gommers, F. G.; Dropkin, V. H. *Phytopathology* **1977**, *67*(7), 869–873.

498. Perera, N. H.; Hartmann, E.; Holaday, A. S. *Plant Sci.* **1995**, *111*(2), 133–143.

499. Zhukovskaya, N. V.; Loladze, L. N. *Tr. Inst. Bot., Akad. Nauk Kaz. SSR* **1972**, *32*, 102–111.

500. Morgan, L. J.; Pridham, J. B. *Phytochemistry* **1986**, *25*(1), 19–22.

501. Scott, P.; Kruger, N. J. *Planta* **1994**, *193*(1), 16–20.

502. Kumar, S.; Batra, V. I. P. *Curr. Sci.* **1989**, *58*(12), 686–687.

503. Bergmann, H. *Biol. Zentralbl.* **1973**, *92*(4), 443–453.

504. Franz, M.; Meier, H. *Planta* **1969**, *85*(2), 202–208.

505. Prasad, B. K.; Sinha, N. P.; Singh, S. P.; Dayal, S.; Kumar, S.; Jallaluddin, ; Prasad, P. *Indian Phytopathol.* **1997**, *50*(3), 362–364.

506. Collis, B. E.; Pollock, C. J. *New Phytol.* **1991**, *119*(4), 483–489.

507. Pollock, C. J.; Cairns, A. J.; Collis, B. E.; Walker, R. P. *J. Plant Physiol.* **1989**, *134*(2), 203–208.

508. Barta, A. L. *Plant Cell Environ.* **1986**, *9*, 127–131.

509. Kuroda, H.; Sagisaka, S. *Biosci. Biotechnol. Biochem.* **1992**, *56*(5), 712–715.

510. Keiller, D. R.; Paul, M. J.; Cockburn, W. *New Phytol.* **1987**, *107*(1), 1–13.

511. Mertens, E.; Marcellin, P.; van Schaftingen, E.; Hers, H.-G. *Eur. J. Biochem.* **1987**, *167*(3), 579–583.

512. Kurasawa, H.; Igaue, I.; Hayakawa, T.; Ogami, H. *Nippon Nogei Kagaku Kaishi* **1959**, *33*(5), 388–392.

513. Reddy, A. R.; Das, V. S. R. *J. Exp. Bot.* **1987**, *38*, 828–833.

514. Mertens, E.; Larondelle, Y.; Hers, H.-G. *Plant Physiol.* **1990**, *93*(2), 584–587.

515. Lange, D. L.; Kader, A. A. *J. Am. Soc. Hortic. Sci.* **1997**, *122*(2), 253–257.

516. Bennett, A. B.; Smith, G. M.; Nichols, B. G. *Plant Physiol.* **1987**, *83*(4), 973–976.

517. Nordal, A.; Benson, A. A. *J. Am. Chem. Soc.* **1954**, *76*(20), 5054–5055.

518. Kesy, J.; Kowalczyk, S. *J. Plant Physiol.* **1987**, *127*(5), 411–421.

519. Seemann, J. R.; Sharkey, T. D. *Plant Physiol.* **1986**, *82*(2), 555–560.

520. Watada, A. E. *Proc. W. Va. Acad. Sci.* **1969**, *41*, 145–149.

521. Botha, F. C.; Small, J. G. C. *Plant Physiol.* **1987**, *83*(4), 772–777.

522. Hampp, R.; Schaeffer, C.; Wallenda, T.; Stülten, C.; Johann, P.; Einig, W. *Can. J. Bot.* **1995**, (Suppl. 1), S548–S556.

523. Bartels, H. *Flora (Jena)* **1968**, *A159*(5), 391–402.

524. Drozdova, I. S.; Bondar, V. V.; Bukhov, N. G. *Fiziol. Rast.* **1992**, *39*(4), 639–644 (Engl. 404–410).

525. Paz, N.; Xu, D.-P.; Black, C. C., Jr. *Plant Physiol.* **1985**, *79*(4), 1133–1136.

526. Lee, R. B.; Ratcliffe, R. G. *J. Exp. Bot.* **1983**, *34*(146), 1222–1244.

527. Lunn, J. E.; ap Rees, T. *Biochem. J.* **1990**, *267*(3), 739–743.

528. Kruger, N. J.; Beevers, H. *Plant Physiol.* **1985**, *77*(2), 358–364.

529. Lingle, S. E.; Smith, R. C. *Crop Sci.* **1991**, *31*(1), 172–177.

530. Egger, B.; Einig, W.; Schlereth, A.; Wallenda, T.; Magel, E.; Loewe, A.; Hampp, R. *Physiol. Plant.* **1996**, *96*(1), 91–100.

531. Ślaski, J. J. *Plant Soil* **1994**, *167*, 165–171.

532. Morrell, S.; ap Rees, T. *Phytochemistry* **1986**, *25*(7), 1579–1585.

533. Merlo, L.; Geigenberger, P.; Hajirezaei, M.; Stitt, M. *J. Plant Physiol.* **1993**, *142*(4), 392–402.

534. Stitt, M.; Gerhardt, R.; Künzel, B.; Heldt, H. W. *Plant Physiol.* **1983**, *72*(4), 1139–1141.

535. Sagisaka, S. *Soil Sci. Plant Nutr.* **1995**, *41*(4), 807–812.

536. Reddy, A. R. *Indian J. Exp. Biol.* **1998**, *36*(4), 403–406.

537. Mahajan, R.; Singh, R. *Indian J. Biochem. Biophys.* **1990**, *27*(5), 324–328.

538. Esposito, S.; Bowsher, C. G.; Emes, M. J.; Tetlow, I. J. *J. Plant Physiol.* **1999**, *154*(1), 24–29.

539. Zagdańska, B. *J. Plant Physiol.* **1989**, *134*(3), 320–326.

540. Ashihara, H.; Matsumura, H. *Int. J. Biochem.* **1977**, *8*(6), 461–471.

541. Ashihara, H.; Sato, F. *Biochim. Biophys. Acta* **1993**, *1156*(2), 123–127.

542. Yazaki, Y.; Asukagawa, N.; Ishikawa, Y.; Ohta, E.; Sakata, M. *Plant Cell Physiol.* **1988**, *29*(6), 919–924.

543. Uchiyama, T.; Sakata, M. *Oyo Butsuri* **1988**, *57*(2), 263–269.

544. Lee, R. B.; Ratcliffe, R. G. *J. Exp. Bot.* **1983**, *34*(146), 1213–1221.

545. Stitt, M.; Heldt, H. W. *Planta* **1985**, *164*(2), 179–188.

546. Tobias, R. B.; Boyer, C. D.; Shannon, J. C. *Plant Physiol.* **1992**, *99*(1), 146–152.

547. Horecker, B. L.; Wood, W. A. *Methods Enzymol.* **1957**, *3*, 152–154.

548. Kumar, R.; Singh, R. *J. Agric. Food Chem.* **1984**, *32*(4), 806–808.

549. Buchanan, J. G.; Cummerson, D. A.; Turner, D. M. *Carbohydr. Res.* **1972**, *21*(2), 283–292.

550. Meyerhof, O.; Lohmann, K.; Schuster, P. *Biochem. Z.* **1936**, *286*(5/6), 319–335.

551. Koerner, T. A. W., Jr.; Voll, R. J.; Cary, L. W.; Younathan, E. S. *Biochemistry* **1980**, *19*(12), 2795–2801.

552. Lefebvre, M. J.; Gonzalez, N. S.; Pontis, H. G. *J. Chromatogr.* **1964**, *15*(4), 495–500.

553. Pogell, B. M. *J. Biol. Chem.* **1953**, *201*(2), 645–649.

554. Tankó, B.; Robison, R. *Biochem. J.* **1935**, *29*, 961–972.

555. Meyerhof, O.; Lohmann, K. *Biochem. Z.* **1927**, *185*(1/3), 113–164.

556. Reynhardt, E. C. *Chem. Phys. Lett.* **1989**, *155*(3), 317–320.

557. Lis, T. *Acta Crystallogr.* **1986**, *C42*(12), 1745–1747.

558. Narendra, N.; Seshadri, T. P.; Viswamitra, M. A. *Acta Crystallogr.* **1985**, *C41*(11), 1612–1614.

559. Neuberg, C.; Lustig, H.; Rothenberg, M. A. *Arch. Biochem.* **1943**, *3*(1), 33–44.

560. Meyerhof, O.; Suranyi, J. *Biochem. Z.* **1926**, *178*(4/6), 427–443.

561. McGilvery, R. W. *Biochemistry* **1965**, *4*(10), 1924–1930.

562. Barrett, T. W. *Spectrochim. Acta* **1981**, *37A*(40), 233–239.

563. Bergmeyer, H. U.; Naeher, G.; Traeger, H.; Brustmann, H. Ger. Patent 2,061,731 to Boehringer Mannheim G.m.b.H. 1972; from *Chem. Abstr.* **1972**, *77*, 34850g.

564. Narendra, N.; Seshadri, T. P.; Viswamitra, M. A. *Acta Crystallogr.* **1985**, *C41*(1), 31–34.

565. Stepanenko, B. N.; Silaeva, E. A. *Biokhimiya* **1949**, *14*(6), 544–551.

566. Neuberg, C.; Sabetay, S. *Biochem. Z.* **1925**, *161*, 240–243.

567. McCready, R. M.; Hassid, W. Z. *J. Am. Chem. Soc.* **1944**, *66*(4), 560–563.

568. McGilvery, R. W. *J. Biol. Chem.* **1953**, *200*(2), 835–838.

569. Voll, E. J.; Ramaprasad, S.; Vargas, D.; Younathan, E. S.; Laban, S.; Koerner, T. A. W. *Carbohydr. Res.* **1990**, *203*(2), 173–182.

570. Uyeda, K.; Furuya, E.; Sherry, A. D. *J. Biol. Chem.* **1981**, *256*(16), 8679–8684.

571. Raymond, A. L.; Levene, P. A. *J. Biol. Chem.* **1929**, *83*(3), 619–630.

572. Leuthardt, F.; Testa, E. *Helv. Chim. Acta* **1951**, *34*(3), 931–938.

573. Hers, H. G.; Beaufays, H.; de Duve, C. *Biochim. Biophys. Acta* **1953**, *11*, 416–426.

574. Mandl, I.; Neuberg, C. *Methods Enzymol.* **1957**, *3*, 162–172.

575. Van Schaftingen, E.; Hers, H. G. *Biochem. Biophys. Res. Commun.* **1980**, *96*(4), 1524–1531.

576. Van Schaftingen, E.; Hers, H.-G. *Eur. J. Biochem.* **1981**, *117*(2), 319–323.

577. Robison, R. *Biochem. J.* **1932**, *26*, 2191–2202.

578. Volkova, N. V.; Semenyuk, I. I.; Yasnikov, A. A. *Ukr. Khim. Zh.* **1967**, *33*(7), 712–716 (Engl. 56–59).

579. Fleury, P.; Courtois, J.; Desjobert, A. *Bull. Soc. Chim. Fr.* **1948**, (5/6), 694–700.

580. Macleod, M.; Robison, R. *Biochem. J.* **1933**, *27*, 286–297.

581. Lee, J. B. *J. Org. Chem.* **1963**, *28*(9), 2473–2475.

582. Midelfort, C. F.; Gupta, R. K.; Rose, I. A. *Biochemistry* **1976**, *15*(10), 2178–2185.

583. Benkovic, S. J.; Eagle, L. L.; Mildvan, A. S. *Biochem. Biophys. Res. Commun.* **1972**, *47*(4), 852–858.

584. Barker, J.; Isherwood, F. A.; Jakes, R.; Solomos, T.; Younis, M. E. *Nature* **1962**, *196*, 1115.

585. Van Schaftlingen, E.; Lederer, B.; Bartrons, R.; Hers, H.-G. *Eur. J. Biochem.* **1982**, *129*(1), 191–195.

586. Van Schaftingen, E.; Hers, H.-G. *FEBS Lett.* **1983**, *164*(1), 195–200.

587. Loughman, B. C.; Ratcliffe, R. G.; Schwabe, J. W. R. *Plant Sci.* **1989**, *59*(1), 11–23.

588. Miller, W. B. *Physiol. Plant.* **1989**, *77*(1), 123–128.

589. Kim, J.; Gross, K. C.; Solomos, T. *Postharvest. Biol. Technol.* **1991**, *1*(1), 67–80.

590. Harris, G. C.; Gibbs, P. B.; Ludwig, G.; Un, A.; Sprengnether, M.; Kolodny, N. *Plant Physiol.* **1986**, *82*(4), 1081–1089.

591. Colowick, S. P. *J. Biol. Chem.* **1938**, *124*(2), 557–558.

592. Kosterlitz, H. W. *Biochem. J.* **1955**, *37*(3), 3211–3212.

593. Hirano, S.; Fukuda, T.; Kondo, S. *Agric. Biol. Chem.* **1974**, *38*(8), 1515–1520.

594. Heidlas, J. E.; Lees, W. J.; Whitesides, G. M. *J. Org. Chem.* **1992**, *57*(1), 152–157.

595. MacDonald, D. L. *Carbohydr. Res.* **1968**, *6*(3), 376–381.

596. Piras, R.; Cabib, E. *Anal. Chem.* **1963**, *35*(6), 755–760.

597. Kosterlitz, H. W. *Biochem. J.* **1939**, *33*(2), 1087–1093.

598. Krishman, R.; Seshadri, t.P. *Acta Crystallogr.* **1990**, *C46*(12), 2299–2302.

599. Sabesan, S.; Neira, S. *Carbohydr. Res.* **1992**, *223*, 169–185.

600. Warren, C. D.; Jeanloz, R. W. *Biochemistry* **1972**, *11*(14), 2565–2572.

601. Foster, A. B.; Overend, W. G.; Stacey, M. *J. Chem. Soc.* **1951**, (4), 980–987.

602. Wenger, W. C.; Anderson, R. L. *Carbohydr. Res.* **1981**, *88*(2), 267–276.

603. Ward, D. N.; Wen, T.; Bousfield, G. R. *J. Chromatogr.* **1987**, *398*, 255–264.

604. Levene, P. A.; Raymond, A. L. *J. Biol. Chem.* **1931**, *92*(3), 765–768.

605. Grant, G. A. *Biochem. J.* **1935**, *29*(2), 1551–1576.

606. Gorin, P. A. J.; Mazurek, M. *Can. J. Chem.* **1974**, *52*(17), 3070–3076.

607. Perchemlides, R.; Osawa, T.; Davidson, E. A.; Jeanloz, R. W. *Carbohydr. Res.* **1967**, *3*(4), 461–477.

608. Warren, C. D.; Jeanloz, R. W. *Biochemistry* **1973**, *12*(25), 5031–5037.

609. Danilov, L. L.; Volkova, L. V.; Bondarenko, V. A.; Evstigneeva, R. P. *Bioorg. Khim.* **1975**, *1*(7), 905–911.

610. Posternak, T.; Rosselet, J. P. *Helv. Chim. Acta* **1953**, *36*(6), 1614–1623.

611. Meldal, M.; Christensen, M. K.; Bock, K. *Carbohydr. Res.* **1992**, *235*, 115–127.

612. Hamdaoui, B.; Dewynter, D.; Capony, F.; Montero, J.-L.; Toiron, C.; Hnach, M.; Rochefort, H. *Bull. Soc. Chim. Fr.* **1994**, *131*(8), 854–864.

613. Pallanca, J. E.; Turner, N. J. *J. Chem. Soc., Perkin Trans. 1* **1993**, (21), 3017–3022.

614. Jephcott, C. M.; Robison, R. *Biochem. J.* **1934**, *28*(2), 1844–1853.

615. Foster, A. B.; Overend, W. G.; Stacey, M. *J. Chem. Soc.* **1951**, (4), 987–991.

616. Bamann, E.; Trapmann, H. *Biochem. Z.* **1955**, *326*(3), 161–163.

617. Danilov, L. L.; Utkina, N. S.; Shibaev, V. S. *Bioorg. Khim.* **1980**, *6*(5), 780–782.

618. Klenow, H.; Emberland, R. *Arch. Biochem. Biophys.* **1955**, *58*(2), 276–287.

619. Göring, H.; Reckin, E.; Kaiser, R. *Flora (Jena)* **1968**, *159A*(1), 82–103.

620. Okuda, T.; Mori, K. *Phytochemistry* **1974**, *13*(6), 961–964.

621. Bean, R. C.; Porter, G. G.; Barr, B. K. *Plant Physiol.* **1963**, *38*(3), 280–284.

622. McFadden, B. A.; Barden, L. L.; Rokke, N. W.; Uyeda, M.; Siek, T. J. *Carbohydr. Res.* **1967**, *4*(3), 254–259.

623. Franke, F. P.; Kapuscinski, M.; MacLeod, J. K.; Williams, J. F. *Carbohydr. Res.* **1984**, *125*(2), 177–184.

624. Horecker, B. L.; Smyrniotis, P. Z.; Hiatt, H. H.; Marks, P. A. *J. Biol. Chem.* **1955**, *212*(2), 827–836.

625. Horecker, B. L. *Methods Enzymol.* **1957**, *3*, 195–198.

626. Taggart, J.; Hallaway, M. *Biochem. J.* **1966**, *101*(2), 4–5.

627. Sokatch, J. R.; McFadden, B. A. *Biochem. Prep.* **1968**, *12*, 1–5.

628. Clark, M. G.; Blackmore, P. F.; Williams, J. F. *J. Label. Compds.* **1972**, *8*(4), 637–652.

629. Schmidt, U.; Stiller, R.; Brade, H.; Thiem, J. *Synlett.* **1998**, (2), 125–126.

630. Tolbert, N. E.; Nystrom, C. W.; Kerr, P. C. *Plant Physiol.* **1957**, *32*(4), 269–274.

631. Ishikawa, H.; Takaichi, K. *Nippon Nogei Kagaku Kaishi* **1957**, *31*(6), 380–383.

632. Holaday, A. S.; Martindale, W.; Alred, R.; Brooks, A. L.; Leegood, R. C. *Plant Physiol.* **1992**, *98*(3), 1105–1114.

633. Kapuscinski, M.; Williams, J. F. *Proc. Aust. Biochem. Soc.* **1984**, *16*, 6.

634. Tarelli, E.; Wheeler, S. F. *Carbohydr. Res.* **1995**, *269*(2), 359–368.

635. Keys, A. J.; Martin, R. V. *Biochem. J.* **1963**, *88*(3), P44–P45.

636. Leloir, L. F.; Cardini, C. E. *J. Biol. Chem.* **1955**, *214*(1), 157–165.

637. Krause, K.-P.; Stitt, M. *Phytochemistry* **1992**, *31*(4), 1143–1146.

638. Mendicino, J. *J. Biol. Chem.* **1960**, *235*(12), 3347–3352.

639. Kerr, R. W., Ed.; *Starch Sugars and Related Compounds*; Academic Press: New York, 1944; pp 156–159.

640. Greenwood, C. T. *Stärke* **1960**, *12*(6), 168–174.

641. Tabata, S.; Hizukuri, S. *Stärke* **1971**, *23*(8), 267–272.

642. Muhrbeck, P.; Tellier, C. *Starch/Stärke* **1991**, *43*(1), 25–27.

643. Muhrbeck, P.; Svensson, E.; Eliasson, A.-C. *Starch/Stärke* **1991**, *43*(12), 466–468.

644. Muhrbeck, P.; Eliasson, A.-C. *J. Sci. Food Agric.* **1991**, *55*(1), 13–18.

645. Tabata, S.; Hizukuri, S.; Nagata, K. *Carbohydr. Res.* **1978**, *67*(1), 189–195.

646. McBurney, L. J.; Smith, M. D. *Cereal Chem.* **1965**, *42*(2), 161–167.

647. Posternak, T. *J. Biol. Chem.* **1951**, *188*(1), 317–325.

648. Parrish, F. W.; Whelan, W. J. *Stärke* **1961**, *13*(6), 231–236.

649. Blennow, A.; Bay-Smidt, A. M.; Olsen, C. E.; Møller, B. L. *J. Chromatogr.* **1998**, *829A*, 385–391.

650. Samec, M.; Blinc, M. *Kolloidchem. Beih.* **1938**, *47*(9/12), 371–472.

651. Lim, S.-T.; Kasemsuwan, T.; Jane, J.-L. *Cereal Chem.* **1994**, *71*(5), 488–493.

652. Edwards, D. W.; Ripperton, J. C. *J. Agric. Res.* **1933**, *47*(3), 179–191.

653. Hizukuri, S.; Tabata, S.; Nikuni, Z. *Stärke* **1970**, *22*(10), 338–343.

654. Blennow, A.; Bay-Smidt, A. M.; Wischmann, B.; Olsen, C. E.; Møller, B. L. *Carbohydr. Res.* **1998**, *307*, 45–54.

655. Fernandez, O. *Ion (Madrid)* **1948**, *8*(790–792), 807.

656. Harris, G.; MacWilliams, I. C. *Stärke* **1963**, *15*(3), 98–101.

657. Topper, Y. J.; Mehler, A. H.; Bloom, B. *Science* **1957**, *126*, 1287.

658. Singh, V.; Ali, S. Z. *Starch/Stärke* **1987**, *39*(8), 277–279.

659. Taylor, T. C.; Walton, R. P. *J. Am. Chem. Soc.* **1929**, *51*(11), 3431–3440.

660. Hizukuri, S.; Shirasaka, K.; Juliano, B. O. *Starch/Stärke* **1983**, *35*(10), 348–350.

661. Reyes, A. C.; Albano, E. L.; Briones, V. P.; Juliano, B. O. *J. Agric. Food Chem.* **1965**, *13*(5), 438–442.

662. Kapustina, V. V.; Dudkin, M. S.; Kalashnikova, N. A. *Prikl. Biokhim. Mikrobiol.* **1972**, *8*(5), 554–558.

663. Goering, K. J.; Schuh, M. *Cereal Chem.* **1967**, *44*(5), 532–538.

664. Chang, S.-M. *J. Food Sci.* **1988**, *3*(3), 837–840.

665. Fortuna, T. *Acta Aliment. Pol.* **1991**, *17*(4), 275–283.

666. Bay-Smidt, A. M.; Wischmann, B.; Olsen, C. E.; Nielsen, T. H. *Starch/Stärke* **1994**, *46*(5), 167–172.

667. Quick, W. A.; Li, P. H. *Potato Res.* **1976**, *19*(4), 305–312.

668. Baldwin, M. E. *J. Am. Chem. Soc.* **1930**, *52*(7), 2907–2919.

669. Lim, S.; Seib, P. A. *Cereal Chem.* **1993**, *70*(2), 145–152.

670. Tabata, S.; Nagata, K.; Hizukuri, S. *Stärke* **1975**, *27*(10), 333–335.

671. Meyer, K. H.; Bernfeld, P.; Wolff, E. *Helv. Chim. Acta* **1940**, *23*(2), 854–864.

672. Posternak, S. *Schweiz. Wochenschr. Chem. Pharm.* **1904**, *42*, 405–406, from *Chem. Zentralbl.* **1904**, *II*, 717.

673. Graf, E. *Phytic Acid: Chemistry & Applications.* Pilatus Press: Minneapolis, MN, 1986; 344 pp. (a) pp 23–42. (b) pp 43–55. (c) pp 119–125.

674. Pfeffer, W. *Jahrb. Wiss. Bot.* **1872**, *8*, 429–574.

675. Winterstein, E. *Ber. Deutsch. Chem. Ges.* **1897**, *30*, 2299–2302.

676. Cosgrove, D. J.; Irving, G. C. J. *Inositol Phosphates: Their Chemistry, Biochemistry and Physiology.* Elsevier Publishing Company: Amsterdam, 1980; xi + 191 pp.

677. Anderson, R. J. *J. Biol. Chem.* **1914**, *17*, 171–190.

678. Anon. *Phytic Acid*; Tsuno Rice Fine Chemicals Co., Ltd: Wakayama, Japan, 7 pp (undated).

679. Johnson, L. F.; Tate, M. E. *Can. J. Chem.* **1969**, *47*(1), 63–73.

680. Patten, A. J.; Hart, E. B. *Am. Chem. J.* **1904**, *31*(5), 564–572.

681. Brigando, C.; Mossoyan, J. C.; Favier, F.; Benlian, D. *J. Chem. Soc., Dalton Trans.* **1995**, (4), 575–578.

682. Barre, R.; Courtois, J. E.; Wormser, G. *Bull. Soc. Chim. Biol.* **1954**, *36*(4/5), 455–474.

683. Inhülsen, D.; Niemeyer, R. *Z. Pflanzenphysiol.* **1978**, *88*, 103–116.

684. Sobolev, A. M. *Fiziol. Rast.* **1966**, *13*(2), 193–200 (Engl. 177–183).

685. Parrish, F. W.; Madacsi, J. P.; Phillippy, B. Q.; Wilfred, A. G.; Buco, S. M. *J. Agric. Food Chem.* **1990**, *38*(2), 407–409.

686. Isbrandt, L. R.; Oertel, R. P. *J. Am. Chem. Soc.* **1980**, *102*(9), 3144–3148.

687. Johansson, C.; Kördel, J.; Drakenberg, T. *Carbohydr. Res.* **1990**, *207*(2), 177–183.

688. Emsley, J.; Niazi, S. *Phosphorus Sulfur* **1981**, *10*(3), 401–408.

689. Johnson, K.; Barrientos, L. G.; Le, L.; Murthy, P. P. N. *Anal. Biochem.* **1995**, *231*(2), 421–431.

690. Sherman, W. R.; Ackermann, K. E.; Berger, R. A.; Gish, B. G.; Zinbo, M. *Biomed. Environ. Mass Spectrom.* **1986**, *13*(7), 133–141.

691. Samuni, A.; Neta, P. *J. Phys. Chem.* **1973**, *77*(20), 2425–2429.

692. Desjobert, A.; Petek, F. *Bull. Soc. Chim. Biol.* **1956**, *38*(5/6), 871–883.

693. Martin, J. K. *Anal. Biochem.* **1970**, *36*(1), 233–237.

694. Asada, K.; Tanaka, K.; Kasai, Z. *Ann. N. Y. Acad. Sci.* **1969**, *165*(2), 801–814.

695. Roberts, R. M.; Loewus, F. *Plant Physiol.* **1968**, *43*(10), 1710–1716.

696. Gerritse, R. G. *J. Sci. Food Agric.* **1978**, *29*(7), 577–586.
697. Phillippy, B. Q.; Johnston, M. R. *J. Food Sci.* **1985**, *50*(2), 341–342.
698. Arnold, P. W. *Biochim. Biophys. Acta* **1956**, *19*(3), 552–554.
699. O'Dell, B. L.; de Boland, A. *J. Agric. Food Chem.* **1976**, *24*(4), 804–808.
700. Blatny, P.; Kvasnicka, F.; Kenndler, E. *J. Agric. Food Chem.* **1995**, *43*(1), 129–133.
701. Fennessey, J. P.; Nowacki, W. *Helv. Chim. Acta* **1968**, *51*(6), 14–75.
702. Posternak, S. *Helv. Chim. Acta* **1921**, *4*(1), 150–165.
703. Blank, G. E.; Pletcher, J.; Sax, M. *Acta Crystallogr.* **1975**, *B31*, 2584–2592.
704. Truter, M. R.; Tate, M. E. *J. Chem. Soc.* **1970**, (1), 70–71.
705. Jackman, R. H.; Black, C. A. *Soil Sci.* **1951**, *72*(3), 179–186.
706. Lott, J. N. A.; Goodchild, D. J.; Craig, S. *Aust. J. Plant Physiol.* **1984**, *11*(6), 459–469.
707. Li, N.; Wahlberg, O.; Puigdomenech, I. *Chem. Scr.* **1989**, *29*(1), 91–95.
708. Anderson, R. J. *J. Biol. Chem.* **1912**, *12*, 97–113.
709. Grynspan, F.; Cheryan, M. *J. Am. Oil Chem. Soc.* **1983**, *60*(10), 1761–1764.
710. Hoff-Jorgensen, E. *Kgl. Danske Vid. Selsk., Mat. Nat. Med.* **1944**, *21*(7), 27.
711. Seligson, N. *Arch. Pharm.* **1930**, *268*, 104–113.
712. Clarke, G. *J. Chem. Soc.* **1914**, *105*, 535–545.
713. Clarke, G. *J. Chem. Soc.* **1915**, *107*, 360.
714. Cosgrove, D. J. *J. Sci. Food Agric.* **1966**, *17*(12), 550–554.
715. Anderson, R. J. *J. Biol. Chem.* **1912**, *11*, 471–488.
716. Angyal, S. J.; Russell, A. F. *Aust. J. Chem.* **1969**, *22*(2), 391–404.
717. Bullock, J. I.; Duffin, P. A.; Nolan, K. B. *J. Sci. Food Agric.* **1993**, *63*(2), 261–263.
718. Cosgrove, D. J. *Biochem. J.* **1963**, *89*(1), 172–175.
719. Cosgrove, D. J. *Ann. N. Y. Acad. Sci.* **1969**, *165*(2), 677–686.
720. Tate, M. E. *Anal. Biochem.* **1968**, *23*(1), 141–149.
721. Schormüller, J.; Bressau, G. *Z. Lebensm. Unters. Forsch.* **1960**, *113*(6), 484–491.
722. Brown, D. M.; Hall, G. E. *J. Chem. Soc.* **1959**, (1), 357–359.
723. Tomlinson, R. V.; Ballou, C. E. *Biochemistry* **1962**, *1*(1), 166–171.
724. Heggen, M. H. M.; Reith, J. F. *Pharm. Weekbl.* **1948**, *83*, 801–805.
725. Paton, G.; Noailly, M.; Mossoyan, J. C. *J. Phys. Org. Clhem.* **1999**, *12*(5), 401–407.
726. Bauman, A. T.; Chateauneuf, G. M.; Boyd, B. R.; Brown, R. E.; Murthy, P. P. N. *Tetrahedron Lett.* **1999**, *40*(24), 4489–4492.
727. Anderson, R. J.; Kulp, W. L. *J. Biol. Chem.* **1920**, *43*(2), 469–475.
728. Fowler, H. D. *J. Sci. Food Agric.* **1956**, *7*(6), 381–386.
729. Van der Kaay, J.; Van Haastert, P. J. M. *Anal. Biochem.* **1995**, *225*(1), 183–185.
730. Hull, S. R.; Montgomery, R. *J. Agric. Food Chem.* **1995**, *43*(6), 1516–1523.
731. Heubner, W.; Stadler, H. *Biochem. Z.* **1914**, *64*, 422–437.
732. Earley, E. B. *Ind. Eng. Chem. Anal. Ed.* **1944**, *16*(6), 389–391.
733. Wheeler, E. L.; Ferrel, R. E. *Cereal Chem.* **1971**, *48*(3), 312–320.

734. Latta, M.; Eskin, M. *J. Agric. Food Chem.* **1980**, *28*, 1313–1315.

735. McCance, R. A.; Widdowson, E. M. *Biochem. J.* **1935**, *29*, 2694–2699.

736. Haug, W.; Lantzsch, H.-J. *J. Sci. Food Agric.* **1983**, *34*(12), 1423–1426.

737. Harland, B. F.; Oberleas, D. *Cereal Chem.* **1977**, *54*(4), 827–832.

738. Harland, B. F.; Oberleas, D. *J. Assoc. Off. Anal. Chem.* **1986**, *69*(4), 667–670.

739. Tangendjaja, B.; Buckle, K. A.; Wooton, M. *J. Chromatogr.* **1980**, *197*, 274–277.

740. Graf, E.; Dintzis, F. R. *Anal. Biochem.* **1982**, *119*, 413–417.

741. Camire, A. I.; Clydesdale, F. M. *J. Food Sci.* **1982**, *47*(2), 575–578.

742. Lee, K.; Abendroth, J. A. *J. Food Sci.* **1983**, *48*(4), 1344–1345 1351.

743. O'Neill, I. K.; Sargent, M.; Trimble, M. L. *Anal. Chem.* **1980**, *52*(8), 1288–1291.

744. Babakhodzhaeva, S. A.; Rizaev, M. Yu.; Gorokhovskaya, A. S. *Khim. Farm. Zh.* **1968**, *2*(4), 42–44 (Engl. 215–217).

745. Uppström, B.; Svensson, R. *J. Sci. Food Agric.* **1980**, *31*(7), 651–656.

746. Kikunaga, S.; Takahashi, M.; Huzisige, H. *Plant Clell Physiol.* **1985**, *26*(7), 1323–1330.

747. Grases, F.; March, P. *Anal. Chim. Acta* **1989**, *219*(1), 89–95.

748. Darbre, A.; Norris, F. W. *Biochem. J.* **1956**, *64*(3), 441–446.

749. Ellis, R.; Morris, E. R. *Cereal Chem.* **1982**, *59*(3), 232–233.

750. Ellis, R.; Morris, E. R.; Philpot, C. *Anal. Biochem.* **1977**, *77*(2), 536–539.

751. Mothes, R.; Schwenke, K. D.; Grossmann, S. *Nahrung* **1990**, *34*(1), 99–101.

752. Plaami, S.; Kumpulainen, J. *J. Assoc. Off. Anal. Chem.* **1991**, *74*(1), 32–36.

753. Young, L. *Biochem. J.* **1936**, *30*, 252–257.

754. Makower, R. U. *Cereal Chem.* **1970**, *47*(5), 288–295.

755. Vaintraub, I. A.; Lapteva, N. A. *Anal. Biochem.* **1988**, *175*(1), 227–230.

756. Doherty, C.; Faubion, J. M.; Rooney, L. W. *Cereal Chem.* **1982**, *59*(5), 373–378.

757. Michel-Durand, E. *Bull. Soc. Chim. Biol.* **1938**, *20*, 413–422.

758. Courtois, J.; Perez, C. *Bull. Soc. Chim. Biol.* **1948**, *30*(3/4), 195–201.

759. Bourdet, A.; Feillet, P. *Cereal Chem.* **1967**, *44*(5), 457–482.

760. Kamaya, M.; Furuki, T.; Nagashima, K.; Ishii, E.; Saito, H. *Phytochem. Anal.* **1995**, *6*(5), 251–254.

761. March, J. G.; Simonet, B. M.; Grases, F. *Analyst (Lond.)* **1999**, *124*(6), 897–900.

762. Kamaya, M.; Furuki, T.; Nagashima, K. *Phytochem. Anal.* **1998**, *9*(3), 119–123.

763. Ellis, R.; Morris, E. R. *Cereal Chem.* **1983**, *60*(2), 121–124.

764. Ellis, R.; Morris, E. R. *Cereal Chem.* **1986**, *63*(1), 58–59.

765. Cilliers, J. J. L.; van Niekerk, P. J. *J. Agric. Food Chem.* **1986**, *34*(4), 680–683.

766. Rounds, M. A.; Nielsen, S. S. *J. Chromatogr.* **1993**, *653*(1), 148–152.

767. Minnear, R. A.; Segars, J. E.; Elwood, J. W.; Mulholland, P. J. *Analyst* **1988**, *113*(4), 645–649.

768. Clarkin, C. M.; Minnear, R. A.; Kim, S.; Elwood, J. W. *Environ. Sci. Technol.* **1992**, *26*(1), 199–204.

769. Talamond, P.; Gallon, G.; Treche, S. *J. Chromatogr.* **1998**, *805A*, 143–147.

770. Kientz, C. E.; Verweij, A.; de Jong, G. J.; Brinkman, U. A. T. *J. High Resolut. Chromatogr.* **1989**, *12*(12), 793–796.

771. Lazar, S.; Lafosse, M.; Herbreteau, B.; Guillaumet, G. *Analusis* **1993**, *21*(10), 415–418.

772. Knuckles, B. E.; Kuzmicky, D. D.; Betschart, A. A. *J. Food Sci.* **1982**, *47*(4), 1257–1258 1262.

773. Sandberg, A.-S.; Ahderinne, R. *J. Food Sci.* **1986**, *51*(3), 547–550.

774. Lehrfeld, J. *Cereal Chem.* **1989**, *66*(6), 510–515.

775. Lehrfeld, J. *J. Agric. Food Chem.* **1994**, *42*(12), 2726–2731.

776. Bos, K. D.; Verbeek, C.; van Eeden, C. H. P.; Slump, P.; Wolters, M. G. E. *J. Agric. Food Chem.* **1991**, *39*(10), 1770–1772.

777. Lehrfeld, J.; Wu, Y. V. *J. Agric. Food Chem.* **1991**, *39*(10), 1820–1824.

778. Champagne, E. T.; Rao, R. M.; Liuzzo, J. A.; Robinson, J. W.; Gale, R. J.; Miller, F. *Cereal Chem.* **1986**, *63*(2), 160–163.

779. Shimizu, N.; Tomoda, M. *Chem. Pharm. Bull.* **1987**, *35*(9), 3918–3921.

780. Le Boever, J. L.; Eeckhout, W.; Boucque, C. V. *Neth. J. Agric. Sci.* **1994**, *42*(4), 357–369.

781. Mazzola, E. P.; Phillippy, B. Q.; Harland, B. F.; Miller, T. H.; Potemra, J. M.; Katsimpiris, E. W. *J. Agric. Food Chem.* **1986**, *34*(1), 60–62.

782. Ersöz, A.; Akgün, H.; Aras, N. K. *J. Agric. Food Chem.* **1990**, *38*(3), 733–735.

783. Asada, K.; Kasai, Z. *Plant Cell Physiol.* **1962**, *3*(4), 397–406.

784. Norris, F. W.; Darbre, A. *Analyst* **1956**, *81*(7), 394–400.

785. de Koning, A. J. *Analyst (Cambridge)* **1994**, *119*(6), 1319–1323.

786. Kon, S.; Olson, A. C.; Frederick, D. P.; Eggling, S. B.; Wagner, J. R. *J. Food Sci.* **1973**, *38*(2), 215–217.

787. Seiffert, U. B.; Agranoff, B. W. *Biochim. Biophys. Acta* **1965**, *98*(3), 574–581.

788. Jackson, J. F.; Jones, G.; Linskens, H. F. *Phytochemistry* **1982**, *21*(6), 1255–1258.

789. Williams, S. G. *Plant Physiol.* **1970**, *45*(4), 376–381.

790. Nardi, A.; Cristalli, M.; Desiderio, C.; Ossicini, L.; Shukla, S. K.; Fanali, S. *J. Microcolumn Sep.* **1992**, *4*(1), 9–11.

791. Schmalfuss, K. *Bodenkd. Pflanzenernähr.* **1941**, *20*(3/4), 151–177.

792. Schmalfuss, K. *Phosphorsäure* **1939/40**, *8/9*, 404–412.

793. Tolibaev, M.; Mukhamedova, Kh.S.; Akramov, S. T. *Khim. Prir. Soedin.* **1975**, *11*(6), 799–800 (Engl. 813).

794. Mukhamedova, Kh.S.; Akramov, S. T. *Khim. Prir. Soedin.* **1977**, (4), 505–507 (Engl. 422–424).

795. Mukhamedova, Kh.S.; Akbarov, R. R.; Akramov, S. T. *Khim. Prir. Soedin.* **1978**, (3), 395–396 (Engl. 330–331).

796. Akbarov, R. R.; Mukhamedova, Kh.S.; Akramov, S. T. *Khim. Prir. Soedin.* **1979**, (2), 223–224 (Engl. 190–191).

797. Suzuki, U.; Yoshimura, K. *Bull. Coll. Agric. Tokyo* **1907**, *7*(4), 495–502, from *J. Chem. Soc. Abstr.* **1908**, *94* (ii), 124 and *Chem. Zentralbl.* **1907**, *II*, 1636.

798. Wenzel, W. *Z. Pflanzenernähr. Düng. Bodenk.* **1956**, *75*(3), 191–203.

799. Campen, W. A. C. *Landbouwk. Tijdschr.* **1958**, *68*, 727–730.

800. Ogawa, M.; Tanaka, K.; Kasai, Z. *Agric. Biol. Chem.* **1975**, *39*(3), 695–700.

801. Ogawa, M.; Tanaka, K.; Kasai, Z. *Cereal Chem.* **1977**, *54*(5), 1029–1034.

802. Fontaine, T. D.; Pons, W. A., Jr.; Irving, G. W., Jr. *J. Biol. Chem.* **1946**, *164*(2), 487–507.

803. Kazakov, A. L.; Samokish, I. I. *Khim. Prir. Soedin.* **1980**, (4), 483–485 (Engl. 343–345).

804. Tombs, M. P. *Plant Physiol.* **1967**, *42*(6), 797–813.

805. Morton, R. K.; Palk, B. A.; Raison, J. K. *Biochem. J.* **1964**, *9*(3), 522–528.

806. Fischler, P.; Kürten, F. H. *Biochem. Z.* **1932**, *254*, 138–147.

807. Posternak, S. *Compt. Rend.* **1903**, *137*, 202–204.

808. Reeves, R. J.; Carroll, R. T.; Gennaro, G. P. *Talanta* **1979**, *26*(11), 1033–1034.

809. Vasiliev, R.; Scintee, V.; Anastasescu, O.; Fruchter, J. *Rev. Chim. (Bucharest)* **1960**, *11*(7), 421.

810. Sobolev, A. M. *Fiziol. Rast.* **1964**, *11*(1), 106–111 (Engl. 89–93).

811. Graf, E. *Anal. Biochem.* **1983**, *131*(2), 351–355.

812. Morris, G. F. I.; Thurman, D. A.; Boulter, D. *Phytochemistry* **1970**, *9*(8), 1707–1714.

813. Organ, M. G.; Greenwood, J. S.; Bewley, J. D. *Planta* **1988**, *174*(4), 513–517.

814. Tewari, K. K.; Singh, M. *Phytochemistry* **1964**, *3*(2), 341–347.

815. DiLollo, A.; Alli, J.; Kermasha, S. *J. Agric. Food Chem.* **1991**, *39*(12), 2128–2130.

816. May, L.; Morris, E. R.; Ellis, R. *J. Agric. Food Chem.* **1980**, *28*(5), 1004–1006.

817. Ambe, S.; Ambe, F.; Nozaki, T. *J. Agric. Food Chem.* **1987**, *35*(3), 292–296.

818. Sowa, S.; Connor, K. F. *Plant Sci.* **1995**, *105*(1), 23–30.

819. Jacobsen, J. V.; Knox, R. B.; Pyliotis, N. A. *Planta* **1971**, *101*(3), 189–209.

820. Yiu, S. H.; Poon, H.; Fulcher, R. G.; Altosaar, I. *Food Microstruct.* **1982**, *1*(2), 135–143.

821. Sundararajan, A. R. *Indian J. Med. Res.* **1938**, *25*(3), 685–691.

822. Ferguson, E. L.; Gibson, R. S.; Thompson, L. U.; Ounpuu, S.; Berry, M. *J. Food Compos. Anal.* **1988**, *1*(4), 316–325.

823. Lott, J. N. A.; Buttrose, M. S. *Aust. J. Plant Physiol.* **1977**, *5*(1), 89–111.

824. Ricardo, C. P. P.; Santos, H. *J. Exp. Bot.* **1990**, *41*(222), 79–87.

825. Balogun, A. M.; Fetuga, B. L. *Biol. Wastes* **1989**, *28*(4), 303–308.

826. Common, R. H. *Analyst* **1940**, *65*(2), 79–83.

827. Balasubramanian, S. C.; Rao, B. S. N.; Ramanamurthy, P. S. V.; Shenolikar, I. S. *Indian J. Med. Res.* **1962**, *50*(5), 779–793.

828. Balogun, A. M.; Fetuga, B. L. *Food Chem.* **1988**, *30*(1), 37–43.

829. Obizoba, I. C.; Anyika, J. U. *Plant Foods Hum. Nutr.* **1994**, *46*(2), 157–165.

830. Batten, G. D. *Cereal Chem.* **1986**, *63*(5), 384–387.

831. Wilson, A. M.; Harris, G. A. *Plant Physiol.* **1966**, *41*(9), 1416–1419.

832. Banzon, E. A.; Gonzalez, O. N.; Villacorte, S. C. *Philipp. J. Sci.* **1965**, *94*(2), 141–152.

833. Chapin, F. S., III; Kedrowski, R. A. *Ecology* **1983**, *64*(2), 376–391.

834. Pedersen, B.; Kalinowski, L. S.; Eggum, B. O. *Plant Foods Hum. Nutr.* **1987**, *36*(4), 309–324.

835. Becker, R.; Wheeler, E. L.; Lorenz, K.; Stafford, A. E.; Grosjean, O. K.; Betschart, A. A.; Saunders, R. M. *J. Food Sci.* **1981**, *46*(4), 1175–1180.

836. Bertoni, M. H.; Gomez, R. G.; Cattaneo, P.; Covas, G. *An. Asoc. Quim. Argent.* **1984**, *72*(6), 597–605.

837. Bawa, S. F.; Yadav, S. P. *J. Sci. Food Agric.* **1986**, *37*(5), 504–506.

838. Lorenz, K.; Wright, B. *Food Chem.* **1984**, *14*(1), 27–34.

839. Singhal, R. S.; Kulkarni, P. R. *J. Sci. Food Agric.* **1988**, *42*(4), 325–331.

840. Gupta, K.; Barat, G. K.; Wagle, D. S.; Chawla, H. K. L. *Food Chem.* **1989**, *31*(2), 105–116.

841. de Arellano, M. L.; Fernandez, S.; Luco, J.; de Sanchez, N. E.; Petenatti, E.; de Mucciarelli, S. I. L. *Sci. Aliments* **1996**, *16*(3), 289–296.

842. Giri, K. V. *Indian J. Med. Res.* **1938**, *25*(4), 869–877.

843. Buttrose, M. S.; Lott, J. N. A. *Can. J. Bot.* **1978**, *56*(17), 2062–2071.

844. Bagaoisan, C. L. *Philipp. Agric.* **1932**, *21*(1), 53–59.

845. Singh, U.; Singh, B.; Smith, O. D. *J. Food Sci.Technol.* **1991**, *28*(6), 345–347.

846. Rao, P. U. *J. Am. Oil Chem. Soc.* **1995**, *72*(4), 477–480.

847. Dieckert, J. W.; Snowden, J. E., Jr.; Moore, A. T.; Heinzelman, D. C.; Altschul, A. M. *J. Food Sci.* **1962**, *27*, 321–325.

848. Seo, A.; Morr, C. V. *J. Food Sci.* **1985**, *50*(1), 262–263.

849. Sharma, C. B.; Dieckert, J. W. *Physiol. Plant.* **1975**, *33*(1), 1–7.

850. Graf, E.; Dintzis, F. R. *J. Agric. Food Chem.* **1982**, *30*(6), 1094–1097.

851. Weber, C. W.; Kohlhepp, E. A.; Idouraine, A.; Ochoa, L. J. *J. Agric. Food Chem.* **1993**, *41*(11), 1931–1935.

852. Mosha, T. C.; Gaga, H. E.; Pace, R. D.; Laswai, H. S.; Mtebe, K. *Plant Foods Hum. Nutr.* **1995**, *47*(4), 361–367.

853. Bertoni, M. H.; Gomez, R. G.; Cattaneo, P. *An. Asoc. Quim. Argent.* **1986**, *74*(4), 415–419.

854. Bertoni, M. H.; Gomez, R. G.; Cattaneo, P. *An. Asoc. Quim. Argent.* **1986**, *74*(3), 223–228.

855. Ravindran, V.; Ravindran, G.; Sivalogan, S. *Food Chem.* **1994**, *50*, 133–136.

856. Lott, J. N. A.; Vollmer, C. M. *Can. J. Bot.* **1979**, *57*(9), 987–992.

857. Kuhn, M.; Grau, H.; Wagner, S. *Getreide Mehl Brot* **1995**, *49*(1), 54–57.

858. Lolas, G. M.; Palamidis, N.; Markakis, P. *Cereal Chem.* **1976**, *53*(6), 867–871.

859. Snook, L. C. *Emp. J. Exp. Agric.* **1938**, *6*(21), 20–24.

860. Saastamoinen, M.; Plaami, S.; Kumpulainen, J. *Acta Agric. Scand.* **1992**, *42B*(1), 6–11.

861. Ashton, W. M.; Williams, P. C. *J. Sci. Food Agric.* **1958**, *9*(8), 505–511.

862. Miller, G. A.; Youngs, V. L.; Oplinger, E. S. *Cereal Chem.* **1980**, *57*(3), 189–191.

863. Miller, G. A.; Youngs, V. L.; Oplinger, E. S. *Cereal Chem.* **1980**, *57*(3), 192–194.

864. Saastamoinen, M.; Heinonen, T. *Ann. Agric. Fenn.* **1985**, *24*(2), 103–105.

865. Schormüller, J.; Würdig, G. *Deut. Lebensm. Rundschau* **1957**, *53*(1), 1–10.

866. Froelich, W.; Nyman, M. *J. Cereal Sci.* **1988**, *7*(1), 73–82.

867. Marrese, R. J.; Duell, R. W.; Sprague, M. A. *Crop Sci.* **1961**, *1*(1), 80–81.

868. Fulcher, R. G.; O'Brien, T. P.; Wong, S. I. *Cereal Chem.* **1981**, *58*(2), 130–135.

869. Yiu, S. H. *Food Microstruct.* **1986**, *5*(2), 219–225.

870. Pomeranz, Y. *Cereal Chem.* **1973**, *50*(4), 504–511.

871. Donhowe, E. T.; Peterson, D. M. *Plant Physiol.* **1983**, *71*(3), 519–523.

872. Buttrose, M. S. *Aust. J. Plant Physiol.* **1978**, *5*(5), 631–639.

873. Frøhlich, W.; Wahlgren, M.; Drakenberg, T. *J. Cereal Sci.* **1988**, *8*(1), 47–53.

874. Takasugi, N. *J. Chem. Soc. Jpn.* **1944**, *65*(4), 333–336.

875. Hall, J. R.; Hodges, T. K. *Plant Physiol.* **1966**, *41*(9), 1459–1464.

876. Fasakin, E. A. *Bioresour. Technol.* **1999**, *69*(2), 185–187.

877. Kuo, J.; Hocking, P. J.; Pate, J. S. *Aust. J. Bot.* **1982**, *30*(2), 231–249.

878. Vijayakumari, K.; Siddhuraju, P.; Janardhanan, K. *J. Sci. Food Agric.* **1997**, *73*(3), 279–286.

879. Lott, J. N. A.; Buttrose, M. S. *Can. J. Bot.* **1978**, *56*(17), 2050–2061.

880. Ahmad, M.; Badshah, A.; Chaudhry, M. A. *Pak. J. Sci. Ind. Res.* **1993**, *36*(1), 23–26.

881. Anjou, K.; Lönnerdal, B.; Uppström, B.; Åman, P. *Swed. J. Agric. Res.* **1977**, *7*(4), 169–178.

882. King, R. D.; Dietz, H. M. *Cereal Chem.* **1987**, *64*(6), 411–413.

883. Serraino, M. R.; Thompson, L. U.; Savoie, L.; Parent, G. *J. Food Sci.* **1985**, *50*(6), 1689–1692.

884. Maheshwari, P. N.; Stanley, D. W.; Beveridge, T. J.; van de Voort, F. R. *J. Food Biochem.* **1982**, *5*(1), 39–61.

885. Khokhar, S.; Pushpanjali, ; Fenwick, G. R. *J. Agric. Food Chem.* **1994**, *42*(11), 2440–2444.

886. Lu, S.-Y.; Kim, H.; Eskin, N. A. M.; Latta, M.; Johnson, S. *J. Food Sci.* **1987**, *52*(1), 173–175.

887. Zhou, B.; He, Z.-Q.; Yu, H.-M.; Mukherjee, K. D. *J. Agric. Food Chem.* **1990**, *38*(3), 690–694.

888. Matthäus, B.; Lösing, R.; Fiebig, H.-J. *J. High Resolut. Chromatogr.* **1995**, *18*(4), 267–268.

889. Lemieux, L.; Amiot, J.; Brisson, G. J. *Can. Inst. Food Sci. Technol. J.* **1985**, *18*(1), 29–33.

890. Gillberg, L. *J. Food Sci.* **1978**, *43*(4), 1219–1223, 1228.

891. Van Steveninck, R. F. M.; Babare, A.; Fernando, D. R.; Van Steveninck, M. E. *Plant Soil* **1993**, *155/6*, 525–528.

892. Naczk, M.; Diosady, L. L.; Rubin, L. J. *Lebensm. Wiss. Technol.* **1986**, *19*, 13–16.

893. Josefsson, E. *J. Sci. Food Agric.* **1975**, *26*(9), 1299–1310.

894. Lajolo, F. M.; Marquez, U. M. L.; Filisetti-Cozzi, T. M. C. C.; McGregor, D. I. *J. Agric. Food Chem.* **1991**, *39*(11), 1933–1937.

895. Maldonado, S.; Lima, C.; Etchart, M.; Lainez, V.; de Lederkremer, R. M. *Acta Bot. Neerl.* **1998**, *47*(3), 285–297.

896. Sharma, Y. K.; Tiwari, A. W.; Rao, K. C.; Mishra, A. *J. Food Sci. Technol. (India)* **1977**, *14*(1), 38–40.

897. Fasidi, I. O. *Nahrung* **1981**, *25*(10), 949–953.

898. Rao, P. U.; Deosthale, Y. G. *J. Food Sci. Technol. (India)* **1983**, *20*(5), 195–197.

899. Chitra, U.; Vimala, V.; Singh, U.; Geervani, P. *Plant Foods Hum. Nutr.* **1993**, *47*(2), 163–172.

900. Oboh, H. A.; Muzquiz, M.; Burbano, C.; Cuadrado, C.; Pedrosa, M. M.; Ayet, G.; Osagie, A. U. *J. Chromatogr.* **1998**, *823A*(1/2), 307–312.

901. Igbedioh, S. O.; Olugbemi, K. T.; Akpapunam, M. A. *Food Chem.* **1994**, *50*, 147–151.

902. Rahim, A.; Ahmad, K. *Pak. J. Biol. Agric. Sci.* **1960**, *3*(1), 46–49.

903. Chauhan, E. M.; Duhan, A.; Bhat, C. M. *J. Food Sci. Technol. (India)* **1986**, *23*(2), 106–108.

904. Spitzer, E.; Lott, J. N. A.; Vollmer, C. M. *Can. J. Bot.* **1980**, *58*(11), 1244–1249.

905. Nolasco, S. N.; Bertoni, M. H.; Malec, L.; Cattaneo, P. *An. Asoc. Quim. Argent.* **1987**, *75*(1), 29–34.

906. Latha, T. S.; Prakash, V. *J. Agric. Food Chem.* **1984**, *32*(6), 1412–1416.

907. Averill, H. P.; King, C. G. *J. Am. Chem. Soc.* **1926**, *48*(3), 724–728.

908. Ramachandra, G.; Monteiro, P. V. *J. Food Compos. Anal.* **1990**, *3*(1), 81–87.

909. Aletor, V. A.; Adeogun, O. A. *Food Chem.* **1995**, *53*(4), 375–379.

910. Nolasco, S. M.; Wiese, B.; Quiroga, O.; Vigo, M. S. *An. Asoc. Quim. Argent.* **1991**, *79*(1), 53–57.

911. Maza, M. P.; Zamora, R.; Alaiz, M.; Hidalgo, F. J.; Millan, F.; Vioque, E. *J. Sci. Food Agric.* **1989**, *46*(4), 495–502.

912. Weimer, W. C.; Armstrong, D. E. *Environ. Sci. Technol.* **1979**, *13*(7), 826–829.

913. Tagliaferri, E. cited by Koziol M.J.*J. Food Compos. Anal.* **1992**, *5*(1), 35–68.
914. Chauhan, G. S.; Eskin, N. A. M.; Tkachuk, R. *Cereal Chem.* **1992**, *69*(1), 85–88.
915. Ruales, J.; Nair, B. M. *Food Chem.* **1993**, *48*(2), 137–143.
916. Valencia, S.; Svanberg, U.; Sandberg, A. S.; Ruales, J. *Int. J. Food Sci. Nutr.* **1999**, *50*(3), 203–211.
917. Ekpa, O. D. *Food Chem.* **1996**, *57*(2), 229–232.
918. Edem, D. O.; Eka, O. U.; Ifon, E. T. *Food Chem.* **1984**, *14*(4), 303–311.
919. El-Tinay, A. H.; Mahgoub, S. O.; Mohamed, B. E.; Hamad, M. A. *J. Food Compos. Anal.* **1989**, *2*(1), 69–78.
920. Duhan, A.; Chauhan, B. M.; Punia, D.; Kapoor, A. C. *J. Sci. Food Agric.* **1989**, *49*(4), 449–455.
921. Hussain, B.; Khan, S.; Ismail, M.; Sattar, A. *Nahrung* **1989**, *33*(4), 345–348.
922. Burbano, C.; Muzquiz, M.; Osagie, A.; Ayet, G.; Cuadrado, C. *Food Chem.* **1995**, *52*(3), 321–325.
923. Khan, N.; Zaman, R.; Elahi, M. *J. Agric. Food Chem.* **1988**, *36*(6), 1274–1276.
924. Naik, M. S.; Narayana, N. *Indian J. Appl. Chem.* **1959**, *22*(5/6), 239–242.
925. Harmuth-Hoene, A.-E.; Bognar, A. E.; Kornemann, U.; Diehl, J. F. *Z. Lebensm. Unters. Forsch.* **1987**, *185*(5), 386–393.
926. Elkowicz, K.; Sosulski, F. W. *J. Food Sci.* **1982**, *47*(4), 1301–1304.
927. Naik, M. S.; Narayana, N. *J. Univ. Poona Sci. Technol.* **1957**, *12*, 26–30.
928. Nolasco, S. M.; Quiroga, O.; Wiese, B.; Vigo, M. S. *Grasas Aceites* **1996**, *47*(6), 377–380.
929. Oke, O. L. *Indian J. Med. Res.* **1965**, *53*(5), 417–420.
930. Vigo, M. S.; Bertoni, M. H.; Cattaneo, P. *An. Asoc. Quim. Argent.* **1973**, *61*(5), 241–249.
931. Alaoui, L.; Essatara, M. *Nutr. Rep. Int.* **1985**, *31*(2), 469–476.
932. Møllgaard, H.; Lorenzen, K.; Hansen, I. G.; Christensen, P. E. *Biochem. J.* **1946**, *40*, 589–603.
933. Harland, B. F.; Prosky, L. *Cereal Foods World* **1979**, *24*(9), 387–394.
934. McKenzie, J. M. *Nutr. Rep. Int.* **1984**, *29*(2), 387–395.
935. Franz, H.; Maier, H. G. *Dtsch. Lebensm. Rundsch.* **1994**, *90*(11), 345–349.
936. Marfo, E. K.; Simpson, B. K.; Idowu, J. S.; Oke, O. L. *J. Agric. Food Chem.* **1990**, *38*(7), 1580–1585.
937. Abdulnabi, M.; Dhia, L. Y. *J. Fac. Med. Baghdad* **1964**, *6*(3), 109–116.
938. Bonvehi, J.; Coll, F. V. *Acta Aliment.* **1997**, *26*(3), 243–253.
939. Lott, J. N. A.; Buttrose, M. S. *Can. J. Bot.* **1978**, *56*(17), 2072–2082.
940. von Hofsten, A. *Physiol. Plant.* **1973**, *29*(1), 76–81.
941. Poux, N. *J. Microsc.* **1964**, *4*(6), 771–782.
942. Lott, J. N. A.; Vollmer, C. M. *Plant Physiol.* **1979**, *63*(2), 307–311.
943. Bolley, D. S.; McCormack, R. H. *J. Am. Oil Chem. Soc.* **1952**, *29*(11), 470–472.
944. Henderson, C. W.; Scheerens, J. C.; Berry, J. W. *J. Agric. Food Chem.* **1986**, *34*(3), 434–436.
945. Lott, J. N. A. *Plant Physiol.* **1975**, *55*(5), 913–916.
946. Lott, J. N. A.; Greenwood, J. S.; Vollmer, C. M.; Buttrose, M. S. *Plant Physiol.* **1978**, *61*(6), 984–988.
947. Lendle, A. *Arch. Pharm.* **1938**, *276*, 45–53.
948. Mansour, E. H.; Dworschák, E.; Lugasi, A.; Barna, Ė.; Gergeli, A. *J. Sci. Food Agric.* **1993**, *61*(1), 73–78.
949. Linssen, J. P. H.; Cozijnsen, J. L.; Pilnik, W. *J. Sci. Food Agric.* **1989**, *49*(3), 291–296.

950. Rincon, M.; Chen, Q.; Boss, W. F. *Plant Physiol.* **1989**, *89*(1), 126–132.

951. Marfo, E. K.; Ole, O. L. *Food Chem.* **1989**, *31*(2), 117–127.

952. Van Steveninck, R. F. M.; Van Steveninck, M. E.; Fernando, D. R.; Horst, W. J.; Marschner, H. *J. Plant Physiol.* **1987**, *131*(3/4), 247–257.

953. Aiyesanmi, A. F.; Oguntokun, M. O. *Riv. Ital. Sostanze Grasse* **1996**, *73*(11), 521–523.

954. Wanasundera, J. P. D.; Ravindran, G. *Food Chem.* **1992**, *45*(4), 247–250.

955. Ferguson, E. L.; Gibson, R. S.; Opare-Obisaw, C.; Osei-Opare, F.; Stephen, A. M.; Lehrfeld, J.; Thompson, L. U. *J. Food Compos. Anal.* **1993**, *6*(1), 87–99.

956. Satoh, T. *Nippon Kagaku Zasshi* **1967**, *88*(9), 985–988.

957. Nolasco, S. M.; Quiroga, O.; Wiese, B. *An. Asoc. Quim. Argent.* **1989**, *77*(5), 393–398.

958. Togashi, M.; Sgarbieri, V. de C. *Cienc. Tecnol. Aliment.* **1994**, *14*(1), 85–95, from *Chem. Abstr.* **1995**, *122*, 54585a.

959. Tadzhibaev, Yu.A.; Mukhamedova, Kh.S.; Akramov, S. T. *Khim. Prir. Soedin.* **1976**, (4), 435–439 (Engl. 383–386).

960. Rao, P. U.; Deosthale, Y. G. *Plant Foods Hum. Nutr.* **1988**, *38*(1), 35–41.

961. Antony, U.; Chandra, T. S. *J. Agric. Food Chem.* **1998**, *46*(7), 2578–2582.

962. Buttrose, M. S.; Lott, J. N. A. *Can. J. Bot.* **1978**, *56*(17), 2083–2091.

963. Wallace, P. A.; Marfo, E. K.; Plahar, W. A. *Food Chem.* **1997**, *61*(3), 287–291.

964. Gupta, K.; Thakral, K. K.; Gupta, V. K.; Arora, S. K. *J. Sci. Food Agric.* **1995**, *68*(1), 73–76.

965. Dosunmu, M. I.; Johnson, E. C. *Food Chem.* **1995**, *54*(1), 67–71.

966. Malec, L. S.; Vigo, M. S.; Cattaneo, P. *An. Asoc. Quim. Argent.* **1986**, *74*(3), 311–320.

967. Ogushi, M.; Yanagisawa, E.; Yoshikawa, A.; Harada, R. *Nippon Eiyo Shokuryo Gakkaishi* **1989**, *42*(3), 245–249.

968. Raboy, V.; Dickinson, D. B.; Beloro, F. E. *Crop Sci.* **1984**, *24*(3), 431–434.

969. Bogoroditskaya, V. P. *Voprosy Pitaniya* **1957**, *16*(1), 55–56.

970. Ranhotra, G. S.; Loewe, R. J.; Puyat, L. V. *J. Food Sci.* **1974**, *39*(5), 1023–1025.

971. Bouzas, J. O.; Bertoni, M. H.; Cattaneo, P. *An. Asoc. Quim. Argent.* **1982**, *70*(2), 237–255.

972. Mohamed, A. I.; Nebrahtu, T.; Rangappa, M. *Plant Foods Hum. Nutr.* **1991**, *41*(1), 89–100.

973. Winters, D. D.; Handel, A. P.; Lohrberg, J. D. *J. Food Sci.* **1984**, *49*(4), 1113–1114.

974. Monma, M.; Sugimoto, T.; Hashizume, K.; Saio, K. *Biosci. Biotechnol. Biochem.* **1992**, *56*(7), 1036–1040.

975. Brooks, J. R.; Morr, C. V. *J. Food Sci.* **1982**, *47*(4), 1280–1282.

976. Brooks, J. R.; Morr, C. V. *J. Agric. Food Chem.* **1984**, *32*(3), 672–674.

977. Honig, D. H.; Rackis, J. J.; Wolf, W. J. *J. Agric. Food Chem.* **1987**, *35*, 967–971.

978. Honig, D. H.; Wolf, W. J.; Rackis, J. J. *Cereal Chem.* **1984**, *61*(6), 523–526.

979. Prattley, C. A.; Stanley, D. W. *J. Food Biochem.* **1982**, *6*(4), 243–253.

980. Ologhobo, A. D.; Fetuga, B. L. *J. Food Sci.* **1984**, *49*(1), 199–201.

981. Udosen, E. O.; Ukpanah, U. M. *Plant Foods Hum. Nutr.* **1993**, *44*(3), 285–289.

982. Taha, F. S.; El-Nockrashy, A. S. *Nahrung* **1981**, *25*(8), 759–764.

983. Stansbury, M. F.; Pons, W. A., Jr.; Hoffpauir, C. L. *J. Agric. Food Chem.* **1953**, *1*(1), 75–78.

984. Ergle, D. R.; Guinn, G. *Plant Physiol.* **1959**, *34*(4), 476–481.
985. Martinez, W. H.; Frampton, V. L.; Cabell, C. A. *J. Agric. Food Chem.* **1961**, *9*(1), 64–66.
986. Lui, N. S. T.; Altschul, A. M. *Arch. Biochem. Biophys.* **1967**, *121*(3), 678–684.
987. Eklund, A. *Upsala J. Med. Sci.* **1975**, *80*(1), 5–6.
988. Saeed, M.; Cheryan, M. *J. Food Sci.* **1988**, *53*(4), 1127–1131, 1143.
989. Rucci, A. O.; Bertoni, M. H. *An Asoc. Quim. Argent.* **1974**, *62*(6), 365–368.
990. Miller, N.; Pretorius, H. E.; du Toit, L. J. *Food Chem.* **1986**, *21*, 205–209.
991. Ravindran, V.; Ravindran, G. *Food Chem.* **1988**, *30*(2), 93–102.
992. El-Adawy, T. A.; Khalil, A. H. *J. Agric. Food Chem.* **1994**, *42*(9), 1896–1900.
993. Lee, W. J. *J. Am. Soc. Brew. Chem.* **1990**, *48*(2), 62–65.
994. Harland, B. F.; Harland, J. *Cereal Chem.* **1980**, *57*(3), 226–229.
995. Saastamoinen, M.; Plaami, S.; Kumpulainen, J.; Rantanen, O. *Cereal Res. Commun.* **1991**, *19*(4), 391–397.
996. Tluczkiewicz, J.; Berendt, W. *Acta Soc. Bot. Pol.* **1977**, *46*(1), 3–14.
997. Alexander, J. C.; Gabriel, H. G.; Reichertz, J. L. *Can. Inst. Food Sci. Technol. J.* **1984**, *17*(4), 224–228.
998. Antai, S. P.; Obong, U. S. *Plant Foods Hum. Nutr.* **1992**, *42*(3), 219–224.
999. Oboh, S.; Ologhobo, A.; Tewe, O. *Food Chem.* **1989**, *31*(1), 9–18.
1000. Dixon, K. W.; Kuo, J.; Pate, J. S. *Aust. J. Bot.* **1983**, *31*(1), 85–103.
1001. Makkar, H. P. S.; Becker, K.; Sporer, F.; Wink, M. *J. Agric. Food Chem.* **1997**, *45*(8), 3152–3157.
1002. de Goldfiem, J. S. *Rev. Méd. Trop.* **1934**, *26*, 207–209.
1003. Adewusi, S. R. A.; Falade, O. S. *Food Sci. Technol. Int. (Lond.)* **1996**, *2*(4), 231–239, from *Chem. Abstr.* **1996**, *125*, 326902a.
1004. Deka, R. K.; Sarkar, C. R. *Food Chem.* **1990**, *38*(4), 239–246.
1005. Halmer, P.; Bewley, J. D.; Thorpe, T. A. *Planta* **1978**, *139*, 1–8.
1006. Troszynska, A.; Honke, J.; Milczak, M.; Kozlowska, H. *Pol. J. Food Nutr. Sci.* **1993**, *2*(3), 49–54.
1007. Van Steveninck, R. F. M.; Van Steveninck, M. E.; Wells, A. J.; Fernando, D. R. *J. Plant Physiol.* **1990**, *137*, 140–146.
1008. El-Mahdy, A. R.; Moharram, Y. G.; Abou-Samaha, O. R. *Z. Lebensm. Unters. Forsch.* **1985**, *181*(4), 318–320.
1009. Khan, M. A.; Rana, I. A.; Ullah, I.; Jaffery, S. *J. Food Compos. Anal.* **1987**, *1*(1), 65–70.
1010. Bhatty, R. S.; Slinkard, A. E. *Can. Inst. Food Sci. Technol. J.* **1989**, *22*(2), 137–142.
1011. Sethi, P.; Kulkarni, P. R. *Int. J. Food Sci. Nutr.* **1994**, *45*(1), 5–13.
1012. Lin, J.-J.; Dickinson, D. B.; Ho, T.-H. D. *Plant Physiol.* **1987**, *83*(2), 408–413.
1013. Bhatty, R. S.; Cherdkiatgumchai, P. *J. Am. Oil Chem. Soc.* **1990**, *67*(2), 79–84.
1014. Bertoni, M. H.; Cattaneo, P. *An. Asoc. Quim. Argent.* **1972**, *60*, 363–374.
1015. Rutledge, W. A.; Common, R. H. *J. Agric. Sci.* **1945**, *35*(3), 123–125.
1016. Li, R.; Volenec, J. J.; Joern, B. C.; Cunningham, S. M. *Crop Sci.* **1996**, *36*(3), 617–623.
1017. Muzquiz, M.; Burbano, C.; Gorospe, M. J.; Rodenas, I. *J. Sci. Food Agric.* **1989**, *47*(2), 205–214.
1018. Gad, S. S.; Mohamed, M. S.; El-Zalaki, M. E.; Mohasseb, S. Z. *Food Chem.* **1982**, *8*(1), 11–19.
1019. Trugo, L. C.; Donangelo, C. M.; Duarte, Y. A.; Tavares, C. L. *Food Chem.* **1993**, *47*(4), 391–394.

1020. de la Cuadra, C.; Muzquiz, M.; Burbano, C.; Ayet, G.; Calvo, R.; Osagie, A.; Cuadra-do, C. *J. Sci. Food Agric.* **1994**, *66*(3), 357–364.

1021. Dagnia, S. G.; Petterson, D. S.; Bell, R. R.; Flanagan, F. V. *J. Sci. Food Agric.* **1992**, *60*(4), 419–423.

1022. Chango, A.; Bau, H. M.; Villaume, C.; Mejean, L.; Nicolas, J.-P. *Sci. Aliments* **1993**, *13*(2), 285–295.

1023. Suvorov, V. I.; Sobolev, A. M. *Fiziol. Rast.* **1972**, *19*(3), 578–581 (Engl. 486–489).

1024. Sobolev, A. M.; Buzulukova, N. P.; Dmitrieva, M. I.; Barbashova, A. K. *Fiziol. Rast.* **1976**, *23*(4), 739–746 (Engl. 621–628).

1025. Spitzer, E.; Lott, J. N. A. *Can. J. Bot.* **1980**, *58*(6), 699–711.

1026. Popov, Kh.; Tenov, R.; Popova, Ya.; Richev, G. *Nauchn. Tr., Nauchnoizsled. Inst. Kon-serv. Prom., Plovdiv* **1965**, *3*, 137–144, from *Chem. Abstr.* **1967**, *67*, 31730a.

1027. Bosse, D.; Köck, M. *Plant Cell Environ.* **1998**, *21*(3), 325–332.

1028. Thorn, K. A.; Tinsley, A. M.; Weber, C. W.; Berry, J. W. *Ecol. Food Nutr.* **1983**, *13*(4), 251–256.

1029. Borade, V. P.; Kadam, S. S.; Salunkhe, D. K. *Plant Foods Hum. Nutr.* **1984**, *34*(2), 151–157.

1030. Sergeeva, K. A.; Duseeva, F. A. *Fiziol. Biokhim. Zimostoik Drev. Rast.* **1974**, 33–38.

1031. Renuka, P.; Andalla, B. *J. Oil Technol. Assoc. India* **1997**, *29*(2), 39–40.

1032. Ravindran, G.; Ravindran, V. *Food Chem.* **1988**, *27*(4), 299–309.

1033. Joseph, A. *Ann. Nutr. Aliment.* **1973**, *27*(3), 125–139.

1034. Pons, W. A., Jr.; Stansbury, M. D.; Hoffpauir, C. H. *J. Assoc. Off. Anal. Chem.* **1953**, *36*(2), 492–504.

1035. Wallace, G. W.; Satterlee, L. D. *J. Food Sci.* **1977**, *42*, 473–478.

1036. Campbell, M.; Dunn, R.; Ditterline, R.; Pickett, S.; Raboy, V. *J. Plant Nutr.* **1991**, *14*(9), 925–937.

1037. Al-Kahtani, H. A. *J. Food Sci.* **1995**, *60*(2), 395–398.

1038. Vijayakumari, K.; Siddhuraju, P.; Janardhanan, K. *Food Chem.* **1996**, *55*(2), 173–177.

1039. Ravindran, V.; Ravindran, G. *J. Sci. Food Agric.* **1988**, *46*(1), 71–79.

1040. O'Dell, B. L.; de Boland, A. R.; Koirtyohann, S. R. *J. Agric. Food Chem.* **1972**, *20*(3), 718–721.

1041. Reddy, N. R.; Salunkhe, D. K. *J. Food Sci.* **1980**, *45*(6), 1708–1712.

1042. Toma, R. B.; Tabehkia, M. M. *J. Food Sci.* **1979**, *44*(2), 619–621.

1043. McCall, E. R.; Jurgens, J. F.; Hoffpauir, C. L.; Pons, W. A., Jr.; Stark, S. M., Jr.; Cucullu, A. F.; Heinzelman, D. C.; Cirino, V. O.; Murray, M. D. *J. Agric. Food Chem.* **1953**, *1*(16), 988–993.

1044. Tabekhia, M. M.; Luh, B. S. *Deutsche Lebensm. Rundsch.* **1979**, *75*(2), 57–62.

1045. Resurreccion, A. P.; Juliano, B. O.; Tanaka, Y. *J. Sci. Food Agric.* **1979**, *30*(5), 475–481.

1046. Hayakawa, T. *Niigata Daigaku Nogakubu Kenkyu Hokoku* **1977**, *29*, 35–42.

1047. Tanaka, K.; Yoshida, T.; Asada, K.; Kasai, Z. *Arch. Biochem. Biophys.* **1973**, *155*(1), 136–143.

1048. Ogawa, M.; Tanaka, K.; Kasai, Z. *Soil Sci. Plant Nutr. (Tokyo)* **1979**, *25*(3), 437–448.

1049. Tanaka, K.; Ogawa, M.; Kasai, Z. *Cereal Chem.* **1977**, *54*(3), 684–689.

1050. Tanaka, K.; Ogawa, M.; Kasai, Z. *Cereal Chem.* **1976**, *53*(5), 643–649.

1051. Sarma, M. L. *J. Indian Chem. Soc.* **1942**, *19*, 308–310.

1052. Fujiwara, A.; Mitsuhashi, N. *Nogaku* **1948**, *2*(11), 522–529.

1053. Lorenz, K. *Cereal Chem.* **1983**, *60*(6), 424–426.

1054. Ravindran, G. *Food Chem.* **1991**, *39*(1), 99–107.

1055. Maza, M. P.; Millán, F.; Alaiz, M.; Zamora, R.; Hidalgo, F. J.; Vioque, E. *Grasas Aceites* **1988**, *39*(2), 102–104.

1056. Uprety, D. C.; Austin, A. *Bull. Grain Technol.* **1971**, *10*(4), 249–255.

1057. Chauhan, B. M.; Suneja, N.; Bhat, C. M. *Bull. Grain Technol.* **1986**, *24*(1), 44–49.

1058. Simwemba, C. G.; Hoseney, R. C.; Varriano-Marston, E.; Zeleznak, K. *J. Agric. Food Chem.* **1984**, *32*(1), 31–34.

1059. Khetarpaul, N.; Chauhan, B. M. *J. Sci. Food Agric.* **1989**, *49*(1), 117–121.

1060. Opoku, A. R.; Ohenhen, S. O.; Ejiofor, N. *J. Agric. Food Chem.* **1981**, *29*(6), 1247–1248.

1061. Talamond, P.; Gallon, G.; Guyot, J. P.; Lape, I. M.; Trèche, S. *Analusis* **1998**, *26*(10), 396–400.

1062. Kingsley, M. O. *J. Sci. Food Agric.* **1995**, *68*(2), 153–158.

1063. Park, J. H.; Yang, C. B. *Han'guk Sikp'um Kwahakhoechi* **1990**, *22*(3), 343–349.

1064. Jackson, J. F.; Kamboj, R. K.; Linskens, H. F. *Theor. Appl. Genet.* **1983**, *64*(3), 259–262.

1065. Kyriakidis, N. B.; Galiotou-Panayotou, M.; Stavropoulou, A.; Athanasopoulos, P. *Biotechnol. Lett.* **1998**, *20*(5), 475–478.

1066. Ologhobo, A. D.; Fetuga, B. L. *Nutr. Rep. Int.* **1982**, *26*(4), 605–611.

1067. Alli, I.; Baker, B. E. *J. Sci. Food Agric.* **1981**, *32*, 588–592.

1068. Deshpande, S. S.; Sathe, S. K.; Salunkhe, D. K.; Cornforth, D. P. *J. Food Sci.* **1982**, *47*(6), 1846–1850.

1069. Lolas, O. M.; Markakis, P. *J. Agric. Food Chem.* **1975**, *23*(1), 13–15.

1070. Reddy, N. R.; Pierson, M. D. *J. Food Sci.* **1987**, *52*(1), 109–112 134.

1071. Makower, R. U. *J. Sci. Food Agric.* **1969**, *20*, 82–84.

1072. Tecklenburg, E.; Zabik, M. E.; Uebersax, M. A.; Dietz, J. C.; Lusas, E. W. *J. Food Sci.* **1984**, *49*(2), 569–572 584.

1073. Deshpande, S. S.; Cheryan, M. *Plant Foods Hum. Nutr.* **1984**, *34*(3), 185–196.

1074. DeMason, D. A.; Stillman, J. I. *Planta* **1986**, *167*(3), 321–329.

1075. Reid, D. A.; Lott, J. N. A.; Attree, S. M.; Fowke, L. C. *Plant Sci.* **1999**, *141*(1), 11–18.

1076. Reid, D. A.; Lott, J. N. A.; Attree, S. M.; Fowke, L. C. *Plant Sci.* **1999**, *141*(1), 19–27.

1077. Pittermann, J.; West, M.; Lott, J. N. A. *Can. J. Forest Res.* **1996**, *26*(9), 1697–1702.

1078. Jackson, J. F.; Linskens, H. F. *Aust. Forest Res.* **1982**, *12*(1), 11–18.

1079. West, M. M.; Lott, J. N. A. *Can. J. Bot.* **1993**, *71*, 577–585.

1080. L'Annunziata, M. F.; Fuller, W. H. *Proc. Soil Sci. Soc. Am.* **1971**, *35*(4), 587–595.

1081. Sattar, A.; Jan, M.; Ahmad, A.; Wahid, M.; Khan, I. *Acta Aliment.* **1989**, *18*(1), 45–52.

1082. Zdunczyk, Z.; Godycka, I.; Amarowicz, R. *Plant Foods Hum. Nutr.* **1997**, *50*(1), 37–45.

1083. Fowler, H. D. *J. Sci. Food Agric.* **1957**, *8*(6), 333–341.

1084. Beal, L.; Mehta, T. *J. Food Sci.* **1985**, *50*(1), 96–100 115.

1085. Rosenbaum, T. M.; Baker, B. E. *J. Sci. Food Agric.* **1969**, *20*(12), 709–712.

1086. Crean, D. E. C.; Haisman, D. R. *J. Sci. Food Agric.* **1963**, *14*(11), 824–833.

1087. Naczk, M.; Rubin, L. J.; Shahidi, F. *J. Food Sci.* **1986**, *51*(5), 1245–1247.

1088. Vose, J. R.; Basterrechea, M. J.; Gorin, P. A. J.; Finlayson, A. J.; Youngs, C. G. *Cereal Chem.* **1976**, *53*(6), 928–936.

1089. Welch, R. M.; House, W. A.; Allaway, W. H. *J. Nutr.* **1974**, *104*, 733–740.

1090. Rosenbaum, T. M.; Henneberry, G. O.; Baker, B. E. *J. Sci. Food Agric.* **1966**, *17*(6), 237–240.

1091. Carnovale, E.; Lugano, E.; Lombardi-Boccia, G. *Cereal Chem.* **1988**, *65*(2), 114–117.

1092. Guzman, M. G. A.; Quijada, M. R. G.; Hernandez, R. M. *Arch. Latinoamer. Nutr.* **1998**, *48*(4), 328–333.

1093. Zolfaghari, R.; Harden, M. *Lebensm. Wiss. Technol.* **1985**, *18*(3), 186–191.

1094. Becker, R.; Grosjean, O. K. *J. Agric. Food Chem.* **1980**, *28*(1), 22–25.

1095. van Staden, J.; Comins, N. R. *Planta* **1976**, *130*(2), 219–221.

1096. El-Adawy, T. A.; Rahma, E. H.; El-Badawey, A. A.; Gomba, M. A.; Lásztity, R.; Sar-kadi, L. *Nahrung* **1994**, *38*(1), 12–20.

1097. Funes, J. A.; Bertoni, M. H.; Cattaneo, P. *An. Asoc. Quim. Argent.* **1978**, *66*, 255–272.

1098. Rahma, E. H.; Abd El Aal, M. H. *Food Chem.* **1988**, *28*(1), 31–43.

1099. Tan, N.-H.; Rahim, Z. H. A.; Khor, H.-T.; Wong, K.-C. *J. Agric. Food Chem.* **1983**, *31*(4), 916–917.

1100. Kotaru, M.; Ikeuchi, R.; Yoshikawa, H.; Ibuki, F. *Food Chem.* **1987**, *24*(4), 279–286.

1101. Kadam, S. S.; Kute, L. S.; Lawande, K. M.; Salunkhe, D. K. *J. Food Sci.* **1982**, *47*(6), 2051–2053 2057.

1102. Kantha, S. S.; Hettiarachchy, N. S.; Erdman, J. W., Jr. *Cereal Chem.* **1986**, *63*(1), 9–13.

1103. Bertoni, M. H.; Cattaneo, P.; Covas, G. *An. Asoc. Quim. Argent.* **1987**, *75*(2), 269–272.

1104. Suvorov, V. I.; Buzulukova, N. P.; Sobolev, A. M.; Sveshnikova, I. N. *Fiziol. Rast.* **1970**, *17*(6), 1223–1231 (Engl. 1020–1027).

1105. Greenwood, J. S.; Bewley, J. D. *Planta* **1984**, *160*(2), 113–120.

1106. Azarkovich, M. I.; Dmitrieva, M. I.; Sobolev, A. M. *Fiziol. Rast.* **1999**, *46*(3), 410–418 (Engl. 349–356).

1107. Malec, L. S.; Civeira, M. E.; Vigo, M. S. *An. Asoc. Quim. Argent.* **1993**, *81*(6), 445–450.

1108. Gomez, R. G.; Malec, L. S.; Vigo, M. S. *An. Asoc. Quim. Argent.* **1993**, *81*(1), 9–14.

1109. Morelec-Coulon, M. J.; Faure, M. *Bull. Soc. Chim. Biol.* **1957**, *39*(7/8), 947–951.

1110. Fretzdorff, B.; Weipert, D. *Z. Lebensm. Unters. Forsch.* **1986**, *182*(4), 287–293.

1111. Pedersen, B.; Eggum, B. O. *Plant Foods Hum. Nutr.* **1983**, *32*(2), 185–196.

1112. Lease, J. G. *Poultry Sci.* **1966**, *45*(2), 237–241.

1113. Taha, F. S.; Fahmy, M.; Sadek, M. A. *J. Agric. Food Chem.* **1987**, *35*(3), 289–292.

1114. Villegas Varela, L. *An. Asoc. Quim. Argent.* **1974**, *62*(5), 309–315.

1115. Rajendran, S.; Prakash, V. *J. Agric. Food Chem.* **1988**, *36*(2), 269–275.

1116. Wiseman, M.-O.; Price, R. L. *Cereal Chem.* **1987**, *64*(2), 91–93.

1117. Wager, H. G. *J. Sci. Food Agric.* **1963**, *14*(8), 583–586.

1118. Samotus, B.; Schwimmer, S. *Nature* **1962**, *194*, 578–579.

1119. Mali, P. C.; Gupta, Y. P. *Indian J. Nutr. Diet.* **1976**, *11*(5), 289–295.

1120. Wang, I. C.; Mitchell, H. L.; Barham, H. N. *Trans. Kansas Acad. Sci.* **1959**, *62*(3), 208–211.

1121. Radhakrishnan, M. R.; Sivaprasad, J. *J. Agric. Food Chem.* **1980**, *28*(1), 55–57.

1122. Agunbiade, S. O.; Longe, O. G. *Nahrung* **1996**, *40*(4), 184–188.

1123. Ene-Obong, H. N. *Plant Foods Hum Nutr.* **1995**, *48*(3), 225–233.

1124. Ishola, M. M.; Agbaji, E. B.; Agbaji, A. S. *J. Sci. Food Agric.* **1990**, *51*(1), 141–143.

1125. Udosen, E. O.; Edet, E. E.; Akpanabiatu, M. I. *Global J. Pure Appl. Sci.* **1998**, *4*(3), 251–253, from *Chem. Abstr.* **1998**, *129*, 315306r.

1126. Bonvehi, J. S.; Jordà, R. E. *Z. Naturforsch.* **1998**, *53C*(9/10), 785–792.

1127. Edet, E. E.; Eka, O. U.; Ifon, E. T. *Food Chem.* **1985**, *17*(1), 41–47.

1128. Jakobsen, H. B.; Martens, H.; Lyshede, O. B. *Ann. Bot.* **1994**, *74*(4), 409–415.

1129. Rather, J. B. *J. Am. Chem. Soc.* **1917**, *39*(11), 2506–2515.

1130. Abdel-Nabey, A. A.; Damir, A. A. *Plant Foods Hum. Nutr.* **1990**, *40*(4), 267–274.

1131. Leung, D. W. M.; Bewley, J. D.; Reid, J. S. G. *Planta* **1981**, *153*(2), 95–100.

1132. Young, S. M.; Greaves, J. E. *Food Res.* **1940**, *5*(1), 103–108.

1133. Hira, C. K.; Kokhhar, A.; Sadana, B. K.; Sharma, K. K. *J. Food Sci. Technol.* **1991**, *28*(2), 112–113.

1134. Bassiri, A.; Nahapetian, A. *J. Agric. Food Chem.* **1979**, *27*(5), 984–989.

1135. Barrier-Guillot, B.; Casado, P.; Maupetit, P.; Jondreville, C.; Gatel, F.; Larbier, M. *J. Sci. Food Agric.* **1996**, *70*(1), 62–68.

1136. Nahapetian, A.; Bassiri, A. *J. Agric. Food Chem.* **1976**, *24*(5), 947–950.

1137. Poux, N. *J. Microsc.* **1963**, *2*(5), 557–568.

1138. Sasaki, K.; Loewus, F. A. *Plant Physiol.* **1980**, *66*(4), 740–745.

1139. Lott, J. N. A.; Spitzer, E. *Plant Physiol.* **1980**, *66*(3), 494–499.

1140. March, J. G.; Villacampa, A. I.; Grases, F. *Anal. Chim. Acta* **1995**, *300*(1/3), 269–272.

1141. Nelson, K. J.; Potter, N. N. *J. Food Sci.* **1979**, *44*, 104–107, 111.

1142. Dintzis, F. R.; Lehrfeld, J.; Nelsen, T. C.; Finney, F. L. *Cereal Chem.* **1992**, *69*(5), 577–581.

1143. Hay, J. G. *Cereal Chem.* **1942**, *19*(3), 326–333.

1144. Singh, B.; Reddy, N. R. *J. Food Sci.* **1977**, *42*, 1077–1083.

1145. Singh, B.; Sedeh, H. G. *Cereal Chem.* **1979**, *54*(4), 267–272.

1146. Niziotek, S. *Acta Physiol. Plant.* **1995**, *17*(3), 269–274.

1147. Bertoni, M. H.; Gonzales, A. P.; Cattaneo, P.; Covas, G. *An. Asoc. Quim. Argent.* **1995**, *83*(3/4), 159–163.

1148. Tabekhia, M. M.; Donnelly, B. J. *Cereal Chem.* **1982**, *59*(2), 105–107.

1149. Andrews, J. S.; Bailey, C. H. *Ind. Eng. Chem.* **1932**, *24*(1), 80–82.

1150. Grau, H.; Wagner, S.; Frischkorn, C.; Barreno, J.-C.; Kuhn, M. *Getreide Mehl Brot* **1994**, *48*(2), 39–42.

1151. Fretzdorff, B. *Getr. Mehl Brot.* **1992**, *46*(6), 180–185.

1152. Ologunde, M. O.; Ayorinde, F. O.; Shepard, R. L. *J. Am. Oil Chem. Soc.* **1990**, *67*(2), 92–95.

1153. Griffiths, D. W.; Thomas, T. A. *J. Sci. Food Agric.* **1981**, *32*(2), 187–192.

1154. Griffiths, D. W. *J. Sci. Food Agric.* **1982**, *33*(9), 847–851.

1155. Vinh, L. T.; Dworschak, E. *Nahrung* **1985**, *29*(2), 161–166.

1156. Shehata, A. M. E.; Abu-Bakr, T. M.; El-Shimi, N. M. *J. Food Process. Preserv.* **1983**, *7*(3), 185–192.

1157. Bednarski, W.; Tomasik, J.; Piatkowska, B. *J. Sci. Food Agric.* **1985**, *36*(8), 745–751.

1158. Pawar, V. D.; Sawate, A. R.; Ingle, U. M. *J. Food Sci. Technol. (India)* **1986**, *23*(1), 36–39.

1159. Pawar, V. D.; Ingle, U. M. *J. Food Sci. Technol. (India)* **1988**, *25*(4), 190–193.

1160. Khokhar, S.; Chauhan, B. M. *J. Food Sci.* **1986**, *51*(3), 591–594.

1161. Marconi, E.; Ruggeri, S.; Carnovale, E. *Food Chem.* **1997**, *59*(2), 203–212.

1162. Chau, C.-F.; Cheung, P. C.-K. *J. Agric. Food Chem.* **1997**, *45*(12), 4773–4776.

1163. Reddy, N. R.; Salunkhe, D. K. *J. Food Sci.* **1981**, *46*(2), 564–567, 570.

1164. Hira, C. K.; Kanwar, J. K.; Gupta, N.; Kochhar, A. *J. Food Sci. Technol. (India)* **1988**, *25*(3), 133–136.

1165. Kataria, A.; Chauhan, B. M.; Punia, D. *Plant Foods Hum. Nutr.* **1989**, *39*(3), 257–266.

1166. Kaur, D.; Kapoor, A. C. *Food Chem.* **1992**, *43*(2), 119–124.

1167. Verma, P.; Mehta, U. *J. Food Sci. Technol. (India)* **1988**, *25*(4), 197–200.

1168. Oluwatosin, O. B. *J. Sci. Food Agric.* **1999**, *79*(2), 269–272.

1169. Farinu, G. O.; Ingrao, G. *J. Sci. Food Agric.* **1991**, *55*(3), 401–410.

1170. Longe, O. G. *J. Food Process. Preserv.* **1983**, *7*(3), 143–150.

1171. Ologhobo, A. D.; Fetuga, B. L. *Trop. Agric.* **1984**, *61*(4), 261–264.

1172. Ogun, P. O.; Markakis, P.; Chenoweth, W. *J. Food Sci.* **1989**, *54*(4), 1084–1085.

1173. Popova, O. L.; Murav'eva, D. A. *Biol. Nauk* **1991**, (9), 112–116.

1174. DeMason, D. A. *Am. J. Bot.* **1986**, *73*(9), 1332–1340.

1175. Siddhuraju, P.; Vijayakumari, K.; Janardhanan, K. *Food Chem.* **1995**, *53*(3), 299–304.

1176. Nyotu, H. G.; Alli, I.; Paquette, G. *J. Food Sci.* **1986**, *51*(5), 1204–1207.

1177. Hamilton, T. S.; Hamilton, B. C.; Johnson, B. C.; Mitchell, H. H. *Cereal Chem.* **1951**, *28*(3), 163–176.

1178. Raboy, V.; Below, F. E.; Dickinson, D. B. *J. Hered.* **1989**, *80*(4), 311–315.

1179. Nelson, T. S.; Ferrara, L. W.; Storer, N. L. *Poultry Sci.* **1968**, *47*(4), 1372–1374.

1180. Barba, I.; Gasparovic, C.; Cabañas, M. E.; Alonso, J.; Murillo, I.; San Segundo, B.; Arus, C. *Cell. Mol. Biol. (Paris)* **1997**, *43*(5), 609–620.

1181. Earley, E. B.; De Turk, E. E. *J. Am. Soc. Agron.* **1944**, *36*(10), 803–814.

1182. De Turk, E. E.; Holbert, J. R.; Howk, B. W. *J. Agric. Res.* **1933**, *46*(2), 121–141.

1183. de Boland, A. R.; Garner, G. B.; O'Dell, B. L. *J. Agric. Food Chem.* **1975**, *23*, 1186–1189.

1184. Bassiri, A.; Nahapetian, A. *J. Agric. Food Chem.* **1977**, *25*(5), 1118–1122.

1185. Tanaka, K.; Yoshida, T.; Kasai, Z. *Plant Cell Physiol.* **1974**, *15*(1), 147–151.

1186. Igaue, I.; Shimizu, M.; Miyauchi, S. *Plant Cell Physiol.* **1980**, *21*(2), 351–356.

1187. Asada, K.; Tanaka, K.; Kasai, Z. *Plant Cell Physiol.* **1968**, *9*(2), 185–193.

1188. Blaicher, F. M.; Mukherjee, K. D. *Z. Naturforsch.* **1981**, *36C*(5/6), 383–384.

1189. Ogawa, M.; Tanaka, K.; Kasai, Z. *Agric. Biol. Chem.* **1979**, *43*(10), 2211–2213.

1190. Nahapetian, A.; Young, V. R. *J. Nutr.* **1980**, *110*(7), 1458–1472.

1191. Singsen, E. F.; Matterson, L. D.; Kozeff, A. *Poultry Sci.* **1950**, *29*(5), 635–639.

1192. Jacobson, H. G. M. *Plant Physiol.* **1948**, *23*(6), 636–637.

1193. Biswas, S.; Biswas, B. B. *Biochim. Biophys. Acta* **1965**, *108*, 710–713.

1194. IUPAC-IUB. *Biochem. J.* **1976**, *153*(1), 23–31.

1195. Irvine, R. F., Ed.; *Methods of Inositide Research*; Raven Press: New York, NY, 1990; pp 65–82, xvii + 219 pp.

1196. Morré, D. J.; Boss, W. F.; Loewus, F. A., Eds.; *Inositol Metabolism in Plants*; *Plant Biology*; Vol. 9; Wiley-Liss: New York, NY, 1990; xviii + 393 pp. (a) p 117. (b) p 163.

1197. Biswas, B. B.; Biswas, S., Eds.; *Subcell. Biochem.* **1996**, *26* xxii + 421 pp.

1198. Matheson, N. K.; Strother, S. *Phytochemistry* **1969**, *8*(8), 1349–1356.

1199. Sethi, U.; Basu, A.; Guha-Mukherjee, S. *Phytochemistry* **1990**, *29*(3), 825–828.

1200. Heim, S.; Wagner, K. G. *Plant Sci.* **1989**, *63*(2), 159–165.

1201. Irvine, R. F.; Letcher, A. J.; Lander, D. J.; Drøbak, B. K.; Dawson, A. P.; Musgrave, A. *Plant Physiol.* **1989**, *89*(3), 888–892.

1202. Munnik, T.; Musgrave, A.; de Vrije, T. *Planta* **1994**, *193*(1), 89–98.

1203. Benson, A. A.; Maruo, B. *Biochim. Biophys. Acta* **1958**, *27*(1), 189–195.

1204. Murthy, P. P. N.; Renders, J. M.; Keranen, L. M. *Plant Physiol.* **1989**, *91*(4), 1266–1269.

1205. Brearley, C. A.; Hanke, D. E. *Biochem. J.* **1992**, *283*(1), 255–260.

1206. Fisher, N.; Broughton, M. E.; Peel, D. J.; Bennett, R. *J. Sci. Food Agric.* **1964**, *15*(5), 325–341.

1207. Lepage, M.; Mumma, R.; Benson, A. A. *J. Am. Chem. Soc.* **1960**, *82*(14), 3713–3715.

1208. Brearly, C. A.; Hanke, D. E. *Biochem. J.* **1996**, *318*(1), 279–286.

1209. Molinari, E.; Hoffmann-Ostenhof, O. *Z. Physiol. Chem.* **1968**, *349*(12), 1797–1799.

1210. Phillippy, B. Q.; White, K. D.; Johnston, M. R.; Tao, S.-H.; Fox, M. R. S. *Anal. Biochem.* **1987**, *162*(1), 115–121.

1211. Posternak, S.; Posternak, T. *Helv. Chim. Acta* **1929**, *12*(2), 1165–1181.

1212. Nagai, Y.; Funahashi, S. *Agric. Biol. Chem.* **1963**, *27*(9), 619–624.

1213. Raison, J. K.; Evans, W. J. *Biochim. Biophys. Acta* **1968**, *170*(2), 448–451.

1214. Lee, C. Y.; Parsons, G. F.; Downing, D. L. *J. Food Sci.* **1982**, *47*(3), 1034–1035.

1215. Desjobert, A. *Bull. Soc. Chim. Biol.* **1954**, *36*(9), 1293–1299.

1216. Bartlett, G. R. *Anal. Biochem.* **1982**, *124*(2), 425–431.

1217. Manthey, A. E.; Dickinson, D. B. *Plant Physiol.* **1978**, *61*(6), 904–908.

1218. Saio, K. *Plant Cell Physiol.* **1964**, *5*(4), 393–400.

1219. Coté, G. G.; Morse, M. J.; Crain, R. C.; Satter, R. L. *Plant Cell Rep.* **1987**, *6*(2), 352–355.

1220. Acharya, M. K.; Dureja-Munjal, I.; Guha-Mukherjee, S. *Phytochemistry* **1991**, *30*(9), 2895–2897.

1221. Strasser, H.; Hoffmann, C.; Grisebach, H.; Matern, U. *Z. Naturforsch.* **1986**, *41C*(7/8), 717–724.

1222. Anderson, R. J. *J. Biol. Chem.* **1914**, *18*(3), 441–446.

1223. Angyal, S. J.; Russell, A. F. *Aust. J. Chem.* **1969**, *22*(2), 383–390.

1224. Bollmann, O.; Pittner, F. *Monatsh. Chem.* **1980**, *111*(3), 557–561.

1225. Steward, J. H.; Tate, M. E. *J. Chromatogr.* **1969**, *45*(3/4), 400–406.

1226. Mihailovic, M. I.; Antic, M.; Hadzijev, D. *Plant Soil* **1965**, *23*(1), 117–128.

1227. Goldschmidt, B. *Carbohydr. Res.* **1990**, *208*, 105–110.

1228. Hull, S. R.; Gray, J. S. S.; Montgomery, R. *Anal. Biochem.* **1999**, *273*(2), 252–260.

1229. Phillippy, B. Q.; Bland, J. M. *Anal. Biochem.* **1988**, *175*(1), 162–166.

1230. Beno-Moualem, D.; Naveh, L.; Jacoby, B. *Plant Physiol. Biochem.* **1995**, *33*(3), 311–318.

1231. Piacentini, M. P.; Ricci, D.; Fraternale, D.; Piatti, E.; Manunta, A.; Accorsi, A. *Biochem. Mol. Biol. Int.* **1997**, *41*(4), 833–840.

1232. Cooke, C. J.; Smith, C. J.; Newton, R. F.; Walton, T. J. *Biochem. Soc. Trans.* **1991**, *19*(4), 359S.

1233. Franklin-Tong, V. E.; Drøbak, B. K.; Allan, A. C.; Watkins, P. A. C.; Trewavas, A. J. *Plant Cell* **1996**, *8*(8), 1305–1321.

1234. Toyoda, K.; Shiraishi, T.; Yamada, T.; Ichinose, Y.; Oku, H. *Plant Cell. Physiol.* **1993**, *34*(5), 729–735.

1235. Jones, D. L.; Kochian, L. V. *Plant Cell* **1995**, *7*(11), 1913–1922.

1236. Aducci, P.; Marra, M. *Biochem. Biophys. Res. Commun.* **1990**, *168*(3), 1041–1046.

1237. Loewus, M. W.; Sasaki, K.; Leavitt, A. L.; Munsell, L.; Sherman, W. R.; Loewus, F. A. *Plant Physiol.* **1982**, *70*(6), 1661–1663.

1238. Brearly, C. A.; Hanke, D. E. *Biochem. J.* **1996**, *314*(1), 227–233.

1239. Cooke, C. J.; Smith, C. J.; Newton, R. P.; Walton, T. J. *Biochem. Soc. Trans.* **1991**, *19*(2), 94S.

1240. Wang, S. Y.; Faust, M. *Physiol. Plant.* **1995**, *93*(3), 471–475.

1241. Skoglund, E.; Carlsson, N.-G.; Sandberg, A.-S. *J. Agric. Food Chem.* **1997**, *45*(12), 4668–4673.

1242. Tanaka, K.; Watanabe, K.; Asada, K.; Kasai, Z. *Agric. Biol. Chem.* **1971**, *35*(3), 314–320.

1243. Brearly, C. A.; Hanke, D. E. *Biochem. J.* **1996**, *314*(1), 215–225.

1244. Pal, S.; Acharya, M. K.; Guha-Mukherjee, S. *Phytochemistry* **1993**, *32*(4), 823–825.

1245. Pizer, F. L.; Ballou, C. E. *J. Am. Chem. Soc.* **1959**, *81*(4), 915–921.

1246. Bieth, H.; Jost, P.; Spiess, B. *J. Inorg. Biochem.* **1990**, *39*(1), 59–73.

1247. Brown, D. M.; Clark, B. F. C.; Letters, R. *J. Chem. Soc.* **1961**, (9), 3774–3779.

1248. Heathers, G. P.; Juehne, T.; Rubin, L. J.; Corr, P. B.; Evers, A. S. *Anal. Biochem.* **1989**, *176*(1), 109–116.

1249. Leavitt, A. L.; Sherman, W. R. *Carbohydr. Res.* **1982**, *103*(2), 203–212.

1250. Wreggett, K. A.; Irvine, R. F. *Biochem. J.* **1989**, *262*(3), 997–1000.

1251. Hsu, F.-F.; Goodman, H. D.; Sherman, W. R. *Biomed. Environ. Mass Spectrom.* **1990**, *19*(10), 597–600.

1252. Mernissi-Arifi, K.; Schmitt, L.; Schlewer, G.; Spiess, B. *Anal. Chem.* **1995**, *67*(15), 2567–2574.

1253. Duthu, B.; Houalla, D.; Wolf, R. *Can. J. Chem.* **1988**, *66*(12), 2965–2974.

1254. Henshall, A.; Harrold, M. P.; Tso, J. M. Y. *J. Chromatogr.* **1992**, *608*(1/2), 413–419.

1255. Akiyama, T.; Takechi, N.; Ozaki, S.; Shiota, K. *Bull. Chem. Soc. Jpn.* **1992**, *65*(2), 366–372.

1256. Brown, D. M.; Clark, B. F. C. *J. Chem. Soc.* **1963**, (3), 1475–1477.

1257. Klyashchitskii, B. A.; Pimenova, V. V.; Bashkatova, A. I.; Zhelvakova, E. G.; Sokolov, S. D.; Shvets, V. I.; Evstigneeva, R. P.; Preobrazhenskii, N. A. *Zh. Obshch. Khim.* **1970**, *40*(11), 2482–2489 (Engl. 2469–2474).

1258. Laumen, K.; Ghisalba, O. *Biosci. Biotechnol. Biochem.* **1994**, *58*(11), 2046–2049.

1259. Pietrusiewicz, K. M.; Salamonczyk, G. M.; Bruzik, K. S.; Wieczorek, W. *Tetrahedron* **1992**, *48*(26), 5523–5542.

1260. Walton, T. J.; Hughes, S. Y. *Rapid Commun. Mass Spectrom.* **1993**, *7*(6), 555–560.

1261. Iselin, B. M. *J. Am. Chem. Soc.* **1949**, *71*(11), 3822–3825.

1262. Liang, C.; Ewig, C. S.; Stouch, T. R.; Hagler, A. T. *J. Am. Chem. Soc.* **1994**, *116*(9), 3904–3911.

1263. Fleury, P.; Desjobert, A.; Lecocq, J. *Bull. Soc. Chim. Biol.* **1954**, *36*(9), 1301–1310.

1264. Cerdan, S.; Hansen, C. A.; Johanson, R.; Inubushi, T.; Williamson, J. R. *J. Biol. Chem.* **1986**, *261*(31), 14676–14680.

1265. Cosgrove, D. J. *Aust. J. Biol. Sci.* **1970**, *23*(6), 1207–1220.

1266. Hokin-Neaverson, M.; Sadeghian, K. *J. Chromatogr.* **1976**, *120*(2), 502–505.

1267. Wreggett, K. A.; Irvine, R. F. *Biochem. J.* **1987**, *245*(3), 655–660.

1268. Binder, H.; Weber, P. C.; Siess, W. *Anal. Biochem.* **1985**, *148*(1), 220–227.

1269. Sherman, W. R.; Goodwin, S. L.; Zinbo, M. *J. Chromatogr. Sci.* **1971**, *9*(6), 363–367.

1270. Buscher, B. A. P.; van der Hoeven, R. A. M.; Tjaden, U. R.; Andersson, E.; van der Greef, J. *J. Chromatogr.* **1995**, *712A*(1), 235–243.

1271. Yoo, C. S.; Blank, G.; Pletcher, J.; Sax, M. *Acta Crystallogr.* **1974**, *B30*(8), 1983–1987.

1272. Ballou, C. E.; Pizer, L. I. *J. Am. Chem. Soc.* **1960**, *82*(13), 3333–3335.

1273. Mercier, D.; Barnett, J. E. G.; Géro, S. D. *Tetrahedron* **1969**, *25*(23), 5681–5687.

1274. Grado, C.; Ballou, C. E. *J. Biol. Chem.* **1961**, *236*(1), 54–60.

1275. Vacca, J. P.; deSolms, S. J.; Hutt, J. R.; Billington, D. C.; Baker, R.; Kulagowski, J. J.; Mawer, I. M. *Tetrahedron* **1989**, *45*(17), 5679–5702, *Errata* **1991**, *47*(6), 907.

1276. Molotkovsky, J. G.; Bergelson, L. D. *Tetrahedron Lett.* **1971**, (50), 4791–4794.

1277. Billington, D. C.; Baker, R.; Kulagowski, J. J.; Mawer, I. M.; Vacca, J. P.; deSolms, S. J.; Huff, J. R. *J. Chem. Soc., Perkin Trans. 1* **1989**, (8), 1423–1429.

1278. McCormick, M. H.; Carter, H. E. *Biochem. Prep.* **1952**, *2*, 65–68.

1279. Chen, I.-W.; Charalampous, F. C. *J. Biol. Chem.* **1966**, *241*(10), 2194–2199.

1280. Iacobucci, S.; d'Alarcao, M. *Synth. Commun.* **1994**, *24*(6), 809–817.

1281. Barrientos, L. G.; Murthy, P. P. N. *Carbohydr. Res.* **1996**, *296*(1), 39–54.

1282. Eisenberg, F., Jr.; Parthasarathy, R. *Methods Enzym. Anal.* **1984**, *6*, 371–375.

1283. Martinoia, E.; Locher, R.; Vogt, E. *Plant Physiol.* **1993**, *102*(1), 101–105.

1284. Lim, P. E.; Tate, M. E. *Biochim. Biophys. Acta* **1973**, *302*(2), 316–328.

1285. Chung, S.-K.; Chang, Y.-T.; Kwon, Y. K. *J. Carbohydr. Chem.* **1998**, *17*(3), 369–384.

1286. Schmitt, L.; Bortmann, P.; Schlewer, G.; Spiess, B. *J. Chem. Soc., Perkin Trans. 2* **1993**, (11), 2257–2263.

1287. Krylova, V. N.; Kobel'kova, N. I.; Oleinik, G. F.; Shvets, V. I. *Zh. Org. Khim.* **1980**, *16*(1), 62–68 (Engl. 59–64).

1288. Shevchenko, S. M.; Semenov, S. G.; Kiselev, G. V. *Teor. Eksp. Khim.* **1984**, *20*(2), 223–228.

1289. Skoglund, E.; Carlsson, N.-G.; Sandberg, A.-S. *J. Agric. Food Chem.* **1997**, *45*(2), 431–436.

1290. Drory, A.; Borochov, A.; Mayak, S. *J. Plant Physiol.* **1992**, *140*(1), 116–120.

1291. Barrientos, L.; Scott, J. J.; Murthy, P. P. N. *Plant Physiol.* **1994**, *106*(4), 1489–1495.

1292. Phillippy, B. Q. *J. Agric. Food Chem.* **1989**, *37*(5), 1261–1265.

1293. Spiers, I. D.; Freeman, S.; Poyner, D. R.; Schwalbe, C. H. *Tetrahedrom Lett.* **1995**, *36*(12), 2125–2128.

1294. Hayakawa, T.; Suzuki, K.; Miura, H.; Ohno, T.; Igaue, I. *Agric. Biol. Chem.* **1990**, *54*(2), 279–286.

1295. Mernissi-Arifi, K.; Ballereau, S.; Schlewer, G.; Spiess, B.; Zenkouar, M. *New J. Chem.* **1996**, *20*(10), 1087–1092.

1296. Lindon, J. C.; Baker, D. H.; Farrant, R. D.; Williams, J. M. *Biochem. J.* **1986**, *233*(1), 275–277.

1297. Ling, L.; Ozaki, S. *Carbohydr. Res.* **1994**, *256*, 49–58.

1298. Pinkston, J. D.; Bowling, D. J.; Delaney, T. E. *J. Chromatogr.* **1989**, *474*(1), 97–111.

1299. Church, W. H.; Chiang, H. T. *J. Capillary Electrophor.* **1997**, *4*(6), 261–268, from *Chem. Abstr.* **1999**, *130*, 32507q.

1300. Salamonczyk, G. M.; Pietrusiewicz, K. M. *Tetrahedron Lett.* **1994**, *35*(24), 4233–4234.

1301. Mills, S. J.; Potter, B. V. L. *J. Org. Chem.* **1996**, *61*(25), 8980–8987.

1302. Watanabe, Y.; Ogasawara, T.; Ozaki, S.; Hirata, M. *Carbohydr. Res.* **1994**, *258*, 87–92.

1303. Ozaki, S.; Kondo, Y.; Shiotani, N.; Ogasawara, T.; Watanabe, Y. *J. Chem. Soc., Perkin Trans. 1* **1992**, (6), 729–737.

1304. Tegge, W.; Ballou, C. E. *Proc. Natl. Acad. Sci. U. S. A.* **1989**, *86*(1), 94–98.

1305. Dreef, C. E.; Tuinman, R. J.; Elie, C. J. J.; van der Marel, G. A.; van Boom, J. H. *Recl. Trav. Chim. Pays-Bas* **1988**, *107*(5), 395–397.

1306. Reese, C. B.; Ward, J. G. *Tetrahedron Lett.* **1987**, *28*(20), 2309–2312.

1307. Stepanov, A. E.; Runova, O. B.; Krylova, V. N.; Shvets, V. I.; Bochkov, V. N.; Schlewer, G.; Spiess, B. *Bioorg. Khim.* **1991**, *17*(2), 258–267.

1308. Maitra, R.; Samanta, S.; Mukherjee, M.; Biswas, S.; Biswas, B. B. *Indian J. Biochem. Biophys.* **1988**, *25*(6), 655–659.

1309. Phillippy, B. Q.; Ullah, A. H. J.; Ehrlich, K. C. *J. Biol. Chem.* **1994**, *259*(45), 28393–28399.

1310. Phillippy, B. Q. *Plant Physiol.* **1998**, *116*(1), 291–297.

1311. Irving, G. C. J.; Cosgrove, D. J. *J. Bacteriol.* **1972**, *112*(1), 434–438.

1312. Guedat, P.; Schlewer, G.; Krempp, E.; Riley, A. M.; Potter, B. V. L.; Spiess, B. *J. Chem. Soc., Chem. Commun.* **1997**, (6), 625–626.

1313. Lindon, J. C.; Baker, D. J.; Williams, J. M.; Irvine, R. F. *Biochem. J.* **1987**, *244*(3), 591–595.

1314. Lauman, K.; Ghisalba, O. *Biosci. Biotechnol. Biochem.* **1999**, *63*(8), 1374–1377.

1315. Ozaki, S.; Kondo, Y.; Nakahira, H.; Yamaoka, S.; Watanabe, Y. *Tetrahdron Lett.* **1987**, *28*(40), 4691–4694.

1316. Baudin, G.; Glänzer, B. I.; Swaminathan, K. S.; Vasella, A. *Helv. Chim. Acta* **1988**, *71*, 1367–1378.

1317. Mayr, G. W.; Dietrich, W. *FEBS Lett.* **1987**, *213*(2), 278–282.

1318. Ozaki, S.; Ling, L.; Ogasawara, T.; Watanabe, Y.; Hirata, M. *Carbohydr. Res.* **1994**, *259*(2), 307–310.

1319. Gou, D.-M.; Liu, Y.-C.; Chen, C. S. *Carbohydr. Res.* **1992**, *234*, 51–64.

1320. Stephens, L. R.; Hawkins, P. T.; Stanley, A. F.; Moore, T.; Poyner, D. R.; Morris, P. J.; Hanley, M. R.; Kay, R. R.; Irvine, R. F. *Biochem. J.* **1991**, *275*(2), 485–499.

1321. Rudolf, M. T.; Kaiser, T.; Guse, A. H.; Mayr, G. W.; Schultz, C. *Liebigs Ann. Recl.* **1997**, (9), 1861–1869.

1322. Johnson, L. F.; Tate, M. E. *Ann. N. Y. Acad. Sci.* **1969**, *165*(2), 526–532.

1323. Christensen, S.; Harbak, H. *J. Chromatogr.* **1990**, *533*, 201–206.

1324. Ozaki, S.; Koga, Y.; Ling, L.; Watanabe, Y.; Kimura, Y.; Hirata, M. *Bull. Chem. Soc. Jpn.* **1994**, *67*(4), 1058–1063.

1325. Matthäus, B. *Fett/Lipid* **1998**, *100*(7), 295–301.

1326. Smolenska-Sym, G.; Kacperska, A. *Physiol. Plant.* **1994**, *91*(1), 1–8.

1327. Morris, E. R.; Hill, A. D. *J. Food Compos. Anal.* **1996**, *9*(1), 2–12.

1328. Kasim, A. B.; Edwards, H. M., Jr. *J. Sci. Food Agric.* **1998**, *76*(1), 1–9.

1329. Mori, H. *Eiyo To Shokuryo* **1969**, *22*(2), 122–125.

1330. Sharma, C. B.; Dieckert, J. W. *Phytochemistry* **1974**, *13*(11), 2529–2533.

1331. Klein, J.; Maier, H. G. *Dtsch. Lebensm. Rundsch.* **1995**, *91*(6), 174–177.

1332. Kozlowska, H.; Honke, J.; Sadowska, J.; Frias, J.; Vidal-Valverde, C. *J. Sci. Food Agric.* **1996**, *71*(3), 367–375.

1333. Honke, J.; Kozlowska, H.; Vidal-Valverde, C.; Frias, J.; Gorecki, R. *Z. Lebensm. Unters. Forsch.* **1998**, *206A*(4), 279–283.

1334. Teare, I. D. *HortScience* **1974**, *9*(3), 236–238.

1335. Mandal, N. C.; Biswas, B. B. *Indian J. Biochem.* **1970**, *7*(1), 63–67.

1336. Reddy, A. S. N.; McFadden, J. J.; Friedmann, M.; Poovaiah, B. W. *Biochem. Biophys. Res. Commun.* **1987**, *149*(2), 334–339.

1337. Maiti, I. B.; Majumder, A. L.; Biswas, B. B. *Phytochemistry* **1974**, *13*(7), 1047–1051.

1338. Brearly, C. A.; Hanke, D. E. *Biochem. J.* **1995**, *311*(3), 1001–1007.

1339. Hawthorne, J. N.; Hübscher, G. *Biochem. J.* **1959**, *71*(1), 195–200.

1340. Cicero, T. J.; Sherman, W. R. *Biochem. Biophys. Res. Commun.* **1971**, *42*(3), 428–433.

1341. Merchant, T. E.; Glonek, T. *J. Lipid Res.* **1990**, *31*(3), 479–486.

1342. Brockerhoff, H.; Ballou, C. E. *J. Biol. Chem.* **1961**, *236*(7), 1907–1911.

1343. Hendrickson, H. S.; Reinertsen, J. L. *Biochemistry* **1969**, *8*(12), 4855–4858.

1344. Hendrickson, H. S.; Fullington, J. G. *Biochemistry* **1965**, *4*(8), 1599–1605.

1345. Brown, D. M.; Hall, G. E.; Letters, R. *J. Chem. Soc.* **1959**, (11), 3547–3552.

1346. Brando, C.; Hoffman, T.; Bonvini, E. *J. Chromatogr.* **1990**, *529*(1), 65–80.

1347. Butt, V. S.; Beevers, H. *Plant Physiol.* **1966**, *4B*, 265–414.

1348. Kates, M. *Adv. Lipid Res.* **1970**, *8*, 225–265.

1349. Kates, M. *Techniques of Lipidology: Isolation, Analysis and Identification of Lipids.* American Elsevier Publ. Co: New York, 1972; pp 269–610. (a) p 401; (b) pp 393–394.

1350. Stumpf, P. K., Ed.; *The Biochemistry of Plants: A Comprehensive Treatise; Lipids: Structure and Function*; Vol. 4; Academic Press: New York, 1980; xv + 693 pp.

1351. Moore, T. S., Jr. *Methods Plant Biochem.* **1990**, *3*, 229–239.

1352. Coté, G. G.; Crain, R. C. *Annu. Rev. Plant Physiol. Plant Mol. Biol.* **1993**, *44*, 333–356.

1353. Schulze, E.; Steiger, E. *Z. Physiol. Chem.* **1889**, *13*(4), 365–384.

1354. Klenk, E.; Sakai, R. *Z. Physiol. Chem.* **1939**, *258*(1), 33–38.

1355. Schulte, K. E.; Krause, H. *Biochem. Z.* **1951**, *322*(3), 168–173.

1356. Kates, M.; Eberhardt, F. M. *Can. J. Bot.* **1957**, *35*(6), 895–905.

1357. Benson, A. A.; Strickland, E. H. *Biochim. Biophys. Acta* **1960**, *41*(2), 328–333.

1358. Bomstein, R. A. *Biochem. Biophys. Res. Commun.* **1965**, *21*(1), 49–54.

1359. Boss, W. F.; Massel, M. O. *Biochem. Biophys. Res. Commun.* **1985**, *132*(3), 1018–1023.

1360. Chapman, K. D.; Moore, T. S., Jr. *Plant Physiol.* **1993**, *102*(3), 761–769.

1361. Chapman, K. D.; Moore, T. S., Jr. *Arch. Biochem. Biophys.* **1993**, *301*(1), 21–33.

1362. Brearley, C. A.; Hanke, D. E. *Biochem. J.* **1993**, *290*(1), 145–150.

1363. Frentzen, M.; Griebau, R. *Plant Physiol.* **1994**, *106*(4), 1527–1532.

1364. Morelec-Coulon, M. J.; Faure, M. *Bull. Soc. Chim. Biol.* **1958**, *40*(9/10), 1315–1317.

1365. Coddington, J. M.; Johns, S. R.; Leslie, D. R.; Willing, R. I.; Bishop, D. G. *Aust. J. Chem.* **1981**, *34*(2), 357–363.

1366. Meneses, P.; Glonek, T. *J. Lipid Res.* **1988**, *29*(5), 679–689.

1367. Ushikusa, T. *Oyo Butsuri* **1990**, *59*(10), 1371–1375.

1368. Hurst, W. J.; Martin, R. A., Jr.; Sheeley, R. M. *Spectroscopy* **1987**, *2*(10), 30–32.

1369. Sherman, W. R.; Ackermann, K. E.; Bateman, R. H.; Green, B. N.; Lewis, I. *Biomed. Mass Spectrom.* **1985**, *12*(8), 409–413.

1370. Marto, J. A.; White, F. M.; Seldomridge, S.; Marshall, A. G. *Anal. Chem.* **1995**, *67*(21), 3979–3984.

1371. Takahashi, T.; Hashimoto, T. *Yukagaku* **1967**, *16*, 572–574.

1372. Colacicco, G.; Rapport, M. M. *J. Lipid Res.* **1967**, *8*(5), 513–515.

1373. Aneja, R.; Chadha, J. S.; Knaggs, J. A. *Biochem. Biophys. Res. Commun.* **1969**, *36*(3), 401–406.

1374. Babaev, M. Yu.; Mukhamedova, Kh. S.; Akramov, S. T. *Khim. Prir. Soedin.* **1975**, *11*(3), 328–334 (Engl. 344–348).

1375. Shustanova, L. A.; Umarov, A. U.; Markman, A. L. *Khim. Prir. Soedin.* **1970**, *6*(3), 292–296 (Engl. 293–296).

1376. Shustanova, L. A.; Umarov, A. U.; Markman, A. L. *Khim. Prir. Soedin.* **1971**, *7*(1), 3–7 (Engl. 1–4).

1377. Tolibaev, I.; Mukhamedova, Kh. S.; Akramov, S. T. *Khim. Prir. Soedin.* **1976**, (3), 289–293 (Engl. 258–261).

1378. Iwata, M. *Proc. Imp. Acad. (Tokyo)* **1930**, *6*(3), 212–215, from *Biochem. Z.* **1930**, *224*, 430–433.

1379. Allen, C. F.; Good, P.; Davis, H. F.; Chisum, P.; Fowler, S. D. *J. Am. Oil Chem. Soc.* **1966**, *43*(4), 223–231.

1380. Allen, C. F.; Good, P.; Davis, H. F.; Fowler, S. D. *Biochem. Biophys. Res. Commun.* **1964**, *15*(5), 424–430.

1381. Ballard, R. E.; Jones, J.; Read, D.; Inchley, A.; Cranmer, M. *Chem. Phys. Lett.* **1987**, *135*(4/5), 468–470.

1382. Coulon-Morelec, M. J.; Faure, M. *Bull. Soc. Chim. Biol.* **1967**, *49*(3), 273–277.

1383. Coulon-Morelec, M. J.; Douce, R. *Bull. Soc. Chim. Biol.* **1968**, *50*(9), 1547–1559.

1384. Kondo, Y.; Masuda, K. *Nippon Nogei Kagaku Kaishi* **1982**, *56*(3), 203–208.

1385. Itabashi, Y.; Kuksis, A. *Anal. Biochem.* **1997**, *254*(1), 49–56.

1386. Folch, J.; Lees, M.; Stanley, G. H. S. *J. Biol. Chem.* **1957**, *226*(1), 497–509.

1387. Bligh, E. G.; Dyer, W. J. *Can. J. Biochem. Physiol.* **1959**, *37*, 911–917.

1388. Roughan, P. G.; Slack, C. R.; Holland, R. *Lipids* **1978**, *13*(7), 497–503.

1389. Bieleski, R. L. *Plant Physiol.* **1972**, *49*(5), 740–745.

1390. Weber, E. J. *J. Am. Oil Chem. Soc.* **1979**, *56*(6), 637–641.

1391. Demopoulos, C. A.; Antonopoulou, S.; Andrikopoulos, N. K.; Kapoulas, V. M. *J. Liq. Chromatogr.* **1996**, *19*(4), 521–535.

1392. Morrison, W. R.; Mann, D. L.; Soon, W.; Coventry, A. M. *J. Sci. Food Agric.* **1975**, *26*(4), 507–521.

1393. Colborne, A. J.; Laidman, D. L. *Phytochemistry* **1975**, *14*(12), 2639–2645.

1394. Fujiwara, S.; Mukohara, T.; Koh, Y. *Shokuhin Sangyo Senta Gijutsu Kenkyu Hokoku* **1991**, *17*, 61–72, from *Chem. Abstr.* **1991**, *115*, 254665u.

1395. Montanari, L.; King, J. W.; List, G. R.; Rennick, K. A. *J. Food Sci.* **1996**, *61*(6), 1230–1233, 1253.

1396. Nichols, B. W. *Biochim. Biophys. Acta* **1963**, *70*(4), 417–422.

1397. Ajana, H.; Perrin, J.-L.; Prevot, A. *Rev. Fr. Corps Gras* **1986**, *33*(1), 19–26.

1398. Blacs, P.; Erdélyi, A.; Kurucz-Lusztig, Ė.; Holló, J. *Acta Aliment. Acad. Sci. Hung.* **1978**, *7*(3), 181–193.

1399. Watanabe, M.; Itoh, T.; Iida, M.; Okabe, A.; Kawaguchi, Y.; Shimbo, K.; Sono, R.; Murui, T.; Kaneko, T. *Yukagaku* **1984**, *35*(12), 1018–1024.

1400. Anon., AOCS Official Method Ja 7-86. *Phospholipids in Lecithin Concentrates by Thin Layer Chromatography*; 4th ed.; American Oil Chemists' Society: Champaign, IL, 1986.

1401. Lam, K. K.; Sequera, D. E. *J. Liq. Chromatogr.* **1990**, *13*(10), 1967–1982.

1402. Roughan, G. *Plant Sci.* **1986**, *43*(1), 57–62.

1403. Pomeroy, M. K.; Mudd, J. B. *Phytochemistry* **1993**, *34*(4), 1015–1020.

1404. Landrath, G.; Bonekamp, A.; Kraus, Lj. *J. Chromatogr.* **1991**, *588*(1/2), 303–305.

1405. Colarow, L. *J. Planar Chromatogr. Mod. TLC* **1989**, *2*(1), 19–23.

1406. Ackman, R. G. *Methods Enzymol.* **1981**, *72*, 205–252.

1407. Ashworth, E. N.; St. John, J. B.; Christiansen, M. N.; Patterson, G. W. *J. Agric. Food Chem.* **1981**, *29*(4), 879–881.

1408. Nasner, A.; Kraus, L. *J. Chromatogr.* **1981**, *216*, 389–394.

1409. Kaitaranta, J. K.; Bessman, S. P. *Anal. Chem.* **1981**, *53*(8), 1232–1235.

1410. Demandre, C.; Tremolieres, A.; Justin, A.-M.; Mazliak, P. *Phytochemistry* **1985**, *24*(3), 481–485.

1411. Beare-Rogers, J. L.; Bonekamp-Nasner, A.; Dieffenbacher, A. *Pure Appl. Chem.* **1992**, *64*(3), 447–454.

1412. Itoh, T.; Murui, T.; Asano, S.; Arakawa, A.; Ushikusa, T.; Sakuma, R.; Sono, R.; Tohdo, H.; Hirata, Y.; Nishide, T.; Yamaguchi, T.; Morita, H. *Yukagaku* **1992**, *41*(12), 1171–1179.

1413. Moreau, R. A.; Asmann, P. T.; Norman, H. A. *Phytochemistry* **1990**, *29*(8), 2461–2466.

1414. van der Meeren, P.; Vanderdeelen, J.; Huys, M.; Baert, L. *J. Chromatogr.* **1988**, *447*(2), 436–442.

1415. Valeur, A.; Michelsen, P.; Odham, G. *Lipids* **1993**, *28*(3), 255–259.

1416. Kondoh, Y.; Yamada, A.; Takano, S. *Bunseki Kagaku* **1991**, *40*(2), 57–63.

1417. Conforti, F. D.; Harris, C. H.; Rinehart, J. T. *J. Chromatogr.* **1993**, *645*, 83–88.

1418. Coté, G. G.; DePass, A. L.; Quarmby, L. M.; Tate, B. F.; Morse, M. J.; Satter, R. L.; Crain, R. C. *Plant Physiol.* **1989**, *90*(4), 1422–1428.

1419. de Meulenaer, B.; van der Meeren, P.; Vanderdeelen, J.; Baert, L. *J. Am. Oil Chem. Soc.* **1995**, *72*(9), 1073–1075.

1420. Hanras, C.; Perrin, J. L. *J. Am. Oil Chem. Soc.* **1991**, *68*(11), 804–808.

1421. Elsner, A.; Lange, R. *Fett Wiss. Technol.* **1993**, *95*(1), 31–34.

1422. Ingvardsen, L.; Michaelsen, S.; Sørensen, H. *J. Am. Oil Chem. Soc.* **1994**, *71*(2), 183–188.

1423. Szücs, R.; Verleysen, K.; Duchateau, G. S. M. J. E.; Sandra, P.; Vandeginste, B. G. M. *J. Chromatogr.* **1996**, *738A*(1), 25–29.

1424. Christie, W. W.; Noble, R. C.; Moore, J. H. *Analyst* **1970**, *95*(11), 940–944.

1425. Morrison, W. R.; Tan, S. L.; Hargin, K. D. *J. Sci. Food Agric.* **1980**, *31*(4), 329–340.

1426. Tsuyuki, H.; Itoh, S.; Nakatsukasa, Y. *Nihon Daigaku No-Juigakubu Gakujutsu Kenkyu Hokoku* **1983**, *40*, 173–179.

1427. Camciuc, M.; Vilarem, G.; Gaset, A. *Ol. Corps Gras Lipides* **1997**, *4*(5), 389–392.

1428. El-Nockrashy, A. S.; Osman, F. *Planta Med.* **1965**, *13*(3), 326–330.

1429. Bhakare, H. A.; Kulkarni, A. S.; Khotpal, R. R. *J. Food Sci. Technol.* **1993**, *30*(1), 54–55.

1430. Pukacka, S. *Physiol. Plant.* **1991**, *82*(2), 306–310.

1431. Pukacka, S.; Czubak, A. *Acta Soc. Bot. Pol.* **1998**, *67*(3/4), 249–252.

1432. Bligny, R.; Douce, R. *Biochim. Biophys. Acta* **1980**, *617*(2), 254–263.

1433. Alban, C.; Joyard, J.; Douce, R. *Plant Physiol.* **1988**, *88*(3), 709–717.

1434. Tavernier, E.; Lê Quôc, D.; Lê Quôc, K. *Biochim. Biophys. Acta* **1993**, *1167*(3), 242–247.

1435. Cmelik, S. H. W. *J. Sci. Food Agric.* **1963**, *14*(5), 287–291.

1436. Künzler, K.; Eichenberger, W. *Phytochemistry* **1997**, *46*(5), 883–892.

1437. Bazier, R.; Burghoffer, E.; Costes, C. *Ann. Amelior. Plant.* **1979**, *29*(6), 665–682.

1438. Ivanov, S.; Zlatanov, M.; Ivanova, E.; Aitzetmüller, K. *Fett/Lipid* **1999**, *101*(8), 307–309.

1439. Chapman, D. J.; de Felice, J.; Barber, J. *Photosynth. Res.* **1986**, *8*(3), 257–265.

1440. Huner, N. P. A.; Williams, J. P.; Maissan, E. E.; Myscich, E. G.; Krol, M.; Laroche, A.; Singh, J. *Plant Physiol.* **1989**, *89*(1), 144–150.

1441. Dorne, A.-J.; Cadel, G.; Douce, R. *Phytochemistry* **1986**, *25*(1), 65–68.

1442. Radunz, A.; He, P.; Schmid, G. H. *Z. Naturforsch.* **1998**, *53C*(5/6), 305–310.

1443. He, P.; Radunz, A.; Bader, K. P.; Schmid, G. H. *Z. Naturforsch.* **1996**, *51C*(11/12), 833–840.

1444. Isamukhamedov, A. Sh.; Ganiev, S. S.; Auelbekov, S. A. *Khim. Prir. Soedin.* **1993**, (3), 465–467 (Engl. 402–404).

1445. Rozentsvet, O. A.; Dembitsky, N. M.; Zhuicova, V. S. *Phytochemistry* **1995**, *38*(5), 1209–1213.

1446. Li, L.; Sun, Q.; Li, F. *Yingyang Xuebao* **1990**, *12*(4), 423–425.

1447. Salama, A. M.; Pearce, R. S. *J. Exp. Bot.* **1993**, *44*(265), 1253–1265.

1448. Cooper, K. M.; Lösel, D. M. *New Phytol.* **1978**, *80*(1), 143–151.

1449. Philipp, E.-I.; Franke, W. W.; Keenan, T. W.; Stadler, J.; Jarasch, E.-D. *J. Cell Biol.* **1976**, *68*(1), 11–29.

1450. Murata, N.; Sato, N.; Takahashi, N. *Biochim. Biophys. Acta* **1984**, *795*(1), 147–150.

1451. Müller, H. *Fette Seifen Anstrichm.* **1977**, *79*(6), 259–261.

1452. Sturbois-Balcerzak, B.; Vincent, P.; Maneta-Peyret, L.; Duvert, M.; Satiat-Jeunemaitre, B.; Cassagne, C.; Moreau, P. *Plant Physiol.* **1999**, *120*(1), 245–256.

1453. Yang, K.-Y.; Shin, H.-S. *Korean J. Food Sci. Technol.* **1982**, *14*(4), 388–393.

1454. Berry, A. M.; Moreau, R. A.; Jones, A. D. *Plant Physiol.* **1991**, *95*(1), 111–115.

1455. Opute, F. I.; Osagie, A. U. *J. Sci. Food Agric.* **1978**, *29*(12), 1002–1006.

1456. Afzal, M.; Ali, M.; Hassan, R. A. H.; Sweedan, N.; Dhami, M. S. I. *Planta Med.* **1991**, *57*(1), 38–40.

1457. Fiad, S. *J. Am. Oil Chem. Soc.* **1991**, *68*(1), 26–28.

1458. Opute, F. I. *J. Exp. Bot.* **1979**, *30*(116), 601–606.

1459. Bajwa, S. S.; Sastry, P. S. *Indian J. Biochem. Biophys.* **1972**, *9*(1), 133–134.

1460. Santos, J. S. *Ciencia* **1992**, *45*(2), 113–119.

1461. Lorenz, K.; Hwang, Y. S. *Nutr. Rep. Internat.* **1985**, *31*(1), 83–89.

1462. Knacker, T.; Schaub, H. *Planta* **1984**, *162*(5), 441–449.

1463. Das, R.; Bagga, S.; Sopory, S. K. *Plant Sci.* **1987**, *53*(1), 45–51.

1464. Lemoine, Y.; Dubacq, J.-P.; Zabulon, G.; Ducruet, J.-M. *Can. J. Bot.* **1986**, *64*(12), 2999–3007.

1465. Pillai, P.; St. John, J. B. *Plant Physiol.* **1981**, *68*(3), 585–587.

1466. Lakshminarayana, G.; Pantulu, A. J.; Rao, K. S. *J. Agric. Food Chem.* **1984**, *32*(6), 1361–1363.

1467. Li, B. S.; Kim, D. H.; Ryu, G. O. *Punsok Hwahak* **1977**, *15*(1), 4–6, from *Chem. Abstr.* **1977**, *87*, 116619j.

1468. Nagaraja, K. V. *Plant Foods Hum. Nutr.* **1987**, *37*(4), 307–311.

1469. Maia, G. A.; Brown, W. H.; Whiting, F. M.; Stull, J. W. *J. Food Sci.* **1976**, *41*(4), 961–962.

1470. Rao, K. S.; Lakshminarayana, G. *J. Agric. Food Chem.* **1988**, *36*(3), 475–478.

1471. Zlatanov, M. *Fett Wiss. Technol.* **1994**, *96*(12), 456–457.

1472. Guo, R.; Lu, L.; Xu, Y. *Zhongguo Zhongyao Zazhi* **1991**, *16*(12), 741–742.

1473. Dembitsky, V. M. *Prog. Lipid Res.* **1993**, *32*(3), 281–356.

1474. Raie, M. Y.; Ahmad, I.; Javed, M. A.; Waheed, I. *Proc. Pak. Acad. Sci.* **1989**, *26*(3), 199–206.

1475. Rolph, C. E.; Goad, L. J. *Physiol. Plant.* **1991**, *83*(4), 605–610.

1476. Goodman, L. A.; Berry, J. W. *J. Am. Oil Chem. Soc.* **1986**, *63*(1), 98–100.

1477. Miquel, M.; Browse, J. *J. Biol. Chem.* **1992**, *267*(3), 1502–1509.

1478. Uemura, M.; Joseph, R. A.; Steponkus, P. L. *Plant Physiol.* **1995**, *109*(1), 15–30.

1479. Browse, J.; Somerville, C. R.; Slack, C. R. *Plant Sci.* **1988**, *56*(1), 15–20.

1480. Bancher, E.; Washüttl, J.; Goller, H. J. *Seifen Öle Fette Wachse* **1972**, *98*(16), 505–507.

1481. Wagner, H.; Wolff, P. *Fette Seifen Anstrichm.* **1964**, *66*(6), 425–429.

1482. Tzen, J. T. C.; Cao, Y.-Z.; Laurent, P.; Ratnayake, C.; Huang, A. H. C. *Plant Physiol.* **1993**, *101*(1), 267–276.

1483. Inanaga, S.; Arima, E.; Nishihara, T. *Plant Soil* **1990**, *122*(1), 91–96.

1484. Sreenivasulu, P.; Nayudu, M. V.; Sastry, P. S. *Indian J. Biochem. Biophys.* **1977**, *14*(3), 255–261.

1485. Sai Gopal, D. V. R.; Sreenivasulu, P.; Nayudu, M. V. *Physiol. Mol. Plant Pathol.* **1990**, *37*(1), 1–8.

1486. Reddy, M. N.; Sujathamma, P. *J. Phytopathol.* **1988**, *123*(1), 1–5.

1487. Zlatanov, M. D. *J. Sci. Food Agric.* **1999**, *79*(12), 1620–1624.

1488. Zarrouk, M.; Cherif, A. *Z. Pflanzenphysiol.* **1983**, *112*(4), 373–380.

1489. Guillot-Salomon, T.; Tuquet, C.; de Lubac, M.; Hallais, M.-F.; Signol, M. *Cytobiology* **1978**, *17*(2), 442–452.

1490. Bahl, J.; Demandre, D.; Chauveau, M.; Alpha, M. J.; Roussaux, J. *Plant Physiol. Biochem.* **1997**, *35*(9), 693–700.

1491. Xu, Y.; Wei, L. *Zhonggyo Zhongyao Zazhi* **1995**, *20*(10), 615–616.

1492. Dembitsky, V. M.; Rezanka, T.; Bycher, I. A.; Afonina, O. M. *Phytochemistry* **1993**, *33*(5), 1009–1014.

1493. Dembitsky, V. M.; Rezanka, T. *Biochem. Syst. Ecol.* **1995**, *23*(1), 71–78.

1494. Hirayama, O.; Mihara, M. *Agric. Biol. Chem.* **1987**, *51*(12), 3215–3221.

1495. Yamada, Y.; Hara, Y.; Senda, M.; Nishihara, M.; Kito, M. *Phytochemistry* **1979**, *18*(3), 423–426.

1496. Sahasrabudhe, M. R. *J. Am. Oil Chem. Soc.* **1979**, *56*(2), 80–84.

1497. Dawson, R. M. C.; Clarke, N.; Quarles, R. H. *Biochem. J.* **1969**, *114*(2), 265–270.

1498. Hack, M. H.; Helmy, F. M. *J. Chromatogr.* **1982**, *250*, 80–85.

1499. Acker, L.; Becker, G. *Stärke* **1971**, *23*(12), 419–424.

1500. Liukkonen, K.; Laakso, S. *Stärke* **1992**, *44*(4), 128–132.

1501. Sandstrom, R. P.; Cleland, R. E. *Plant Physiol.* **1989**, *90*(3), 1207–1213.

1502. Uemura, M.; Steponkus, P. L. *Plant Physiol.* **1994**, *104*(2), 479–496.

1503. Fuchs, R.; Haas, R.; Wrage, K.; Heinz, E. *Z. Physiol. Chem.* **1981**, *362*(8), 1069–1078.

1504. Jusaitis, M.; Paleg, L. G.; Aspinall, D. *Phytochemistry* **1981**, *20*(3), 417–423.

1505. Cooke, D. T.; Munkonge, F. M.; Burden, R. S.; James, C. S. *Biochim. Biophys. Acta* **1991**, *1061*(2), 156–162.

1506. Kuiper, P. J. C.; Kähr, M.; Stuiver, C. E. E.; Kylin, A. *Physiol. Plant.* **1974**, *32*(1), 33–36.

1507. Keenan, T. W.; Leonard, R. T.; Hodges, T. K. *Cytobios* **1973**, *7*(26/7), 103–112.

1508. Liljenberg, C.; Kates, M. *Can. J. Biochem. Cell Biol.* **1985**, *63*(2), 77–84.

1509. Markley, J. L.; McMillan, C.; Thompson, G. A., Jr. *Can. J. Bot.* **1982**, *60*(12), 2704–2715.

1510. Ognyanov, I. V.; Elamin, M. E.; Taranjiska, R. B.; Ivanova, B. S. *Dokl. Bolg. Akad. Nauk* **1977**, *30*(8), 1121–1124.

1511. Ksandopulo, S. Yu.; Kornena, E. P.; Moskvina, E. N. *Izv. Vyssh. Uchebn. Zaved. Pishch. Tekhnol.* **1989**, (2), 124–126.

1512. Tsuyuki, H.; Itoh, S.; Nakatsukasa, Y. *Nihon Daigaku No-Juigakubu Gakujutsu Kenkyu Hokoku* **1982**, *39*, 75–81.

1513. Rodionov, V. S.; Zakharova, L. S. *Fiziol. Biokhim. Kul't. Rast.* **1987**, *19*(5), 472–478.

1514. Moreau, R. A. *Phytochemistry* **1987**, *26*(7), 1899–1902.

1515. Beiss, U. *Landwirt. Forsch. Sonderh.* **1969**, *23*(1), 198–206.

1516. Sallé, G.; Le Coz, S.; Tuquet, C. *Physiol. Vég.* **1986**, *24*(1), 73–83.

1517. Wintermans, J. F. G. M. *Biochim. Biophys. Acta* **1960**, *44*(1), 49–54.

1518. Kolesnik, A. A.; Kapustina, V. V.; Drobot, I. V. *Khim. Prir. Soedin.* **1988**, (3), 357–360 (Engl. 299–301).

1519. Stuiver, C. E. E.; Kuiper, P. J. C.; Marschner, H. *Physiol. Plant.* **1978**, *42*(1), 124–128.

1520. Marty, F.; Branton, D. *J. Cell Biol.* **1980**, *87*(1), 72–83.

1521. Yahia, A.; Liljenberg, C.; Nilsson, R.; Lindberg, S.; Banas, A. *J. Plant Physiol.* **1995**, *146*(1/2), 81–87.

1522. Chernobrovkina, N. P.; Il'inova, M. A. *Lipidnyi Obmen Drev. Rast. Usloviyakh Sev.* **1983**, 112–118, from *Chem. Abstr.* **1984**, *101*, 188076k.

1523. Roughan, P. G.; Batt, R. D. *Phytochemistry* **1969**, *8*(2), 363–369.

1524. Omogbai, F. E. *J. Sci. Food Agric.* **1990**, *50*(2), 253–255.

1525. Burke, J. J.; Wilson, R. F.; Swafford, J. R. *Plant Physiol.* **1982**, *70*(1), 24–29.

1526. McKillican, M. E.; Larose, J. A. G. *Lipids* **1974**, *9*(7), 455–460.

1527. Arbenz, E. *Mitt. Geb. Lebensm. Hyg.* **1919**, *10*(3/4), 93–98.

1528. Rafal'son, A. B.; Koifman, T. Sh.; Maiorova, N. N.; Kornena, E. P.; Ksandopulo, S. Yu. *Maslo-Zhir. Prom-st* **1986**, (5), 12–13.

1529. Goraj-Moszora, I.; Drozdowski, B. *Ann. Acad. Med. Silesiensis* **1990**, *20*, 7–12.

1530. Dakhma, W. S.; Zarrouk, M.; Cherif, A. *Phytochemisry* **1995**, *40*(5), 1383–1386.

1531. Svenningsson, H.; Liljenberg, C. *Physiol. Plant.* **1986**, *68*(1), 53–58.

1532. Wu, S. S. H.; Moreau, R. A.; Whitaker, B. D.; Huang, A. H. C. *Lipids* **1999**, *34*(5), 517–523.

1533. Evans, D. E.; Sang, J. P.; Cominos, X.; Rothnie, N. E.; Knox, R. B. *Plant Physiol.* **1990**, *93*(2), 418–424.

1534. Aziz, A.; Lather, F. *J. Plant Physiol.* **1998**, *158*(5/6), 754–762.

1535. Smolenska, G.; Kuiper, P. J. C. *Physiol. Plant.* **1977**, *41*(1), 29–35.

1536. Shinozaki, Y.; Ohara, S.; Kondo, H. *Nippon Nogei Kagaku Kaishi* **1963**, *37*(9) Part I 546–552, Part II 553–557.

1537. Moreau, F.; Dupont, J.; Lance, C. *Biochim. Biophys. Acta* **1974**, *345*(3), 294–304.

1538. Journet, E.-P.; Douce, R. *Plant Physiol.* **1985**, *79*(2), 458–467.

1539. Wright, L. C.; McMurchie, E. J.; Pomeroy, M. K.; Raison, J. K. *Plant Physiol.* **1982**, *69*(6), 1356–1360.

1540. Reut, O. V.; Kolesnik, A. A.; Golubev, V. N. *Khim. Prir. Soedin.* **1989**, (2), 180–185 (Engl. 152–157).

1541. Rakhimov, D. A.; Tolibaev, I. *Khim. Prir. Soedin.* **1994**, (5), 612–615 (Engl. 563–565).

1542. Wheeldon, L. W. *J. Lipid Res.* **1960**, *1*(5), 439–445.

1543. Lepage, M. *Lipids* **1967**, *2*(3), 244–250.

1544. Kavua, M. E.; Duval, E.; Laval-Martin, D. *Z. Naturforsch.* **1991**, *46C*(7/8), 569–574.

1545. Panosyan, A. G.; Avetisyan, G. M.; Nikishchenko, M. N.; Mnatsakanyan, V. A. *Khim. Prir. Soedin.* **1980**, (6), 781–785 (Engl. 554–557).

1546. Hemavathy, J.; Prabhakar, J. V. *J. Food Compos. Anal.* **1988**, *1*(4), 366–370.

1547. Lin, H.; Wu, L. *Environ. Exp. Bot.* **1996**, *36*(3), 239–254.

1548. Prasad, R. B. N.; Rao, Y. N.; Rao, S. V. *J. Am. Oil Chem. Soc.* **1987**, *64*(10), 1424–1427.

1549. Badifu, G. I. O. *J. Food Compos. Anal.* **1989**, *2*(3), 238–244.

1550. Kulkarni, A. S.; Khotpal, R. R.; Bhakare, H. A. *J. Oil Technol. Assoc. India* **1992**, *24*(4), 117–122.

1551. Dembitsky, V. M.; Rezanka, T. *Phytochemistry* **1995**, *40*(1), 93–97.

1552. Spener, F.; Mangold, H. K. *Phytochemistry* **1975**, *14*(5/6), 1369–1373.

1553. Hemavathy, J.; Prabhakar, J. V. *J. Am. Oil Chem. Soc.* **1990**, *67*(12), 955–957.

1554. Lakshminarayana, G.; Rao, K. S.; Pantulu, A. J.; Gupta, D. R. *Fett. Wiss. Technol.* **1988**, *90*(2), 65–68.

1555. Dembitsky, V. M.; Rozentsvet, O. A. *Phytochemistry* **1996**, *41*(2), 483–488.

1556. Roberts, G. R. *J. Sci. Food Agric.* **1974**, *25*(5), 473–475.

1557. Anan, T.; Nakagawa, M. *Nippon Shokuhin Kogyo Gakkaishi* **1977**, *24*(6), 305–310.

1558. Sekiya, J.; Yamashita, K.; Nakagawa, S.; Shibata, Y.; Hatanaka, A. *Agric. Biol. Chem.* **1988**, *52*(1), 243–247.

1559. Lynd, J. Q.; Ansman, R. R. *Plant Cell Physiol.* **1988**, *29*(8), 1289–1295.

1560. Kuhn, M.; Jatsch, C.; von Buttlar, H.-B.; Slansky, I. *Agribiol. Res.* **1997**, *50*(1), 26–31.

1561. Tolibaev, I.; Glushenkova, A. I. *Khim. Prir. Soedin.* **1995**, *31*(3), 498–499 (Engl. 412–413).

1562. Whitaker, B. D. *Planta* **1992**, *187*(2), 261–265.

1563. Kinsella, J. E. *J. Food Sci.* **1971**, *36*(6), 865–866.

1564. Whitaker, B. D. *J. Am. Soc. Hortic. Sci.* **1991**, *116*(3), 528–533.

1565. Whitaker, B. D. *Physiol. Plant.* **1995**, *93*(4), 683–688.

1566. Roughan, P. G. *Plant Physiol.* **1985**, *77*(3), 740–746.

1567. Bhatia, I. S.; Madaan, T. R. *Plant Biochem. J.* **1978**, *5*(2), 118–126.

1568. Burkhardt, H. J. *J. Am. Oil Chem. Soc.* **1970**, *47*(2), 69–72.

1569. Osman, F.; Ashour, A. E.; Gad, A. M. *Fette Seifen Anstrichm.* **1969**, *71*(4), 264–266.

1570. Slack, R.; Roughan, G. *Dev. Plant Biol.* **1980**, *6*, 219–222.

1571. Millichip, M.; Tatham, A. S.; Jackson, F.; Griffiths, G.; Shewry, P. R.; Stobart, A. K. *Biochem. J.* **1996**, *314*(1), 333–337.

1572. Ahmad, I.; Raie, M. Y.; Akhtar, M. W. *Pak. J. Sci. Ind. Res.* **1994**, *37*(4), 150–152.

1573. Thompson, T. E.; Senter, S. D.; Grauke, L. J. *HortScience* **1993**, *28*(12), 1191–1193.

1574. Dresen, H.; Prasad, R. B. N.; Gülz, P.-G. *Z. Naturforsch.* **1989**, *44C*(9/10), 739–742.

1575. Zaka, S.; Akhtar, M. W.; Khan, S. A. *Pak. J. Sci. Ind. Res.* **1988**, *31*(2), 106–113.

1576. Niranjan, G. S.; Katiyar, S. K. *J. Indian Chem. Soc.* **1979**, *56*(7), 722–725.

1577. Imbs, A. B.; Pham, L. Q. *J. Am. Oil Chem. Soc.* **1995**, *72*(8), 957–961.

1578. Rhee, C.-O.; Kim, E.-S.; Kim, D.-Y. *Han'guk Nonghwa Hakhoe Chi* **1983**, *26*(1), 19–27.

1579. Somersalo, S.; Karunen, P.; Aro, E.-M. *Physiol. Plant.* **1986**, *68*(3), 467–470.

1580. Przybylski, E.; Chauhan, G. S.; Eskin, N. A. M. *Food Chem.* **1994**, *51*(2), 187–192.

1581. Radwan, S. S.; Mangold, H. K.; Barz, W.; Hüsemann, W. *Chem. Phys. Lipids* **1979**, *25*(1), 101–109.

1582. Attia, R. S.; Aman, M. E.; Shehata, A. M. E.; Hamza, M. A. *Food Chem.* **1996**, *56*(2), 123–129.

1583. Pranis, A. K. *Izv. Akad. Nauk Az. SSR, Ser. Biol. Nauk* **1987**, (2), 69–73.

1584. Katiyar, S. K.; Bhatia, A. K.; Kumar, N.; Atal, C. K. *J. Sci. Food Agric.* **1985**, *36*(6), 442–444.

1585. Krollmann, P.; Gülz, P.-G. *Z. Pflanzenphysiol.* **1983**, *110*(5), 469–474.

1586. Gleizes, M.; Marpeau, A.; Pauly, G.; Belingheri, L. *Dev. Plant Biol.* **1984**, *9*, 343–346.

1587. Bhatia, I. S.; Grewal, S. S.; Sukhija, P. S. *J. Sci. Food Agric.* **1971**, *22*(12), 638–640.

1588. Tsuyuki, H.; Chaya, H.; Takenaga, F.; Itoh, S. *Nihon Daigaku No-Juigakubu Gakujutsu Kenkyu Hokoku* **1986**, *43*, 63–66.

1589. Javed, M. A.; Raie, M. Y.; Ali, H. *Pak. J. Sci. Ind. Res.* **1994**, *37*(12), 544–546.

1590. Jiao, H. J.; Wang, S. Y.; Civerolo, E. L. *J. Phytopathol.* **1992**, *135*(1), 48–56.

1591. Nordby, H. E.; Yelenovsky, G. *Plant Physiol.* **1982**, *70*(1), 132–135.

1592. Tsuyuki, H.; Ito, S.; Tsujimura, Y. *Nihon Daigaku No-Juigakubu Gakujutsu Kenkyu Hokoku* **1984**, *41*, 166–174.

1593. Takenaga, F.; Kobayashi, M.; Itoh, S.; Tsuyuki, H. *Nihon Daigaku No-Juigakubu Ga-kujutsu Kenkyu Hokoku* **1991**, *48*, 17–25.

1594. Sekiya, J.; Nagai, K.; Shimose, N. *Agric. Biol. Chem.* **1990**, *54*(11), 2777–2782.

1595. Sulaberidze, K. V.; Tushishvili, L. Sh.; Paseshnichenko, V. A. *Fiziol. Rast.* **1989**, *36*(6), 1192–1198 (Engl. 959–964).

1596. Vandercook, C. E.; Guerrero, H. C.; Price, R. L. *J. Agric. Food Chem.* **1970**, *18*(5), 905–907.

1597. Fuh, B.-S.; Ichii, T.; Kawai, Y.; Nakanishi, T. *Engei Gakkai Zasshi* **1988**, *57*(3), 529–537.

1598. Tanaka, H.; Mizuta, Y.; Ichii, T. *Engei Gakkai Zasshi* **1997**, *66*(2), 297–306.

1599. Goldschmidt, E. E. *Phytochemistry* **1977**, *16*(7), 1046–1047.

1600. Gemeinhardt, F.; Knobloch, K.; Ziegler, E.; Brandauer, H. *Z. Lebensm. Unters. Forsch.* **1978**, *166*(6), 368.

1601. Kadota, R.; Tokunaga, Y.; Miura, M. *Miyazaki Daigaku Nogakubu, Kenkyo Hokoku* **1982**, *29*(1), 213–222.

1602. Kolesnik, A. A.; Golubev, V. N.; Meteshkin, Yu.V.; Sprinsyan, T. V.; Papunidze, G. R. *Khim. Prir. Soedin.* **1989**, (3), 342–348 (Engl. 293–299).

1603. Dobson, H. E. M. *Am. J. Bot.* **1988**, *75*(2), 170–182.

1604. Prud'homme, M.-P.; Le Saos, J.; Boucaud, J.; Oursel, A.; Trémolières, A. *Plant Physiol. Biochem.* **1990**, *28*(1), 71–78.

1605. Krishnamurthy, M. N.; Chandrasekhara, N. *J. Food Sci. Technol. (India)* **1983**, *20*(5), 206–209.

1606. Pereira, A., Jr.; Pereira, M. M. A.; Pereira, J. *Garcia Orta* **1967**, *15*(3), 349–360, from *Chem. Abstr.* **1970**, *72*, 120197c.

1607. Tanaka, Y.; Nishida, I.; Higashi, S.; Beppu, T.; Murata, N. *Plant Cell Physiol.* **1990**, *31*(4), 545–550.

1608. Han, J. S.; Rhee, S. H.; Cheigh, H. S. *Han'guk Yongyang Siklyong Hakhoechi* **1987**, *16*(1), 29–35.

1609. Hussain, M.; Norton, G.; Neale, R. J. *J. Sci. Food Agric.* **1984**, *35*(10), 1112–1119.

1610. Lehoczki, E.; Pölös, E.; Laskay, G.; Farkas, T. *Plant Sci.* **1985**, *42*(1), 19–24.

1611. Cahoon, E. B.; Ohlrogge, J. B. *Plant Physiol.* **1994**, *104*(3), 845–855.

1612. Shewry, P. R.; Pinfield, N. J.; Stobart, A. K. *J. Exp. Bot.* **1973**, *24*(83), 1100–1105.

1613. Parcerisa, J.; Codony, R.; Boatella, J.; Rafecas, M. *J. Agric. Food Chem.* **1999**, *47*(4), 1410–1415.

1614. Shinozaki, Y.; Ishii, S. *Nippon Nogei Kagaku Kaishi* **1968**, *42*(4), 207–210.

1615. Tadzhibaev, Yu. A.; Mukhamedova, Kh. S.; Akramov, S. T. *Khim. Prir. Soedin.* **1977**, (4), 502–505 (Engl. 420–422).

1616. Tadzhibaev, Yu. A.; Mukhamedova, Kh. S.; Akramov, S. T. *Khim. Prir. Soedin.* **1978**, (1), 53–57 (Engl. 40–43).

1617. Tadzhibaev, Yu. A.; Mukhamedova, Kh. S.; Akramov, S. T. *Khim. Prir. Soedin.* **1977**, (1), 109–110 (Engl. 95–96).

1618. Gazizov, F. Yu.; Glushenkova, A. I. *Khim. Prir. Soedin.* **1996**, *32*(3), 347–349 (Engl. 335–337).

1619. Gazizov, F. Yu.; Glushenkova, A. I. *Khim. Prir. Soedin.* **1995**, *31*(4), 562–563 (Engl. 465–466).

1620. Malik, C. P.; Chawla, J. *J. Palynol.* **1981**, *17*(1/2), 42–54.

1621. Jain, A.; Shivanna, K. R. *Phytochemistry* **1989**, *28*(4), 999–1002.

1622. Gümüş, Ş. *Fac. Sci. Univ. Ankara* **1979**, *B25*(4), 29–36.

1623. Hemavathy, J. *J. Food Compos. Anal.* **1992**, *5*(1), 90–95.

1624. Horváth, I.; Vigh, L.; van Hasselt, P. R.; Woltjes, J.; Kuiper, P. J. C. *Plant Physiol.* **1983**, *57*(4), 532–536.

1625. Chen, Z.; Lin, S.; Xu, C.; Tang, C.; Chen, M.; Kuang, T. *Zhiwu Shengli Xuebao* **1993**, *19*(3), 293–298.

1626. Fishwick, M. J.; Wright, A. J.; Galliard, T. *J. Sci. Food Agric.* **1977**, *28*(4), 394–398.

1627. Tachibana, S. *Engei Gakkai Zasshi* **1987**, *56*(2), 180–186.

1628. Bulder, H. A. M.; den Nijs, A. P. M.; Speek, E. J.; van Hasselt, P. R.; Kuiper, P. J. C. *J. Plant Physiol.* **1991**, *138*(6), 661–666.

1629. Tsuyuki, H.; Itoh, S.; Yamagata, K. *Nippon Shokuhin Kogyo Gakkaishi* **1985**, *32*(1), 7–15.

1630. Hancock, J. F.; Antonio, T. M.; Dalgarn, D. S.; Newman, D. W. *Ohio J. Sci.* **1983**, *83*(1), 50–54.

1631. Murata, N.; Yamaya, J. *Plant Physiol.* **1984**, *74*(4), 1016–1024.

1632. Xu, Y.; Siegenthaler, P. A. *Plant Cell Physiol.* **1996**, *37*(4), 471–479.

1633. Mamedova, T. K.; Akhmerova, E. S. *Fiziol. Rast.* **1979**, *26*(3), 613–619 (Engl. 492–497).

1634. Tevini, M. *Planta* **1976**, *128*(2), 167–171.

1635. Coulomb, C.; Coulomb, P. J.; Michel, G. *Histochem. J.* **1982**, *14*(2), 175–196.

1636. Wang, C. Y.; Kramer, G. F.; Whitaker, B. D.; Lusby, W. R. *J. Plant Physiol.* **1992**, *140*, 229–235.

1637. Hemavathy, J.; Prabhakar, J. V. *J. Food Sci.* **1988**, *53*(5), 1578–1579.

1638. Sharma, S.; Khanna, R.; Sanwal, G. G. *Physiol. Plant.* **1985**, *63*(3), 315–321.

1639. Zlatanov, M.; Ivanov, S.; Antonova, G.; Kouleva, I. *Riv. Ital. Sostanze Grasse* **1997**, *74*(9), 409–413.

1640. Ul'chenko, N. T.; Nazarova, I. P.; Glushenkova, A. I.; Fatkhiev, F. F.; Tolstikov, G. A. *Khim. Prir. Soedin.* **1991**, (6), 758–762 (Engl. 668–671).

1641. Oderinde, R. A.; Tairu, A. O. *Food Chem.* **1992**, *45*(4), 279–282.

1642. Yoshida, S.; Uemura, M. *Plant Physiol.* **1984**, *75*(1), 31–37.

1643. Manoharan, K.; Prasad, R.; Guha-Mukherjee, S. *Phytochemistry* **1987**, *26*(2), 407–410.

1644. Hölzl, J. *Biochem. Z.* **1965**, *341*(2), 168–173.

1645. Soimajärvi, J.; Linko, R. R. *J. Agric. Food Chem.* **1979**, *27*(6), 1277–1279.

1646. Picchioni, G. A.; Watada, A. E.; Roy, S.; Whitaker, B. D.; Wergin, W. P. *J. Food Sci.* **1994**, *59*(3), 597–601, 605.

1647. Kleinig, H.; Kopp, C. *Planta* **1978**, *139*(1), 61–65.

1648. Gregor, H.-D. *Chem. Phys. Lipids* **1977**, *20*(1), 77–85.

1649. Styer, E. H.; Whitaker, B. D.; Kitto, S. L.; Hoover, D. G. *Phytochemistry* **1996**, *41*(1), 187–191.

1650. Zuñiga, G. E.; Alberdi, M.; Fernandez, J.; Montiel, P.; Corcuera, L. J. *Phytochemistry* **1994**, *37*(3), 669–672.

1651. Engelmann-Sylvestre, I.; Bureau, J.-M.; Trémolières, A.; Paulin, A. *Plant Physiol. Biochem.* **1989**, *27*(6), 931–937.

1652. Brown, J. H.; Chambers, J. A.; Thompson, J. E. *Plant Physiol.* **1991**, *95*(3), 909–916.

1653. Sewon, P.; Mikola, H.; Lehtinen, T.; Kallio, P. *Phytochemistry* **1997**, *46*(8), 1339–1347.

1654. Dembitsky, V. M.; Rezanka, T.; Bycher, I. A.; Afonina, O. M. *Phytochemistry* **1993**, *33*(5), 1021–1027.

1655. Chung, H.-Y. *Han'guk Nonghwa Hakhoechi* **1994**, *37*(6), 509–515.

1656. Osagie, A. U.; Opute, F. I. *J. Agric. Food Chem.* **1982**, *30*(5), 993–996.

1657. Wang, S.; Lu, J.; Xu, Y.; Ye, D. *Zhongguo Zhongyao Zazhi* **1993**, *18*(6), 340–341.

1658. Suzuki, K.; Itoh, S.; Tsuyuki, H. *Nippon Shokuhin Kogyo Gakkaishi* **1982**, *29*(5), 310–315.

1659. Kolesnik, A. A.; Golubev, V. N.; Kostinskaya, L. I.; Khalilov, M. A. *Khim. Prir. Soedin.* **1987**, (4), 501–505 (Engl. 415–418).

1660. Tadzhibaev, Yu. A.; Mukhamedova, Kh. S.; Akramov, S. T. *Khim. Prir. Soedin.* **1977**, (2), 195–198 (Engl. 168–170).

1661. Radunz, A. *Z. Physiol. Chem.* **1968**, *349*(3), 303–309.

1662. Sridhar, R.; Lakshminarayana, G. *J. Agric. Food Chem.* **1992**, *40*(11), 2131–2134.

1663. Lakshminarayana, G.; Rao, K. S.; Pantulu, A. J.; Thyagarajan, G. *Aqu. Bot.* **1984**, *20*(3/4), 219–227.

1664. Banerjee, A.; Dubey, V.; Banerjee, K. *J. Oil Technol. Assoc. India* **1996**, *28*(4), 107–108.

1665. Goncharova, N. P.; Isamukhamedov, A. Sh.; Glushenkova, A. I. *Khim. Prir. Soedin.* **1993**, (5), 646–651 (Engl. 569–573).

1666. Opute, F. I. *Ann. Bot.* **1975**, *39*(164), 1057–1061.

1667. Opute, F. I. *Phytochemistry* **1975**, *14*(4), 1023–1026.

1668. Goh, S. H.; Khor, H. T.; Gee, P. T. *J. Am. Oil Chem. Soc.* **1982**, *59*(7), 296–299.

1669. Kataoka, E.; Tokue, C.; Tanimura, W. *Nogaku Shuho Tokyo (Nogyo Daigaku)* **1987**, *31*(3), 189–196.

1670. Kurbanova, I. M.; Kozlov, Yu. P.; Gasanov, R. A. *Izv. Akad. Nauk Az. SSR, Ser Biol. Nauk* **1974**, (2), 29–34.

1671. Tsuyuki, H.; Itoh, S.; Nakatsukasa, Y. *Nihon Daigaku No-Juigakubu Gakujutsu Kenkyu Hokoku* **1977**, *34*, 155–161.

1672. Sukhija, P. S.; Bhatia, I. S. *Indian J. Biochem.* **1970**, *7*(4), 271–274.

1673. Tadzhibaev, Yu. A.; Mukhamedova, Kh. S.; Akramov, S. T. *Khim. Prir. Soedin.* **1977**, (5), 623–627 (Engl. 517–520).

1674. Béranger-Novat, N.; Davidowicz-Grzegorzewska, A. *Ann. Bot.* **1992**, *70*(1), 93–100.

1675. Obara, T.; Yussa, K. *Nippon Nogei Kagaku Kaishi* **1969**, *43*(2), 91–94.

1676. Mano, Y. *Denpun Kagaku* **1977**, *24*(1), 15–18.

1677. Prasad, R. B. N.; Gülz, P.-G. *Z. Naturforsch.* **1989**, *44C*(9/10), 735–738.

1678. Kolesnik, A. A.; Golubev, V. N.; Gadzhieva, A. A. *Khim. Prir. Soedin.* **1991**, (4), 464–467 (Engl. 404–407).

1679. Harwood, J. L.; Jones, A. V. H. M.; Thomas, H. *Planta* **1982**, *156*(2), 152–157.

1680. Kolesnik, A. A.; Kakhniashvili, T. A.; Zherebin, Yu. L.; Golubev, V. N.; Pilipenko, L. N. *Khim. Prir. Soedin.* **1986**, (4), 423–427 (Engl. 394–397).

1681. Kulkarni, A. S.; Khotpal, R. R.; Gupta, R.; Bhakare, H. A. *Acta Cienc. Indica Chem.* **1992**, *18*(4), 353–356.

1682. Wang, S. Y.; Maas, J. L. *Acta Hortic.* **1997**, *439*, 863–867.

1683. Gazizov, F. Yu.; Glushenkova, A. I. *Khim. Prir. Soedin.* **1997**, *33*(6), 800–802 (Engl. 627–629).

1684. Tsuyuki, H.; Itoh, S.; Nakatsukasa, Y. *Nihon Daigaku No-Juigakubu Gakujutsu Kenkyu Hokoku* **1979**, *36*, 156–162.

1685. Urakami, C.; Oka, S.; Han, J. S. *J. Am. Oil Chem. Soc.* **1976**, *53*(8), 525–529.

1686. Tsululidze, L. A.; Dzhmukhadze, V. A.; Shvedov, I. V. *Izv. Akad. Nauk Gruz. SSR Ser. Khim.* **1979**, *5*(2), 124–127.

1687. Vishnepol'skaya, F. A.; Al'perina, Ts.Ya. *Tr. Vses. Nauchno-Issled. Inst. Zhirov* **1967**, *26*, 142–146, from *Chem. Abstr.* **1968**, *69*, 88169v.

1688. Raison, J. K.; Orr, G. R. *Plant Physiol.* **1986**, *81*(3), 807–811.

1689. Martin, B. A.; Wilson, R. F. *Agrochimica* **1985**, *29*(2/4), 89–96.

1690. Nishihara, M.; Kito, M. *Biochim. Biophys. Acta* **1978**, *531*(1), 25–31.

1691. Yoshida, H.; Tagaki, S.; Ienaga, H.; Tsuchiya, C. *J. Am. Oil Chem. Soc.* **1998**, *75*(7), 767–774.

1692. Chapman, G. W., Jr. *J. Am. Oil Chem. Soc.* **1980**, *57*(9), 299–302.

1693. Kojima, M.; Nagasawa, T.; Furukawa, N.; Mouri, H.; Ohnishi, M.; Ito, S. *Nippon Shokuhin Kogyo Gakkaishi* **1991**, *38*(12), 1076–1085.

1694. Stearns, E. M., Jr.; Morton, W. T. *Lipids* **1977**, *12*(5), 451–454.

1695. Simpson, T. D.; Nakamura, L. R. *J. Am. Oil Chem. Soc.* **1989**, *66*(8), 1093–1096.

1696. Boatright, W. L.; Snyder, H. E. *J. Am. Oil Chem. Soc.* **1993**, *70*(6), 623–628.

1697. Katiyar, S. K.; Bhatia, A. K.; Atal, C. K. *Trop. Sci.* **1988**, *28*(1), 13–17.

1698. Hirayama, O.; Suzuki, T. *Agric. Biol. Chem.* **1968**, *32*(5), 549–554.

1699. Rossignol, M. *Phytochemistry* **1976**, *15*(12), 1893–1896.

1700. Whitman, C. E.; Travis, R. L. *Plant Physiol.* **1985**, *79*(2), 494–498.

1701. Wilson, R. F. *Crop Sci.* **1981**, *21*(1), 69–75.

1702. Nakayama, Y.; Kito, M. *Agric. Biol. Chem.* **1981**, *45*(9), 2155–2156.

1703. Karshiev, Kh.; Mukhamedova, Kh.S.; Akramov, S. T. *Khim. Prir. Soedin.* **1979**, (5), 615–618 (Engl. 536–538).

1704. El-Nockrashy, A. S.; Frampton, V. L. *Seifen Öle Fette Wachse* **1970**, *96*(21), 747–750.

1705. El-Nockrashy, A. S.; Mostafa, H. M.; El-Fouly, M. M.; El-Shattory, Y. *Nahrung* **1976**, *20*(2), 125–131.

1706. Karshiev, Kh.; Akramov, S. T. *Khim. Prir. Soedin.* **1978**, (3), 307–310 (Engl. 254–256).

1707. El-Nockrashy, A. S.; Mostafa, H. M.; El-Shattory, Y.; Abbassy, M. H. *Nahrong* **1974**, *18*(3), 285–293.

1708. Karshiev, Kh.; Shustanova, L. A.; Akramov, S. T. *Khim. Prir. Soedin.* **1976**, (1), 19–21 (Engl. 15–17).

1709. Karshiev, Kh.; Shustanova, L. A.; Akramov, S. T. *Khim. Prir. Soedin.* **1975**, *11*(6), 693–697 (Engl. 728–731).

1710. Reine, A. H.; Stegink, S. J. *J. Chromatogr.* **1988**, *437*(1), 211–219.

1711. Gazizov, F. Yu.; Isamukhamedov, A. Sh.; Akramov, S. T. *Khim Prir. Soedin.* **1982**, (1), 120 (Engl. 113–114).

1712. Kaplunova, T. S.; Mukhamedova, Kh. S.; Akramov, S. T. *Khim. Prir. Soedin.* **1978**, (1), 41–44 (Engl. 31–33).

1713. Kaplunova, T. S.; Mukhamedova, Kh. S.; Akramov, S. T. *Khim. Prir. Soedin.* **1979**, (1), 8–11 (Engl. 6–9).

1714. Babaev, M. Yu.; Mukhamedova, Kh. S.; Akramov, S. T. *Khim. Prir. Soedin.* **1974**, *10*(3), 417–418 (Engl. 428–429).

1715. Thompson, A. C.; Henson, R. D.; Minyard, J. P.; Hedin, P. A. *Lipids* **1968**, *3*(4), 373–374.

1716. Gazizov, F. Yu.; Isamukhamedov, A. Sh.; Akramov, S. T. *Khim Prir. Soedin.* **1981**, (6), 706–710 (Engl. 507–510).

1717. Chapman, K. D.; Trelease, R. N. *J. Cell Biol.* **1991**, *115*(4), 995–1007.

1718. Stone, T. B.; Thompson, A. C.; Pitre, H. N. *J. Entomol. Sci.* **1985**, *20*(4), 422–428.

1719. Zikiryaev, A. *Dokl. Akad. Nauk Resp. Uzb.* **1994**, (1), 55–56.

1720. Chapman, K. D.; Sprinkle, W. B. *J. Plant Physiol.* **1996**, *149*(3/4), 277–284.

1721. Gottlieb, O. R.; Machline, C. *Quimica (Brazil)* **1945**, *1*, 142, from *Chem. Abstr.* **1946**, *40*, 2667[9].

1722. Grewal, S. S.; Sukhija, P. S.; Bhatia, I. S. *Plant Foods Hum. Nutr.* **1978**, *28*(3), 211–218.

1723. El-Sebaiy, L. A.; El-Mahdy, A. R.; Moustafa, E. K.; Mohamed, M. S. *Food Chem.* **1980**, *5*(3), 217–229.

1724. Sperling, P.; Hammer, U.; Friedt, W.; Heinz, E. *Z. Naturforsch.* **1990**, *45C*(3/4), 166–172.

1725. Carelli, A. A.; Brevedan, M. I. V.; Crapiste, G. H. *J. Am. Oil Chem. Soc.* **1997**, *74*(5), 511–514.

1726. Justin, A.-M.; Robert, D.; Mazliak, P. *Compt. Rend.* **1979**, *288D*(1), 81–84.

1727. Navari-Izzo, F.; Quartacci, M. F.; Melfi, D.; Izzo, R. *Physiol. Plant.* **1993**, *87*(4), 508–514.

1728. Poincelot, R. P. *Plant Physiol.* **1976**, *58*(4), 595–598.

1729. Navari-Izzo, F.; Vangioni, N.; Quartacci, M. F. *Phytochemistry* **1990**, *29*(7), 2119–2123.

1730. Lin, S.; Mullin, C. A. *J. Agric. Food Chem.* **1999**, *47*(3), 1223–1229.

1731. Belver, A.; Donaire, J. P. *Phytochemistry* **1987**, *26*(11), 2923–2927.

1732. Uemura, M.; Yoshida, S. *Plant Physiol.* **1986**, *80*(1), 187–195.

1733. Tolibaev, I.; Glushenkova, A. I. *Khim. Prir. Soedin.* **1996**, *32*(1), 10–14 (Engl. 7–10).

1734. Hasma, H.; Subramaniam, A. *J. Nat. Rubber Res.* **1986**, *1*(1), 30–40.

1735. Smith, R. H. *Biochem. J.* **1954**, *57*(1), 130–139.

1736. Mohamed, A.; Bhardwaj, H.; Hamama, A.; Webber, C., III *Ind. Crops Prod.* **1995**, *4*(3), 157–165.

1737. Tolibaev, I.; Mukhamedova, Kh. S.; Glushenkova, A. I. *Khim. Prir. Soedin.* **1986**, (2), 158–160 (Engl. 146–149).

1738. Tolibaev, I.; Mukhamedova, Kh. S.; Akramov, S. T. *Khim. Prir. Soedin.* **1981**, (2), 157–160 (Engl. 119–121).

1739. Isamukhamedov, A. Sh.; Akramov, S. T. *Khim Prir. Soedin.* **1982**, (3), 395–396 (Engl. 365–366).

1740. Goncharova, N. P.; Glushenkova, A. I. *Khim. Prir. Soedin.* **1995**, *31*(5), 667–670 (Engl. 562–564).

1741. Lagazidze, D. S.; Murav'eva, D. A.; Bostoganashvili, V. S. *Khim. Farm. Zh.* **1984**, *18*(6), 713–717.

1742. Selldén, G.; Selstam, E. *Physiol. Plant.* **1976**, *37*(1), 35–41.

1743. Parsons, J. G.; Price, P. B. *J. Agric. Food Chem.* **1979**, *27*(4), 913–915.

1744. Shin, H.-S.; Gray, J. I. *Korean J. Food Sci. Technol.* **1983**, *15*(2), 195–201.

1745. Kim, H. G.; Cheigh, H. S. *Han'guk Yongyang Siklyong Hakhoechi* **1989**, *21*(4), 515–520.

1746. Benson, A. A.; Wintermans, J. F. G. M.; Wiser, R. *Plant Physiol.* **1959**, *34*(3), 315–317.

1747. Navari-Izzo, F.; Quartacci, M. F.; Izzo, R.; Pinzino, C. *Physiol. Plant.* **1992**, *84*(1), 73–79.

1748. Henry, L. E. A.; Mikkelsen, J. D.; Møller, B. L. *Carlsberg Res. Commun.* **1983**, *48*(2), 131–148.

1749. Haschke, H.-P.; Kaiser, G.; Martinoia, E.; Hammer, U.; Teucher, T.; Dorne, A. J.; Heinz, E. *Bot. Acta* **1990**, *103*(1), 32–38.

1750. Rochester, C. P.; Kjellbom, P.; Andersson, B.; Larsson, C. *Arch. Biochem. Biophys.* **1987**, *255*(2), 385–391.

1751. Arnalte, M.-E.; Cornejo, M. J.; Bush, D. S.; Jones, R. L. *Plant Sci.* **1991**, *77*, 223–232.

1752. Tevini, M. *Z. Pflanzenphysiol.* **1971**, *65*(3), 266–272.

1753. Brown, D. B.; DuPont, F. M. *Plant Physiol.* **1989**, *90*(3), 955–961.

1754. Gellerman, J. L.; Anderson, W. H.; Richardson, D. G.; Schlenk, H. *Biochim. Biophys. Acta* **1975**, *388*(2), 277–290.

1755. Nichols, B. W. *Phytochemistry* **1965**, *4*(5), 769–772.

1756. Tevini, M. *Z. Pflanzenphysiol.* **1971**, *66*(1), 64–72.

1757. Bhakare, H. A.; Khotpal, R. R.; Kulkarni, A. S. *J. Food Sci. Technol.* **1993**, *30*(5), 382–384.

1758. Walter, W. M., Jr.; Hansen, A. P.; Purcell, A. E. *J. Food Sci.* **1971**, *36*(5), 795–797.

1759. Boggess, T. S., Jr.; Marion, J. E.; Dempsey, A. H. *J. Food Sci.* **1970**, *35*(3), 306–309.

1760. Yamaki, S.; Uritani, I. *Plant Cell Physiol.* **1972**, *13*(1), 67–79.

1761. Filipenko, L. N.; Kolesnik, A. A. *Khim. Prir. Soedin.* **1991**, (2), 199–205 (Engl. 160–166).

1762. Davies, H.; Jordan, B. R.; Harwood, J. L. *Plant Physiol. Biochem.* **1991**, *29*(3), 281–288.

1763. Levin, E. D.; Rubchevskaya, L. P.; Vol, E. V. *Khim. Drev.* **1983**, (4), 97–100.

1764. Billa, S. P.; Bahadur, K. *Z. Naturforsch.* **1971**, *26B*(12), 1376.

1765. Watanabe, K.; Kito, M.; Ueno, S.; Mitsuda, H. *Agric. Biol. Chem.* **1986**, *50*(7), 1791–1796.

1766. Grenier, G.; Proteau, L.; Marier, J.-P.; Beaumont, G. *Plant Physiol. Biochem.* **1987**, *25*(4), 409–413.

1767. Grenier, G.; Marier, J.-P.; Beaumont, G. *Physiol. Vég.* **1982**, *20*(2), 179–185.

1768. Kim, H. R.; Koh, M. S.; Yang, H. C. *Han'guk Yongyang Siklyong Hakhoechi* **1987**, *16*(3), 75–84, from *Chem. Abstr.* **1988**, *108*, 93301k.

1769. Sekiya, J.; Yamashita, K.; Shimose, N. *Agric. Biol. Chem.* **1990**, *54*(1), 191–195.

1770. Lardans, A.; Trémolières, A. *Phytochemistry* **1991**, *30*(12), 3955–3961.

1771. Wanasundara, P. K. J. P. D.; Wanasundara, U. N.; Shahidi, F. *J. Am. Oil Chem. Soc.* **1999**, *76*(1), 41–48.

1772. Slack, C. R.; Roughan, P. G. *Biochem. J.* **1978**, *170*(2), 437–439.

1773. Hubac, C.; Guerrier, D.; Ferran, J.; Trémolières, A. *Plant Physiol. Biochem.* **1989**, *27*(5), 737–744.

1774. Meyer, S.; Hung, S. P. N.; Trémolières, A.; de Kouchkovsky, Y. *Photosynth. Res.* **1992**, *32*(2), 95–107.

1775. Rossignol, M.; Grignon, N.; Grignon, C. *Biochimie* **1982**, *64*(4), 263–270.

1776. Oursel, A.; Lamant, A.; Salsac, L.; Mazliak, P. *Phytochemistry* **1973**, *12*(8), 1865–1874.

1777. Isamukhamedov, A. Sh.; Akramov, S. T. *Khim Prir. Soedin.* **1982**, (4), 513–514 (Engl. 481–482).

1778. Soulie, M.-C.; Bompeix, G.; Laval-Martin, D. *Phytochemistry* **1992**, *31*(6), 1961–1967.

1779. Minamide, T.; Ueda, Y.; Ogata, K.; Kamata, H. *Nippon Shokuhin Kogyo Gakkaishi* **1970**, *17*(3), 104–109.

1780. Whitaker, B. D. *Phytochemistry* **1993**, *32*(2), 265–271.

1781. Kalra, S. K.; Brooks, J. L. *Phytochemistry* **1973**, *12*(3), 487–492.

1782. Whitaker, B. D. *Planta* **1986**, *169*(3), 313–319.

1783. Krupa, Z.; Baszynski, T. *Acta Physiol. Plant.* **1985**, *7*(2), 55–64.

1784. Güçlü, J.; Paulin, A.; Soudain, P. *Physiol. Plant.* **1989**, *77*(3), 413–419.

1785. Gazizov, F. Yu.; Isamukhamedov, A. Sh.; Akramov, S. T. *Khim Prir. Soedin.* **1984**, (2), 168–173 (Engl. 156–160).

1786. Rodionov, V. S.; Zakharova, L. S. *Fiziol. Biokhim. Kul't. Rast.* **1985**, *17*(6), 581–588.

1787. Kihara, M.; Xu, L.; Konishi, K.; Kida, K.; Nagao, Y.; Kobayashi, S.; Shingu, T. *Chem. Pharm. Bull.* **1994**, *42*(2), 289–292.

1788. Kihara, M.; Konishi, K.; Xu, L.; Kobayashi, S. *Chem. Pharm. Bull.* **1991**, *39*(7), 1849–1853.

1789. Rozentsvet, O. A.; Ponomareva, E. R.; Mazepova, Ya.N.; Koneva, N. V. *Khim. Prir. Soedin.* **1995**, *31*(2), 206–209 (Engl. 169–171).

1790. Wang, S. Y.; Sun, T.; Whitaker, B. D.; Faust, M. *Physiol. Plant.* **1988**, *73*(4), 560–564.

1791. Wang, S. Y.; Faust, M. *J. Am. Soc. Hortic. Sci.* **1990**, *115*(3), 462–467.

1792. Galliard, T. *Phytochemistry* **1968**, *7*(11), 1915–1922.

1793. Kimura, S.; Kanno, M.; Yamada, Y.; Takahashi, K.; Murashige, H.; Okamoto, T. *Agric. Biol. Chem.* **1982**, *46*(12), 2895–2902.

1794. Mazliak, P.; Justin, A. M. *Fruits* **1967**, *22*(9), 413–432.

1795. Prabha, T. N.; Raina, P. L.; Patwardhan, M. V. *J. Biosci. (India)* **1988**, *13*(1), 33–38.

1796. Wang, S. Y.; Faust, M. *J. Plant Growth Regul.* **1989**, *8*(2), 153–161.

1797. Ketchie, D. O.; Bervaes, J. C. A. M.; Kuiper, P. J. C. *Physiol. Plant.* **1987**, *71*(4), 419–424.

1798. Asilbekova, D. T.; Glushenkova, A. I. *Khim. Prir. Soedin.* **1998**, *34*(6), 741–745 (Engl. 664–667).

1799. El-Khalafy, H. M.; Aly, F. M. *Grasas Aceites* **1971**, *22*(2), 86–88.

1800. Ali, M. A.; Gafur, M. A.; Rahman, M. S.; Ahmed, G. M. *J. Am. Oil Chem. Soc.* **1985**, *62*(3), 520–523.

1801. Hemavathy, J.; Prabhakar, J. V.; Sen, D. P. *J. Food Sci.* **1987**, *52*(3), 833–834.

1802. Khor, H.-T.; Tan, H.-L. *J. Sci. Food Agric.* **1981**, *32*(4), 399–402.

1803. Lalaguna, F.; Agudo, M. *J. Am. Oil Chem. Soc.* **1988**, *65*(11), 1808–1811.

1804. Marion, D.; Douillard, R.; Gandemer, G. *Rev. Fr. Corps Gras* **1988**, *35*(5), 229–234.

1805. Harzallah-Skhiri, F.; Guillot-Salomon, T.; Signol, M. *Dev. Plant Biol.* **1982**, *8*, 423–426.

1806. Panikkar, K. R.; Majella, V. L.; Pillai, P. M. *Planta Med.* **1987**, *53*(5), 503.

1807. Hemavathy, J. *J. Am. Oil Chem. Soc.* **1991**, *68*(9), 651–652.

1808. Hegazy, H. S.; Hassan, N. S.; Nassar, A. H. *Delta J. Sci.* **1991**, *15*(2), 266–282, from *Chem. Abstr.* **1993**, *119*, 135685y.

1809. Takenaga, F.; Itoh, S.; Tsuyuki, H. *Nippon Shokuhin Kogyo Gakkaishi* **1985**, *32*(4), 247–254.

1810. Pollero, R. J. *Aquat. Bot.* **1986**, *26*(1/2), 103–111.

1811. Nichols, B. W.; James, A. T. *Fette, Seifen, Anstrichm.* **1964**, *66*(12), 1003–1006.

1812. Meir, S.; Philosoph-Hadas, S. *Plant Physiol. Biochem.* **1995**, *33*(2), 241–249.

1813. Lakshminarayana, G.; Rao, R. S.; Pantulu, A. J.; Gupta, D. R. *J. Oil Technol. Assoc. India* **1987**, *19*(2), 35–37.

1814. Orr, G. R.; Raison, J. K. *Plant Physiol.* **1987**, *84*(1), 88–92.

1815. Ladygina, M. E.; Grishkova, V. P. *S.-Kh. Biol.* **1979**, *14*(1), 50–55.

1816. Beiss, U. *Phytopathol. Z.* **1963**, *49*(1), 29–40.

1817. Matsuzaki, T.; Koiwai, A.; Nagao, T.; Sato, F.; Yamada, Y. *Agric. Biol. Chem.* **1984**, *48*(7), 1699–1706.

1818. Dorne, A.-J.; Kappler, R.; Kristen, U.; Heinz, E. *Phytochemistry* **1988**, *27*(7), 2027–2031.

1819. Matsuzaki, T.; Koiwai, A.; Kawashima, N. *Plant Cell Physiol.* **1983**, *24*(2), 199–206.

1820. Wassef, M. K.; Hendrix, J. W. *J. Agric. Food Chem.* **1975**, *23*(4), 688–690.

1821. Zeitoun, M. A. M.; Neff, W. E. *Ol. Corps Gras Lipides* **1995**, *2*(3), 245–248.

1822. Alberdi, M.; Fernandez, J.; Cristi, R.; Romero, M.; Rios, D. *Phytochemistry* **1990**, *29*(8), 2467–2471.

1823. Lag, M. A.; Meza-Basso, L.; Fernandez, J.; Cristi, R.; Romero, M. *Phytochemistry* **1991**, *30*(3), 763–768.

1824. Read, H. *Lipids* **1985**, *20*(8), 510–515.

1825. Lotti, G.; Quartacci, M. F. *Agrochimica* **1990**, *34*(3), 243–250.

1826. Mukherjee, K. D.; Kiewitt, I. *J. Agric. Food Chem.* **1987**, *35*(6), 1009–1012.

1827. Vioque, E.; Maza, M. P. *Grasas Aceites* **1973**, *24*(6), 323–326.

1828. Mancha, M. *Grasas Aceites* **1974**, *25*(3), 159–164.

1829. Rodriguez-Rosales, M. P.; Donaire, J. P. *New Phytol.* **1988**, *108*(4), 509–514.

1830. Zarrouk, M.; Cherif, A. *Dev. Plant Biol.* **1984**, *9*, 595–599.

1831. Alter, M.; Gutfinger, T. *Riv. Ital. Sostanze Grasse* **1982**, *59*(1), 14–18.

1832. Donaire, J. P.; Belver, A. *Rev. Esp. Fisiol.* **1987**, *43*(2), 253–257.

1833. Williams, M.; Sanchez, J.; Hann, A. C.; Harwood, J. L. *J. Exp. Bot.* **1993**, *44*(268), 1717–1723.

1834. Komaitis, M. E.; Revinthi-Moraiti, K.; Evangelatos, G. *Food Chem.* **1988**, *27*(1), 25–32.

1835. Mostafa, S. A.; El Gammal, S. M. A.; Attia, S. H. A. *Zentralbl. Mikr.* **1982**, *137*(8), 651–660.

1836. Ito, S.; Naito, M.; Fujino, Y. *Obihiro Chikusan Daigaku Gakujutsu Kenkyu Hokoku, Dai-1-Bu* **1980**, *11*(4), 591–596.

1837. Obara, T.; Miyata, N. *Nippon Shokuhin Kogyo Gakkaishi* **1969**, *16*(7), 304–309.

1838. Bradbury, J. H.; Collins, J. G. *Cereal Chem.* **1982**, *59*(3), 159–162.

1839. Ito, S.; Sato, S.; Fujino, Y. *Stärke* **1979**, *31*(7), 217–221.

1840. Brown, D. J.; Beevers, H. *Plant Physiol.* **1987**, *84*(2), 555–559.

1841. Jung, J.; Kim, Y.-K.; Park, S.-G. *Han'guk Nonghwa Hakhoechi* **1983**, *26*(1), 58–64.

1842. Toriyama, S. I.; Hinata, K.; Nishida, I.; Murata, N. *Plant Cell Physiol.* **1988**, *29*(4), 615–621.

1843. Hemavathy, J.; Prabhakar, J. V. *J. Am. Oil Chem. Soc.* **1987**, *64*(7), 1016–1019.

1844. Ahn, T.-H.; Rhee, C.-O.; Kim, D. Y. *Korean J. Food Sci. Technol.* **1984**, *16*(2), 192–200.

1845. Miyazawa, T.; Yoshino, Y.; Fujino, Y. *J. Sci. Food Agric.* **1977**, *28*(10), 889–894.

1846. Takano, K.; Kamoi, I.; Obara, T. *Nippon Shokuhin Kogyo Gakkaishi* **1989**, *36*(6), 468–474.

1847. Mano, Y.; Ohnishi, M.; Ito, S. *Denpun Kagaku* **1988**, *35*(1), 49–55.

1848. Ros, R.; Cooke, D. T.; Martinez-Cortina, C.; Picazo, I. *J. Exp. Bot.* **1992**, *43*(256), 1475–1481.

1849. Bao, F.; Xu, Y.; Zhang, K.; Wen, D. *Shengwu Huaxue Yu Shengwu Wuli Xiebao* **1985**, *17*(1), 50–55.

1850. Kasamo, K.; Kagita, F.; Yamanishi, H.; Sakaki, T. *Plant Cell Physiol.* **1992**, *33*(5), 609–616.

1851. Dermaku, S.; Popovic, M. *Rad Jugosl. Akad. Znan. Umjet* **1985**, *416*, 11–16, from *Chem. Abstr.* **1986**, *105*, 112057s.

1852. Knoth, R.; Hansmann, P.; Sitte, P. *Planta* **1986**, *168*(2), 167–174.

1853. Kim, W. K.; Lee, C. S.; Jeong, B. K. *Koryo Insam Hakhoechi* **1988**, *12*(2), 93–103.

1854. Seit-Ablaeva, S. K.; Nechaev, A. P.; Denisenko, Ya.I. *Vop. Pitan.* **1973**, (4), 75–76.

1855. Lorenz, K.; Hwang, Y. S. *Cereal Chem.* **1986**, *63*(5), 387–390.

1856. Hoekstra, F. A.; Van Roekel, T. *Plant Physiol.* **1988**, *88*(3), 626–632.

1857. Jabbar, A.; Hossein, S. A.; Ahmed, R. *Pak. J. Biol. Agric. Sci.* **1966**, *9*(1), 7–12.

1858. Natekar, S.; Malse, P. T. *Natl. Acad. Sci. Lett. (India)* **1985**, *8*(6), 179–180.

1859. Passaquet, C.; Teodorescu-Ionescu, N.; Zuily-Fodil, Y.; Pham Thi, A. T. *Physiol. Plant.* **1986**, *67*(2), 211–216.

1860. Osagie, A. U.; Kates, M. *Lipids* **1984**, *19*(12), 958–965.

1861. Hoekstra, F. A.; Crowe, L. M.; Crowe, J. H. *Plant Cell Environ.* **1989**, *12*(1), 83–91.

1862. Min, Y.-K.; Kim, Z.-U. *Korean J. Food Sci. Technol.* **1992**, *35*(3), 146–151.

1863. Shin, H.-S.; Kim, S.-W. *J. Am. Oil Chem. Soc.* **1994**, *71*(6), 619–622.

1864. Tsuyuki, H.; Itoh, S.; Nakatsukasa, Y. *Nihon Daigaku No-Juigakubu Gakujutsu Kenkyu Hokoku* **1978**, *35*, 224–230.

1865. Schweriner, H. A.; Biale, J. B. *J. Lipid Res.* **1973**, *14*(2), 235–242.

1866. Ellenbracht, F.; Barz, W.; Mangold, H. K. *Planta* **1980**, *150*(2), 114–119.

1867. Nouchi, I.; Toyama, S. *Plant Physiol.* **1988**, *87*(3), 638–646.

1868. Skorzynska-Polit, E.; Tukendorf, A.; Selstam, E.; Baszynski, T. *Environ. Exp. Bot.* **1998**, *40*(3), 275–286.

1869. Skórzyńska, E.; Urbanik-Sypniewska, T.; Russa, R.; Baszyński, T. *J. Plant Physiol.* **1991**, *138*(4), 454–459.

1870. Macsymiec, W.; Russa, R.; Urbanik-Sypniewska, T.; Baszyński, T. *Physiol. Plant.* **1994**, *91*(4), 715–721.

1871. Bartholomew, L.; Mace, K. D. *Cytobios* **1972**, *5*(2), 241–247.

1872. Katiyar, S. K.; Bhatia, A. K. *J. Food Sci. Technol. (India)* **1988**, *25*(3), 165–166.

1873. Drumm, T. D.; Gray, J. I.; Hosfield, G. L. *J. Sci. Food Agric.* **1990**, *50*(4), 485–497.

1874. Allen, C. F.; Good, P.; Mollenhauer, H. H.; Totten, C. *J. Cell Biol.* **1971**, *48*(3), 542–546.

1875. Hartmann, E.; Jeck, U.; Grasmück, I. *Z. Pflanzenphysiol.* **1981**, *103*(5), 427–436.

1876. Mukhamedova, K. S.; Akramov, S. T. *Khim. Prir. Soedin.* **1982**, (6), 688–691 (Engl. 651–654).

1877. Hoppe, H. H.; Heitefuss, R. *Physiol. Plant Pathol.* **1975**, *5*(3), 263–271.

1878. Ivanova, A.; Stefanov, K.; Yordanov, I. *Dokl. Bulg. Akad. Nauk* **1995**, *48*(9/10), 119–121.

1879. Stefanov, K.; Popova, I.; Nikolova-Damyanova, B.; Kimenov, G.; Popov, S. *Phytochemistry* **1992**, *31*(11), 3745–3748.

1880. Sednina, G. V. *Biol. Nauki (Moscow)* **1980**, (10), 110–113.

1881. Wang, H.; Zhang, C. *Zhiwu Xuebao* **1993**, *35*(7), 533–540.

1882. Khotimchenko, S. V. *Phytochemistry* **1993**, *33*(2), 369–372.

1883. Kozukue, E.; Kozukue, N. *J. Food Sci.* **1981**, *46*(3), 751–755.

1884. Senser, M.; Beck, E. *Z. Pflanzenphysiol.* **1982**, *105*(3), 241–253.

1885. Lemoine, Y.; Dubacq, J.-P.; Zabulon, G. *Physiol. Vég.* **1982**, *20*(3), 487–503.

1886. Zwiazek, J. J.; Blake, T. J. *Physiol. Plant.* **1990**, *79*(1), 71–77.

1887. Rais, M. Y.; Ahmad, I.; Javed, M. A.; Waheed, A. *Fett Wiss. Technol.* **1995**, *97*(12), 455–457.

1888. Hirata, Y.; Sekiguchi, R.; Saitoh, M.; Kubota, K.; Kayama, M. *Yukagaku* **1994**, *43*(7), 579–582.

1889. Zwiazek, J. J.; Shay, J. M. *Can. J. Bot.* **1988**, *66*(3), 535–541.

1890. Andrikopoulos, N. K.; Siafaka-Kapadai, A.; Demopoulos, C. A.; Kapoulas, V. M. *Phytochemistry* **1985**, *24*(12), 2953–2957.

1891. Kim, M.; Rhee, S. H.; Cheigh, H. S. *Han'guk Yongyang Siklyong Hakhoechi* **1984**, *13*(4), 406–412.

1892. Kojima, M.; Shiraki, H.; Ohnishi, M.; Ito, S. *Obihiro Chikusan Daigaku Gakujutsu Kenkyu Hokoku, Dai-1-Bu* **1990**, *17*(1), 13–19.

1893. Alaiz, E. P.; Velez, J.; Alsasua, M. T. *Dev. Plant Biol.* **1982**, *8*, 445–450.

1894. McIlwain, D. L.; Ballou, C. E. *Biochemistry* **1966**, *5*(12), 4054–4061.

1895. Donskaya, L. I.; Muzalevskaya, O. V. *Rastit. Resur.* **1994**, *30*(3), 64–66.

1896. Fuksman, I. L.; Novitskaya, Yu.E.; Tikhova, M. A.; Sulimova, G. M. *Lipidnyi Obmen Drev. Rast. Usloviyakh Sev.* **1983**, 97–105, from *Chem. Abstr.* **1984**, *101*, 188139h.

1897. Selstam, E.; Őquist, G. *Plant Sci.* **1985**, *42*(1), 41–48.

1898. Wingsle, G.; Mattson, A.; Ekblad, A.; Hällgren, J.-E.; Selstam, E. *Plant Sci.* **1992**, *82*(2), 167–178.

1899. Fuksman, I. L.; Ponkina, N. A. *Khim Drev.* **1984**, (5), 85–88.

1900. Kashani, G. G.; Valadon, L. R. G. *J. Food Technol.* **1983**, *18*(4), 461–467.

1901. Haydar, M.; Hadziyev, D. *J. Sci. Food Agric.* **1974**, *25*(10), 1285–1305.

1902. Thomas, W. *Stärke* **1979**, *31*(2), 54–57.

1903. El-Refai, A. A.; Gouda, M. S.; Ammar, K. A. *Food Chem.* **1987**, *23*(2), 117–127.

1904. Miyazawa, T.; Fujino, Y. *Nippon Nogei Kagaku Kaishi* **1976**, *50*(4), 169–174.

1905. Hoover, R.; Cloutier, L.; Dalton, S.; Sosulski, F. W. *Stärke* **1988**, *40*(9), 326–342.

1906. Kaufmann, H. P.; Hamza, Y.; Mangold, H. K. *Chem. Phys. Lipids* **1971**, *6*(4), 325–332.

1907. Bolton, P.; Harwood, J. L. *Biochem. J.* **1977**, *168*(2), 261–269.

1908. Chapman, D. J.; De-Felice, J.; Barber, J. *Planta* **1983**, *157*(3), 218–223.

1909. Dorne, A.-J.; Heinz, E. *Plant Sci.* **1989**, *60*(1), 39–46.

1910. Cline, K.; Andrews, J.; Mersey, B.; Newcomb, E. H.; Keegstra, K. *Proc. Natl. Acad. Sci. U. S. A.* **1981**, *78*(6), 3595–3599.

1911. Hernandez, L. E.; Cooke, D. T. *Phytochemistry* **1996**, *42*(2), 341–346.

1912. Xue, L.; McCune, L. M.; Kleppinger-Sparace, K. F.; Brown, M. J.; Pomeroy, M. K.; Sparace, S. A. *Plant Physiol.* **1997**, *113*(2), 549–557.

1913. Chapman, D. J.; De-Felice, J.; Barber, J. *Planta* **1985**, *166*(2), 280–285.

1914. Lösel, D. M. *New Phytol.* **1978**, *80*(1), 167–174.

1915. Latsague, M.; Acevedo, H.; Fernandez, J.; Romero, M.; Cristi, R.; Alberdi, M. *Phytochemistry* **1992**, *31*(10), 3419–3426.

1916. Rao, Y. N.; Prasad, R. B. N.; Rao, S. V. *Fette Seifen Anstrichm.* **1984**, *86*(3), 107–109.

1917. Yoshida, S.; Sakai, A. *Plant Cell Physiol.* **1973**, *14*(2), 353–359.

1918. Dorne, A.-J.; Joyard, J.; Douce, R. *Can. J. Bot.* **1987**, *65*(11), 2368–2372.

1919. Zlatanov, M.; Janakieva, I. *Fett/Lipid* **1998**, *100*(7), 312–315.

1920. Takenaga, F.; Miyata, M.; Itoh, S.; Tsuyuki, H. *Nihon Daigaku No-Juigakubu Gaku-jutsu Kenkyu Hokoku* **1990**, *47*, 210–217.

1921. Schmidt, E. R.; Feucht, W. *Sci. Hortic. (Amsterdam)* **1993**, *55*(3/4), 273–282.

1922. Yuri, A.; Schmitt, E.; Feucht, W.; Treutter, T. *J. Plant Physiol.* **1990**, *135*(6), 692–697.

1923. Trusova, V. M.; Molchanov, M. I.; Balaur, N. S. *Dokl. Akad. Nauk SSSR* **1991**, *316*(2), 494–497 (Engl. *Biochem.* 10–13).

1924. Ogihara, H.; Itoh, S.; Tsuyuki, H. *Nihon Daigaku No-Juigakubu Gakujutsu Kenkyu Hokoku* **1982**, *39*, 97–106.

1925. Takenaga, F.; Itoh, S.; Tsuyuki, H. *Nippon Shokuhin Kogyo Gakkaishi* **1983**, *30*(1), 345–349, from *Chem. Abstr.* **1983**, *99*, 102309u.

1926. Monge, E.; Pérez, C.; Pequerul, A.; Madero, P.; Val, J. *Plant Soil* **1993**, *154*(1), 97–102.

1927. Khor, H.-T.; Chan, S.-L. *Phytochemistry* **1988**, *27*(7), 2041–2044.

1928. Ohnishi, M.; Obata, S.; Ito, S.; Fujino, Y. *Obihiro Chikusan Daigaku Gakujutsu Kenkyu Hokoku, Dai-1-Bu* **1989**, *16*(2), 109–114.

1929. Mukhamedova, Kh.S.; Akramov, S. T. *Khim. Prir. Soedin.* **1979**, (1), 12–14 (Engl. 9–11).

1930. Jarvis, M. C.; Duncan, J. J. *Phytochemistry* **1974**, *13*, 979–981.

1931. Isamukhamedov, A. Sh.; Akramov, S. T. *Khim Prir. Soedin.* **1982**, (3), 396–397 (Engl. 367–368).

1932. Tsuyuki, H.; Itoh, S.; Nakatsukasa, Y. *Nihon Daigaku No-Juigakubu Gakujutsu Kenkyu Hokoku* **1981**, *38*, 141–148.

1933. Kolesnik, A. A.; Golubev, V. N.; Farzaliev, E. B. *Khim. Prir. Soedin.* **1989**, (2), 175–180 (Engl. 148–152).

1934. Tsuyuki, H.; Shiozawa, T.; Takenaga, F.; Itoh, S. *Nihon Daigaku No-Juigakubu Gaku-jutsu Kenkyu Hokoku* **1989**, *46*, 210–218.

1935. Bayer, M. H. *Physiol. Plant.* **1991**, *81*(3), 313–318.

1936. Bayer, M. H. *Plant Physiol.* **1983**, *73*(1), 179–181.

1937. Stepanovic, B.; Thu, P. T. A.; de Paula, F. M.; da Silva, J. V. *Can. J. Bot.* **1992**, *70*(1), 107–113.

1938. Hudson, B. J. F.; Karis, I. G. *J. Sci. Food Agric.* **1974**, *25*(12), 1491–1502.

1939. Krupa, Z. *Physiol. Plant.* **1987**, *73*(4), 518–524.

1940. Novitskii, Yu.A.; Novitstkaya, G. V.; Sokolova, I. A. *Fiziol. Rast.* **1990**, *37*(1), 54–63 (Engl. 39–46).

1941. Opute, F. I. *J. Sci. Food Agric.* **1978**, *29*(2), 115–120.

1942. Loubaresse, M.; Paulin, A.; Dereuddre, J. *Compt. Rend. Acad. Sci., Ser 3* **1991**, *313*(10), 453–459.

1943. Kondoh, H.; Kawabe, S. *Agric. Biol. Chem.* **1975**, *39*(3), 745–746.

1944. Donaldson, R. P.; Beevers, H. *Plant Physiol.* **1977**, *59*(2), 259–263.

1945. Moreau, R. A.; Liu, K. D. F.; Huang, A. H. C. *Plant Physiol.* **1980**, *65*(5), 1176–1180.

1946. Sakaki, T.; Tanaka, K.; Yamada, M. *Plant Cell Physiol.* **1994**, *35*(1), 53–62.

1947. Singh, J.; de la Roche, I. A.; Siminovich, D. *Nature* **1975**, *257*, 669–670.

1948. Wuttke, H.-G. *Z. Naturforsch.* **1976**, *31C*(7/8), 456–460.

1949. Czech-Kozlowska, M.; Krzywanski, Z. *Acta Physiol. Plant.* **1983**, *5*(1/2), 21–27.

1950. Asilbekova, D. T.; Gusakova, S. D.; Glushenkova, A. I.; Bessonova, I. A. *Khim. Prir. Soedin.* **1995**, *31*(2), 215–219 (Engl. 175–178).

1951. Cerbón, J.; Villegas, T. *Plant Cell, Tissue Organ Cult.* **1983**, *2*(4), 317–326.

1952. Garg, A. *Int. Sugar J.* **1979**, *81*(967), 200–205.

1953. Prasad, R. B. N.; Rao, S. V. *J. Oil Technol. Assoc. India* **1980**, *11*(2), 35–38, from *Chem. Abstr.* **1980**, *92*, 90963t.

1954. Bidatze, M. A.; Dalakishvili, Ts.M.; Kemertelidze, E. P.; Mukhamedova, Kh.S. *Khim. Prir. Soedin.* **1991**, (5), 619–621 (Engl. 545–547).

1955. Mazza, G.; Biliaderis, C. G.; Przybylski, R.; Oomah, B. D. *J. Agric. Food Chem.* **1992**, *40*(9), 1520–1523.

1956. Biliaderis, C. G.; Mazza, G.; Przybylski, R. *Starch/Stärke* **1993**, *45*(4), 121–127.

1957. Mano, Y.; Nishiyama, S.; Kojima, M.; Ohnishi, M.; Ito, S. *Cereal Chem.* **1991**, *68*(3), 280–284.

1958. Vecher, A. S.; Nenadovich, R. A.; Reshetnikov, V. N. *Dokl. Akad. Nauk BSSR* **1985**, *29*(1), 80–82.

1959. Nenadovich, R. A.; Reshetnikov, V. N. *Dokl. Akad. Nauk BSSR* **1988**, *32*(5), 460–463.

1960. Nenadovich, R. A. *Vestsi Akad. Navuk BSSR, Ser. Biyal. Navuk* **1990**, (3), 14–17.

1961. Novitskaya, G. V.; Zvereva, O. A.; Sokolova, I. A. *Fiziol. Rast.* **1986**, *33*(5), 997–1004 (Engl. 768–774).

1962. Lynch, D. V.; Steponkus, P. L. *Plant Physiol.* **1987**, *83*(4), 761–767.

1963. Krupa, Z.; Huner, N. P. A.; Williams, J. P.; Maissan, E.; James, D. R. *Plant Physiol.* **1987**, *84*(1), 19–24.

1964. Uemura, M.; Steponkus, P. *Plant Physiol.* **1997**, *114*(4), 1493–1500.

1965. Clarkson, D. T.; Hall, K. C.; Roberts, J. K. M. *Planta* **1980**, *149*(5), 464–471.

1966. Yoshida, H.; Shigezaki, J.; Takagi, S.; Kajimoto, G. *J. Sci. Food Agric.* **1995**, *68*(4), 407–415.

1967. Takano, K.; Matsumoto, S.; Kamoi, I.; Obara, T. *Tokyo Nogyo Daigaku Nogaku Shuho* **1982**, *27*(2), 141–144.

1968. Gülz, P.-G.; Eich, C. *Z. Naturforsch.* **1982**, *37C*(11/12), 1286–1287.

1969. Guillot-Salomon, T.; Bahl, J.; Ben-Rais, L.; Alpha, M.-J.; Cantrel, C.; Dubacq, J.-P. *Plant Physiol. Biochem.* **1991**, *29*(6), 667–679.

1970. Cantrel, C.; Guillot-Salomon, T.; Brown, S. C.; Marie, D.; Dubacq, J. P. *Plant Cell Physiol.* **1991**, *32*(7), 959–967.

1971. Chrétien, D.; Guillot-Salomon, T.; Bahl, J.; Hirsch, A.-M.; Alpha, M.-J.; Cantrel, C.; Dubacq, J.-P. *Plant Physiol. Biochem.* **1990**, *28*(1), 79–86.

1972. Rodionov, V. S.; Zakharova, L. S.; Nyuppieva, K. A. *Fiziol. Biokhim. Kul't. Rast.* **1981**, *13*(4), 411–415.

1973. Lucini, E. I.; Grosso, N. R.; Lamarque, A. L.; Maestri, D. M.; Zygadlo, J. A.; Guzman, C. A. *J. Agric. Food Chem.* **1994**, *42*(12), 2743–2745.

1974. Palta, J. W.; Whitaker, B. D.; Weiss, L. S. *Plant Physiol.* **1993**, *103*(3), 793–803.

1975. Barrientos, M.; Contreras, A.; Alberdi, M. *Phyton* **1990**, *51*(2), 133–138.

1976. Cocallemen, S.; Farines, M.; Faill, H.; Soulier, J.; Morin, O. *Rev. Fr. Corps Gras* **1988**, *35*(3), 105–109.

1977. Rodionov, V. S.; Nyupieva, K. A.; Zakharova, L. S. *Biokhimiya* **1974**, *39*(1), 215–224 (Engl. 178–185).

1978. Rodionov, V. S.; Kozubov, G. M.; Tikhova, M. A.; Sulimova, G. M. *Fiziol. Rast.* **1975**, *22*(2), 333–337 (Engl. 275–279).

1979. Galliard, T. *Phytochemistry* **1968**, *7*(11), 1907–1914.

1980. Lepage, M. *Lipids* **1968**, *3*(6), 477–481.

1981. Lee, S.-Y.; Shin, H.-S. *Korean J. Food Sci. Technol.* **1979**, *11*(4), 304–313.

1982. Meunier, D.; Mazliak, P. *Compt. Rend.* **1972**, *275D*(2), 213–216.

1983. Justin, A. M.; Demandre, C.; Mazliak, P. *Biochim. Biophys. Acta* **1989**, *1005*(1), 51–55.

1984. Pun, W. H.; Khan, A. A.; Chung, I.; Haydar, M.; Hadziyev, D. *Potato Res.* **1980**, *23*(1), 57–74.

1985. Korableva, N. P.; Bagdasaryan, S. G.; Ladyzhenskaya, E. P.; Protsenko, M. O. *Ukr. Bot. Zh.* **1983**, *40*(2), 75–78.

1986. Mano, Y.; Wada, T.; Fujino, Y. *Obihiro Otani Tanki Daigaku Kiyo* **1978**, *15*, 47–53, from *Chem. Abstr.* **1980**, *93*, 44267s.

1987. Mukhamedova, Kh.S.; Glushenkova, A. I. *Khim. Prir. Soedin.* **1985**, (3), 281–284 (Engl. 259–262).

1988. Mukhamedova, Kh.S.; Glushenkova, A. I. *Khim. Prir. Soedin.* **1997**, *33*(4), 578–582 (Engl. 445–448).

1989. Osagie, A. U. *J. Agric. Food Chem.* **1987**, *35*(4), 601–604.

1990. Gehring, R. Z.; Todd, G. W. *Proc. Ark. Acad. Sci.* **1985**, *39*, 44–49.

1991. Boissy, M. C.; Perlès, R. *Bull. Soc. Chim. Biol.* **1965**, *47*(5), 859–866.

1992. Wu, J.; Seliskar, D. M.; Gallagher, J. L. *Physiol. Plant.* **1998**, *102*(2), 307–317.

1993. Rochester, C. P.; Kjellbom, P.; Larsson, C. *Physiol. Plant.* **1987**, *71*(3), 257–263.

1994. Edman, K.; Ericson, I. *Biochem. J.* **1987**, *243*(2), 575–578.

1995. Poincelot, R. P. *Arch. Biochem. Biophys.* **1973**, *159*(1), 134–142.

1996. Farineau, N.; Guillot-Salomon, T.; Tuquet, C.; Farineau, J. *Photochem. Photobiol.* **1984**, *40*(3), 387–390.

1997. Dorne, A.-J.; Joyard, J.; Douce, R. *Proc. Natl. Acad. Sci. U. S. A.* **1990**, *87*(1), 71–74.

1998. Block, M. A.; Dorne, A.-J.; Joyard, J.; Douce, R. *J. Biol. Chem.* **1983**, *258*(21), 13281–13286.

1999. Heemskerk, J. W. M.; Bögemann, G.; Scheijen, M. A. M.; Wintermans, J. F. G. M. *Anal. Biochem.* **1986**, *154*(1), 85–91.

2000. Sukhija, P. J.; Sital, J. S.; Raheja, R. K.; Bhatia, I. S. *Physiol. Plant.* **1976**, *38*(3), 221–223.

2001. Lechevallier, D. *Physiol. Vég.* **1977**, *15*(3), 387–402.

2002. Lechevallier, D. *Physiol. Vég.* **1977**, *15*(1), 95–119.

2003. Menetrez, M. L.; Fites, R. C.; Wilson, R. F. *J. Am. Oil Chem. Soc.* **1988**, *65*(4), 634–637.

2004. Tuquet, C.; Farineau, N.; Sallé, G. *Physiol. Plant.* **1989**, *78*(4), 574–582.

2005. Leach, R. P.; Wheeler, K. P.; Flowers, T. J.; Yeo, A. R. *J. Exp. Bot.* **1990**, *41*(230), 1089–1094.

2006. Génard, H.; Le Saos, J.; Boucaud, J. *Compt. Rend. Acad. Sci., Ser. 3* **1988**, *306*(2), 75–80.

2007. Génard, H.; Le Saos, J.; Billard, J.-P.; Trémolières, A.; Boucaud, J. *Plant Physiol. Biochem.* **1991**, *29*(5), 421–427.

2008. Pitke, P. M.; Singh, P. P.; Srivastava, H. C. *J. Am. Oil Chem. Soc.* **1979**, *56*(5), 559.

2009. Tuquet, C.; Sallé, G. *Plant Physiol. Biochem.* **1996**, *34*(2), 283–292.

2010. Kimura, S.; Kai, T.; Mitsuhashi, T. *Tokyo Gakugai Daigaku Kiyo [4]* **1969**, *21*, 31–37.

2011. Parsons, J. G.; Keeney, P. G.; Patton, S. *J. Food Sci.* **1969**, *34*(6), 497–499.

2012. Arruda, D. H.; Dimick, P. S. *J. Am. Oil Chem. Soc.* **1991**, *68*(6), 385–390.

2013. Bhatia, I. S.; Lal, S.; Sukhija, P. S.; Pruthi, T. D. *J. Sci. Food Agric.* **1970**, *21*(10), 517–519.

2014. Shipova, E. V.; Pankrushina, A. N.; Likholat, T. V. *S.-Kh. Biol.* **1986**, (1), 90–93.

2015. Weenink, R. O. *Biochem. J.* **1964**, *93*(3), 606–611.

2016. Hemavathy, J.; Prabhakar, J. V. *Food Chem.* **1989**, *31*(1), 1–7.

2017. Radwan, S. S. *J. Chromatogr. Sci.* **1978**, *16*(11), 538–542.

2018. Ferretti, R.; Filesni, C.; Serlapi-Crescenzi, G.; Tassi-Micco, C. *Riv. Soc. Ital. Sci. Aliment.* **1982**, *11*(4), 219–222.

2019. Pomeranz, Y.; Chung, O.; Robinson, R. J. *J. Am. Oil Chem. Soc.* **1966**, *43*(2), 511–514.

2020. MacMurray, T. A.; Morrison, W. R. *J. Sci. Food Agric.* **1970**, *21*(10), 520–528.

2021. Prieto, J. A.; Ebri, A.; Collar, C. *J. Chromatogr. Sci.* **1993**, *31*(2), 55–60.

2022. Bedi, G. S.; Sandhu, S. S.; Nath, N. *J. Food Sci. Technol.* **1973**, *10*(4), 164–167.

2023. Arunga, R. O.; Morrison, W. R. *Lipids* **1971**, *6*(10), 768–776.

2024. Meredith, P.; Dengate, H. N.; Morrison, W. R. *Stärke* **1978**, *30*(4), 119–125.

2025. Shin, M. G.; Min, B. K.; Chang, M. S. *Anal. Sci. Technol.* **1992**, *5*(3), 339–343, from *Chem. Abstr.* **1993**, *119*, 70700c.

2026. de la Roche, I. A.; Andrews, C. J.; Kates, M. *Plant Physiol.* **1973**, *51*(3), 468–473.

2027. Hargin, K. D.; Morrison, W. R. *J. Sci. Food Agric.* **1980**, *31*(9), 877–888.

2028. Jelsema, C. L.; Morré, D. J.; Ruddat, M.; Turner, C. *Bot. Gaz.* **1977**, *138*(2), 138–149.

2029. Vardevanian, P. O.; Boyadzhian, B. G.; Tiratsuian, S. G.; Vardevanian, A. O.; Panosian, G. A. *Fiziol. Rast.* **1996**, *43*(4), 616–619 (Engl. 541–543).

2030. Belova, T. E.; Dubtsova, G. N.; Kolesnik, G. B.; Doronina, O. D.; Nechaev, A. P. *Prikl. Biokhim. Mikrobiol.* **1981**, *17*(5), 734–738 (Engl. 549–553).

2031. Novitskaya, G. V. *Prikl. Biokhim. Mikrobiol.* **1983**, *19*(2), 261–266 (Engl. 214–219).

2032. Horvath, I.; Vigh, L.; Belea, A.; Farkas, T. *Physiol. Plant.* **1980**, *49*(2), 117–120.

2033. Prabha, C.; Arora, Y. K.; Wagle, D. S. *Plant Sci.* **1985**, *38*(1), 13–16.

2034. Bahl, J.; Francke, B.; Monegér, R. *Planta* **1976**, *129*(3), 193–201.

2035. Hudson, B. J. F.; Karis, I. G. *J. Sci. Food Agric.* **1973**, *24*(12), 1541–1550.

2036. Zhang, G.; Slaski, J. J.; Archambault, D. H.; Taylor, G. J. *Physiol. Plant.* **1996**, *96*(4), 683–691.

2037. Mansour, M. M. F.; van Hasselt, P. R.; Kuiper, P. J. C. *Physiol. Plant.* **1994**, *92*(3), 473–478.

2038. Pankratova, S. I.; Khokhlova, L. P. *Fiziol. Biokhim. Kul't. Rast.* **1984**, *9*(2), 129–135.

2039. Khokhlova, L. P.; Belogub, O. V.; Abdrakhimova, J. R.; Zabotin, A. I.; Khala-buda, L. N. *Fiziol. Biokhim. Kul't. Rast.* **1988**, *20*(3), 246–254.

2040. Alam, A.; Nechaev, A. P.; Eremenko, T. V.; Doronina, O. D. *Izv. Vyssh. Uchebn. Zaved. Pishch. Tekhnol.* **1980**, (1), 130–131.

2041. Horvath, I.; Vigh, L. *Dev. Plant Biol.* **1984**, *9*, 535–538.

2042. Lösel, D. M.; Lewis, D. H. *New Phytol.* **1974**, *73*(6), 1157–1169.

2043. Caffrey, M.; Werner, B. G.; Priestley, D. A. *Biochim. Biophys. Acta* **1987**, *921*(1), 124–134.

2044. van Bilsen, D. G. J. L.; Hoekstra, F. A. *Plant Physiol.* **1993**, *101*(2), 675–682.

2045. Antonopoulou, S.; Demopoulos, C. A.; Andrikopoulos, N. K. *J. Agric. Food Chem.* **1996**, *44*(10), 3052–3056.

2046. Croteau, R.; Fagerson, J. S. *Phytochemistry* **1969**, *8*(11), 2219–2222.

2047. Harwood, J. L.; James, A. T. *Eur. J. Biochem.* **1975**, *50*(2), 325–334.

2048. Webb, M. S.; Williams, J. P. *Plant Cell Physiol.* **1984**, *25*(8), 1541–1550.

2049. Sasaki, T.; Satoh, A.; Tanaka, K.; Omasa, K.; Shimazaki, K. *Phytochemistry* **1995**, *40*(4), 1065–1070.

2050. Mackender, R. O.; Leech, R. M. *Plant Physiol.* **1974**, *53*(3), 496–502.

2051. Kojima, M.; Ohnishi, M.; Mano, Y.; Sasaki, S.; Ito, S.; Fujino, Y. *Yukagaku* **1989**, *38*(8), 619–625.

2052. Kojima, M.; Suzuki, H.; Ohnishi, M.; Ito, S. *Obihiro Chikusan Daigaku Gakujutsu Kenkyu Hokoku, Dai-1-Bu* **1990**, *17*(1), 21–29.

2053. Urakami, C.; Hirosawa, N. *Arch. Int. Physiol. Biochim.* **1968**, *76*(4), 635–645.

2054. Akbarov, R. R.; Mukhamedova, K. S.; Akramov, S. T. *Khim. Prir. Soedin.* **1984**, (2), 177–179 (Engl. 163–166).

2055. McCarty, R. E.; Douce, R.; Benson, A. A. *Biochim. Biophys. Acta* **1973**, *316*(2), 266–270.

2056. Um, S. H.; Cheigh, H. S. *Han'guk Yongyang Siklyong Hakhoechi* **1988**, *17*(2), 164–171.

2057. Goldberg, R.; Prat, R.; Dubacq, J.-P. *Plant Cell Physiol.* **1983**, *24*(3), 541–549.

2058. Yoshida, S.; Uemura, M. *Plant Physiol.* **1986**, *82*(3), 807–812.

2059. Stalleart, V. M.; Geuns, J. M. C. *Phytochemistry* **1994**, *36*(5), 1177–1180.

2060. Munshi, S. K.; Kinger, A. K.; Bhatia, N. K.; Sukhija, P. S. *J. Plant Sci. Res.* **1991**, *7*(1/4), 68–73.

2061. Farhangi, M.; Valadon, L. R. G. *J. Food Sci.* **1982**, *47*(4), 1158–1163.

2062. Kasamo, K.; Nouchi, I. *Plant Physiol.* **1987**, *83*(2), 323–328.

2063. Trémolières, A.; Laurent, P.; Connan, A. *Plant Physiol. Biochem.* **1988**, *26*(3), 339–346.

2064. de Paula, F. M.; Pham Thi, A. T.; da Silva, J. V.; Justin, A. M.; Demandre, C. *Plant Sci.* **1990**, *66*(2), 185–193.

2065. Huner, N. P. A.; Krol, M.; Williams, J. P.; Maissan, E. *Plant Physiol.* **1988**, *87*(3), 721–726.

2066. Quartacci, M. F.; Izzo, R.; Muratore, G. *Riv. Ital. Sostanze Grasse* **1995**, *72*(7), 299–302.

2067. Kansal, V. K.; Lal, B. M. *Indian J. Exp. Biol.* **1976**, *14*(4), 514–516.

2068. Mas, A.; Navarro-Pedreño, J.; Cooke, D. T.; James, C. S. *Phytochemistry* **1994**, *35*(5), 1249–1253.

2069. Zherebin, Yu.L.; Kolesnik, A. A. *Khim Prir. Soedin.* **1984**, (2), 165–168 (Engl. 152–155).

2070. Kuiper, P. J. C. *Plant Physiol.* **1968**, *43*(9), 1367–1371.

2071. Bhakuni, D. S.; Srivastava, S. N.; Sharma, V. N.; Kaul, K. N. *Indian J. Appl. Chem.* **1961**, *24*(5/6), 197–201.

2072. Melo, N.; Tavares, R. M.; Morais, F.; Barroso, J. G.; Pais, M. S. S. *Phytochemistry* **1995**, *40*(5), 1367–1371.

2073. Gospodinova, V.; Tevekelev, D. *Izv. Inst. Khranene, Bulg. Akad. Nauk* **1968**, *7*, 33–36, from *Chem. Abstr.* **1969**, *70*, 17549d.

2074. Tan, S. L.; Morrison, W. R. *J. Am. Oil Chem. Soc.* **1979**, *56*(4), 531–535.

2075. Gardner, H. W.; Miernyk, J. A.; Christianson, D. D.; Khoo, U. *Physiol. Plant.* **1987**, *69*(3), 541–549.

2076. Cowan, D. S. C.; Cooke, D. T.; Clarkson, D. T.; Hall, J. L. *J. Exp. Bot.* **1993**, *44*(262), 991–994.

2077. Acker, L.; Schmitz, H. J. *Stärke* **1967**, *19*(9), 275–280.

2078. Tuquet, C.; Guillot-Salomon, T.; de Lubac, M.; Signol, M. *Plant Sci. Lett.* **1977**, *8*(1), 59–64.

2079. Chicha, A.; Justin, A.-M.; Jolliot, A.; Demandre, C.; Mazliak, P. *Plant Physiol. Biochem.* **1993**, *31*(4), 507–513.

2080. Grandmougin, A.; Bouvier-Navé, P.; Ullmann, P.; Benveniste, P.; Hartmann, M.-A. *Plant Physiol.* **1989**, *90*(2), 591–597.

2081. Nasyrova, G. F.; Palladina, T. A. *Fiziol. Rast.* **1984**, *31*(2), 351–355 (Engl. 282–285).

2082. Kumar, A.; Pant, R. C. *Indian J. Exp. Bot.* **1980**, *18*(2), 192–196.

2083. Ashworth, D. J.; Adams, D. O.; Giang, B. Y.; Cheng, M. T.; Lee, R. Y. *Anal. Chem.* **1985**, *57*(3), 710–715.

2084. Ahn, D. H.; Ha, B. S. *Han'guk Yongyang Siklyong Hakhoechi* **1987**, *16*(4), 350–363, from *Chem. Abstr.* **1989**, *111*, 114018c.

2085. Kim, D. J.; Jeon, Y. M. *Han'guk Yongyang Siklyong Hakhoechi* **1991**, *20*(5), 473–478, from *Chem. Abstr.* **1992**, *116*, 191183h.

2086. Ohnishi, M.; Yasui, Y.; Mano, Y.; Ito, S.; Fujino, Y. *Agric. Biol. Chem.* **1989**, *53*(2), 565–567.

2087. Singh, I. P.; Jogi, B. S.; Dua, H. S.; Gupta, M. L. *Indian J. Agric. Sci.* **1975**, *45*(11/12), 545–549.

2088. Goncharova, N. P.; Isamukhamedov, A. Sh.; Glushenkova, A. I. *Khim. Prir. Soedin.* **1990**, (1), 21–24 (Engl. 16–18).

2089. Dall, W.; Smith, D. M.; Moore, L. E. *Aquaculture* **1992**, *101*(1/2), 75–83.

2090. Dembitsky, V. M.; Rozentsvet, O. A. *Phytochemistry* **1989**, *28*(12), 3341–3343.

2091. Hammond, E. W. *Chromatography for the Analysis of Lipids*. CRC Press: Boca Raton, FL, 1993; viii + 188 pp.

2092. Murata, N.; Sato, N.; Takahashi, N.; Hamazaki, Y. *Plant Cell Physiol.* **1982**, *23*(8), 1071–1079.

2093. Harwood, J. L. *Phytochemistry* **1976**, *15*(10), 1459–1463.

2094. Noda, M.; Obata, T. *Nippon Nogei Kagaku Kaishi* **1975**, *49*(5), 251–256.

2095. Haverkate, F.; Van Deenen, L. L. M. *Biochim. Biophys. Acta* **1965**, *106*(1), 78–92.

2096. Devor, K. A.; Mudd, J. B. *J. Lipid Res.* **1971**, *12*(4), 396–402.

2097. Kimura, S.; Sano, Y.; Shibasaki, K. *Nippon Shokuhin Kogyo Gakkaishi* **1970**, *17*(4), 123–128.

2098. Isamukhamedov, A. Sh.; Shustanova, L. A.; Akramov, S. T. *Khim. Prir. Soedin.* **1977**, (3), 324–330 (Engl. 277–282).

2099. Weber, N.; Benning, H. *Eur. J. Biochem.* **1985**, *146*(2), 323–329.

2100. Abidi, S. L.; Mounts, T. L. *J. Chromatogr.* **1992**, *598*(2), 209–218.

2101. Abidi, S. L.; Mounts, T. L.; Rennick, K. A. *J. Liq. Chromatogr.* **1991**, *14*(3), 573–588.

2102. Abidi, S. L.; Mounts, T. L. *J. Chromatogr. Sci.* **1993**, *31*(6), 231–236.

2103. Abidi, S. L.; Mounts, T. L. *J. Chromatogr.* **1995**, *694A*(2), 365–373.

2104. Pike, C. S.; Norman, H. A.; Kemmerer, E. C.; Wessner, D. R.; Greenberg, C. M.; Kaplan, L. J.; Brodsky, N. M.; Ellis, A. A. *Plant Sci.* **1990**, *68*(2), 189–196.

2105. Norman, H. A.; St. John, J. B. *J. Lipid Res.* **1986**, *27*(10), 1104–1107.

2106. Kim, H.-Y.; Salem, N., Jr. *Anal. Chem.* **1987**, *59*(5), 722–726.

2107. Porter, N. A.; Wolf, R. A.; Nixon, J. R. *Lipids* **1979**, *14*(1), 20–24.

2108. Abidi, S. L. *J. Chromatogr.* **1991**, *587*(2), 193–203.

2109. Kesselmeier, J.; Heinz, E. *Anal. Biochem.* **1985**, *144*(2), 319–328.

2110. Hullin, F.; Kim, H.-Y.; Salem, N., Jr. *J. Lipid Res.* **1989**, *30*(12), 1963–1975.

2111. Myher, J. J.; Kuksis, A. *Can. J. Biochem.* **1982**, *60*(6), 638–650.

2112. Murata, N. *Plant Cell Physiol.* **1983**, *24*(1), 81–86.

2113. Norberg, P.; Engström, L.; Nilsson, R.; Liljenberg, C. *Biochim. Biophys. Acta* **1992**, *1112*(1), 52–56.

2114. Myher, J. J.; Kuksis, A. *Biochim. Biophys. Acta* **1984**, *795*(1), 85–90.

2115. Olsson, M.; Nilsson, R.; Norberg, P.; von Arnold, S.; Liljenberg, C. *Plant Physiol. Biochem.* **1994**, *32*(2), 225–232.

2116. Singleton, J. A.; Stikeleather, L. F. *J. Am. Oil Chem. Soc.* **1995**, *72*(4), 485–488.

2117. Aid, F.; Demandre, C.; Kesir-Benhassaine, G.; Mazliak, P. *Plant Physiol. Biochem.* **1995**, *33*(2), 235–240.

2118. Privett, O. S.; Nutter, L. J. *Lipids* **1967**, *2*(2), 149–154.

2119. Tadzhibaev, Yu.A.; Mukhamedova, Kh. S.; Akramov, S. T. *Khim. Prir. Soedin.* **1977**, (6), 780–783 (Engl. 657–660).

2120. Williams, J. P.; Khan, M. U. *Plant Physiol. Biochem.* **1996**, *34*(1), 93–100.

2121. Yokoyama, Y.; Suzuki, B. *Proc. Imp. Acad. (Tokyo)* **1931**, *7*(1), 12–14.

2122. Catlow, D. A. *Int. J. Mass Spectrom. Ion Phys.* **1983**, *46*, 387–390.

2123. Crawford, C. G.; Plattner, R. D. *J. Lipid Res.* **1984**, *25*(5), 518–522.

2124. Wang, X.-M.; Norman, H. A.; St. John, J. B.; Yin, T.; Hildebrand, D. F. *Phytochemistry* **1989**, *28*(2), 411–414.

2125. Justin, A. M.; Demandre, C.; Mazliak, P. *Biochim. Biophys. Acta* **1987**, *922*(3), 364–371.

2126. Shustanova, L. A.; Umarov, A. U.; Markman, A. L. *Khim. Prir. Soedin.* **1971**, *7*(2), 137–141 (Engl. 129–133).

2127. Gut, H.; Matile, P. *Bot. Acta* **1989**, *102*(1), 31–36.

2128. L'Heureux, G. P.; Bergevin, M.; Thompson, J. E.; Willemot, C. *J. Plant Physiol.* **1994**, *143*(6), 699–703.

2129. Nguyen, X. V.; Côme, D.; Justin, A.-M.; Lewak, S.; Mazliak, P. *Plant Physiol. Biochem.* **1987**, *25*(4), 697–703.

2130. Willemot, C.; Verret, G. *Lipids* **1973**, *8*(10), 588–591.

2131. Kojima, M.; Hotsubo, K.; Kinoshita, M.; Ohnishi, M.; Ito, S. *Obihiro Chikusan Daigaku Gakujutsu Kenkyu Hokoku, Dai-1-Bu* **1991**, *17*(2), 135–142.

2132. Justin, A. M.; Demandre, C.; Trémolières, A.; Mazliak, P. *Biochim. Biophys. Acta* **1985**, *836*(1), 1–7.

2133. Serghini-Caid, H.; Demandre, C.; Justin, A.-M.; Mazliak, P. *Phytochemistry* **1988**, *27*(8), 2543–2548.

2134. Lehmann, W. D.; Kessler, M. *Chem. Phys. Lipids* **1983**, *32*(2), 123–135.

2135. Chen, S.; Benfenati, E.; Fanelli, R.; Kirschner, G.; Pregnolato, F. *Biomed Environ. Mass Spectrom.* **1989**, *18*(12), 1051–1056.

2136. Jensen, J. J.; Tomer, K. B.; Gross, M. L. *Lipids* **1987**, *22*(7), 480–489.

2137. Varenne, P.; Das, B. C.; Polonsky, J.; Tencé, M. *Biomed. Mass Spectrom.* **1985**, *12*(1), 6–10.

2138. Norman, H. A.; McMillan, C.; Thompson, G. A., Jr. *Plant Cell Physiol.* **1984**, *25*(8), 1437–1444.

2139. Ohashi, Y. *Biomed. Mass Spectrom.* **1984**, *11*(8), 383–385.

2140. Demirev, P. A. *Biomed. Environ. Mass Spectrom.* **1987**, *14*(5), 241–246.

2141. Siebertz, H. P.; Heinz, E.; Linscheid, M.; Joyard, J.; Douce, R. *Eur. J. Biochem.* **1979**, *101*(2), 429–438.

2142. Levene, P. A.; Rolf, I. P. *J. Biol. Chem.* **1923**, *55*(4), 743–749.

2143. Hirao, S. *Nippon Nogei Kagaku Kaishi* **1931**, *7*, 364–367.

2144. Morrison, W. R. *J. Cereal Sci.* **1988**, *8*(1), 1–15.

2145. Albon, N. *J. Cryst. Growth* **1976**, *35*(1), 105–109.

2146. Sakurai, I.; Sakurai, S.; Sakurai, T.; Seto, T.; Ikegami, A.; Iwayanagi, S. *Chem. Phys. Lipids* **1980**, *26*(1), 41–48.

2147. Michelsen, P.; Gronowitz, S.; Åkesson, B.; Herslöf, B. *Chem. Phys. Lipids* **1983**, *32*(2), 137–143.

2148. Brosio, E.; Conti, F.; Di Nola, A.; Kovacs, A. L. *J. Theor. Biol.* **1977**, *67*(2), 319–334.

2149. Ehlert, G.; Heintze, W. *Kiel. Milchwirtsch. Forschungsber.* **1969**, *21*(4), 425–446.

2150. Asher, S. A.; Stearns, R.; Urabe, T.; Pershan, P. S. *Mol. Cryst. Liq. Cryst.* **1981**, *63*(1/4), 193–204.

2151. Klein, R. A. *J. Lipid Res.* **1971**, *12*(2), 123–131.

2152. Aberth, W.; Straub, K. M.; Burlingame, A. L. *Anal. Chem.* **1982**, *54*(12), 2029–2034.

2153. Crawford, C. G.; Plattner, R. D. *J. Lipid Res.* **1983**, *24*(4), 456–460.

2154. Kuwabara, H.; Viden, I.; Tsuchiya, M.; Sugatani, J.; Saito, K. *J. Biochem.* **1986**, *100*(2), 477–484.

2155. Hayashi, A.; Matsubara, T.; Morita, M.; Kinoshita, T.; Nakamura, T. *J. Biochem.* **1989**, *106*(2), 264–269.

2156. Eckard, P. R.; Taylor, L. T.; Slack, G. C. *J. Chromatogr.* **1998**, *826A*(2), 241–247.

2157. Baer, E.; Maurukas, J. *J. Am. Chem. Soc.* **1952**, *74*(1), 158–160.

2158. Baer, E.; Kates, M. *J. Am. Chem. Soc.* **1950**, *72*(2), 942–949.

2159. Collins, F. D. *Chem. Phys. Lipids* **1967**, *1*(2), 91–99.

2160. Lewis, R. N. A. H.; Mak, N.; McElhaney, R. N. *Biochemistry* **1987**, *26*(19), 6118–6126.

2161. Baer, E. *J. Am. Chem. Soc.* **1953**, *75*(3), 621–623.

2162. Tuchtenhagen, J.; Ziegler, W.; Blume, A. *Eur. Biophys. J.* **1994**, *23*(5), 323–335.

2163. Cullis, P. R.; de Kruyff, B.; Richards, R. E. *Biochim. Biophys. Acta* **1976**, *426*(3), 433–446.

2164. Rothgeb, T. M.; Oldfield, E. *J. Biol. Chem.* **1981**, *256*(12), 6004–6009.

2165. Marinetti, G. V.; Stotz, E. *J. Am. Chem. Soc.* **1955**, *77*(24), 6668–6670.

2166. Touchstone, J. C.; Levin, S. S.; Dobbins, M. F.; Carter, P. J. *J. High Resolut. Chromatogr.* **1981**, *4*(8), 423–424.

2167. Albon, N.; Doucet, J. *Chem. Phys. Lipids* **1983**, *33*(4), 375–382.

2168. Mushayakarara, W.; Albon, N.; Levin, I. W. *Biochim. Biophys. Acta* **1982**, *686*(2), 153–159.

2169. Bruzik, K. S.; Harwood, J. S. *J. Am. Chem. Soc.* **1996**, *119*(28), 6629–6637.

2170. Hanahan, D. J.; Brockerhoff, H. *Arch. Biochem. Biophys.* **1960**, *91*(2), 326–328.

2171. Chia, N.-C.; Mendelsohn, R. *Biochim. Biophys. Acta* **1996**, *1283*(2), 141–150.

2172. Bousquet, C.; Masson, M.; Harrand, M. *J. Raman Spectrosc.* **1995**, *26*(4), 273–282, from *Chem. Abstr.* **1995**, *122*, 301668d.

2173. Santaren, J. F.; Rico, M.; Ribera, A. *Org. Magn. Reson.* **1983**, *21*(4), 238–242.

2174. Urbina, J. A.; Moreno, B.; Arnold, W.; Taron, C. H.; Orlean, P.; Oldfield, E. *Biophys. J.* **1998**, *75*(3), 1372–1383.

2175. Dessort, D.; van Dorsselaer, A.; Tian, S. J.; Vincendon, G. *Tetrahedron Lett.* **1982**, *23*(13), 1395–1398.

2176. Jensen, N. J.; Tomer, K. B.; Gross, M. L. *Lipids* **1986**, *21*(9), 580–588.

2177. Wood, G. W.; Lau, P.-Y.; Morrow, G.; Rao, G. S. N.; Schmidt, D. E., Jr.; Tuebner, J. *Chem. Phys. Lipids* **1977**, *18*(3), 316–333.

2178. Harvey, D. J. *J. Mass Spectrom.* **1995**, *30*(9), 1333–1346.

2179. Kim, Y. H.; Yoo, J. S.; Kim, M. S. *Bull. Korean Chem. Soc.* **1997**, *18*(8), 874–880 Errata *19* (4), no page number.

2180. Nilsson, R.; Liljenberg, C. *Phytochem. Anal.* **1996**, *7*(5), 228–232.

2181. Raith, K.; Wolf, R.; Wagner, J.; Neubert, R. H. H. *J. Chromatogr.* **1998**, *802A*(1), 185–188.

2182. Hermetter, A.; Stütz, H.; Franzmair, R.; Paltauf, F. *Chem. Phys. Lipids* **1989**, *50*(1), 57–62.

2183. Davis, P. J.; Fleming, B. D.; Coolbear, K. P.; Keough, K. M. W. *Biochemistry* **1981**, *20*(12), 3633–3636.

2184. Ge, M.; Budil, D. E.; Freed, J. H. *Biophys. J.* **1994**, *66*(5), 1515–1521.

2185. Ohashi, M.; Iino, H.; Tsujimoto, K.; Ohashi, Y.; Shida, Y. *Org. Mass Spectrom.* **1985**, *20*(10), 642–643.

2186. Xia, J.; Hui, Y.-Z. *Tetrahedron: Asymmetry* **1997**, *8*(3), 451–458.

2187. Kim, H.-Y.; Wang, T.-C. L.; Ma, Y.-C. *Anal. Chem.* **1994**, *66*(22), 3977–3982.

2188. de Haas, G. H.; van Deenen, L. L. M. *Rec. Trav. Chim.* **1961**, *80*(9/10), 951–970.

2189. Chapman, D.; Williams, R. M.; Ladbrooke, B. D. *Chem. Phys. Lipids* **1967**, *1*, 445–475.

2190. Phillips, M. C.; Hauser, H.; Paltauf, F. *Chem. Phys. Lipids* **1972**, *8*, 127–133.

2191. Stoffel, W.; Pruss, H.-D. *Z. Physiol. Chem.* **1969**, *350*(11), 1385–1393.

2192. Stoffel, W.; Tunggal, B. D.; Zierenberg, O.; Schreiber, E.; Binczak, E. *Z. Physiol. Chem.* **1974**, *355*(11), 1367–1380.

2193. Shvets, V. I.; Dorofeeva, L. T.; Grum-Grzhimailo, M. A.; Shmidt, I. S.; Volkova, L. V.; Preobrazhenskii, N. A. *Zh. Obshch. Khim.* **1964**, *34*(12), 3983–3986 (Engl. 4042–4045).

2194. Coolbear, K. P.; Berde, C. C. B.; Keough, K. M. W. *Biochemistry* **1983**, *22*(6), 1466–1473.

2195. Santarén, J. F.; Rico, M.; Guilleme, J.; Ribera, A. *Org. Magn. Reson.* **1982**, *18*(2), 98–103.

2196. Molotkovskii, Yu.G.; Lazurkina, T. Yu.; Bergel'son, L. D. *Izv. Akad. Nauk SSSR, Ser. Khim.* **1969**, (8), 1784–1789 (Engl. 1650–1654).

2197. Kögl, F.; de Haas, G. H.; van Deenen, L. L. M. *Rec. Trav. Chim.* **1966**, *79*(7), 551–574.

2198. Robles, E. C.; van den Berg, D. *Biochim. Biophys. Acta* **1969**, *187*(4), 520–526.

2199. Koyama, Y.; Ikeda, K.-I. *Chem. Phys. Lipids* **1980**, *26*(2), 149–172.

2200. Münster, H.; Stein, J.; Budzikiewicz, H. *Biomed. Environ. Mass Spectrom.* **1986**, *13*(8), 423–427.

2201. Lammers, J. G.; Liefkens, T. J.; Bus, J.; van der Meer, J. *Chem. Phys. Lipids* **1978**, *22*(4), 293–305.

2202. Baer, E.; Buchnea, D.; Newcombe, A. G. *J. Am. Chem. Soc.* **1956**, *78*(1), 232–237.

2203. Hauser, H.; Guyer, W.; Paltauf, F. *Chem. Phys. Lipids* **1981**, *29*(2), 103–120.

2204. Eriksson, P.-O.; Lindblom, G.; Arvidson, G. *J. Phys. Chem.* **1987**, *91*(4), 846–853.

2205. Okazaki, M.; Hara, I.; Tsutsui, Y. K.; Fujiyama, T. *Bull. Chem. Soc. Jpn.* **1981**, *54*(8), 2399–2407.

2206. Shvets, V. I.; Antal, L.; Volkova, L. V.; Preobrazhenskii, N. A. *Zh. Obshch. Khim.* **1964**, *34*(6), 1908–1911 (Engl. 1921–1923).

2207. Keough, K. M. W.; Kariel, N. *Biochim. Biophys. Acta* **1987**, *902*(1), 11–18.

2208. Baba, N.; Akiyama, T.; Tahara, S.; Nakajima, S. *Biosci. Biotchnol. Biochem.* **1995**, *59*(2), 353–354.

2209. Warner, T. G.; Benson, A. A. *J. Lipid Res.* **1977**, *18*(4), 548–552.

2210. Bruzik, K. S.; Salamonczyk, G.; Stec, W. J. *J. Org. Chem.* **1986**, *51*(12), 2368–2370.

2211. Lindh, I.; Stawinski, J. *J. Org. Chem.* **1989**, *54*(6), 1338–1342.

2212. Tattrie, N. H.; McArthur, C. S. *Can. J. Biochem. Physiol.* **1957**, *35*(12), 1165–1170.

2213. Hermetter, A.; Paltauf, F. *Chem. Phys. Lipids* **1981**, *28*(1), 111–115.

2214. Hirt, R.; Berchtold, R. *Pharm. Acta Helv.* **1958**, *33*, 349–356, from *Chem. Abstr.* **1959**, *52*, 9289g.

2215. Phuong, N. H.; Thuong, N. T.; Chabrier, P. *Compt. Rend.* **1976**, *283C*(5), 229–231.

2216. Nicholas, A. W.; Khouri, L. G.; Ellington, J. C., Jr.; Porter, N. A. *Lipids* **1983**, *18*(6), 434–438.

2217. Woolley, P.; Eibl, H. *Chem. Phys. Lipids* **1988**, *47*(1), 55–62.

2218. Ghosh, D.; Lyman, R. L.; Tinoco, J. *Chem. Phys. Lipids* **1971**, *7*(3), 173–184.

2219. Baer, E.; Buchnea, D. *Can. J. Biochem. Physiol.* **1959**, *37*(8), 953–959.

2220. Robles, E. C.; de Iongh, H. *Rec. Trav. Chim.* **1967**, *86*(8), 762–764.

2221. Patel, K. M.; Morrisett, J. D.; Sparrow, J. T. *J. Lipid Res.* **1979**, *20*(5), 674–677.

2222. Baer, E.; Kindler, A. *Biochemistry* **1962**, *1*(3), 518–521.

2223. Roodsari, F. S.; Wu, D.; Pum, G. S.; Hajdu, H. *J. Org. Chem.* **1999**, *64*(21), 7727–7737.

2224. Shvets, V. I.; Klopova, V. A.; Preobrazhenskii, N. A. *Khim. Prir. Soedin., Akad. Nauk Uz. SSR* **1966**, *2*(4), 225–229, from *Chem. Abstr.* **1967**, *67*, 29325d.

2225. Gordon, D. T.; Jensen, R. G. *Lipids* **1972**, *7*(4), 261–262.

2226. Ghosh, D.; Williams, M. A.; Tinoco, J. *Biochim. Biophys. Acta* **1973**, *291*(2), 351–362.

2227. Decker, C. J.; van Kessel, W. S. M. G.; Klomp, J. P. G.; Pieters, J.; de Kruijff, B. *Chem. Phys. Lipids* **1983**, *33*(1), 93–106.

2228. Zuidam, N. J.; Crommelin, D. J. A. *J. Pharm. Sci.* **1995**, *84*(9), 1113–1119.

2229. Van Deenen, L. L. M.; de Haas, G. H. *Biochim. Biophys. Acta* **1963**, *70*(5), 538–553.

2230. de Haas, G. H.; van Deenen, L. L. M. *Tetrahedron Lett.* **1960**, (22), 7–11.

2231. Hanahan, D. J.; Rodbell, M.; Turner, L. D. *J. Biol. Chem.* **1954**, *206*(1), 431–441.

2232. Slotboom, A. J.; de Haas, G. H.; Bonsen, P. P. M.; Burbach-Westerhuis, G. J.; van Deenen, L. L. M. *Chem. Phys. Lipids* **1970**, *4*(1), 15–29.

2233. Slotboom, A. J.; de Haas, G. H.; Burbach-Westerhuis, G. J.; van Deenen, L. L. M. *Chem. Phys. Lipids* **1970**, *4*(1), 30–36.

2234. Alvarez, J. G.; Lopez, I.; Storey, B. T.; Levin, S. S.; Touchstone, J. C. *J. Liq. Chromatogr.* **1986**, *9*(16), 3495–3504.

2235. Stafford, R. E.; Fanni, T.; Dennis, E. A. *Biochemistry* **1989**, *28*(12), 5113–5120.

2236. Bunow, M. R.; Levin, I. W. *Biochim. Biophys. Acta* **1977**, *487*(2), 388–394.

2237. Plückthun, A.; Dennis, E. A. *Biochemistry* **1982**, *21*(8), 1743–1750.

2238. Tokumura, A.; Handa, Y.; Yoshioka, Y.; Tsukatani, H. *Chem. Pharm. Bull.* **1983**, *31*(12), 4425–4435.

2239. Tinsley, P. W.; Maerker, G. *J. Am. Oil Chem. Soc.* **1993**, *70*(2), 187–191.

2240. Eibl, H.; Woolley, P. *Chem. Phys. Lipids* **1988**, *47*(1), 63–68.

2241. Kabashima, I. *Berichte* **1938**, *71B*(5), 1073–1076.

2242. Bush, S. F.; Levin, H.; Levin, I. W. *Chem. Phys. Lipids* **1980**, *27*(2), 101–111.

2243. Han, X.; Gross, R. W. *J. Am. Chem. Soc.* **1996**, *118*(2), 451–457.

2244. Chakrabarti, P.; Khorana, H. G. *Biochemistry* **1975**, *14*(23), 5021–5033.

2245. Plattner, R. D.; Stack, R. J.; Al-Hassan, J. M.; Summers, B.; Criddle, R. S. *Org. Mass Spectrom.* **1988**, *23*(12), 834–840.

2246. Hanahan, D. J. *J. Biol. Chem.* **1954**, *211*(1), 321–325.

2247. Kumar, V. V.; Baumann, W. J. *Biophys. J.* **1991**, *59*(1), 103–107.

2248. Adkins, J. A.; Risby, T. H.; Scocca, J. J.; Yasbin, R. E.; Ezzell, J. W. *J. Anal. Appl. Pyrolysis* **1984**, *7*(1), 35–51.

2249. Fenwick, G. R.; Eagles, J.; Self, R. *Biomed. Mass Spectrom.* **1983**, *10*(6), 382–386.

2250. Bird, P. R.; de Haas, G. H.; Heemskerk, C. H. T.; van Deenen, L. L. M. *Biochim. Biophys. Acta* **1965**, *98*(3), 566–573.

2251. Rouet-Mayer, M.-A.; Valentova, O.; Simond-Côte, E.; Daussant, J.; Thévenot, C. *Lipids* **1995**, *30*(8), 739–746.

2252. Morimoto, T.; Murakami, N.; Nagatsu, A.; Sakakibara, J. *Tetrahedron Lett.* **1993**, *34*(15), 2487–2490.

2253. Smith, N. B.; Kuksis, A. *Can. J. Biochem.* **1978**, *56*(12), 1149–1153.

2254. Sugatani, J.; Kawasaki, N.; Saito, K. *Biochim. Biophys. Acta* **1978**, *529*(1), 29–37.

2255. Baba, N.; Yoneda, K.; Iwasa, J.; Tahara, S.; Kaneko, T.; Matsuo, M. *Indian J. Chem.* **1992**, *31B*(12), 824–827.

2256. Saito, K.; Kates, M. *Biochim. Biophys. Acta* **1974**, *369*(2), 245–253.

2257. Ohno, Y.; Okazaki, M.; Hara, I. *Biochem. Cell Biol.* **1990**, *68*(1), 376–379.

2258. Baer, E.; Maurukas, J.; Russell, M. *J. Am. Chem. Soc.* **1952**, *74*(1), 152–157.

2259. Tenchov, B. G.; Boyanov, A. I.; Koynova, R. D. *Biochemistry* **1984**, *23*(15), 3553–3558.

2260. Akutsu, H.; Kyogoku, Y.; Nakahara, H.; Fukuda, K. *Chem. Phys. Lipids* **1975**, *15*(2), 222–242.

2261. Davies, M. A.; Hubner, W.; Blume, A.; Mendelsohn, R. *Biophys. J.* **1992**, *63*(4), 1059–1062.

2262. Bicknell-Brown, E.; Brown, K. G.; Person, W. B. *J. Raman Spectrosc.* **1981**, *11*(5), 356–362.

2263. Evans, N.; Games, D. E.; Harwood, J. L.; Jackson, A. H. *Biochem. Trans.* **1974**, *2*(5), 1091–1093.

2264. Ayanoglu, E.; Wegmann, A.; Pilet, O.; Marbury, G. D.; Hass, J. R.; Djerassi, C. *J. Am. Chem. Soc.* **1984**, *106*(18), 5246–5251.

2265. Chen, S.; Li, K. W. *J. Biochem.* **1994**, *116*(4), 811–817.

2266. Kaplun, A. P.; Kabanova, M. A.; Lyutik, A. I.; Shvets, V. I.; Evstigneeva, R. P. *Zh. Obshch. Khim.* **1973**, *43*(8), 1839–1844 (Engl. 1823–1827).

2267. Nyström, E.; Sjövall, J. *Anal. Biochem.* **1965**, *12*(2), 235–248.

2268. Billimoria, J. D.; Owen, J. S.; Scott, G. H. *Chem. Phys. Lipids* **1974**, *12*(4), 327–343.

2269. Daemen, F. J. M. *Chem. Phys. Lipids* **1967**, *1*(6), 476–481.

2270. Senak, L.; Davies, M. A.; Mendelsohn, R. *J. Phys. Chem.* **1991**, *95*(6), 2565–2571.

2271. Hong, M.; Schmidt-Rohr, K.; Nanz, D. *Biophys. J.* **1995**, *69*(5), 1939–1950.

2272. Münster, H.; Budzikiewicz, H. *Biol. Chem. Hoppe-Seyler* **1988**, *369*(4), 303–308.

2273. Kim, H.-Y.; Salem, N., Jr. *Anal. Chem.* **1986**, *58*(1), 9–14.

2274. Martin, S. F.; Josey, J. A.; Wong, Y.-L.; Dean, D. W. *J. Org. Chem.* **1994**, *59*(17), 4805–4820.

2275. Daemen, F. J. M.; de Haas, G. H.; van Deenen, L. L. M. *Rec. Trav. Chim.* **1963**, *82*(5), 487–496.

2276. de Haas, G. H.; Daemen, F. J. M.; van Deenen, L. L. M. *Biochim. Biophys. Acta* **1962**, *65*(2), 260–270.

2277. Baer, E.; Buchnea, D. *Can. J. Biochem. Physiol.* **1961**, *39*(9), 1471–1478.

2278. Hermetter, A.; Stütz, H.; Lohner, K.; Paltauf, F. *Chem. Phys. Lipids* **1987**, *43*(1), 69–77.

2279. Chapman, D.; Byrne, P.; Shipley, G. G. *Proc. R. Soc. Lond.* **1966**, *290A*(1420), 115–142.

2280. Chapman, D.; Salsbury, N. J. *Trans. Far. Soc.* **1966**, *62*(9), 2607–2621.

2281. Li, S.; Wang, G.; Lin, H.; Huang, C. H. *J. Biol. Chem.* **1998**, *273*(30), 19009–19018.

2282. Dorofeeva, L. T.; Zharova, T. V.; Volkova, L. V.; Tolkachev, O. N.; Preobrazhen-skii, N. A. *Zh. Obshch. Khim.* **1964**, *34*(9), 2935–2939 (Engl. 2970–2973).

2283. Baer, E.; Buchnea, D. *J. Am. Chem. Soc.* **1959**, *81*(7), 1758–1762.

2284. Lewis, R. N. A. H.; Mannock, D. A.; McElhaney, R. N.; Turner, D. C.; Gruner, S. M. *Biochemistry* **1989**, *288*(2), 541–548.

2285. Lammers, J. G.; van Boom, J. H. *Rec. Trav. Chim.* **1979**, *98*(4), 243–250.

2286. Cullis, P. R.; de Kruyff, B. *Biochim. Biophys. Acta* **1976**, *436*(3), 523–540.

2287. Volkova, L. V.; Morozova, S. F.; Preobrazhenskii, N. A. *Zh. Obshch. Khim.* **1965**, *35*(1), 84–85 (Engl. 82–83).

2288. Shvets, V. I.; Volkova, L. V.; Ryzhenkova, S. F.; Lukashenko, E. E.; Preobrazhen-skii, N. A. *Zh. Obshch. Khim.* **1963**, *33*(9), 2876–2879 (Engl. 2803–2806).

2289. Bova, L. M.; Shvets, V. I.; Shutova, F. Ya.; Preobrazhenskii, N. A. *Zh. Org. Khim.* **1965**, *1*(10), 1720–1723.

2290. Dorofeeva, L. T.; Tolkachev, O. N.; Preobrazhenskii, N. A. *Zh. Obshch. Khim.* **1963**, *33*(9), 2880–2883 (Engl. 2807–2809).

2291. Pfeiffer, F. R.; Cohen, S. R.; Weisbach, J. A. *J. Org. Chem.* **1969**, *34*(9), 2795–2796.

2292. Rakhit, S.; Bagli, J. F.; Deghenghi, R. *Can. J. Chem.* **1969**, *47*(15), 2906–2910.

2293. Baer, E.; Suzuki, Y.; Blackwell, J. *Biochemistry* **1963**, *2*(6), 1227–1232.

2294. Sukhanov, V. A.; Sergovskaya, N. L.; Shvets, V. I.; Evstigneeva, R. P. *Zh. Obshch. Khim.* **1977**, *47*(9), 2130–2136 (Engl. 1943–1947).

2295. Tomoi, M.; Kimura, Y.; Tokuyama, S. *Synth. Commun.* **1990**, *20*(9), 1363–1371.

2296. Chadha, J. S. *Chem. Phys. Lipids* **1968**, *2*(6), 415–418.

2297. Billimoria, J. D.; Lewis, K. O. *J. Chem. Soc. C* **1968**, (12), 1404–1412.

2298. Molotkovsky, J. G.; Nikulina, L. F.; Bergelson, L. D. *Chem. Phys. Lipids* **1976**, *17*(2), 108–110.

2299. Eibl, H.; Nicksch, A. *Chem. Phys. Lipids* **1978**, *22*(1), 1–8.

2300. Hayashi, K.; Nakamura, K.; Hashimoto, M.; Shimizu, M.; Hatano, N.; Takeyama, K. Jpn. Patent 74/126624 to Dainippon Pharmaceutical Co., Ltd., December 4, 1974; from *Chem. Abstr.* **1975**, *82*, 170036b.

2301. Baer, E.; Blackwell, J. *Biochemistry* **1964**, *3*(7), 975–976.

2302. Aneja, R.; Chadha, J. S.; Robles, E. C.; Van Daal, R. *Biochim. Biophys. Acta* **1969**, *187*(3), 439–441.

2303. Shvets, V. I.; Klopova, V. A.; Preobrazhenskii, N. A. *Khim. Prir. Soedin., Akad. Nauk Uz. SSR* **1966**, *2*(2), 80–83, from *Chem. Abstr.* **1966**, *65*, 13534e.

2304. Slater, J. L.; Huang, C.; Adams, R. G.; Levin, I. W. *Biophys. J.* **1989**, *56*(2), 243–252.

2305. Tomoi, M.; Kimura, Y.; Tokuyama, S. *React. Polym.* **1991**, *15*, 63–69.

2306. Pascher, I.; Sundell, S.; Hauser, H. *J. Mol. Biol.* **1981**, *153*(3), 807–824.

2307. Daemen, F. J. M.; de Haas, G. H.; van Deenen, L. L. M. *Rec. Trav. Chim.* **1962**, *81*(4), 348–358.

2308. Tilcock, C. P. S.; Cullis, P. R.; Hope, M. J.; Gruner, S. M. *Biochemistry* **1986**, *25*(4), 816–822.

2309. Åkesson, B.; Arner, A.; Sundler, R. *Biochim. Biophys. Acta* **1976**, *441*(3), 453–464.

2310. Chapman, K. D.; Lin, I.; DeSouza, A. D. *Arch. Biochem. Biophys.* **1995**, *318*(2), 401–407.

2311. Sandoval, J. A.; Huang, Z.-H.; Garrett, D. C.; Gage, D. A.; Chapman, K. D. *Plant Physiol.* **1995**, *109*(1), 269–275.

2312. Miyazawa, T.; Fujino, Y. *J. Food Sci.* **1978**, *43*(1), 98–101.

2313. Shvets, V. I.; Lyutik, A. I.; Kabanova, M. A.; Alekseeva, L. D.; Tkachenko, A. S.; Evstigneeva, R. P. *Bioorg. Khim.* **1975**, *1*(6), 758–764.

2314. Domingo, J. C.; Mora, M.; de Madariaga, M. A. *Chem. Phys. Lipids* **1995**, *75*(1), 15–25.

2315. Lafrance, D.; Marion, D.; Pézolet, M. *Biochemistry* **1990**, *29*(19), 4592–4599.

2316. Domingo, J. C.; de Madariaga, M. A. *Chem. Phys. Lipids* **1996**, *84*(2), 147–153.

2317. Lafrance, C. P.; Blochet, J. E.; Pezolet, M. *Biophys. J.* **1997**, *72*(6), 2559–2568.

2318. Wren, J. J.; Merryfield, D. S. *J. Chem. Soc.* **1964**, (Suppl. 2), 6251–6252.

2319. Wu, J.; Browse, J. *Plant Cell* **1995**, *7*(1), 17–27.

2320. Li, T.; Lynch, D. V.; Steponkus, P. L. *Cryo-Lett.* **1987**, *8*(6), 314–321.

2321. Wohlgemuth, R.; Waespe-Šarčevic, N.; Seelig, J. *Biochemistry* **1980**, *19*(14), 3315–3321.

2322. Watts, A.; Harlos, K.; Maschke, W.; Marsh, D. *Biochim. Biophys. Acta* **1978**, *510*(1), 63–74.

2323. Zhang, Y.-P.; Lewis, R. N. A. H.; McElhaney, R. N. *Biophys. J.* **1997**, *72*(2), 779–793.

2324. Haddadian, F.; Shamsi, S. A.; Schaeper, J. P.; Danielson, N. D. *J. Chromatogr. Sci.* **1998**, *36*(8), 395–400.

2325. Abe, M.; Hasegawa, M.; Ogino, K.; Nakano, Y.; Tsubaki, N. *Yukagaku* **1995**, *44*(2), 134–139.

2326. Stepanov, A. E.; Makarova, I. M.; Shvets, V. I. *Zh. Org. Khim.* **1984**, *20*(5), 985–988 (Engl. 896–899).

2327. Bonsen, P. P. M.; de Haas, G. H.; van Deenen, L. L. M. *Chem. Phys. Lipids* **1966**, *1*(1), 33–40.

2328. Smaal, E. B.; Nicolay, K.; Mandersloot, J. G.; de Gier, J.; de Kruijff, B. *Biochim. Biophys. Acta* **1987**, *897*(2), 453–466.

2329. Wang, P.; Schuster, M.; Wang, Y.-F.; Wong, C.-H. *J. Am. Chem. Soc.* **1993**, *115*(23), 10487–10491.

2330. Wittels, B. *J. Biol. Chem.* **1973**, *248*(8), 2906–2911.

2331. Redelmeier, T. E.; Hope, M. J.; Cullis, P. R. *Biochemistry* **1990**, *29*(12), 3046–3053.

2332. Ramirez, F.; Ioannou, P. V.; Marecek, J. F.; Dodd, G. H.; Golding, B. T. *Tetrahedron* **1977**, *33*(6), 599–608.

2333. Rainier, S.; Jain, M. K.; Ramirez, F.; Ioannou, P. V.; Marecek, J. F.; Wagner, R. *Biochim. Biophys. Acta* **1979**, *558*(2), 187–198.

2334. de Haas, G. H.; Bonsen, P. P. M.; van Deenen, L. L. M. *Biochim. Biophys. Acta* **1966**, *116*, 114–124.

2335. Sankaram, M. B.; Powell, G. L.; Marsh, D. *Biochim. Biophys. Acta* **1989**, *980*(3), 389–392.

2336. de Haas, G. H.; van Deenen, L. L. M. *Rec. Trav. Chim.* **1965**, *84*(3), 436–438.

2337. Spooner, P. J. R.; Duralski, A. A.; Rankin, S. E.; Pinheiro, T. J. T.; Watts, A. *Biophys. J.* **1993**, *65*(1), 106–112.

2338. Keana, J. F. W.; Shimizu, M.; Jernstedt, K. K. *J. Org. Chem.* **1986**, *51*(12), 2297–2299.

2339. Inoue, K.; Nojima, S. *Chem. Pharm. Bull.* **1968**, *16*(1), 76–81.

2340. Mishina, I. M.; Vasilenko, I. A.; Stepanov, A. E.; Shvets, V. I. *Bioorg. Khim.* **1987**, *13*(8), 1110–1115, from *Chem. Abstr.* **1988**, *109*, 55123s.

2341. Salamonczyk, G. M.; Bruzik, K. S. *Tetrahedron Lett.* **1990**, *31*(14), 2015–2016.

2342. Kume, T.; Taguchi, R.; Tomita, M.; Tokuyama, S.; Morizawa, K.; Nakachi, O.; Hirano, J.; Ikezawa, H. *Chem. Pharm. Bull.* **1992**, *40*(8), 2133–2137.

2343. Jones, M.; Rana, K. K.; Ward, J. G.; Young, R. C. *Tetrahedron Lett.* **1989**, *30*(39), 5353–5356.

2344. Ward, J. G.; Young, R. C. *Tetrahedron Lett.* **1988**, *29*(46), 6013–6016.

2345. Klyashchitskii, B. A.; Zhelvakova, E. G.; Pimenova, V. V.; Shvets, V. I.; Evstigneeva, R. P.; Preobrazhenskii, N. A. *Zh. Obshch. Khim.* **1971**, *41*(6), 1386–1393 (Engl. 1391–1397).

2346. Molotkovsky, J. G.; Bergelson, L. D. *Chem. Phys. Lipids* **1973**, *11*(2), 135–147.

2347. Lyutik, A. I.; Sukhanov, V. A.; Shvets, V. I.; Evstigneeva, R. P. *Bioorg. Khim.* **1975**, *1*(5), 684–687.

2348. Nakagawa, Y. *Lipids* **1993**, *28*(11), 1033–1035.

2349. Lin, G.; Liu, S.-H.; Wu, F.-C. *J. Chin. Chem. Soc. (Taipei)* **1993**, *40*(1), 81–86.

2350. Murakami, S.; Tokoyama, S.; Ozawa, K.; Kyota, R.; Nakachi, O. Jpn. Patent 63,175,993 to Nippon Oils and Fats Co., Ltd., December 23, 1988; from *Chem. Abstr.***1989**, *111*, 134685w.

2351. Filthuth, E.; Eibl, H. *Chem. Phys. Lipids* **1992**, *60*(3), 253–261.

2352. van Breemen, R. B.; Wheeler, J. J.; Boss, W. F. *Lipids* **1990**, *25*(6), 328–334.

2353. Wheeler, J. J.; Boss, W. F. *Biochim. Biophys. Acta* **1989**, *984*(1), 33–40.

2354. VanMiddlesworth, F.; Lopez, M.; Zweerink, M.; Edison, A. M.; Wilson, K. *J. Org. Chem.* **1992**, *57*, 4753–4754.

2355. Ettlinger, C.; Lehle, L. *Nature* **1988**, *331*(6152), 176–178.

2356. Heim, S.; Wagner, K. G. *Biochem. Biophys. Res. Commun.* **1986**, *134*(3), 1175–1181.

2357. Wheeler, J. J.; Boss, W. F. *Plant Physiol.* **1987**, *85*(2), 389–392.

2358. Sandelius, A. S.; Morré, D. J. *Plant Physiol.* **1987**, *84*(4), 1022–1027.

2359. Memon, A. R.; Boss, W. F. *J. Biol. Chem.* **1990**, *265*(25), 14817–14821.

2360. Helsper, J. P. F. G.; de Groot, P. F. M.; Linskens, H. F.; Jackson, J. F. *Phytochemistry* **1986**, *25*(9), 2193–2199.

2361. Wagh, S. S.; Menon, K. K. G.; Natarajan, V. *Biochim. Biophys. Acta* **1988**, *962*(2), 178–185.

2362. Drøbak, B. K.; Ferguson, I. B.; Dawson, A. P.; Irvine, R. F. *Plant Physiol.* **1988**, *87*(1), 217–222.

2363. Helsper, J. P. F. G.; Linskens, H. F.; Jackson, J. F. *Phytochemistry* **1984**, *23*(9), 1841–1845.

2364. Piacentini, M. P.; Ricci, D.; Fraternale, D.; Piatti, E.; Manunta, A.; Accorsi, A. *Comp. Biochem. Physiol.* **1999**, *122B*(3), 293–299.

2365. Chicha, A.; Demandre, C.; Justin, A.-M.; Mazliak, P. *Compt. Rend. III* **1994**, *317*(5), 419–423.

2366. Hegewald, H. *Anal. Biochem.* **1996**, *242*(1), 152–155.

2367. Gumber, S. C.; Lowenstein, J. M. *Biochem. J.* **1986**, *235*(2), 617–619.

2368. Painter, G. F.; Grove, S. J. A.; Gilbert, I. H.; Holmes, A. B.; Raithby, P. R.; Hill, M. L.; Hawkins, P. T.; Stephens, L. R. *J. Chem. Soc., Perkin Trans. 1* **1999**, (8), 923–935.

2369. Bruzik, K. S.; Kubiak, R. J. *Tetrahedron Lett.* **1995**, *36*(14), 2415–2418.

2370. Wang, D. S.; Chen, C. S. *J. Org. Chem.* **1996**, *61*(17), 5905–5910.

2371. Sadovnikova, M. S.; Shvets, V. I.; Evstigneeva, R. P. *Bioorg. Khim.* **1975**, *1*(1), 17–24 (Engl. 9–14).

2372. Thum, O.; Chen, J.; Prestwich, G. D. *Tetrahedron Lett.* **1996**, *37*(50), 9017–9020.

2373. Gu, Q. M.; Prestwich, G. D. *J. Org. Chem.* **1996**, *61*(24), 8642–8647.

2374. Dreef, C. E.; Elie, C. J. J.; Hoogerhout, P.; van der Marel, G. A.; van Boom, H. H. *Tetrahedron Lett.* **1988**, *29*(49), 6513–6516.

2375. Chen, J.; Feng, L.; Prestwich, G. D. *J. Org. Chem.* **1998**, *63*(19), 6511–6522.

2376. Reddy, K. K.; Riz, J.; Falck, J. R. *Tetrahedron Lett.* **1997**, *38*(27), 4729–4730.

2377. Kabanova, M. A.; Suleimanyan, N. E.; Loginova, N. F.; Mairanovskii, V. G.; Shvets, V. I.; Evstigneeva, R. P. *Zh. Obshch. Khim.* **1973**, *43*(1), 193–199 (Engl. 189–194).

2378. Luna, E. J.; McConnell, H. M. *Biochim. Biophys. Acta* **1977**, *470*(2), 303–316.

2379. Browning, J.; Seelig, J. *Chem. Phys. Lipids* **1979**, *24*(2), 103–118.

2380. MacDonald, R. C.; Simon, S. A.; Baer, E. *Biochemistry* **1976**, *15*(4), 885–891.

2381. Gomez-Fernandez, J. C.; Villalain, J. *Chem. Phys. Lipids* **1998**, *96*(1/2), 41–52.

2382. Hübner, W.; Mantsch, H. H.; Paltauf, F.; Hauser, H. *Biochemistry* **1994**, *33*(1), 320–326.

2383. Chen, S.; Kirschner, G.; Traldi, P. *Anal. Biochem.* **1990**, *191*(1), 100–105.

2384. Casal, H. L.; Mantsch, H. H.; Paltauf, F.; Hauser, H. *Biochim. Biophys. Acta* **1987**, *919*(3), 275–286.

2385. Chen, N. R.; Ho, J. T. *Biochem. Biophys. Res. Commun.* **1985**, *127*(1), 220–225.

2386. Chen, S.; Menon, G.; Traldi, P. *Org. Mass Spectrom.* **1992**, *27*(3), 215–218.

2387. Shvets, V. I.; Petrova, M. K.; Kazenova, G. A.; Preobrazhenskii, N. A. *Zh. Obshch. Khim.* **1967**, *37*(7), 1454–1458 (Engl. 1378–1381).

2388. Tilcock, C. P. S.; Bally, M. P.; Farren, S. B.; Cullis, P. R.; Gruner, S. M. *Biochemistry* **1984**, *23*(12), 2696–2703.

2389. Turner, D. L.; Silver, M. H.; Baczynski, E. *Lipids* **1966**, *1*(6), 439–443.

2390. Shvets, V. I.; Volkova, L. V.; Miroshnikov, A. I.; Morozova, S. F.; Grineva, V. G.; Polyanskaya, V. A.; Preobrazhenskii, N. A. *Zh. Obshch. Khim.* **1966**, *36*(1), 49–54 (Engl. 52–56).

2391. Rezanka, T.; Podojil, M. *J. Chromatogr.* **1989**, *463*(2), 397–408.

2392. van Dieijen, G.; Tans, G.; van Rijn, J.; Zwaal, R. F. A.; Rosing, J. *Biochemistry* **1981**, *20*(25), 7096–7101.

2393. Benfenati, E.; de Bellis, G.; Chen, S.; Bettazzoli, L.; Fanelli, R.; Tacconi, M. T.; Kirschner, G.; Pege, G. *J. Lipid Res.* **1989**, *30*(12), 1983–1986.

2394. Lee, M.-H.; Bell, R. M. *J. Biol. Chem.* **1989**, *264*(25), 14797–14805.

2395. Roux, M.; Huynh-Dinh, T.; Igolen, J.; Prigent, Y. *Chem. Phys. Lipids* **1983**, *33*(1), 41–45.

2396. Ramirez, F.; Marecek, J. F. *Synthesis* **1984**, (11), 917–919.

2397. Baer, E. *J. Biol. Chem.* **1951**, *189*(1), 235–247.

2398. Jacobson, K.; Papahadjopoulos, D. *Biochemistry* **1975**, *14*(1), 152–161.

2399. Kaplun, A. P.; Kabanova, M. A.; Shvets, V. I.; Evstigneeva, R. P. *Zh. Obshch. Khim.* **1973**, *43*(7), 1617–1619 (Engl. 1598–1600).

2400. Cameron, D. G.; Gudgin, E. F.; Mantsch, J. H. *Biochemistry* **1981**, *20*(15), 4496–4500.

2401. Aurengo, A.; Masson, M.; Dupeyrat, R.; Levy, Y.; Hasmonay, H.; Barbillat, J. *Biochem. Biophys. Res. Commun.* **1979**, *89*(2), 559–564.

2402. Batrakov, S. G.; Panosyan, A. G.; Kogan, G. A.; Bergel'son, L. D. *Bioorg. Khim.* **1975**, *1*(11), 1593–1599.

2403. Wood, G. W.; Perkins, S. E. *Anal. Biochem.* **1982**, *122*(2), 368–371.

2404. Lapidot, Y.; Barzilay, I.; Hajdu, J. *Chem. Phys. Lipids* **1969**, *3*(2), 125–134.

2405. Bevan, T. H.; Brown, D. A.; Malkin, T. *J. Chem. Soc.* **1962**, (9), 3495–3498.

2406. Eibl, H.; Blume, A. *Biochim. Biophys. Acta* **1979**, *553*(3), 476–488.

2407. Sadovnikova, M. S.; Golovina, L. E.; Shvets, V. I.; Evstigneeva, R. P. *Zh. Obshch. Khim.* **1976**, *46*(6), 1389–1394 (Engl. 1365–1370).

2408. Demel, R. A.; Yin, C. C.; Lin, B. Z.; Hauser, H. *Chem. Phys. Lipids* **1992**, *60*(2), 209–223.

2409. Baer, E.; Buchnea, D. *Arch. Biochem. Biophys.* **1958**, *78*(2), 294–305.

2410. Hong, C. I.; An, S.-H.; Schliselfeld, L.; Buchheit, D. J.; Nechaev, A.; Kirisits, A. J.; West, C. R. *J. Med. Chem.* **1988**, *31*(9), 1793–1798.

2411. Stanacev, N. Z.; Kates, M. *Can. J. Biochem. Physiol.* **1960**, *38*, 297–300.

2412. Predvoditelev, D. A.; Fursenko, I. V. *Bioorg. Khim.* **1981**, *7*(9), 1426–1428.

2413. Gupta, C. M.; Radhakrishnan, R.; Khorana, H. G. *Proc. Natl. Acad. Sci. U. S. A.* **1977**, *74*, 4315–4319.

2414. Borle, F.; Seelig, J. *Biochemistry* **1983**, *22*(24), 5536–5544.

2415. Longmuir, K. J.; Martin, O. C.; Pagano, R. E. *Chem. Phys. Lipids* **1985**, *36*(3), 197–207.

2416. Shvets, V. I.; Dorofeeva, L. T.; Klyatshchitskii, B. A.; Preobrazhenskii, N. A. *Sint. Prir. Soedin. Ikh Analogov Fragm.* **1965**, 23–27, from *Chem. Abstr.* **1966**, *65*, 20000a.

2417. Imagawa, W.; Bandyopadhyay, G. K.; Wallace, D.; Nandi, S. *Proc. Natl. Acad. Sci. U. S. A.* **1989**, *86*(11), 4122–4126.

2418. Uhlenbroek, J. H.; Verkade, P. E. *Rec. Trav. Chim.* **1953**, *72*(6), 558–562.

2419. Stoffel, W.; Wolf, G. D. *Z. Physiol. Chem.* **1966**, *347*(1/3), 94–101.

2420. Tokumura, A.; Handa, Y.; Yoshioka, Y.; Higashimoto, M.; Tsukatani, H. *Chem. Pharm. Bull.* **1982**, *30*(6), 3119–3126.

2421. Nachippan, V.; Rajasekharan, R. *Anal. Biochem.* **1994**, *222*(1), 283–285.

2422. Kennerly, D. A. *J. Chromatogr.* **1987**, *409*, 291–297.

2423. Slotboom, A. J.; de Haas, G. H.; van Deenen, L. L. M. *Chem. Phys. Lipids* **1967**, *1*(4), 317–336.

2424. van der Bend, R. L.; de Widt, J.; van Corven, E. J.; Moolenaar, W. H.; van Blitter-swijk, W. J. *Biochim. Biophys. Acta* **1992**, *1125*(1), 110–112.

2425. Pieringer, R. A.; Hokin, L. E. *J. Biol. Chem.* **1962**, *237*(3), 659–663.

2426. Feingold, D. S. *Encycl. Plant Physiol.* **1982**, *13A*, 3–76.

2427. Pradet, A.; Raymond, P. *Annu. Rev. Plant Physiol.* **1983**, *34*, 199–204.

2428. Wasternack, C. *Encycl. Plant Physiol.* **1982**, *14B*, 263–301.

2429. Brown, E. G. *Methods Plant Biochem.* **1981**, *5*, 53–90.

2430. Townsend, L. B., Ed.; *Chemistry aof Nucleosides and Nucleotides*, Vol. 1; Plenum Press: New York, 1988; Vol. 1, x + 394 pp, Vol. 2, ix + 468 pp (1991).

2431. Albaum, H. G.; Ogur, M.; Hirshfeld, A. *Arch. Biochem.* **1950**, *27*(1), 130–142.

2432. Bergkvist, R. *Acta Chem. Scand.* **1956**, *10*(8), 1303–1316.

2433. Ginsburg, V.; Stumpf, P. K.; Hassid, W. Z. *J. Biol. Chem.* **1956**, *223*(2), 977–983.

2434. Nygaard, P. *Physiol. Plant.* **1972**, *26*(1), 29–33.

2435. Dutta, I.; Dutta, P. K.; Smith, D. W.; O'Donovan, G. A. *J. Chromatogr.* **1991**, *536*(1/2), 237–243.

2436. Ingle, J. *Phytochemistry* **1963**, *2*(4), 353–370.

2437. Brown, E. G. *Biochem. J.* **1962**, *85*(3), 633–640.

2438. Nieman, R. H.; Pap, D. L.; Clark, R. A. *J. Chromatogr.* **1978**, *161*, 137–146.

2439. Guinn, G. *Anal. Biochem.* **1973**, *54*(1), 276–282.

2440. Meyer, R.; Wagner, K. G. *Anal. Biochem.* **1985**, *148*(2), 269–276.

2441. Brown, E. G.; Davies, D. *Anal. Biochem.* **1989**, *183*(2), 312–319.

2442. Ching, T. M.; Ranzoni, M. W.; Ching, K. K. *Plant Sci. Lett.* **1975**, *4*(5), 331–333.

2443. Chung, M. J.; Kim, Y. S.; Lee, I. S.; Jo, J. S.; Sung, N. J. *Han'guk Yongyang Siklyong Hakhoechi* **1995**, *24*(6), 911–916, from *Chem. Abstr.* **1996**, *124*, 226520k.

2444. Kluge, M.; Ziegler, H. *Planta* **1964**, *61*(2), 167–177.

2445. Szczotka, Z.; Tomaszewska, E. *Bull. Acad. Polon. Sci. Ser., Sci. Biol.* **1980**, *28*(8/9), 549–553.

2446. Fromm, J.; Eschrich, W. *Trees (Berlin)* **1986**, *1*(1), 42–46.

2447. Brown, E. G.; Short, K. C. *Phytochemistry* **1969**, *8*(8), 1365–1372.

2448. Glas, R.; Robinson, D. G. *Protoplasma* **1982**, *113*(2), 97–102.

2449. Shimizu, T.; Clifton, A.; Komamine, A.; Fowler, M. W. *Physiol. Plant.* **1977**, *40*(2), 125–129.

2450. Simmonds, J. A.; Dumbroff, E. B. *Plant Physiol.* **1974**, *53*(1), 91–95.

2451. Chung, M. J.; Kim, Y. S.; Lee, I. S.; Jo, J. S.; Sung, N. J. *Han'guk Yongyang Siklyong Hakhoechi* **1995**, *24*(5), 727–733, from *Chem. Abstr.* **1996**, *124*, 144958x.

2452. Xu, B.; Gu, Z. *Zhongguo Nongye Kexue (Beijing)* **1984**, (3), 51–55, from *Chem. Abstr.* **1984**, *101*, 186054c.

2453. Kosanke, R.; Bernhardt, D.; Köhler, K. H.; Voigt, B. *Biol. Plant.* **1990**, *32*(2), 145–152.

2454. Mazor, L.; Nagbi, M.; Perl, M. *J. Exp. Bot.* **1984**, *35*(157), 1128–1135.

2455. Wilson, C.; Gross, J. W.; Lucas, W. J. *Planta* **1985**, *164*(2), 227–240.

2456. Hashida, W.; Mouri, T.; Shiga, I.; Nishikawa, S.; Teramoto, S. *Hakko Kogaku Zasshi* **1963**, *41*(8), 420–426.

2457. Wieser, U.; Pankow, W.; Wiemken, A. *J. Plant Physiol.* **1986**, *124*(1/2), 181–186.

2458. Bray, C. M.; Davison, P. A.; Ashraf, M.; Taylor, R. M. *Ann. Bot.* **1989**, *63*, 185–193.

2459. Noguchi, K.; Terashima, I. *Physiol. Plant.* **1997**, *101*(1), 1–7.

2460. Grossi, L. *Biochem. Physiol. Pflanz.* **1985**, *180*(4), 277–283.

2461. Gill, R. K.; Malik, C. P. *J. Palynol.* **1981**, *17*(1/2), 37–41.

2462. Chen, R.; Cai, D.; Fu, J. *Zhongshan Daxue Xuebao, Ziran Kexueban* **1987**, (1), 117–120, from *Chem. Abstr.* **1987**, *107*, 72745x.

2463. Zheng, X.; Fu, J. *Zhongshan Daxue Xuebao, Ziran Kexueban* **1987**, (1), 21–29, from *Chem. Abstr.* **1987**, *107*, 74390b.

2464. Sreenivasulu, P.; Nayudu, M. V. *Acta Phytopathol. Acad. Sci. Hung.* **1978**, *13*(3/4), 255–261.

2465. Crompton, C.; Wynne, J. C.; Patterson, R. P. *Crop Sci.* **1978**, *18*(5), 736–739.

2466. Mouri, T.; Hashida, W.; Shiga, I.; Teramoto, S. *Hakko Kogaku Zasshi* **1966**, *44*(5), 237–247.

2467. Urushizaki, S.; Nagashima, N.; Takeda, Y.; Ota, Y. *Jpn. Agric. Res. Quart.* **1987**, *20*(3), 196–201.

2468. Nguyen, T. T.; Palcic, M. M.; Hadziyev, D. , Bioflavour '87, Proceedings of the International Conference on Würzburg; Schreier, P., Ed.; Walter de Gruyter: Berlin, 1988; pp 275–286.

2469. Chow, P. N. P.; LaBerge, D. E. *J. Agric. Food Chem.* **1978**, *26*(5), 1134–1137.

2470. Ladonin, V. F. *Agrokhimiya* **1967**, (2), 85–95.

2471. Bergkvist, R. *Acta Chem. Scand.* **1957**, *11*(9), 1457–1464.

2472. Hampp, R.; Goller, M.; Ziegler, H. *Plant Physiol.* **1982**, *69*(2), 448–455.

2473. Baumann, G.; Günther, G. *Biochem. Physiol. Pflanz.* **1975**, *167*(5), 399–410.

2474. Sisler, E. C.; Klein, W. H. *Physiol. Plant.* **1961**, *14*, 115–123.

2475. Elnaghy, M. A.; Nordin, P. *Arch. Biochem. Biophys.* **1966**, *113*(1), 72–76.

2476. McKee, K. L.; Mendelssohn, I. A. *Environ. Exp. Bot.* **1987**, *27*(2), 147–156.

2477. Lunn, G.; Madsen, E. *Physiol. Plant.* **1981**, *53*(2), 164–169.

2478. Gendraud, M.; Lafleuriel, L. *Z. Pflanzenphysiol.* **1980**, *96*(5), 435–444.

2479. Pavlinova, O. A.; Afanas'eva, T. P. *Fiziol. Rast.* **1962**, *9*(2), 133–141 (Engl. 106–111).

2480. Lucas, W. J.; Wilson, C. *Plant Physiol.* **1987**, *85*(2), 423–429.

2481. Santarius, K. A.; Heber, U. *Biochim. Biophys. Acta* **1965**, *102*(1), 39–54.

2482. Rao, I. M.; Fredeen, A. L.; Terry, N. *Plant Physiol. Biochem.* **1993**, *31*(2), 223–231.

2483. Volovik, O. I.; Grigora, M. Yu.; Dmitrieva, V. V.; Polischuk, A. I.; Ostrovskaya, L. K. *Fiziol. Biokhim. Kul't Rast.* **1985**, *17*(6), 549–556.

2484. Petraglia, T.; Poole, R. J. *Plant Physiol.* **1980**, *65*(5), 969–972.

2485. Suzuki, M.; Okazaki, S.; Masuda, H.; Sugawara, S. *Obihiro Chikusan Daigaku Gakujutsu Kenkyu Hokoku, Dai-1-Bu* **1973**, *7*(4), 641–649.

2486. Silvius, J. E.; Snyder, F. W.; Kremer, D. F. *Plant Physiol.* **1982**, *70*(1), 316–317.

2487. Sagisaka, S. *Plant Cell Physiol.* **1982**, *23*(8), 1337–1346.

2488. Tripepi, R. R.; Mitchell, C. A. *Plant Physiol.* **1984**, *76*(1), 31–35.

2489. Desbiez, M.-O. *Z. Pflanzenphysiol.* **1973**, *69*(2), 174–180.

2490. Ling, J.; Lin, Z. *Zhiwu Shenglixue Tongxun* **1993**, *29*(2), 90–93, from *Chem. Abstr.* **1993**, *119*, 221839h.

2491. Xu, B.; Gu, Z. *Zhiwu Shenglixue Tongxun* **1984**, (3), 39–42, from *Chem. Abstr.* **1985**, *102*, 3326w.

2492. Sobczyk, E. A.; Kacperska-Palacz, A. *Plant Physiol.* **1978**, *62*(6), 875–878.

2493. Barbaro, A.; Filek, W.; Pienkowski, S.; Randak, G. *Acta Agrar. Silvestria, Ser. Agrar.* **1985**, *24*, 3–16.

2494. Börner, H. *Z. Pflanzenkr. Pflanzensch.* **1974**, *81*(10), 584–590.

2495. Raymond, P.; Al-Ani, A.; Pradet, A. *Plant Physiol.* **1985**, *79*(3), 879–884.

2496. Tashiro, T.; Fujita, E. *Nippon Shokuhin Kogyo Gakkaishi* **1981**, *28*(11), 588–593.

2497. Shen, Q. G.; Liu, C. D.; Zhang, J. Y.; Yan, T.; Yan, L. F.; Tang, P. S. *Zhiwu Xuebao* **1982**, *24*(6), 531–539.

2498. Yuan, H.-F. *Bot. Bull. Acad. Sin.* **1977**, *18*(1), 32–38.

2499. Takino, Y.; Imagawa, H.; Shishido, K. *Nippon Shokuhin Kogyo Gakkaishi* **1972**, *19*(5), 213–218.

2500. Selvendran, R. R.; Isherwood, F. A. *Phytochemistry* **1971**, *10*(3), 525–529.

2501. Slonov, L. Kh.; Petinov, N. S. *Fiziol. Rast.* **1980**, *27*(5), 1095–1100 (Engl. 811–815).

2502. Turner, L. B.; Wellburn, A. R. *J. Plant Physiol.* **1985**, *120*(2), 111–122.

2503. Nieman, E. H.; Clark, R. A.; Pap, D.; Ogata, G.; Maas, E. V. *J. Exp. Bot.* **1988**, *39*(200), 301–309.

2504. Shimazaki, A.; Hirose, F.; Ashihara, H. *Z. Pflanzenphysiol.* **1982**, *106*(3), 191–198.

2505. Ashihara, H.; Li, X.-N.; Ukaji, T. *Ann. Bot.* **1988**, *61*(2), 225–232.

2506. Ukaji, T.; Ashihara, H. *Z. Naturforsch.* **1986**, *41C*(11/12), 1045–1051.

2507. Brodelius, P.; Vogel, H. J. *J. Biol. Chem.* **1985**, *260*(6), 3556–3560.

2508. Wagner, E.; Tetzner, J.; Haertlé, U.; Deitzer, G. F. *Ber. Dtsch. Bot. Ges.* **1974**, *87*(2), 291–302.

2509. Hatzfeld, W.-D.; Dancer, J.; Stitt, M. *Planta* **1990**, *180*(2), 205–211.

2510. Singh, B. P.; Misra, A. K. *Proc. Indian Acad. Sci.* **1974**, *B80*(2), 97–104.

2511. Vasseur, J. *Compt. Rend.* **1975**, *280D*(1), 33–36.

2512. Al-Baldawi, N. F. *Iraqi J. Sci.* **1986**, *27*(3/4), 323–328, from *Chem. Abstr.* **1987**, *107*, 130868c.

2513. Buslig, B. S.; Attaway, J. A. *Proc. Florida State Hortic. Soc.* **1969**, *82*, 206–208.

2514. Abou-Zamzam, A. M.; Wallace, A.; Motoyama, E. *J. Am. Soc. Hortic. Sci.* **1970**, *95*(2), 203–206.

2515. Avtomonova, L. V.; Egamberdiev, A. R. *Subtrop. Kul't.* **1974**, (4), 58–63.

2516. Akao, S.; Tsukahara, S. *Shikoku Nogyo Shikenjo Hokoku* **1979**, *34*, 23–29.

2517. Barmore, C. R.; Biggs, R. H. *J. Food Sci.* **1972**, *37*(5), 712–714.

2518. Bennett, M. C. *J. Agric. Food Chem.* **1977**, *25*(2), 219–221.

2519. Biggs, R. H. *Proc. Florida State Hort. Soc.* **1972**, *85*, 204–206.

2520. Bruemmer, J. H.; Roe, B. *Phytochemistry* **1985**, *24*(9), 2105–2106.

2521. Palit, P. *Indian J. Exp. Bot.* **1987**, *25*(2), 135–138.

2522. Bradbeer, J. W.; Jarvis, B. C. *Biochem. J.* **1967**, *104*, 39–40.

2523. Li, Z.; Attias, J.; Marin, B. *Compt. Rend., Ser. 3* **1992**, *315*(3), 127–132.

2524. Morohashi, Y.; Sugimoto, M. *Plant Cell Physiol.* **1988**, *29*(5), 893–896.

2525. Decleire, M.; de Cat, W. *Weed Res.* **1981**, *21*(2), 93–97.

2526. Yamaya, T.; Tanigawa, M.; Kawasaki, T.; Matsumoto, H. *Soil Sci. Plant Nutr.* **1982**, *28*(3), 349–357.

2527. Stanchev, L. B.; Mashev, N. P. *Pochvozn. Agrokhim.* **1972**, *7*(2), 43–50.

2528. Rasi-Caldogno, F.; de Michelis, M. I. *Plant Physiol.* **1978**, *61*(1), 85–88.

2529. Kutis, T. L.; Kutis, S. D. *Fiziol. Biokhim. Kul't Rast.* **1984**, *16*(6), 608–610.

2530. Mamedov, T. K.; Akhmerova, E. S. *Dokl. Akad. Nauk Azerb. SSR* **1975**, *31*(4), 42–48.

2531. Lehmann, J. *Z. Pflanzenphysiol.* **1979**, *94*(4), 331–338.

2532. Vyskrebentseva, E. I.; Krasavina, M. S. *Fiziol. Rast.* **1966**, *13*(3), 433–445 (Engl. 388–398).

2533. Mamedova, T. Kh.; Rasulov, F. A. *Dokl. Akad. Nauk Azerb. SSR* **1970**, *26*(10/11), 71–76.

2534. Chao, C. C.; Lin, T. P. *Bot. Bull. Acad. Sin.* **1996**, *37*(4), 229–237, from *Chem. Abstr.* **1997**, *126*, 44971s.

2535. Mitsui, K.; Ashihara, H. *Plant Cell. Physiol.* **1988**, *29*(7), 1177–1183.

2536. Ohtsuka, H.; Take, T. *J. Food Sci.* **1969**, *34*(5), 392–394.

2537. Ashihara, H.; Nygaard, P. *Physiol. Plant.* **1989**, *75*(1), 31–36.

2538. Kanabue, J.; Bressan, R. A.; Carpita, N. C. *Plant Physiol.* **1986**, *82*(2), 363–368.

2539. Sloan, M. E.; Camper, N. D. *Pestic. Biochem. Physiol.* **1981**, *15*(3), 201–208.

2540. Stuhlfauth, T.; Beckedahl, J.; Fock, H. P. *Photosynthetica* **1991**, *25*(1), 11–15.

2541. Joung, S. Y.; Lee, S. J.; Sung, N. J.; Jo, J. S.; Kang, S. K. *Han'guk Yongyang Siklyong Hakhoechi* **1995**, *24*(5), 720–726, from *Chem. Abstr.* **1996**, *124*, 144138x.

2542. McKee, K. L.; Mendelssohn, I. A. *Physiol. Plant.* **1984**, *62*(1), 1–7.

2543. Bogacheva, I. I.; Shlykova, I. M. *Fiziol. Rast.* **1973**, *20*(4), 708–713 (Engl. 595–599).

2544. Zungsontiporn, S.; Kusanagi, T.; Sugiyama, H.; Murata, Y. *Zasso Kenkyu* **1989**, *34*(4), 280–284, from *Chem. Abstr.* **1990**, *113*, 75034z.

2545. Schwink, L. *Planta* **1956**, *47*, 165–218.

2546. Wilson, J. M. *New Phytol.* **1978**, *80*(2), 325–334.

2547. Sung, M. W.; Kwon, B. K. *Sikmul Hakhoechi* **1980**, *23*(2), 37–44.

2548. Gorianova, T. M. *Uch. Zap. Gor'k. Gos. Univ.* **1972**, (195), 97–102, from *Chem. Abstr.* **1973**, *79*, 15104s.

2549. Chernyad'ev, I. I.; Kozlovskikh, A. L. *Fiziol. Rast.* **1990**, *37*(2), 335–341 (Engl. 251–256).

2550. Augsten, H.; Felsberg, I. *Flora* **1969**, *160A*(5), 474–478.

2551. Lavarenne, S.; Champciaux, M.; Barnola, P.; Gendraud, M. *Physiol. Vég.* **1982**, *20*(3), 371–376.

2552. Galitz, D. S.; Howell, R. W. *Physiol. Plant.* **1968**, *21*(6), 1296–1300.

2553. Yokoyama, Z.; Mori, T.; Matsushita, S. *Kyoto Daigaku Shokuryo Kagaku Kenkyusho Hokoku* **1970**, (33), 52–60.

2554. Quebedaux, B. *Plant Physiol.* **1981**, *68*(1), 23–27.

2555. Sung, M. W.; Yang, W. J. *Sikmul Hakhoechi* **1980**, *23*(1), 11–15.

2556. Buchvarov, P. Z.; Alekhina, N. D. *Dokl. Bulg. Akad. Nauk* **1984**, *37*(7), 919–922.

2557. Yaklich, R. W.; Kulik, M. M.; Anderson, J. D. *Crop Sci.* **1979**, *19*(6), 806–810.

2558. Weissman, G. S. *Plant Physiol.* **1972**, *49*(1), 142–145.

2559. Saka, H.; Chisaka, H.; Uezono, T. *Nippon Sakumotsu Gakkai Kiji* **1983**, *52*(2), 183–189.

2560. Key, J. L.; Hanson, J. B. *Plant Physiol.* **1961**, *36*(2), 145–152.

2561. Fader, G. M.; Koller, H. R. *Plant Physiol.* **1984**, *75*(3), 694–699.

2562. Tajima, S.; Kouzai, K. *Plant Cell Physiol.* **1989**, *30*(4), 589–593.

2563. Mitsumori, F.; Yoneyama, T.; Ito, O. *Plant Sci.* **1985**, *38*(2), 87–92.

2564. Sa, T. M.; Lim, S. U.; Israel, D. W. *Han'guk Nonghwa Hakhoechi* **1992**, *35*(6), 449–456.

2565. Oresnik, I. J.; Layzell, D. B. *Plant Physiol.* **1994**, *104*(1), 217–225.

2566. Franz, G. *Phytochemistry* **1969**, *8*(4), 737–741.

2567. Basra, A. S.; Malik, C. P. *Proc. Indian Natl. Sci. Acad.* **1982**, *B48*(2), 229–232.

2568. Yang, W.; Qiu, J. *Sepu* **1990**, *8*(2), 126.

2569. Stewart, J. Mc D.; Guinn, G. *Plant Physiol.* **1971**, *48*(2), 166–170.

2570. Sagdullaev, I. N.; Valikhanov, M. N. *Sb. Nauchn. Tr. Tashk. Gos. Univ.* **1982**, (687), 20–23, from *Chem. Abstr.* **1983**, *99*, 155380s.

2571. Carpita, N. C.; Delmer, D. P. *J. Biol. Chem.* **1981**, *256*(1), 308–315.

2572. Ibragimov, A. P.; Plekanova, L. S. *Fiziol. Rast.* **1969**, *16*(4), 696–701 (Engl. 580–583).

2573. Warman, T. W.; Solomon, T. *J. Exp. Bot.* **1988**, *39*(203), 685–694.

2574. Jacob, J.; Lawlor, D. W. *Plant Cell Environ.* **1993**, *16*(7), 785–795.

2575. Gendraud, M.; Prevôt, J.-C. *Compt. Rend.* **1972**, *274D*(14), 2049–2052.

2576. Taniguchi, H.; Umemura, Y.; Nakamura, M. *Agric. Biol. Chem.* **1967**, *31*(2), 231–239.

2577. Pavlinova, O. A. *Fiziol. Rast.* **1965**, *12*(4), 606–617 (Engl. 527–535).

2578. McMullen, A. L. *Biochim. Biophys. Acta* **1960**, *41*(2), 341–343.

2579. Amalou, Z.; Bangratz, J.; Chrestin, H. *Plant Physiol.* **1992**, *98*(4), 1270–1276.

2580. Xue, J.; Yang, Y.; Sun, X. *Zhongguo Nongye Kexue (Beijing)* **1986**, (3), 51–58, from *Chem. Abstr.* **1986**, *105*, 208060p.

2581. Bergmann, H. *Kulturpflanze* **1972**, *20*, 327–349.

2582. Andrews, C. J.; Pomeroy, M. K. *Plant Physiol.* **1989**, *91*(3), 1063–1068.

2583. Parker, D. K.; Proudlove, M. O. *J. Inst. Brewing* **1993**, *99*(4), 327–330.

2584. Fric, F.; Haspelová-Horvatovicová, A. *Phytochemistry* **1967**, *6*(5), 633–640.

2585. Sawert, A.; Backer, A.; Plank-Schumacher, K.-H.; Wagner, K. G. *J. Plant Physiol.* **1987**, *127*(1/2), 183–186.

2586. Loughman, B. C.; Russell, R. S. *J. Exp. Bot.* **1957**, *8*(23), 280–293.

2587. Oji, Y.; Morikawa, K.; Wakiuchi, N.; Okamoto, S. *Kobe Daigaku Nogakubu Kenkyu Hokoku* **1984**, *16*(1), 297–302.

2588. Akiyama, Y.; Shibasaka, M.; Kawasaki, T. *Okayama Daigaku Shigen Seibutsu Kagaku Kenkyusho Hokoku* **1995**, *3*(2), 137–143.

2589. Jackson, P. C.; Pfeffer, P. E.; Gerasimowicz, W. V. *Plant Physiol.* **1986**, *81*(4), 1130–1133.

2590. Bewley, J. D.; Gwoźdź, E. A. *Plant Physiol.* **1975**, *55*(6), 1110–1114.

2591. Ashihara, H. *Phytochemistry* **1993**, *33*(6), 1427–1430.

2592. Murata, T. *Nippon Nogei Kagaku Kaishi* **1970**, *44*(9), 412–421.

2593. Takamura, T.; Uritani, I. *Agric. Biol. Chem.* **1973**, *37*(7), 1511–1515.

2594. Pamatmat, M. M.; Skjoldal, H. R. *Estuarine Coastal Mar. Sci.* **1979**, *9*(1), 79–90.

2595. Dedonder, A.; Dirix, F.; Rethy, R.; Fredericq, H.; de Greef, J. A. *Isr. J. Bot.* **1990**, *39*(4/6), 337–346.

2596. Pistelli, L.; Marigo, G.; Ball, E.; Lüttge, U. *Planta* **1987**, *172*(4), 479–486.

2597. Marigo, G.; Lüttge, U.; Smith, J. A. C. *Physiol. Vég.* **1981**, *19*(2), 123–133.

2598. Standard, S. A.; Vaux, P.; Bray, C. M. *J. Chromatogr.* **1985**, *318*(2), 433–439.

2599. Raymond, P.; Al-Ani, A.; Pradet, A. *Physiol. Vég.* **1983**, *21*(4), 677–687.

2600. Ching, T. M.; Danielson, R. *Proc. Assoc. Off. Seed Anal.* **1972**, *62*, 116–124.

2601. Cumming, D. F. *Biochem. J.* **1970**, *116*(2), 189–198.

2602. Su, J. C. *J. Chin. Agric. Chem. Soc.* **1965**, (Spec. Issue), 45–53, from Chen,[3802] table 2.

2603. Kondo, T.; Nakashima, H. *Physiol. Plant.* **1979**, *45*(4), 357–362.

2604. Gower, R. A.; Posner, H. B. *Plant Physiol.* **1979**, *63*(3), 548–551.

2605. Zhou, C.; Xie, M.; Shi, H. *Sepu* **1989**, *7*(2), 108–110.

2606. Linskens, H. F. *Acta Bot. Neerl.* **1958**, *7*(1), 61–68.

2607. Abdul-Wahab, M. F. *J. Chem. U.A.R.* **1964**, *7*(1), 1–12.

2608. Titok, V. V.; Rusinova, O. V.; Khotyleva, L. V. *Dokl. Akad. Nauk Belarusi* **1994**, *38*(3), 63–66.

2609. Wellburn, A. R.; Higginson, C.; Robinson, D.; Walmsley, C. *New Phytol.* **1981**, *88*(2), 223–237.

2610. Dziegielewski, T.; Kedzierski, W.; Pawelkiewicz, J. *Biochim. Biophys. Acta* **1979**, *564*(1), 37–42.

2611. Yoshimura, Y.; Take, T.; Otsuka, H. *Kaseigaku Zasshi* **1969**, *20*(7), 481–484.

2612. Roux, L. *Ann. Physiol. Veg.* **1963**, *5*(3), 203–215.

2613. McCune, D. C.; Weinstein, L. H.; Mancini, J. F. *Contrib. Boyce Thompson Inst.* **1970**, *24*(19), 213–225.

2614. Brown, J.; Nordin, P. *J. Agric. Food Chem.* **1969**, *17*(2), 341–343.

2615. Bueschler, R. W.; Dostal, H. C. *J. Food Sci.* **1974**, *39*(4), 774–776.

2616. Ogura, N.; Ishii, T.; Nakagawa, H. *Chiba Daigaku Engeigakubu Gakujutsu Hokoku* **1980**, (27), 9–14.

2617. Inaba, A.; Yamamoto, T.; Ito, T.; Nakamura, R. *Engei Gakkai Zasshi* **1980**, *49*(3), 435–441.

2618. Dutta, I.; O'Donovan, G. A.; Smith, D. W. *J. Chromatogr.* **1988**, *444*, 183–189.

2619. Lin, T.-P.; Chen, M.-H. *Ann. Bot.* **1995**, *76*(4), 381–387.

2620. Han, Z. N.; Wang, Q.; Shen, T. *J. Plant Nutr.* **1994**, *17*(7), 1257–1264.

2621. Thévenot, C.; Simond-Côte, E.; Gendraud, M. *J. Plant Physiol.* **1989**, *135*(2), 138–143.

2622. Schmidt, S. *Flora* **1965**, *155*(4), 580–585.

2623. Kollar, A.; Seemüller, E. *J. Phytopathol.* **1990**, *128*(2), 99–111.

2624. Yamashita, T. *Plant Cell Physiol.* **1983**, *24*(6), 1151–1155.

2625. Yoshioka, H.; Ueda, Y.; Iwata, T. *Nippon Shokuhin Kogyo Gakkaishi* **1982**, *29*(6), 333–339.

2626. Sunderland, D. W.; Merrett, M. J. *Physiol. Plant.* **1967**, *20*(2), 368–372.

2627. Kakie, T. *Hatano Tabako Shikenjo Hokohu* **1969**, 43–50.

2628. Meyer, R.; Wagner, K. G. *Physiol. Plant.* **1986**, *67*(4), 666–672.

2629. Wray, V.; Schiel, O.; Berlin, J. *Z. Pflanzenphysiol.* **1983**, *112*(3), 215–220.

2630. Brown, D. C. W.; Thorpe, T. A. *Plant Physiol.* **1980**, *65*(4), 587–590.

2631. Banerjee, A.; Choudhuri, M. M.; Ghosh, B. *Z. Pflanzenphysiol.* **1981**, *102*(1), 33–36.

2632. Kukanaga, S.; Katolh, Y.; Takahashi, M. *J. Sci. Food Agric.* **1991**, *56*(3), 335–343.

2633. Murakami, T. *Nogyo Gijutsu Kenkyusho Hokoku* **1975**, *26D*, 223–262.

2634. Barua, K.; Choudhuri, M. M.; Ghosh, B. *Plant Sci. Lett.* **1981**, *21*(4), 357–360.

2635. Asada, K.; Kasai, Z. *Soil Sci. Plant Nutr. (Tokyo)* **1963**, *9*(1), 13–17.

2636. Murakami, T. *Nippon Sakumotsu Gakkai Kiji* **1970**, *39*(3), 287–294.

2637. Hayashi, H.; Chino, M. *Plant Cell Physiol.* **1990**, *31*(2), 247–251.

2638. Menegus, F.; Cattaruzza, L.; Mattana, M.; Beffagna, N.; Ragg, E. *Plant Physiol.* **1991**, *95*(3), 760–767.

2639. Mori, S.; Nishizawa, N.; Hayashi, H.; Chino, M.; Yoshimura, E.; Ishihara, J. *Plant Soil* **1991**, *130*(1/2), 143–156.

2640. Zhukova, T. M.; Chirkova, T. V.; Tret'yakov, A. V. *Vestn. Leningr. Univ., Biol.* **1984**, (3), 74–81, from *Chem. Abstr.* **1994**, *101*, 227008m (but article not in the cited issue).

2641. Mitsui, T.; Sado, T.; Terada, A.; Igaue, I. *Agric. Biol. Chem.* **1989**, *53*(7), 1991–1993.

2642. Koch, M. S.; Mendelssohn, I. A.; McKee, K. L. *Limnol. Oceanogr.* **1990**, *35*(2), 399–408.

2643. Ohnishi, J.; Kanai, R. *Plant Cell Physiol.* **1987**, *28*(2), 243–251.

2644. Espen, L.; Morgutti, S.; Alisi, C.; Pirovano, L.; Ragg, E.; Cocucci, S. M. *Physiol. Plant.* **1995**, *93*(4), 577–583.

2645. Thigpen, S. P.; Sachs, R. M. *Physiol. Plant.* **1985**, *65*(2), 156–162.

2646. Jin, S.-P.; Jin, S.-M.; Lian, H. K.; Ren, X. C. *Zhiwu Xuebao* **1986**, *28*(4), 381–386.

2647. Colling, M.; Wolfram, G. *Z. Lebensm. Unters. Forsch.* **1987**, *185*(4), 288–291.

2648. Fiussello, N. *Allionia* **1970**, *16*, 53–58.

2649. Weinstein, L. H.; McCune, D. C.; Mancini, J. F.; van Leuken, P. *Plant Physiol.* **1969**, *44*(11), 1499–1510.

2650. Wallis, M. E.; Bradbeer, J. W. *J. Exp. Bot.* **1970**, *21*(69), 1039–1047.

2651. Nieman, R. H.; Clark, R. A. *J. Chromatogr.* **1984**, *317*, 271–281.

2652. Wilson, C.; Clark, R. A.; Nieman, R. H. *J. Exp. Bot.* **1992**, *43*(253), 1009–1014.

2653. Brown, E. G.; Mangat, B. S. *Phytochemistry* **1970**, *9*(8), 1859–1868.

2654. Mikityuk, O. D.; Chkanikov, D. I. *Agrokhimiya* **1967**, (6), 128–137.

2655. Mangat, B. S. *J. Exp. Bot.* **1982**, *33*(133), 269–278.

2656. Mangat, B. S. *Pestic. Biochem. Physiol.* **1979**, *10*(3), 251–258.

2657. Šebesta, K.; Šorm, F. *Coll. Czech. Chem. Commun.* **1959**, *24*, 2781–2789.

2658. Rychter, A. M.; Chauveau, M.; Bomsel, J.-L.; Lance, C. *Physiol. Plant.* **1992**, *84*(1), 80–86.

2659. Markowski, A.; Augustyniak, G.; Janowiak, F. *Acta Physiol. Plant.* **1990**, *12*(2), 167–173.

2660. Fürtig, K.; Pavelic, D.; Brunold, C.; Brändle, R. *Limnologica* **1999**, *29*(1), 60–63.

2661. Chen, C.-S.; Su, J.-C. *Bot. Bull. Acad. Sin.* **1968**, *9*(1), 27–33.

2662. Kicinski, H. G.; Kettrup, A.; Boos, K.-S.; Masuch, G. *Int. J. Environ. Anal. Chem.* **1988**, *32*(3/4), 213–241.

2663. Hampp, R.; Einig, W.; Egger, B. *Environ. Pollut.* **1990**, *68*(3/4), 305–318.

2664. Bourgeois, J.; Malek, L. *Tree Physiol.* **1991**, *8*(4), 407–413.

2665. Durzan, O. J.; Pitel, J.; Ramaiah, P. K. *Can. J. Forest Res.* **1972**, *2*(3), 206–216.

2666. Harvey, G. W.; Legge, A. H. *Can. J. Bot.* **1979**, *57*(7), 759–764.

2667. Murphy, J. B.; Noland, T. L. *Plant Physiol.* **1982**, *70*(5), 1410–1412.

2668. Noland, T. L.; Murphy, J. B. *Seed Sci. Technol.* **1984**, *12*(3), 777–787.

2669. Nygaard, P. *Physiol. Plant.* **1973**, *28*(1), 361–371.

2670. Ching, T. M.; Ching, K. K. *Plant Physiol.* **1972**, *50*(5), 536–540.

2671. Johnson, M. A.; Carlson, J. A.; Conkey, J. H.; Noland, T. L. *J. Exp. Bot.* **1987**, *38*(188), 518–524.

2672. Lorenc-Plucińska, G.; Ziegler, H. *Photosynthetica* **1996**, *32*(1), 141–144.

2673. Lorenc-Plucińska, G.; Ziegler, H. *Acta Physiol. Plant.* **1995**, *17*(3), 225–232.

2674. Kazaryan, V. V.; Vardanyan, Zh.A. *Byull. Gl. Bot. Sada* **1989**, *152*, 37–41.

2675. Okuntsov, M. M.; Vrublevskaya, K. G.; Zaitseva, T. A. *Vop. Fotosin.* **1970**, (2), 32–46.

2676. Déjardin, A.; Rochat, C.; Mangenest, S.; Boutin, J.-P. *Planta* **1997**, *201*(2), 128–137.

2677. Pronina, N. B.; Kholodenko, N. Ya *Fiziol. Biokhim. Kul't Rast.* **1974**, *6*(6), 653–657.

2678. Vecher, A. S.; Kurbatova, S. I.; Raitsina, G. I.; Reshetnikov, V. N. *Dokl. Akad. Nauk BSSR* **1966**, *10*(4), 265–268.

2679. Brown, A. P.; Wray, J. L. *Biochem. J.* **1968**, *108*(3), 437–444.

2680. Miettinen, J. K.; Savioja, T. *Acta Chem. Scand.* **1959**, *13*(8), 1693–1698.

2681. Rowan, K. S. *J. Exp. Bot.* **1957**, *8*(23), 256–271.

2682. Sofrová, D.; Pačes, V.; Leblová, S.; Koštiř, J. *Naturwissenschaften* **1967**, *54*(2), 45–46.

2683. Brown, E. G.; Cassells, A. C. *Phytochemistry* **1971**, *10*(6), 1251–1260.

2684. Hasson-Porath, E.; Poljakoff-Mayber, A. *Plant Physiol.* **1971**, *47*(1), 109–113.

2685. Sagisaka, S. *Plant Cell Physiol.* **1981**, *22*(7), 1287–1292.

2686. Endress, R. *Phytochemistry* **1977**, *16*(10), 1549–1554.

2687. Vakis, N.; Soule, J.; Ahmed, E. M.; Grierson, W. *Proc. Florida State Hortic. Soc.* **1970**, *83*, 285–290.

2688. Balander, P.; Rageau, R.; Gendraud, M.; Bonhomme, M.; Parisot, E. *Acta Hortic.* **1995**, *409*, 39–45.

2689. Duncan, H. J. *J. Sci. Food Agric.* **1974**, *25*(2), 211–219.

2690. Tonelli, D.; Girotti, S.; Budini, R. *Atti Accad. Sci. Ist. Bologna, Cl. Sci. Fis., Rend.* **1980**, *7*(1), 239–247.

2691. Moreland, D. E.; Hussey, G. G.; Shriner, C. R.; Farmer, F. S. *Plant Physiol.* **1974**, *54*(4), 560–563.

2692. Cocucci, S. M.; Morgutti, S.; Abruzzese, A.; Alisi, C.; Yahie, A. H.; Espen, L. *Physiol. Plant.* **1993**, *87*(4), 609–615.

2693. Hall, S. M.; Baker, D. A. *Planta* **1972**, *106*(2), 131–140.

2694. Mengel, K.; Haeder, H.-E. *Plant Physiol.* **1977**, *59*(2), 282–284.

2695. Murphy, T. M.; Matson, G. B.; Morrison, S. L. *Plant Physiol.* **1983**, *73*(1), 20–24.

2696. Alexander, A. G. *J. Agric. Univ. Puerto Rico* **1967**, *51*(3), 210–227.

2697. Gardner, D. C. J.; Peel, A. H. *Nature* **1969**, *222*, 774.

2698. Hess, C. M.; Meeuse, B. J. D. *Proc. Kon. Ned. Akad. Wet.* **1968**, *71C*(5), 443–455.

2699. Grzesiuk, S.; Tluczkiewicz, J.; Gorecki, R. J. *Biochem. Physiol. Pflanz.* **1983**, *178*(6/7), 487–491.

2700. Bon, M.-C. *Physiol. Plant.* **1988**, *72*(4), 796–800.

2701. Monteunis, O.; Gendraud, M. *Tree Physiol.* **1987**, *3*(3), 257–263.

2702. Friederich, K. E.; Mohr, H. *Photochem. Photobiol.* **1975**, *22*(1/2), 49–53.

2703. Bodson, M. *Planta* **1985**, *163*(1), 34–37.

2704. Buri, R.; Signer, V.; Solms, J. *Schweiz. Landwirt. Monatsh.* **1970**, *48*(10), 346–350.

2705. Buri, R.; Solms, J. *Naturwissenchaften* **1971**, *58*(1), 56.

2706. Maga, J. A.; McNeill, J. G. *Lebensm. Wiss. Technol.* **1986**, *19*(1), 31–33.

2707. Amir, J.; Kahn, V.; Unterman, M. *Phytochemistry* **1977**, *16*(10), 1495–1498.

2708. Roberts, J. K. M.; Aubert, S.; Gout, E.; Bligny, R.; Douce, R. *Plant Physiol.* **1997**, *113*(1), 191–199.

2709. Rees, W. R.; Duncan, H. J. *J. Sci. Food Agric.* **1972**, *23*(3), 337–343.

2710. Akatsu, M.; Hosoi, Y.; Sasamoto, H.; Ashihara, H. *J. Plant Physiol.* **1996**, *149*(1/2), 133–137.

2711. Mendelssohn, I. A.; McKee, K. L. *Aquat. Bot.* **1981**, *11*(1), 37–55.

2712. Dancer, J.; Neuhaus, H. E.; Stitt, M. *Plant Physiol.* **1990**, *92*(3), 637–641.

2713. Bonzon, M.; Hug, M.; Wagner, E.; Greppin, H. *Planta* **1981**, *152*, 189–194.

2714. Knauer, G. A.; Ayers, A. V. *Limnol. Oceanogr.* **1977**, *22*(3), 408–414.

2715. Cherry, J. H.; Hageman, R. H. *Plant Physiol.* **1961**, *36*(2), 163–168.

2716. Sioufi, A.; Percheron, F.; Courtois, J. E. *Phytochemistry* **1970**, *9*(5), 991–999.

2717. Ching, T. M.; Poklemba, C. J.; Metzger, R. H. *Plant Physiol.* **1983**, *79*(3), 652–657.

2718. Briggs, K. G.; Horak, A. *Can. J. Plant Sci.* **1980**, *60*(4), 1455–1457.

2719. Borschel, H.; Bringezu, T.; Müller, G. *Biochem. Physiol. Pflanz.* **1986**, *181*(3), 165–176.

2720. Krause, C. *Wiss. Z. Humboldt-Univ. Berlin, Math. Naturwiss. Reihe* **1976**, *25*(6), 769–775.

2721. Murakami, T. *Nippon Sakumotsu Gakkai Kiji* **1970**, *39*(4), 409–417.

2722. Sangwan, R. S.; Popli, S.; Singh, R. *Natl. Acad. Sci. Lett. (India)* **1982**, *5*(10), 327–330.

2723. Jenner, C. F. *Plant Physiol.* **1968**, *43*(1), 41–49.

2724. Collins, G. G.; Jenner, C. F.; Paleg, L. G. *Plant Physiol.* **1972**, *49*(3), 398–403.

2725. Standard, S. A.; Perret, D.; Bray, C. M. *J. Exp. Bot.* **1983**, *34*(145), 1047–1054.

2726. Cheung, C. P.; Suhadolnik, R. J. *Anal. Biochem.* **1977**, *83*(1), 52–56.

2727. Blekhman, G. I. *Fiziol. Rast.* **1979**, *26*(4), 779–787 (Engl. 627–634).

2728. Savitch, L. V.; Gray, G. R.; Huner, N. P. A. *Planta* **1997**, *201*(1), 18–26.

2729. Sellami, A. *Biochim. Biophys. Acta* **1976**, *423*(3), 524–539.

2730. Stitt, M.; Lilley, R. McC; Heldt, H. W. *Plant Physiol.* **1982**, *70*(4), 971–977.

2731. Bomsel, J.-L.; Pradet, A. *Physiol. Vég.* **1967**, *5*(3), 223–236.

2732. Ladonin, V. F. *Fiziol. Rast.* **1970**, *17*(4), 744–750 (Engl. 622–627).

2733. Rybka, Z. *J. Plant Physiol.* **1990**, *136*(5), 635–637.

2734. Cheung, C. P.; Schadolnik, R. J. *Nature* **1978**, *271*, 357–358.

2735. Bergmann, H.; Müntz, R.; Rammelt, R.; Müller-Stoll, W. R. *Flora* **1967**, *158A*(4), 384–406.

2736. Quick, W. A.; Shaw, M. *Can. J. Bot.* **1966**, *44*(4), 777–788.

2737. Grass, L.; Burris, J. S. *Can. J. Plant Sci.* **1995**, *75*(4), 831–839.

2738. Abdel-Wahab, M. F.; El-Kinawi, S. A. *Acta Chem. Scand.* **1960**, *14*(7), 1667–1668.

2739. Michalke, B.; Schnabl, H. *J. Plant Physiol.* **1987**, *130*(2/3), 243–253.

2740. Toko, K.; Iiyama, S.; Tanaka, C.; Hayashi, K.; Yamafuji, K.; Yamafuji, K. *Biophys. Chem.* **1987**, *27*(1), 39–58.

2741. Lahiri, K.; Chattopadhyay, S.; Chatterjee, S.; Ghosh, B. *Ann. Bot.* **1993**, *71*(6), 485–488.

2742. Gregoire, J.; Gregoire, J.; Limozin, N.; Van, L. V. *Bull. Soc. Chim. Biol.* **1965**, *47*(2), 195–212.

2743. Yoshida, S.; Matsuura, C.; Etani, S. *Plant Physiol.* **1989**, *89*(2), 634–642.

2744. Hayasahi, T.; Kusunose, H.; Oshibuchi, T. *Nippon Nogei Kagaku Kaishi* **1978**, *52*(8), 301–308.

2745. Ali-Zade, M. A.; Abdullaev, F. I. *Izv. Akad. Nauk Az. SSR, Ser. Biol. Nauk* **1974**, (3), 33–37.

2746. Marutyan, S. A. *Dokl. Akad. Nauk Arm. SSR* **1976**, *62*(1), 59–63.

2747. Tesnière, C.; Romieu, C.; Dugelay, I.; Nicol, M. Z.; Flanzy, C.; Robin, J. P. *J. Exp. Bot.* **1994**, *45*(270), 145–151.

2748. Tesnière, C.; Raymond, P.; Abbal, P. *Sci. Aliments* **1987**, *7*(3), 447–458.

2749. Eichhorn, M.; Augsten, H. *Z. Pflanzenphysiol.* **1977**, *84*(1), 37–48.

2750. Esashi, Y.; Fuwa, M.; Kurota, A.; Oota, H.; Abe, M. *Plant Cell Physiol.* **1987**, *28*(1), 141–150.

2751. Foley, M. E.; Nafziger, E. D.; Slife, F. W.; Wax, L. M. *Weed Sci.* **1983**, *31*(1), 76–80.

2752. Cohn, M. A.; Obendorf, R. L. *Crop Sci.* **1976**, *16*(4), 449–452.

2753. Zeleneva, I. V. *Ontogenez* **1977**, *8*(3), 289–296.

2754. Christianson, D. D.; Paulis, J. W.; Wall, J. S. *Anal. Biochem.* **1968**, *22*(1), 35–46.

2755. Dankert, M.; Passeron, S.; Recondo, E. *An. Asoc. Quim. Argent.* **1972**, *60*(3/4), 257–271.

2756. Barbaro, A.; Mesjasz-Przybylowicz, J.; Parczewska, J.; Radwan-Kaźnika, M. *Acta Physiol. Plant.* **1981**, *3*(3), 103–116.

2757. Ohshima, T.; Hayashi, H.; Chino, M. *Plant Cell Physiol.* **1990**, *31*(5), 735–737.

2758. Roeckel-Drevet, P.; Digonnet, C.; Matthys-Rochon, E.; Champiat, D.; Dumas, C. *Plant Physiol. Biochem.* **1995**, *33*(3), 289–294.

2759. Peterson, T. A.; Nieman, R. H.; Clark, R. A. *Plant Physiol.* **1987**, *85*(4), 984–989.

2760. Schell, L. P.; Danehower, D. A.; Anderson, J. R., Jr.; Patterson, R. P. *Crop Sci.* **1991**, *31*(2), 425–430.

2761. Chang, C. W. *Plant Physiol.* **1968**, *43*(5), 669–674.

2762. Lin, W.; Hanson, J. B. *Plant Physiol.* **1974**, *54*(3), 250–256.

2763. Delistraty, D. A.; Hershner, C. *J. Appl. Biochem.* **1983**, *5*(6), 404–419.

2764. Lowry, O. H.; Passonneau, J. V. *A Flexible System of Enzymatic Analysis.* Academic Press: New York, 1972; xii + 291 pp.

2765. Hurlbert, R. B.; Schmitz, H.; Brumm, A. F.; Potter, V. R. *J. Biol. Chem.* **1954**, *209*(1), 23–39.

2766. Pontis, H. G.; Blumson, N. L. *Biochim. Biophys. Acta* **1958**, *27*, 618–624.

2767. Mazurova, T. A. *Vestn. Leningr. Univ., Biol.* **1978**, (2), 112–119.

2768. Keys, A. J. *J. Exp. Bot.* **1963**, *14*(40), 14–28.

2769. Messinger, A.; Powell, J. W.; Weidlich, T.; Genzel, L. *J. Biomol. Struct. Dyn.* **1993**, *10*(5), 841–852.

2770. Huhn, G. F.; Jensen, J. H.; Pierce, M. E.; Schlabach, E. M. *Sep. Sci. Technol.* **1993**, *28*(11–12), 1959–1970.

2771. Hooks, M. A.; Shearer, G. C.; Roberts, J. K. M. *Plant Physiol.* **1994**, *104*(2), 581–589.

2772. Lutz, N. W.; Yahi, N.; Fantini, J.; Cozzone, P. J. *Magn. Reson. Med.* **1996**, *36*(5), 788–795.

2773. Stroinski, A.; Szczotka, Z. *Rocz. Akad. Roln. Poznaniu* **1994**, *264*, 79–85.

2774. Raymond, P.; Narayanan, A.; Pradet, A. *Biochem. Biophys. Res. Commun.* **1973**, *53*(4), 1115–1121.

2775. Ashton, A. R.; Polya, G. M. *Biochem. J.* **1977**, *165*(1), 27–32.

2776. Wilson, T.; Moustafa, E.; Renwick, A. G. C. *Biochem. J.* **1978**, *175*(3), 931–936.

2777. Brewin, N. J.; Northcote, D. H. *J. Exp. Bot.* **1973**, *24*(82), 881–888.

2778. Amrhein, N. *Planta* **1974**, *118*(3), 241–258.

2779. Mascarenhas, J. P.; Salomon, D. *Proceedings of the 8th International Conference on Plant Growth Substances*1984 pp 1218–1224, from *Chem. Abstr.* **1977**, *87*, 148683s.

2780. Wellburn, A. R.; Ashby, J. P.; Wellburn, F. A. M. *Biochim. Biophys. Acta* **1973**, *320*(2), 363–371.

2781. Bressan, R. A.; Ross, C. W.; Vandepeute, J. *Plant Physiol.* **1976**, *57*(1), 29–37.

2782. Narayanan, A.; Pradet, A. *Indian J. Exp. Bot.* **1974**, *12*(3), 287–289.

2783. Suh, G.-R.; Park, I. *Chayon Kwahak Taehak Nonmunjip* **1976**, *1*(1), 101–106.

2784. Drlica, K. A.; Gardner, J. M.; Kado, C. I.; Vijay, I. K.; Troy, F. A. *Biochem. Biophys. Res. Commun.* **1974**, *56*(3), 753–759.

2785. Francko, D. A.; Wetzel, R. G. *Physiol. Plant.* **1981**, *52*(1), 33–36.

2786. Srivastava, A. K.; Azhar, S.; Krishna Murti, C. R. *FEBS Lett.* **1974**, *47*(2), 330–332.

2787. Kessler, B.; Levensein, R. *Biochim. Biophys. Acta* **1974**, *343*(1), 156–166.

2788. Johnson, L. P.; MacLeod, J. K.; Parker, C. W.; Letham, D. S.; Hunt, N. H. *Planta* **1981**, *152*(3), 195–201.

2789. Tanchuco, J. Q.; Ramirez, D. A.; Mujer, C. V. *Kalikasan* **1981**, *10*(2/3), 351–354.

2790. Kurosaki, F.; Tsurusawa, Y.; Nishi, A. *Phytochemistry* **1987**, *26*(7), 1919–1923.

2791. Kaul, R.; Sachar, R. C. *Biochem. Biophys. Res. Commun.* **1982**, *104*(1), 126–132.

2792. Cyong, J. C.; Takahashi, M. *Chem. Pharm. Bull.* **1982**, *30*(7), 2463–2466.

2793. Handa, A. K.; Johri, M. M. *Plant Physiol.* **1977**, *59*(3), 490–496.

2794. Hahn, M. G.; Grisebach, H. *Z. Naturforsch.* **1983**, *38C*(7/8), 578–582.

2795. Spiteri, A.; Viratelle, O. M.; Raymond, P.; Rancillac, M.; Labouesse, J.; Pradet, A. *Plant Physiol.* **1989**, *91*(2), 624–628.

2796. Truelsen, T. A.; Wyndaele, R. *Physiol. Plant.* **1978**, *42*(3), 324–330.

2797. Giannattasio, M.; Mandato, E.; Macchia, V. *Biochem. Biophys. Res. Commun.* **1974**, *57*(2), 365–371.

2798. Alvarez, R.; Moore, T. C.; Vandepeute, J. *Plant Physiol.* **1974**, *53*(2), 144–148.

2799. Miller, J. J.; Galsky, A. G. *Phytochemistry* **1974**, *13*(8), 1295–1296.

2800. Stroiński, A.; Floryszak-Wieczorek, J. *Acta Physiol. Plant.* **1985**, *7*(2), 35–43.

2801. Stroiński, A.; Floryszak-Wieczorek, J. *Plant Sci.* **1985**, *42*(1), 1–4.

2802. Kull, U.; Kühn, B. *Z. Naturforsch.* **1975**, *30C*(1), 69–72.

2803. Gangwani, L.; Tamot, B. K.; Khurana, J. P.; Maheshwari, S. C. *Biochem. Biophys. Res. Commun.* **1991**, *178*(3), 1113–1119.

2804. Kato, R.; Fujii, T. *Plant Cell Physiol.* **1988**, *29*(1), 85–88.

2805. Hasunuma, K.; Funadera, K.; Furukawa, K.; Miyamoto-Shinohara, Y. *Photochem. Photobiol.* **1988**, *48*(1), 89–92.

2806. Legendre, L.; Derckel, J. P.; Wrisez, F.; Corrèze, C.; Audran, J. C.; Haye, B.; Lambert, B. *Phytochemistry* **1997**, *44*(5), 769–774.

2807. Ashton, A. R.; Polya, G. M. *Plant Physiol.* **1978**, *61*(5), 718–722.

2808. Jayaswal, R. K. *Cytobios* **1991**, *67*(270/1), 153–159.

2809. Cooke, C. J.; Smith, C. J.; Walton, T. J.; Newton, R. P. *Phytochemistry* **1994**, *35*(4), 889–895.

2810. Francko, D. A.; Wetzel, R. G. *Aquat. Bot.* **1984**, *19*(1/2), 13–22.

2811. Rosenberg, N.; Pines, M.; Sela, I. *FEBS Lett.* **1982**, *137*(1), 105–107.

2812. Ames, I. H. *Plant Cell Physiol.* **1976**, *17*(3), 1059–1066.

2813. Roef, L.; Witters, E.; Gadeyne, J.; Marcussen, J.; Newton, R. P.; Van Onckelen, H. A. *Anal. Biochem.* **1996**, *233*(2), 188–196.

2814. Witters, E.; Roef, L.; Newton, R. P.; Van Dongen, W.; Esmans, E. L.; Van Onckelen, H. A. *Rapid Commun. Mass Spectrom.* **1996**, *10*(2), 225–231.

2815. Witters, E.; Vanhoutte, K.; Dewitte, W.; Machackova, I.; Benkova, E.; Van Dongen, W.; Esmans, E. L.; Van Onckelen, H. A. *Phytochem. Anal.* **1999**, *10*(3), 143–151.

2816. Lundeen, C. V.; Wood, H. N.; Braun, A. C. *Differentiation* **1973**, *1*(4), 255–260.

2817. Handa, A. K.; Bressan, R. A. *Anal. Biochem.* **1980**, *102*(2), 332–339.

2818. Lee, C.-H.; Lim, B. T.; Ong, K. K. *Biochem. Int.* **1982**, *5*(5), 643–649.

2819. Renelt, A.; Colling, C.; Hahlbrock, K.; Nürnberger, T.; Parker, J. E.; Sacks, W. R.; Scheel, D. *J. Exp. Bot.* **1993**, *44*(Suppl.), 257–268.

2820. Newton, R. P.; Gibbs, N.; Moyse, C. D.; Wiebers, J. L.; Brown, E. G. *Phytochemistry* **1980**, *19*(9), 1909–1911.

2821. Kingston, E. E.; Beynon, J. E.; Newton, R. P. *Biomed. Mass Spectrom.* **1984**, *11*(7), 367–374.

2822. Karinova, F. G.; Tarchevskaya, O. I.; Levnova, S. A.; Zhukov, S. N. *Fiziol. Biokhim. Kul't. Rast.* **1990**, *22*(6), 532–537.

2823. Brown, E. G.; Newton, R. P. *Phytochemistry* **1983**, *12*(11), 2683–2685.

2824. Brown, E. G.; Al-Najafi, T.; Newton, R. P. *Phytochemistry* **1979**, *18*(1), 9–14.

2825. Van Onckelen, H. A.; Dupon, M.; De Greef, J. A. *Physiol. Plant.* **1982**, *55*(2), 93–97.

2826. Brown, E. G.; Newton, R. P.; Evans, D. E.; Walton, T. J.; Younis, L. M.; Vaughan, J. M. *Phytochemistry* **1989**, *28*(10), 2559–2563.

2827. Takahashi, N.; Ejiri, S.; Katsumata, T. *Agric. Biol. Chem.* **1978**, *42*(8), 1605–1606.

2828. Smeltzer, R. H.; Johnson, M. A. *Can. J. Forest Res.* **1977**, *7*(1), 68–75.

2829. Newton, R. P.; Chiatante, D.; Ghosh, D.; Brenton, A. G.; Walton, T. J.; Harris, F. M.; Brown, E. G. *Phytochemistry* **1989**, *28*(9), 2243–2254.

2830. Becker, D.; Ziegler, H. *Planta* **1973**, *110*(1), 85–89.

2831. Yavorskaya, V. K.; Dragovoz, I. V.; Savinsky, S. V.; Markova, V. E.; Kofuran, I. Sh. *Phytochem. Anal.* **1991**, *2*(5), 204–207.

2832. Janistyn, B.; Drumm, H. *Planta* **1975**, *125*(1), 81–85.

2833. Isherwood, F. A.; Ring, S. G. *Phytochemistry* **1977**, *16*(3), 309–310.

2834. Ocheretina, O. Yu.; Il'inskaya, L. I.; Karavaeva, K. A.; Sokolovskii, V. Yu.; Ozerets-kovskaya, O. L.; Kritskii, M. S. *Prikl. Biokhim. Mikrobiol.* **1991**, *27*(6), 890–897 (Engl. 675–681).

2835. Tu, J. C. *Protoplasma* **1979**, *99*(1/2), 139–146.

2836. Niles, R. M.; Mount, M. S. *Plant Physiol.* **1974**, *54*(3), 372–373.

2837. Lee, C. H. *Biochem. Int.* **1985**, *10*(1), 123–128.

2838. Janistyn, B. *Planta* **1983**, *159*(4), 382–385.

2839. Janistyn, B. *Z. Naturforsch.* **1981**, *36C*(3/4), 193–196.

2840. Edlich, W.; Gräser, H. *Biochem. Physiol. Pflanz.* **1978**, *173*(2), 114–122.

2841. Bachofen, R. *Plant Sci. Lett.* **1973**, *1*(11), 447–450.

2842. Sim, W.-S.; Kim, H.-R. *Plant Cell Physiol.* **1987**, *28*(3), 415–420.

2843. Sen, S.; Pilet, P. E. *Experientia* **1981**, *37*(12), 1279–1280.

2844. Cyong, J.-C.; Hanabusa, K. *Phytochemistry* **1980**, *19*(2), 2747–2748.

2845. Cyong, J.-C.; Takahashi, M. *Phytochemistry* **1982**, *21*(8), 1871–1874.

2846. Hoffmann, J.; Pahlich, E.; Steubing, L. *Int. J. Environ. Anal. Chem.* **1976**, *4*(3), 183–196.

2847. Pradet, A. *Physiol. Vég.* **1967**, *5*(3), 209–221.

2848. Moyer, J. D.; Henderson, J. F. *Anal. Biochem.* **1983**, *131*(1), 187–189.

2849. Gendraud, M. *Physiol. Vég.* **1977**, *15*, 121–132.

2850. Kaimakov, E. A. *Biofizika* **1977**, *22*(6), 1107–1108.

2851. Thannhauser, S. J. *Z. Physiol. Chem.* **1919**, *107*(2/3), 157–164.

2852. Embden, G.; Schmidt, O. *Z. Physiol. Chem.* **1929**, *181*, 130–139.

2853. Lipkin, D.; Cook, W. H.; Markham, R. *J. Am. Chem. Soc.* **1959**, *81*(23), 6198–6203.

2854. Beleznay, F.; Biazo, G.; Ladik, J. *Acta Phys. Acad. Sci. Hung.* **1965**, *18*(3), 213–219.

2855. Skolachev, V. P.; Denisovich, L. I. *Biokhimiya* **1966**, *31*(1), 132–136.

2856. Garel, J. P.; Filliol, D.; Mandel, P. *J. Chromatogr.* **1973**, *78*(2), 381–391.

2857. Zaslavskii, B. Yu.; Shchukina, L. G.; Rogozhin, S. V. *Mol. Biol. (Moscow)* **1981**, *15*(6), 1315–1320.

2858. Duggan, D. E.; Bowman, R. L.; Brodie, B. B.; Udenfriend, S. *Arch. Biochem. Biophys.* **1957**, *68*(1), 1–14.

2859. Steele, R. H.; Szent-Györgyi, A. *Proc. Natl. Acad. Sci. U. S. A.* **1957**, *43*, 477–491.

2860. Kiessling, M.; Winsel, K.; Freytag, B. *Isotopenpraxis* **1990**, *26*(7), 330–331.

2861. Levene, P. A.; Simms, H. S. *J. Biol. Chem.* **1925**, *65*, 519–534.

2862. Bock, R. M.; Ling, N.-S.; Morell, S. A.; Lipton, S. H. *Arch. Biochem. Biophys.* **1956**, *62*(2), 253–264.

2863. Danyluk, S. S.; Hruska, F. E. *Biochemistry* **1968**, *7*(3), 1038–1043.

2864. Anon. Pabst Laboratories, *Circular* OR-10, 1956.

2865. Lenormant, H.; Blout, E. R. *Compt. Rend.* **1954**, *239*(20), 1281–1283.

2866. Tajmir-Riahi, H.-A. *Biochim. Biophys. Acta* **1990**, *1087*(1), 49–54.

2867. Aylward, N. N.; Koenig, J. L. *Macromolecules* **1970**, *3*(5), 590–596.

2868. Lin, C. Y.; Urry, D. W.; Eyring, H. *Biochem. Biophys. Res. Commun.* **1964**, *17*(6), 642–647.

2869. Follmann, H.; Kuntz, I.; Zacharias, W. *Eur. J. Biochem.* **1975**, *58*(1), 31–41.

2870. Sukhov, D. A.; Dodonova, N. Ya.; Vilesov, F. I. *Biofizika* **1976**, *21*(5), 817–819.

2871. Shulman, R. G.; Rahn, R. O. *J. Chem. Phys.* **1966**, *41*(8), 2940–2946.

2872. Simonsen, M. G.; Coleman, R. V.; Hansma, P. K. *J. Chem. Phys.* **1974**, *51*(9), 3789–3799.

2873. Buckin, V. A.; Sarvazyan, A. P.; Dudchenko, E. I.; Hemmes, P. *J. Phys. Chem.* **1980**, *84*(7), 696–699.

2874. Nishizawa, S.; Kuramoto, N.; Huang, H.; Jordan, F. *J. Phys. Chem.* **1999**, *103B*(18), 3754–3757.

2875. Feldman, I.; Agarwal, R. P. *J. Am. Chem. Soc.* **1968**, *90*(26), 7329–7333.

2876. Bain, A. D.; Bell, R. A.; Everett, J. R.; Hughes, D. W. *Can. J. Chem.* **1980**, *58*(18), 1947–1956.

2877. Allerhand, A.; Doddrell, D.; Komoroski, R. *J. Chem. Phys.* **1971**, *55*(1), 189–198.

2878. Richards, R. E.; Thomas, N. A. *J. Chem. Soc., Perkin Trans. 2* **1980**, (4), 368–374.

2879. Markowski, V.; Sullivan, G. R.; Roberts, J. D. *J. Am. Chem. Soc.* **1977**, *99*(3), 714–718.

2880. Gerlt, J. A.; Demon, P. C.; Mehdi, S. *J. Am. Chem. Soc.* **1982**, *104*(10), 2848–2856.

2881. Sun, B. Q.; Rienstra, C. M.; Costa, P. R.; Williamson, J. R.; Griffin, R. G. *J. Am. Chem. Soc.* **1997**, *119*(36), 8540–8546.

2882. Bast, P.; Berger, S.; Günther, H. *Magn. Reson. Chem.* **1992**, *30*(7), 587–594.

2883. Lawson, A. M.; Stillwell, R. N.; Tacker, M. M.; Tsuboyama, R.; McCloskey, J. A. *J. Am. Chem. Soc.* **1971**, *93*(4), 1014–1023.

2884. Schulten, H. R.; Beckey, H. D. *Org. Mass Spectrom.* **1973**, *7*(7), 861–867.

2885. Schram, K. H.; Slowikowski, D. L. *Biomed Environ. Mass Spectrom.* **1986**, *13*(5), 263–264.

2886. Eicke, A.; Sichtermann, W.; Benninghoven, A. *Org. Mass Spectrom.* **1980**, *15*(6), 289–294.

2887. Edmonds, C. G.; Loo, J. A.; Barinaga, C. H.; Udseth, H. R.; Smith, R. D. *J. Chromatogr.* **1989**, *474*(1), 21–37.

2888. Lai, S. T. F.; Evans, C. A., Jr. *Biomed. Mass Spectrom.* **1979**, *6*(1), 10–14.

2889. Van Vaeck, L.; Van Espen, P.; Adams, F.; Gijbels, R.; Lauwers, W.; Esmans, E. *Biomed. Environ. Mass Spectrom.* **1989**, *18*(8), 581–591.

2890. Ghosh, D.; Newton, R. P.; Brenton, A. G.; Harris, F. M.; Donovan, M. P.; Brown, E. G.; Walton, T. J. *Anal. Chim. Acta* **1991**, *247*(2), 187–199.

2891. Jost, W.; Hauck, H. E. *Anal. Biochem.* **1983**, *135*(1), 120–127.

2892. Hashizume, T.; Sasaki, Y. *Anal. Biochem.* **1966**, *15*(1), 199–203.

2893. Gelotte, B. *J. Chromatogr.* **1960**, *3*, 330–342.

2894. Krauss, G.; Wasternack, C.; Reinbothe, H. *J. Chromatogr.* **1973**, *76*(1), 248–251.

2895. Greenhut, J.; Rudolph, F. B. *J. Chromatogr.* **1985**, *319*(3), 461–466.

2896. Sanno, Y.; Honjo, M.; Tanaka, K. *Chem. Pharm. Bull.* **1961**, *9*(8), 657–659.

2897. Beckers, J. L.; Everaerts, F. M. *J. Chromatogr.* **1972**, *71*(2), 380–385.

2898. Bock, R. M.; Alberty, R. A. *J. Biol. Chem.* **1951**, *193*(1), 435–442.

2899. Lucy, C. A.; Yeung, K. K.-C.; Peng, X.; Chen, D. D. Y. *LC-GC* **1998**, *16*(1), 26–32.

2900. Gross, L.; Young, E. S. *J. Chromatogr.* **1989**, *480*, 169–178.

2901. Elisabeth, P.; Yoshioka, M.; Sasaki, T.; Senda, M. *J. Chromatogr.* **1998**, *806A*(1), 199–207.

2902. Berthod, H.; Pullman, B. *FEBS Lett.* **1973**, *33*(1), 147–150.

2903. Zink, S.; Beckel, C. L.; Larson, A. C. *J. Theor. Biol.* **1980**, *85*(2), 293–303.

2904. Alving, R. E.; Laki, K. *J. Theor. Biol.* **1972**, *34*(2), 199–214.

2905. Alberty, R. A.; Goldberg, R. N. *Biochemistry* **1992**, *31*(43), 10610–10615.

2906. Subbanaidu, D. V.; Sagar, N. L. S. V.; Murthy, V. R. *Asian Chem. Lett.* **1997**, *1*(1), 5–8, from *Chem. Abstr.* **1999**, *131*, 130214g.

2907. Levene, P. A. *J. Biol. Chem.* **1920**, *41*(4), 483–492.

2908. Kraut, J.; Jensen, L. H. *Acta Crystallogr.* **1963**, *16*(2), 79–88.

2909. Tajmir-Riahi, H. A.; Messaoudi, S. *J. Biomol. Struct. Dyn.* **1992**, *10*(2), 345–365.

2910. Lord, R. C.; Thomas, G. J., Jr. *Spectrochim. Acta* **1967**, *23A*(9), 2551–2591.

2911. Eagles, J.; Javanaud, C.; Self, R. *Biomed. Mass Spectrom.* **1984**, *11*(1), 41–46.

2912. Laudát, J.; Laudát, F. *Eur. Biophys. J.* **1992**, *21*(3), 233–239.

2913. Sugawara, Y.; Urabe, H.; Kobayashi, K.; Iimura, Y.; Iwasaki, M. *Mol. Cryst. Liq. Cryst. Sci. Technol.* **1996**, *277A*, 615–618.

2914. Escobar, R.; Carmona, P.; Molina, M.; Lasagabaster, A. *Analyst* **1994**, *119*(6), 1115–1118.

2915. Sternglanz, H.; Subramanian, E.; Lacey, J. C., Jr.; Bugg, C. E. *Biochemistry* **1976**, *15*(22), 4797–4802.

2916. Grigoratou, A.; Katsaros, N. *J. Inorg. Biochem.* **1985**, *24*(2), 147–154.

2917. Wagner-Jauregg, T. *Z. Physiol. Chem.* **1936**, *239*(4/6), 188–194.

2918. Lohmann, K. *Biochem. Z.* **1935**, *282*(1/2), 109–119.

2919. Khym, J. X. *Biochemistry* **1963**, *2*(3), 401–406.

2920. Ross, J. B. A.; Rousslang, K. W.; Motten, A. G.; Kwiram, A. L. *Biochemistry* **1979**, *18*(9), 1808–1813.

2921. Alberty, R. A.; Smith, R. M.; Bock, R. M. *J. Biol. Chem.* **1951**, *193*(1), 425–434.

2922. Matthies, M.; Zundel, G. *J. Chem. Soc., Perkin Trans. 2* **1977**, (14), 1824–1830.

2923. Lemieux, R. U.; Lown, J. W. *Can. J. Chem.* **1963**, *41*(4), 889–897.

2924. Rimai, L.; Cole, T.; Parsons, J. L.; Hickmott, J. T., Jr.; Carew, E. B. *Biophys. J.* **1969**, *9*(3), 320–329.

2925. Isenberg, I.; Szent-Györgyi, A. *Proc. Natl. Acad. Sci. U. S. A.* **1959**, *45*(8), 1232–1233.

2926. Breitmaier, E.; Voelter, W. *Eur. J. Biochem.* **1972**, *31*(2), 234–238.

2927. Gerlt, J. A.; Reynolds, M. A.; Demon, P. C.; Kenyon, G. I. *J. Am. Chem. Soc.* **1983**, *105*(21), 6469–6474.

2928. André, F.; Demassier, V.; Bloch, G.; Neumann, J.-M. *J. Am. Chem. Soc.* **1990**, *112*(19), 6784–6789.

2929. John, C. M.; Odom, R. W. *Int. J. Mass Spectrom. Ion. Proc.* **1997**, *161*(1/3), 47–67.

2930. Wada, H. *J. Chromatogr.* **1985**, *322*, 255–263.

2931. Zink, S.; Beckel, C. L.; Larson, A. C. *J. Theor. Biol.* **1980**, *85*(2), 305–323.

2932. Beaucamp, K.; Daiser, G. *Biochem. Z.* **1966**, *344*(1), 121–122.

2933. Hosur, M. V.; Viswamitra, M. A. *Curr. Sci.* **1979**, *48*(23), 1027–1028.

2934. Huang, C. M.; Chuang, P. C.; Chiang, T. T.; Yu, K. F. *Wu Li Hsueh Pao* **1966**, *22*(8), 911–918, from *Chem. Abstr.* **1967**, *67*, 77206n.

2935. Reynhardt, E. C.; Jurga, K.; Andrew, E. R. *J. Magn. Reson.* **1987**, *74*(3), 480–502.

2936. Swaminathan, P.; Sundaralingam, M. *Acta Crystallogr.* **1980**, *B36*(11), 2590–2597.

2937. Katti, S. K.; Viswamitra, M. A. *Curr. Sci.* **1979**, *48*(22), 989–990.

2938. Baddiley, J.; Todd, A. R. *J. Chem. Soc.* **1947**, (5), 648–651.

2939. Mizutani, M. *Ôsaku Daigaku Igaku Zassi* **1956**, *8*, 1334–1340.

2940. Svetlova, I. E.; Smirnova, N. S.; Dobrinina, N. A.; Martinenko, L. I.; Evseev, A. M.; Spitsyn, V. I. *Dokl. Akad. Nauk SSSR* **1987**, *294*(4), 890–893 (Engl. 266–269).

2941. Kiseleva, M. N.; Zarochentseva, E. P.; Dodonova, N. Ya. *Biofisika* **1975**, *20*(4), 561–565.

2942. Khalil, F. L.; Brown, T. L. *J. Am. Chem. Soc.* **1964**, *86*(23), 5113–5117.

2943. Levedahl, B. H.; James, T. W. *Biochim. Biophys. Acta* **1956**, *21*(2), 298–302 *Errata***1957**, *23* (2), 442–443.

2944. Gilligan, T. J., III; Schwartz, G. *Biophys. Chem.* **1976**, *4*(1), 55–63.

2945. Artykov, A. A.; Valaeva, S. R.; Zinov'ev, O. I.; Sarvazyan, A. P.; Shakhparonov, M. I. *Akust. Zh.* **1983**, *29*(5), 585–589 (Engl. 347–350).

2946. Dorman, D. E.; Roberts, J. D. *Proc. Natl. Acad. Sci. U. S. A.* **1970**, *65*(1), 19–26.

2947. Fujii, I.; Isobe, R.; Kanematsu, K. *J. Chem. Soc., Chem. Commun* **1985**, (7), 405–406.

2948. Liébecq, C.; Degueldre-Guillaume, M.-J. *Bull. Soc. Chim. Biol.* **1968**, *50*(10), 1615–1623.

2949. Berger, L. *Biochim. Biophys. Acta* **1956**, *20*(1), 23–26.

2950. Kennard, O.; Isaacs, N. W.; Motherwell, W. D. S.; Coppola, J. C.; Wampler, D. L.; Larson, A. C.; Watson, D. G. *Proc. R. Soc. Ser. A* **1971**, *325*(1562), 401–436.

2951. Katsaros, N.; Vrachnou-Astra, E.; Konstantatos, J. *J. Inorg. Biochem.* **1982**, *16*(3), 227–235.

2952. Urabe, H.; Sugawara, Y.; Kasuya, T. *Phys. Rev. B Condens. Matter* **1995**, *51*(9), 5666–5672.

2953. Sandström, A.; Chattopadhyaya, J. *J. Chem. Soc., Chem. Commun.* **1987**, (11), 862–864.

2954. Ballantine, J. A.; Games, D. E.; Slater, P. S. *Rapid Commun. Mass Spectrom.* **1997**, *11*(6), 624–629.

2955. Kerr, S. E. *J. Biol. Chem.* **1941**, *139*(1), 121–130.

2956. Nerheim, A. G.; Shaw, E. H., Jr. *Proc. S. Dakota Acad. Sci.* **1951**, *30*, 65–67.

2957. Baddiley, J.; Michelson, A. M.; Todd, A. R. *J. Chem. Soc.* **1949**, (3), 582–586.

2958. Evans, F. E.; Sarma, R. H. *J. Biol. Chem.* **1974**, *249*(15), 4754–4759.

2959. Son, T.-D.; Chachaty, C. *Biochim. Biophys. Acta* **1977**, *500*(2), 405–418.

2960. Fazakerley, G. V.; Reid, D. G. *Eur. J. Biochem.* **1979**, *93*(3), 535–543.

2961. Tanswell, P.; Thornton, J. M.; Korda, A. V.; Williams, R. J. P. *Eur. J. Biochem.* **1975**, *57*(1), 135–145.

2962. Perahia, D.; Pullman, B.; Saran, A. *Biochem. Biophys. Res. Commun.* **1972**, *47*(6), 1284–1289.

2963. Millner, O. E., Jr.; Andersen, J. A. *Biopolymers* **1975**, *14*(10), 2159–2179.

2964. Yoshikawa, M.; Kato, T. *Bull. Chem. Soc. Jpn.* **1967**, *40*(12), 2849–2853.

2965. Koransky, W.; Grunze, H.; Münch, G. *Z. Naturforsch.* **1962**, *17B*(5), 291–295.

2966. Ikehara, M.; Ohtsuka, E. *Chem. Pharm. Bull.* **1963**, *11*(4), 435–440.

2967. Ikehara, M.; Ohtsuka, E.; Ishikawa, F. *Chem. Pharm. Bull.* **1961**, *9*(3), 173–178.

2968. Khwaja, T. A.; Reese, C. B. *Tetrahedron* **1971**, *27*(24), 6189–6199.

2969. Hayakawa, Y.; Wakabayashi, S.; Nobori, T.; Noyori, R. *Tetrahedron Lett.* **1987**, *28*(20), 2259–2262.

2970. Engels, J.; Krahmer, U. *Synthesis* **1981**, (6), 485–486.

2971. Takaku, H. *Chem. Pharm. Bull.* **1977**, *25*(8), 2121–2123.

2972. Hata, T.; Chong, K.-J. *Bull. Soc. Chim. Jpn.* **1972**, *45*(2), 654.

2973. Naruse, M.; Fujimoto, Y. *Yakugaku Zasshi* **1966**, *86*(3), 192–195.

2974. Brown, D. M.; Haynes, L. J.; Todd, A. R. *J. Chem. Soc.* **1950**, (11), 3299–3304.

2975. Brown, D. M.; Fasman, G. D.; Magrath, D. I.; Todd, A. R. *J. Chem. Soc.* **1954**, (5), 1448–1455.

2976. Sowa, T.; Ouchi, S. *Bull. Chem. Soc. Jpn.* **1978**, *48*(7), 2084–2090.

2977. Hes, J.; Mertes, M. P. *J. Org. Chem.* **1974**, *39*(25), 3767–3769.

2978. Takaku, H.; Shimada, Y. *Tetrahedron Lett.* **1974**, (14), 1279–1282.

2979. Asai, M.; Miyaki, M.; Shimizu, B. *Chem. Pharm. Bull.* **1967**, *15*(12), 1856–1862.

2980. Shimizu, B.; Asai, M.; Nishimura, T. *Chem. Pharm. Bull.* **1967**, *15*(12), 1847–1855.

2981. Davisson, V. J.; Davis, D. R.; Dixit, V. M.; Poulter, C. D. *J. Org. Chem.* **1987**, *52*(9), 1794–1801.

2982. Khorana, H. G. *J. Am. Chem. Soc.* **1954**, *76*(13), 3517–3522.

2983. Kozarich, J. W.; Chinault, A. C.; Hecht, S. M. *Biochemistry* **1973**, *12*(22), 4458–4463.

2984. Chase, B. H.; Kenner, G. W.; Todd, A. R.; Webb, R. F. *J. Chem. Soc.* **1956**, (6), 1371–1376.

2985. Clark, V. M.; Kirby, G. W.; Todd, A. *J. Chem. Soc.* **1957**, (4), 1497–1501.

2986. Michelson, A. M. *J. Chem. Soc.* **1958**, (5), 1957–1963.

2987. Sawai, H.; Inaba, Y.; Hirano, A.; Wakai, H.; Shimazu, M. *Tetrahedron Lett.* **1993**, *34*(30), 4801–4804.

2988. Scheit, K.-H.; Kampe, W. *Chem. Ber.* **1965**, *98*(4), 1045–1048.

2989. Furusawa, K.; Sekine, M.; Hata, T. *J. Chem. Soc., Perkin Trans. 1* **1976**, (16), 1711–1716.

2990. Shumyantzeva, V. V.; Sokolova, N. I.; Shabarova, Z. A. *Nucleic Acids Res.* **1976**, *3*(4), 903–916.

2991. Chambers, R. W.; Khorana, H. G. *J. Am. Chem. Soc.* **1958**, *80*(14), 3749–3752.

2992. Goldman, L.; Marsico, L. W.; Anderson, G. W. *J. Am. Chem. Soc.* **1960**, *82*(11), 2969–2970.

2993. Cramer, F.; Neunhoeffer, N. *Chem. Ber.* **1962**, *95*(7), 1664–1669.

2994. Purygin, P. P.; Kolodkina, I. I.; Kon'kova, E. P.; Yurkevich, A. M. *Khim-Farm. Zh.* **1983**, *17*(10), 1235–1237.

2995. Kolodkina, I. I.; Varshavskaya, L. S.; Yurkevich, A. M.; Preobrazhenskii, N. A. *Zh. Obshch. Khim.* **1967**, *37*(9), 1996–2002 (Engl. 1894–1898).

2996. den Hartog, J. A. J.; van Boom, J. H. *Recl. J. R. Neth. Chem. Soc.* **1981**, *100*(7/8), 285–290.

2997. Cerletti, P.; Ipata, P. L.; Siliprandi, N. *Boll. Soc. Ital. Biol. Sper.* **1956**, *32*, 1113–1116.

2998. Dawson, R. M. C.; Ford, M.; Eichberg, J. *Biochem. J.* **1965**, *95*(1), 104–106.

2999. Mishra, N. C.; Broom, A. D. *J. Chem. Soc., Chem. Commun.* **1991**, *18*, 1276–1277.

3000. Ludwig, J. *Acta Biochim. Biophys. Sci. Hung.* **1981**, *16*(3/4), 131–133.

3001. Fukuoka, K.; Suda, F.; Ishikawa, M.; Hata, T. *Nucleosides Nucleotides* **1995**, *14*(3/5), 693–694.

3002. Michelson, A. M.; Todd, A. R. *J. Chem. Soc.* **1949**, (10), 2487–2490.

3003. Maeda, M.; Patel, A. D.; Hampton, A. *Nucleic Acids Res.* **1977**, *4*(8), 2843–2853.

3004. Crans, D. C.; Whitesides, G. M. *J. Org. Chem.* **1983**, *48*(18), 3130–3132.

3005. Baughn, R. L.; Adalsteinsson, Ö.; Whitesides, G. M. *J. Am. Chem. Soc.* **1978**, *100*(1), 304–306.

3006. Tanaka, K.; Honjo, M.; Sanno, Y.; Moriyama, H. *Chem. Pharm. Bull.* **1962**, *10*(3), 220–225.

3007. Moffatt, J. G.; Khorana, H. G. *J. Am. Chem. Soc.* **1961**, *83*(3), 649–658.

3008. Kitani, A.; Tsunetsugu, S.; Sasaki, K. *J. Chem. Soc., Perkin 2* **1991**, *2*(2), 329–331.

3009. Hosseini, M. W.; Lehn, J.-M. *J. Chem. Soc., Chem. Commun.* **1988**, (5), 397–399.

3010. Michelson, A. M. *Biochim. Biophys. Acta* **1964**, *91*(1), 1–13.

3011. Hata, T.; Furusawa, K.; Sekine, M. *J. Chem. Soc., Chem. Commun.* **1975**, (6), 196–197.

3012. Takaku, H.; Konishi, T.; Hata, T. *Chem. Lett.* **1977**, (6), 655–658.

3013. Shaoul, O.; Sporns, P. *J. Food Sci.* **1987**, *52*(3), 810–812.

3014. Hock, A.; Huber, G. *Biochem. Z.* **1956**, *328*(1), 44–55.

3015. Trapmann, H.; Devani, N. *Arch. Pharm.* **1965**, *298*(4), 253–257.

3016. Zhu, B.; Wu, Y. J.; Zhao, D. Q.; Ni, J. Z. *BioMetals* **1999**, *12*(1), 11–17, from *Chem. Abstr.* **1999**, *131*, 195902z.

3017. Liébecq, C. *Arch. Int. Physiol. Biochim.* **1957**, *65*(1), 141–142.

3018. Ishikawa, H. *J. Biochem.* **1935**, *22*(3), 385–395.

3019. Holbrook, K. A.; Ouellet, L. *Can. J. Chem.* **1957**, *35*(12), 1496–1503.

3020. Tetas, M.; Lowenstein, J. M. *Biochemistry* **1963**, *2*(2), 350–357.

3021. Yohannes, P. G.; Bowman-James, K. *Inorg. Chim. Acta* **1993**, *209*(2), 115–117.

3022. Shivakumar, K.; Appukutan, P. S.; Kartha, C. C. *Biochem. Int.* **1989**, *19*(4), 845–853.

3023. Hosseini, M. W.; Lehn, J.-M.; Mertes, M. P. *Helv. Chim. Acta* **1983**, *66*(8), 2454–2466. *Errata* **1985**, *68*, 818.

3024. Hutchings, G. J.; Banks, B. E. C.; Mruzek, M.; Ridd, J. H.; Vernon, C. A. *Biochemistry* **1981**, *20*(20), 5809–5816.

3025. Vanderheiden, B. S. *Anal. Biochem.* **1968**, *22*(2), 231–237.

3026. Sigel, H.; Hofstetter, F.; Martin, R. B.; Milburn, R. M.; Scheller-Krattiger, V.; Scheller, K. H. *J. Am. Chem. Soc.* **1984**, *106*(25), 7935–7946.

3027. Ramirez, F.; Marecek, J. F.; Szamosi, J. *J. Org. Chem.* **1980**, *45*(23), 4748–4752.

3028. Tsujimoto, T.; Yamaba, T. *Wakayama Med. Rep.* **1956**, *3*, 77–83, from *Chem. Abstr.* **1957**, *51*, 4457^1.

3029. Tsuhako, M.; Kumitomi, R.; Baba, Y.; Miyajima, T. *Bull. Chem. Soc. Jpn.* **1991**, *64*(2), 490–497.

3030. Lohrmann, R. *J. Mol. Evol.* **1975**, *6*(4), 237–252.

3031. Schoffstall, A. M.; Barto, R. J.; Ramos, D. L. *Origins of Life* **1982**, *12*(2), 143–151.

3032. Udenfriend, S.; Zaltzman, P. *Anal. Biochem.* **1962**, *3*(1), 49–59.

3033. Hug, O.; Kronski, D.; Kroedel, W.; Steenken, S. *Stud. Biophys.* **1968**, *7*, 161–168.

3034. Kai, M.; Ohkura, Y.; Yonekura, S.; Iwasaki, M. *Anal. Chim. Acta* **1994**, *287*(1/2), 75–81.

3035. Sohr, H.; Wienhold, K. *Anal. Chim. Acta* **1972**, *60*(2), 413–420.

3036. Lee, G. C. Y.; Chan, S. I. *J. Am. Chem. Soc.* **1972**, *94*(9), 3218–3229.

3037. Pearlman, D. A.; Kim, S. H. *J. Mol. Biol.* **1990**, *211*(1), 171–187.

3038. Gottarelli, G.; Proni, G.; Spada, G. P. *Liq. Cryst.* **1997**, *22*(5), 563–566, from *Chem. Abstr.* **1997**, *127*, 26509p.

3039. Ogawa, M.; Sakaguchi, T. *Yakugaku Zasshi* **1971**, *91*(7), 750–754.

3040. Navarro, R.; Morcillo, J. *J. Mol. Struct.* **1988**, *175*, 49–54.

3041. Nishimura, Y.; Tsuboi, M.; Sato, T.; Aoki, K. *J. Mol. Struct.* **1986**, *146*, 123–153.

3042. Yang, J. T.; Samejima, T.; Sarkar, P. K. *Biopolymers* **1966**, *4*(6), 623–636.

3043. Sprecher, C. A.; Johnson, W. C. *Biopolymers* **1977**, *16*(10), 2243–2264.

3044. Son, T. D.; Guschlbauer, W. *Nucleic Acids Res.* **1975**, *2*(6), 873–886.

3045. Norton, R. S.; Allerhand, A. *J. Am. Chem. Soc.* **1976**, *98*(4), 1007–1014.

3046. West, R. T.; Garza, L. A., II; Winchester, W. R.; Walmsley, J. A. *Nucleic Acids Res.* **1994**, *22*(23), 5128–5134.

3047. Lang, J.; Sturm, J.; Zana, R. *J. Phys. Chem.* **1973**, *77*(19), 2329–2335.

3048. Bertrand, M. J.; Benham, V.; St.-Louis, R.; Evans, M. J. *Can. J. Chem.* **1989**, *67*(5), 910–920.

3049. Randerath, K.; Randerath, E. *J. Chromatogr.* **1964**, *16*(1), 111–125.

3050. Hansske, F.; Cramer, F. *Carbohydr. Res.* **1977**, *54*(1), 75–84.

3051. Berthod, H.; Pullman, B. *FEBS Lett.* **1973**, *30*(2), 231–235.

3052. Starikov, E. B. *Mol. Phys.* **1993**, *80*(6), 1459–1467.

3053. Buell, M. V.; Perkins, M. E. *J. Biol. Chem.* **1927**, *72*(1), 21–26.

3054. Murayama, W.; Nagashima, N.; Shimizu, Y. *Acta Crystallogr.* **1969**, *B25*(11), 2236–2245.

3055. Emerson, J.; Sundaralingam, M. *Acta Crystallogr.* **1980**, *B36*(6), 1510–1513.

3056. Kamio, H.; Yakamachi, H. *Yakugaku Zasshi* **1967**, *87*(12), 1436–1441.

3057. Katti, S. K.; Seshadri, T. P.; Viswamitra, M. A. *Acta Crystallogr.* **1981**, *B37*(10), 1825–1831.

3058. Nishimura, Y.; Tsuboi, M.; Sato, T. *Nucleic Acids Res.* **1984**, *12*(17), 6901–6908.

3059. Paris, A.; Laszlo, P. *ACS Symp. Ser.* **1976**, *34*, 418–435.

3060. Churchill, M. E. A.; Pritchard, R. G.; Helliwell, M. *Acta Crystallogr.* **1992**, *C48*(7), 1223–1227.

3061. Levene, P. A. *J. Biol. Chem.* **1919**, *40*(1), 171–174.

3062. Phillips, R.; Eisenberg, P.; George, P.; Rutman, R. J. *J. Biol. Chem.* **1965**, *240*(11), 4393–4397.

3063. Sutherland, J. C.; Griffin, K. *Biopolymers* **1984**, *23*(12), 2715–2724.

3064. Sigel, H. *Eur. J. Biochem.* **1968**, *3*(4), 530–537.

3065. Lee, C.-H.; Evans, F. E.; Sarma, R. H. *J. Biol. Chem.* **1975**, *250*(4), 1290–1296.

3066. Tener, G. M. *J. Am. Chem. Soc.* **1961**, *83*(1), 159–168.

3067. Chambers, R. W.; Moffatt, J. G.; Khorana, H. G. *J. Am. Chem. Soc.* **1957**, *79*(14), 3747–3752.

3068. Imai, K.-I.; Fujii, S.; Takanohashi, K.; Furukawa, Y.; Masuda, T.; Honjo, M. *J. Org. Chem.* **1969**, *34*(6), 1547–1550.

3069. Chambers, R. W.; Khorana, H. G. *J. Am. Chem. Soc.* **1957**, *79*(14), 3752–3755.

3070. Ueda, T.; Ohtsuka, E. *Chem. Pharm. Bull.* **1959**, *7*(6), 740–743.

3071. Chambers, R. W.; Shapiro, P.; Kurkov, V. *J. Am. Chem. Soc.* **1960**, *82*(4), 970–975.

3072. Simon, E. S.; Grabowski, S.; Whitesides, G. M. *J. Org. Chem.* **1990**, *55*(6), 1834–1841.

3073. Lomakina, T. S.; Grineva, N. I. *Khim. Prir. Soedin. Akad. Nauk Uz. SSR* **1965**, (4), 275–283, from *Chem. Abstr.* **1966**, *64*, 6736h.

3074. Sigel, H.; Buisson, D. H.; Prijs, B. *Bioinorg. Chem.* **1975**, *5*(1), 1–20.

3075. Thannhauser, S. J.; Dorfmüller, G. *Z. Physiol. Chem.* **1919**, *104*(2/3), 64–72.

3076. Bredereck, H.; Richter, G. *Berichte* **1938**, *71B*(4), 718–720.

3077. Guéron, M.; Eisinger, J.; Shulman, R. G. *J. Chem. Phys.* **1967**, *47*(10), 4077–4091.

3078. Janik, B.; Paleček, E. *Arch. Biochem. Biophys.* **1964**, *105*(2), 225–236.

3079. Angell, C. L. *J. Chem. Soc.* **1961**, (2), 504–515.

3080. Reetz, C. P., Jr.; Ascarelli, G. *Biopolymers* **1982**, *21*(8), 1569–1586.

3081. Seuvre, A.-M.; Mathlouthi, M. *Carbohydr. Res.* **1987**, *169*, 83–103.

3082. Aylward, N. N.; Koenig, J. L. *Macromolecules* **1970**, *3*(5), 583–589.

3083. Takaku, O.; Haraguchi, H.; Toda, S.; Fuwa, K. *Agric. Biol. Chem.* **1975**, *39*(12), 2373–2381.

3084. Bolton, P. H. *J. Magn. Reson.* **1981**, *45*(2), 239–253.

3085. Budzikiewicz, H.; Linscheid, M. *Biomed. Mass Spectrom.* **1977**, *4*(3), 103–106.

3086. Read, M.; Duchesne, J. *Compt. Rend.* **1963**, *257*(10), 1724–1727.

3087. Schram, K. H.; Taniguchi, Y.; McCloskey, J. A. *J. Chromatogr.* **1978**, *155*(2), 355–361.

3088. Shimizu, B.; Saito, A.; Nishimura, T.; Miyaki, M. *Chem. Pharm. Bull.* **1967**, *15*(12), 2011–2014.

3089. Reddy, B. S.; Viswamitra, M. A. *Curr. Sci.* **1971**, *40*(13), 344.

3090. Miles, H. T. *Biochim. Biophys. Acta* **1956**, *22*(2), 247–253.

3091. Hogle, J.; Sundaralingam, M.; Lin, G. H. Y. *Acta Crystallogr.* **1980**, *36B*(3), 564–570.

3092. Levene, P. A. *J. Biol. Chem.* **1919**, *39*(1), 77–81.

3093. Braginskaya, F. I.; El'piner, I. E. *Biofizika* **1973**, *18*(1), 18–24.

3094. Viswamitra, M. A.; Seshadri, T. P.; Post, M. L.; Kennard, O. *Nature* **1975**, *258*(5535), 497–501.

3095. Ahmad, N.; Subhan, M.; Roellgen, F. W. *J. Chem. Soc. Pak.* **1991**, *13*(1), 24–28, from *Chem. Abstr.* **1991**, *115*, 169738a.

3096. Moffatt, J. G. *Can. J. Chem.* **1964**, *42*(3), 599–604.

3097. Davies, D. B.; Danyluk, S. S. *Biochemistry* **1974**, *13*(21), 4417–4434.

3098. Davies, D. B. *Stud. Biophys.* **1976**, *55*(1), 29–38.

3099. Kenner, G. W.; Reese, C. G.; Todd, A. *J. Chem. Soc.* **1958**, (2), 546–551.

3100. Kuriyama, C.; Fujizaki, F.; Murata, K. *Eiyo to Shokuryo* **1965**, *17*(5), 337–341, from *Chem. Abstr.* **1966**, *64*, 9988b.

3101. Levene, P. A. *J. Biol. Chem.* **1920**, *41*(1), 1–2.

3102. Levene, P. A.; Tipson, R. S. *J. Biol. Chem.* **1964**, *106*(1), 113–124.

3103. Ogawa, M.; Sakaguchi, T. *Chem. Pharm. Bull.* **1972**, *20*(1), 190–193.

3104. This reference for page 324, lines 25–26 is lost.

3105. Mantsch, H. H.; Smith, I. C. P. *Biochem. Biophys. Res. Commun.* **1972**, *46*(2), 808–815.

3106. Tsuda, T.; Takagi, K.; Watanabe, T.; Satake, T. *J. High Resolut. Chromatogr.* **1988**, *11*(10), 721–723.

3107. Seshadri, T. P.; Viswamitra, M. A.; Kartha, G. *Acta Crystallogr.* **1980**, *B36*(4), 925–927.

3108. Carmona, P.; Rodriguez-Casado, A.; Molina, M.; Escobar, R. *J. Raman Spectrosc.* **1999**, *30*(8), 631–636.

3109. Escobar, R.; Carmona, P.; Molina, M. *Z. Anal. Chem.* **1991**, *340*(4), 213–216.

3110. Levene, P. A. *J. Biol. Chem.* **1918**, *33*, 229–234.

3111. Shefter, E.; Trueblood, K. N. *Acta Crystallogr.* **1965**, *18*(6), 1067–1077.

3112. Levene, P. A. *J. Biol. Chem.* **1919**, *40*(2), 395–398.

3113. Viswamitra, M. A.; Post, M. L.; Kennard, O. *Acta Crystallogr.* **1979**, *B35*(5), 1089–1094.

3114. Khan, M. M. T.; Reddy, P. R. *J. Inorg. Nucl. Chem.* **1976**, *38*(6), 1234–1236.

3115. Chinsky, L.; Turpin, P. Y.; Duquesne, M.; Brahms, J. *Biopolymers* **1978**, *17*(5), 1347–1359.

3116. Malakhov, K. V.; Turkina, M. Ya.; Dobychin, S. L. *Zh. Anal. Khim.* **1986**, *41*(3), 511–518 (Engl. 406–412).

3117. Baczynskij, L.; Bronson, G. E. *Biochem. Environ. Mass Spectrom.* **1988**, *16*, 253–254.

3118. Yathindra, N.; Sundaralingam, M. *Biopolymers* **1973**, *12*(10), 2261–2277.

3119. Takaku, H.; Shimada, Y.; Arai, K. *Bull. Chem. Soc. Jpn.* **1974**, *47*(3), 779–780.

3120. Gulland, J. M.; Smith, H. *J. Chem. Soc.* **1947**, (3), 338–342.

3121. Smith, M. *Biochem. Prep.* **1961**, *8*, 130–133.

3122. Bredereck, H. *Z. Physiol. Chem.* **1934**, *224*(1/2), 79–85.

3123. Wasserman, H. H.; Cohen, D. *J. Org. Chem.* **1964**, *29*(7), 1817–1820.

3124. Chambers, R. W. *J. Am. Chem. Soc.* **1959**, *81*(12), 3032–3035.

3125. Kenner, G. W.; Todd, A. R.; Weymouth, F. J. *J. Chem. Soc.* **1952**, (10), 3675–3681.

3126. Hall, R. H.; Khorana, H. G. *J. Am. Chem. Soc.* **1954**, *76*(20), 5056–5060.

3127. Kenner, G. W.; Todd, A. R.; Webb, R. F.; Weymouth, F. J. *J. Chem. Soc.* **1954**, (7), 2288–2293.

3128. Sigel, H.; Hofstetter, F. *Eur. J. Biochem.* **1983**, *132*(3), 569–577.

3129. Odmark, G.; Kihlman, B. A. *Mutat. Res.* **1965**, *2*(3), 274–286.

3130. Lapper, R. D.; Mantsch, H. H.; Smith, I. C. P. *J. Am. Chem. Soc.* **1973**, *95*(9), 2878–2880.

3131. Klein, W.; Thannhauser, S. J. *Z. Physiol. Chem.* **1934**, *224*, 252–260.

3132. Volkin, E.; Khym, J. X.; Cohn, W. E. *J. Am. Chem. Soc.* **1951**, *73*(4), 1533–1536.

3133. Szerenyi, P.; Dearman, H. H. *Chem. Phys. Lett.* **1972**, *15*(1), 81–85.

3134. Zimmerman, S. B.; Murphy, L. D. *Biopolymers* **1992**, *32*(10), 1365–1373.

3135. Klukanova, H.; Studnickova, M.; Kovar, J.; Turanek, J.; Kahle, V. *Bioelectrochem. Bioenerg.* **1986**, *15*(3), 317–328.

3136. Kinoshita, H.; Christian, S. D.; Dryhurst, G. *J. Electroanal. Chem.* **1977**, *83*(1), 151–166.

3137. Hurst, R. O.; Marko, A. M.; Butler, G. C. *J. Biol. Chem.* **1953**, *204*(2), 847–856.

3138. Clauwaert, J.; Stockx, J. *Z. Naturforsch.* **1968**, *23B*(1), 25–30.

3139. Phillips, R.; Eisenberg, P.; George, P.; Rutman, R. J. *J. Biol. Chem.* **1965**, *240*(11), 4393–4397.

3140. Rawitscher, M.; Sturtevant, J. M. *J. Am. Chem. Soc.* **1960**, *82*(14), 3739–3740.

3141. Espinosa-Müller, A. W.; Bravo, A. N. *J. Mol. Struct.* **1985**, *133*, 179–191.

3142. Benevides, J. M.; Lemeur, D.; Thomas, G. J., Jr. *Biopolymers* **1984**, *23*(6), 1011–1024.

3143. Fodor, S. P. A.; Rava, R. P.; Hays, T. R.; Spiro, T. G. *J. Am. Chem. Soc.* **1985**, *107*(6), 1520–1529.

3144. Cozzone, P. M.; Jardetzky, O. *Biochemistry* **1976**, *15*(22), 4853–4859.

3145. Wiebers, J. L. *Anal. Biochem.* **1973**, *51*(2), 542–556.

3146. Cerny, R. L.; Gross, M. L.; Grotjahn, L. *Anal. Biochem.* **1986**, *156*(2), 424–435.

3147. Hernández, H. *Rapid Commun. Mass Spectrom.* **1996**, *10*(12), 1543–1550.

3148. Viari, A.; Ballini, J. P.; Meleard, P.; Vigny, P.; Dousset, P.; Blonski, C.; Shire, D. *Biomed. Environ. Mass Spectrom.* **1987**, *16*(1/12), 225–228.

3149. Ho, Y.; Kebarle, P. *Int. J. Mass Spectrom. Ion Process.* **1997**, *165/6*, 433–455.

3150. Sato, K.; Seki, T.; Inukai, S.; Shimizu, S. *Anal. Biochem.* **1981**, *116*(1), 185–188.

3151. Cadet, J.; Voituriez, L.; Berger, M. *J. Chromatogr.* **1983**, *259*(1), 111–119.

3152. Dizdaroglu, M.; Hermes, B. *J. Chromatogr.* **1979**, *171*, 321–330.

3153. Caronia, J. P.; Crowther, J. B.; Hartwick, R. A. *J. Liq. Chromatogr.* **1983**, *6*(9), 1673–1691.

3154. Whitehouse, R. P.; Greenstock, C. L. *J. Liq. Chromatogr.* **1982**, *5*(11), 2085–2093.

3155. Wakizaka, A.; Kurosaka, K.; Okuhara, E. *J. Chromatogr.* **1979**, *162*(3), 319–326.

3156. Hancock, R. L. *J. Gas Chromatogr.* **1966**, *4*(10), 363–366.

3157. Lerner, J.; Schepartz, A. I. *J. Chromatogr.* **1969**, *39*, 132–139.

3158. Row, K. H.; Griest, W. H.; Maskarinec, M. P. *J. Chromatogr.* **1987**, *409*, 193–203.

3159. Kathawala, F.; Cramer, F. *Ann. Chem.* **1968**, *712*, 195–200.

3160. Thornburg, W. W.; Werum, L. N.; Gordon, H. T. *J. Chromatogr.* **1961**, *6*, 131–141.

3161. Keck, K.; Hagen, U. *Biochim. Biophys. Acta* **1964**, *87*(4), 685–687.

3162. Row, K. H.; Raw, J. I. *Sep. Sci. Technol.* **1990**, *25*(3), 323–333.

3163. Britz-McKibbin, P.; Chen, D. D. Y. *J. Chromatogr.* **1997**, *781A*(1/2), 23–34.

3164. Cornillon, R.; Lespinasse, J. N.; Broch, H.; Vasilescu, D. *Stud. Biophys.* **1975**, *48*(2), 139–143.

3165. Starikov, E. B. *Int. J. Quantum Chem.* **1996**, *58*(5), 497–515.

3166. Starikov, E. B. *Int. J. Quantum Chem.* **1998**, *69*(2), 209–217.

3167. Kim, N. S.; Jiang, Q.; Lebreton, P. H. *Int. J. Quantum Chem.* **1996**, *60*(8), 1735–1743.

3168. Reddy, B. S.; Viswamitra, M. A. *Acta Crystallogr.* **1975**, *B31*(1), 19–26.

3169. Tabeta, R.; Aida, M.; Saitô, H. *Bull. Chem. Soc. Jpn.* **1986**, *59*(6), 1957–1966.

3170. Kirtley, S. M.; Mullins, O. C.; Chen, J.; van Elp, J.; George, S. J.; Chen, C. T.; O'Halloran, T.; Cramer, S. P. *Biochim. Biophys. Acta* **1992**, *1132*(3), 249–254.

3171. Klein, W.; Thannhauser, S. J. *Z. Physiol. Chem.* **1935**, *231*, 96–103.

3172. Katahira, M.; Nishimura, Y.; Tsuboi, M.; Sato, T.; Mitsui, Y.; Iitaka, Y. *Biochim. Biophys. Acta* **1986**, *867*(4), 256–267.

3173. Sato, T. *Acta Crystallogr.* **1984**, *C40*(5), 738–740.

3174. Schwartz, P. M.; Drach, J. C. *J. Chromatogr.* **1975**, *106*(1), 200–203.

3175. Perrett, D. *Chromatographia* **1982**, *16*, 211–213.

3176. Yu, C.; Ha, D.; Morris, P. W.; Puttemans, M. L.; Venton, D. L. *Anal. Biochem.* **1985**, *147*(1), 180–185.

3177. Di Pierro, D.; Tavazzi, B.; Perno, C. F.; Bartolini, M.; Balestra, E.; Calió, R.; Giardina, B.; Lazzarino, G. *Anal. Biochem.* **1995**, *231*(2), 407–412.

3178. Shalamberidze, M. B.; Sokolova, N. I.; Shabanova, Z. A. *Izv. Akad. Nauk Gruz. SSR, Ser. Biol.* **1978**, *4*(3), 254–259, from *Chem. Abstr.* **1979**, *90*, 23498r.

3179. Ramirez, F.; Marecek, J. F.; Szamosi, J. *Phorphorus Sulfur* **1982**, *13*(2), 249–257.

3180. Bogachev, V. S. *Bioorg. Khim.* **1996**, *22*(9), 699–705.

3181. Garyaev, P. P.; Gorelik, V. S.; Moiseenko, V. N.; Popnin, V. P.; Chudin, V. I.; Shcheglov, V. A. *Kratk. Soobshch. Fiz.* **1992**, (1/2), 33–36, from *Chem. Abstr.* **1993**, *119*, 3250t.

3182. Labotka, R. J.; Glonek, T.; Myers, T. C. *J. Am. Chem. Soc.* **1976**, *98*(12), 2699–2704.

3183. O'Hair, R. A. J.; McLuckey, S. A. *Int. J. Mass Spectrom. Ion Proc.* **1997**, *162*(1/3), 183–202.

3184. Burgess, K.; Russell, D. R.; Shitangkoon, A.; Zhang, A. J. *Nucleosides Nucleotides* **1996**, *15*(11/12), 1719–1723.

3185. Orlov, V. M.; Pustobaev, V. M.; Beschetnikova, Zh.A.; Maslova, R. N. *Mol. Biol. (Moscow)* **1994**, *28*(3), 708–713.

3186. Takigiku, R.; Schneider, R. E. *J. Chromatogr.* **1991**, *559*(1/2), 247–256.

3187. Hayes, D. H.; Michelson, A. M.; Todd, A. R. *J. Chem. Soc.* **1955**, (3), 808–815.

3188. Jankowska, J.; Stawiński, J. *Bull. Pol. Acad. Sci. Chem.* **1983**, *31*(1/2), 17–22.

3189. Dawson, W. H.; Cargill, R. L.; Dunlap, R. B. *J. Carbohydr. Nucleosides Nucleotides* **1977**, *4*(6), 363–375.

3190. Philipp, M.; Seliger, H. *Naturwisseschaften* **1977**, *64*(5), 273.

3191. Tanaka, T.; Yamada, Y.; Ikehara, M. *Tetrahedron Lett.* **1986**, *27*(46), 5641–5644.

3192. Pal, S.; Nair, V. *Biotechnol. Lett.* **1998**, *20*(12), 1149–1151.

3193. Hoard, D. E.; Ott, D. G. *J. Am. Chem. Soc.* **1965**, *87*(8), 1785–1788.

3194. Ladner, W. E.; Whitesides, G. M. *J. Org. Chem.* **1985**, *50*(7), 1076–1079.

3195. Etaix, E.; Orgel, L. E. *J. Carbohydr. Nucleosides Nucleotides* **1978**, *5*(2), 91–110.

3196. Inagaki, T. *J. Biochem.* **1940**, *32*, 63–79, from *Chem. Abstr.* **1941**, *35*, 762^9.

3197. Marinello, E.; Quadri, G. *Giorn. Biochim.* **1964**, *13*(3), 178–187.

3198. Oivanen, M.; Darzynkiewicz, E.; Lönnberg, H. *Acta Chem. Scand.* **1988**, *B42*(4), 250–253.

3199. Miyama, S.; Asanuma, H.; Komiyama, M. *J. Chem. Soc., Perkin Trans. 2* **1997**, (9), 1685–1688.

3200. Yashiro, M.; Miyama, S.; Takarada, T.; Komiyama, M. *J. Inclusion Phenom. Mol. Recognit. Chem.* **1994**, *17*(4), 393–397, from *Chem. Abstr.* **1995**, *123*, 144412s.

3201. Klein, W.; Thannhauser, S. J. *Z. Physiol. Chem.* **1933**, *218*, 173–180.

3202. Hirota, K.; Inoue, Y.; Chûjô, R. *Bull. Soc. Chem. Jpn.* **1984**, *57*(1), 247–250.

3203. Yuan, C.; Chang, C.; Chen, D.; Huang, M. *Bopuxue Zazhi* **1984**, *1*(5), 401–408, from *Chem. Abstr.* **1986**, *104*, 221551y.

3204. Andrievskii, G. V.; Nikolaeva, M. I.; Serebryanyi, A. M.; Tantsyrev, G. D.; Sharova, O. L. *Zh. Anal. Khim.* **1988**, *43*(10), 1850–1856 (Engl. 1498–1503).

3205. Murthy, M. R. V.; Ranjekar, P. K. *J. Chromatogr.* **1973**, *76*(2), 337–343.

3206. Young, D. W.; Tollin, P.; Wilson, H. R. *Acta Crystallogr.* **1974**, *B30*(8), 2812–2818.

3207. Viswamita, M. A.; Seshadri, T. P. *Nature* **1974**, *252*(5479), 176–177.

3208. Tajmir-Riahi, H.-A. *J. Biomol. Struct. Dyn.* **1990**, *8*(2), 303–313.

3209. Aberth, W.; Straub, K. M.; Burlingame, A. L. *Anal. Chem.* **1982**, *54*(12), 2029–2034.

3210. Topal, M. D.; Warshaw, M. M. *Biopolymers* **1976**, *15*(9), 1755–1773.

3211. Michelson, A. M.; Todd, A. R. *J. Chem. Soc.* **1954**, (1), 34–40.

3212. Kulinska, K.; Laaksonen, A. *NATO ASI Ser. C* **1994**, *435*, 509–512.

3213. Jardetzky, C. D. *J. Am. Chem. Soc.* **1961**, *83*(13), 2919–2920.

3214. Dizdaroglu, M. *J. Chromatogr.* **1984**, *295*(1), 103–121.

3215. Mullay, J. *J. Comput. Chem.* **1991**, *12*(3), 369–375.

3216. Lespinasse, J. N.; Broch, H.; Cormillon, R.; Vasilescu, D. *Stud. Biophys.* **1975**, *49*(3), 227–232.

3217. Tasaki, K.; Yang, X.; Urano, S.; Fetzer, S.; LeBreton, P. R. *J. Am. Chem. Soc.* **1990**, *112*(2), 538–548.

3218. Viswamitra, M. A.; Reddy, B. S.; Lin, G. H.-Y.; Sundaralingam, M. *J. Am. Chem. Soc.* **1971**, *93*(18), 4565–4573.

3219. Close, D. M.; Bernhard, W. A. *J. Chem. Phys.* **1979**, *70*(1), 210–215.

3220. Pandit, J.; Seshadri, T. P.; Viswamitra, M. A. *Acta Crystallogr.* **1983**, *C39*(3), 342–345.

3221. Aoki, K.; Saenger, W. *J. Inorg. Biochem.* **1984**, *20*(3), 225–245.

3222. Kudelin, B. K.; Kaminskii, Yu.L.; Ivanova, I. F.; Gavrilin, S. S. *Biokhimiya* **1979**, *44*(2), 368–371.

3223. Tyagi, A. K.; Jayaram, J. A.; Anandaraj, S.; Taylor, B.; Cooney, D. A. *J. Biochem. Biophys. Methods* **1979**, *1*(4), 221–226.

3224. Shapiro, R.; Danzig, M. *Biochim. Biophys. Acta* **1973**, *319*(1), 5–10.

3225. Ihara, T.; Shimura, H.; Ohmuri, K.; Tsuji, H.; Takeuchi, J.; Tagaki, M. *Chem. Lett.* **1996**, (6), 687–688.

3226. Smrt, J.; Šorm, F. *Coll. Czech. Chem. Commun.* **1960**, *25*(2), 553–558.

3227. Davies, D. B.; Rajani, P.; Sadikot, H. *J. Chem. Soc., Perkin Trans. 2* **1985**, (2), 279–285.

3228. Liu, L.-K.; Li, W.-S.; Su, C.-Y. *Proc. Natl. Sci. Counc. Repub. China* **1986**, *10A*(4), 359–367.

3229. Edelson, E. H.; Lawless, J. G.; Wehr, C. T.; Abbott, S. R. *J. Chromatogr.* **1979**, *174*(2), 409–419.

3230. Svoboda, V.; Vockova, J. *J. Chromatogr.* **1986**, *353*, 139–152.

3231. Lim, C. K.; Peters, T. J. *J. Chromatogr.* **1989**, *461*, 259–266.

3232. Viswamitra, M. A.; Seshadri, T. P.; Post, M. L. *Acta Crystallogr.* **1980**, *B36*(9), 2019–2024.

3233. Aoki, K.; Saenger, W. *J. Chem. Soc., Dalton Trans.* **1984**, (7), 1401–1409.

3234. Agarwal, S.; Garcia, A. A.; Miles, D. *Sep. Sci. Technol.* **1998**, *33*(1), 1–18, from *Chem. Abstr.* **1998**, *128*, 179836g.

3235. Ludwig, J.; Tomasz, J. *Synthesis* **1982**, (1), 32–34.

3236. Scarano, E.; Talarico, M. *Boll. Soc. Ital. Biol. Sper.* **1959**, *35*, 747–749.

3237. Samanta, S. R.; Lytle, F. E. *Appl. Spectrosc.* **1982**, *36*(3), 306–309.

3238. Carmichael, I.; Hug, G. L. *J. Phys. Chem. Ref. Data* **1986**, *15*(1), 1–250.

3239. Bours, J.; Dempewolf, J.; Schaake, S.; Rink, H. *J. Chromatogr.* **1987**, *403*, 336–342.

3240. Sato, T. *Acta Crystallogr.* **1984**, *C40*(5), 736–738.

3241. Trueblood, K. N.; Horn, P.; Luzzati, V. *Acta Crystallogr.* **1961**, *14*(9), 965–982. *Errata* **1962**, *15* (6), 613.

3242. Tanaka, M.; Tanaka, J. *Bull. Chem. Soc. Jpn.* **1971**, *44*(4), 938–944.

3243. Michelson, A. M.; Todd, A. R. *J. Chem. Soc.* **1953**, (3), 951–959.

3244. Griffin, B. E.; Todd, A. *J. Chem. Soc.* **1958**, (4), 1389–1393.

3245. Agarwal, K. L.; Dhar, M. M. *Experientia* **1965**, *21*(8), 432–433.

3246. Kimura, J.; Fujisawa, Y.; Yoshizawa, T.; Fukuda, K.; Mitsunobu, O. *Bull. Chem. Soc. Jpn.* **1979**, *52*(4), 1191–1196.

3247. Kellner, H. A.; Schneiderwind, R. G. K.; Eckert, H.; Ugi, I. K. *Angew. Chem.* **1981**, *93*(6/7), 581–582.

3248. Hata, T.; Sekine, M. *Tetrahedron Lett.* **1974**, (45), 3943–3946.

3249. Tanaka, T.; Yamada, Y.; Uesugi, S.; Ikehara, M. *Tetrahedron* **1989**, *45*(3), 651–660.

3250. Wreesmann, C. T. J.; Fidder, A.; Veeneman, G. H.; van der Marel, C. A.; van Boom, J. H. *Tetrahedron Lett.* **1985**, *26*(7), 933–936.

3251. Ikehara, M.; Ohtsuka, E. *Chem. Pharm. Bull.* **1963**, *11*(11), 1353–1358.

3252. Ponnamperuma, C.; Mack, R. *Science* **1965**, *148*(3674), 1221–1223.

3253. Slabaugh, M. R.; Harvey, A. J.; Nagyvary, J. *J. Mol. Evol.* **1974**, *3*(4), 317–321.

3254. Simakov, M. B.; Kuzicheva, E. A.; Malko, I. L. *Adv. Space Res.* **1997**, *19*(7), 1059–1062.

3255. Pollard, C. J. *Biochim. Biophys. Acta* **1970**, *201*(3), 511–512.

3256. Narayanan, A.; Vermeersch, J.; Pradet, P. *Compt. Rend.* **1970**, *271D*(25), 2406–2407.

3257. Amrhein, N. *Annu. Rev. Plant Physiol.* **1977**, *28*, 123–132.

3258. Brown, E. G.; Newton, R. P. *Phytochemistry* **1981**, *20*(11), 2453–2463.

3259. Volker, T. T.; Viratelle, O. M.; Delaage, M. A.; Labouesse, J. *Anal. Biochem.* **1985**, *144*(2), 347–355.

3260. Watenpaugh, K.; Dow, J.; Jensen, L. H.; Furberg, S. *Science* **1968**, *159*(3811), 206–207.

3261. Bryan, A. M.; Olafsson, P. G. *Biopolymers* **1973**, *12*(2), 229–235.

3262. Dworkin, M.; Keller, K. H. *J. Biol. Chem.* **1977**, *252*(3), 864–865.

3263. de Juan, A.; Izquierdo-Ridorsa, A.; Tauler, R.; Fonrodona, G.; Casassas, E. *Biophys. J.* **1997**, *73*(6), 2937–2948.

3264. Tate, M. E. *Biochem. J.* **1981**, *195*(2), 419–426.

3265. Smith, M.; Drummond, G. I.; Khorana, H. G. *J. Am. Chem. Soc.* **1961**, *83*(3), 698–706.

3266. Lipkin, D.; Markham, R.; Cook, W. H. *J. Am. Chem. Soc.* **1959**, *81*(22), 6075–6080.

3267. Nishimura, Y.; Tsuboi, M.; Kubasek, W. L.; Bajdor, K.; Peticolas, W. L. *J. Raman Spectrosc.* **1987**, *18*(3), 221–227.

3268. Klee, W. A.; Mudd, S. H. *Biochemistry* **1967**, *6*(4), 988–998.

3269. Brunner, W. C.; Maestre, M. F. *Biopolymers* **1975**, *14*(3), 555–565.

3270. Jardetzky, C. D. *J. Am. Chem. Soc.* **1962**, *84*(1), 62–66.

3271. Barry, C. D.; Martin, D. R.; Williams, R. J. P.; Xavier, A. V. *J. Mol. Biol.* **1974**, *84*(4), 491-502.

3272. Saran, A.; Berthod, H.; Pullman, B. *Biochim. Biophys. Acta* **1973**, *331*(2), 154–164.

3273. Wang, C.; Wang, Y.; Cao, K.; Zhang, L.; Ma, L.; Wang, X. *Huaxue Xuebao* **1953**, *41*(5), 435–440, from *Chem. Abstr.* **1983**, *99*, 66079h.

3274. Hunt, D. F.; Shabanowitz, J.; Botz, F. K.; Brent, D. A. *Anal. Chem.* **1977**, *49*(8), 1160–1163.

3275. Yamashita, M.; Fenn, J. B. *Iyo Masu Kenkyukai Koenshu* **1984**, *9*, 203–206, from *Chem. Abstr.* **1985**, *102*, 92637a.

3276. Woods, W. D.; Waitzman, M. B. *J. Chromatogr.* **1970**, *47*(3), 536–542.

3277. Lepri, L.; Coas, V.; Desideri, P. G.; Pettini, L.; Checchini, L. *J. Planar Chromatogr. Mod. TLC* **1989**, *2*(6), 456–460.

3278. Aoyagi, S.; Hirayanagi, K.; Yoshimura, T.; Ishikawa, T. *J. Chromatogr.* **1982**, *253*(1), 133–137.

3279. White, A. A.; Zeaser, T. V. *Anal. Biochem.* **1971**, *41*(2), 372–396.

3280. Uematsu, T.; Sasaki, T. *J. Chromatogr.* **1979**, *179*(1), 229–232.

3281. Morales-Vallarta, M.; Castafieda, M. *J. Chromatogr.* **1972**, *68*(1), 298–301.

3282. Marsh, F. J.; Weiner, P.; Douglas, J. E.; Kollman, P. A.; Kenyon, G. L.; Gerit, J. A. *J. Am. Chem. Soc.* **1980**, *102*(5), 1660–1665.

3283. Lespinasse, J. N.; Vasilescu, D. *Biopolymers* **1974**, *13*(1), 63–75.

3284. Jordan, F. *J. Theor. Biol.* **1973**, *41*(1), 23–40.

3285. Varughese, K. I.; Lu, C. T.; Kartha, G. *J. Am. Chem. Soc.* **1982**, *104*(12), 3398–3401.

3286. Druyan, M. E.; Sparagana, M.; Peterson, S. W. *J. Cyclic Nucleotide Res.* **1976**, *2*(5), 373–377.

3287. Kettani, A.; Guéron, M.; Leroy, J.-L. *J. Am. Chem. Soc.* **1997**, *119*(5), 1108–1115.

3288. Forrest, G.; Millis, A. J. T.; Whipple, A. P. *Arch. Biochem. Biophys.* **1977**, *178*(1), 51–57.

3289. Bertoluzza, A.; Fini, G.; Tosi, M. R.; Tugnoli, V. *J. Raman Spectrosc.* **1994**, *25*(7/8), 709–712.

3290. Williams, A. L., Jr.; Moore, D. S. *Biopolymers* **1983**, *22*(2), 755–786.

3291. Lee, C. H.; Sarma, R. H. *J. Am. Chem. Soc.* **1976**, *98*(12), 3541–3548.

3292. Gorenstein, D. G.; Kar, D. *Biochem. Biophys. Res. Commun.* **1975**, *65*(3), 1073–1080.

3293. Krauss, F.; Schmidt, A. *J. Chromatogr.* **1983**, *264*(1), 111–118.

3294. Eto, M.; Iio, M.; Kobayashi, Y.; Omura, H.; Eto, M. *Agric. Biol. Chem.* **1974**, *38*(11), 2081–2092.

3295. Kim, H.; Budzinski, E. E.; Box, H. C. *J. Chem. Phys.* **1989**, *90*(3), 1448–1451.

3296. Chwang, A. K.; Sundaralingam, M. *Acta Crystallogr.* **1974**, *30B*(5), 1233–1240.

3297. Gargallo, R.; Cuesta-Sanchez, F.; Izquierdo-Ridosa, A.; Massart, D. L. *Anal. Chem.* **1996**, *68*(13), 2241–2247.

3298. Newton, R. P.; Salih, S. G.; Salvage, B. J.; Kingston, E. E. *Biochem. J.* **1984**, *221*(3), 665–673.

3299. Markham, A. F.; Uesugi, S.; Ohtsuka, E.; Ikehara, M. *Biochemistry* **1979**, *18*(22), 4936–4942.

3300. Tomasz, J. *J. Chromatogr* **1979**, *169*, 466–468.

3301. Dutta, S. P.; Mittelman, A.; Chhedá, G. B. *Anal. Biochem.* **1976**, *75*(2), 649–651.

3302. Parthasarathy, R.; Srikrishnan, T. *Nucleosides Nucleotides* **1997**, *16*(10/11), 1967–1981.

3303. Zheng, Q.; Dou, S.; Gu, Y. *Wuli Xuebao* **1981**, *30*(10), 1369–1375, from *Chem. Abstr.* **1981**, *95*, 229895e.

3304. Markham, A. F.; Nakagawa, E.; Ohtsuka, E.; Ikehara, M. *Biopolymers* **1980**, *19*(2), 285–296.

3305. Smith, M.; Jardetzky, O. *J. Mol. Spectrosc.* **1968**, *28*(1), 70–80.

3306. Engels, J.; Hoftiezer, H. *Chem. Ber.* **1977**, *110*(6), 2019–2027.

3307. Coulter, C. L. *Acta Crystallogr.* **1969**, *25B*(10), 2055–2065. *Errata* **1970**, *26B* (4), 441.

3308. Tener, G. M.; Khorana, H. G.; Markham, R.; Poi, E. H. *J. Am. Chem. Soc.* **1958**, *80*(23), 6223–6230.

3309. Blackburn, R. J.; Lapper, R. D.; Smith, I. C. P. *J. Am. Chem. Soc.* **1973**, *95*(9), 2873–2878.

3310. Bols, N. C.; Bowen, B. W. M.; Khor, K. G.; Boliska, S. A. *Anal. Biochem.* **1980**, *106*(1), 230–237.

3311. Ramos, D. L.; Schoffstall, A. M. *J. Chromatogr.* **1983**, *261*(1), 83–93.

3312. Bentrude, W. G.; Sopchik, A. E.; Gajda, T. *J. Am. Chem. Soc.* **1989**, *111*(11), 3981–3987.

3313. Hemmes, P.; Oppenheimer, L.; Jordan, F. *J. Chem. Soc., Chem. Commun.* **1976**, (22), 929–930.

3314. Lavallee, D. K.; Zeltmann, A. H. *J. Am. Chem. Soc.* **1974**, *96*(17), 5552–5556.

3315. Yathindra, N.; Sundaralingam, M. *Biochem. Biophys. Res. Commun.* **1974**, *56*(1), 119–126.

3316. Kirschner, K. N.; Sherer, E. C.; Shields, G. C. *J. Phys. Chem.* **1996**, *100*(8), 3293–3298.

3317. Hayashi, F.; Akasaka, K.; Hatano, H. *J. Magn. Reson.* **1977**, *27*(3), 419–439.

3318. Topiol, S.; Morgan, T. K., Jr.; Sabio, M.; Lumma, W. C., Jr. *J. Am. Chem. Soc.* **1990**, *112*(4), 1452–1459.

3319. Jiang, S.; Wu, W.; Luo, S.; Huang, T. *Fenzi Kexue Xuebao* **1982**, *2*(1), 49–54 (in English).

3320. Schweizer, M. P.; Robins, R. K. *Jerusalem Symp. Quantum Chem. Biochem.* **1973**, *5*, 329–343, from *Chem. Abstr.* **1975**, *83*, 159352b.

3321. Inagaki, F.; Takahashi, S.; Tasumi, M.; Miyazawa, T. *Bull. Chem. Soc. Jpn.* **1976**, *49*(3), 611–616.

3322. Genieser, H.-G.; Butt, E.; Bottin, U.; Dostmann, W.; Jastorff, B. *Synthesis* **1989**, (1), 53–54.

3323. Mukaiyama, T.; Hashimoto, M. *J. Am. Chem. Soc.* **1972**, *84*(24), 8528–8532.

3324. Taguchi, Y.; Mushika, Y. *Bull. Chem. Soc. Jpn.* **1975**, *48*(5), 1528–1532.

3325. Takaku, H.; Kato, M.; Hata, T. *Chem. Lett.* **1978**, (2), 181–184.

3326. Borden, R. K.; Smith, M. *J. Org. Chem.* **1966**, *31*(10), 3247–3253.

3327. van Boom, J. H.; Burgers, P. M. J.; Van Deursen, P.; Reese, C. B. *J. Chem. Soc., Chem. Commun.* **1974**, (15), 618–619.

3328. Marumoto, R.; Nishimura, T.; Honjo, M. *Chem. Pharm. Bull.* **1975**, *23*(10), 2295–2300.

3329. Yamagata, Y.; Kojima, H.; Ejiri, K.; Inomata, K. *Origins Life* **1982**, *12*, 333–337.

3330. Oivanen, M.; Rajamäki, M.; Varila, J.; Hovinen, J.; Mikhailov, S. *J. Chem. Soc., Perkin Trans. 2* **1994**, (2), 309–314.

3331. Yonezawa, K.; Matsumoto, Y.; Komiyama, M. *Chem. Expr.* **1991**, *6*(12), 965–968.

3332. Li, X.; Zhu, B.; Zhao, D.; Ni, J. *Indian J. Chem.* **1996**, *35B*(11), 1144–1147.

3333. Zhu, B.; Zhao, D.; Ni, J. *Indian J. Chem.* **1998**, *37A*(9), 755–758.

3334. Drummond, G. I.; Gilgan, M. W.; Reiner, E. J.; Smith, M. *J. Am. Chem. Soc.* **1964**, *86*(8), 1626–1630.

3335. Sumaoka, J.; Yashiro, M.; Komiyama, M. *J. Chem. Soc., Chem. Commun.* **1992**, (23), 1707–1708.

3336. Hayaishi, O.; Greengard, P.; Colowick, S. *J. Biol. Chem.* **1971**, *246*, 5840–5843.

3337. Greengard, P.; Rudolph, S. A.; Sturtevant, J. M. *J. Biol. Chem.* **1969**, *244*(17), 4798–4800.

3338. Gerlt, J. A.; Westheimer, F. H.; Sturtevant, J. M. *J. Biol. Chem.* **1975**, *250*(13), 5059–5067.

3339. Scheffers-Sap, M. M. E.; Buck, H. M. *Recl. Trav. Chim.* **1978**, *97*(10), 252–256.

3340. Rudolph, S. A.; Johnson, E. M.; Greengard, P. *J. Biol. Chem.* **1971**, *246*(5), 1271–1273.

3341. Roby, C.; Martin, J.-B.; Bligny, R.; Douce, R. *J. Biol. Chem.* **1987**, *262*(11), 5005–5007.

3342. Murata, T. *Agric. Biol. Chem.* **1975**, *39*(7), 1401–1406.

3343. ap Rees, T.; Leja, M.; Macdonald, F. D.; Green, J. H. *Phytochemistry* **1984**, *23*(11), 2463–2468.

3344. Buchanan, J. G.; Lynch, V. H.; Benson, A. A.; Bradley, D. F.; Calvin, M. *J. Biol. Chem.* **1953**, *203*(2), 935–945.

3345. Katan, R.; Avigad, G. *Israel J. Chem.* **1965**, *3*, 110.

3346. Kluge, M. *Planta* **1969**, *85*(2), 160–170.

3347. Amino, S.; Takeuchi, Y.; Komamine, A. *Physiol. Plant.* **1985**, *64*(2), 197–201.

3348. Kim, T. Y.; Yoon, Y. J.; Lee, K. W. *Han'guk Sikp'um Kwahakhoe Chi* **1978**, *10*(2), 258–262, from *Chem. Abstr.* **1979**, *90*, 4705q.

3349. Takeuchi, Y.; Amino, S. *Plant Cell Physiol.* **1984**, *25*(8), 1589–1593.

3350. Gonzalez, N. S.; Pontis, H. G. *Biochim. Biophys. Acta* **1963**, *69*(1), 179–181.

3351. Meyer, R.; Wagner, K. G. *Planta* **1985**, *166*(4), 439–445.

3352. Franz, G.; Meier, H. *Biochim. Biophys. Acta* **1969**, *184*, 658–659.

3353. Kim, T. Y.; Shin, H. J. *Taehan Hwahakhoe Chi* **1966**, *10*(2), 59–61, from *Chem. Abstr.* **1967**, *67*, 61602c.

3354. Hayashi, T.; Matsuda, K. *Agric. Biol. Chem.* **1981**, *45*(12), 2907–2908.

3355. Mamedova, T. K.; Sherstnev, B. A. *Fiziol. Rast.* **1965**, *12*(4), 618–621 (Engl. 536–539).

3356. Umemura, Y.; Nakamura, M.; Funahashi, S. *Arch. Biochem. Biophys.* **1967**, *199*, 240–252.

3357. Tluczkiewicz, J. *Bull. Acad. Polon. Sci., Ser. Sci. Biol.* **1974**, *22*(6), 421–424.

3358. Sicher, R. C.; Kremer, D. F.; Harris, W. G. *Plant Physiol.* **1986**, *82*(1), 15–18.

3359. Brown, E. G.; Mangat, B. S. *Biochim. Biophys. Acta* **1967**, *148*(2), 350–355.

3360. McPharlin, I. R.; Bieleski, R. L. *Physiol. Plant.* **1989**, *76*(1), 95–99.

3361. Rowan, K. S. *Biochim. Biophys. Acta* **1959**, *34*(1), 270–271.

3362. Schlüpmann, H.; Bacic, A.; Read, S. M. *Plant Physiol.* **1994**, *105*(2), 459–470.

3363. Golebski, W.; Meyer, R.; Wagner, K. G. *Physiol. Plant.* **1988**, *72*(1), 29–35.

3364. Palmer, C. E. *Z. Pflanzenphysiol.* **1976**, *77*, 345–349.

3365. Murata, T.; Minamikawa, T.; Akazawa, T.; Sugiyama, T. *Arch. Biochem. Biophys.* **1964**, *106*, 371–378.

3366. Sandermann, H., Jr.; Grisebach, H. *Eur. J. Biochem.* **1968**, *6*(3), 404–410.

3367. Hampe, M. M. V. *Rev. Brasil. Farm.* **1970**, *51*(3), 145–152, from *Chem. Abstr.* **1971**, *74*, 20379m.

3368. Smith, A. M.; Neuhaus, H. E.; Stitt, M. *Planta* **1990**, *181*(3), 310–315.

3369. Bhatia, I. S.; Uppal, D. *Plant Sci. Lett.* **1979**, *16*(1), 59–66.

3370. Sharma, K. P.; Bhatia, I. S. *Indian J. Biochem. Biophys.* **1978**, *15*(2), 133–135.

3371. Jenner, C. F. *Aust. J. Plant Physiol.* **1991**, *18*(2), 165–177.

3372. Cheung, C. P.; Suhadolnik, R. J. *Plant Physiol.* **1979**, *63*(1), 146–148.

3373. Stitt, M.; Wirtz, W.; Gerhardt, R.; Heldt, H. W.; Spencer, C.; Walker, D.; Foyer, C. *Planta* **1985**, *166*(3), 354–364.

3374. Sawert, A.; Aust, H.-J.; Wagner, K. G. *Physiol. Mol. Plant Pathol.* **1988**, *33*(3), 497–506.

3375. Tolbert, N. E.; Gailey, F. B. *Plant Physiol.* **1955**, *30*(6), 491–499.

3376. Gilkes, N. R.; Herman, E. M.; Chrispeels, M. J. *Plant Physiol.* **1979**, *64*(1), 38–42.

3377. Solms, J.; Hassid, W. Z. *J. Biol. Chem.* **1957**, *228*(1), 357–364.

3378. Recondo, E.; Dankert, M.; Leloir, L. F. *Biochem. Biophys. Res. Commun.* **1963**, *12*(3), 204–207.

3379. Shannon, J. C.; Pien, F.-M.; Liu, K.-C. *Plant Physiol.* **1996**, *110*(3), 835–843.

3380. Hooks, M. A.; Clark, R. A.; Nieman, R. H.; Roberts, J. K. M. *Plant Physiol.* **1989**, *89*(3), 963–969.

3381. Roberts, J. K. M.; Testa, M. P. *Plant Physiol.* **1988**, *86*(4), 1127–1130.

3382. Paladini, A. C.; Leloir, L. F. *Biochem. J.* **1952**, *51*(3), 426–430.

3383. Carminatti, H.; Passeron, S.; Dankert, M.; Recondo, E. *J. Chromatogr.* **1965**, *18*(2), 342–348.

3384. Hirano, S. *Biochem. Biophys. Res. Commun.* **1971**, *43*(6), 1219–1222.

3385. Wolucka, B. A.; de Hoffmann, E.; Rush, J. S.; Waechter, C. J. *J. Am. Soc. Mass Spectrom.* **1996**, *7*(6), 541–549.

3386. Budovskii, E. I.; Shibaev, V. N. *Vop. Med. Khim.* **1967**, *13*(5), 554–556.

3387. Xu, G.; Lehmann, R.; Schleider, E.; Häring, H. U.; Liebich, H. *Biomed. Chromatogr.* **1998**, *12*(3), 113–115.

3388. Ii, T.; Okuda, S.; Hirano, T.; Tsujimoto, K.; Ohashi, M. *Org. Mass Spectrom.* **1993**, *28*(2), 127–131.

3389. Sugawara, Y.; Iwasaki, H.; Tsukakoshi, M.; Kasuga, T. *Reza Kagaku Kenkyu* **1984**, *6*, 31–33, from *Chem. Abstr.* **1985**, *102*, 140062w.

3390. Moffatt, J. G. *Biochem. Prep.* **1961**, *8*, 125–129.

3391. London, R. E.; Sherry, A. D. *Biochemistry* **1978**, *17*(17), 3663–3666.

3392. Kim, K. B.; Behrman, E. C.; Behrman, E. J. *Nucleosides Nucheotides* **1999**, *18*(4/5), 1055–1056.

3393. Kenner, G. W.; Todd, A. R.; Webb, R. F. *J. Chem. Soc.* **1954**, (8), 2843–2847.

3394. Kampe, W. *Chem. Ber.* **1966**, *99*(2), 593–601.

3395. Ito, M.; Yano, K.; Eto, M.; Omura, H.; Eto, M. *Agric. Biol. Chem.* **1977**, *41*(1), 155–160.

3396. Roseman, S.; Distler, J. J.; Moffatt, J. G.; Khorana, H. G. *J. Am. Chem. Soc.* **1961**, *83*(3), 659–663.

3397. Belousova, Z. P.; Purygin, P. P.; Danilov, L. L.; Shibaev, V. N. *Bioorg. Khim.* **1986**, *12*(11), 1522–1529.

3398. Michelson, A. M.; Todd, A. *J. Chem. Soc.* **1956**, (9), 3459–3463.

3399. Nakagawa, I.; Hata, T. *Bull. Chem. Soc. Jpn.* **1973**, *46*(10), 3275–3277.

3400. Nunez, H. A.; Barker, R. *Biochemistry* **1976**, *15*(17), 3843–3847.

3401. Budovskii, E. I.; Shibaev, V. N. *Khim. Prir. Soedin.* **1968**, *4*(4), 233–236 (Engl. 199–201).

3402. Frey, P. A.; Arabshahi, A. *Biochemistry* **1995**, *34*(36), 11307–11310.

3403. Dixon, M. published simultaneously in *Science* **1960**, *132*, 1548–1550 and *Nature* **1960**, *188*, 464–466 .

3404. Alberty, R. A. *Arch. Biochem. Biophys.* **1993**, *307*(1), 8–14.

3405. von Euler, H.; Nordenfelt, E. *Arkiv Kemi Mineral. Geol.* **1926**, *9*(35), 1–6.

3406. Whatley, F. R. *New Phytol.* **1951**, *50*(2), 244–251 corr. *Aust. J. Sci.* **1953**, *15* (4), 173–174.

3407. Marré, E.; Bianchetti, R. *Biochim. Biophys. Acta* **1961**, *48*(3), 583–585.

3408. Fritz, G. J.; Stout, E. R.; Leister, D. E. *Plant Physiol.* **1963**, *38*(6), 642–648.

3409. Sunderland, D. W.; Merrett, M. J. *Nature* **1963**, *200*(4909), 921.

3410. Sunderland, D. W.; Merrett, M. J. *Nature* **1963**, *199*(4898), 1116–1117.

3411. Yamamoto, Y. *Plant Physiol.* **1963**, *38*(1), 45–54.

3412. Kosuge, T.; Gilchrist, D. G. *Encycl. Plant Physiol.* **1976**, *4*(679–702), 695.

3413. Wigge, B.; Krömer, S.; Gardeström, P. *Physiol. Plant.* **1993**, *88*(1), 10–18.

3414. Johnson, R. D.; Waller, G. R. *Phytochemistry* **1974**, *13*(8), 1493–1500.

3415. Griffiths, M. M.; Bernofsky, C. *J. Biol. Chem.* **1972**, *247*(5), 1473–1478.

3416. Peine, G.; Hoffmann, P.; Seifert, G.; Schilling, G. *Biochem. Physiol. Pflanz.* **1985**, *180*(1), 1–14.

3417. Aso, Y.; Gotoh, S.; Yamasaki, N. *Agric. Biol. Chem.* **1989**, *53*(6), 1635–1639.

3418. Lowry, O. H.; Passoneau, J. V.; Rock, M. K. *J. Biol. Chem.* **1961**, *236*(10), 2756–2759.

3419. Tezuka, T.; Yamaguchi, F.; Ando, Y. *J. Photochem. Photobiol.* **1994**, *24B*(1), 33–40.

3420. Bevilacqua, L. R.; Scuncia, P. E. *Boll. Soc. Ital. Biol. Sper.* **1959**, *35*(16), 1017–1019.

3421. Bevilacqua, L. R. *Atti Accad. Ligure Sci. Lett. Genoa* **1954**, *11*(1), 145–154.

3422. Lieckfeldt, E.; Jung, S.-J.; Peine, G.; Hoffmann, P. *Biochem. Physiol. Pflanz.* **1987**, *182*(5), 393–405.

3423. Bevilacqua, L. R.; Scotti, R. *Atti Accad. Ligure Sci. Lett. Genoa* **1954**, *10*, 196–203.

3424. Gallais, S.; Pou de Crescenzo, M. A.; Laval-Martin, D. L. *J. Plant Physiol.* **1998**, *153*(5/6), 664–669.

3425. Fujii, T.; Kondo, N. *Embryologia* **1969**, *11*(1), 40–45.

3426. Rao, I. M.; Arulanantham, A. R.; Terry, N. *Photosynth. Res.* **1990**, *23*, 205–212.

3427. Heber, U. W.; Santarius, K. A. *Biochim. Biophys. Acta* **1965**, *109*(2), 390–408.

3428. von Euler, H.; Ahlström, L.; Högberg, B.; Pettersson, I. *Arkiv Kemi Mineral. Geol.* **1944**, *19A*(4), 1–15.

3429. Maciejewska, U.; Kakperska, A. *Physiol. Plant.* **1987**, *69*(4), 687–691.

3430. Okunuki, K. *Acta Phytochim. (Jpn.)* **1943**, *13*(2), 93–98.

3431. Li, X.-N.; Ashihara, H. *Phytochemistry* **1990**, *29*(2), 497–500.

3432. Asali, E. C.; Mutharasan, R.; Humphrey, A. E. *J. Biotechnol.* **1992**, *23*(1), 83–94.

3433. Tesnière, C.; Soudain, P.; de Coninck, B. *Physiol. Vég.* **1984**, *22*(4), 447–459.

3434. Bruemmer, J. H. *J. Agric. Food Chem.* **1969**, *17*(6), 1312–1315.

3435. Davis, P. L.; Roe, B.; Bruemmer, J. H. *J. Food Sci.* **1973**, *38*(2), 225–229.

3436. Wakiuchi, N.; Matsumoto, H.; Kondo, S.; Takahashi, E. *Physiol. Plant.* **1972**, *26*(2), 230–232.

3437. Whyte, B. J.; Griffiths, W. T. *Biochem. J.* **1993**, *291*(3), 939–944.

3438. Chermnykh, L. N.; Chugunova, N. G.; Kosobryukov, A. A. *Sel'skokhoz. Biol.* **1974**, *9*(2), 238–243.

3439. Zhao, Z.; Hu, X.; Ross, C. W. *Plant Physiol.* **1987**, *84*(4), 987–988.

3440. Shinkle, J. R.; Swoap, S. J.; Simon, P.; Jones, R. L. *Plant Physiol.* **1992**, *98*(4), 1336–1341.

3441. Morohashi, Y.; Sasanuma, S.-I.; Seto, T. *Physiol. Plant.* **1993**, *89*(2), 295–300.

3442. Rong, G. S.; Nada, K.; Tachibana, S. *J. Jpn. Soc. Hortic. Sci.* **1998**, *67*(4), 613–618.

3443. Takeuchi, Y.; Nihira, J.; Kondo, N.; Tezuka, T. *Plant Cell Physiol.* **1985**, *26*(6), 1027–1035.

3444. Yamasue, Y.; Hasegawa, R.; Ueki, K. *Zasso Kenkyu* **1988**, *33*(3), 172–179, from *Chem. Abstr.* **1989**, *111*, 4367h.

3445. Dalton, D. A.; Langeberg, L.; Treneman, N. C. *Physiol. Plant.* **1993**, *87*(3), 365–370.

3446. Kanayama, Y.; Yamamoto, Y. *Plant Cell Physiol.* **1990**, *31*(6), 893–895.

3447. Ryrie, I. J.; Scott, K. J. *Plant Physiol.* **1968**, *43*(5), 687–692.

3448. Schmidt, G.; Boyle, C.; Aust, H. J.; Wagner, K. G. *J. Phytopathol.* **1994**, *141*(4), 402–414.

3449. Goto, K. *Z. Naturforsch.* **1984**, *39C*(1/2), 73–84.

3450. Taguchi, H.; Kashimoto, A.; Nishitani, H.; Shimabayashi, Y.; Iwai, K. *Agric. Biol. Chem.* **1988**, *52*(1), 85–89.

3451. Dubrovo, P. N. *Fiziol. Rast.* **1978**, *25*(3), 615–622 (Engl. 481–487).

3452. Pashchenko, V. N.; Murel, I. A.; Nichiporovich, A. A. *Fiziol. Rast.* **1971**, *18*(6), 1134–1140 (Engl. 961–966).

3453. Roux, L. *Ann. Physiol. Vég.* **1964**, *6*(2), 141–147.

3454. Graham, J. S. D.; Hawker, J. S. *Plant Physiol.* **1962**, *37*(6), 765–767.

3455. Rhodes, M. J. C.; Wooltorton, L. S. C. *Phytochemistry* **1968**, *7*(3), 337–353.

3456. Mukherji, S.; Dey, B.; Sircar, S. M. *Physiol. Plant.* **1968**, *21*(2), 360–368.

3457. Ke, D.; Yahia, E.; Hess, B.; Zhou, L.; Kader, A. A. *J. Am. Soc. Hortic. Sci.* **1995**, *120*(3), 481–490.

3458. Anderson, D. G.; Vennesland, B. *J. Biol. Chem.* **1954**, *207*(2), 613–620.

3459. Tezuka, T.; Yamamoto, Y. *Plant Physiol.* **1972**, *50*(4), 458–462.

3460. Walter, G.; Peine, G.; Dujardin, E. C.; Hoffmann, P. *Photosynthetica* **1988**, *22*(4), 502–510.

3461. Juszczuk, S. M.; Rychter, A. M. *J. Plant Physiol.* **1997**, *151*(4), 399–404.

3462. Ogren, W. L.; Krogmann, D. W. *J. Biol. Chem.* **1965**, *240*(12), 4603–4608.

3463. Simons, P. C.; van den Briel, W.; Bienfait, H. F. *Plant Physiol.* **1984**, *75*(1), 219–221.

3464. Lorenc-Plucińska, G.; Karolewski, P. *Acta Soc. Bot. Pol.* **1994**, *63*(2), 167–171.

3465. Zelles, L. *Environ. Exp. Bot.* **1978**, *18*(1), 39–45.

3466. van Herk, A. W. H. *Arkiv Kemi Mineral. Geol.* **1935**, *11A*(22), 1–10.

3467. Greenbaum, A. L.; Clark, J. B.; McLean, P. *Biochem. J.* **1965**, *95*(1), 161–166.

3468. Harvey, M. J.; Brown, A. P. *Biochim. Biophys. Acta* **1969**, *172*(1), 116–125.

3469. Hason-Porath, E.; Poljakoff-Mayber, A. *J. Exp. Bot.* **1970**, *21*(67), 300–303.

3470. Escuredo, P. R.; Minchin, F. R.; Gogorcana, Y.; Iturbe-Ormaetxe, I.; Klucas, R. V.; Becana, M. *Plant Physiol.* **1996**, *110*(4), 1187–1195.

3471. Brown, E. G. *Biochem. J.* **1965**, *95*(2), 509–514.

3472. Muto, S.; Miyachi, S.; Usuda, H.; Edwards, G. E.; Bassham, J. A. *Plant Physiol.* **1981**, *68*(2), 324–328.

3473. Schmidt, W.; Schuck, C. *Physiol. Plant.* **1996**, *98*(1), 215–221.

3474. Gupta, A. S.; Alscher, R. G.; McCune, D. *Plant Physiol.* **1991**, *96*(2), 650–655.

3475. Tanaka, Y.; Sato, J. *Aichi-ken Nogyo Sogo Shikenjo Kenkyu Hokoku* **1979**, (11), 130–136, from *Chem. Abstr.* **1981**, *94*, 101566k.

3476. Bovalini, L.; Ferri, S.; Martelli, P.; Franchi, G. G. *Boll. Soc. Ital. Biol. Sper.* **1984**, *60*(10), 1891–1895.

3477. Donaldson, R. P. *Arch. Biochem. Biophys.* **1982**, *215*(1), 274–279.

3478. Frosch, S.; Wagner, E.; Mohr, H. *Z. Naturforsch.* **1974**, *29C*(7/8), 392–398.

3479. Brinkman, F. G.; van der Plas, L. H. W.; Verleur, J. D. *Z. Pflanzenphysiol.* **1973**, *68*(4), 364–372.

3480. Cassells, A. C.; Harmey, M. A. *Arch. Biochem. Biophys.* **1968**, *126*(2), 486–491.

3481. Kikuta, Y.; Harada, T.; Akemine, T.; Tagawa, T. *Plant Cell Physiol.* **1977**, *18*(2), 361–370.

3482. Corcuera, L. J.; Hintz, M.; Pahlich, E. *J. Plant Physiol.* **1989**, *134*(3), 290–293.

3483. Heineke, D.; Riens, B.; Grosse, H.; Hoferichter, P.; Peter, U.; Flügge, U.-I.; Heldt, H. W. *Plant Physiol.* **1991**, *95*(4), 1131–1137.

3484. Shibata, H.; Sawa, Y.; Ochiai, H.; Kawashima, T.; Yamane, K. *Plant Sci.* **1987**, *51*(2/3), 129–133.

3485. Lechevallier, D.; Vermeersch, J.; Monéger, R. *Physiol. Vég.* **1977**, *15*(1), 63–93.

3486. Ishikawa, H. A.; Ishikawa, K.; Tateyama, M.; Takizawa, N.; Urami, S. *Plant Cell Physiol.* **1977**, *18*(3), 527–537.

3487. Pao, L. Y.; Godavari, H. R.; Waygood, E. R. *Plant Biochem. J.* **1974**, *1*(1), 16–25.

3488. Krishnaraj, S.; Thorpe, T. A. *Int. J. Plant Sci.* **1996**, *157*(1), 110–117.

3489. Slaski, J. J. *J. Plant Physiol.* **1989**, *133*(6), 696–701.

3490. Waygood, E. R. *Plant Physiol.* **1965**, *40*(6), 1242–1247.

3491. Dahse, I.; Schnabl, H.; Hampp, R.; Ziegler, H.; Müller, E.; Liebermann, B. *Planta* **1988**, *173*(3), 391–398.

3492. Graham, D.; Cooper, J. E. *Aust. J. Biol. Sci.* **1967**, *20*(2), 319–327.

3493. Chakrabarti, S.; Chakrabarti, A.; Chatterjee, N.; Banerjee, S. K. *Int. J. Environ. Stud.* **1986**, *28*(2/3), 197–205.

3494. Wall, J. S.; Young, M. R.; Carpenter, K. J. *J. Agric. Food Chem.* **1987**, *35*(5), 752–758.

3495. Carrier, J. M.; Neve, N. *Photosynthetica* **1979**, *13*(3), 323–331.

3496. Usuda, H. *Plant Physiol.* **1988**, *88*(4), 1461–1468.

3497. Cherry, J. H.; Hageman, R. H. *Plant Physiol.* **1960**, *35*(3), 343–352.

3498. Razumovich, A. N.; Rusinova, O. V. *Vestsi Akad. Navuk BSSR, Ser. Biyal. Navuk* **1985**, (3), 38–42.

3499. Qiu, Z.-S.; Rubinstein, B.; Stern, A. I. *Planta* **1985**, *165*(3), 383–391.

3500. Bevilacqua, L. R. *Boll. Soc. Ital. Biol. Sper.* **1959**, *35*(24 bis), 1854–1855, from *Chem. Abstr.* **1962**, *56*, 13253h.

3501. Robinson, J.; Levitas, N.; Rosen, R.; Perlzweig, W. A. *J. Biol. Chem.* **1947**, *170*(2), 653–659.

3502. Ciotti, M. M.; Kaplan, N. O. *Methods Enzymol.* **1957**, *3*, 890–899.

3503. Colowick, S. P.; Kaplan, N. O.; Ciotti, M. M. *J. Biol. Chem.* **1951**, *191*(2), 447–459.

3504. Cheung, C. P.; Suhadolnik, R. J. *Anal. Biochem.* **1979**, *97*(2), 309–314.

3505. Sarma, R. H.; Mynott, R. J. *J. Am. Chem. Soc.* **1973**, *95*(22), 7470–7480.

3506. Sarma, R. H.; Mynott, R. J. *Org. Magn. Reson.* **1972**, *4*(4), 577–584.

3507. Gol'dshtein, I. I.; Roshchupkin, D. I.; Skulachev, V. P. *Biokhimiya* **1966**, *31*(5), 927–931 (Engl. 799–802).

3508. Gallego, E.; Peral, F. *J. Mol. Struct.* **1993**, *298*, 87–94.

3509. Haid, E.; Lehmann, P.; Ziegenhorn, J. *Clin. Chem.* **1975**, *21*(7), 884–887.

3510. Takahashi, F.; Sakai, Y.; Yoshida, T.; Nakazawa, Y. *Nippon Gagaku Kaishi* **1995**, (11), 922–924, from *Chem. Abstr.* **1996**, *124*, 3764s.

3511. Pfleidlerer, G.; Schulz, W. *Ann. Chem.* **1953**, *580*(3), 237–246.

3512. Hogeboom, G. H.; Barry, G. T. *J. Biol. Chem.* **1948**, *176*(2), 935–948.

3513. Yang, P.-H.; Rupley, J. A. *J. Chem. Thermodyn.* **1979**, *11*(3), 301–302.

3514. Cunningham, A. J.; Underwood, A. L. *Biochemistry* **1967**, *6*(1), 266–271.

3515. Carruthers, C.; Tech, J. *Arch. Biochem. Biophys.* **1955**, *56*(2), 441–447.

3516. Zieger, G.; Konrat, R.; Sterk, H. *Magn. Reson. Chem.* **1992**, *30*(Special Issue), S3–S6, from *Chem. Abstr.* **1993**, *118*, 60035e.

3517. Moore, C. E., Jr.; Underwood, A. L. *Anal. Biochem.* **1969**, *29*(1), 149–153.

3518. Johnson, R. W.; Marschner, T. M.; Oppenheimer, N. J. *J. Am. Chem. Soc.* **1988**, *110*(7), 2257–2263.

3519. Siegel, J. M.; Montgomery, G. A.; Bock, R. M. *Arch. Biochem. Biophys.* **1959**, *82*(2), 288–299.

3520. Nadolny, C.; Zundel, G. *J. Mol. Struct.* **1996**, *385*(2), 81–87.

3521. Yue, K. T.; Martin, C. L.; Chen, D.; Nelson, P.; Sloan, D. L.; Callender, R. *Biochemistry* **1986**, *25*(17), 4941–4947.

3522. Rodgers, E. G.; Peticolas, W. L. *J. Raman Spectrosc.* **1980**, *9*(6), 372–375.

3523. Jardetzky, O.; Wade-Jardetzky, N. G. *J. Biol. Chem.* **1966**, *241*(1), 85–91.

3524. Sarma, R. H.; Kaplan, N. O. *J. Biol. Chem.* **1969**, *244*(3), 771–774.

3525. Oppenheimer, N. J.; Davidson, R. M. *Org. Magn. Reson.* **1980**, *13*(1), 14–16.

3526. Reisbig, R. R.; Woody, R. W. *Biochemistry* **1978**, *17*(10), 1974–1984.

3527. Xiao, Y. J.; Chen, Y. F.; Gao, X. X. *Spectrochim. Acta* **1999**, *55A*(6), 1209–1218.

3528. Schulten, H.-R.; Schiebel, H. M. *Z. Anal. Chem.* **1985**, *321*(6), 531–537.

3529. Schueler, B.; Krueger, F. R. *Org. Mass Spectrom.* **1980**, *15*(6), 295–301.

3530. Verachtert, H.; Bass, S. T.; Wilder, J.; Hansen, R. G. *Anal. Biochem.* **1965**, *11*(3), 497–509.

3531. Mezzasoma, I. *Boll. Soc. Ital. Biol. Sper.* **1967**, *43*(16), 974–976, from *Chem. Abstr.* **1968**, *68*, 44612h.

3532. Hansson, K.-A. *J. Chromatogr.* **1983**, *266*, 395–399.

3533. Pastore, E. J.; Friedkin, M. *J. Biol. Chem.* **1961**, *236*(8), 2314–2316.

3534. Senda, M.; Sasaki, T.; Nakamoto, H.; Hiyama, T. *Anal. Sci.* **1991**, *7*(Suppl.), 1545–1548.

3535. Buskov, S.; Møller, P.; Sørensen, H.; Sørensen, J. C.; Sørensen, S. *J. Chromatogr.* **1998**, *802A*, 233–241.

3536. Pullman, B.; Pullman, A. *Proc. Natl. Acad. Sci. U. S. A.* **1959**, *45*(2), 136–144.

3537. Grabe, B. *Arkiv Fys.* **1959**, *17*(5), 97–111.

3538. Miller, S. I.; Smith-Magowan, D. *J. Phys. Chem. Ref. Data* **1990**, *19*(4), 1049–1053.

3539. Winer, A. D. *J. Biol. Chem.* **1964**, *239*(10), 3598–3600.

3540. Parthasarathy, R.; Fridey, S. M. *Science* **1984**, *226*, 969–971.

3541. Saenger, W.; Reddy, B. S.; Mühlegger, K.; Weimann, G. *Nature* **1977**, *267*, 225–229.

3542. Reddy, B. S.; Saenger, W.; Mühlegger, K.; Weimann, G. *J. Am. Chem. Soc.* **1981**, *103*(4), 907–914.

3543. Wallenfels, K.; Christian, W. *Angew. Chem.* **1952**, *64*(15), 419–420.

3544. Newton, C. J.; Faynor, S. M.; Northrop, D. B. *Anal. Biochem.* **1983**, *132*(1), 50–53.

3545. Applegarth, D. A.; Eden, J. *Clin. Chim. Acta* **1966**, *13*(6), 720–724.

3546. Gerhardt, W.; Kofoed, B.; Westlund, L.; Pavlu, B. *Scand. J. Clin. Lab. Invest. Suppl.* **1975**, *33*(139), 51, from *Chem. Abstr.* **1975**, *82*, 120802c.

3547. Scott, T. G.; Spencer, R. D.; Leonard, N. J.; Weber, G. *J. Am. Chem. Soc.* **1970**, *92*(3), 687–698.

3548. Ziegenhorn, J.; Senn, M.; Bücher, T. *Clin. Chem.* **1976**, *22*(2), 151–160.

3549. Sovago, I.; Martin, R. B. *FEBS Lett.* **1979**, *106*(1), 132–134.

3550. Miles, D. W.; Urry, D. W. *J. Biol. Chem.* **1968**, *243*(16), 4181–4188.

3551. Boldridge, D. W.; Morton, T. H.; Scott, G. W.; Clark, J. H.; Philips, L. A.; Webb, S. P.; Yeh, S. M.; Van Eikeren, B. *Springer Ser. Chem. Phys.* **1984**, *38*, 477–480, from *Chem. Abstr.* **1985**, *102*, 91936y.

3552. Lehninger, A. L. *Biochem. Prep.* **1952**, *2*, 92–96.

3553. Briggs, T. N.; Stuehr, J. E. *Anal. Chem.* **1975**, *47*(12), 1916–1920.

3554. Mas, M. T.; Colman, R. F. *Biochemistry* **1984**, *23*(8), 1675–1683.

3555. Wasternack, C. *Pharmazie* **1970**, *25*(12), 740–748.

3556. Nikandrov, V. V.; Krasnovskii, A. A. *Biofizika* **1978**, *23*(4), 721–723, from *Chem. Abstr.* **1978**, *89*, 159391n.

3557. Schauenstein, E.; Saenger, W.; Schaur, J. J.; Desoye, G.; Schreibmayer, W. *Z. Naturforsch.* **1980**, *36C*(1/2), 76–79.

3558. Oppenheimer, N. J.; Arnold, L. J.; Kaplan, N. O. *Proc. Natl. Acad. Sci. U. S. A.* **1971**, *68*(12), 3200–3205.

3559. Blumenstein, M.; Raftery, M. A. *Biochemistry* **1973**, *12*(19), 3585–3590.

3560. Thornton, J. M.; Bayley, P. M. *Biopolymers* **1977**, *16*(9), 1971–1986.

3561. Wu, Y.-D.; Houk, K. N. *J. Am. Chem. Soc.* **1991**, *113*(7), 2353–2358.

3562. Smith, P. E.; Tanner, J. J. *J. Am. Chem. Soc.* **1999**, *121*(37), 8637–8644.

3563. Ragg, E.; Scaglioni, L.; Mondelli, R.; Carelli, I.; Casini, A.; Tortorella, S. *Biochim. Biophys. Acta* **1991**, *1076*(1), 49–60.

3564. Hughes, N. A.; Kenner, G. W.; Todd, A. *J. Chem. Soc.* **1957**, (9), 3733–3738.

3565. Mel'nikova, L. M.; Berezovskii, V. M. *Zh. Obshch. Khim.* **1970**, *40*(4), 918–923 (Engl. 897–901).

3566. Shuster, L.; Kaplan, N. O.; Stolzenbuch, F. E. *J. Biol. Chem.* **1955**, *215*(1), 195–209.

3567. Lee, J.; Churchill, H.; Choi, W.-B.; Lynch, J. E.; Roberts, F. E.; Volante, R. P.; Reider, P. J. *J. Chem. Soc., Chem. Commun.* **1999**, (8), 729–730.

3568. Honjo, M.; Furukawa, Y.; Moriyama, H.; Tanaka, K. *Chem. Pharm. Bull.* **1963**, *11*(6), 712–720.

3569. Walt, D. R.; Findeis, M. A.; Rios-Mercadillo, V. M.; Augé, J.; Whitesides, G. M. *J. Am. Chem. Soc.* **1984**, *106*(1), 234–239.

3570. Avigad, G. *Biochim. Biophys. Acta* **1979**, *571*(1), 171–174.

3571. Powning, R. F.; Keatzing, C. C. *Arch. Biochem. Biophys.* **1957**, *66*(1), 249–251.

3572. Chaykin, S.; King, L.; Watson, J. G. *Biochim. Biophys. Acta* **1966**, *124*(1), 13–25.

3573. Kaplan, N. O.; Colowick, S. P.; Barnes, C. C. *J. Biol. Chem.* **1951**, *191*(2), 461–472.

3574. Guilbert, C. C.; Johnson, S. L. *Biochemistry* **1971**, *10*(12), 2313–2316.

3575. Berti, P. J.; Schramm, V. L. *J. Am. Chem. Soc.* **1997**, *119*(50), 12069–12078.

3576. Oppenheimer, N. J. In: *Pyridine Nucleotide Coenzymes;* Everse, J.; Anderson, B.; You, K., Eds.; Academic Press: New York, 1982; pp 51–89 Review with 112 references.

3577. Liu, R.; Visscher, J. *Nucleosides Nucleotides* **1994**, *13*(5), 1215–1216.

3578. Johnson, S. I.; Tuazon, P. T. *Biochemistry* **1977**, *16*(6), 1175–1183.

3579. Alivisatos, S. G. A.; Ungar, F.; Abraham, G. J. *Biochemistry* **1965**, *4*(12), 2616–2630.

3580. Lindsay, R. H.; Paik, W. K.; Cohen, P. P. *Biochim. Biophys. Acta* **1962**, *58*(3), 585–587.

3581. Ben-Hayyim, G.; Hochman, A.; Avron, M. *J. Biol. Chem.* **1967**, *242*(12), 2837–2839.

3582. Bernofsky, C. *Arch. Biochem. Biophys.* **1975**, *166*(2), 645–650.

3583. Fels, I. G. *Nature* **1962**, *195*, 704–705.

3584. Bevilacqua, L. R. *Atti Accad. Ligure Sci. Lett. Genoa* **1955**, *12*, 102–109.

3585. Appaji Rao, N.; Cama, J. R.; Kumar, S. A. *J. Indian Inst. Sci.* **1961**, *43*(1), 1–9.

3586. Giri, K. V.; Krishnaswamy, P. R.; Appaji Rao, N. *Biochem. J.* **1958**, *70*(1), 66–71.

3587. Giri, K. V.; Appaji Rao, N.; Cama, H. R.; Kumar, S. A. *Biochem. J.* **1960**, *75*(2), 381–386.

3588. Koop, W.; Barz, W. *Phytochemistry* **1976**, *15*(11), 1581–1583.

3589. Ravindranath, S. D.; Appaji Rao, N. *Arch. Biochem. Biophys.* **1969**, *133*(1), 54–59.

3590. Begley, T. P.; Kinsland, C.; Taylor, S.; Tandon, M.; Nicewonger, R.; Wu, M.; Chiu, H.-J.; Kelleher, N.; Campobasso, N.; Zhang, Y. *Polyketides Vitamins* **1998**, *195*, 93–142.

3591. Russell, L. F.; Vanderslice, J. T. *J. Micronutr. Anal.* **1990**, *8*, 257–310.

3592. Swaraj, K.; Toponov, A. F.; Golubeva, L. I.; Kretovich, V. L. *Fiziol. Rast.* **1986**, *33*(1), 87–92 (Engl. 70–74).

3593. Koziol, J. *Zesz. Nauk Wyzsza Szk. Ekon. Poznaniu, Ser. I* **1966**, *26*, 79–89, from *Chem. Abstr.* **1967**, *66*, 54328u.

3594. Treadwell, G. E., Jr.; Metzler, D. E. *Anal. Biochem.* **1972**, *46*(2), 261–275.

3595. Warburg, O.; Christian, W. *Biochem. Z.* **1938**, *298*, 150–168.

3596. Sanwal, B. D.; Waygood, E. R. *Can. J. Bot.* **1963**, *41*(1), 55–62.

3597. Koziolova, A. *Zesz. Nauk Wyzsza Szk. Ekon. Poznaniu, Ser.* **1966**, *25*, 41–73, from *Chem. Abstr.* **1967**, *66*, 84797p.

3598. Van Gestelen, P.; Asard, H.; Caubergs, R. J. *Physiol. Plant.* **1996**, *98*(2), 389–398.

3599. Lüttje, S.; Van Gestelen, P.; Córdoba-Pedregosa, M. C.; González-Reyes, J. A.; Asard, J.; Villalba, J. M.; Böttger, M. *Protoplasma* **1998**, *205*(1/4), 43–51.

3600. Vecher, A. S.; Bekman, L. S.; Bykova, L. N. *Dokl. Akad. Nauk BSSR* **1977**, *21*(3), 266–268.

3601. Bykova, L. N.; Vecher, A. S. *Vestsi Akad. Navuk BSSR, Ser. Biyal. Navuk* **1969**, (3), 52–57.

3602. Asomaning, E. J. A.; Galston, A. W. *Plant Physiol.* **1961**, *36*(4), 453–464.

3603. Takiguchi, K.; Hotta, K. *Bitamin* **1960**, *21*(4), 503–505.

3604. Rubin, B. A.; Chernavina, I. A.; Krendeleva, T. E. *Fiziol. Rast.* **1965**, *12*(2), 204–208 (Engl. 176–179).

3605. Mitsuda, H.; Kawai, F.; Miyoshi, T. *Nippon Nogei Kagaku Kaishi* **1958**, *32*, 847–851.

3606. Barr, R.; Sandelius, A. S.; Crane, F. L.; Morré, D. J. *Biochim. Biophys. Acta* **1986**, *852*(2), 254–261.

3607. Tanaka, K. *Eiyo To Shokuryo* **1955/6**, *8*, 143–144, from *Chem. Abstr.* **1960**, *54*, 12404c.

3608. Welkie, G. W.; Miller, G. W. *Plant Physiol.* **1960**, *35*(4), 516–520.

3609. Gvamichava, N. E.; Kezeli, T. A.; Tarasashvili, K. M.; Piranishvili, N. S. *Soobshch. Akad. Nauk Gruz. SSR* **1982**, *107*(1), 133–136, from *Chem. Abstr.* **1983**, *98*, 48483t.

3610. Kuczmak, M.; Tolbert, N. E. *Plant Physiol.* **1962**, *37*(6), 729–734.

3611. Kezeli, T. A.; Gvamicheva, N. E.; Tarasashvili, K. M.; Piranishvili, N. S.; Kotaeva, D. V.; Chkhubianishvili, E. A. *Fiziol. Moroz. Vinograd. Lozy, Akad. Nauk Gruz. SSR* **1986**, 48–63.

3612. Bessey, O. A.; Lowry, O. H.; Love, R. H. *J. Biol. Chem.* **1949**, *180*(2), 755–769.

3613. Faeder, E. J.; Siegel, L. M. *Anal. Biochem.* **1973**, *53*(1), 332–336.

3614. Draper, R. D.; Ingraham, L. L. *Arch. Biochem. Biophys.* **1968**, *125*(3), 802–808.

3615. Lowe, H. J.; Clark, W. M. *J. Biol. Chem.* **1956**, *221*(2), 983–992.

3616. Fazekar, A. G.; Kokai, K. *Methods Enzymol.* **1971**, *18B*, 385–398.

3617. Koziol, J. *Methods Enzymol.* **1971**, *18B*, 253–285.

3618. Rauen, H. M.; Waldmann, H. *Z. Physiol. Chem.* **1951**, *287*, 216–221.

3619. Brunk, G. R.; Martin, K. A.; Nishimura, A. M. *Biophys. J.* **1976**, *16*(16), 1373–1384.

3620. Gibson, Q. H.; Massey, V.; Atherton, N. M. *Biochem. J.* **1962**, *85*(2), 369–383.

3621. Tatsumi, H.; Nakase, H.; Kano, K.; Ikeda, T. *J. Electroanal. Chem.* **1998**, *443*(2), 236–242.

3622. Ke, B. *Arch. Biochem. Biophys.* **1957**, *68*(2), 330–340.

3623. Cerletti, P. *Anal. Chim. Acta* **1959**, *20*(3), 243–250.

3624. Copeland, R. A.; Spiro, T. G. *J. Phys. Chem.* **1986**, *90*(25), 6648–6654.

3625. Listowsky, I.; England, S.; Betheil, J. J.; Seifter, S. *Biochemistry* **1966**, *5*(8), 2548–2552.

3626. Miles, D. W.; Urry, D. W. *Biochemistry* **1968**, *7*(8), 2791–2799.

3627. Tollin, G. *Biochemistry* **1968**, *7*(5), 1720–1727.

3628. Kainosho, M.; Kyogoku, Y. *Biochemistry* **1972**, *11*(5), 741–752.

3629. Yagi, K.; Ohishi, N.; Takai, A.; Kawano, K.; Kyogoku, Y. *Biochemistry* **1976**, *15*(13), 2877–2880.

3630. Shiga, T.; Piette, L. H. *Photochem. Photobiol.* **1964**, *3*(3), 213–222.

3631. Kurreck, H.; Bock, M.; Bretz, N.; Elsner, M.; Kraus, H.; Lubitz, W.; Müller, F.; Geissler, J.; Kroneck, P. M. H. *J. Am. Chem. Soc.* **1984**, *106*(3), 737–746.

3632. Xu, J.; Birke, R. L.; Lombardi, J. R. *J. Am. Chem. Soc.* **1987**, *109*(19), 5645–5649.

3633. Drabent, R.; Frąckowiak, D.; Jadżyn, Cz.; Hotchandani, S.; Leblanc, R. M. *Biochim. Biophys. Acta* **1985**, *843*(1/2), 25–28.

3634. Abián, J.; Susín, S.; Abadía, J.; Gelpí, E. *Anal. Chim. Acta* **1995**, *302*(2/3), 215–223.

3635. Kilgour, G. L.; Felton, S. P.; Huennekens, F. M. *J. Am. Chem. Soc.* **1957**, *79*(9), 2254–2256.

3636. Giri, K. V.; Krishnaswamy, P. R. *J. Indian Inst. Sci.* **1956**, *38A*(4), 232–244.

3637. Nielsen, P.; Rauschenbach, P.; Backer, A. *Anal. Biochem.* **1983**, *130*(2), 359–368.

3638. Ukita, T.; Irie, M. *Chem. Pharm. Bull.* **1960**, *8*, 436–440.

3639. Johnson, R. D.; LaJohn, L.; Carrico, R. J. *Anal. Biochem.* **1978**, *86*(2), 526–530.

3640. Cerletti, P.; Siliprandi, N. *Biochem. J.* **1955**, *61*(2), 324–328.

3641. Kenndler, E.; Schwer, C.; Kaniansky, D. *J. Chromatogr.* **1990**, *508*(1), 203–207.

3642. Budavari, S.; et al., *The Merck Index*, 12th ed., CRC Press, 1966.

3643. Fujisawa, K. *J. Biochem.* **1958**, *45*(5), 359–361.

3644. Ghisla, S.; Massey, V.; Lhoste, J. M.; Mayhew, S. G. *Biochemistry* **1974**, *13*(3), 589–597.

3645. Moonen, C. T. W.; Verwoort, J.; Müller, F. *Biochemistry* **1984**, *23*(21), 4859–4867.

3646. Verwoort, J.; Müller, F.; Mayhew, S. G.; van den Berg, W. A. M.; Moonen, C. T. W.; Bacher, A. *Biochemistry* **1986**, *25*(22), 6789–6799.

3647. Enescu, M.; Lindqvist, L.; Soep, B. *Photochem. Photobiol.* **1998**, *68*(2), 150–156.

3648. Cerletti, P.; Rossi-Fanelli, A. *J. Vitaminol.* **1958**, *4*(1), 71–76.

3649. Mitra, C.; Torstensson, A. *Bioelectrochem. Bioenerg.* **1980**, *7*(4), 749–756.

3650. Honeychurch, M. J.; Ridd, M. J. *Electroanalysis* **1996**, *8*(4), 362–369.

3651. Birss, V. I.; Hinman, A. S.; McGarvey, C. E.; Segal, J. *Electrochim. Acta* **1994**, *39*(16), 2449–2454.

3652. Templeton, D. M.; Hollebone, B. R.; Tsai, C. S. *Biochemistry* **1980**, *19*(16), 3868–3873.

3653. Sarma, R. H.; Dannies, P.; Kaplan, N. O. *Biochemistry* **1968**, *7*(12), 4359–4367.

3654. Ulrich, E. L.; Westler, W. M.; Markley, J. L. *Tetrahedron Lett.* **1983**, *24*(5), 473–476.

3655. Stob, S.; Kemmink, J.; Kaptein, R. *J. Am. Chem. Soc.* **1989**, *111*(18), 7036–7042.

3656. Russell, L. F.; Vanderslice, J. T. *Food Chem.* **1992**, *43*(2), 151–162.

3657. Izumiya, M. *J. Vitaminol.* **1960**, *6*(2), 87–93.

3658. Song, P.-S. *J. Am. Chem. Soc.* **1969**, *91*(7), 1850–1851.

3659. Govil, G.; Phadke, R. S.; Sonawat, H. M. *THEOCHEM* **1988**, *179*, 165–175.

3660. Raszka, M.; Kaplan, N. O. *Proc. Natl. Acad. Sci. U. S. A.* **1974**, *71*(11), 4546–4550.

3661. Visser, A. J. W. G. *Photochem. Photobiol.* **1984**, *40*(6), 703–706.

3662. Kuhn, R.; Rudy, H.; Weygand, F. *Berichte* **1936**, *69B*(6), 1543–1547.

3663. Viscontini, M.; Ebnöther, C.; Karrer, P. *Helv. Chim. Acta* **1952**, *35*(2), 457–459.

3664. Berezovskii, V. M.; Artemkina, R. V.; Khomutova, E. D. *Zh. Obshch. Khim.* **1963**, *34*(8), 2791–2796 (Engl. 2813–2817).

3665. Artemkina, R. V.; Berezovskii, V. M. *Zh. Obshch. Khim.* **1966**, *36*(5), 823–828 (Engl. 841–844).

3666. Sato, T.; Yoshimura, J.; Takaoka, T. *Proc. Jpn. Acad. Sci.* **1953**, *29*, 260–261, from *Chem. Abstr.* **1955**, *49*, 1063i.

3667. Tanaka, T. *Yakugaku Zasshi* **1960**, *80*, 1053–1056.

3668. Forrest, H. S.; Todd, A. R. *J. Chem. Soc.* **1950**, (11), 3295–3299.

3669. Strehler, B. L.; Harvey, E. N.; Chang, J. J.; Cormier, M. J. *Proc. Natl. Acad. Sci. USA* **1954**, *40*(1), 10–12.

3670. Flexser, L. A.; Farkas, W. G. U.S. Patent 2,604,470 to Hoffmann-LaRoche, Inc., July 22, 1952; from *Chem. Abstr.* **1953**, *47*, 3354b.

3671. Huennekens, F. M.; Kilgour, G. L. *J. Am. Chem. Soc.* **1955**, *77*(24), 6716–6717.

3672. DeLuca, D.; Kaplan, N. O. *J. Biol. Chem.* **1956**, *223*(1), 569–576.

3673. Moffatt, J. G.; Khorana, H. G. *J. Am. Chem. Soc.* **1958**, *80*(14), 3756–3761.

3674. Kuroda, T. *Bitamin* **1964**, *29*(5), 437–444, from *Chem. Abstr.* **1965**, *62*, 7851g.

3675. Christie, S. M. H.; Kenner, G. W.; Todd, A. R. *J. Chem. Soc.* **1954**, (1), 46–52.

3676. Nielsen, P.; Harksen, J.; Bacher, A. *Eur. J. Biochem.* **1985**, *152*(2), 465–473.

3677. Baddiley, J.; Buchanan, J. G.; Cares, B. *J. Chem. Soc.* **1957**, (9), 4058–4063.

3678. Bamann, E.; Trapmann, H. *Arch. Pharm.* **1956**, *289*, 338–343.

3679. Hotta, K.; Terao, M. *Bitamin* **1959**, *17*(3), 534–538.

3680. Kilgour, G. L.; Huennekens, F. M. *J. Am. Chem. Soc.* **1957**, *79*(9), 2256–2258.

3681. Sebrell, Jr., W. H.; Harris, R. S., Eds.; *The Vitamins: Chemistry, Physiology, Pathology, Methods*, Vol. 5; Academic Press: New York, 1972; pp 114–115, xviii + 468 pp.

3682. de la Fuente, G.; Diaz-Cadavisco, R. *Nature* **1954**, *174*, 1014.

3683. Krampitz, L. O. *Annu. Rev. Biochem.* **1969**, *38*, 213–240.

3684. Tauber, H. *Proc. Soc. Exp. Biol. Med.* **1937**, *37*(3), 541–543.

3685. Rindi, G.; de Giuseppe, L. *Int. Z. Vitaminforsch.* **1961**, *31*, 321–326.

3686. Wenz, A.; Göttmann, G.; Koop, H. *Ann. Chem.* **1958**, *618*, 210–220.

3687. Leichssenring, G.; Schmidt, J. *Chem. Ber.* **1962**, *95*(3), 767–772.

3688. Lohmann, K.; Schuster, P. *Biochem. Z.* **1937**, *294*(3/4), 188–214.

3689. de Bernard, B.; Cartier, P. H. *Giorn. Biochim.* **1958**, *7*, 1–7.

3690. Watanabe, A.; Asahi, Y.; Hori, M. *Yakugaku Zasshi* **1957**, *77*(2), 157–160.

3691. Okamoto, K. *Bull. Chem. Soc. Jpn.* **1963**, *36*(4), 366–371.

3692. Rossi-Fanelli, A.; Segre, A.; Siliprandi, D.; Siliprandi, N. *Farmaco. Ed. Sci.* **1955**, *10*(4), 173–178.

3693. Gibson, E. P.; Turnbull, J. H. *J. Chem. Soc., Perkin Trans. 2* **1980**, (9), 1288–1292.

3694. Goncharenko, L. S. *Biofizika* **1977**, *12*(1), 155–157 (Engl. 176–179).

3695. Chauvet-Monges, A. M.; Briand, C.; Crevat, A.; Surzur, J. M. *Ann. Phys. Biol. Med.* **1971**, *5*(3), 165–172.

3696. Gallo, A. A.; Sable, H. Z. *J. Biol. Chem.* **1974**, *249*(5), 1382–1389.

3697. Siliprandi, D.; Siliprandi, N. *Biochim. Biophys. Acta* **1954**, *14*, 52–61.

3698. Kiessling, K.-H. *Acta Chem. Scand.* **1956**, *10*(8), 1356–1358.

3699. David, S.; Hirshfeld, H. *Bull. Soc. Chim. Fr.* **1963**, 1011–1013.

3700. Levorato, C.; Cima, L. *J. Chromatogr.* **1968**, *32*(4), 771–773.

3701. Parkhomenko, Yu.M.; Rybina, A. A.; Pol'shchak, R. B. *Ukr. Biokhim. Zh.* **1976**, *48*(3), 384–387, from *Chem. Abstr.* **1976**, *85*, 16543v.

3702. Hilker, D. M.; Clifford, A. J. *J. Chromatogr.* **1982**, *231*(2), 433–438.

3703. Davidek, J.; Pudil, F.; Seifert, J. *J. Chromatogr.* **1977**, *140*, 316–318.

3704. Rossi-Fanelli, A.; Mondovi, B.; Boffi, V. *Boll. Soc. Ital. Biol. Sper.* **1953**, *29*(7), 1330–1333.

3705. Crevat, A.; Jacquet-Francillon, M. L.; Bernard, P. *Ann. Phys. Biol. Med.* **1968**, *2*(2), 73–80.

3706. Gudbjarnason, S. *Biochim. Biophys. Acta* **1967**, *148*(1), 22–36.

3707. Khan, B. T.; Rao, P. N.; Khan, M. M. T. *Indian J. Chem.* **1981**, *20A*(8), 857–859.

3708. Pletcher, J.; Wood, M.; Blank, G.; Shin, W.; Sax, M. *Acta Crystallogr.* **1977**, *B33*(10), 3349–3359.

3709. Carlisle, C. H.; Cook, D. S. *Acta Crystallogr.* **1969**, *B25*(5), 1359–1367.

3710. Berezovskii, V. M.; Koltunova, V. I.; Pekel', N. D.; Shlimovich, E. A. *Zh. Obshch. Khim.* **1963**, *33*(1), 49–55 (Engl. 44–49).

3711. Wenz, A.; Göttmann, G.; Koop, H. *Ann. Chem.* **1964**, *680*, 82–87.

3712. Baran, E. J.; Etcheverry, S. B. *Z. Naturforsch.* **1974**, *34B*(11), 1615–1616.

3713. Pletcher, J.; Blank, G.; Wood, M.; Sax, M. *Acta Crystallogr.* **1979**, *B35*(7), 1633–1637.

3714. Pletcher, J.; Sax, M. *J. Am. Chem. Soc.* **1972**, *94*(11), 3998–4005.

3715. Katz, H. B.; Kustin, K. *Biochim. Biophys. Acta* **1973**, *313*(2), 235–248.

3716. Taglioni, J.-P.; Fournier, J. *Compt. Rend.* **1979**, *288C*, 141–144.

3717. Fedorova, R. V.; Fedorov, P. I. *Zh. Prikl. Khim.* **1969**, *42*(4), 876–880, from *Chem. Abstr.* **1969**, *71*, 85067r.

3718. Melnick, J. L. *J. Biol. Chem.* **1939**, *131*(2), 615–620.

3719. Lenormant, H.; Hérisson, M. *Compt. Rend.* **1951**, *232*, 815–816.

3720. Kotera, K. *Chem. Pharm. Bull.* **1965**, *13*(4), 440–443.

3721. White, W. D.; Drago, R. S. *Inorg. Chem.* **1971**, *10*(12), 2727–2735.

3722. Lachmann, H.; Freund, A.; Schnackertz, K. D. *Z. Anal. Chem.* **1984**, *317*, 697–698.

3723. Peix, E. S. *Cienc. Ind. Farm.* **1974**, *6*(11), 341–362, from *Chem. Abstr.* **1975**, *82*, 160189q.

3724. Jordan, F. *J. Am. Chem. Soc.* **1974**, *96*(11), 3623–3630.

3725. Friedemann, R.; Uslar, W. *Biol. Zentralbl.* **1988**, *107*(1), 31–37.

3726. Lewin, L. M.; Wei, R. *Anal. Biochem.* **1966**, *16*(1), 29–35.

3727. Crevat, A.; Bernard, P. M.; Rogeret, C. *Bull. Soc. Pharm. Marseille* **1969**, *18*(70), 83–90, from *Chem. Abstr.* **1970**, *73*, 28825d.

3728. Kimura, M.; Fujita, T.; Nishida, S.; Itakawa, Y. *J. Chromatogr.* **1980**, *188*, 417–419.

3729. Wildemann, L. *Z. Klin. Chem. Klin. Biochem.* **1969**, *7*(5), 509–513.

3730. Hilker, D. M.; Mee, J. M. L. *J. Chromatogr.* **1973**, *76*, 239–241.

3731. Vergara, T.; Marin, D.; Vera, J. *Anal. Chim. Acta* **1980**, *120*, 347–351.

3732. Vergara, T.; Marin, D.; Vera, J.; Sarabia, A. *Electrochim. Acta* **1980**, *25*, 1641–1645, from *Chem. Abstr.* **1981**, *94*, 79077e.

3733. Pleticha, R. *Anal. Chim. Acta* **1958**, *18*, 146–154.

3734. Ciszewski, A.; Studnickova, M.; Fischer, O. *Bioelectrochem. Bioenerg.* **1984**, *13*, 25–38.

3735. Budz, J.; Karpinski, P. H.; Mydlarz, J.; Nyolt, J. *Ind. Eng. Chem. Prod. Res. Dev.* **1986**, *25*(4), 657–664.

3736. Penttinen, H. K. *Finn. Chem. Lett.* **1976**, (1), 1–2, from *Chem. Abstr.* **1976**, *84*, 164720v.

3737. Gallo, A. A.; Hansen, I. L.; Sable, H. Z.; Swift, T. J. *J. Biol. Chem.* **1972**, *247*(18), 5913–5920.

3738. Viscontini, M.; Bonetti, G.; Ebnöther, C.; Karrer, P. *Helv. Chim. Acta* **1951**, *34*(5), 1384–1387.

3739. Mosolov, N. N.; Stepuro, I. I.; Ostrovsky, Yu.M. *Biokhimiya* **1972**, *37*(6), 1272–1275.

3740. Yusa, T. *J. Biochem.* **1959**, *46*(3), 391–393.

3741. Rossi-Fanelli, A.; Siliprandi, N.; Siliprandi, D.; Ciccarone, P. *Arch. Biochem.* **1955**, *58*(1), 237–243.

3742. Weijland, J. *J. Am. Chem. Soc.* **1942**, *64*(10), 2279–2282.

3743. Tanaka, T. *Yakugaku Zasshi* **1956**, *76*(11), 1314–1318.

3744. Karrer, P.; Viscontini, M. *Helv. Chim. Acta* **1946**, *29*(3), 711–718.

3745. Greiling, H.; Kiesow, L. *Z. Naturforsch.* **1957**, *12B*(10), 605–609.

3746. Viscontini, M.; Bonetti, G.; Ebnöther, C.; Karrer, P. *Helv. Chim. Acta* **1951**, *34*(5), 1388–1392.

3747. Velluz, L.; Bartos, J. *Bull. Soc. Chim. Fr.* **1952**, 115–118.

3748. Viscontini, M.; Bonetti, G.; Karrer, P. *Helv. Chim. Acta* **1949**, *32*(5), 1478–1484.

3749. Weijland, J.; Tauber, H. *J. Am. Chem. Soc.* **1938**, *60*(9), 2263–2264.

3750. Weijland, J. *J. Am. Chem. Soc.* **1941**, *63*(4), 1160–1161 *Errata*.[519].

3751. Galamon, T.; Filipowicz, B. *Rocz. Chem.* **1967**, *41*(9), 1513–1519.

3752. Weil-Malherbe, H. *Biochem. J.* **1940**, *34*, 980–990.

3753. Velluz, L.; Amiard, G.; Bartos, J. *Bull. Soc. Chim. Fr.* **1948**, (7/8), 871–878.

3754. Karle, I. L.; Britta, K. *Acta Crystallogr.* **1966**, *20*(1), 118–124.

3755. Aoki, K.; Yamazaki, H.; Wanagai, K.; Itokawa, H. *Acta Crystallogr., Sect. C Cryst. Struct. Commun.* **1968**, *C44*(11), 1945–1956.

3756. Okumura, K. *Bitamin* **1961**, *24*, 158–163, from *Chem. Abstr.* **1964**, *60*, 7068b.

3757. Ebert, M.; Swallow, A. J. *Radiat. Res.* **1957**, *7*(3), 229–235.

3758. Gregolin, C.; Rossi, C. S.; Siliprandi, N. *Radiat. Res.* **1960**, *13*(2), 185–191.

3759. Riva, F.; Giarnieri, D.; Hammady, J. *Ital. J. Biochem.* **1962**, *11*(5), 346–351.

3760. Bamann, E.; Schuegraf, A. *Biochem. Z.* **1955**, *326*(7), 503–510.

3761. Khan, B. T.; Rao, P. N. *Anal. Chim. Acta* **1985**, *106*, 97–100.

3762. Mulley, E. A.; Stumbo, C. R.; Hunting, W. M. *J. Food Sci.* **1975**, *40*(5), 989–992.

3763. Khan, B. T.; Rao, P. N. *Anal. Chim. Acta* **1982**, *67*, 79–82.

3764. Farrer, K. T. H. *Aust. J. Exp. Biol.* **1949**, *27*(6), 591–596.

3765. Chernykevich, I. P.; Gritsenko, E. A.; Luchko, T. A.; Zabrodskaya, S. V. *Dokl. Akad. Nauk. BSSR* **1990**, *34*(3), 274–278, from *Chem. Abstr.* **1991**, *114*, 42354w.

3766. Koltunova, V. I.; Berezovskii, V. M. *Khim. Farm. Zh.* **1967**, *1*(2), 10–14 (Engl. 68–71).

3767. Avison, A. W. D. *J. Chem. Soc.* **1955**, 732–738.

3768. Kuroda, T.; Masaki, M. *Bitamin* **1964**, *29*(2), 109–115, from *Chem. Abstr.* **1965**, *62*, 7851f.

3769. Gallo, A. A.; Sable, H. Z. *J. Biol. Chem.* **1975**, *250*(13), 4986–4991.

3770. Monti, J. P.; Dell'Amico, M.; Chauvet-Monges, A. M.; Fournier, J.; Crevat, A. *C. R. Hebd. Seances Acad. Sci., Ser. C* **1977**, *284*(7), 285–288.

3771. Aoki, K.; Yamazaki, H. *J. Am. Chem. Soc.* **1985**, *107*(22), 6242–6249.

3772. Grande, H. J.; Houghton, R. L.; Veeger, C. *Eur. J. Biochem.* **1973**, *37*, 563–569.

3773. Jordan, F.; Mariam, Y. H. *Microchem. J.* **1977**, *22*(2), 182–200.

3774. Aoki, K.; Yamakazi, H. *J. Am. Chem. Soc.* **1980**, *102*(22), 6878–6890.

3775. Kozik, A.; Rapala-Kozik, R. *Plant Physiol. Biochem.* **1996**, *34*(6), 779–786.

3776. McCormick, D. B.; Wright, L. D., Eds.; *Methods Enzymol.* **1979**, *62*, 63–68 xxv + 616 pp.

3777. Chytil, F.; McCormick, D. B., Eds.; *Methods Enzymol.* **1986**, *122* xxx + 456 pp.

3778. Korytnyk, W.; Ahrens, H. *Methods Enzymol.* **1970**, *18A*, 475–480.

3779. Teysseire, Y. *C. R. Soc. Biol.* **1945**, *139*, 822.

3780. Yusa, T. *Plant Cell Physiol.* **1961**, *2*(4), 471–474.

3781. Vanderslice, J. T.; Huang, M. H. A. *J. Micronutr. Anal.* **1986**, *2*(3), 189–199.

3782. Roux, H.; Chevalier, C.; Teysseire, Y. *Compt. Rend. Soc. Biol.* **1944**, *138*, 638–639.

3783. Kochibe, N.; Yusa, T.; Hayashi, K. *Plant Cell Physiol.* **1963**, *4*(2), 239–242.

3784. Hieda, T. *Nagoya Igakkai Zasshi* **1944**, *59*, 403–410, from *Chem. Abstr.* **1948**, *42*, 6891.

3785. Dougherty, R. H.; Cobb, W. Y. *J. Agric. Food Chem.* **1970**, *18*(5), 921–925.

3786. Sen, N. P.; Robinson, A. D. *Can. J. Biochem. Physiol.* **1963**, *41*(1), 97–100.

3787. Zeijlemaker, F. C. J. *Proc. Acad. Sci. Amsterdam* **1939**, *42*(2), 187–194.

3788. Siesto, A. J. *Quad. Nutr.* **1959**, *19*(3/4), 129–135.

3789. Molin, W. T.; Wilkerson, C. G., Jr.; Fites, R. C. *Plant Physiol.* **1980**, *66*(2), 313–315.

3790. Weakley, F. B.; Eldridge, A. C.; McKinney, L. L. *J. Agric. Food Chem.* **1961**, *9*(6), 435–439.

3791. Bunting, A. H.; James, W. O. *New Phytol.* **1941**, *40*(4), 262–267.

3792. Melnick, D.; Field, H., Jr. *J. Biol. Chem.* **1939**, *127*(3), 531–540.

3793. Bos, T. J.; Romani, R. J. *J. Food Biochem.* **1981**, *5*(4), 245–261.

3794. Chikubu, S.; Horiuchi, H.; Furuya, M.; Niizeki, S. *Nippon Nogei Kagaku Kaishi* **1955**, *29*(3), 225–229.

3795. Kulesza, J.; Moszczynski, P.; Kroh, J. *Biol. Plant.* **1967**, *9*(1), 15–19.

3796. Booth, R. G. *J. Soc. Chem. Ind.* **1940**, *59*(9), 181–184.

3797. Hennessy, D. H.; Cerecedo, L. R. *J. Am. Chem. Soc.* **1939**, *61*(1), 179–183.

3798. Kulesza, J.; Moszczynski, P.; Kroh, J. *Radiobiologiya* **1967**, *7*(2), 267–270.

3799. Tsutusi, T.; Adachi, T. *Kyoto-fu Eisei Kogai Kenkyusho Nenpo* **1994**, *39*, 50–55, from *Chem. Abstr.* **1995**, *123*, 168113w.

3800. Herbert, E. W., Jr.; Vanderslice, J. T.; Huang, M.-H.; Higgs, D. J. *Apidologie* **1987**, *18*(2), 129–136.

3801. Obermeyer, H. G.; Fulmer, W. C.; Young, J. M. *J. Biol. Chem.* **1944**, *154*(3), 557–559.

3802. Gale, E. F.; Epps, H. M. R. *Biochem. J.* **1944**, *38*(3), 250–256.

3803. Rabinowitz, J. C.; Snell, E. E. *J. Biol. Chem.* **1947**, *169*(3), 643–650.

3804. Yasumoto, K.; Iwami, K.; Tsuji, H.; Okada, J.; Mitsuda, H. *Vitamins* **1976**, *50*(8), 327–333.

3805. Lee, Y.-P. *Biochim. Biophys. Acta* **1960**, *43*, 25–30.

3806. Shimomura, S.; Emman, K.; Fukui, T. *J. Biochem.* **1980**, *87*(4), 1043–1052.

3807. Matheson, N. K.; Richardson, R. H. *Phytochemistry* **1978**, *17*(2), 195–200.

3808. Akopyan, T. N.; Braunstein, A. E.; Goryachenkova, E. V. *Proc. Natl. Acad. Sci. U. S. A.* **1975**, *72*(4), 1617–1621.

3809. Masada, M.; Fukushima, K.; Tamura, G. *J. Biochem.* **1975**, *77*(5), 1107–1115.

3810. Mazelis, M.; Creveling, R. K. *Biochem. J.* **1975**, *147*, 485–491.

3811. Ramakrishna, S.; Adiga, P. R. *Eur. J. Biochem.* **1975**, *59*, 377–386.

3812. Schramm, W.; Bitsch, R. *Ber. Bundesforschungsanst. Ernähr.* **1993**, *93*(2), 417–421, from *Chem. Abstr.* **1995**, *122*, 185997j.

3813. Cane, D. E.; Du, S.; Robinson, J. K.; Hsiung, Y.; Spenser, I. D. *J. Am. Chem. Soc.* **1999**, *121*(33), 7722–7723.

3814. Morrison, A. L.; Long, R. F. *J. Chem. Soc.* **1958**, (1), 211–215.

3815. Argoudelis, C. J. *J. Agric. Food Chem.* **1986**, *34*(6), 995–998.

3816. Peterson, E. A.; Sober, H. A. *J. Am. Chem. Soc.* **1954**, *76*(1), 169–175.

3817. Bazhulina, N. P.; Morozov, Yu.V.; Karpeiskii, M. Ya.; Ivanov, V. I.; Kuklin, A. I. *Biofizika* **1966**, *11*(1), 42–47.

3818. Hedin, P. A. *J. Agric. Food Chem.* **1963**, *11*(4), 343–345.

3819. Vanderslice, J. T.; Stewart, K. K.; Yarman, M. M. *J. Chromatogr.* **1979**, *176*(2), 280–285.

3820. Korytnyk, W.; Fricke, G.; Paul, B. *Anal. Biochem.* **1966**, *17*(1), 66–75.

3821. Abe, M.; Nishimune, T.; Hayashi, R. *Vitamins* **1974**, *48*(12), 545–549, from *Chem. Abstr.* **1975**, *82*, 77142y.

3822. Takeuchi, Y.; Kawabata, A.; Iwasa, Y. *Bull. Univ. Osaka Prefect.* **1960**, *10B*, 15–24, from *Biol. Abstr.* **1961**, *36*, 29698.

3823. Colombini, C. E.; McCoy, E. E. *Anal. Biochem.* **1970**, *34*(2), 451–458.

3824. Kurosu, Y.; Satou, Y.; Udagawa, K.; Kurioka, S. *Bitamin* **1992**, *66*(2), 91–100.

3825. Peterson, E. A.; Sober, H. A.; Meister, A. *Biochem. Prep.* **1953**, *3*, 34–39.

3826. Kuroda, T. *J. Vitaminol.* **1964**, *10*, 252–263, from *Chem. Abstr.* **1965**, *63*, 5734.

3827. Kozlov, E. I.; L'vova, M. Sh. *Khim. Farm. Zh.* **1977**, *11*(11), 109–115 (Engl. 1543–1549).

3828. Manousek, O.; Zuman, P. *Coll. Czech. Chem. Commun.* **1964**, *29*(6), 1432–1457.

3829. Anderson, F. J.; Martell, A. E. *J. Am. Chem. Soc.* **1964**, *86*(4), 715–720.

3830. Martinez-Carrion, M. *Eur. J. Biochem.* **1975**, *54*(1), 39–43.

3831. Witherup, T. H.; Abbott, E. H. *J. Org. Chem.* **1975**, *40*(15), 2229–2234.

3832. Farago, M. E.; McMillan, M. M.; Sabir, S. S. *Inorg. Chim. Acta* **1975**, *14*, 207–214.

3833. Heyl, D.; Luz, E.; Harris, S. A.; Folkers, K. *J. Am. Chem. Soc.* **1951**, *73*(7), 3430–3433.

3834. Umbreit, W. W.; O'Kane, D. J.; Gunsalus, I. C. *J. Biol. Chem.* **1948**, *176*(2), 629–637.

3835. Bridges, J. W.; Davies, D. S.; Williams, R. T. *Biochem. J.* **1966**, *98*, 451–468.

3836. Wampler, J. E.; Churchich, J. E. *J. Biol. Chem.* **1969**, *244*(6), 1477–1480.

3837. Korytnyk, W.; Singh, R. P. *J. Am. Chem. Soc.* **1963**, *85*(18), 2813–2817.

3838. Lapper, R. D.; Mantsch, H. H.; Smith, I. C. P. *Can. J. Chem.* **1975**, *53*(16), 2406–2412.

3839. Sammons, M. C.; Bursey, M. M.; Brent, D. A. *Biomed. Mass Spectrom.* **1974**, *1*(3), 169–171.

3840. Fujiwara, T. *Bull. Chem. Soc. Jpn.* **1973**, *46*(3), 863–871.

3841. Yamada, M.; Saito, A. *J. Vitaminol.* **1965**, *11*(3), 192–198, from *Chem. Abstr.* **1966**, *64*, 6955h.

3842. Mahuren, J. D.; Coburn, S. P. *Anal. Biochem.* **1977**, *82*(1), 246–249.

3843. Gregory, J. F., III; Feldstein, D. *J. Agric. Food Chem.* **1985**, *33*(3), 359–363.

3844. Siliprandi, N.; Siliprandi, D.; Lis, H. *Biochim. Biophys. Acta* **1954**, *14*, 212–218.

3845. Llor, J.; Cortijo, M. *J. Chem. Soc., Perkin Trans. 2* **1977**, (13), 1715–1719.

3846. Viscontini, M.; Ebnöther, C.; Karrer, P. *Helv. Chim. Acta* **1951**, *34*(6), 1834–1842.

3847. Viscontini, M.; Karrer, P. *Helv. Chim. Acta* **1952**, *35*(6), 1924–1925.

3848. Peterson, E. A.; Sober, H. A.; Meister, A. *Biochem. Prep.* **1953**, *3*, 29–33.

3849. Chen, R. F. *Science* **1965**, *150*, 1593–1595.

3850. Williams, V. R.; Neilands, J. B. *Arch. Biochem. Biophys.* **1954**, *53*(1), 56–70.

3851. Churchich, J. F. *Biochim. Biophys. Acta* **1965**, *102*, 280–288.

3852. Churchich, J. F. *Biochim. Biophys. Acta* **1964**, *79*(3), 643–646.

3853. Viscontini, M.; Ebnöther, C.; Karrer, P. *Helv. Chim. Acta* **1951**, *34*(7), 2199–2202.

3854. Giordano, F.; Mazzarella, L. *Acta Crystallogr.* **1971**, *27B*, 128–134.

3855. Tanaka, T. *Yakugaku Zasshi* **1959**, *79*, 1301–1305.

3856. Baddiley, J.; Mathias, A. P. *J. Chem. Soc.* **1952**, 2583–2591.

3857. Long, R. F. Br. Patent 749,800 to Roche Products Ltd., May 30, 1956; from *Chem. Abstr.* **1957**, *51*, 1297d.

3858. L'vova, S. D.; Yakovleva, L. N.; Cherkasova, I. S.; Balyakina, M. V.; Gunar, V. I. *Khim. Farm. Zh.* **1975**, *9*(11), 23–25 (Engl. 705–706).

3859. Iwanami, M.; Numata, T.; Murakami, M. *Bull. Chem. Soc. Jpn.* **1968**, *41*(1), 161–165.

3860. Nakagawa, K.; Yoshimura, I.; Sueda, N.; Fukawa, H. *Agric. Biol. Chem.* **1977**, *41*(8), 1431–1433.

3861. Heyl, D.; Luz, E.; Harris, S. A.; Folkers, K. *J. Am. Chem. Soc.* **1951**, *73*(7), 3436–3437.

3862. Zuman, P.; Manousek, O. *Coll. Czech. Chem. Commun.* **1961**, *26*(9), 2134–2143.

3863. Heyl, D.; Harris, S. A.; Folkers, K. *J. Am. Chem. Soc.* **1953**, *75*(3), 653–655.

3864. Cunningham, W. C.; Thanassi, J. W. *Experientia* **1979**, *35*(4), 451–452.

3865. Asso, L.; Asso, M.; Mossoyan, J.; Benlian, D. *J. Chim. Phys.* **1978**, *75*(5), 561–567.

3866. de Obieta, C. A.; Munoz, M. S. *An. Edafol. Agrobiol.* **1981**, *40*(9/10), 1825–1831.

3867. Soda, K.; Yorifuji, T.; Misono, H.; Moriguchi, M. *Biochem. J.* **1969**, *114*, 629–633.

3868. Ekanayake, A.; Nelson, P. E. *J. Micronutr. Anal.* **1988**, *4*(1), 1–15.

3869. Toukairin-Oda, T.; Sakamoto, E.; Hirose, H.; Mori, M.; Itoh, T.; Tsuge, H. *J. Nutr. Sci. Vitaminol.* **1989**, *35*(3), 171–180.

3870. Tadera, K.; Naka, Y. *Agric. Biol. Chem.* **1991**, *55*(2), 563–564.

3871. Gregory, J. F., III; Sartain, D. B. *J. Agric. Food Chem.* **1991**, *39*, 899–905.

3872. Gregory, J. F., III; Ink, S. L. *J. Agric. Food Chem.* **1987**, *35*(1), 76–82.

3873. Zavenyagina, T. N.; Bukin, Yu.V. *Biokhimiya* **1971**, *36*(1), 205–215 (Engl. 172–180).

3874. Wang, X. *Sepu* **1993**, *11*(6), 364–366, from *Chem. Abstr.* **1994**, *120*, 53027s.

3875. Addo, C.; Augustin, J. *J. Food Sci.* **1988**, *53*(3), 749–752.

3876. Toepfer, E. W.; Polansky, M. M.; Richardson, L. R.; Wilkes, S. *J. Agric. Food Chem.* **1963**, *11*(6), 523–525.

3877. Ivanoff, L. *Ber. Deutsch. Botan. Ges.* **1902**, *20*, 366–372, from *Chem. Zentralb.***1902** (ii), 1001–1002.

3878. Bernardini, L. *Atti Accad. Real Lincei* **1912**, *21*(1), 283–289, from *Chem. Zentralbl.***1912** (i), 1126 and *J. Chem. Soc. Abstr.***1912**, *102* (ii), 380–381.

3879. Zlataroff, A. *Biochem. Z.* **1916**, *75*, 200–210.

3880. Kursanov, A. *Biokhimiya* **1938**, *3*(4), 467–480.

3881. Klein, R. M. *Plant Physiol.* **1952**, *27*, 335–354.

3882. Aimi, R.; Konno, S. *Nippon Sakumotsu Gakkai Kiji* **1958**, *26*(3), 228–230.

3883. Ergle, D. R.; Eaton, F. M. *Plant Physiol.* **1957**, *32*(2), 106–113.

3884. Äyräpää, T. *Acta Chem. Scand.* **1957**, *11*(9), 1565–1575.

3885. Tluczkiewicz, J.; Berendt, W. *Acta Soc. Bot. Pol.* **1977**, *46*(1), 15–30.

3886. Zuev, L. A.; Golubeva, P. F. *Dokl. Akad. Nauk SSSR* **1955**, *104*(6), 929–931.

3887. Zuev, L. A.; Poruchikova, V. I. *Dokl. Akad. Nauk SSSR* **1950**, *70*(3), 469–472.

3888. Michalik, I. *Acta Fytotech.* **1978**, *33*, 303–318.

3889. Pons, W. A., Jr.; Guthrie, J. D. *Ind. Eng. Chem. Anal. Ed.* **1946**, *18*(3), 184–186.

3890. Verma, S. C.; Lal, B. M. *J. Sci. Food Agric.* **1966**, *17*(1), 43–46.

3891. Lal, B. M.; Verma, S. C. *J. Sci. Food Agric.* **1968**, *19*(2), 113–116.

3892. Martin, E. M.; Morton, R. K. *Biochem. J.* **1956**, *53*, 221–235.

3893. Jennings, A. C.; Morton, R. K. *Aust. J. Biol. Sci.* **1963**, *16*(2), 332–341.

3894. Jennings, A. C.; Watt, W. B. *J. Sci. Food Agric.* **1967**, *18*(11), 527–535.

3895. Chapin, F. S., III; Follett, J. M.; O'Connor, K. F. *J. Ecol.* **1982**, *70*(1), 305–321.

3896. Weiss, R. *Arch. Inst. Grand-Ducal. Luxemb., Sect. Sci. Nat. Phys. Math.* **1953**, *20*, 127–157.

3897. Nigmatov, M. M.; Zakhar'yants, I. L. *Uzb. Biol. Zh.* **1980**, (6), 24–27.

3898. Hewitt, E. J.; Notton, B. J. *J. Sci. Food Agric.* **1960**, *11*(11), 653–658.

3899. Charles, D. J.; Singh, M.; Sanwal, G. G. *Physiol. Plant.* **1980**, *49*(4), 336–340.

3900. Charles, D. J.; Singh, M.; Sanwal, G. G. *Physiol. Plant.* **1982**, *56*(2), 211–216.

3901. Earle, F. R.; Milner, R. T. *Oil Soap* **1938**, *15*(2), 41–42.

3902. Shcheglova, V. F.; Rachkov, L. D. *Tr. Vses. Nauch. Tekh. Konf., Fiziol. Rast. Agro-khim. Pochvoved* **1957**, 257–262, from *Chem. Abstr.* **1961**, *55*, 22498i.

3903. Sokolova, N. A. *Fiziol. Biokhim. Khlop.* **1972**, 8–12, from *Chem. Abstr.***1974**, *80*, 130450i.

3904. Popov, P. S. *Sb. Nauchno-Issled. Rab. Maslichn. Efirovmaslichn. Kul't.* **1960**, 227–242, from *Chem. Abstr.***1962**, *56*, 716c.

3905. Kurmies, B. *Phosphorsäure* **1960**, *20*, 126–145.

3906. Michalik, I. *Pol'nohospodarstvo* **1986**, *32*(5), 396–407.

3907. Javillier, M.; Colin, Y. *Bull. Soc. Chim. Biol.* **1933**, *15*, 1552–1562.

3908. Kosheleva, L. L.; Terent'ev, V. M. *Fotosin. Pitan. Rast.* **1969**, 139–146.

3909. Kakie, T. *Hatano Tabako Shikenjo Hokoku* **1970**, (67), 91–96.

3910. Holden, M. *Biochem. J.* **1952**, *51*, 433–442.

3911. Guardiola, J. L.; Sutcliffe, J. F. *Ann. Bot.* **1971**, *35*(9), 809–823.

3912. Greenwood, J. S.; Gifford, D. J.; Bewley, J. D. *Can. J. Bot.* **1984**, *62*(2), 255–261.

3913. Yarosh, N. P.; Antonova, O. G. *Tr. Prikl. Bot., Genet. Selek.* **1972**, *48*(1), 158–167.

3914. Knowles, F.; Watkin, J. E. *J. Agric. Sci.* **1932**, *22*, 755–766.

3915. Webster, J. E. *J. Agric. Res.* **1928**, *37*(2), 123–125.

3916. Michalik, I. *Acta Fytotech.* **1971**, *22*, 25–35.

3917. Kedrowski, R. A. *J. Plant Nutr.* **1983**, *6*(1), 989–1011.

3918. Sugino, Y.; Miyoshi, Y. *J. Biol. Chem.* **1964**, *239*(7), 2360–2364.

3919. Sokolov, A. V. *Chem. Soc. Agric. (USSR)* **1940**, *10*, 36–38, from *Chem. Abstr.* **1943**, *37*, 4986^9.

3920. Nilova, V. P. *Tr. Vses. Nauchn.-Issled. Inst. Zashchity Rast.* **1964**, *21*(2), 71–80, from *Chem. Abstr.* **1966**, *65*, 7630g.

3921. Pucacka, S. *Acta Physiol. Plant.* **1986**, *8*(4), 219–227.

3922. de Ruiz, A. S. C.; Bressani, R. *Cereal Chem.* **1990**, *67*(7), 519–522.

3923. Wagner, S.; Kuhn, M. *Getreide Mehl Brot* **1994**, *48*(4), 6–10.

3924. Perl, M. *J. Exp. Bot.* **1982**, *33*(134), 449–455.

3925. Gopal, G. R.; Rao, G. R. *J. Nucl. Agric. Biol.* **1983**, *12*(2), 37–41.

3926. Bartnik, M.; Szafranska, I. *J. Cereal Sci.* **1987**, *5*(1), 23–28.

3927. Bevilacqua, L. R. *Atti Accad. Naz. Lincei, Cl. Sci. Fis. Mat. Nat., Rend.* **1953**, *14*(6), 827–833, from *Chem. Abstr.* **1954**, *48*, 4640e and *Chem. Zentralbl.* **1954**, *125*, 7204.

3928. Gendraud, M.; Lafleuriel, J. *Phytochemistry* **1977**, *16*(8), 1289–1290.

3929. Weinberger, P.; Vladut, R. *Can. J. Forest Res.* **1981**, *11*(4), 796–804.

3930. Gambhir, P. N.; Pande, P. C.; Ratchiffe, R. G. *Magn. Reson. Chem.* **1997**, *35*, S125–S132.

3931. Rotkiewicz, D.; Kozlowska, H.; Zadernowski, R.; Budzynski, W.; Horodyski, A. *Acta Acad. Agric. Tech. Olstenensis Technol. Aliment.* **1987**, *21*, 153–163, from *Chem. Abstr.* **1988**, *109*, 51825z.

3932. Thompson, L. U.; Serraino, M. R. *J. Food Sci.* **1985**, *59*(4), 1200.

3933. Perry, H. J.; Harwood, J. L. *Phytochemistry* **1993**, *33*(2), 329–333.

3934. Perry, H. J.; Harwood, J. L. *Phytochemistry* **1993**, *34*(1), 69–73.

3935. Norton, G.; Harris, J. F. *Planta* **1975**, *123*(2), 163–174.

3936. Ching, T. M.; Crane, J. M.; Stamp, D. L. *Plant Physiol.* **1974**, *54*(5), 748–751.

3937. Belavady, B.; Banerjee, S. *Food Res.* **1953**, *18*, 223–226.

3938. Akpapunam, M. A.; Sefa-Dedeh, S. *Food Chem.* **1997**, *59*(1), 121–125.

3939. Makarevich-Gal'perin, L. M. *Ukr. Biokhim. Zh.* **1953**, *25*, 219–226, from *Chem. Abstr.* **1954**, *48*, 7122c.

3940. Ichihara, K.; Noda, M. *Phytochemistry* **1981**, *20*(5), 1023–1030.

3941. Ahmad, I.; Raie, M. Y.; Akhtar, M. W. *Pak. J. Sci. Ind. Res.* **1994**, *37*(5), 194–197.

3942. Banerjee, S.; Nandi, N. *Proc. Soc. Exp. Biol. Med.* **1949**, *71*(4), 588–589.

3943. Rao, B. S. N.; Prabhavathi, T. *J. Sci. Food Agric.* **1982**, *33*, 89–96.

3944. Prabhavathi, T.; Rao, B. S. N. *J. Sci. Food Agric.* **1979**, *30*(6), 597–602.

3945. Bhosale, S. B. *Biovigyanam* **1974**, *2*(2), 179–182.

3946. Gupta, K.; Thakral, K. K.; Arora, S. K.; Wagle, D. S. *J. Sci. Food Agric.* **1991**, *54*(1), 43–46.

3947. Bafor, M.; Jonsson, L.; Stobart, A. K.; Stymne, S. *Biochem. J.* **1990**, *272*(1), 31–38.

3948. Wiberg, E.; Bafor, M. *Phytochemistry* **1995**, *39*(6), 1325–1327.

3949. Suparmo, ; Markakis, P. *J. Food Sci.* **1987**, *52*(6), 1736–1737.

3950. Raboy, V.; Hudson, S. J.; Dickinson, D. B. *Plant Physiol.* **1985**, *79*(1), 323–325.

3951. Lee, C. K.; Karunanithy, R. *J. Sci. Food Agric.* **1990**, *51*(4), 437–441.

3952. Ahn, B.; Yang, C. B. *Han'guk Sikp'um Kwahakhoechi* **1985**, *17*(6), 516–521, from *Chem. Abstr.* **1986**, *104*, 106278c.

3953. Harwood, J. L. *Phytochemistry* **1975**, *14*(9), 1985–1990.

3954. Wagner, H.; Hölzl, J.; Schmid, A.; Hörhammer, L. *Naturwissenschaften* **1963**, *50*(6), 230.

3955. Wilson, R. F.; Rinne, R. W. *Plant Physiol.* **1974**, *54*(5), 744–747.

3956. Yoshida, H.; Kajimoto, G. *Agric. Biol. Chem.* **1978**, *42*(7), 1323–1330.

3957. Yoshida, H.; Kajimoto, G. *Agric. Biol. Chem.* **1977**, *41*(10), 1857–1863.

3958. Negishi, T. *Obihiro Chikusan Daigaku Gakujutsu Kenkyu Hokoku, Dai-1-Bu* **1974**, *9*(1), 211–217.

3959. Asamov, D. K. *Nauchn. Tr., Tashkent Gos. Univ.* No. 398, 65–68, 1972.

3960. Valikhanov, M. N. *Sb. Nauchn. Tr. Tashk. Gos. Univ.* **1982**, *687*, 3–13, from *Chem. Abstr.* **1983**, *99*, 119445h.

3961. Igamnazarov, R. P.; Asamov, D. K.; Valikhanov, M. N. *Uzb. Biol. Zh.* **1985**, (6), 16–18.

3962. Rakhimov, M. M.; Abdullaeva, M. M.; Valikhanov, M. N.; Babaev, M. U. *Fiziol. Rast.* **1986**, *33*(6), 1121–1129 (Engl. 858–865).

3963. Kajimoto, G.; Tsutsui, Y.; Yoshida, H. *Nippon Nogei Kagaku Kaishi* **1981**, *55*(1), 31–36.

3964. Gazizov, F. Yu.; Glushenkova, A. I. *Khim. Prir. Soedin.* **1995**, *31*(4), 564–567 (Engl. 467–470).

3965. Valikhanov, M. N. *Nauchn. Tr. Tashk. Gos. Univ.* **1968**, (334), 15–19.

3966. Maxwell, W. *Am. Chem. J.* **1891**, *13*(1), 16–24.

3967. Sobolev, A. M. *Fiziol. Rast.* **1962**, *9*(3), 334–341 (Engl. 263–269).

3968. Lotti, G.; Navari-Izzo, F.; Izzo, R.; Quartacci, M. F. *Agrochimica* **1987**, *31*(3), 275–286.

3969. Rodriguez-Rosales, M. P.; Kerbeb, L.; Ferrol, N.; Donaire, J. P. *Plant Physiol. Biochem.* **1998**, *36*(4), 285–291.

3970. Shcherbakov, V. G.; Malyshev, A. M. *Biol. Nauki* **1968**, (8), 80–85, from *Chem. Abstr.* **1968**, *69*, 93605p.

3971. Baisted, D. J. *Phytochemistry* **1981**, *20*(5), 985–988.

3972. Becker, G.; Acker, L. *Fette Seifen Anstrichm.* **1972**, *74*(6), 324–327.

3973. Katayama, N.; Suzuki, H. *Plant Cell Physiol.* **1980**, *21*(1), 115–129.

3974. Bog-Hansen, T. C. *Bios (Nancy)* **1973**, *4*(7/8), 349–353.

3975. Mayer, A. M. *Enzymologia* **1958**, *19*(1), 1–8.

3976. Pradet, A.; Narayanan, A.; Vermeersch, J. *Physiol. Vég.* **1968**, *14*(1), 107–114.

3977. Huang, L.-S.; Grunwald, C. *Phytochemistry* **1988**, *27*(7), 2049–2053.

3978. Huang, L.-S.; Grunwald, C. *Phytochemistry* **1990**, *29*(5), 1441–1445.

3979. Koiwai, A.; Matsuzaki, T. *Phytochemistry* **1990**, *29*(1), 73–76.

3980. Koiwai, A.; Suzuki, F.; Matsuzaki, T.; Kawashima, N. *Phytochemistry* **1982**, *21*(2), 305–308.

3981. Footitt, S.; Cohn, M. A. *Plant Physiol.* **1995**, *107*(4), 1365–1370.

3982. Giri, K. V.; Sreenivasan, A. *Biochem. Z.* **1938**, *296*(5/6), 428–433.

3983. Nakamura, A.; Kôno, T.; Funahashi, S. *Bull. Agric. Chem. Soc. Jpn.* **1958**, *22*(5), 320–324.

3984. Seo, S. W.; Ota, Y. *Nippon Sakumotsu Gakkai Kiji* **1981**, *50*(4), 514–518.

3985. Seo, S. W.; Ota, Y. *Nippon Sakumotsu Gakkai Kiji* **1981**, *50*(4), 519–524.

3986. Vorob'ev, N. V.; Aleshin, E. P.; Pozharskaya, E. A. *Sel'skokhoz. Biol.* **1974**, *9*(1), 39–45.

3987. Heim, S.; Bauleke, A.; Wylegalla, C.; Wagner, K. G. *Plant Sci.* **1987**, *49*(3), 159–165.

3988. Kumar, A.; Chauhan, B. M. *Cereal Chem.* **1993**, *70*(5), 504–506.

3989. Jackson, J. F.; Linskens, H. F. *Acta Bot. Neerl.* **1982**, *31*(5/6), 441–447.

3990. Dogras, C. C.; Dilley, D. R.; Herner, R. C. *Plant Physiol.* **1977**, *60*(6), 897–902.

3991. Sathe, S. K.; Deshpande, S. S.; Reddy, N. R.; Goll, D. E.; Salunkhe, D. K. *J. Food Sci.* **1983**, *48*(6), 1796–1800.

3992. Walker, K. A. *Planta* **1974**, *116*(2), 91–98.

3993. Tabekhia, M. M.; Luh, B. S. *J. Food Sci.* **1980**, *45*(2), 406–408.

3994. Crans, D. C.; Mikus, M.; Friehauf, R. B. *J. Plant Physiol.* **1995**, *145*(1/2), 101–107.

3995. Pretorius, J. C.; Small, J. C. C.; Fagerstedt, K. V. *Seed Sci. Res.* **1998**, *8*(1), 17–28.

3996. Bevilacqua, L. R. *Atti Accad. Naz. Lincei, Cl. Sci. Fis. Mat. Nat., Rend.* **1955**, *18*(2), 214–222.

3997. Lee, S. J.; Chung, D. H. *Han'guk Nonghwa Hakhoechi* **1961**, *2*, 35, from *Chem. Abstr.* **1963**, *58*, 8241f.

3998. Chikubu, S.; Horiuchi, H.; Furuya, M.; Niizeki, S. *Nippon Nogei Kagaku Kaishi* **1956**, *30*(1), 7–11.

3999. Neumann, G.; Preissler, M.; Azaizeh, H. A.; Romheld, V. *J. Plant Nutr. Soil Sci.* **1999**, *162*(3), 295–300, from *Chem. Abstr.* **1999**, *131*, 56497h.

4000. Ferguson, I. B.; Bollard, E. G. *Ann. Bot.* **1976**, *40*, 1047–1055.

4001. Di Nola, L.; Mayer, A. M. *Phytochemistry* **1986**, *25*(10), 2255–2259.

4002. Di Nola, L.; Mayer, A. M. *Phytochemistry* **1986**, *25*(12), 2725–2731.

4003. Daftyan, L. V. *Biol. Zh. Arm.* **1973**, *26*(12), 50–54.

4004. Murcia, M. A.; Rincón, F. *Food Chem.* **1992**, *44*(1), 29–35.

4005. Sung, H. Y.; Su, J. C. *Chung Kuo Nung Yeh Hua Hsueh Hui Chih* **1973**, (Special Issue), 1–13, from *Chem. Abstr.* **1975**, *82*, 54296n.

4006. Maffei-Faccioli, A. *Boll. Soc. Ital. Biol. Sper.* **1959**, *35*, 2166–2171, from *Chem. Abstr.* **1962**, *57*, 10230h.

4007. Munshi, S. K.; Sukhija, P. S. *J. Sci. Food Agric.* **1984**, *35*(6), 689–697.

4008. Ching, T. M. *Plant Physiol.* **1966**, *41*(8), 1313–1319.

4009. Prokof'ev, A. A.; Sveshnikova, I. N.; Sobolev, A. M. *Fiziol. Rast.* **1967**, *14*(5), 889–897 (Engl. 746–753).

4010. Ohnuma, N.; Yamada, M. *Sci. Pap. Coll. Gen. Educ., Univ. Tokyo* **1971**, *21*(2), 137–145.

4011. Greiner, R.; Konietzny, U.; Jany, K. D. *J. Food Biochem.* **1998**, *22*(2), 143–161.

4012. Mukherjee, K. D.; Kiewitt, I. *Phytochemistry* **1984**, *23*(2), 349–352.

4013. Knowles, L. O.; Knowles, N. R. *Ann. Bot.* **1990**, *65*(2), 217–223.

4014. Youssef, A. M.; Abu-Foul, N. S. *Alexandria Sci. Exch.* **1998**, *19*(2), 353–368, from *Chem. Abstr.* **1999**, *130*, 94719u.

4015. Mahgoub, S. E. O.; Elhag, S. E. *Food Chem.* **1998**, *61*(1/2), 77–80.

4016. Dumazert, C.; Vovan, L. *Compt. Rend. Soc. Biol.* **1963**, *157*(2), 329–331.

4017. El-Shimi, N. M.; Damir, A. A.; Ragab, M. *Food Chem.* **1984**, *14*(1), 11–19.

4018. Saad, M. K.; Nahed, M. E. S.; Shakib, L. *Alexandria Sci. Exch.* **1992**, *13*(2), 279–284, from *Chem. Abstr.* **1992**, *117*, 230230p.

4019. El-Sebaiy, L. A.; El-Mahdy, A. R. *Food Chem.* **1983**, *10*(4), 309–319.

4020. Nahapetian, A.; Bassiri, A. *J. Agric. Food Chem.* **1975**, *23*(6), 1179–1182.

4021. Abernethy, R. H.; Paulsen, G. M.; Ellis, R., Jr. *J. Agric. Food Chem.* **1973**, *21*(2), 282–284.

4022. Batten, G. D.; Wardlaw, I. F. *Aust. J. Plant Physiol.* **1987**, *14*(3), 267–275.

4023. Eastwood, D.; Laidman, D. L. *Phytochemistry* **1971**, *10*(6), 1275–1284.

4024. Mihailovic, M. L.; Antic, M.; Hadzijev, D. *Bull. Soc. Chim. Beograd* **1962**, *27*(2/3), 135–155.

4025. Vakharia, D. N.; Brearley, C. A.; Wilkinson, M. C.; Galliard, T.; Laidman, D. L. *Planta* **1987**, *172*(4), 502–507.

4026. Varty, K.; Laidman, D. L. *J. Exp. Bot.* **1976**, *27*(99), 748–758.

4027. Stokes, D. N.; Galliard, T.; Harwood, J. L. *Phytochemistry* **1986**, *25*(4), 811–815.

4028. Bahl, J.; Bougain, G.; Monéger, R. *Physiol. Veg.* **1983**, *21*(3), 475–483.

4029. Likholat, T. V.; Konstantinova, O. B. *Fiziol. Rostovykh Protsessov* **1980**, 69–74, from *Chem. Abstr.* **1982**, *96*, 214455h.

4030. Obendorf, R. L.; Marcus, A. *Plant Physiol.* **1974**, *53*(5), 779–781.

4031. Ching, T. M.; Kronstad, W. E. *Crop Sci.* **1972**, *12*(6), 785–789.

4032. Kumar, R.; Singh, R. *Ann. Bot.* **1984**, *53*(5), 749–752.

4033. Dani, H. M.; Saini, H. S.; Allag, I. S.; Sareen, K. *Curr. Sci.* **1970**, *39*(3), 55–57.

4034. Kikuchi, M.; Yamane, Y. *Vitamins* **1972**, *46*(6), 291–298, from *Chem. Abstr.* **1973**, *78*, 94920k.

4035. Batten, G. D.; Wardlaw, J. F. *Aust. J. Plant Physiol.* **1987**, *14*, 253–265.

4036. Niziotek, S.; Bartoszewicz, K.; Kaczkowski, J. *Acta Physiol. Plant.* **1994**, *16*(3), 171–176.

4037. Sartirana, M. L.; Bianchetti, R. *Physiol. Plant.* **1967**, *20*(4), 1066–1075.

4038. Bernardini, L.; Chiarulli, G. *Staz. Sper. Agrar. Ital.* **1909**, *42*, 97–115, from *Chem. Abstr.* **1910**, *4*, 1870a and *Chem. Zentralbl.* **1909**, *1*, 1249–1250.

4039. Eskin, N. A. M.; Wiebe, S. *J. Food Sci.* **1983**, *48*(1), 270–271.

4040. Sharma, A.; Sehgal, S. *Food Chem.* **1992**, *43*(5), 383–385.

4041. Borowska, J.; Zadernowski, R.; Wierzbicka, B.; Roszko, J. *Acta Acad. Agric. Tech. Olstenensis Technol. Aliment.* **1994**, *472*(26), 71–80.

4042. Abdul-Wahab, M. F.; El-Kinawi, S. A. *J. Chem. U.A.R.* **1963**, *6*(1), 43–54.

4043. Kikuchi, M.; Hayashi, K. *Shokubutsugaku Zasshi* **1969**, *82*(967), 28–31, from *Chem. Abstr.* **1969**, *71*, 846w.

4044. Kojima, M.; Sasaki, S.; Ohnishi, M.; Mano, Y.; Ito, S. *Phytochemistry* **1990**, *29*(7), 2091–2096.

4045. Kataria, A.; Chauhan, B. M.; Gandhi, S. *Food Chem.* **1988**, *30*(2), 149–156.

4046. Srinivasan, ; Kayastha, A. M.; Malhotra, O. P. *Plant Physiol.* **1992**, *100*(4), 2109–2112.

4047. Kataria, A.; Chauhan, B. M.; Punia, D. *Food Chem.* **1989**, *32*(1), 9–17.

4048. Sattar, A.; Durrani, S. K.; Nahmood, F.; Ahmad, A.; Khan, J. *Food Chem.* **1989**, *34*(2), 111–120.

4049. Kaur, D.; Kapoor, A. C. *Food Chem.* **1990**, *37*(3), 171–179.

4050. Majumder, A. L.; Biswas, B. B. *Indian J. Exp. Biol.* **1973**, *11*(2), 120–123.

4051. Vo-Van, L.; Grégoire, J. *Compt. Rend. Soc. Biol.* **1958**, *152*(7), 1260–1262.

4052. Vovan, L.; Dumazert, C.; Grégoire, J. *Compt. Rend. Soc. Biol.* **1962**, *156*(3), 708–710.

4053. Akpapunam, M. A.; Achinewho, S. C. *Plant Foods Hum. Nutr.* **1985**, *35*(4), 353–358.

4054. Esashi, Y.; Nagayama, T.; Satoh, M.; Saijoh, K. *Plant Cell Environ.* **1983**, *6*(1), 55–63.

4055. Horvatic, M.; Balint, L. *J. Agron. Crop Sci.* **1996**, *176*(2), 73–77, from *Chem. Abstr.* **1996**, *124*, 338070v.

4056. Weber, E. J. *J. Am. Oil Chem. Soc.* **1970**, *47*(9), 340–343.

4057. Sharma, M. L.; Mishra, D. P. *Int. J. Trop. Agric.* **1989**, *6*(1/2), 117–124.

4058. Tan, S. L.; Morrison, W. R. *J. Am. Oil Chem. Soc.* **1979**, *56*(8), 759–764.

4059. Guba, F. *Agrokémia és Talajtan* **1951**, *1*(1), 11–14.

4060. van de Venter, H. A.; Hoffman, R.; de Meillon, S. *S. Afr. J. Plant Soil* **1988**, *5*(4), 193–196.

4061. Cho, Y. D.; Cho, B. G.; Lee, S. H.; Kang, Y. H. *Hanguk Saenghwa Hakhoe Chi* **1982**, *15*(2), 125–133, from *Chem. Abstr.* **1982**, *97*, 159670t.

4062. Ingle, J.; Beevers, L.; Hageman, R. H. *Plant Physiol.* **1964**, *39*(5), 735–740.

4063. Kikuchi, M. *Vitamins* **1973**, *47*(7), 351–356, from *Chem. Abstr.* **1973**, *79*, 89607e.

4064. Bevilacqua, L. R.; Castelli, L. *Boll. Soc. Ital. Biol. Sper.* **1955**, *31*(7/8), 890–891, from *Chem. Zentralbl.* **1957**, *128*, 3556.

4065. de Silva, N. S.; Fowler, M. W. *Phytochemisry* **1976**, *15*(11), 1735–1740.

4066. Clair-Maczulajtys, D.; Bory, G. *Physiol. Vég.* **1985**, *23*(3), 275–282.

4067. Horiguchi, G.; Kodama, H.; Nishimura, M.; Iba, K. *Planta* **1996**, *199*(3), 439–442.

4068. Shugaev, A. G.; Bukhov, N. G. *J. Plant Physiol.* **1997**, *150*(1/2), 53–56.

4069. Einig, W.; Lauxmann, U.; Hauch, B.; Hampp, R.; Landolt, W.; Maurer, S.; Matyssek, R. *New Phytol.* **1997**, *137*(4), 673–680.

4070. Rodionov, V. S.; Il'inova, M. K. *Biol. Nauki (Moscow)* **1988**, (2), 74–79, from *Chem. Abstr.* **1988**, *109*, 51857m.

4071. Hobbs, D. H.; Hume, J. H.; Rolph, C. E.; Cooke, D. T. *Phytochemistry* **1996**, *42*(2), 335–339.

4072. Markowski, A.; Filek, M. *Acta Physiol. Plant.* **1986**, *8*(2), 123–153.

4073. Hoffmann, P.; Meinl, G. *Wiss. Z. Humboldt-Univ. Berlin, Math.-Naturwiss. Reihe* **1978**, *27*(5), 517–520.

4074. Rao, G. G.; Rao, G. R. *Indian J. Exp. Biol.* **1978**, *16*(2), 270–271.

4075. Nielsen, T. H.; Skjaerbaek, H. C.; Karlsen, P. *Physiol. Plant.* **1991**, *82*(2), 311–319.

4076. Kubota, K.; Ashihara, H. *Phytochemistry* **1993**, *34*(6), 1509–1513.

4077. Lecharny, A.; Trémolières, A.; Wagner, E. *Planta* **1990**, *182*(2), 211–215.

4078. Pranis, A. K. *Fiziol. Biokhim. Kul't. Rast.* **1988**, *20*(5), 480–487.

4079. Parra, M.; Ortuño, A. *An. Edafol. Agrobiol.* **1981**, *40*(3/4), 609–614.

4080. Parra, M.; Parra, T. *An. Edafol. Agrobiol.* **1982**, *41*(3/4), 705–712.

4081. Fuh, B.-S.; Ichii, T.; Nakanishi, T.; Kawai, Y. *Engei Gakkai Zasshi* **1988**, *57*(1), 109–115.

4082. Yelenovsky, G. *Proc. Florida State Hortic. Soc.* **1982**, *95*, 7–10.

4083. Ferguson, C. H. R.; Simon, E. W. *J. Exp. Bot.* **1973**, *24*(79), 307–316.

4084. Bafor, M. E.; Osagie, A. U. *Oléagineux* **1988**, *43*(6), 261–266.

4085. Zombré, G.; Naudy-De Serres, M.; Viala, G.; Calmes, J.; Cavalié, G. *Compt. Rend. III* **1990**, *310*(6), 253–258.

4086. Raboy, V.; Dickinson, D. B. *Plant Physiol.* **1987**, *85*(3), 841–844.

4087. Abdurakhmanova, Z. N.; Abdullaev, A.; Dzhumaev, B. B.; Gorenkova, L. G. *Fiziol. Rast.* **1995**, *42*(2), 201–205 (Engl. 176–179).

4088. Gazizov, F. Yu.; Isamukhamedov, A. Sh.; Akramov, S. T. *Khim Prir. Soedin.* **1980**, (4), 485–488 (Engl. 345–348).

4089. Avtomonova, L. E. *Dokl. Akad. Tadzh. SSR* **1986**, *29*(7), 439–442.

4090. Meshcheryakova, E. A. *Dokl. Akad. Nauk Tadzh. SSR* **1973**, *16*(1), 59–63.

4091. Pétel, G.; Sueldo, R.; Coudret, A.; Gendraud, M. *Biol. Plant.* **1992**, *34*(5/6), 373–380.

4092. Chetal, S.; Wagle, D. S.; Nainawatee, H. S. *Z. Pflanzenphysiol.* **1982**, *107*(2), 187–191.

4093. Thomson, L. W.; Zalik, S. *Plant Physiol.* **1981**, *67*(4), 646–654.

4094. Guillot-Salomon, T.; Farineau, N.; Cantrel, C.; Oursel, A.; Tuquet, C. *Physiol. Plant.* **1987**, *69*(1), 113–122.

4095. Sawert, A.; Backer, A. I.; Wagner, K. G. *Plant Cell Physiol.* **1988**, *29*(1), 61–65.

4096. Song, M.; Tattrie, N. *Can. J. Bot.* **1973**, *51*(10), 1893–1897.

4097. Rubchevskaya, L. P.; Ignatova, E. V.; Repyakh, S. M. *Khim. Prir. Soedin.* **1998**, *34*(4), 549–550 (Engl. 504–505).

4098. Dobrovo, P. N.; Il'yasova, V. B.; Shirinskaya, M. G.; Yagodin, B. A. *Fiziol. Rast.* **1970**, *26*(3), 599–605 (Engl. 480–486).

4099. Nguyen, X. V.; Mazliak, P. *Plant Physiol. Biochem.* **1990**, *28*(2), 283–291.

4100. Yoshioka, H.; Harata, J.; Okamoto, T. *Hirosaki Daigaku Nogakubu Gakujutsu* **1973**, *21*, 71–81.

4101. Wang, S. Y.; Faust, M. *J. Am. Soc. Hortic. Sci.* **1990**, *115*(5), 803–808.

4102. Yamashita, T. *Ann. Bot.* **1984**, *54*(2), 283–288.

4103. Yamashita, T. *Tree Physiol.* **1990**, *6*(2), 191–200.

4104. Chouteau, J. *Ann. Inst. Exp. Tabac Bergerac* **1960**, *3*, 379–512.

4105. Koiwai, A.; Matsuzaki, T.; Suzuki, F.; Kawashima, N. *Plant Cell Physiol.* **1981**, *22*(6), 1059–1065.

4106. Meyer, R.; Wagner, K. G. *Physiol. Plant.* **1985**, *65*(4), 439–445.

4107. Murakami, T. *Nippon Sakumotsu Gakkai Kiji* **1968**, *37*(4), 539–550.

4108. Fong, F.; Heath, R. L. *Phytochemistry* **1977**, *16*(2), 215–217.

4109. Liljenberg, C.; von Arnold, S. *J. Plant Physiol.* **1987**, *130*(2/3), 255–265.

4110. Chen, K.-S.; Zhang, C.-L.; Liang, H.-G. *Zhiwu Shengli Xuebao* **1991**, *17*(2), 139–144.

4111. Gniazdowska, A.; Mikulska, M.; Rychter, A. M. *Biol. Plant.* **1998**, *41*(2), 217–226.

4112. Loewe, A.; Einig, W.; Hampp, R. *Plant Physiol.* **1996**, *112*(2), 641–649.

4113. Hampp, R.; Egger, B.; Effenberger, S.; Einig, W. *Physiol. Plant.* **1994**, *90*(2), 299–306.

4114. Olsson, M. *Plant Physiol. Biochem.* **1995**, *33*(5), 547–553.

4115. Sagisaka, S. *Plant Physiol.* **1974**, *54*(4), 544–549.

4116. Kennedy, R. A.; Laetsch, W. M. *Planta* **1973**, *115*(2), 113–124.

4117. Parra, M.; Parra, T. *An. Edafol. Agrobiol.* **1983**, *42*(11/12), 2185–2191.

4118. Goto, M.; Minamide, T.; Iwata, T. *Engei Gakkai Zasshi* **1988**, *56*(4), 479–485.

4119. Parra, M.; Ortuño, A.; Armero, T. *An. Edafol. Agrobiol.* **1977**, *36*(7/8), 719–724.

4120. Parra, M. *An. Edafol. Agrobiol.* **1981**, *40*(3/4), 615–620.

4121. Yoshida, S. *Teion Kagaku Seibutsu Hen* **1969**, *27*, 119–123.

4122. Hatch, M. D.; Slack, C. R.; Johnson, H. S. *Biochem. J.* **1967**, *102*(2), 417–422.

4123. Peng, C. C.; Tzen, J. T. C. *Plant Cell Physiol.* **1998**, *39*(1), 35–42.

4124. Samotus, B.; Schwimmer, S. *Plant Physiol.* **1962**, *37*(4), 519–522.

4125. Quick, W. A.; Snyder, J. C.; Desborough, S. L. *Potato Res.* **1979**, *22*(1), 1–17.

4126. Berkeley, H. D.; Galliard, T. *J. Sci. Food Agric.* **1974**, *25*(7), 861–867.

4127. Anich, M.; Fanta, N.; Mancilla, M.; Kettlur, A. M.; Valenzuela, M. A.; Traverso-Cori, A. *Phytochemistry* **1990**, *29*(5), 1411–1415.

4128. Khanna, R.; Sinha, S. K. *Biochem. Biophys. Res. Commun.* **1973**, *52*(1), 121–124.

4129. Robinson, J. M.; Rowland, R. A. *Photosynth. Res.* **1996**, *50*(2), 103–115.

4130. Suzuki, S.; Nakamoto, H.; Ku, M. S. B.; Edwards, G. E. *Plant Physiol.* **1987**, *84*(4), 1244–1248.

4131. Galey, S.; Francke, B.; Bahl, J. *Planta* **1980**, *149*(5), 433–439.

4132. Lechowicz, W.; Maternicka, K.; Faltynowicz, M.; Poskuta, J. *Photosynth. Res.* **1986**, *8*(2), 133–147.

4133. Bolling, H.; el Bayâ, A. W. *Chem. Phys. Lipids* **1972**, *8*(2), 102–111.

4134. Sawert, A.; Aust, H.-J.; Wagner, K. G. *Physiol. Mol. Plant Pathol.* **1988**, *33*(3), 507–514.

4135. Perepadya, Yu.G. *Tr. Prikl. Bot. Genet. Selek.* **1970**, *43*(1), 41–50.

4136. Kondo, H.; Sumi, M. *Kagaku (Science)* **1956**, *26*, 254–255.

4137. Savithramma, N.; Swamy, P. M. *Proc. Indian Acad. Sci., Plant Sci.* **1989**, *99*(2), 139–145.

4138. Savithramma, N.; Swamy, P. M. *J. Indian Bot. Soc.* **1995**, *74*(1/4), 165–169.

4139. Zherebin, Yu.L.; Kolesnik, A. A.; Bogatskii, A. V. *Fiziol. Biokhim. Kul't. Rast.* **1984**, *16*(3), 243–248.

4140. Leech, R. M.; Rumsby, M. G.; Thomson, W. W. *Plant Physiol.* **1973**, *52*(3), 240–245.

4141. Georgieva, M.; Kudrev, T. *Fiziol. Rast. (Sofia)* **1982**, *8*(3), 12–17.

4142. Kudrev, T.; Georgieva, M.; Penev, S. *Fiziol. Rast. (Sofia)* **1978**, *4*(2), 52–57.

4143. Helal, H. M.; Ragab, A. S.; Monem, M. A.; Schnug, E. *Commun. Soil Sci. Plant Anal.* **1996**, *27*(5/8), 1169–1175.

4144. Bobodzhanova, S. A. *Izv. Akad. Nauk Tadzh. SSR, Otd. Biol. Nauk* **1978**, (3), 79–84.

4145. Anderson, G. *J. Sci. Food Agric.* **1956**, *7*(6), 437–444.

4146. Smith, D. H.; Clark, F. E. *Soil Sci. Soc. Am. Proc.* **1952**, *16*, 170–172.

4147. Reddy, N. R.; Sathe, S. K.; Salunkhe, D. K. *Adv. Food Res.* **1982**, *28*, 1–92.

4148. Reddy, N. R.; Balakrishnan, C. V.; Salunkhe, D. K. *J. Food Sci.* **1978**, *43*(2), 540–543.

4149. Kennedy, B. M.; Schelstraete, M. *Cereal Chem.* **1975**, *52*(2), 173–182.

4150. Ranhotra, G. S.; Hepburn, F. N.; Bradley, W. B. *Cereal Chem.* **1971**, *48*, 699–706.

4151. Ranhotra, G. S.; Loewe, R. J. *J. Food Sci.* **1975**, *40*, 940–942.

4152. Pringle, W. J. S.; Moran, T. *J. Soc. Chem. Ind.* **1942**, *61*, 108–110.

4153. Chen, L. H.; Pan, S. H. *Nutr. Rep. Int.* **1977**, *16*(1), 125–131.

4154. Toma, R. B.; Tabekhia, M. M.; Williams, J. D. *Nutr. Rep. Int.* **1979**, *20*(1), 25–31.

4155. Frølich, W.; Drakenberg, T.; Asp, N.-G. *J. Cereal Sci.* **1986**, *4*(4), 325–334.

4156. Plimmer, R. H. A. *Biochem. J.* **1913**, *7*, 72–80.

4157. Aremu, C. Y. *Food Chem.* **1988**, *27*(1), 77–82.

4158. Reddy, N. R.; Sathe, S. K.; Pierson, M. D. *J. Food Sci.* **1988**, *53*(1), 107–110, 147.

4159. Peng, A. C. *J. Food Sci.* **1982**, *47*(3), 1036–1037.

4160. Abdullah, A.; Baldwin, R. A.; Minor, H. *J. Food Prot.* **1984**, *47*(6), 441–444.

4161. Maga, J. A. *J. Agric. Food Chem.* **1982**, *30*(1), 1–9.

4162. Morris, E. R.; Ellis, R. *J. Nutr.* **1980**, *110*(10), 2000–2010.

4163. Samotus, B.; Schwimmer, S. *Biochim. Biophys. Acta* **1962**, *57*(2), 389–391.

4164. Samotus, B.; Schwimmer, S. *Anal. Biochem.* **1962**, *3*(3), 215–220.

4165. Anderson, G. *J. Sci. Food Agric.* **1963**, *14*(5), 352–359.

4166. Pedersen, J. G. A. *Beretn.* Forsøgslab No. 193, 1–223, 1940, cited in ref. 263; *Chem. Zentralbl.* **1941**, *II*, 1037 and *Chem. Abstr.* **1944**, *38*, 1776.

4167. Serraino, M. R.; Thompson, L. U. *J. Agric. Food Chem.* **1984**, *32*(1), 38–40.

4168. Gillberg, L.; Törnell, B. *J. Food Sci.* **1976**, *41*(5), 1063–1069.

4169. McKinney, L. L.; Sollars, W. F.; Setzkorn, E. A. *J. Biol. Chem.* **1949**, *178*(1), 117–132.

4170. Kawatra, B. L.; Kaur, S. *Nahrung* **1989**, *33*(4), 311–314.

4171. Tangkongchitr, U.; Seib, P. A.; Hoseney, R. C. *Cereal Chem.* **1981**, *58*(3), 229–234.

4172. Holt, R. *J. Sci. Food Agric.* **1955**, *6*(3), 136–142.

4173. Summers, J. D.; Slinger, S. L.; Cisneros, G. *Cereal Chem.* **1967**, *44*(3), 318–323.

4174. Henrici, M. *Verhandl. Natf. Ges. Basel* **1927**, *38*, 316–326.

4175. McCance, R. A.; Widdowson, E. M.; Moran, T.; Pringle, W. J. S.; Macrae, T. F. *Biochem. J.* **1945**, *39*, 213–222.

4176. Glass, R. L.; Geddes, W. F. *Cereal Chem.* **1959**, *36*(2), 186–190.

4177. Deshpande, S. S.; Damodaran, S. *J. Food Sci.* **1989**, *54*(3), 695–699.

4178. Deshpande, S. S.; Cheryan, M. *J. Food Sci.* **1984**, *49*(2), 516–519, 524.

4179. Khetarpaul, N.; Chauhan, B. M. *Int. J. Food Sci. Technol.* **1989**, *24*(3), 327–331.

4180. Jones, J. D. *Proceedings of the International Symposium on Rapeseed and Mustard, Mysore, India*, 1976, p 167, cited in *J. Agric. Food Chem.* **1984**, *32* (3), 493–498.

4181. Weinstein, L. H.; Bozarth, R. F.; Porter, C. A.; Mandl, R. H.; Tweedy, B. G. *Contrib. Boyce Thompson Inst.* **1964**, *22*(7), 389–397. *Errata*, p vi.

4182. Arbenz, E. *Mitt. Geb. Lebensmitt. Hyg.* **1922**, *13*(1/2), 45–52.

4183. Rucci, A. O.; Bertoni, M. H. *An Asoc. Quim. Argent.* **1973**, *61*(4), 165–182.

4184. Bodalski, T. *Wiadomosci Farm.* **1931**, *58*, 683–684, 697–699, from *Chem. Abstr.* **1932**, *26*, 2767.

4185. Wildflush, E. T. *Tr. Beloruss. S-kh. Akad.* **1939**, *8*(30), 49 see *Chem. Zentralbl.* **1941**, *II*, 2828, *Herbage Abstr.* **1939**, *9* (4), 328, or *Chem. Abstr.* **1941**, *35*, 5160; also **1944**, *38*, 1848.

4186. Neuberg, C.; Grauer, A. *Biochim. Biophys. Acta* **1953**, *12*, 265–272.

4187. Karvanek, M.; Pokorny, J.; Kozlowska, H.; Rutkowski, A. *Nahrung* **1964**, *8*(8), 675–680.

4188. McKenzie-Parnell, J. M.; Guthrie, B. E. *Biol. Trace Elem. Res.* **1986**, *10*(2), 107–121.

4189. Lott, J. N. A.; Greenwood, J. S.; Vollmer, C. M. *Can. J. Chem.* **1978**, *56*(19), 2408–2414.

4190. Saio, K.; Watanabe, T. *Agric. Biol. Chem.* **1966**, *30*(11), 1133–1138.

4191. Shah, F. H.; Shah, W. H.; Bano, S. *Plant Foods Hum. Nutr.* **1984**, *34*(2), 75–79.

4192. Alli, I.; Baker, B. E. *J. Sci. Food Agric.* **1980**, *31*, 1316–1322.

4193. Rogozinski, F. *Bull. Acad. Sci. Cracovie B* **1915**, (5), 87–98.

4194. El-Mahdy, A. R.; El-Sebaiy, L. A. *Food Chem.* **1982**, *9*(3), 149–158.

4195. Rendleman, J. A. *Cereal Chem.* **1982**, *59*(4), 302–309.

4196. Tryfiates, G. P.; Smith, R. R.; Bishop, R. *J. Nutr. Growth Cancer* **1983**, *1*(2), 85–89.

4197. Barker, J.; Solomos, T. *Nature* **1962**, *196*, 189.

4198. Barker, J. *New Phytol.* **1968**, *67*(1), 213–217.

4199. Schoch, T. J. *J. Am. Chem. Soc.* **1942**, *64*(12), 2954–2956.

4200. Schoch, T. J. *J. Am. Chem. Soc.* **1942**, *64*(12), 2957–2961.

4201. Singal, H. R.; Sheoran, I. S.; Singh, R. *Physiol. Plant.* **1986**, *66*(3), 457–462.

4202. Sassenrath, G. F.; Ort, D. R.; Portis, A. R., Jr. *Arch. Biochem. Biophys.* **1990**, *282*(2), 302–308.

4203. Krall, A. R. *Plant Physiol.* **1955**, *30*(3), 269–271.

4204. Benedict, C. R.; Beevers, H. *Plant Physiol.* **1961**, *36*(5), 540–544.

4205. Tolbert, N. E.; Wiebe, H. *Plant Physiol.* **1955**, *30*(6), 499–504.

4206. Ball, K. L.; Green, J. H.; ap Rees, T. *Eur. J. Biochem.* **1991**, *197*(1), 265–269.

4207. Burma, D. P.; Mortimer, D. C. *Can. J. Biochem. Physiol.* **1957**, *35*(10), 835–843.

4208. Hoad, G. V.; Peel, A. J. *J. Exp. Bot.* **1965**, *16*(49), 742–758.

4209. Biddulph, G.; Cory, R. *Plant Physiol.* **1957**, *32*(6), 608–619.

4210. Edwards, J.; ap Rees, T.; Wilson, P. M.; Morrell, S. *Planta* **1984**, *162*(2), 188–191.

4211. Dixon, W. L.; ap Rees, T. *Phytochemistry* **1980**, *19*(7), 1297–1301.

4212. Stitt, M.; Wirtz, W.; Heldt, H. W. *Biochim. Biophys. Acta* **1980**, *593*(1), 85–102.

4213. Hatch, M. D.; Slack, C. R. *Biochem. J.* **1966**, *101*(1), 103–111.

4214. Hassid, W. Z. *Plant Physiol.* **1938**, *13*(3), 641–647.

4215. Loughman, B. C.; Martin, R. P. *J. Exp. Bot.* **1957**, *8*(23), 272–279.

4216. Liu, T.-T. Y.; Shannon, J. C. *Plant Physiol.* **1981**, *67*(3), 518–524.

4217. Weiss, D.; Shomer-Ilan, A.; Vainstein, A.; Halevy, A. H. *Physiol. Plant.* **1990**, *78*(3), 345–350.

4218. Johnson, H. S.; Hatch, M. D. *Biochem. J.* **1969**, *114*(1), 127–134.

4219. Maitra, P. K.; Estabrook, R. W. *Anal. Biochem.* **1964**, *7*, 472–484.

4220. Patthy, M.; Balla, T.; Arányi, P. *J. Chromatogr.* **1990**, *523*, 201–216.

4221. Barnett, J. E. G.; Brice, R. E.; Corina, D. L. *Biochem. J.* **1970**, *119*(2), 183–186.

4222. Kiely, D. E.; Abruscato, G. J.; Baburso, V. *Carbohydr. Res.* **1974**, *34*(2), 307–313.

4223. Posternak, T. *Helv. Chim. Acta* **1958**, *41*(6), 1891–1898.

4224. Chen, I. W.; Charalampous, F. C. *Arch. Biochem. Biophys.* **1966**, *117*(1), 154–157.

4225. Memon, A. R.; Rincon, M.; Boss, W. F. *Plant Physiol.* **1989**, *91*(2), 477–480.

4226. Drøbak, B. K.; Watkins, P. A. C.; Chattaway, J. A.; Roberts, K.; Dawson, A. P. *Plant Physiol.* **1991**, *95*(2), 412–419.

4227. Joseph, S. K.; Esch, T.; Bonner, W. D., Jr. *Biochem. J.* **1989**, *264*(3), 851–856.

4228. Tanaka, T.; Yoshida, T.; Kasai, Z. *Agric. Biol. Chem.* **1976**, *40*(7), 1319–1325.

4229. Connett, R. J. A.; Hanke, D. E. *Planta* **1986**, *169*(2), 216–221.

4230. Chattaway, J. A.; Drøbak, B. K.; Watkins, P. A. C.; Dawson, A. P.; Letcher, A. J.; Stephens, L. R.; Irvine, R. F. *Planta* **1992**, *187*(4), 542–545.

4231. Mandal, N. C.; Burman, S.; Biswas, B. B. *Phytochemistry* **1972**, *11*(2), 495–502.

4232. Majumder, A. N. L.; Mandal, N. C.; Biswas, B. B. *Phytochemistry* **1972**, *11*(2), 503–508.

4233. Meek, J. L.; Davidson, F.; Hobbs, F. W., Jr. *J. Am. Chem. Soc.* **1988**, *110*(7), 2317–2318.

4234. Dureja-Munjal, I.; Acharya, M. K.; Guha-Mukherjee, S. *Phytochemistry* **1992**, *31*(4), 1161–1163.

4235. Kerr, S. E.; Kfoury, G. A. *Arch. Biochem. Biophys.* **1962**, *96*(2), 347–353.

4236. Bartlett, G. R. *J. Biol. Chem.* **1959**, *234*(3), 466–468.

4237. Shayman, J. A.; BeMent, D. M. *Biochim. Biophys. Res. Commun.* **1988**, *151*(1), 114–122.

4238. Irth, H.; Lamoree, M.; de Jong, G. J.; Brinkman, U. A. T.; Frei, R. W.; Kornfeldt, R. A.; Persson, L. *J. Chromatogr.* **1990**, *499*, 617–625.

4239. Lehrfeld, J.; Morris, E. R. *J. Agric. Food Chem.* **1992**, *40*(11), 2208–2210.

4240. Woodcock, E. A.; Tanner, J. K. *J. Chromatogr.* **1992**, *581*(1), 134–138.

4241. De, B. P.; Biswas, B. B. *J. Biol. Chem.* **1979**, *254*(18), 8717–8719.

4242. Lewin, L. M. *Anal. Biochem.* **1962**, *3*(2), 124–130.

4243. Newman, D. W. *Ohio J. Sci.* **1964**, *64*(4), 258–263.

4244. Privett, O. S.; Dougherty, K. A.; Erdahl, W. L.; Stolyhwo, A. *J. Am. Oil Chem. Soc.* **1973**, *50*(12), 516–520.

4245. Donaldson, R. P. *Plant Physiol.* **1976**, *57*(4), 510–515.

4246. Mazliak, P. *Phytochemistry* **1967**, *6*(5), 687–702.

4247. Beiss, U. *Fette Seifen Anstrichm.* **1969**, *71*(5), 363–365.

4248. Wilson, R. F.; Rinne, R. W. *Plant Physiol.* **1976**, *57*(3), 375–381.

4249. Singh, H.; Privett, O. S. *Biochim. Biophys. Acta* **1970**, *202*(1), 200–202.

4250. Vijayalakshmi, B.; Rao, V.; Achaya, K. T. *Fette Seifen Anstrichm.* **1969**, *71*(9), 757–761.

4251. Trémolières, A.; Mazliak, P. *Compt. Rend.* **1968**, *267D*(12), 1039–1042.

4252. Abdelkader, A. B.; Mazliak, P. *Compt. Rend.* **1967**, *264D*(9), 1162–1165.

4253. Abdelkader, A. B. *Compt. Rend.* **1969**, *268D*(20), 2406–2409.

4254. Hölzl, J.; Wagner, H. *Z. Naturforsch.* **1971**, *26B*(5), 425–434.

4255. Wilson, R. F.; Rinne, R. W. *Plant Physiol.* **1976**, *57*(4), 556–559.

4256. Willemot, C. *Plant Physiol.* **1975**, *55*(2), 356–359.

4257. Radunz, A. *Z. Naturforsch.* **1980**, *35C*(11/12), 1024–1031.

4258. Slack, C. R.; Roughan, P. G.; Balasingham, N. *Biochem. J.* **1978**, *170*(2), 421–433.

4259. Carter, H. E.; Strobach, D. R.; Hawthorne, J. N. *Biochemistry* **1969**, *8*(1), 383–388.

4260. Wagner, H.; Zofcsik, W.; Heng, I. *Z. Naturforsch.* **1969**, *24B*(7), 922–927.

4261. de Kok, L. J.; Kuiper, P. J. C. *Physiol. Plant.* **1977**, *39*(2), 123–128.

4262. Haydar, M.; Hadziyev, D. *J. Food Sci.* **1973**, *38*(5), 772–778.

4263. Wilson, A. C.; Kates, M.; de la Roche, A. I. *Lipids* **1978**, *13*(7), 504–510.

4264. Pruthi, T. D.; Bhatia, I. S. *J. Sci. Food Agric.* **1970**, *21*(8), 419–422.

4265. Galliard, T.; Matthew, J. A.; Wright, A. J.; Fishwick, M. J. *J. Sci. Food Agric.* **1977**, *28*(9), 863–868.

4266. Yapa, P. A. J.; Kawasaki, T.; Matsumoto, H. *Plant Cell Physiol.* **1986**, *27*(2), 223–232.

4267. Camara, B.; Bardat, F.; Dogbo, O.; Brangeon, J.; Monéger, R. *Plant Physiol.* **1983**, *73*(1), 94–99.

4268. Browse, J.; Warwick, N.; Somerville, C. R.; Slack, C. R. *Biochem. J.* **1986**, *235*(1), 25–31.

4269. Pun, W. H.; Hadziyev, D. *Can. Inst. Food Sci. Technol. J.* **1978**, *11*(3), 134–141.

4270. Sosulki, F.; Zadernowski, R.; Babuchowski, K. *J. Am. Oil Chem. Soc.* **1981**, *58*(4), 561–564.

4271. Bhandal, I. S.; Malik, C. P. *J. Exp. Bot.* **1980**, *31*(123), 931–935.

4272. Thi, A. T. P.; Borrel-Flood, C.; da Silva, J. V.; Justin, A. M.; Mazliak, P. *Phytochemistry* **1985**, *24*(4), 723–727.

4273. Slack, C. R.; Campbell, L. C.; Browse, J. A.; Roughan, P. G. *Biochim. Biophys. Acta* **1983**, *754*(1), 10–20.

4274. El-Aal, M. H. A.; Khalil, M. K. M.; Rahma, E. H. *Food Chem.* **1986**, *19*(4), 287–298.

4275. Batra, A.; Mehta, B. K.; Bokadia, M. M. *Fette Seifen Anstrichm.* **1986**, *88*(5), 189–191.

4276. Franzke, C.; Göbel, R.; Kroll, J.; Klein, S. *Nahrung* **1986**, *30*(8), 851–853.

4277. Revinthi-Moraiti, K.; Komaitis, M. E.; Evangelatos, G.; Kapoulas, B. B. *Food Chem.* **1985**, *16*(1), 15–24.

4278. Daza, L. M.; Donaire, J. P. *Physiol. Plant.* **1982**, *54*(2), 207–212.

4279. Zhukov, A. V.; Stefanov, K. L.; Vereshchagin, A. E. *Fiziol. Rast.* **1987**, *34*(3), 518–527 (Engl. 419–427).

4280. Mukherjee, K. D. *Plant Physiol.* **1983**, *73*(4), 929–934.

4281. Whitman, C. E.; Travis, R. L. *Phytochemistry* **1987**, *26*(5), 1311–1315.

4282. Simpson, T. D.; Nakamura, L. K. *J. Am. Oil Chem. Soc.* **1989**, *66*(9), 1309–1311.

4283. Khan, A. A.; Malhotra, S. S. *Phytochemistry* **1987**, *26*(6), 1627–1630.

4284. Malhotra, S. S.; Khan, A. A. *Phytochemistry* **1978**, *17*(2), 241–244.

4285. Weihrauch, J. L.; Son, Y.-S. *J. Am. Oil Chem. Soc.* **1983**, *60*(12), 1971–1978.

4286. Rodionov, V. S. *Fiziol. Rast.* **1978**, *25*(4), 773–777 (Engl. 603–606).

4287. Harlos, K.; Eibl, H.; Pascher, I.; Sundell, S. *Chem. Phys. Lipids* **1984**, *34*(2), 115–126.

4288. Shibata, T.; Uzawa, J.; Sugiura, Y.; Hayashi, K.; Takizawa, T. *Chem. Phys. Lipids* **1984**, *34*(2), 107–113.

4289. Sotirhos, N.; Ho, C.-T.; Chang, S. S. *Fette Seifen Anstrichm.* **1986**, *88*(1), 6–8.

4290. Zuñiga, G. E.; Fernandez, J.; Cristi, R.; Alberdi, M.; Corcuera, L. H. *Phytochemistry* **1990**, *29*(10), 3087–3090.

4291. Paliyath, G.; Lynch, D. V.; Thompson, J. E. *Physiol. Plant.* **1987**, *71*(4), 503–511.

4292. Mancha, M. *Grasas Aceites* **1978**, *29*(4), 263–268.

4293. Kalra, S. K.; Brooks, J. L. *Plant Biochem. J.* **1978**, *5*(1), 14–21.

4294. Luthra, R.; Sharma, B. N.; Paul, Y. *Plant Biochem. J.* **1980**, *7*(2), 159–166.

4295. Obara, T.; Kitamura, M. *J. Agric. Chem. Soc. Jpn.* **1955**, *29*(1), 37–39.

4296. Bevan, T. H.; Gregory, G. I.; Malkin, T.; Poole, A. G. *J. Chem. Soc.* **1951**, 841–842.

4297. Cevc, G., Ed.; *Phospholipids Handbook*; Marcel Dekker: New York, 1993; xi + 988 pp.

4298. Tani, T.; Chikubu, S.; Arisake, S.; Shikano, T. *Rept. Food Res. Inst. (Jpn.)* **1950**, *3*, 11–26.

4299. Westheimer, F. H. *ACS Symp. Ser.* **1992**, *486*, 1–17.

4300. Grzesiuk, S.; Tluczkiewicz, J.; Gorecki, R. J. *Acta Physiol. Plant.* **1983**, *5*(3), 61–69.

4301. Spiers, I. D.; Freeman, S.; Schwalbe, K. H. *J. Chem. Soc., Chem. Commun.* **1995**, (21), 2219–2220.

4302. Mills, S. J.; Potter, B. V. L. *J. Org. Chem.* **1996**, *61*(25), 8980–8987.

4303. Bartl, F.; Urjasz, H.; Brzezinski, B. *J. Mol. Struct.* **1998**, *441*(1), 77–82.

4304. Buskov, S.; Møller, P.; Sørensen, H.; Sørensen, J. C.; Sørensen, S. *J. Chromatogr.* **1998**, *802A*(1), 233–241.

4305. Sidorova, N. N.; Usmanov, R. A.; Kochetov, G. A. *Biochem. Mol. Biol. Int.* **1995**, *35*(1), 111–115.

4306. Um, S. H.; Song, Y. O.; Cheigh, H. S. *Han'guk Yongyang Siklyong Hakhoechi* **1990**, *19*(1), 87–93, from *Chem. Abstr.* **1991**, *114*, 162608d.

4307. Popova, O. I.; Muravyeva, D. A. *Farmatsiya (Moscow)* **1995**, *44*(6), 12–15, from *Chem. Abstr.* **1996**, *124*, 337833j.

4308. Chernenko, T. V.; Khodzhaeva, M. A.; Glushenkova, A. I.; Turakhozhaev, M. T. *Chem. Nat. Compd.* **1998**, *33*(6), 624–626, from *Chem. Abstr.* **1998**, *129*, 214119b.

4309. Whitby, L. G. *Biochem. J.* **1953**, *54*(3), 437–442.

4310. Komatsu, T. *Seikagaku* **1960**, *32*, 351–355, from *Chem. Abstr.* **1964**, *60*, 3220h.

4311. Komatsu, T. *Seikagaku* **1960**, *32*, 355–358, from *Chem. Abstr.* **1964**, *60*, 3221a.

4312. Huennekens, F. M.; Felton, S. P. *Methods Enzymol.* **1957**, *3*, 950–959.

4313. Wu, J.; Takayama, S.; Wong, H.; Siuzdak, G. *Chem. Biol.* **1997**, *4*(9), 653–657.

4314. Sevilla, J. M.; Pineda, T.; Roman, A. J.; Blazquez, M. *J. Electroanal. Chem.* **1997**, *428*(1/2), 91–95, from *Chem. Abstr.* **1997**, *127*, 141902y.

4315. Bergmeyer, H.U. *Methods of Enzymatic Analysis.* Academic Press: New York, 2nd edition, 1974; p. 358.

Index

Note: Page numbers followed by *f* indicate figures and *t* indicate tables.

B

G

I

J

K

L

O

X

Y

Z

Edwards Brothers Malloy
Ann Arbor MI. USA
June 13, 2013